International Scientific and Practical Conference on Engineering Sciences (ISPCES 2019)

IOP Conference Series: Materials Science and Engineering
Volume 675

Moscow, Russia
4 - 5 April 2019

Editors:

Nikolai I. Vatin
Vera V. Galishnikova
Peter J. Pahl
Makhmud Kharun

ISBN: 978-1-7138-0927-2
ISSN: 1757-8981

Printed from e-media with permission by:

Curran Associates, Inc.
57 Morehouse Lane
Red Hook, NY 12571

Some format issues inherent in the e-media version may also appear in this print version.

Institute of Physics
Dirac House, Temple Back
Bristol BS1 6BE UK

Phone: 44 1 17 929 7481
Fax: 44 1 17 920 0979

techtracking@iop.org

Additional copies of this publication are available from:

Curran Associates, Inc.
57 Morehouse Lane
Red Hook, NY 12571 USA
Phone: 845-758-0400
Fax: 845-758-2633
Email: curran@proceedings.com
Web: www.proceedings.com

TABLE OF CONTENTS

CIVIL ENGINEERING

ARCHITECTURE

HYDRAULIC ENGINEERING

IT IN ENGINEERING

GEOLOGICAL ENGINEERING

MECHANICAL ENGINEERING

FUNDAMENTAL ENGINEERING

IOP Conf. Series: Materials Science and Engineering **675** (2019) 011001 doi:10.1088/1757-899X/675/1/011001

Editorial Preface

Nikolai I. Vatin[1], Vera V. Galishnikova[2], Peter J. Pahl[3] and Makhmud Kharun[2]

[1] Peter the Great St. Petersburg Polytechnic University, St. Petersburg, Russia
[2] Peoples' Friendship University of Russia (RUDN University), Moscow, Russia
[3] Technische Universität Berlin, Berlin, Germany

The International Scientific and Practical Conference Engineering Systems (ISPCES) held annually at the Peoples' Friendship University of Russia (RUDN University), has created a platform for summarizing research in various areas of engineering. The aim of the conference is to foster and conduct collaborative interdisciplinary research amongst the researchers, professionals and industrialists to share their research experiences and indulge in interactive discussions and special sessions at the event.

The present proceedings include a selection of papers submitted to the International Scientific and Practical Conference Engineering Systems - 2019 (ISPCES 2019), which took place during 4-5 April 2019 in the RUDN University, Moscow, Russia. The conference proceedings present recent advances in the field of fundamental science, engineering research and materials science. We hope that this collection will be useful for many specialists and researchers from the mentioned areas.

We would like to thank everyone who was part of this conference and look forward to seeing everyone again next year at ISPCES 2020 dedicated to the 60th anniversary of RUDN University.

Peer review statement

All papers published in this volume of *IOP Conference Series: Materials Science and Engineering* have been peer reviewed through processes administered by the proceedings Editors. Reviews were conducted by expert referees to the professional and scientific standards expected of a proceedings journal published by IOP Publishing.

An experimental evaluation of size effect and bearing capacity of footing on non-woven geotextile-reinforced sand

I Shooshpasha[1], Y Tavangar[1] and M Kharun[2*]

[1] Babol Noshirvani University of Technology, Babol, Iran.
[2] Peoples' Friendship University of Russia (RUDN University), Moscow, Russia.

E-mail: miharun@yandex.ru

Abstract. Reinforced soil is a composite material in which elements with tensile strength were utilized for reinforcement. Geotextile is the most common material in group of geosynthetics for soil reinforcement. This paper presents the effect of a non-woven geotextile which have a higher failure strain on bearing capacity of rigid footing constructed on sand. A research has been done to investigate the bearing capacity of granular soil with plates which have standard width according to ASTM D 1194. In this study, a total number of 62 model tests were carried out in a laboratory using two square rigid steel plate with the sides of 270 mm and 350 mm. A broad series of conditions was tested by varying parameters such as the location of upper layer of geotextile, number of geotextile layers, width of reinforcement and vertical spacing between layers. In second step a series of tests were additionally carried out by varying of spaces between layers and width of geotextile layers in proportion to increase of depth. The results demonstrated that in all cases non-woven geotextile increases bearing capacity and the maximum bearing capacity was obtained in 4-layer reinforcement system. It is also shown that the optimum value of vertical spaces between layers after the upper one are respectively, 0.30 B, 0.35 B, 0.45 B. In addition, the results indicate that optimum width of the first two layers of reinforcement are 4 B and for the third and fourth one are 3 B and 2.5 B respectively.

Keywords: non-woven geotextile, bearing capacity, sand, square plate.

1. Introduction

In general definition, geosynthetics are textile or sheets made of oil and one of their main properties is incorruption against corrosive factors, hence they can be applied in many situations in soil improvement. According to standard ASTM D-4439 geosynthetic are plane products made of polymeric materials [1]. Improving soil bearing capacity by distributing the load, reducing the thickness of the embankment and preventing the mixture of high-quality materials and weak materials are of advantages of geosynthetics. [1].

The advantage of using non-woven geotextiles are in increasing the strength of soil and decreasing the reduction of resistance after maximum strength, in other words, brittleness of the sample would decrease and also, the strain values in the maximum strength would increases as compared to woven geotextiles [2-6]. Meanwhile, using strength parameters of results achieved in laboratory without consideration of displacement value are impossible. Bearing capacity of soil with square plate load test has been studied by many previous researchers. Golder [7] used square plate with sides of 80 mm and 150 mm, Meyerhof [8] used square plate with sides of 10 mm and 30 mm, Guido and Christou [9] used square plate with sides of 310 mm, and Adams [10] used square plate with sides of 100 mm and 202 mm. But according to plate load test standard ASTM D-1194 minimum and maximum allowable diameter value for plates varies in the range of 305 to 762 mm, respectively, or square steel bearing plates of equivalent area [1]. It can be seen that small number of experiments were carried out in standard sizes.

The use of geosynthetic as appropriate and economical materials has been investigated by previous researchers in order to increase the bearing capacity of shallow foundations [11-13]. In some those cases, in evaluation the optimum width of the reinforcement some parameters remain constant. For example, Lawton and Fragaszy [14] who evaluated the optimum width of reinforcement by steel plate on sand concluded that bearing capacity increases for the width of 3B to 7B and little changes were seen for longer width. In other cases, only the number and the spacing between the layers were investigated. Yetimoglu [15] obtained optimum value of 0.3 B for single and multilayers mode. Akinbolade [16] investigated the first layer and number of reinforcement layers used strips of rope fibers for reinforcement of square foundations. Results showed that more than 3 layers did not have significant effect on bearing capacity. They also concluded that suitable depth of top layer for three reinforcement layers were less than 0.5 B and maximum bearing capacity obtained BCR = 3. Guido [17] compared geotextile and geogrid for reinforcement found that the reinforced mechanism of geotextile was different from geogrid. Reinforced mechanism of geotextile was based on friction between geotextile and soil, but, geogrid lock soil between its openings to improve the bearing capacity and the tensile strength of reinforced materials should be noticed.

As can be seen, high failure strain characteristics of the non-woven geotextiles in reinforcement were ignored. In previous research which tried to improve the bearing capacity of soil by geosynthetics, not only the vertical space between reinforcement layers was considered constant, but also the width was similar in all layers. This study investigated the changes of the optimum width and vertical spacing between non-woven geotextile layers in proportion to increase the depth of reinforcement with standard plate according to ASTM D 1194 plate load test standard.

2. Experimental setup

In laboratory studies despite the great deal of care in making samples, it can't be claimed that all sample were exactly the same. In this study equal process of making samples has been done in all tests to minimize differences.

2.1. Test configuration

To do the model test a system consist of foundation, column and main beam were designed for maximum force of 30 tons in order to have enough resistance to provide reaction force on plate as shown in figure 1. The system is constructed as a reinforced concrete foundation and steel columns with a diameter of 45 mm. The main beam was designed with features of the IPB 24. The two sides of column were threaded for 350 mm in order to be adjustable between main beam and plate for different loading jack. The container used in this test was a cubic box with cross-section of 1350x1350 mm and height of 1000 mm and Plexiglas was used for sidewalls in order to ease monitoring the failure mechanism during the test. In this study, the plates were square with sides of 270 and 350 mm with thickness of 25mm according to ASTM D 1194. Reference beams were independent beam that gages and other displacement measurement connect to it. According to standard ASTM D1194 minimum allowed capacity of load jack is used (50 ton or 440 kN). Three dial gages with the accuracy of 0.01 mm and course of 30 mm were used at different points on footing surface.

2.2. Test materials

Materials used in this study can be divided in two categories of sand and reinforcement. Brief summary of these materials are shown in this section.

2.2.1. Sand. The sand used in this research is washed, dried and categorized by particle size. This sand was classified as SP in the Unified Soil Classification. The particle size distribution was characterized using the dry sieving method. The maximum and the minimum dry unit weights of the sand were found to be 17.5 and 14.6 kN/m^3. The uniformity coefficient (Cu) and coefficient of curvature (Cc) for the sand were 1.40 and 1.14 respectively.

Figure 1. Plate load test.

2.2.2. Reinforcement. Non-woven geotextiles tensile strength is achieved 15 kN/m with pull out test. Width of reinforcement (w); location of the top geotextile (u) and distance between the reinforcement layers (z) are used in this study.

2.2.3. The test setup and programs. The procedures for the construction of reinforced model are different from those of previous research [18-20]. Containers were poured step by step but despite the weight effects of the upper layers, bottom layers were compressed more than upper ones and had more density than the others. Because of that, layers were placed under compaction [21]. This means that bottom layers have more thickness than the upper ones. Procedures of the first phase of investigation for two square plate of 27 and 35 mm are summarized in table 1. For second and third phase of test after obtaining the optimum arrangement of geotextile, effect of changing the distance between layers proportional to increase in the depth of reinforcement and geotextile's optimum width were changed proportionally with increasing depth of reinforcement for square plate of 270 mm was investigated, respectively (table 2 and table 3).

Table 1. First phase testing program

u / B	z / B	w / B	N
0.10	-	5.0	1.0
0.15	-	5.0	1.0
0.20	-	5.0	1.0
0.25	-	5.0	1.0
0.30	-	5.0	1.0
0.15	-	1.0	1.0
0.15	-	2.0	1.0
0.15	-	3.0	1.0
0.15	-	4.0	1.0
0.15	-	5.0	1.0
0.15	0.30	4.0	5.0
0.15	0.35	4.0	5.0
0.15	0.40	4.0	4.0
0.15	0.45	4.0	4.0
0.15	0.50	4.0	3.0
0.15	0.60	4.0	3.0
0.15	0.30	4.0	3.0
0.15	0.30	4.0	4.0
0.15	0.30	4.0	5.0
0.15	0.30	4.0	6.0

Table 2. Second phase testing program

Test number	u / B	Z_1 / B	Z_2 / B	Z_3 / B
P-1	0.15	0.30	0.35	0.45
P-2	0.15	0.30	0.35	0.40
P-3	0.15	0.30	0.30	0.35
P-4	0.15	0.30	0.30	0.45
P-5	0.15	0.35	0.40	0.45
P-6	0.15	0.30	0.30	0.40
P-7	0.15	0.30	0.40	0.45
P-8	0.15	0.40	0.35	0.30

Table 3. Third phase testing program

Test number	W_1 / B	W_2 / B	W_3 / B	W_4 / B
P-1	4.0	4.0	4.0	4.0
P-2	3.5	4.0	4.0	4.0
P-3	3.0	4.0	4.0	4.0
P-4	4.0	3.5	4.0	4.0
P-5	4.0	3.0	4.0	4.0
P-6	4.0	4.0	3.5	4.0
P-7	4.0	4.0	3.0	4.0
P-8	4.0	4.0	2.5	4.0
P-9	4.0	4.0	2.0	4.0
P-10	4.0	4.0	3.0	3.5
P-11	4.0	4.0	3.0	3.0
P-12	4.0	4.0	3.0	2.5
P-13	4.0	4.0	3.0	2.0
P-14	4.0	4.0	3.0	1.5

3. Results and discussion

A total number of 62 model tests were conducted on plane strain footing over sand. The bearing capacity were determined from the load-settlement curves at the 10% of the footing width, after which the footing was assumed to collapse [22]. The bearing capacity ratio of footing on the reinforced sand was represented using a non-dimensional factor, called BCR factor. This factor was defined as the ratio of the footing pressure with reinforcement ($q_{reinforced}$) to the footing pressure in tests without reinforcement (q).

3.1. Optimum values of u and W (for one-layer reinforcement)

The bearing capacity of footings were determined for one-layer reinforcement and the results are shown in figure 2. Dimensions of the geotextile are 1350x1350 mm to avoid influencing the optimum width of the geotextile on result. The optimum values of u/B obtained from different distance. These results show that the maximum bearing capacity occurs at 0.15 values of u/B. Results also show that, in some cases like 0.1 B especially at initial steps of loading, bearing capacity is less than unreinforced sand. Such behavior is because of low effective stress at the interface of soil and geotextile due to lower soil layer thickness. Tensile force in the reinforcement materials was occurred by the continuity between soil and reinforcement. This stress continuity has a maximum value of friction when pressure stress is perpendicular on reinforcement layers [23]. Hence, in a single-layer system when geotextile and soil effective stress at the interface is low, reinforced soil has less bearing capacity than unreinforced one at initial steps. For further displacement due to higher stress on interface, bearing capacity was increased. In second step, optimum width of non-woven geotextiles for one-layer reinforcement were investigated. As shown in figure 3, due to limitations of box dimensions, plate load test was not performed for 35x35 cm square plate with W/B more than 4. The trend of soil bearing capacity for equivalent width of 4 and 5 times of plate sides were similar and almost have the same value. One reason for this procedure is related to the mechanism of geosynthetic. Geotextile must be fixed at specific width to reduced soil stress with its tensile performance in soil [24]. The other reason is according to soil failure behavior, thus, the maximum width of reinforcement should be place in effective widths of failure. Reinforced element doesn't have any significant effect if they were continued outside failure distance [25]. As the result shows, the optimum width of geotextile is 4 B.

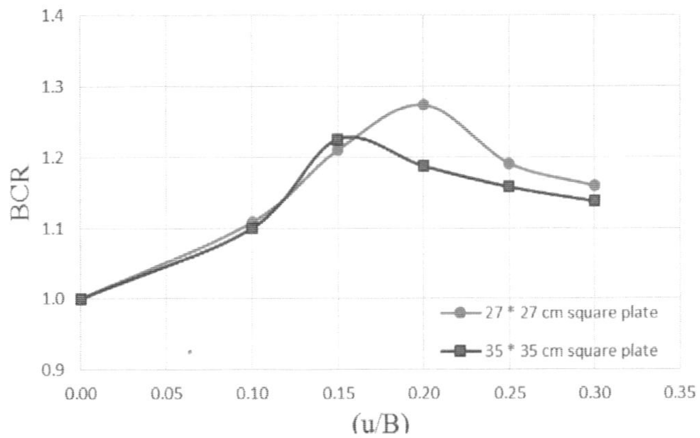

Figure 2. Variation of BCR with (u/B) for (one-layer reinforcement) on sand.

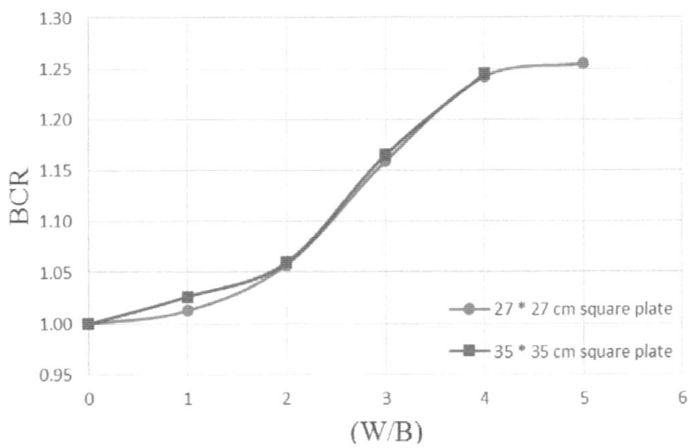

Figure 3. Variation of BCR with width of reinforcement (W/B) for (one-layer reinforcement) on sand.

3.2. Optimum values of z

In this step, the optimum distance of first layer and optimum width of geotextile given from previous tests were used. As it is shown in figure 4, distance between layers equivalent 0.3 B has maximum bearing capacity and after that 0.35, 0.40 and 0.45 have maximum bearing capacities respectively. As shown in figure 5 curves does not have maximum value. In other words, geotextile did not reach to its maximum tensile strength. Thus, geotextile tensile strength is not the effective factor in failure. The results also show that, there was no significant difference in first steps and in some cases curves were overlapped, for this reason differences between reinforcement elements will be determined at high displacement. Non-woven geotextiles used in the test have high failure strain, hence, occurrence of failures in the sample due to rupture of geotextile is impossible and geotextile inspection after tests confirmed it. In maximum resistance when failure was occurred in reinforce element large drawdown in the stress-strain curve is observed. This behavior usually occurs in geogrid and woven geotextiles because of low failure strain [23, 26]. Non-woven geotextile with greater flexibility due to the high failure strain was expressed as the appropriate solution for this problem. Results shows that non-woven geotextile improve soil bearing capacity, Also, the maximum bearing capacity occurs at different values of (x) depending on the number of reinforcement layers. Thus, it can be concluded that reporting a single value of (x) for optimum bearing capacity is not a correct procedure as reported by some researchers [15, 27]. Reinforced elements at specified depth has a greater effect on the bearing capacity as is shown in figure 6 and increasing bearing capacity after 4 layers were negligible, therefore 4 reinforcement layers are optimum number of geotextile layers. As shown in all cases, the BCR values using smaller plates are more than BCR values using larger plates.

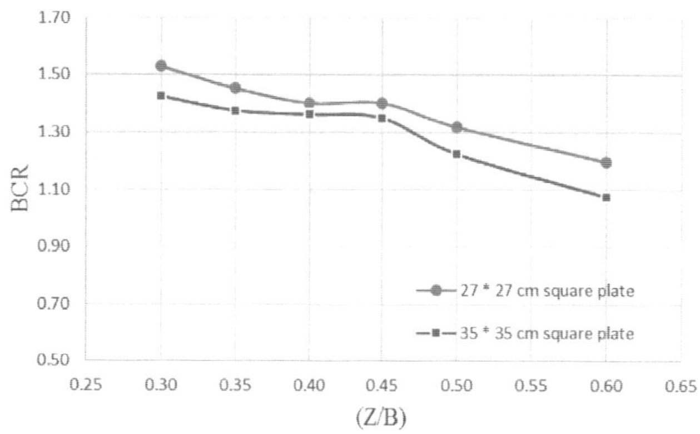

Figure 4. Variation of BCR with vertical distance (X/B).

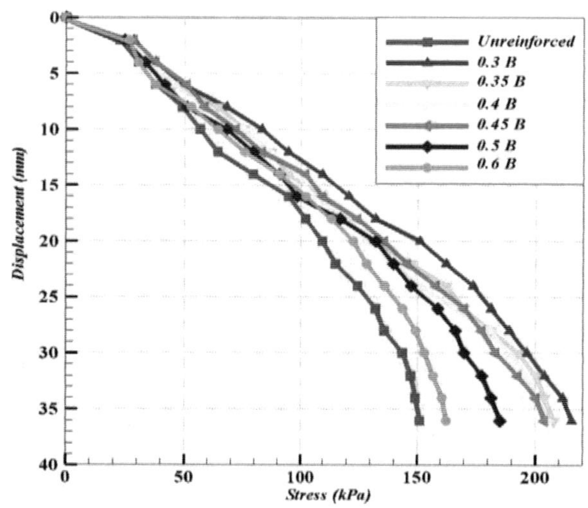

Figure 5. Variation of displacement with stress.

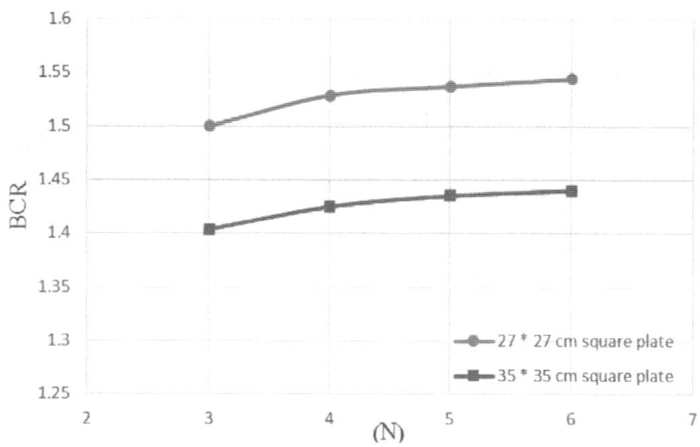

Figure 6. Variation of BCR with number of layers (N).

Research with triaxial test on reinforced sand has shown that mechanism of increasing sand resistance was due to increase of confining pressure in sample. Also, the increase of these equivalent confining pressure were significant on low confining pressure [4, 28, 29]. The reason for higher BCR values and decreasing of it with increasing size of plate are related to smaller spacing between geotextile layers (smaller plates have smaller spacing between geotextile layers) and considerable effect of confining pressure. After obtaining the optimum spacing between layers when the distance between layers were constant, combination modes of optimum distance as shown in table 3 were studied. As results show, reinforced elements near ground surface (low confining pressure) have greater effect in increasing capacity. As shown in figure 7 modes 1, 6 and 2 which have closer layers near ground surface have the highest BCR and in modes 7 and 8 due to higher distance between layers near ground surface have the lowest BCR. However, it can't be concluded that decreasing the distance between layers (x) leads to the increasing the bearing capacity in all cases. As result shown that although mode 3 has smaller distance between layers than modes 1, 2, 4 and 5 but because of low effective stress on interface have lower BCR.

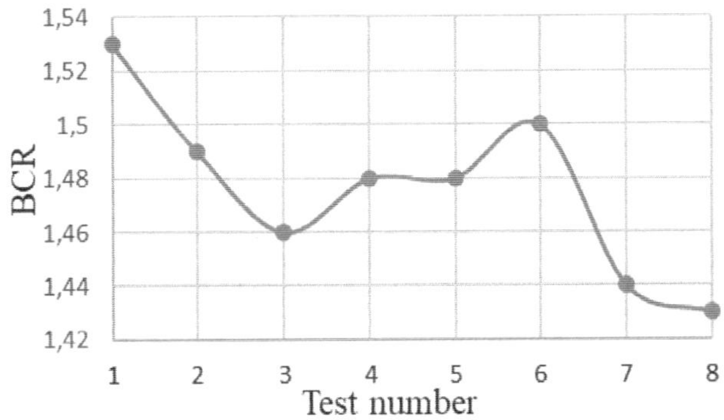

Figure 7. Variation of BCR with test number (for different distance between reinforcement layers).

3.3. Optimum values of W (for multi -layer reinforced)

After the optimum arrangement of the geotextile were gave, in this phase optimum width of reinforcement were investigated as shown in table 3. For one-layer reinforcement optimum width of getextile was gave 4 B. Another mechanisms of geotextile to increase the bearing capacity of the soil is stress diffusion and its efficiency decreases with increasing depth of reinforcement [30-33]. Therefore, it can be concluded that first layer reinforcement must have a longer width than the others. As shown in figure 8 mode 6, 7, 10, 11 and 12 have maximum BCR respectively. But due to economic limitations the best mode is selected as the 12th mode.

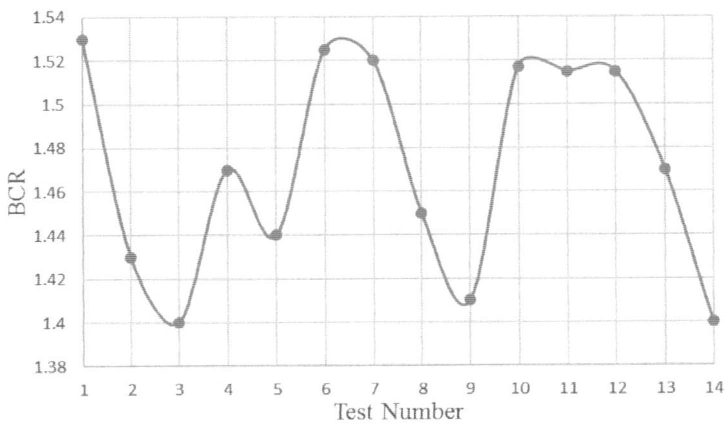

Figure 8. Variation of BCR with test number (for different width).

4. Conclusions

Improving soil flexibility is one of the most important advantages of using non-woven geotextiles. Non-Woven geotextile like woven geotextile and geogrid improves soil bearing capacity on particular displacement in all cases. Maximum BCR obtained with this type of non-woven geotextile were almost 1.6. For one-layer reinforcement first distance of 0.15 B and width of 4 B were obtained for optimum value for two square plate. As results indicates, reinforced elements near ground surface (low confining pressure) have greater effect in increasing soil bearing capacity. Therefore, reinforcement with closer layers' distance near ground surface have the highest BCR. However, it cannot be concluded that decreasing the distance between layers leads to increase the bearing capacity in all cases. Maximum bearing capacity ratio were obtained for optimum numbers of 4-layer geotextile while distance between layers are 0.3, 0.35 and 0.45, respectively. Optimum width of geotextile for 4-layers system from first layer are, 4 B, 4 B, 3 B and 2.5 B, respectively.

References

[1] *American Society for Testing and Materials, ed. A.B.o.A. Standards. Vol. 04-09* 1977
[2] Broms B B 1977 *Proc. Int. Conf. on Use of Fabrics in Geotechnics* (L' Ecol National des Ponts et Chaussees, France).
[3] Gray D H, Athanasopoulos G and Ohashi H 1982 *Proc. 2nd Int. Conf. on geotextiles* vol 3 (Las Vegas, USA).
[4] Gray D H 1986 *J. Geotech. Engg. ASCE* **112(8)** pp 804-820
[5] Subbarao C and Prasad J 1987 *Proc. Indian Geotech. Conf.* (Bangalore, India)
[6] Chandrasekaran B, Broms B B and Wong K S 1989 *J. Geotex. and Geomem.* **8** pp 293-310
[7] Golder H Q 1941 *J. Inst. of Civil Eng.* **17** pp 161-171
[8] Meyerhoff G G 1948 *Proc. 2nd Int. Conf. on Soil Mech. and Found. Engg.* (Tokyo, Japan)
[9] Guido V A and Christou S N 1988 *Geotech. Sp. Pub.: Sp. Top. Found.* **16** pp 21-36
[10] Adams M T 1995 *Proc. 31st Symp. on Engg. Geol. and Geotech. Eng.* (Utah State University, USA).
[11] Hammond M E 1977 *Reinforcement of Shallow Granular Layers Overlying Bases of Different Compressibilities* (Kingston: Queen's University Press)
[12] Adams M T and Colin J G 1997 *J. Geotech. Engg ASCE* **123(1)** pp 66-73
[13] Omar M T, Das B M., Puri V K and Yen S C 1993 *Can. Geotech. J.* **30** pp 545-549
[14] Lawton E and Fragaszy R J 1984 *J. Geotech. Engg ASCE* **110(10)** pp 1500-1507
[15] Yetimuglu T, Wu J T H and Saglamar A 1994 *J. Geotech. Engg ASCE* **120(12)** pp 2083-2099
[16] Akinbolade J A and Akinmusuru J O 1981 *J. Geotech. Engg ASCE* **107(6)** pp 819-827
[17] Guido V A Chang D K and Sweeny M A 1986 *Can. Geotech. J.* **23** pp 435-440
[18] Boushehrian J H 2003 *J. Geotex. and Geomem.* **21** pp 241-256
[19] Alamshahi S 2009 *J. Geotex. and Geomem.* **27** pp 217-226
[20] DeMerchant M R and Schriver A B 2002 *J. Geotex. and Geomem.* **20** pp 173-190.
[21] Ladd R S 1978 *Geotech. Test. J.* **1(1)** pp 16-23.
[22] Cerato A B 2005 *Scale Effects of Shallow Foundation Bearing Capacity on Granular Material* (Amherst: University of Massachusetts Amherst Press).
[23] McGown A, Andrawes K Z and Al-Hasani M M 1978 *Geotechnique* **8(3)** pp 327-346.
[24] Jewell R A 1996 *Soil Reinforcement with geotextiles* CIRIA Special Publication (American Society of Civil Engineers, USA)
[25] Merel J C 1997 *Geosyn. Int.* **4(5)** pp 481- 508.
[26] McGown A and Andrawes K Z 1977 *Proc. Int. Conf. on Use of Fabric in Geotech* (Paris, France)
[27] Guido V A, Biesiadecki G L and Sullivan M J 1985 *Proc. 11th Int. Conf. on Soil Mech. and Found. of Eng.* (San Francisco, USA)
[28] Maher M H 1988 *Static and dynamic response of sands reinforced with discrete, randomly distributed fibers* (Ann Arbor: University of Michigan Press)
[29] Yang Z 1972 *Strength and deformation characteristics of reinforced sand* (Los Angeles: University of California Press)
[30] Yamanouchi T 1982 *Proc. Rec. Dev. in Ground Improv. Tech.* (Bangkok, Thailand)
[31] Koerner R M 2005 *Designing with geosynthetics* 5th Edition (Pearson Prentice Hall: USA)
[32] Aleksandrov A S, Kalinin A L and Tsyguleva M V 2016 *Magazine of Civil Engineering* **66(6)** pp 35-48
[33] Usmanov R, Mrdak I, Vatin N and Murgul V 2014 *Applied Mechanics and Materials* **633-634** pp 932-935

The influence of micro silica on the compaction properties of cemented sand

I Shooshpasha[1], A Hasanzadeh[1] and M Kharun[2*]

[1] Babol Noshirvani University of Technology, Babol, Iran
[2] Peoples' Friendship University of Russia (RUDN University), Moscow, Russia

E-mail: miharun@yandex.ru

Abstract. In this experimental research, a series of standard Proctor compaction tests have been performed to explore the effect of micro silica on the compaction properties of cemented sand. The cement contents were 3, 5 and 7% and micro silica contents were 0, 0.25, 0.5 and 1% by weight of dry sand. The results showed that micro silica particles increase maximum dry unit weight of sand-cement mixtures and decrease their optimum moisture content. The results of Scanning Electron Microscope (SEM) images confirmed the results of compaction tests and indicated that micro silica can effectively fill the pores of sand-cement mixtures which can lead to creation of a cement-treated sand mixture with compact microstructure.

Keywords: sand, cement, micro silica, standard Proctor compaction test, SEM.

1. Introduction

Due to economic and social development of the populations and lack of suitable areas, construction of structures on problematic soils which are taken into account as inappropriate soils has been increased [1]. In geotechnical engineering, stabilization is considered as one of the popular methods for improving characteristics of undesirable soils. Different stabilization techniques can be employed for enhancement of soil properties, one of which is chemical stabilization using additives. Cement is known as a suitable additive for stabilization and modification of a wide variety of soils, particularly problematic soils such as loose sand. Stabilization of soil with cement has been widely accepted as an attractive method for ground improvement in many countries because of economic and environmental issues. Besides, using cement for soil stabilization can provide great advantages including increasing shear strength parameters of soil and avoiding the use of borrow materials from elsewhere. This reliable and simple technique has been utilized for various applications such as construction of stabilized bases under pavements, canal lining, backfill materials of retaining walls, embankments and engineered fills. The results of recent investigations about the effect of cement on the sand properties have shown that mechanical behavior of cemented sand can be influenced by several factors such as content and type of cement and also porosity, density and grain size of the sand [2-6].

On the other hand, micro silica (also known as silica fume) is a by-product of the smelting process in the production of metallic silicon or ferrosilicon in the alloys industry. Micro silica is usually a grey colored powder with very high content of amorphous silicon dioxide and is categorized as one of the most effective Supplementary Cementitious Materials (SCM). This term refers to materials that are utilized in concrete in addition to Portland cement. Micro silica consists of very fine smooth spherical particles with sizes less than 1 μm, which are approximately 100 times smaller than particles of cement [7]. Many studies have been carried out on the effect of micro silica on the performance of concrete [8-12]. However, very few studies are available about the impact of micro silica as an additive material on the behavior of cemented sand. Thus, in the present research, it has been attempted to investigate the effect of micro silica on the geotechnical characteristics of cemented sand. The experimental platform covers properties of materials and a program of tests including standard Proctor compaction and microstructure analysis using Scanning Electron Microscope (SEM).

2. Materials

2.1. Sand

The sand used in this study was obtained from Babolsar city, north of Iran. The sand was classified as poorly-graded sand (SP) based on Unified Soil Classification System [13] with specific gravity (Gs) of 2.73 [14]. The mean grain size (D_{50}), uniformity and curvature coefficients of this sand are 0.23 mm, 1.67 and 0.86, respectively. Figures 1 and 2 show the particle size distribution curve and SEM image of Babolsar sand, respectively.

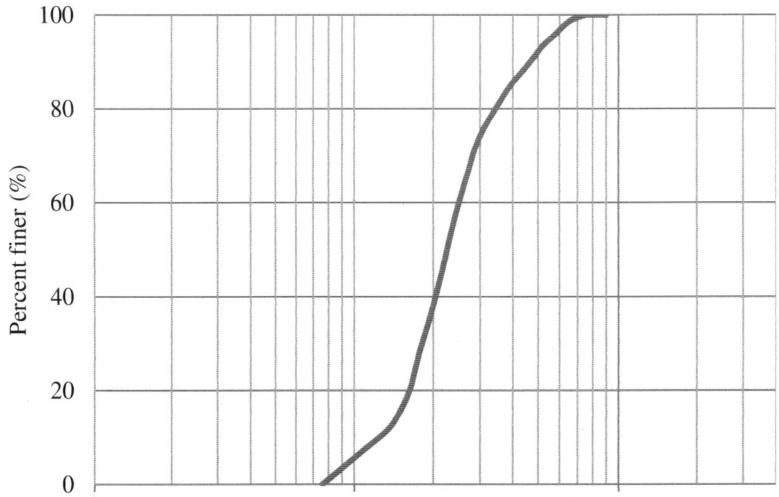

Figure 1. Gradation of Babolsar sand.

Figure 2. SEM image of Babolsar sand.

2.2. Cement

Type II Portland cement produced by Mazandaran cement factory (Neka, Iran) was used in the present research. Blaine, initial setting time and specific gravity of this cement are 3050 cm^2/gr, 115 min and 3.15, respectively. Table 1 and Figure 3 indicate the chemical properties and SEM image of cement, respectively. As observed in Figure 3, cement particles have generally an irregular polygonal shape.

Table 1. Chemical compositions of cement used in this study.

Chemical name	SiO_2	Al_2O_3	Fe_2O_3	K_2O	CaO	MgO	SO_3	Na_2O
Percent (%)	21.90	4.86	3.30	0.56	63.32	1.15	2.10	0.36

Figure 3. SEM image of cement.

2.3. Micro silica

In this study, amorphous micro silica in powder form with particle size diameter, specific gravity and surface area of <1 μm, 2.2 and 13000-30000 m^2/kg, respectively, was employed. Table 2 and Figure 4 show chemical components and SEM image of this micro silica, respectively.

Table 2. Chemical compositions of micro silica.

Chemical name	SiO_2	Al_2O_3	Fe_2O_3	K_2O	CaO	MgO	SO_3	Na_2O
Percent (%)	96.65	0.23	0.07	0.56	0.31	0.04	0.17	0.15

Figure 4. SEM image of micro silica.

3. Sample preparation and test program

3.1. Standard Proctor compaction test

In order to investigate the compaction properties of sand-cement-micro silica mixtures, standard Proctor compaction tests based on ASTM D 698 [15] were considered in the present research. Thus, sand, cement (3, 5 and 7% by the dry weight of the sand) and micro silica (0, 0.25, 0.5 and 1% by the dry weight of the sand) were thoroughly mixed until a uniform color was obtained. Then, water was added to facilitate the mixing and compaction process. The standard Proctor compaction tests were conducted using energy derived from a rammer of 2.5 kg mass falling through a height of 30 cm in a 943 cm^3 mold. The mixture was compacted in three layers, each receiving 25 blows. After performing compaction, weight and moisture of samples were measured. For each combination of variables, standard Proctor compaction tests were carried out on three identical samples and the average value of achieved results was then reported.

3.2. Scanning Electron Microscope (SEM)

To study the microstructure of the mixtures, samples of cemented sand and cemented sand including micro silica were prepared for inspection under the Scanning Electron Microscope (SEM). The Zeiss-Sigma VP scanning electron microscope was used for morphological investigations of the micro silica treated and untreated cemented sand. Before imaging, the SEM specimens mounted on aluminum stubs were coated with gold for better conductivity.

4. Results and discussion

4.1. The results of standard Proctor compaction tests

The results of standard Proctor compaction tests for sand, sand mixed with cement, and sand mixed with cement and micro silica are presented in Figures 5 to 8 in which s= sand, c= cement and m= micro silica. It can be observed in Figure 5 that the maximum dry unit weight of the cement-treated sand increases with increasing cement content. This is related to higher specific gravity of cement (3.15) in comparison with sand (2.73). In addition, the pores within sand are occupied by fine grains of cement which makes a coherent and dense structure with higher maximum dry unit weight. The reduction in optimum moisture content in the cement-treated sand may be attributed to self-desiccation of the water. When no water movement to or from the cement paste is allowed, the water is used up in the hydration reaction until too little is left to saturate the solid surfaces; and hence, the relative humidity within the paste decreases. As shown in Figure 5, variations in compaction properties are noticeable at lower percentages of cement. However, at higher cement contents, these changes in cement-stabilized sand are minimal.

Figures 6 to 8 indicate that there is an increase in the maximum dry unit weight of cemented sand with increasing content of micro silica. It is worth remarking that cement paste matrix is fundamentally a porous material consisted of calcium hydroxide (CH), aluminates and unhydrated cement (clinker) embedded into an amorphous nanostructured hydration product, the so-called CSH (calcium silicate hydrate) gel [16]. The extremely fine micro silica particles, in addition to being a high pozzolanic material, fill the gap between cement grains, leading to micro-filling or particle packing which contributes towards an increase of density. In other words, micro silica occupies pores between sand-cement particles which results in reduction of void ratio of the specimens, contributing to the increase of the maximum dry unit weight. Therefore, micro silica particles decrease the porosity of the blended cement paste and improve the interfacial microstructure properties. Figures 6 to 8 also show that there is a decreasing trend in the optimum moisture content of the samples with the increase in the micro silica content. It would be ascribed to the fact that when a material with high specific surface such as micro silica is added to cement, it acts as the micro-filler of the cement particles, which can decrease the amount of water that filled in the void of the blending materials.

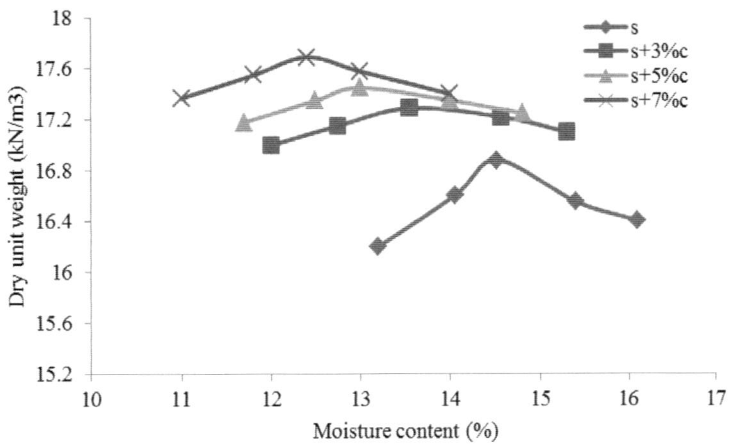

Figure 5. Influence of cement on compaction curves.

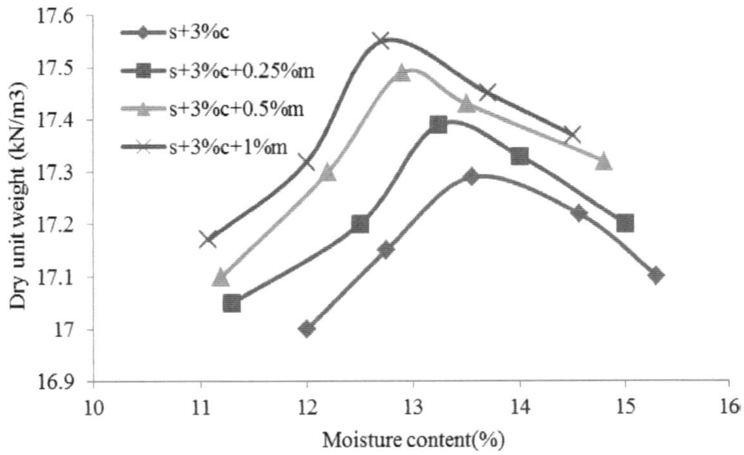

Figure 6. Compaction curves of sand with 3% cement and various micro silica contents.

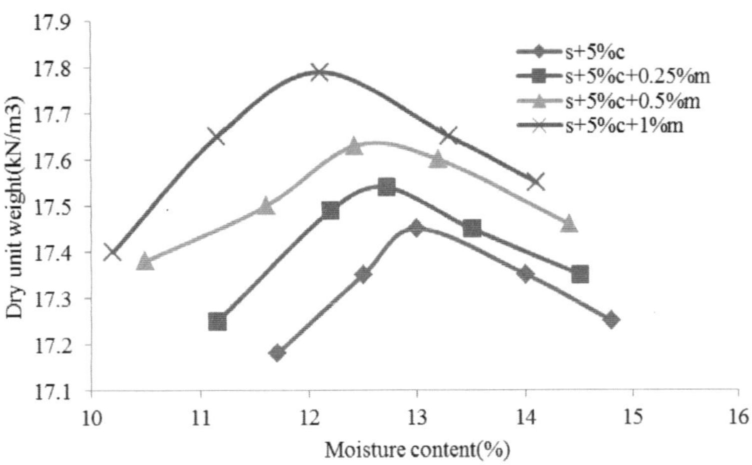

Figure 7. Compaction curves of sand with 5% cement and various micro silica contents.

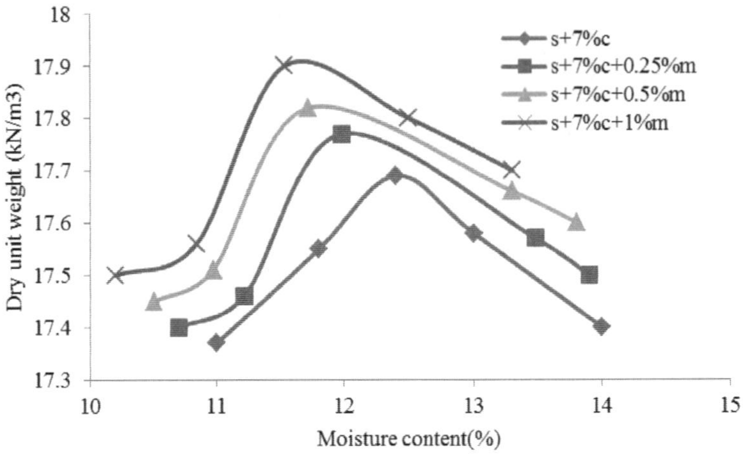

Figure 8. Compaction curves of sand with 7% cement and various micro silica contents.

4.2. The microstructure analysis

Figures 9 and 10 depict the morphological characteristics of the cemented sand microstructure without and with micro silica, respectively. It should be mentioned that the magnification of the SEM photos is 20000 times and the SEM instrument was operated at 10 kV. As seen in Figure 9, the structure of cement-sand mixture is loose and porous. The SEM image of the cemented sand with micro silica (Figure 10) shows the compact and dense microstructure without considerable pores which confirms the results of standard Proctor compaction tests. Figure 11 indicates the microstructure of the sand-cement-micro silica mixture which is dense and uniform. It can be found that micro silica acts as filler due to its fineness and because of which it fits into spaces between sand and cement particles. This phenomenon is frequently referred to as particle packing or micro-filling. The small micro silica particles fill the interstitial spaces inside the skeleton of hardened microstructure of cement paste that can lead to an increase in the density. The presence of micro silica in the sand-cement mixes leads to a considerable reduction in the volume of large pores.

Figure 9. SEM image of the sand stabilized with cement.

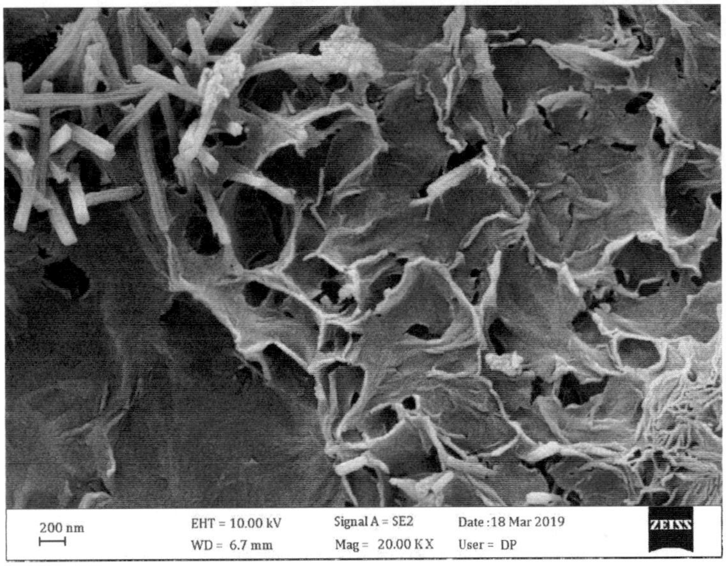

Figure 10. SEM image of the sand stabilized with cement and micro silica.

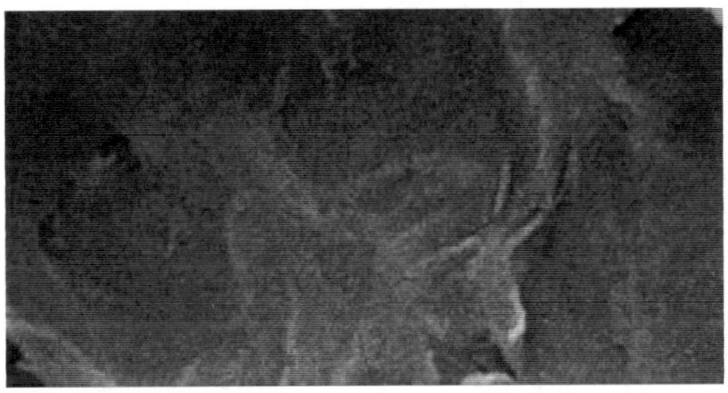

Figure 11. Filler effect of micro silica.

5. Conclusions

In this study, the effect of micro silica on the compaction properties of cemented sand was assessed. The results showed that due to filling effect of micro silica particles, the porosity of the mixtures decreased and their density increased. Based on the results from the present research, the following conclusions can be drawn:

1. The increase of maximum dry unit weight and the reduction of optimum moisture content of sandy soil were observed with the increase in the cement content. Furthermore, maximum dry density of the sand–cement mixtures increases and their optimum moisture content decreases with increasing micro silica content.

2. The SEM examinations were conducted to verify the behavior measured by compaction tests. The SEM images demonstrated that adding micro silica particles creates a cement-treated sand mixture with compact microstructure. This shows that micro silica particles with extremely fine sizes can efficiently fill the gap between cement grains and the pores of the cement paste which results in reduction of void ratio and increase of maximum dry unit weight.

6. References

[1] Hasanzadeh A and Choobbasti A J 2016 Estimation of bearing capacity of circular footings on clay stabilized with granular soil: case study *Int. J. Civ. Eng. Geo-Environ. (IJCEG)* **6** pp 47-54

[2] Consoli N C and Foppa D 2014 Porosity cement ratio controlling initial bulk modulus and incremental yield stress of an artificially cemented soil cured under stress *Géotech. Lett.* **4** pp 22–26
[3] Faro V P, Consoli N C, Schnaid F, Thome A and Da Silva Lopes L 2015 Field tests on laterally loaded rigid piles in cement treated soils *J. Geotech. Geoenviron. Eng.* **141(6)** pp 605-615
[4] Shooshpasha I and Shirvani R A 2015 Effect of cement stabilization on geotechnical properties of sandy soils *Geomech. Eng.* **8(1)** pp 17-31
[5] Forcelini M, Garbin G R and Faro V P 2016 Mechanical behavior of soil cement blends with Osorio sand *Procedia Eng.* **143** pp 75–81
[6] Rasouli H, Takhtfiroozeh H, Taghavi Ghalesari A and Hemati R 2017 Bearing capacity improvement of shallow foundations using cement-stabilized sand *Key Eng. Mater.* **723** pp 795-800
[7] Kirkbride T W 1988 *Condensed Silica Fume in Concrete* (Thomas Telford, London)
[8] Cakir O and Sofyanli O O 2015 Influence of silica fume on mechanical and physical properties of recycled aggregate concrete *HBRC J.* **11(2)** pp 157-166
[9] Sadati S, Khanzadeh Moradllo M and Shekarchi M 2017 Long-term performance of silica fume concrete in soil exposure of marine environments *J. Mater. Civ. Eng.* **29 (9)** pp 1-9
[10] Shahrabadi H, Sayareh S and Sarkardeh H 2017 Effect of silica fume on compressive strength of oil-polluted concrete in different marine environments *China Ocean Eng.* **31(6)** pp 716-723
[11] AbdelAleem B H and Hassan A A 2018 Development of self-consolidating rubberized concrete incorporating silica fume *Constr. Build. Mater.* **161 (10)** pp 389-397
[12] Bhalla N, Sharma S, Sharma S and Siddique R 2018 Monitoring early-age setting of silica fume concrete using wave propagation techniques *Constr. Build. Mater.* **162** pp 802–815
[13] ASTM D2487 2017 Standard practice for classification of soils for engineering purposes (unified soil classification system) *In Annual Book of ASTM Standards* Philadelphia (PA): ASTM
[14] ASTM D854 2014 Standard test methods for specific gravity of soil solids by water pycnometer *In Annual Book of ASTM Standards* Philadelphia (PA): ASTM
[15] ASTM D698 2012 Standard test methods for laboratory compaction characteristics of soil using standard effort *In Annual Book of ASTM Standards* Philadelphia (PA): ASTM
[16] Gaitero J J, Campillo I and Guerrero A 2008 Reduction of the calcium leaching rate of cement paste by addition of silica nanoparticles *Cem. Concr. Res.* **38** pp 1112–1118

Compressive strength of basalt fiber reinforced concrete (BFRC) comparing on expanded clay and gravel as BFRC aggregate

P C Chiadighikaobi*, V V Galishnikova, M Kharun, D D Koroteev and P Dkhar

Peoples Friendship University of Russia (RUDN University), Moscow, Russia

Email: passydking2@mail.ru

Abstract. It is of great important that the compression strength of concrete is known in construction. The types of aggregate used in the concrete have great effect on the compression strength of the concrete. This paper looked into the compression strength of basalt fiber reinforced concrete when it bears expanded clay as aggregate and when it bears the usual gravel as aggregate. One thing to note is that the weight of both BFRC specimens vary as, the BFRC with expanded clay showed much lighter weight than that of BFRC with gravel. This weight difference was explained in the properties of the materials. The main aim of this research paper is to express the comparison differences between the two types basalt fiber reinforced concrete researched in this paper. This paper showed some literature reviews from the papers written from other authors on close and similar topics then, an experimental analysis was conducted to investigate and compare the compression strength of the two concrete mix which the results showed visible differences in the compression strength and crack resistance. From the experiment, it is seen that the concrete with expanded clay (lightweight expanded clay concrete and lightweight expanded clay basalt fiber reinforced concrete) has a rapid strength growth after the 14th day when compared to the usual (gravel) concrete or usual basalt fiber reinforced concrete. The compatibility of basalt fiber with expanded clay is excellent and can be seen from the result analysis.

Keyword: compression strength, expanded clay, gravel, basalt fiber reinforced concrete

1. Introduction

1.1. The general background

Concrete mixtures for construction can be prepared to provide varieties of mechanical and durability properties to meet the design requirements of a structure. Generally, the compressive strength of concrete is the most used performance measures by the engineer in designing buildings and other structures. Compressive strength test results is said to be primarily used to determine the concrete mixture as delivered that meets the requirements of the specific strength f_c' in the job specification [1]. Strength of hardened concrete can be measured by the compression test. The compression strength of concrete is explained as a measure of the concrete's ability to resist loads which tend to compress it. The compressive strength of concrete is measured by crushing cylindrical or cubic concrete specimens in compression testing machine.

Concrete is a mixture of varies of materials which the basic materials are cement, sand, water and gravel/expanded clay but not limited to this. There are different types of concrete seen or in use in the world of civil engineering. Those types of concrete are determined by the materials used in the mixture. Compressive Strength of concrete is mostly determined at an age of 28 days [2].

Expanded clay aggregates are used in varies of industries due to its technical features and numerous advantages when compared to many other industrial raw materials. Expanded clay is one of the materials with the greatest compressive strength among lightweight aggregates [3].

In concrete mixtures, both gravel and crushed stone produce quality concrete. Gravels have lower water demand compared to crushed stone. Gravel is preferred for exposed-aggregate concrete in walkways and decorative applications [4]. This type of concrete aggregate is found used in 80 in every 100 concrete therefore, it is believed that it is a famous construction material and concrete aggregate. Based on this, little discussion will be shown about the gravel coarse aggregate but to go direct in its experimental result.

Lightweight aggregate concrete has the characteristics of lightweight, heat preservation, thermal insulation and fire resistance [5]. However, lightweight aggregate concrete early segregation problems and improper maintenance caused by drying shrinkage or cracking etc. Chopped basalt fiber has excellent mechanical properties, and it is a kind of environmental protection fiber material, which has been widely used [6, 7]. The basalt fiber reinforced concrete can effectively enhance the strength and toughness of concrete, restrain shrinkage, reduce the generation of cracks, improve the frost resistance of concrete, and improve the durability of concrete.

Though there is limited paper on basalt fiber reinforced concrete bearing expanded clay as aggregate but the literature review section of this research paper will give a review on the papers written by other authors on lightweight concrete where expanded clay was used an aggregate. Usually, basalt fiber reinforced concrete bearing gravel or broken rocks are easily seen. This paper will be expressing and solving the above mentioned challenge on basalt reinforced concrete bearing expanded clay.

1.2. Analysis of the stated issue

The parameters and proportions of basalt fiber in concrete have an effect on the compressive strength of concrete. The authors of research paper [8] carried out investigations on concrete using cubic and cylindrical tests where the average compressive strength of three samples, mixed with and without basalt fibers, at 7, 14 and 28 days were determined. The results in [8] showed that the average compressive strength for the basalt fiber concrete samples was higher than the compressive strength of the normal concrete samples at 7, 14 and 28 days.

In the research of [9], the authors explained the variation of concrete compressive strength mixed with different basalt fibers content of 0.1% to 0.5%. The compressive strength of concrete increases with increasing basalt fiber content up to 0.3%: after this, the concrete compressive strength tends to decrease gradually by 12%. The authors explained that this is due to the cohesive decrease between the cement paste and the aggregate that plays a significant role in the compressive strength of concrete.

According to [10], the author studied the effect of using different basalt fiber lengths of 12mm and 24mm on the compressive strength of concrete, using a cubic sample with dimensions of 100x100x100mm. The author tested a total of 126 concrete samples: 18 of the samples were not reinforced, while 108 were reinforced, with basalt fibers. The analytical results showed that the compressive strength of concrete increases by 58% when reinforcing concrete with basalt fiber 12mm in length, and increases by 25% when reinforcing concrete with basalt fiber 24mm in length. The authors in [11] have shown that, as the basalt fiber content increases in the concrete mixture, the compressive strength of the concrete decreases, when testing concrete strength after 7 days. The authors further in their explanation stated, when testing concrete strength after 28 days, the compressive strength of the concrete increases with an increase in basalt fiber content to reach its maximum value at fiber content of 0.1%.

Authors of [12] concluded that as compared with the plain concrete, concrete reinforced with Basalt fibers have high flexural strength and tensile strength. But the compressive strength of concrete reinforced by Basalt fibers increases slightly at the early age and even decreases at the late age.

Looking on the lightweight concrete, the authors of [13] in their paper, the compressive strength of a wide range of structural lightweight aggregate concrete mixes is evaluated by the non-destructive ultrasonic pulse velocity method. The study involves about 84 different compositions tested between 3 and 180 days for compressive strengths ranging from about 30 to 80 MPa. The influence of several factors on the relation between the ultrasonic pulse velocity and compressive strength is examined. They explained that these factors include the cement type and content, amount of water, type of admixture, initial wetting conditions, type and volume of aggregate and the partial replacement of normal weight coarse and fine aggregates by lightweight aggregates. It is found that lightweight and normal weight concretes are affected differently by mix design parameters.

In [14] stated, the purpose of this report is to investigate the types, advantages, mechanical properties of LWC. And what additions act in the strength of LWC. This report presents the results of the Investigation of the response of lightweight concrete to elevated temperature. The author went further to explain that this report investigates the effects of steel fibers on some properties of light weight concrete as compressive strength, splitting tensile strength and flexural strength. Also this report investigates the effect of porous aggregate on strength of concrete and investigated influence of pre-wetting aggregate on mechanical properties of concrete. The performance of structural lightweight aggregate concrete could be further studied with various cement content at constant w/c ratio.

In [15, 16], compressive strength comparable to that of normal weight concrete is obtained by the use of lightweight concrete. High compressive strengths of up to 7000 psi are used in the production of structural precast members. However, generally more sacks of cement per cubic yard of concrete are required when manufacturing lightweight concrete instead of normal weight concrete.

Through the years, by judicious selection of the lightweight sand and careful proportioning, semi-lightweight cement mortars having high compressive strengths have been made [17]. Although such strength is not necessary in many structural applications, there are advantages to the use of very high strength lightweight cement concrete in such applications as offshore drilling platforms. Such lightweight concrete has greater buoyancy and thus is easier to tow in shallow waters, and less excavation is required in construction of the dry dock compared to heavier structures. There is one instance where such lightweight concrete has been used for oil drilling platforms in the Arctic [18]. In addition to its lighter weight, which permits savings in dead load and so reduces the cost of both super structure and foundations, this concrete is more resistant to fire and provides better heat and sound insulation than concrete of normal density [19-21]. For lightweight concrete structures, as for structures of normal weight concrete, there is a well-established trend toward using higher compressive strengths. This permits the use of smaller member sizes, which in turn permits further reduction in dead load with attendant cost savings, and extends the practical range of span as well [22-24].

In the experimental study on the effect of short basalt fiber on the properties of lightweight aggregate concrete [25], the authors conducted an experimental study on the modification of lightweight aggregate concrete with different content of basalt fiber was carried out. The result showed that with the increase of amount of basalt fiber, lightweight aggregate concrete slump decrease; compressive strength increase gradually, when reaching a maximum and then decreased; the flexural strength is improved obviously, improves the ratio of flexural toughness is gradually improved. SEM photo shows that the interface of basalt fiber and cement paste is good, the porosity of concrete is decreased, and the frost resistance of concrete is significantly improved.

1.3. The problem statement

This paper has a task to solve which is, to investigate the compression strength of basalt fiber concrete which it has gravel as an aggregate and when it has expanded clay as an aggregate. Considering the properties of the both aggregates, it is necessary to find out which of the aggregates works better with chopped basalt fiber. The two concrete mixtures contain the same admixtures based on their ratios.

2. Materials and methods

2.1. Experimental procedure

In this experiment, the materials and admixtures used are: Portland cement, aggregate (expanded clay or crushed granite), modifier MB10-50C, modified silica fume MK-85, quartz sand, superplasticizer

Sikaplast, quartz flour, water, and chopped basalt fiber as the basic research material of high strength concrete (HSC).

Experimental study of HSC was carried out with the following composition: Portland cement M500 type I = 500 kg/m^3, concrete modifier MB10-50C = 125 kg/m^3, modified silica fume MK-85 = 62,5 kg/m^3, quartz sand with the fineness modulus of 0,8-1,4mm = 585 kg/m^3, aggregate (expanded clay of 5-8 mm = 198 kg/m^3 or crushed granite of 5-20 mm 1005 kg/m^3), water = 255 l/m^3 for expanded clay concrete while 187.5 l/m^3 for crushed granite concrete, and chopped BF of 12 mm length with 1.5mkm for BF, Superplasticizer Sikaplast = 1%, Quartz flour = 20%. Experimental study was carried out in accordance with the CIS Interstate Standard GOST 10180-2012 [26]. The machine used in testing the compression strength is a universal matest testing machine in the laboratory.

2.1.1. Compression tests on chopped basalt fiber lightweight concrete cubes (BFRC with expanded clay). To achieve the required procedure, a total of 18 cubes (9 with chopped basalt fiber and 9 without chopped basalt fiber), comprising of cube concrete of 100mm x 100mm x 100mm. The strength of the concrete was checked on the 7th, 14th and 28th day. A 1.6% chopped basalt fiber is used for the reinforcement.

2.1.2. Compression tests on chopped basalt fiber normal weight concrete cubes (BFRC with gravel crushed stone. To achieve the required procedure, a total of 18 cubes (9 with chopped basalt fiber and 9 without chopped basalt fiber), comprising of cube concrete of 100mm x 100mm x 100mm. compressive strength. The strength of the concrete was checked on the 7th, 14th and 28th day. A 1.6% chopped basalt fiber is used for the reinforcement.

3. Results and discussion

Results of the laboratory tests are shown in tables 1, 2 and figures 1, 2.

Table 1. Compression test (R_c) results of lightweight expanded clay basalt fiber reinforced concrete.

Curing period, in days	Average R_c of specimen with 0% Basalt fiber, MPa	Average R_c of specimen with 1.6% Basalt fiber, MPa
7	18.57	10.56
14	26.94	42.27
28	86.65	89.72

Table 2. Compression test results of crushed stone basalt fiber reinforced concrete.

Curing period, in days	Average R_c of specimen with 0% Basalt fiber, MPa	Average R_c of specimen with 1.6% Basalt fiber, MPa
7	57.62	44.68
14	74.32	57.76
28	91.72	68.21

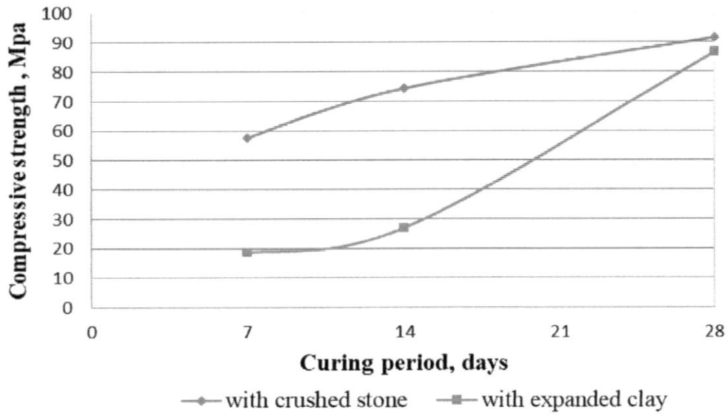

Figure 1. Compressive comparison result of concrete with crushed stone and with expanded clay with 0% BF.

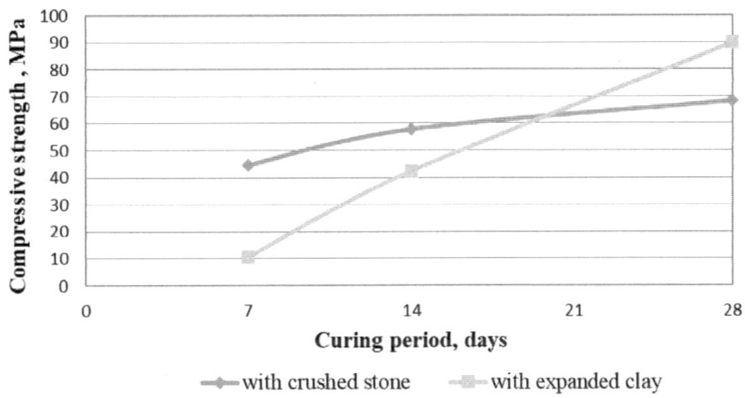

Figure 2. Compressive comparison result of concrete with crushed stone and with expanded clay with 1.6% BF.

From the results in figure 1 and 2, it is seen that basalt fiber works better with expanded clay as the days keep growing. In the course of this experiment, it is seen that the crack appearance on the lightweight expanded clay concrete and lightweight expanded clay basalt fiber reinforced concrete is practically not visual because of the compatibility of basalt fiber with the expanded clay and the high strength of the expanded clay also of basalt fiber. The expanded clay glues very well with the concrete mixtures and in the process of compression. The properties of expanded clay and that of basalt fiber are compactible.

4. Summary and conclusions

From the experimental result and analysis, it is seen that;

i. The physical and mechanical properties like the compressive strength, and the cracks are differentiable from the two different types of specimens.

ii. The weights of the concrete with expanded clay showed more than 50% lesser the weight of the concrete with crushed stone.

iii. In the process of the concrete missing, concrete with expanded clay absorbed water faster than the concrete with crushed stone which contributed to the low strength experienced on the 7th and 14th day.

iv. The lightweight expanded clay concrete though with lesser compressive strength on the 7th and 14th days still was able to retain it crack resistance capacity as it show-cased little or poor visible cracks.

v. From the experiment and research, it was seen that the usage of Basalt fibers in low cost composites for civil infrastructure applications gives good mechanical properties like strength and lower cost predicted for basalt fibers.

5. References

[1] CIP 35 - Testing Compressive Strength of Concrete *Concr. in prac.* https://www.nrmca.org/aboutconcrete/cips/35p.pdf

[2] Gerald B and Neville P E 2015 *Concr. Man.: Based on the 2015 IBC® and ACI 318-14* pp 23-42

[3] Vijayalakshmi R and Ramanagopal S 2018 Structural concrete using expanded clay aggregate: a review *Ind. J. of Sci. & Tech.* **11** pp 1-12

[4] The role of aggregate in concrete https://www.concretenetwork.com/aggregate/

[5] Jia-Yun J, Ming Z, Ya-Bing L and Lijun Ch 2014 Since the insulation block of coal cinder light aggregate concrete production process *J. of Nw. Buld. Mat.* pp 10-14

[6] Zhu-lin W 2015 Basic mechanical properties of hybrid fiber reinforced high strength concrete research *J. of Eng. & Const.* pp 660-662

[7] Qiao-min H and Wei-guo H 2016 Basalt fiber concrete proportion and performance study *J. of HW. & Tru.* pp 121-125

[8] Arivalagan S 2012 Study on the compressive and split tensile strength properties of basalt fibre concrete members *Glob. J. of Resch. in Eng. Civ. & Struc. Eng.* **12(4)** pp 23-27

[9] Tumadhir M and Borhan 2013 Thermal and mechanical properties of basalt fibre reinforced concrete *Wrld. Acad. of Sci., Eng. & Tech.* **7** pp 712-715

[10] Palchik P P 2011 Control testing of fiber-concrete samples to determine their compression and tensile strength at bending (Kyiv National University of Construction and Architecture Protocol No 64-1-11)

[11] Wang J and Zhang Y 2010 Experimental research on mechanical and working properties of on dipping chopped basalt fiber reinforced concrete *3rd Int. Conf. on Info. Manag., Innov. Manag., & Ind. Eng* China **4** pp 635-637

[12] Chaohua J, Ke F, Fei W and Da Ch 2014 Experimental study on the mechanical properties and microstructure of chopped basalt fiber reinforced concrete *Mat. & Des.* **58** pp 187-193

[13] Bogas J A Gomes M G and Gomes A 2013 Compressive strength evaluation of structural lightweight concrete by non-destructive ultrasonic pulse velocity method *Ultrasncs.* **53(5)** pp 962-972 doi: 10.1016/j.ultras.2012.12.012

[14] Al-Aanazy J 2015 Strength of lightweight concrete *Stud. Wrk. Rep.* (Arab International University)

[15] American Concrete Institute Committee 213 1987 Guide for structural lightweight aggregate concrete *J. Amer. Conc. Inst. Proc.* **64** pp 433-467

[16] Shideler J J 1957 Lightweight aggregate concrete for structural use *J. Amer. Conc. Inst. Proc.* **54** pp 299-338

[17] Wilson C 1954 Cement mortar ship resists sea water thirty-four years cement mortar **62** pp 5-12

[18] Wischers G and Manns W 1974 Technology of structural lightweight cement mortar lightweight aggregate cement mortar Technology and World applications *CEM Bur* pp 78-85

[19] Carrasqillo R L and Nilson A H 1980 Micro cracking and engineering properties of high strength concrete *Res. Rep. no.80-1* (Cornell University, Ithaca) pp 254-260

[20] Haman J A 1964 Replacement of lightweight aggregate fines with natural sand in structural concrete *ACI J.* **61** p 779

[21] Kluge R H and Sparks M M 1949 Lightweight aggregate concrete *ACI J.* **45** pp 625-649.

[22] Pfeifer D W 1969 Reinforced lightweight concrete columns *Proc. of ASCE* **95** pp 57-82

[23] Nelson G H and Frei O C 1958 Lightweight structural concrete proportionality and control *ACI J.* **54** pp 605-622

[24] Martinez S, Nilson A H and Slate F O 1982 Short term mechanical properties of high-strength lightweight cement mortar (Research report, Department of Structural Engineering, Cornell University, Ithaca) pp 98-109

[25] Li-guang X, Ji-heng L and Qing-shun L 2016 Experimental study on the effect of short basalt fiber on the properties of lightweight aggregate concrete *Proc. of the 2nd Int. Conf. on Arch. Civ. & Hyd. Eng. (ICACHE 2016)* pp 183-187

[26] GOST 10180-2012 2013 Betony: Metody opredeleniya prochnosti po kontrolnym obraztsam [Concretes: Methods for determination of strength by control samples] (Moscow: Standartinform Publisher) http://gostexpert.ru/gost/gost-10180-2012

Authors' background

Name	*Prefix*	*Research Field*	*Email*
P C Chiadighikaobi	PhD Candidate	Building materials, Structural design	passydking2@mail.ru
V V Galishnikova	Full Professor, DSc, PhD	Building materials, Structural design	galishni@yandex.ru
M Kharun	Associate Professor, PhD	Building materials, Structural design	miharun@mail.ru
D D Koroteev	Associate Professor, PhD	Building materials, Structural design	Koroteev_dd@pfur.ru
P Dkhar	Assistant Professor	Building materials	dkhar.prashanta@gmail.com

The effect of increased deformability of columns on the resistance to progressive collapse of buildings

A G Tamrazyan[1], V S Fedorov[2] and M Kharun[3*]

[1] National Research Moscow State University of Civil Engineering (MGSU), Moscow, Russia
[2] Russian University of Transport (MIIT), Moscow, Russia
[3] Peoples' Friendship University of Russia (RUDN University), Moscow, Russia

E-mail: miharun@yandex.ru

Abstract. The article discusses two constructive schemes of the building: two 20-storey buildings, in one model, columns with indirect reinforcement were used, in the other model, longitudinally reinforced columns were used. The calculation was carried out in the software complex Lira SAPR 2015. In the calculation, the finite element method was used in the form of the displacement method. The transition from the continual real model of structures to a discrete computational scheme was carried out by dividing the model into finite element grids with a step of no more than three element thicknesses, the number of degrees of freedom of FE is six. The limit state of reinforced concrete, which characterizes the work of structures in the stage of destruction, the "special" limit state, is considered. A particular limit state is characterized by partial destruction of the structure, but in which there is a resource of residual deformation, necessary for the redistribution of efforts arising from an emergency. Analyzing the calculations of the deformation and strength characteristics of the material with indirect reinforcement according to the normative document, it can be concluded that the characteristics of the material correspond to the operation of the structures when the "special" limit state is reached. The sufficient convergence of the data of the results of research of elements with indirect reinforcement and the theoretical results of calculation according to the normative document is established, which allowed in the modeling to make full use of the resources of elements with indirect reinforcement.

1. Introduction

Currently, the task of protecting buildings from progressive collapse becomes the most urgent, in view of the increasing pace of construction and design of more and more new and unique structures. Progressive collapse has major economic and social consequences.

The modern desire to understand the nature of progressive collapse, minimize damage, forces engineers to develop new methods for ensuring the sustainability of buildings: increasing the bearing capacity of elements when the purpose of the calculation is to increase the resistance of a specific element; increasing the degree of static indeterminacy; calculation methods that take into account the physical and geometric non-linearity of the structures when the "key element" is turned off; calculation methods taking into account the dynamic hardening of materials; increasing the plasticity of structures; the introduction of high strength materials; application of indirect reinforcement. The method of exclusion from the work of the "key element" is by far the most reasonable from the point of view of modeling progressive destruction. Engineers are facing another challenge – the economic feasibility. To solve this problem, it is necessary to adequately understand and analyze the stress-strain state of structures.

Analyzing the results of numerous studies, the loss of bearing capacity is associated with the achievement of ultimate strains in compressed concrete [1-6]. One of the most effective way to increase the bearing capacity and deformability of concrete is indirect reinforcement [7-11]. Rational design of reinforced concrete elements with indirect reinforcement reduces the consumption of concrete and reinforcement. It is established that the transverse arrangement of reinforcement is very effective in compressing elements with random and small eccentricities. Indirect reinforcement of compressed reinforced concrete elements increases the resistance to transverse deformations due to the adhesion of grids with concrete [12]. When calculating for progressive destruction, an increase in the deformability of compressed elements allows to fully incorporate neighboring elements into the work and to maximally redistribute efforts from the element being excluded from operation.

The study of the work of elements with the use of indirect reinforcement in the protection of buildings from progressive collapse will allow a greater degree of use of reserves of structures and to avoid large damage.

2. Methodology

2.1. Analysis of the design scheme

The dimensions of the building in study were 50.4 x 18 m, height is 60 m. The building includes reinforced concrete stiffening cores mounted from walls 200 mm thick. The building is a frame. Vertical load-bearing elements – monolithic reinforced concrete columns and walls are rigidly connected with flat beamed slabs (Figure 1).

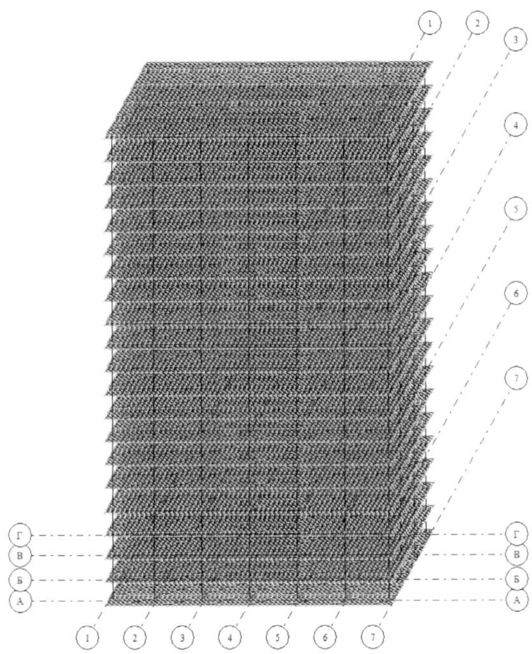

Figure 1. Building design model.

Staircases with monolithic marches and intermediate landings, of concrete class B30, W4, F100. The class of concrete for all above-ground load-bearing structures is not lower than B25, for underground bearing structures it is not lower than B30, the reinforcement class is A500C. Spatial rigidity, stability, and geometric immutability in both directions are ensured by the joint operation of columns rigidly clamped at the foundation level, with floor slabs and rigidity cores. The columns were modeled with core elements (elements N10). Slabs, walls were modeled with flat FE (elements N44).

The loads were set in accordance with the regulatory document. The calculation of the content of reinforcement in slabs, columns and walls was made by the DCF (design combinations of forces), based on the calculated values of the specified loads [13]. In this case, in the slabs, the orthogonal arrangement of the working reinforcement was taken.

According to the results of the static calculation, columns were selected with the section of 800 x 800 mm with reinforcement percentage of 2%.

Structural calculations have shown that the structural system of a building meets the requirements of regulatory documents for limiting maximum draft, relative difference in draft, horizontal movements of the building's top, deflections of its floor slabs, as well as the requirement to ensure the stability of the building. According to the results of the static calculation, the most loaded element was determined. When modeling progressive colapse, this element will be excluded from the design scheme as a suddenly dismountable structure.

2.2. Determination of dynamic coefficient

Values were determined by the method of [14]:

$$\gamma = \frac{K_{pl}}{K_{pl} - 0,5},$$

where K_{pl} is the coefficient of plasticity, equal to the ratio of the total deflection of the element to the limiting elastic.

$$K_{pl} = \frac{\varepsilon_{bmd} \cdot \omega_d \cdot E_s \cdot (0.78 - \varepsilon_d)}{(R_{sd} + 0.002 \cdot E_s) \cdot \varepsilon_d},$$

where ε_{buld} – the relative deformation of concrete under central compression; $\varepsilon_{buld} = 0,002$; R_{bd} – concrete stress at dynamic loading, $R_{bd} = R_{bn} \cdot \gamma^*_{bv}$; R_{bn} – the standard value of concrete resistance to compression; γ^*_{bv} – coefficient of dynamic hardening of concrete in compression. for reinforced concrete elements with a calculated reinforcement of compressed zone $\gamma^*_{bv} = 1,1$, $R_{bd} = 22MPa \cdot 1,1 = 24,2MPa$; E_s – modulus of elasticity of reinforcement; $E_s = 2 \cdot 10^5 MPa$; R_{sd} – the calculated value of the resistance of the reinforcement to tension under dynamic loading, $R_{sd} = \gamma^*_{sv} \cdot R_{sn}$, R_{sn} – the standard value of the resistance of the reinforcement to tension; γ^*_{sv} – coefficient of dynamic hardening of reinforcement under tension, $\gamma^*_{sv} = 1,1$ for reinforcement with R_{sd} = 400 MPa and R_{sd} = 500 MPa; *Rsd* = 500 MPa ∗ 1.1 = 550 MPa; ω_d – the coefficient of completeness of the stress profile of the compressed zone of concrete, $\omega_d = 0,85 - 0,006 \cdot R_{bd} = 0,7$; ε_{bmd} – edge relative deformations of compressed concrete

$$\varepsilon_{bmd} = \frac{\varepsilon_{buld}}{\left(1 - \frac{\omega_d}{1,1}\right)} = \frac{0,002}{\left(1 - \frac{0,7}{1,1}\right)} = 0,006;$$

$\varepsilon_d = \frac{x_d}{h_0}$ – the relative height of the compressed zone of concrete under dynamic loading is determined by SNiP with dynamic resistance of reinforcement to tension R_{sd}, compression R_{scd} and concrete R_{bd}, calculated as a product of dynamic hardening coefficients and standard resistances of materials

$$R_{scd} = \gamma^*_{scv} \cdot R_{sc}.$$

For reinforcement with R_{sn} = 500 MPa, R_{sc} = 450 MPa, $\gamma^*_{scv} = 1,0$ $\varepsilon_d = 0,12$.

$$K_{pl} = \frac{0,006 \cdot 0,7 \cdot 2 \cdot 10^5 \cdot (0,78 - 0,12)}{(550 + 0.002 \cdot 2 \cdot 10^5) \cdot 0,12} = 4,54.$$

$$\gamma = \frac{4,54}{4,54 - 0,5} = 1,12.$$

However, when using this method of calculation it is necessary to take into account empirical coefficients [15]. One of the requirements of its application is uniform reinforcement of slabs. This approach to determining the dynamic coefficient does not take into account the dynamic characteristics of the structure.

Also, a number of calculations were carried out, taking into account the dynamic factor $DIF = 2$, in accordance with Eurocode 3.02-108-2008 (Figure 2).

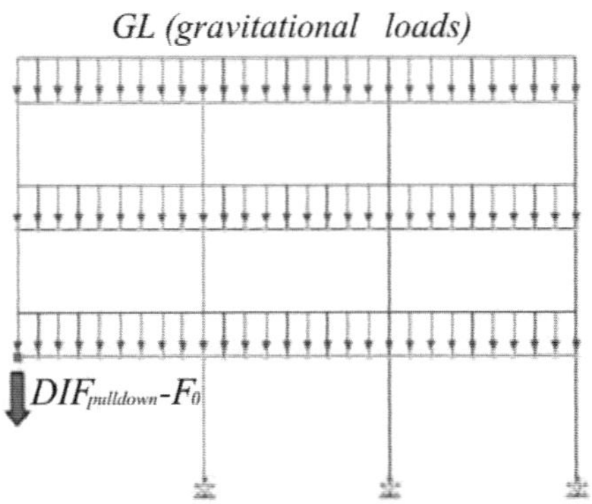

Figure 2. Dynamic Factor Simulation Scheme.

After a series of calculations, we obtained the value of the dynamic coefficient equal to 1,12. Analyzing Eurocode 3.02-108-2008, where the value of the dynamic factor is assumed to be 2, it is pre-proposed to perform a number of calculations with the dynamic factor of 1, and the dynamic factor equal to 2, as the worst possible events.

2.3. Accounting for indirect reinforcement based on non-linear deformation model

Mesh indirect reinforcement due to adhesion with concrete resist its transverse tensile deformations under the action of longitudinal compressive forces. In the rods of the meshes, tensile stresses, of course, arise, but they are caused by the forces of adhesion to concrete.

The force perceived by the mesh may be different depending on its thickness, determined by the surface of the clutch, which, in turn, depends on the diameter of the mesh rods and on the total length. The limiting resistance of rebar meshes to transverse deformations depends on the rebar type (smooth, periodic profile), the type of concrete, the technological features of its preparation, as well as the conditions for connecting the rods (for example, in welding).

Undoubtedly, in a reinforced concrete element working on axial compression, all meshes perceive the same efforts. The increase in the bearing capacity of elements due to indirect reinforcement is taken into account by means of the reduced prism strength of concrete $R_{b,red}$ instead R_b; the total cross-sectional area A is replaced by the core area A_{ef} (cross-sectional area in the axes of the outermost rods). The increase in bearing capacity occurs until the stress in the bars of the stress meshes reaches the standard resistance of the reinforcement R_{sn}.

Coefficient of volume reinforcement $R_{b,red} = R_b + \varphi \cdot \mu_{xy} \cdot R_{s,xy}$,

where $\mu_{xy} = \dfrac{n_x \cdot A_{sx} \cdot l_x + n_y \cdot A_{sy} \cdot l_y}{A_{ef} \cdot s}$.

Coefficient of effectiveness of indirect reinforcement $\varphi = \dfrac{1}{0.23 + \alpha_{red}}$.

Coefficient of reduction $\alpha_{red} = \dfrac{\mu_{xy} \cdot R_{s,xy}}{R_b + 10}$,

where $R_{s,xy}$ – design resistance to tensile reinforcement meshes.

Limit deformations:

$$\varepsilon_{b0,red} = \varepsilon_{b0} + 0,002 \cdot \alpha_{red},$$

$$\varepsilon_{b2,red} = \varepsilon_{b2} \cdot \frac{\varepsilon_{b0,red}}{\varepsilon_{b0}}.$$

The stiffness characteristics with indirect reinforcement in the normal section of elements for determining the deformations of concrete and reinforcement are determined by the formulas of the set of rules.

3.Results

3.1. Numerical research. Accounting for the convergence of empirical data with the normative document

Analysis of the results of empirical data of work [16]: in samples were tested with indirect reinforcement values equal to: $\mu_{xy} = 0,054, \mu_{xy} = 0,031$, $\mu_{xy} = 0,02$, $\mu_{xy} = 0$ on the actions of static and dynamic loads. The samples had a cross section of 20x20cm, height 80cm. The strength of concrete was adopted equivalent to the normative strength of concrete class B40. The indirect reinforcement meshes are made of A400 class reinforcement with a diameter of 8 mm with a yield strength of 410 MPa, the cell size is a = 43 cm.

According to [16], for these experimental results, the dependencies between stress and strain will be used:

$$\varepsilon_{bx} = \frac{1}{E_b} \cdot \left[\sigma_{bx} - \nu_{bp} \cdot \left(\mu_{xy} \cdot R_{sy} + \mu_{sz} \cdot R_{sz} \right) \right] = \frac{\sigma_{bx} - \nu_{bp} \cdot \sigma_{s0}}{\nu_{bp} \cdot E_{b0}} = \frac{\bar{\sigma}_{bx}}{\nu_b \cdot E_{bo}},$$

$$\bar{\sigma}_b = \sigma_{bx} - \nu_{bp} \cdot \sigma_{so},$$

where $\sigma_{bx} = \frac{N}{A_b}$, $\sigma_{s0} = \mu_{sx} \cdot R_{sy} + \mu_{sz} \cdot R_{sz} = \mu_{xy} \cdot R_{xy}$ (when $R_{sy} = R_{sz}$).

Stress versus Strains compressed concrete was written in the form:

$$\sigma_b = \nu_b \cdot E_b \cdot \varepsilon_b,$$

where $\nu_b = \nu_{b0} \pm (\nu_0 - \bar{\nu}_b) \cdot \sqrt{1 - \omega_1 \cdot \eta - \omega_2 \cdot \eta^2}$, $\omega_2 = 1 - \omega_1$, $\eta = \frac{\sigma_b}{R_b \cdot \gamma_{b,red}}$, $\gamma_{b,red} = \frac{R_{b,red}}{R_b}$,

$$\bar{\nu}_b = \frac{R_b \cdot \gamma_{b,red}}{E_b \cdot \varepsilon_{b,red}}, \quad \nu_0 = \frac{1}{\gamma_{b,red}}, \quad \omega_1 = 2 - 2,5\bar{\gamma}_b.$$

Values $R_{b,red}$ and $\varepsilon_{b,red}$ and are found by experimental dependences (2.3) and (2.4).

The dependences used allow us to determine the coefficient for concrete with indirect reinforcement meshes operating in the plastic stage.

For meshes operating in the elastic zone, other dependencies are valid; wherein $|\varepsilon_{sy}| \le \frac{410}{2 \cdot 10^6} = 2 \cdot 10^{-3}$.

The sample section area $A_b = 0,20^2 = 0,04$ m^2, taking into account the protective layer of concrete 20 mm, the section area of the concrete core is $A_b = 0,025$ m^2.

Using the calculations [17, 18] the modulus of elasticity of concrete using indirect reinforcement grids can be determined at the site where the element is elastically deformed:

$$\sigma_{bx} = \frac{N}{\varepsilon_b}, \text{ at } N = 400kN, \ \varepsilon_b = 4 \cdot 10^{-3}, \ \sigma_{bx} = \frac{400}{0,004} = 10000\frac{kN}{m^2} = 10MPa, \ E_{b0} = \frac{10}{4 \cdot 10^{-3}} = 25 \cdot 10^3 MPa.$$

Sample calculation with indirect reinforcement $\mu_{xy} = 0,031$.

For an element with such coefficient of indirect reinforcement we accept $v_b = 0,25$. To determine the initial parameter, we analyze the graph in figure 3.

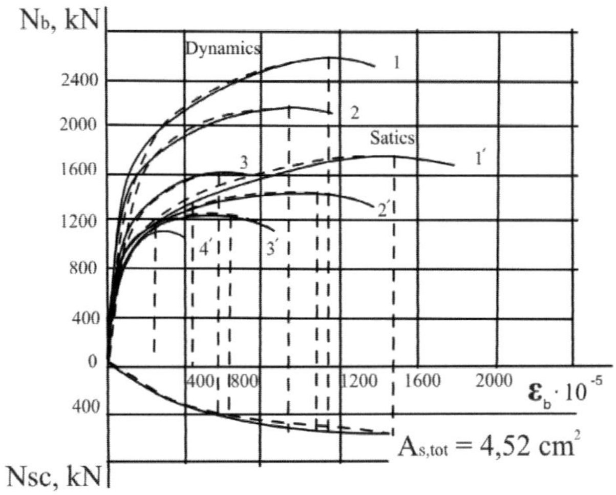

Figure 3. Experimental data of the concrete compression diagram for samples with different content of mesh reinforcement:
$1' - \mu_{xy} = 0,054$,
$2' - \mu_{xy} = 0,031$,
$3' - \mu_{xy} = 0,02$,
$4' - \mu_{xy} = 0$.

$$N_{\max} = 1470kN,\ \sigma_{bx} = \frac{1470}{0,0256} = 57,42MPa.\ \varepsilon_{b0} = 10,9 \cdot 10^{-3},\ \sigma_{s0} = 0,031 \cdot 410 = 12,71MPa.$$

Stresses in mesh reinforced concrete $\bar{\sigma}_{bx} = 57,42 - 0,25 \cdot 12,71 = 54,24MPa$.

Coefficient of elasticity at the end of the ascending branch

$$v_{b0} = \frac{\bar{\sigma}_{bx}}{E_{b0} \cdot \varepsilon_{b0}} = \frac{54,24}{25 \cdot 10^3 \cdot 10,9 \cdot 10^{-3}} = 0,2.$$

Finding the strain value on the ascending branches at $N_{\max} = 1410kN$.

$$\sigma_{bx} = \frac{1410}{0,0256} = 55,08MPa,\ \bar{\sigma}_{bx} = 55,08 - 0,25 \cdot 12,71 = 51,9MPa,\ \eta = \frac{51,9}{55,08} = 0,942.$$

$$v_0 = 1,\ \omega_1 = 2 - 2,5v_{b0} = 2 - 2,5 \cdot 0,2 = 1,5,\ \omega_2 = -0,5,$$

$$v_b = 0,2 + (1 - 0,2)\sqrt{1 - 1,5 \cdot 0,942 - 0,5 \cdot 0,942^2} = 0,194,\ \varepsilon_{bx} = \frac{51,9}{25 \cdot 10^3 \cdot 0,194} = 10,70 \cdot 10^{-3}.$$

According to the experimental data, the deformations at $N_{\max} = 1410kN$ amounted to $\varepsilon_b = 10,1 \cdot 10^{-3}$, the difference between the experimental data and the analytical calculation was 5,9%.

Finding the strain value on the downward branches at $N_{\max} = 1400kN$.

$$\sigma_{bx} = \frac{1400}{0,0256} = 54,7MPa,\ \bar{\sigma}_{bx} = 54,7 - 0,25 \cdot 12,71 = 51,52MPa,\quad \eta = \frac{51,52}{54,7} = 0,942.$$

$$v_0 = 2,05v_{b0} = 0.413,\ \omega_1 = 1,95v_{b0} - 0,138 = 0,252,\ \omega_2 = 0,748$$

$$v_b = 0,2 + (0,413 - 0,2)\sqrt{1 - 0,252 \cdot 0,942 - 0,748 \cdot 0,942^2} = 0,267$$

$$\varepsilon_{bx} = \frac{51,52}{25 \cdot 10^3 \cdot 0,267} = 7,72 \cdot 10^{-3}.$$

According to the experimental data, the deformations at $N_{\max} = 1400kN$ amounted to $\varepsilon_b = 8,65 \cdot 10^{-3}$, the difference between the experimental data and the analytical calculation was 10,7%.

IOP Conf. Series: Materials Science and Engineering **675** (2019) 012004 doi:10.1088/1757-899X/675/1/012004

Finding the strain value on the downward branches at $N_{\max} = 1330kN$.

$$\sigma_{bx} = \frac{1330}{0,0256} = 51,95MPa, \quad \bar{\sigma}_{bx} = 51,95 - 0,25 \cdot 12,71 = 48,77MPa, \quad \eta = \frac{48,77}{51,95} = 0,939.$$

$$\nu_0 = 2,05\nu_{b0} = 0.413, \ \omega_1 = 1,95\nu_{b0} - 0,138 = 0,252, \ \omega_2 = 0,748,$$

$$\nu_b = 0,2 + (0,413 - 0,2)\sqrt{1 - 0,252 \cdot 0,939 - 0,748 \cdot 0,939^2} = 0,232,$$

$$\varepsilon_{bx} = \frac{48,77}{25 \cdot 10^3 \cdot 0,232} = 8,41 \cdot 10^{-3}.$$

According to the experimental data, the deformations at $N_{\max} = 1330kN$ amounted to $\varepsilon_b = 8,11 \cdot 10^{-3}$ the difference between the experimental data and the analytical calculation was 3,7%.

Analysis of the calculation of the characteristics of the element with indirect reinforcement according to normative document: $\mu_{xy} = 0,031$,

$$\alpha_{red} = \frac{0,031 \cdot 410}{22 + 10} = 0,397,$$

$$\varphi = \frac{1}{0,23 + 0,397} = 1,595.$$

$$R_{b,red} = 22 + 1,595 \cdot 0,031 \cdot 410 = 42,27MPa.$$

$$\varepsilon_{b0,red} = 0,002 + 0,002 \cdot 0,326 = 0,00852, \ \varepsilon_{b2,red} = 0,0035 \cdot \frac{0,00852}{0,002} = 0,01491.$$

According to the results of the calculation for the normative document, a three-line diagram of concrete operation with indirect reinforcement was constructed (Figure 4).

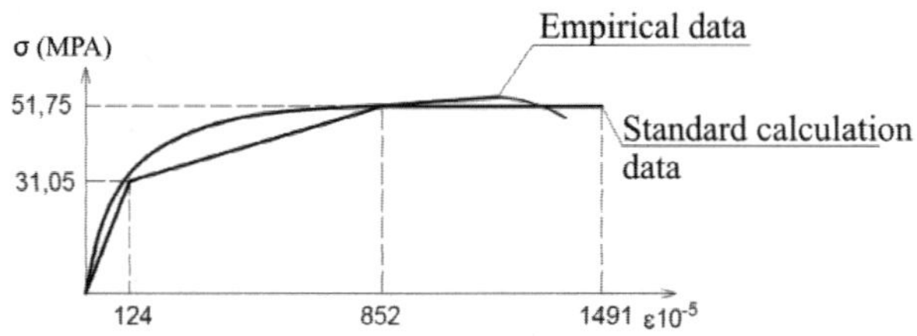

Figure 4. Trilinear diagram of concrete with indirect reinforcement.

3.2. Stress analysis

We find the stress value at the end of the ascending branch at $N_{\max} = 1470kN$.

Stresses in reinforced concrete $\delta_{b,\max} = 57,42 - 0,25 \cdot 12,71 = 54,24MPa.$

According to the calculations for normative document, the limiting stresses are $R_{b,red} = 51,75MPa.$

The results of numerical studies have shown sufficient convergence with the calculations according to the normative document.

4. Conclusions

1. The difference between the results of the calculation of deformations on the normative document and experimental data is valid.

2. Numerous test results of reinforced concrete elements with indirect reinforcement prove that indirect reinforcement meshes create a volumetric stress state in concrete, which increases the ultimate deformations and the carrier item ability. It is established that the destruction occurs due to the destruction of the protective layer of concrete, and the achievement of indirect reinforcement of the yield strength in the meshes.

3. When calculating the resistance to progressive collapse, it was proved that due to the increased deformability of the element with indirect reinforcement meshes, the element has the necessary residual deformation for complete redistribution effort. Due to the possibility of redistribution of forces, the elements in the zone of impact of the emergency load are not destroyed, which allows the rest of the circuit structures to remain in the design position.

4. The method of calculating columns with increased deformability is to use modern methods of calculation, which allow to take into account the non-linear operation of the structure, the correct modeling of the structure with indirect reinforcement; empirical comparison data with theoretical studies led to the conclusion that in the method of calculation it is necessary to use the dependencies of the characteristics of the element with indirect reinforcement, described in the normative document.

References:

[1] Tamrazyan A and Avetisyan L 2016 *MATEC Web of Conferences* **86** 01029.
[2] Tamrazyan A G and Avetisyan L A 2016 *Procedia Engineering* **153** pp 721-725
[3] Tamrazyan A and Avetisyan L 2014 *Applied Mechanics and Materials* **638** pp 62-65
[4] Tamrazyan A 2014 *Applied Mechanics and Materials* **475** pp 1563-1566
[5] Kabantsev O V and Tamrazian A G 2014 *Magazine of Civil Engineering* **49(5)** pp 15-26
[6] Rastorguev B S and Vanus D S 2014 *Construction and Renovation* **6(56)** pp 83-89
[7] Long Y and Cai J 2013 *Journal of Constructional Steel Research* **88** pp 1-14
[8] Attard M and Samani A K 2012 *Eng. Struct.* **41** pp 335-349
[9] Krishan A L, Troshkina E A, Rimshin V I, Rahmanov V A and Kurbatov V L 2016 *Research Journal of Pharmaceutical, Biological and Chemical Sciences* **7(3)** pp 2518-2529
[10] Binici B 2005 *Eng. Struct.* **27** pp 1040-1051
[11] Sfer D, Gettu R and Etse G 2002 *J. Eng. Mech.* **128** p156-163
[12] Vanus D S 2011 *The use of indirect mesh reinforcement to increase the rigidity and crack resistance of reinforced concrete elements. Dissertation of the candidate of technical sciences* PhD Thesis (Moscow: Moscow State University of Civil Engineering) p 186
[13] Perelmuter A V, Kriskunov E Z and Mosina N V 2009 *Magazine of Civil Engineering* **2** pp13-18
[14] Popov N N, Rastorguev B S and Zabegaev A V 1992 *Design calculation for dynamic special loads* (Moscow: Vysshaya Shkola) p 319
[15] Trekin H H, Popov H H and Matkov N G 1986 *Beton I Zhelezobeton* [*Concrete and Reinforced Concrete*] **11** pp 33-34
[16] Vasiliev A P, Matkov N G and Filippov B N 1973 *Beton I Zhelezobeton* [*Concrete and Reinforced Concrete*] **4** pp 101-111
[17] Gvozdev A A 1949 *Calculation of the bearing capacity of structures according to the method of limiting equilibrium* (Moscow: Gosstroyizdat) p 280
[18] Kolchunov V I and Savin S Yu 2018 Survivability criteria for reinforced concrete frame at loss of stability *Magazine of Civil Engineering* **80(4)** pp 73-80

Authors' background

Name	*Prefix*	*Research Field*	*Email*
Tamrazyan A.G.	DSc in Engg., Full Professor	Reinforced concrete structures, Reliability assessment, and Engineering safety of buildings	tamrazian@mail.ru
Fedorov V. S.	DSc in Engg., Full Professor	Reinforced concrete structures, Fire resistance	vs_skzs@mail.ru
Kharun M.	PhD in Engg., Associate Professor	Construction materials, Reinforced concrete structures	miharun@mail.ru

Nonlinear finite element analysis of stress-strain state of plate-rod reinforced concrete structures under the action of bending and torsion

A G Tamrazyan[1] and M Kharun[2*]

[1] National Research Moscow State University of Civil Engineering (MGSU), Moscow, Russia
[2] Peoples' Friendship University of Russia (RUDN University), Moscow, Russia

E-mail: miharun@yandex.ru

Abstract. In this study, loadings of plate-rod structures were considered, causing a complex stress state, which are exposed to significant external forces applied outside the plane, causing torsion of the structure as a whole. An analysis of the stress-strain state of a plate-rod reinforced concrete truss with regard to concrete creep is given. The core and volumetric models and the calculation of the plate-rod trusses in the ANSYS software package taking into account the creep of concrete are created, the results of computer calculations and the comparative characteristics of the results obtained are presented. The forces in the truss elements from the load located in its plane are determined by any method known from structural mechanics for the calculation of plate-rod systems. External torques that act on the structure are balanced in the amount of reactive support torques. The calculation of the strength of elements of reinforced concrete plate-rod trusses is carried out for sections that are normal to their longitudinal axis, and also for sections of the most dangerous direction that are inclined to it. Calculation of the plate-rod trusses under the action of bending and torque is associated with the determination of the twist angle of the truss sections or with the determination of its torsional stiffness.

1. Introduction

Compared to the traditional rod reinforced concrete structures the plate-rod reinforced concrete structures are able to perceive loads acting outside their plane and have their own stability. The use of plate-rod structures is advisable where loads are applied outside the plane of the structure. In such loading conditions, flat lattice structures are adopted as structural formations that consist entirely of plate elements that can work on bending from the plane of the structure, or in combination with linear rods that receive axial forces.

2. Methodology

2.1. Determination of internal forces and bending moments in elements of plate-rod trusses

The calculation of the strength of elements of reinforced concrete plate-rod trusses is carried out for sections that are normal to their longitudinal axis, and also for sections of the most dangerous direction that are inclined to it. Since the lower belt is experiencing only axial stretching, its strength, in the general case having pre-tensioned reinforcement in the cross section, with cross-sectional area Asp and unstressed with cross-sectional area A_s, is calculated by the condition [1]:

$$N \le \gamma_{sp} \cdot R_s \cdot A_{sp} + R_s A_s \tag{1}$$

2.2. Calculation of movements in the stage without cracks

Calculation of the plate-rod trusses, which are under the action of bending and torque, is associated with determining the angle of twist of sections of the truss φ or determining its torsional stiffness.

The angle of twisting can be determined from the condition of equality to zero of the sum of the work of external forces and the internal forces and bending moments of the structure caused by them [2]. By analogy with the formula for displacements of flat rod systems, the twist angle for plate-rod trusses can be determined from the expression:

$$\varphi = \sum_{i=1}^{i=n} \int_0^l \frac{M_{pi} \cdot \overline{M_i}}{E \cdot J_i} dx, \tag{2}$$

where M_{pi} – bending moment in the i-th member of the truss from external torque; M_i – the bending moment in the *i*-th element of the truss from a single torque applied in a section for which the twist angle is determined; l – the length of each element.

For the upper belt, the product $E{\cdot}J_i$ in the case of no cracks is constant, therefore the fraction of the twisting angle of the deformations of the upper belt [3]:

$$\varphi_m^b = \frac{l^b}{3EJ_i} \cdot \sum_{i=1}^{i=n} M_{pi\max}^b \cdot \overline{M}_{i\max}^b \tag{3}$$

For diagonals, the proportion of the twisting angle is given by the following formula:

$$\varphi_m^g = B \cdot \sum_{i=1}^{i=n} M_{pi\max}^g \cdot \overline{M}_{i\max'}^g \tag{4}$$

where
$$B = \frac{12 \cdot l^g}{E \cdot b \cdot \left(h^T - h^l\right)^3} \cdot \left[\ln\left|\frac{h^T}{h^l}\right| + \frac{h^l\left(4h^T - h^l\right)}{2\left(h^T\right)^2} - 1.5 \right] \tag{5}$$

The total twisting angle of the plate-rod truss from the action of external torques applied at the nodes of the upper belt is

$$\varphi_m = \frac{l^b}{3EJ^b} \cdot \sum_{i=1}^{i=n} M_{pi\max}^b \cdot \overline{M}_{i\max}^b + B \cdot \sum_{i=1}^{i=n} M_{pi\max}^g \cdot \overline{M}_{i\max}^g \tag{6}$$

Farm torsional stiffness in the crack-free stage:

$$C = \frac{T}{\varphi_m}, \tag{7}$$

where l^b – top bar length; J^b – the moment of inertia of the cross section of the upper belt relative to the vertical axis of symmetry; $M_{pi\max}^b$ and $M_{pi\max}^g$ – maximum bending moments respectively in the panels of the upper belt and bracing from the action of external torque; $\overline{M}_{i\max}^b$ and $\overline{M}_{i\max}^g$ – maximum bending moments, respectively, in the panels of the upper belt and diagonals of the action of a single torque applied in the section for which the twist angle is determined.

Simulation of the lamellar rod truss was performed in the certified ANSYS software package.

2.3. Calculation by the first group of limit states

The initial data for the calculation (Figure 1) are the data in [4]:

- Applied load $P = 20$ kN with eccentricity $e_0 = 0.4$ m;
- longitudinal reinforcement classA400;
- reinforcement transverse class B500;
- Concrete is heavy on small aggregate of class B40.

The design scheme of the truss and the diagram of the moment vectors are shown in figure 2.

The posts on the diagram are not indicated, since no effort arises in them at this load.

Axial forces are found by constructing a Cremona-Maxwell diagram.

2.4. Construction of the core and volume model of a plate-rod truss in the ANSYS software package

A truss model built in a software package corresponds to a manually calculated truss: $l = 3100$ mm, $h^{в}$ = 200 mm, $h^{н.п.}$ = 70 mm, b = 50 mm.

IOP Conf. Series: Materials Science and Engineering **675** (2019) 012005 doi:10.1088/1757-899X/675/1/012005

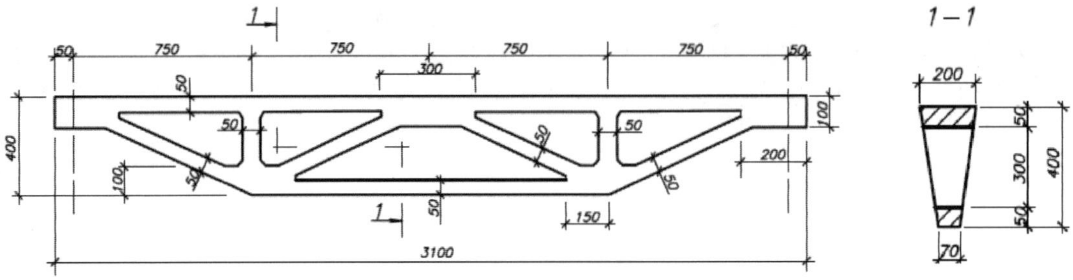

Figure 1. General view of the plate-rod truss.

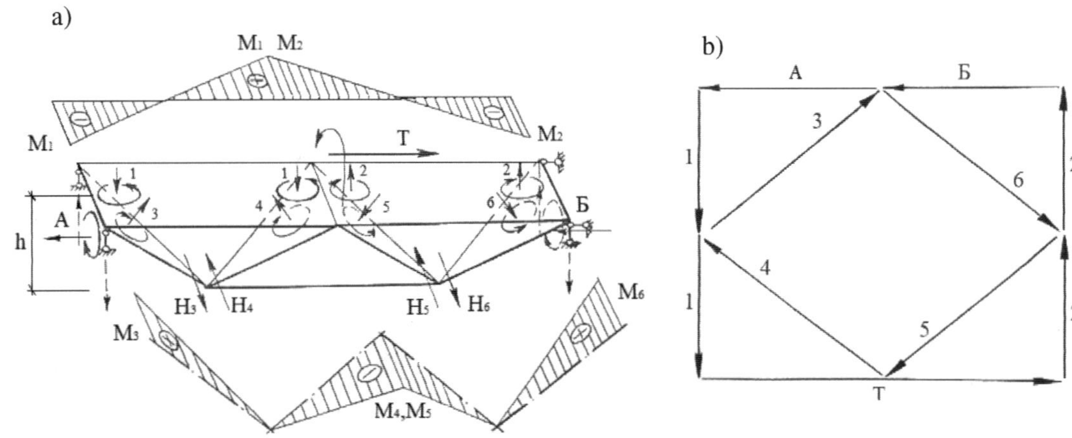

Figure 2. Design scheme plate-rod truss
a) diagrams of bending moments; b) moment vector diagram.

Variable cross sections in the core model can only be defined by splitting the cross section into several parts and specifying each part of a different cross section size, as shown in figure 3.

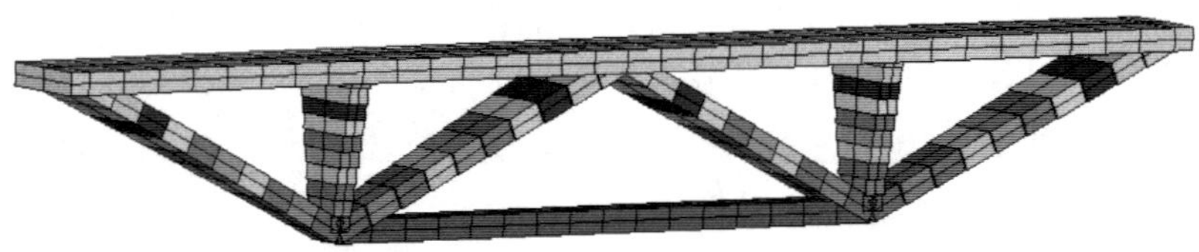

Figure 3. General view of the truss rod model.

Fastening the hinge truss, the load is applied to the middle node with eccentricity $e = 0.4$ m, $P = 20$ kN. The results of the linear calculation of the core model are shown in figure 4.

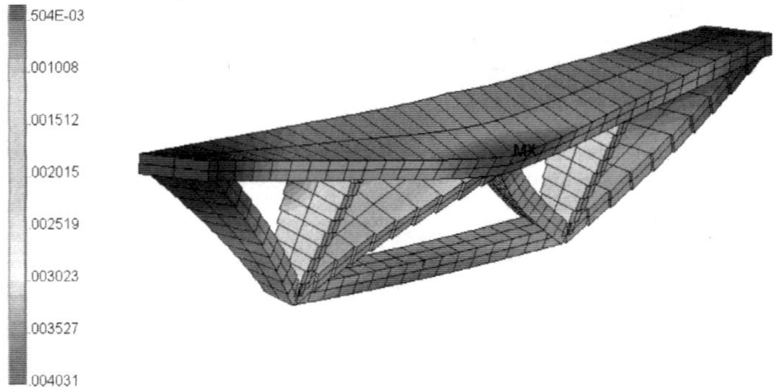

Figure 4. Graph of the deformed model with display of total displacements.

Considering the efforts arising in the elements of the frame are shown in figure 5-7.

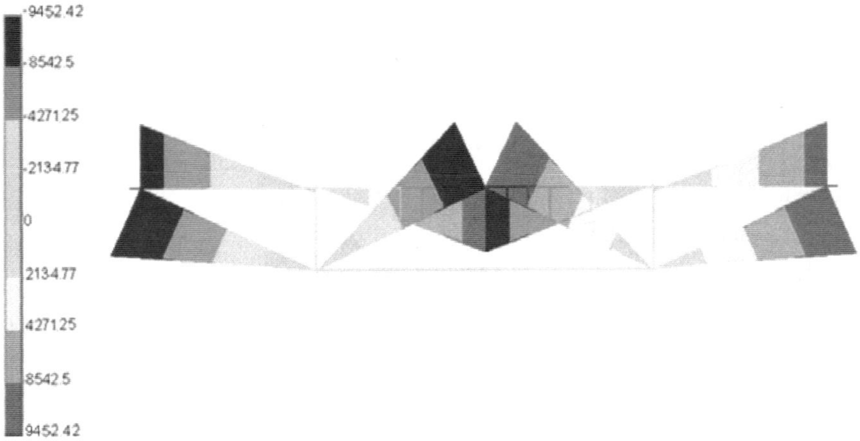

Figure 5. The diagram of bending moments relative to the axis Y (My).

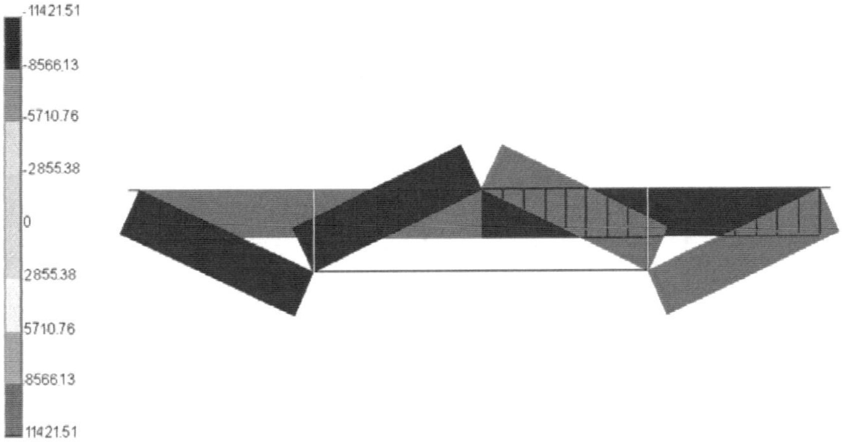

Figure 6. The diagram of transverse forces along the axis Z (Qz).

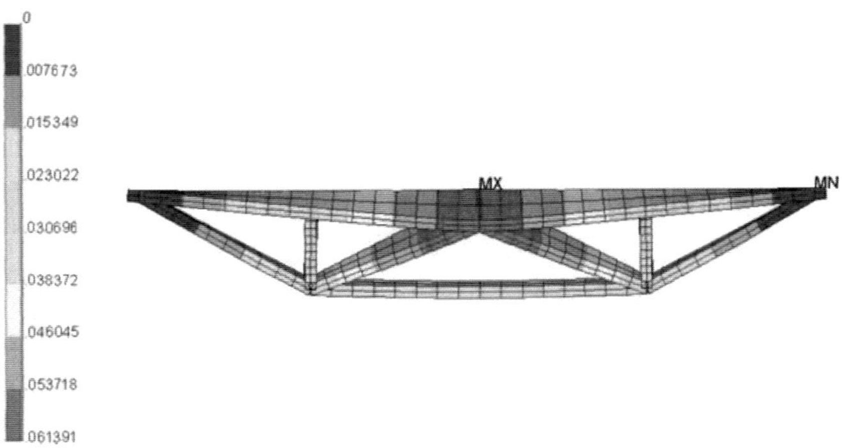

Figure 7. The angle of twist farm relative to the axis X (Ux).

3. Results

3.1. Results with plots and isofields for convenience

The results with plots and isofields for convenience are summarized in table 1.

Table 1. Summary table of the calculation results.

	Upper belt	Lower belt	Stretched brace	Compressed brace
Longitudinal force	23.8 kN	42.8 kN	25.3 kN	21.2 kN
Bending moment (max)	8.54 kNm	-	9.45 kNm	9.45 kNm
Lateral force	11.42 кN	-	11.42 kN	11.42 kN
Total displacement (max)	4 mm	-	-	-
Torsion angle (max)	0.06139 rad	-	-	-

3.2. Volume model

The *volume* model consists of bulk finite elements. First, the volume of the truss is created in the Ansys Space Claim subroutine, then using the triangulation network, the model is transformed into a set of three-dimensional elements [5]. All properties of materials are set the same as in the manual calculation. Fastening the hinge truss, as shown in figure 8.

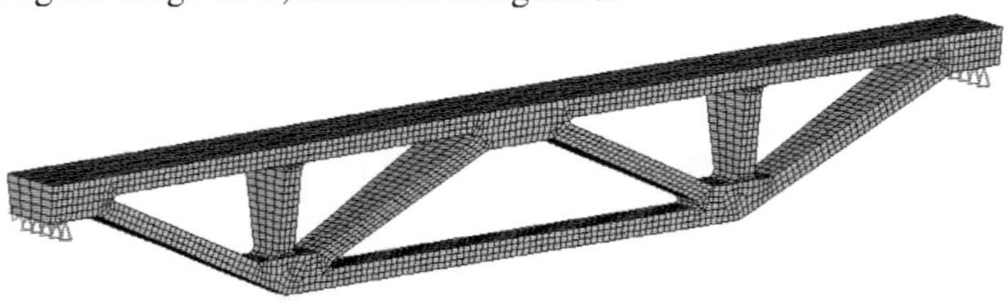

Figure 8. General view of the volume model of the truss.

The volumetric model does not show a plot of effort in braces and belts. But one can compare the total displacements and the general scheme of deformation of the volume model of a plate-rod truss with a rod model (Figure 9).

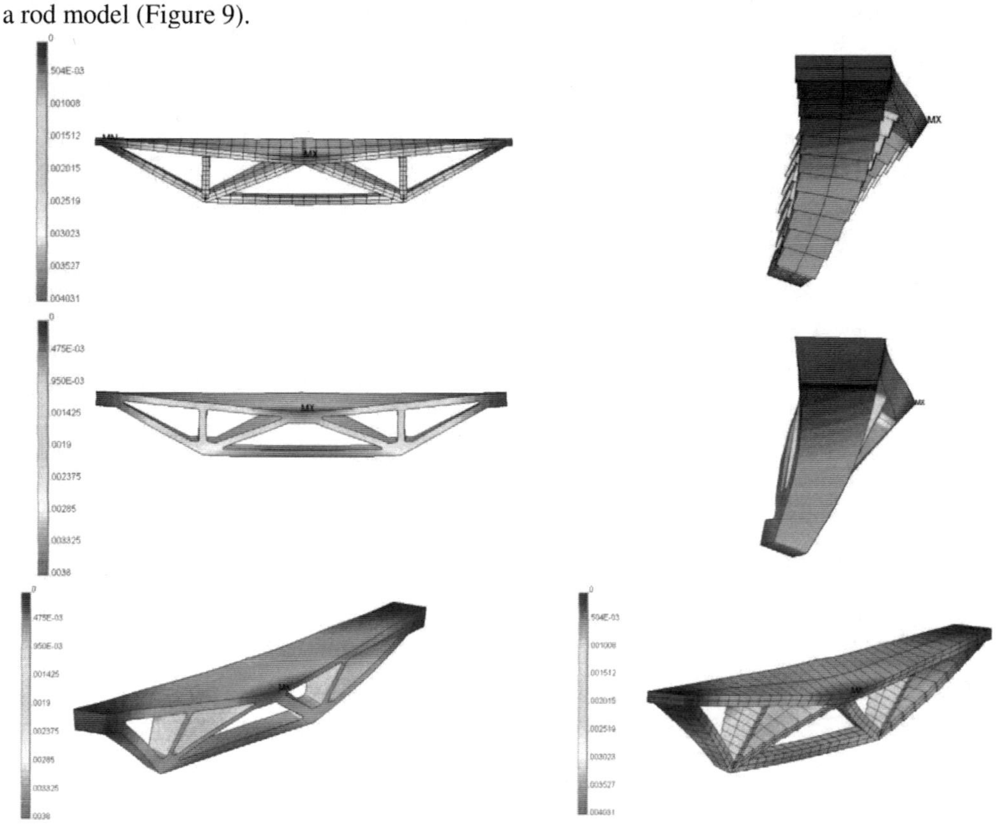

Figure 9. Comparative diagram of the core and volume models.

3.3. Accounting for the creep of concrete in the software package

The creep of concrete in the software package is taken into account according to the laws of nonlinear deformation. To take into account creep, it is necessary to calculate the coefficients φ_0 and β_H according to [6]:

Conditional creep factor:

$$\varphi_0 = \varphi_{RH} \cdot \beta_{(fcm)} \cdot \beta_{(t0)} \quad (8)$$

For $f_{cm} \leq 35$ MPa:

$$\varphi_{RH} = 1 + \frac{1 - \frac{RH}{100} T}{0.1 \cdot \sqrt[3]{h_0}} \quad (9)$$

$$h_0 = \frac{2A_c}{u} \quad (10)$$

$$\beta_{(fcm)} = \frac{16.8}{\sqrt{f_{cm}}} \quad (11)$$

$$\beta_{(t_0)} = \frac{1}{0.1 + (t_0)^{0.2}} \quad (12)$$

$$\beta_H = 1.5 \cdot [1 + (0.012 \cdot RH)^{18}] \cdot h_0 + 250 \leq 1500 \quad (13)$$

where A_c – total cross-sectional area of concrete; u – the perimeter of the element in contact environment; RH – relative air humidity of the environment, %; f_{cm} – average compressive strength of concrete at the age of 28 days, MPa; t_0 – concrete age at the time of application of the load, days; β_H – coefficient taking into account the relative humidity (RH, %) and the conditional size of the element ($h_{0,mm}$).

If execute coefficient calculation:

$$\varphi_{RH} = 1 + \frac{1 - \frac{73}{100}}{0.1 \cdot \sqrt[3]{40}} = 1 + \frac{0.27}{0.342} = 1.789$$

$$h_0 = \frac{2 \cdot 10{,}000}{500} = 40 \text{ mm}$$

$$\beta_{(fcm)} = \frac{16.8}{\sqrt{19.27}} - 3.827$$

$$\beta_{(t_0)} = \frac{1}{0.1 + (28)^{0.2}} = \frac{1}{2.047} = 0.4885$$

$$\varphi_0 = 1.789 \cdot 3.827 \cdot 0.4885 = 3.34$$

$$\beta_H = 1.5 \cdot [1 + (0.012 \cdot 73)^{18}] \cdot 40 + 250 = 1.5 \cdot 1.092 \cdot 40 + 250 = 315.54$$

3.4. Modeling and non-linear calculation of plate-rod trusses, taking into account the concrete creep of compressed elements in the ANSYS software package by the finite element method

For analyzing the physically and geometrically nonlinear behavior of concrete in ANSYS, special volumetric hexagonal (eight-node) elements of the SOLID 65 type with three degrees of freedom in the node were used.

A concrete material model, CONCRETE, was used to model the concrete, which supports the consideration of cracking and crushing (compression damage). Before the onset of cracking, the tensile behavior is assumed to be linearly elastic. To take into account the nonlinear behavior in compression before the onset of failure, a material diagram is additionally introduced (exponential law of deformation) [7, 8].

The volume-stressed state of the SOLID 65 elements with a non-linear calculation is estimated in three orthogonal directions. Upon reaching any major surface stress fracture for stretching, a crack appears. A crack may occur in one or several directions [9].

For reinforcement, a plasticity diagram is introduced (exponential law of deformation).

Additionally, creep is taken into account by specifying the dependence of additional creep deformations on time and on the existing stress-strain state (using the power law of creep) [10].

Consider the resulting farm model. In order to prevent local destruction of the truss when setting the load, it is necessary to provide steel holders on the trusses of the truss and in the center, simulating an experimental installation with a jack (Figure 10).

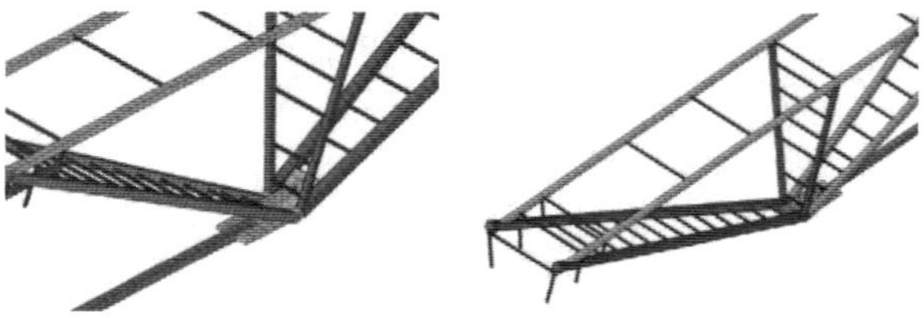

Figure 10. Fragment of a volume model for non-linear calculation.

Since nonlinear calculation is more complicated and time consuming than linear, in order to accelerate the calculation can be used the symmetry, thereby reducing the number of bulk finite elements by half [11-14].

Consider the results obtained after the calculation (Figure 11).

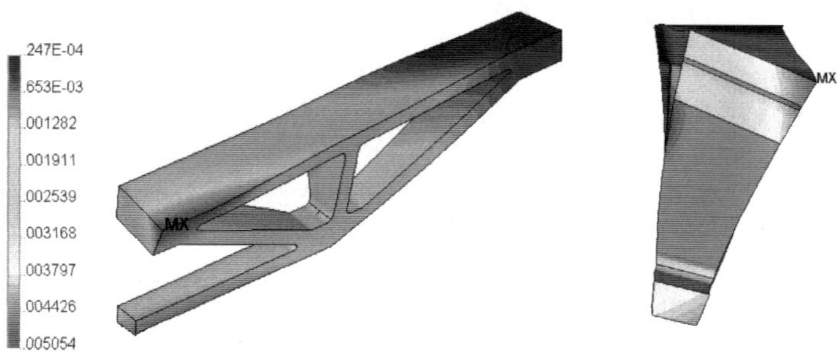

Figure 11. Graph of the deformed model with the display of the total displacements when taking into account creep.

By taking into account the creep of concrete in a non-linear calculation, we obtain maximum deformations of 0.5 cm. (what is a valid value $f_{\max} = \frac{l}{150} = \frac{300}{150} = 2.0$ cm) . In the linear calculation, a value of 0.38 cm was obtained. Consequently, when taking into account the creep of concrete when performing non-linear calculation, we obtain greater deformations than in the linear calculation.

Also, after conducting a non-linear calculation, it is possible to estimate the stresses in the reinforcement and the appearance of cracks in the truss elements (Figure 12).

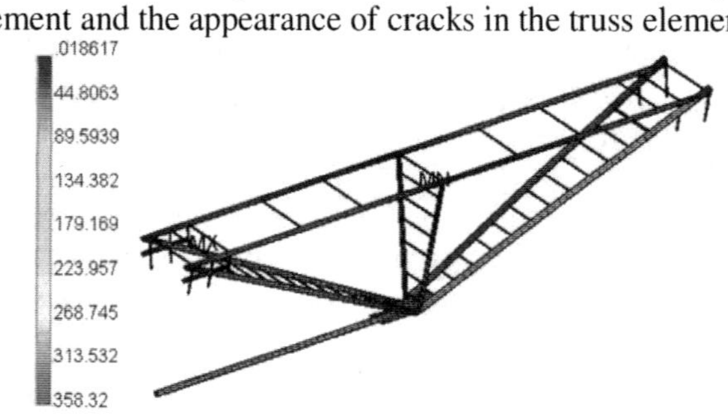

Figure 12. Equivalent stresses in reinforcement, MPa.

3.5. Comparison and analysis of the results
The obtained results for ease of comparison are summarized in table 2.

Table 2. Comparison of results by force and angles of twist

Stress	ANSYS	Manual account	Experiment
N^b	23.8 kN	21.4 kN	20.49 kN
N^g	21.2 kN	23.6 kN	22.79 kN
$N^{нп}$	42.8 kN	42.8 kN	
M^b_{max}	8.54 kNm	8.56 kNm	
M^g_{max}	9.45 kNm	9.44 kN	
Q	11.42 kN	11.42 kN	
φ	0.06139 rad	0.0614 rad	0.0617 rad

The results in the table clearly show good convergence. The error between the manual count, the experiment and the software complex is minimal. This suggests that the method of determining the angle of twist is correct and it is convenient to design such structures as lamellar-rod trusses in a software package.

4. Conclusions

1. Analyzing the data obtained after updating the calculation for the second group of limit states, the calculation of the Ansys software package and the experiment, we obtain good convergence of the obtained results with an accuracy of 0,5%.
2. The core and volume models of a plate-rod reinforced concrete truss after carrying out a linear calculation give a good convergence of the obtained results.
3. In a lamellar rod farm under the action of a torsional load, the total deformations increase with consideration of concrete creep.

References:

[1] *SP 63.13330.2012 Concrete and Reinforced Concrete Structures* (Moscow: Ministry of Construction of Russia)
[2] Baikov V N and Fomichev V I 1975 *University News: Constr. and Arch.* **2** pp 19-25
[3] Topilin A N 2016 *Industrial and Civil Engg.* **7** pp 61-64
[4] Topilin A фтв Sannikova N 2018 *Bulletin of Building Technology* 2 pp 58-59
[5] Basov K A 2005 *ANSYS. User Reference* (Moscow: DMK Press) p 640
[6] *EN 1992-1-1 Eurocode 2: Design of concrete structures Part 1-1: General rules and rules for buildings*
[7] Tamrazyan A G and Avetisyan L A 2016 *Procedia Engineering* **153** pp 721-725
[8] Tamrazyan A and Avetisyan L 2016 *MATEC Web of Conferences* **86**
[9] Parvin A, Khandel O, Crozier D and Raad J 2017 *Proc. 4th Int. Conf. on Smart Monitoring, Assessment and Rehabilitation of Civil Structures* (Zurich, Switzerland)
[10] Tamrazyan A G and Yesayan S G 2012 *Concrete creep mechanics* (Moscow: Moscow State University of Civil Engineering Press) p 524
[11] Kabantsev O V and Tamrazian A G 2014 *Magazine of Civil Engineering* **49(5)** p 15-26
[12] Tyukalov Yu Ya 2018 *Magazine of Civil Engineering* **79(3)** pp 54-65
[13] Lukashevich A A 2018 *Magazine of Civil Engineering* **81(5)** pp 149-159
[14] Benin A V, Semenov A S, Semenov S G, Beliaev M O and Modestov V S 2017 *Magazine of Civil Engineering* **76(8)** pp 279-297

IOP Conf. Series: Materials Science and Engineering **675** (2019) 012005 doi:10.1088/1757-899X/675/1/012005

Authors' background

Name	*Prefix*	*Research Field*	*Email*
Tamrazyan A.G.	DSc in Engg., Full Professor	Reinforced concrete structures, Reliability assessment, and Engineering safety of buildings	tamrazian@mail.ru
Kharun M.	PhD in Engg., Associate Professor	Construction materials, Reinforced concrete structures	miharun@mail.ru

Hyper plasticizers and basalt fibers as the resistance of concrete that can be affected by vegetable oils

T S F Gamal and P C Chiadighikaobi*

Peoples' Friendship University of Russia (RUDN University), Moscow, Russia

Email: passydking2@mail.ru

Abstract. With the development of the agro-industrial complex and the construction of enterprises for processing oleaginous crops, the issue of protecting construction structures from the effects of oil arises. The negative effect of the oils on concrete and reinforced concrete leads to the reduction of reliability and durability of constructions in general, and to different accidents or unscheduled repairs at processing enterprises. In the experimental study of the effective ways to increase the durability of concrete and concrete structures due to the use of modern plasticizing additives, basalt fiber and finely dispersed fillers has been determined. The obtained results of the research can be used in the design of reinforced concrete structures operated with the growth of vegetable or mineral oils, as well as for practical purposes when carrying out a technical survey at industrial plants where vegetable or mineral oil is produced. Basalt fiber reinforced concrete (BFRC) is suggested to be a solution to the effect of oil on concrete. The properties and characteristics of basalt fibers give the fiber advantages when compared to other ways to curing or solving the effects of oil on concrete.

Keywords: basalt fiber, basalt reinforced concrete, vegetable oil, permeability, oil resistance

1. Introduction

1.1. General review

Prevention of accidents and emergencies are one of the most important scientific and technical purposes of design, construction and operation of industrial buildings. At industrial engineering enterprises, metal- working, power, machine tools oil-containing liquids are widely used: mineral oils, oil emulsions, kerosene oil, fuel oil which get on foundations under equipment, on load-bearing structures, and gradually impregnate them. All these kinds of oil also impregnate reinforced concrete tanks, railway sleepers, span structures of bridges, and many other structures.

Oil has negative impact on the strength and reformative properties of impregnated concrete and reinforced concrete elements of building constructions. This decreases the compressive strength of concrete, adhesion to reinforcement deteriorates; all this causes the occurrence of destructive processes in load-bearing concrete and reinforced concrete structures of industrial buildings and destructions of varying degrees of severity [1].

The use of some admixtures and reinforcement materials help in fighting the degradation caused by oil on concrete. Considering the use of hyper-plasticizers and basalt fiber in concrete which have excellent properties as quality materials and admixtures for not only fighting oil effect on concrete but increase the strength of concrete [2, 3].

Basalt fiber reinforced concrete has the capacities to withstand so many environmental factors including the ability to resist corrosion, chemical destructions, and absorption of oil into concrete that will prevent the impregnation of concrete [3].

The purpose of this research is to determine the effectiveness and ways of increasing the durability of concrete by use of hyper-plasticizers on a polycarboxylate basis. Experimenting by checking different variants of hyper-plasticizers and suggesting ways the use of basalt fiber in concrete and the use of basalt mesh as confinement on the walls and floors can increase the strength of oil affected concrete.

1.2. Analysis of the stated issue

The durability of concrete and reinforced concrete products mean the ability to maintain operational characteristics under the influence of aggressive factors. In this case the most important physical and mechanical indicators are water resistance and cold resistance. The pores size plays an important role and more attention given to the size and quantity of capillary pores [4].

To reduce the permeability of concrete, special binding materials to increase concrete density are used in constructions, and special coatings based on hydrophobization composites are applied [5].

One of the widely used ways to increase concrete resistance is the use of special modifying agents or the additional screed on the surface with basalt reinforced concrete or confinement with basalt grid/mesh. The chopped basalt fiber (mineral fiber) is good fiber due to its excellent properties as potential reinforcement of composite materials [6]. This way is more effective than previous ones. The use of complex additives helps to increase the mobility while maintaining or reducing the water-binding ratio of which leads to the improvement of concrete stone density.

According to author [7], according to Concrete and Reinforced Concrete Research Institute of Moscow "out of the total amount of reinforced constructions about 25% is exploited in aggressive environment. While 25% out was exploited in weakly aggressive environment, in which their secondary corrosion protection is not provided, as a rule. That is why constructions should be designed and manufactured to provide their durability by means of self-endurance".

The use of concrete mixture with fine dispersed additives and hyper-plasticizers helps to increase strength characteristics of concrete with a decrease in its porosity. The water-binding ratio must not be less than 0.4. The positive effect of plasticizing additives is in a high water-reducing and dispersing effect – the reduction of the necessary amount of mixing water and more even distribution of individual particles of binder and micro filler. Dispersion helps to exclude clumping of particles and evenly distributes the centers of crystallization throughout the volume of the cement stone. However, we should take into account that when the maximum possible concentration of micro fillers is reached, the strength characteristics sharply decrease due to the deterioration of the bond between the cement stone and the aggregate [8].

A study on the effect of Kerosene, gas oil and crude oil on the compressive and tensile strength for ordinary strength concrete and high performance concrete at 30, 60, 90 and 120 days exposure after 28 days normal water curing was done in [9]. The test results showed that the loss in compressive strength and splitting tensile strength resulting from exposure to oil was relatively smaller for high performance concrete compared to normal strength concrete and the difference about 10%. They also found that the reduction in the compressive strength increases with the decrease in viscosity of oil. The decrease in compressive strength was 25, 19 and 15% for normal concrete and 12, 8 and 6% for high performance concrete exposed to kerosene, gas oil and crude respectively after 120 days of exposure.

An investigation conducted by authors on the experiment about the bonding properties of basalt fiber with cementitious material as well as the effect of fiber orientation on the tensile strength of basalt fiber for evaluating basalt fiber's suitability as a reinforcing fiber. In the experiments, the authors performed single fiber pullout test and tensile strength of fiber measured according to fiber orientation. Their test results showed that basalt fiber has a strong chemical bond with the cementitious matrix, 1.88 times higher than that of polyvinyl alcohol fibers with it. Theoretical fiber-bridging curves showed that the basalt fiber reinforcing system has a higher cracking strength than the polyvinyl alcohol fiber reinforcing system, but the reinforcing system showed softening behavior after cracking [10].

In the research paper [11, 12], the authors presented the results of studies of the effect of petroleum products in the impregnation of concrete and its deformation properties. The results they obtained from their experiment allow assessing changes in deformation characteristics of load-bearing concrete and reinforced concrete of industrial buildings.

Petroleum products, impregnating in concrete and reinforced concrete structures, have a negative impact on their strength and deformation characteristics. The negative impact of petroleum products on concrete and reinforced concrete is associated with changes in the hydration process of cement, as well as changes in the structure of the concrete. Strength and deformation characteristics of concrete change due to hydraulic pressure of petroleum products in the pores exerted on the skeleton of cement stone. In this aspect, the crucial point is the porosity of concrete as a permeability factor for petroleum products [13].

One of the most important factors affecting the physical and mechanical characteristics of oil-impregnated concrete is their viscosity [14].

Basalt fiber used in fiber reinforced polymers FRPs and structural composites has high potential and is expected to grow at a significant rate due to its high temperature resistance, abrasion resistance, ultraviolet resistance and high-energy electromagnetic radiation. Basalt fibers are ideally suited for demanding applications requiring high temperatures, chemical resistance, durability, mechanical strength and low water absorption [15-17]. Such structures combine the best properties of each component to possess enhanced mechanical & superconducting properties for advanced applications.

From some of the above stated characteristics and properties of basalt fibers, it is suggestible that in industries or areas where vegetable oil is produced or highly used, basalt fiber concrete is needed. This will contribute to the solving the damages or cure the negative effect of vegetable oil on concrete.

1.3 Problem statement

This paper has a task on the efficiency of increasing the density of concrete to improve the resistance to the effect of vegetable oils on concrete using hyper-plasticizers on a polycarboxylate basis and the use of basalt fiber.

2. Method of experimental research

The research is based on, analysis, generalization and evaluation of experimental data on the effect of vegetable oil on the physical and mechanical properties of concrete.

The optimum concentration of mineral additives was experimental selected in order to maintain the basalt fiber cement stone strength characteristics at the first stage of this study.

After optimization of the mineral additive concentration, a choice of a hyper-plasticizer took place, which significantly reduces the water requirement of the concrete mix. The effect of the plasticizing additive was evaluated by the diameter of the spreading of the standard cone (Hagermann cone), while there should be no water separation of the mixture. The optimum concentration of plasticizer is 0.3% of the cement mass, with W/C = 0.3.

Absorption of oil samples is determined by the procedure of GOST 12730.5-84 [18], which replaces the water with vegetable (sunflower) oil. The saturation of the samples with oil occurred when the samples were completely immersed oil (the oil level was about 50 mm higher than the samples) at room temperature.

Saturation was assessed by weighing the samples (every 24 hours) until the difference between two consecutive weighs was equal to 0.1%. At the same time, the mass of oil that leaked from the sample was included in the sample mass. The parameters of the basalt fiber used: length 20mm and diameter 1.5μm

2.1 Manufacture and storage of samples

The concrete mixer used in the preparation of fine-grained basalt fiber concrete is BL-10. In the 3PK70.7 forms was placed resulting mixture. Concrete mix was compacted using vibrating table VM-6.4. The samples hardening and the set of strength of the samples were carried out in a chamber of normal hardening. At the end of the period of normal hardening, control samples placed in containers filled with vegetable (sunflower) oil. In accordance with the research plan, after a certain amount of time, the density control checks were done. For each composition, six control samples-cubes with a rib size of 7 cm were made. The values of oil saturation were determined from the series of three samples.

2.2 Study of the influence of vegetable oil on physico-mechanical properties of impregnated concrete
During the work, the compositions of fine-grained basalt fiber cement concrete with the following plasticizing additives were examined (Table 1):

Table 1. Compositions of fine-grained basalt fiber cement concrete with plasticizing agents.

№	Name of plasticizer	Manufacturer country	Appearance	Basis of plasticizer
1	C-3	Russia	Powder of light brown color	Polynaphthalene-sodium methylene sulphonate
2	Sika ViskoCrete®-20 HE	Switzerland	Yellowish liquid, density 1.085kg / l	Aqueous solution of modified polycarboxylate
3	Sika® ViskoCrete®-5 Neu	Switzerland	Unclear colorless liquid, density 1.084 kg/l	Aqueous solution of modified polycarboxylate

As a binder, Portland cement M400 PC D20 Volsky was used while, as a finely dispersed additive; limestone flour grade A was used in accordance with GOST 14050-93 in the amount of 10% of the mass of the cement, 0.45% basalt fiber (BF) was added in the mixture. The content of the additive in the control compositions is given in table 2.

Table 2. The content of the additive in the control compositions.

№	Content of additive,% of cement+ BF mass			Average density,kg / m^3	Cement-water ration
	C-3	Sika 20HE	Sika 5Neu		
1	-	-	-	2180	0,45
2		0,8		2340	0,32
3		1,0		2350	0,31
4			0,8	2370	0,31
5			1,0	2370	0,31
6	0,8			2260	0,37

3. Results

Figure 1 shows the tests results for the saturation of samples of fine-grained basalt fiber concrete. Analysis of the data obtained shows that the application of plasticizing additives has a significant effect on the compaction of concrete (more than 8% in comparison with the control composition).

Dynamics of the increase in the density of the samples under study is regular with a decrease in the water-cement ratio, and consequently, the magnitude of the water-reducing effect of plasticizing additives.

Analysis of the curves shows that when using compacting additives in a concrete mixture, its oil conductivity reduces significantly. The introduction of an additive of plasticizers based on polycarboxylates in the amount of 1% of the weight of cement leads to a decrease in oil conductivity by an average of more than 3 times, and when a superplasticizer C-3 is introduced at a dosage of 0.8%, the oil conductivity decreases on average by 30% compared to concrete without additives.

Figure 2 shows the results of electron microscopy on TESCAN VEGA 3 equipment saturated with vegetable oil control samples of fine-grained basalt fiber concrete. The structure of oil-saturated concrete of increased density (with the use of a hyper plasticizing additive) is distinguished by smaller new growths, destruction along the interface of filler-cement stone phases and a smaller amount of impregnated vegetable oil.

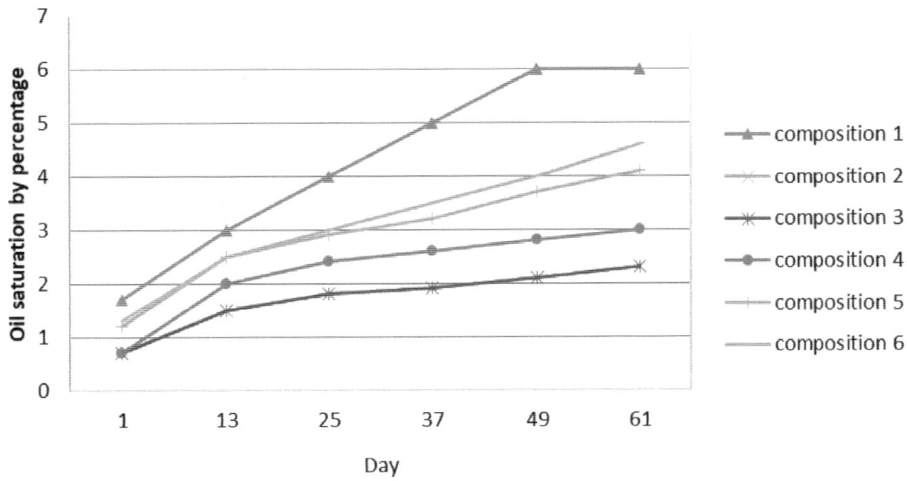

Figure 1. Saturation of vegetable oil samples of fine-grained basalt fiber concrete, depending on the composition.

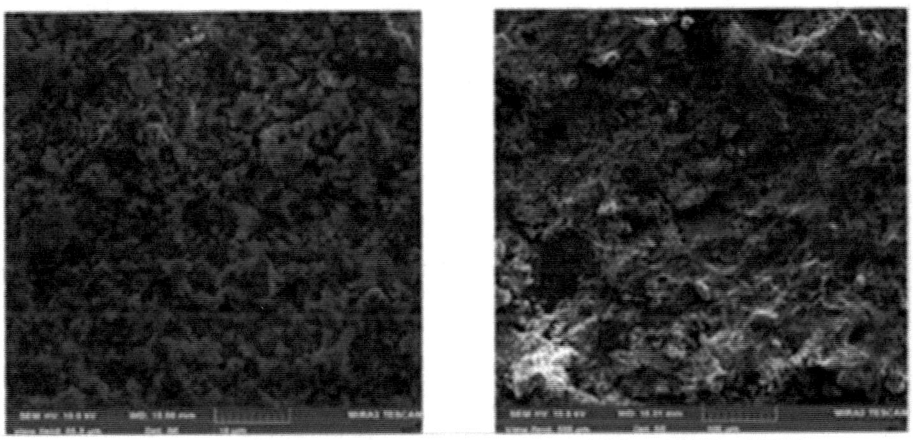

a. without additives b. with the addition of Sika 20NE

Figure 2. Microphotographs of the structure of concrete saturated with vegetable oil.

An analysis of the results of an experimental study shows that an increase in the density of concrete due to the introduction of modern plasticizing additives into its composition is an effective method of preventing oil permeability.

4. Conclusion

It is advisable to protect concrete from the negative effects of vegetable oils during the design phase of the composition. Hyper plasticizing, basalt fiber and finely dispersed additives make it possible to increase the resistance of concrete to oil permeability by more than 3 times.

In the designing of reinforced concrete structures operated with the influence of vegetable or mineral oils, the results obtained from the research are used, as well as for practical purposes when carrying out a technical inspection at industrial facilities where the production of vegetable or mineral oil are done.

The use of basalt fiber in concrete will be of great advantage to concrete industrial users. With the characteristics of basalt fiber, it is of sure and certainty that basalt fiber is a solution to the negative effect of oil on concrete.

References

[1] Svintsov A P and Shambina S L 2018 Influence of viscosity of vegetable and mineral oil on deformation properties of concrete and cement-sand mortar *Construction and Building Materials* **190** pp 964-974

[2] *Polycarboxilate based Hyperplasticizer Powder* https://www.dostkimya.com/en/products/superplasticizers//polycarboxilate-based-hyperplasticizer-powder

[3] Zongwen Li, Jianxun Ma, Hongmin Ma and Xin Xu 2018 Properties and Applications of Basalt Fiber and Its Composites *IOP Conf. Series Erth and Environ. Sci.* **186(2)**:012052

[4] Izotov V S 2008 *Organization of a quality control system and improving the durability of concrete and reinforced concrete structures* (Kazan: Publishing House Model) p 300

[5] Svintsov A P and Shambina S L 2018 The dataset about the deformations under axial compression of concrete and cement-sand mortar impregnated with oil *Data in Brief* **21** pp 1363-1369

[6] Rataram P, Sandeshkumar J and Subhas C S 2015 Development of basalt fiber-reinforced thermoplastic composites and effect of PE-g-MA on composites *Plym. Comp.* **38(11)** 2798-2809

[7] Alekseev S N and Rosenthal N K 1976 *Corrosion resistance of reinforced concrete structures in an aggressive industrial environment* (Moscow: Stroyizdat*)* p 205

[8] Izotov V S and Sokolova Y A 2006 *Chemical additives for concrete modification* (Moscow: Paleotyp) pp 64-65

[9] Jasim A T and Jawad F A 2010 Effect of oil on strength of normal and high performance concrete *Al-Qadisiyah J. for Eng. Sci.* **3(1)** pp 24-32

[10] Jeong-Il C and Bang Yu L 2015 Bonding properties of basalt fiber and strength reduction according to fiber orientation *Materials* **8** pp 6719-6727

[11] Svintsov A P, Nikolenko Yu V, Kharun M I and Kazakov A S 2014 Effect of viscosity of petroleum products on deformation properties of concrete *Magazine of Civil Engineering* **51(7)** pp 16-22

[12] Svintsov A P, Nikolenko Y V and Kazakov A S 2014 Endurance Assessment of the Concrete and Reinforced Concrete Load-Bearing Structures, Impregnated by Petroleum *RUDN Journal of Engineering Researches* **4** pp 35-40

[13] Svintsov A P 2019 Effect of petroleum products on physical and mechanical properties of concrete and the reliability of load-bearing structures *Arab. J. for Sci. & Eng.* **44(5)** pp 4277-4287

[14] Diab H M 2012 Compressive strength performance of low- and high-strength concrete soaked in mineral oil *Construction and Building Materials* **33** pp 25-31

[15] Kameche Z A, Ghomari F, Choinska M and Khelidj A 2014 Assessment of liquid water and gas permeabilities of partially saturated ordinary concrete *Construction and Building Materials* **65** pp 551-565

[16] Sim J, Heolwoo P C and Moon D Y 2005 Characteristics of basalt fiber as a strengthening material for concrete structures *Comp. B.* **36** pp 504-512

[17] Subramanian R V and Austin H F 1980 Silane coupling agents in basalt reinforced polyester composites *Int. J. of Adh. & Adhsiv* **1(1)** pp 50-54

[18] GOST 12730.5-84 2007 *Concretes: Methods for determination of watertightness* (Moscow: Standartinform, Moscow) URL: http://docs.cntd.ru/document/901707639

Authors' background

Name	*Prefix*	*Research Field*	*Email*
T S F Gamal	PhD Candidate	Building materials, Structural design	dktareqjamal@gmail.com

IOP Conf. Series: Materials Science and Engineering **675** (2019) 012006 doi:10.1088/1757-899X/675/1/012006

P C Chiadighikaobi	PhD Candidate	Building materials, Structural design	passydking2@mail.ru

IOP Conf. Series: Materials Science and Engineering **675** (2019) 012007 doi:10.1088/1757-899X/675/1/012007

Translucent cover design of solar energy equipment for manufacturing of prefabricated concrete components

D D Koroteev*, M Kharun, M Istrakova and M E Romero Barrenechea

Peoples' Friendship University of Russia (RUDN University), Moscow, Russia

Email: koroteev_dd@pfur.ru

Abstract. Translucent cover is one the main design features of solar energy equipment like solar collector. Its material and layers number influence on overall translucency of the cover and, as a result, on efficiency of the equipment. Formwork, equipped by translucent cover, can be considered as solar energy equipment like solar collector due to similar thermal-physical processes, taking place in it. Such equipment can be used during manufacturing of prefabricated concrete components to speed up the curing process by heating the structures using solar energy. The aim of our research work is determination of optimal layers' number of translucent cover. The research was made in various climatic conditions. The obtained results testify to certain influence of translucent cover design on efficiency of solar energy equipment. However, this influence depends on climatic conditions.

Keywords: Translucent cover, solar energy equipment, precast concrete structures, heat treatment, renewable energy resources.

1. Introduction

The research on development of various types of solar energy equipment is made in different countries [1-3]. At present, the most common type is a solar collector. Solar collectors are widely used in different fields, such as household equipment systems, hot water production [4-7].

Concrete, as the most common construction material in the world, and prefabricated concrete components have considerable energy requirements for their manufacturing, including energy expenses for heat treatment of concrete to speed up the curing process of its curing. Fossil fuels, such as natural gas, oil, coal are required for this purpose [8]. Possibility to obtain temperature of heater about 60-80 °C in the solar energy equipment allows using solar energy for heat treatment of concrete during manufacturing of prefabricated concrete components [9].

Formwork, which is used during manufacturing of prefabricated concrete components, can be equipped by translucent cover and considered as solar energy equipment. Such equipment is comparable with solar collector for thermo-physical processes, taking place in it.

Such method of heat treatment of the concrete components can be made on condition of placement of the formwork with translucent cover on open air. Concrete is direct heated by solar energy in the day time and stores heat in the night time.

Our research results show that this technology can be used for manufacturing of prefabricated concrete elements using solar energy within 5 months for Moscow region (φ=56° N) and 7 months for South of Russia (φ=45° N). The manufacturing period does not exceed 24 hours during this time [10].

However, the efficiency of heat treatment of the concrete components in the solar energy equipment depends on design features of such equipment. Design of translucent cover as one the main design features influences on amount of solar radiation, which can be absorbed by heater during the daytime, and heat loss to environment [11-13]. Taking into account the above, the aim of this research work is the determination of optimal layers' number of translucent cover.

IOP Conf. Series: Materials Science and Engineering **675** (2019) 012007 doi:10.1088/1757-899X/675/1/012007

2. Materials and methods of research

The increase of the layers' number reduces the amount of heat, consumed by concrete, due to reduction of coefficient of the solar radiation transmission and the growth of the shaded area of the concrete component (Equation 1).

$$C_{tr} = C_{shade} + C_{dust} \cdot C_{cover} - 1 \tag{1}$$

where C_{tr} – coefficient of the solar radiation transmission; C_{shade} – coefficient of shading; C_{dust} – coefficient of dusting; C_{cover} – integral coefficient of the translucent cover transmission (it depends on incidence angle of direct solar radiation, number of layer and material of the cover).

At the same time, the increase of the layers' number reduces heat loss from concrete in the environment through the tranlucent cover (Equation 2).

$$U_{loss} = \left(\frac{1}{h^{e}_{con} + h^{e}_{rad}} + \frac{1}{h^{t_1-t_2}_{con} + h^{t_1-t_2}_{rad}} + \ldots + \frac{1}{h^{t_{n-1}-t_n}_{con} + h^{t_{n-1}-t_n}_{rad}} + \frac{1}{h^{c}_{con} + h^{c}_{rad}} \right)^{-1} \tag{2}$$

where U_{loss} – coefficient of heat loss through the translucent cover, W/(°C·m²); h^{e}_{con} – convective coefficient of heat transfer between the translucent cover and the environment, W/(°C·m²); h^{e}_{rad} – radiation coefficient of heat transfer between the translucent cover and the environment, W/(°C·m²); $h^{t_1-t_2}_{con}$, $h^{t_{n-1}-t_n}_{con}$ – convective coefficients of heat transfer between layers of the translucent cover, W/(°C·m²); $h^{t_1-t_2}_{rad}$, $h^{t_{n-1}-t_n}_{rad}$ – radiation coefficients of heat transfer between layers of the translucent cover, W/(°C·m²); h^{c}_{con} – convective coefficient of heat transfer between the concrete component and the translucent cover, W/(°C·m²); h^{c}_{rad} – radiation coefficient of heat transfer between the concrete component and the translucent cover, W/(°C·m²).

The calculation and experimental research were carried out to determine temperature and strength of concrete, curing in formwork, equipped by translucent cover, in different climate conditions under various design of the cover.

The experimental research in laboratory conditions was made to verify the above calculation model. The temperature conditions of humid subtropical climate were simulated in climatic camera. Walls, floor and roof of the camera were insulated. The electric reflector lamps were used for imitation of solar radiation intensity. They were placed on a special panel on the roof of the camera. The passage of the airflow above the lamps allowed keeping the necessary air temperature. The air temperature in the camera was controlled by digital potentiometer with help of thermocouple. The air humidity was controlled by digital hygrometer.

The intensity of solar radiation changed according to parabolic law with the maximum value of 1300 W/m² in the camera. The air temperature changed according to sine law in the range of 24-50 °C, the air humidity – 6-39% in the camera.

The research of the concrete strength was made for standard methodic with the employment of series laboratory equipment. Concrete B25 was used in the research. The reference standard to compare the obtained concrete strength was concrete, curing within 28 days in normal conditions.

We made the research of materials of translucent cover. The main materials were the following: glass, polymer foil, polycarbonate, acrylic glass. Polymer foils had a number of advantages as compared with other materials, the main are: the price and usability.

Polyethylene-terephthalate foil with integral coefficient of the transmission of 0.88-0.85 and polyethylene foil with integral coefficient of the transmission of 0.92-0.94 were chosen as the most suitable types after the comparison of the polymer foils characteristics.

Single layer and double layer design of translucent cover was chosen in the research because the increase layers' number more than two layers cause considerable reduction of amount of solar radiation, which can be consumed by heater [10].

IOP Conf. Series: Materials Science and Engineering **675** (2019) 012007 doi:10.1088/1757-899X/675/1/012007

3. Results and discussion

The calculation of concrete temperature and strength was carried out for various climatic conditions of Russian Federation in summertime (July): humid subtropical climate (Krasnodar region) and humid continental climate (Moscow region). The results are shown in table 1 and table 2.

Table 1. Concrete curing in formwork, equipped by single and double layer translucent cover, in humid subtropical climate conditions: Krasnodar region, Russia.

Single layer translucent cover					Double layer translucent cover				
Time of day	Q_{abs} [kJ]	U_{loss} [W/(°C·m²)]	Q_{loss} [kJ]	$t_{concrete}$ [°C]	Time of day	Q_{abs} [kJ]	U_{loss} [W/(°C·m²)]	Q_{loss} [kJ]	$t_{concrete}$ [°C]
10:00	93.7	1.15	8.7	21	10:00	68.7	0.39	7.5	21
11:00	109	1.18	9.9	23.3	11:00	93.3	0.4	7.8	23
12:00	118	1.2	12.3	25.9	12:00	112	0.4	9.2	25.4
13:00	121	1.22	16	28.8	13:00	127	0.45	11.7	28.2
14:00	126	11.04	37.8	31.7	14:00	100	2.96	68.4	31.2
15:00	123	9.27	107.7	34.5	15:00	79.5	3.26	63	33.2
16:00	107	8.97	113.3	36.5	16:00	54.2	3.42	56.3	35.1
17:00	82	8.85	116.7	38.8	17:00	20.8	3.57	46.7	37.4
18:00	49.5	8.82	116.6	40.7	18:00	-	3.69	44.1	39.5
19:00	20.7	8.81	117.5	42.4	19:00	-	3.79	53.6	41.4
20:00	3.4	8.82	123.9	43.6	20:00	-	3.91	66.3	43.6
21:00	-	8.77	133.2	44.4	21:00	-	3.96	71	45.1
22:00	-	8.83	142.7	45.2	22:00	-	4.01	77	46.2
23:00	-	8.83	145.7	44.5	23:00	-	4.03	81	46.4
24:00	-	8.83	146.9	43.7	24:00	-	4.05	84.2	46.6
1:00	-	8.82	146.2	42.9	1:00	-	4.07	86.2	46.7
2:00	-	8.81	143.5	42.2	2:00	-	4.07	87.1	46.8
3:00	-	8.81	138.9	41.4	3:00	-	4.07	86.8	46.9
4:00	-	8.8	132.5	40.7	4:00	-	4.06	85.4	47
5:00	-	8.8	124.7	40.1	5:00	-	4.05	83.1	47.2
6:00	3.4	8.8	116.6	39.6	6:00	-	4.03	80.1	47.3
7:00	20.2	8.8	114.9	39.2	7:00	-	4.01	76.6	47.5
8:00	49.5	8.81	120.7	39.1	8:00	1.1	3.98	73.5	47.8
9:00	82.4	8.81	130.4	39.3	9:00	37.4	3.96	89.8	48.1
Maturity of concrete [°C-hours]				909	Maturity of concrete [°C-hours]				968
Relative age of concrete [days]				4	Relative age of concrete [days]				4.2
Strength of concrete [% from reference standard]				61	Strength of concrete [% from reference standard]				64

Note: size of concrete sample 30x30x30(h) cm; temperature ($t_{concrete}$) is in the middle of concrete sample; reference standard is concrete C25 (compressive strength in the age of 28 days is 29.8 MPa); material of transparent cover is polyethylene-terephthalate foil.

Q_{abs} and Q_{loss} in the tables are the amount of heat, absorbed by concrete during the day time and lost through the translucent cover.

The analysis of the research results shows that concrete is heated more intensively under single layer translucent cover during the day time. Its maximum temperature is 1.5-2 °C more than under double layer translucent cover. However, concrete temperature under single layer cover starts decreasing after sundown, while it continues increasing under double layer cover.

The coefficient of heat loss (U_{loss}) of double layer cover decreases almost in two times as compared with single layer cover. The temperature of concrete decreases slower under double layer cover at night time. The maximum excess of the concrete temperature is from 3-5 °C (table 1) until 8-10 °C (table 2) as compared with single layer cover.

Table 2. Concrete curing in formwork, equipped by single and double layer translucent cover, in humid continental climate conditions: Moscow region, Russia.

Single layer translucent cover					Double layer translucent cover				
Time of day	Q_{abs} [kJ]	U_{loss} [W/(°C·m²)]	Q_{loss} [kJ]	$t_{concrete}$ [°C]	Time of day	Q_{abs} [kJ]	U_{loss} [W/(°C·m²)]	Q_{loss} [kJ]	$t_{concrete}$ [°C]
10:00	36.7	1.13	7.5	21	10:00	28.6	0.42	5.8	21
11:00	44.1	1.15	8.2	23.9	11:00	40.1	0.42	6	22.8
12:00	53.6	15.2	20.5	27.1	12:00	43.4	2.62	27.6	25.3
13:00	60.2	11.2	49.9	30.2	13:00	51.1	3.38	33.2	27.3
14:00	57.9	10.5	51.2	32.3	14:00	42.8	3.61	31.5	29.4
15:00	52.7	10.1	52.9	34.2	15:00	32.9	3.74	29.1	31.4
16:00	44.8	9.93	53.7	36.1	16:00	21.9	3.84	26.6	33.2
17:00	33.7	9.83	52.9	37.4	17:00	7.4	3.94	22.7	35.3
18:00	20.4	9.77	50.7	38.3	18:00	4.8	4.03	21.7	37
19:00	7.2	9.74	48.1	38.8	19:00	3.5	4.11	25.3	38.2
20:00	2.6	9.72	49.6	38.9	20:00	2.1	4.18	28.7	38.9
21:00	0.7	9.7	52.1	38.3	21:00	0.6	4.23	31.6	39.6
22:00	-	9.68	52.9	37.5	22:00	-	4.26	33.7	40.6
23:00	-	9.68	52.5	35.8	23:00	-	4.29	35.2	40.3
24:00	-	9.68	51.5	34.3	24:00	-	4.3	36.2	39.9
1:00	-	9.68	49.6	32.6	1:00	-	4.31	36.6	39.4
2:00	-	9.69	47	31.2	2:00	-	4.3	36.1	39
3:00	-	9.72	43.8	29.7	3:00	-	4.28	35	38.5
4:00	-	9.77	39.9	28.5	4:00	-	4.26	33.7	38.2
5:00	0.7	9.85	35.5	27.3	5:00	0.6	4.24	32	37.9
6:00	2.4	9.99	31	26.4	6:00	2.1	4.21	29.9	37.5
7:00	6.9	10.21	28.2	25.8	7:00	3.6	4.17	27.7	37.3
8:00	19.7	10.57	30.8	25.8	8:00	5	4.13	25.4	37.1
9:00	32.5	11.06	34.7	26.3	9:00	11.9	4.08	27.8	37.1
Maturity of concrete [°C-hours]				757.7	Maturity of concrete [°C-hours]				842
Relative age of concrete [days]				2.2	Relative age of concrete [days]				2.7
Strength of concrete [% from reference standard]				50	Strength of concrete [% from reference standard]				53

Note: size of concrete sample 20x20x20(h) cm; temperature ($t_{concrete}$) is in the middle of concrete sample; reference standard is concrete C25 (compressive strength in the age of 28 days is 29.8 MPa); material of transparent cover is polyethylene foil.

Maturity of concrete (3), which is total sum of numerical values of concrete temperatures over the period of concrete curing, is almost the same in both climatic conditions, but under double layer cover it is more than under single layer cover.

$$M_{concrete} = \sum_{i=1}^{n} t_{concrete} \Delta\tau_i \tag{3}$$

where $\Delta\tau_i$ – time between the measurements of the concrete temperature.

Nevertheless, it is more for the concrete samples, curing under double layer cover in humid continental climate conditions (table 2). The same situation is for relative age of concrete, which depends on the average concrete temperature and intervals between its measurements.

The results of calculation and experimental research of the concrete temperature in conditions of subtropical climate are shown in fig. 1. The similarity of graphs, obtained by the calculation and the experimental research, testifies to adequacy of chosen calculation model and the assessment of the process, taking place during heat treatment of the concrete components using solar energy.

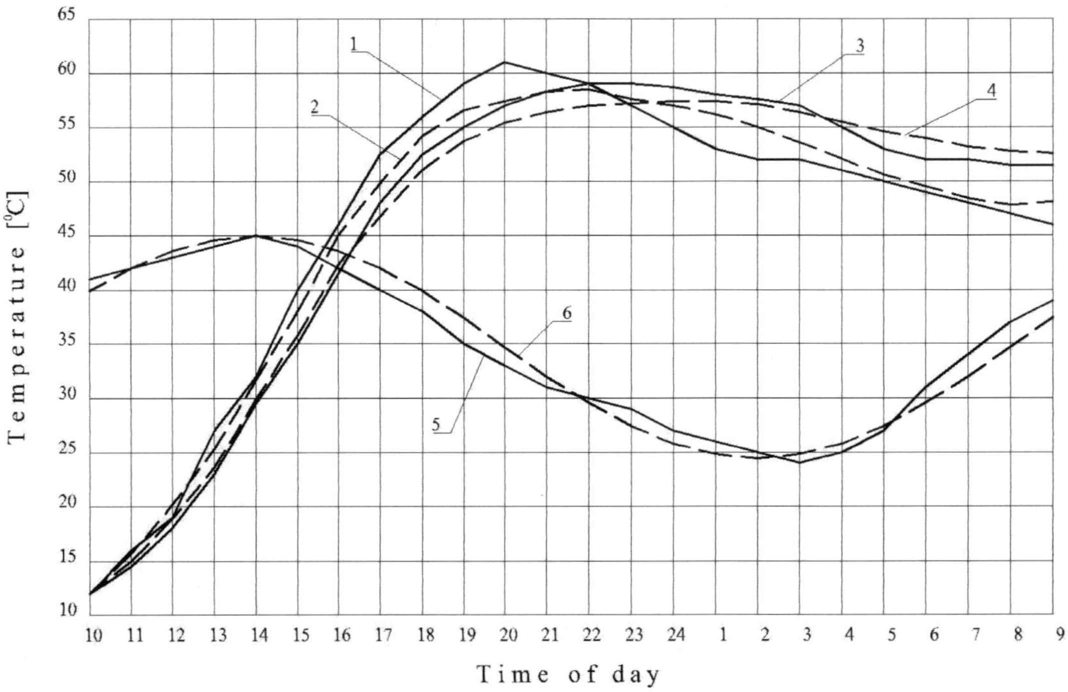

Figure 1. Temperature of concrete sample, curing in formwork, equipped by single and double layer translucent cover in subtropical climate conditions: $\varphi = 44°$ N, in July.

1 – experimental concrete temperature under single layer cover (climatic camera); 2 – calculated concrete temperature under single layer cover; 3 – experimental concrete temperature under double layer cover (climatic camera); 4 – calculated concrete temperature under double layer cover; 5 – actual air temperature during the experiment (climatic camera); 6 – calculated air temperature.

Strength of concrete, curing in formwork, equipped by double layer translucent cover, is better in both climate conditions. However, the difference in the values of the concrete strength decreases with increasing the average air temperature and intensity of solar radiation.

4. Conclusion

The obtained research results testify to the increase of translucent cover layers in humid subtropical climate (Krasnodar region) does not improve the values of strength, maturity and relative age of concrete. It testifies to inefficiency of double layer cover in such climate conditions.

In spite of better values of strength, maturity and relative age of concrete, curing in formwork, equipped by double layer translucent cover in humid continental climate (Moscow region), we have to admit its inefficiency. This is because we have to take into account the economic component, connected with the equipment costs. The increase of the concrete strength by 1-4% does not justify the increase of the equipment price, connected with addition of one more layer of translucent cover at the time of manufacturing and operation of the equipment.

References

[1] Farjana S H, Huda N, Parvez Mahmud M A and Saidur R 2018 Solar process heat in industrial systems – A global review *Renewable and Sustainable Energy Reviews* **82(3)** pp 2270-2286

[2] Abdullah M A, Muttaqi K M and Agalgaonkar A P 2015 Sustainable energy system design with distributed renewable resources considering economic, environmental and uncertainty aspects *Renewable Energy* **78** pp 165-172

[3] Foster E, Contestabile M, Blazquez J, Manzano B, Workman M and Shah N 2017 The unstudied barriers to widespread renewable energy deployment: Fossil fuel price responses *Energy Policy* **103** pp 258-264

[4] Fudholi A and Sopian K 2019 A review of solar air flat plate collector for drying application Renewable and Sustainable *Energy Reviews* **102** pp 333-345

[5] Hu J and Zhang G 2019 Performance improvement of solar air collector based on airflow reorganization: A review *Applied Thermal Engineering* **155** pp 592-611

[6] Shafieian A, Khiadani M and Nosrati A 2018 A review of latest developments, progress, and applications of heat pipe solar collectors *Renewable and Sustainable Energy Reviews* **95** pp 273-304

[7] Kalogirou S A, Karellas S, Braimakis K, Stanciuc C and Badescu V 2016 Exergy analysis of solar thermal collectors and processes *Progress in Energy and Combustion Science* **26** pp 106-137

[8] Li Z, Shen G Q and Xue X 2014 Critical review of the research on the management of prefabricated construction *Habitat International* **43** pp 240-249

[9] Benammar B, Mezghiche B and Guettala S 2013 Influence of atmospheric steam curing by solar energy on the compressive and flexural strength of concretes *Construction and Building Materials* **49** pp 511-518

[10] Koroteev D D, Kharun M and Stashevskaya N A 2017 Manufacturing of concrete elements on mobile polygons with the using of solar energy *International Journal of Advanced and Applied Sciences* **4(10)** pp 10-14

[11] Egolf P W, Amacker N, Gottschalk G, Courret G, Noume A and Hutter K 2018 A translucent honeycomb solar collector and thermal storage module for building facades *International Journal of Heat and Mass Transfer* **127(A)** pp 781-795

[12] Dowson M, Pegg I, Harrison D and Dehouche Z 2012 Predicted and in situ performance of a solar air collector incorporating a translucent granular aerogel cover *Energy and Buildings* **49** pp 173-187

[13] Paneri A, Wong I L and Burek S 2019 Transparent insulation materials: An overview on past, present and future developments *Solar Energy* **184** pp 59-83

Acknowledgments

This paper is financially supported by the Ministry of Education and Science of the Russian Federation on the program to improve the competitiveness of Peoples' Friendship University of Russia (RUDN University) among the world's leading research and education centers in the 2016-2020.

Authors' background

Name	*Prefix*	*Research Field*	*Email*	*Personal website*
Dmitry D. Koroteev	Associate Professor	Construction materials	koroteev_dd@pfur.ru	-
Makhmud Kharun	Associate Professor	Construction materials	miharun@yandex.ru	-
Moises Esau Romero Barrenechea	Associate Professor	Construction materials	moises0555@mail.ru	-

IOP Conf. Series: Materials Science and Engineering **675** (2019) 012008 doi:10.1088/1757-899X/675/1/012008

Features of concrete creep numerical modeling in the calculation of building structures

V N Sidorov[1,2,3,4,5*] and K Nowak[3]

[1] Russian University of Transport, Moscow, Russia
[2] Peoples' Friendship University of Russia (RUDN University), Moscow, Russia
[3] Kielce University of Technology, Kielce, Poland
[4] Perm National Research Polytechnic University, Perm, Russia
[5] Moscow Institute of Architecture, Moscow, Russia

Email: sidorov.vladimir@gmail.com

Abstract. The paper presents a study and justification of the concrete creep computational model for use in computer calculations of concrete and reinforced concrete structures of arbitrary complexity, with a possible change in the intensity of the current load over time. The computational model is based on the generalized Maxwell-Wiechert model of the viscoelastic material. The model was verified on the basis of data from experimental studies of concrete creep, was justified and calibrated in accordance with the requirements of regulatory documents for calculations of concrete and reinforced concrete structures, taking into account the material creep. The paper focuses on the situation where the creep of the material leads to a dangerous change in the structural scheme of a building, which in turn can cause its destruction. Verification of the creep computational model and calculation of structures using the selected computational model was carried out in the SIMULIA Abaqus software environment.

1. Introduction

The relevance of the calculation of concrete and reinforced concrete structures, taking into account the creep of the material, is considered in the paper in the aspect of a possible dangerous change in time of the structural scheme of a building due to material creep, which often leads to the destruction of the structure. The considered computational model of concrete creep was verified on the data of experimental studies of concrete rheological manifestations over time under load, and was also substantiated and calibrated in accordance with the requirements of regulatory documents for calculations of concrete and reinforced concrete structures, taking into account the material creep.

2. Materials and methods

2.1. Requirements for creep calculational models

The requirements for computational creep models are built on the experience of comprehensive experimental and theoretical studies of the processes of creep, relaxation and shrinkage of concrete over time [1 - 3] etc. These requirements are clearly stated in international and national regulatory documents. In this paper, they are based on the recommendations of Eurocode [4] (EC2), the Pre-norma fib Model Code [5] (PN), as well as the NIIZhB Recommendations [6]. In turn, the Recommendations [8] are based on the results of the relevant basic research [1, 2] and oth. and underlie the requirements of the SNiP [7].

The complete deformation of concrete ε is usually represented by its three components:

$$\varepsilon = \varepsilon_0 + \varepsilon_{cr} + \varepsilon_{shr}, \tag{1}$$

where ε_0 is the so-called "instantaneous" deformation which manifests itself at the moment of loading the structure, ε_{cr} is the creep deformation that develops over time, ε_{shr} is the shrinkage deformation.

The key parameter characterizing the creep of concrete in EC2 and PN is the creep coefficient (creep factor) $\varphi(t,t_0)$ established between the moment of application of the load t_0 and the age of concrete t:

$$\varphi(t,t_0)=\frac{\varepsilon_{cr}}{\varepsilon_0}. \tag{2}$$

At the same time, the concept of creep measure $C(t,t_0)$ is also used in these regulatory documents:

$$\varphi(t,t_0)=\mathrm{E_b}(t_0)\cdot C(t,t_0), \tag{3}$$

where $\mathrm{E_b}(t_0)$ is the elasticity modulus of concrete at the time of loading t_0.

In turn, the creep measure is the initial key parameter in the NIIZhB Recommendations:

$$C(t,t_0)=\frac{1}{\mathrm{E_b}(t_0)}-\frac{1}{\mathrm{E_b}(t)}+C(\infty,t_0), \tag{4}$$

where $C(\infty,t_0=28)$ is the limit creep measure.

In this case the concept of creep characteristic (creep coefficient) $\varphi(t,t_0)$ is also used:

$$C(t,t_0)=\frac{\varphi(t,t_0)}{E(t_0)}. \tag{3a}$$

An important requirement for the calculation models of concrete creep is the justified determination of the limits of the use of linear or non-linear dependencies of creep deformations on the stress values arising in the structure. EC2 permits the use of linear dependencies on σ_c stresses arising in structures, based on (2), subject to

$$\sigma_c \leq 0.45\cdot f_{cm}(t_0), \tag{5}$$

where f_{cm} is the average compressive strength of concrete. And the PN condition in this case is more stringent:

$$\sigma_c \leq 0.40\cdot f_{cm}(t_0). \tag{5a}$$

The NIIZhB Recommendations establish the threshold for eligibility of a linear creep model by the condition

$$\sigma_c \leq k_\sigma\cdot R_b, \tag{6}$$

where the coefficient k_σ varies in the range of $0.55 \div 1.00$ depending on the class of concrete (strength) and class of buildings (liability); R_b is the prism strength of concrete.

EC2 recommends using the nonlinear exponential proportion of the change in the creep coefficient if the limit (5) is exceeded:

$$\varphi_{nl}(t,t_0)=\varphi(t,t_0)\cdot e^{(1.5(k_\sigma-0.45))} \quad \text{for} \quad k_\sigma \geq 0.45, \tag{7}$$

where k_σ is the ratio of the compressive stress σ_c and the average compressive strength of concrete f_{cm} at the moment of loading, and the PN document here adds the condition:

$$\varphi_\sigma(t,t_0)=\varphi(t,t_0)\cdot e^{(1.5(k_\sigma-0.4))} \quad \text{for} \quad 0.4 < k_\sigma \leq 0.6. \tag{8}$$

The NIIZhB recommendations in case of non-observance of condition (6) establish the non-linear dependence of the change in the limiting creep measure:

$$C(\infty,t_0)=C(\infty,t_0)\cdot\left(1+\mathrm{v_c}\cdot\eta(t_0)^{\mathrm{m_c}}\right), \tag{9}$$

wherein the values of v_c and m_c depending on the concrete class are taken from the table from [8], and the parameter $\eta(t_0)$ is calculated by the formula $\eta(t_0) = 0.78 \cdot k_\sigma$, where $k_\sigma = \frac{\sigma_b(t_0)}{R_{bn}(t_0)}$, $0.55 \le k_\sigma \le 1.0$ - relative level of stress, $\sigma_b(t_0)$ is the compressive stress acting in concrete at the time of loading t_0, $R_{bn}(t_0)$ is the strength of concrete at the time of loading t_0.

3. Results

3.1. Calculative rheological model of concrete

The rheological model based on the generalized model of a viscoelastic material – the generalized Maxwell model (Wiechert model) [8, 9] was chosen to study and apply to the modeling of concrete creep. Elements of the mechanical interpretation of this model are n springs (i = 1 ÷ n) with "temporary" elastic moduli G_i, and n dampers with viscosity coefficients η_i connected in series with each other in pairs, as well as an elastic element with "long-term" stiffness G_∞ (figure 1*a*).

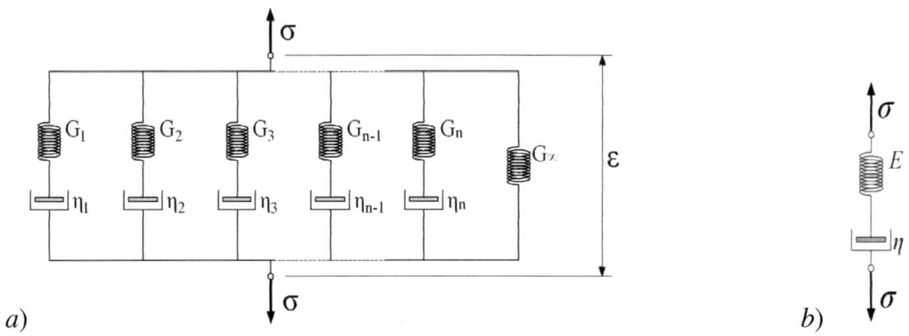

Figure 1. Mechanical representation of the Maxwell model
a) the generalized model; *b*) the elementary model.

The basis of the generalized Maxwell model is the elementary model of the viscoelastic Maxwell material. According to this model, the material resistance is proportional to the speed of its strain

$$\dot{\varepsilon} = \dot{\varepsilon}_e + \dot{\varepsilon}_v, \tag{10}$$

developing from the rate of elastic strain $\dot{\varepsilon}_e$ and the rate of viscous strain $\dot{\varepsilon}_v$. In turn, the elastic strain of the material is interpreted by the strain of the elastic spring with the stiffness E (figure 1*b*)

$$\varepsilon_e = \frac{\sigma_y}{E}, \quad \text{from where} \quad \dot{\varepsilon}_e = \frac{\dot{\sigma}_e}{E}, \tag{11}$$

and the rate of viscous strain in this model is represented by the properties of a damper with a coefficient of viscosity η:

$$\dot{\varepsilon}_v = \frac{\sigma_v}{\eta}. \tag{12}$$

The serial connection of the spring and damper in the Maxwell model means that the stresses σ in both its elements are the same, i.e. $\sigma = \sigma_e = \sigma_v$, but the strains in the elastic and viscous elements are different, i.e. $\varepsilon_e \neq \varepsilon_v$ ($\varepsilon = \varepsilon_e + \varepsilon_v$). Thus, according to (10) - (12), the strain rate of the material, represented by the Maxwell elementary model, will vary in proportion to time t

$$\dot{\varepsilon}(t) = \frac{\sigma(t)}{\eta} + \frac{\dot{\sigma}(t)}{E}, \tag{10a}$$

herewith at a constant stress σ_0 the strain will increase linearly (figure 2*a*).

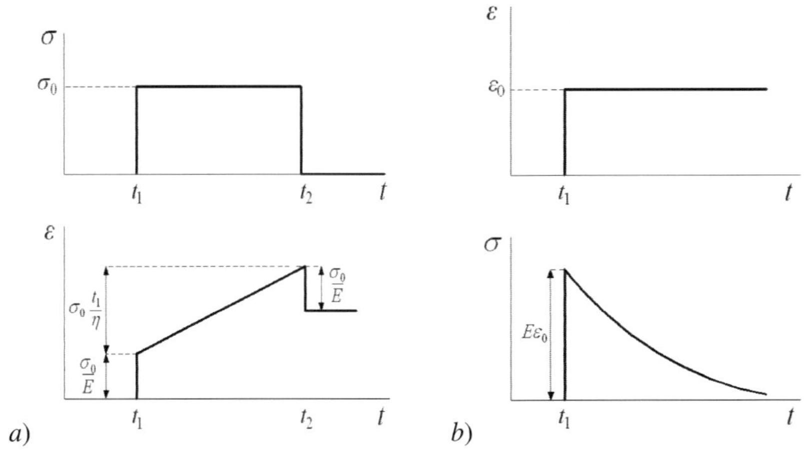

Figure 2. The interdependencies of stresses and strains in the Maxwell elementary model *a*) strain at a constant stress, *b*) stress at constant strain.

In turn, the solution (10*a*) with respect to $\sigma(t)$:

$$\sigma(t) = \sigma_0 \cdot e^{\left(\frac{-t}{\tau}\right)} = E \cdot \varepsilon_0 \cdot e^{\left(\frac{-t}{\tau}\right)} \tag{13}$$

shows that according to the elementary Maxwell model, when the strain ε_0 is fixed, the stress $\sigma(t)$ decreases in time t according to the exponential law (figure 2*b*), which corresponds to the nature of the manifestation of relaxation in a deformable material. In (13) $\tau = \frac{\eta}{E}$ is the so-called relaxation time.

In the transition to the generalized Maxwell model, dependence (13) takes the form:

$$\sigma(t) = E_\infty \cdot \varepsilon_0 + \sum_{i=1}^{n} E_i \cdot \varepsilon_0 \cdot e^{\left(\frac{-t}{\tau_i}\right)} = \varepsilon_0 \cdot \left[E_\infty + \sum_{i=1}^{n} E_i \cdot e^{\left(\frac{-t}{\tau_i}\right)} \right], \tag{14}$$

where $\tau_i = \frac{\eta_i}{E_i}$ is the relaxation time of the i -th damper ($i = 1 \div n$) in the generalized Maxwell model.

Thus, in the generalized Maxwell model, the viscoelasticity of a material is characterized by the generalized relaxation modulus depending on time (the so-called relaxation function):

$$E_R(t) = \frac{\sigma(t)}{\varepsilon_0} = E_\infty + \sum_{i=1}^{n} E_i \cdot e^{\left(-\frac{t}{\tau_i}\right)}. \tag{15}$$

It can be seen that the relaxation function $E_R(t)$ of a material is actually represented in (14), (15) in the form of an exponential series (Prony series).

3.2. Computational implementation of the creep model

The generalized Maxwell model of viscoelastic properties of a material is implemented in the SIMULIA Abaqus software environment with the "viscoelastic" option [13]. Here it is represented by two relaxation functions, having the form of an exponential Proni series. This is the shear relaxation function:

$$G(t) = G_\infty + \sum_{i=1}^{n} G_i \cdot e^{\left(-\frac{t}{\tau_i}\right)}, \tag{16}$$

where $G = \dfrac{E}{2 \cdot (1 + v_0)}$ is the modulus of elasticity in shear, v_0 is the Poisson coefficient, and also the function of bulk relaxation:

$$K(t) = K_{\infty} + \sum_{i=1}^{n} K_i \cdot e^{\left(-\frac{t}{\tau_i}\right)}, \tag{17}$$

where $K = \dfrac{E}{3 \cdot (1 - 2 \cdot v_0)}$ is the coefficient of volume stiffness.

In the SIMULIA Abaqus program, the parameters of the "viscoelastic" material model are set by the dimensionless shear relaxation modulus:

$$g_R(t) = \frac{G(t)}{G_0} = 1 - \sum_{i=1}^{n} g_i \cdot \left(1 - e^{\left(-\frac{t}{\tau_i}\right)}\right), \tag{16a}$$

where $G_0 = E_0/(2 \cdot (1 + v_0))$ is the instantaneous modulus of elasticity in shear, as well as the dimensionless modulus of bulk relaxation:

$$k_R(t) = \frac{K(t)}{K_0} = 1 - \sum_{i=1}^{n} k_i \cdot \left(1 - e^{\left(-\frac{t}{\tau_i}\right)}\right), \tag{17a}$$

where $K_0 = E_0/ \cdot (1 - 2 \cdot v_0)$ is the coefficient of instantaneous volume stiffness.

Obvious complexity here can be the assignment of values of the parameters of the material G_i, K_i, τ_i, which can be obtained by static and dynamic tests of the material for creep and relaxation. However, under the condition of relatively small changes in stresses over time in the simulated structure, you can use the simple case of specifying the modules $g_r(t)$ (16*a*) and $k_r(t)$ (17*a*) with the introduction of the concept of effective modulus of elasticity [6], [8]:

$$E_{c,eff} = \frac{E_0}{1 + \varphi(t, t_0)}. \tag{18}$$

Then the expression of the dimensionless shear relaxation modulus is considerably simplified:

$$g_r(t) = \frac{G_R(t)}{G_0} = \frac{\dfrac{E_0}{2 \cdot (1 + v_0) \cdot (1 + \ (t, t_0))}}{\dfrac{E_0}{2(1 + v_0)}}, \quad \text{or} \quad g_r(t) = \frac{1}{1 + \varphi(t, t_0)}, \tag{16b}$$

where $\varphi(t, t_0)$ is the creep coefficient of the material (2). In this case, the values of the time-varying coefficient $\varphi(t, t_0)$ in the computational model of creep can be set both on the basis of the material test results, and in accordance with the regulatory recommendations [6] - [8] and others. Similarly, the view of the dimensionless bulk relaxation modulus is simplified:

$$k_r(t) = \frac{K_R(t)}{K_0} = \frac{1}{1 + \varphi(t, t_0)} \tag{17b}$$

4. Discussion

4.1. Verification of the computational creep model

To test the "viscoelastic" model, we used the results of experimental studies of two groups of concrete samples [9, 10] (figure 3).

Samples of the 1-st group were loaded ($\sigma_I = 10.67$ [MPa]) at the age of $t_0 = 28$ days, having on this day the modulus of elasticity $E_{(28)} = 36200$ [MPa]. Samples of the 2-nd group were loaded ($\sigma_{II} = 2.47$ [MPa]) 1 day after their manufacture ($t_0 = 1$), having the elastic modulus $E_{(1)} = 22201$ [MPa].

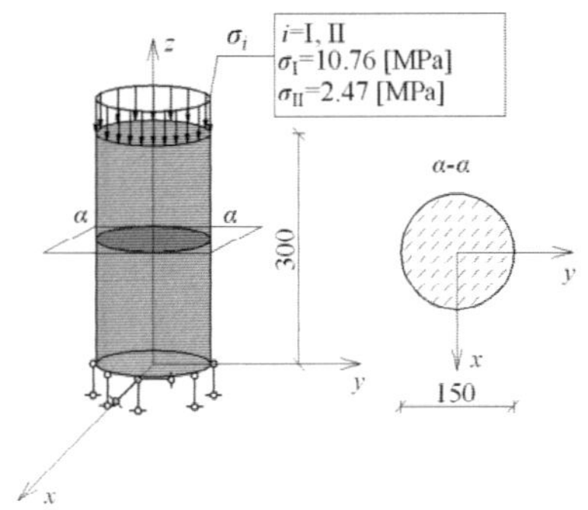

Figure 3. Parameters of the tested samples.

When specifying the dimensionless relaxation moduli of the material using formulas (16*b*) and (17*b*) in the "viscoelastic" model, the concrete creep factor $\varphi(t,\ t_0)$ was calculated based on the experimental results. The figures 4 and 5 show the results of a comparison of the dependences of the creep strain in the samples obtained in the experiment and in numerical simulation using the "viscoelastic" model.

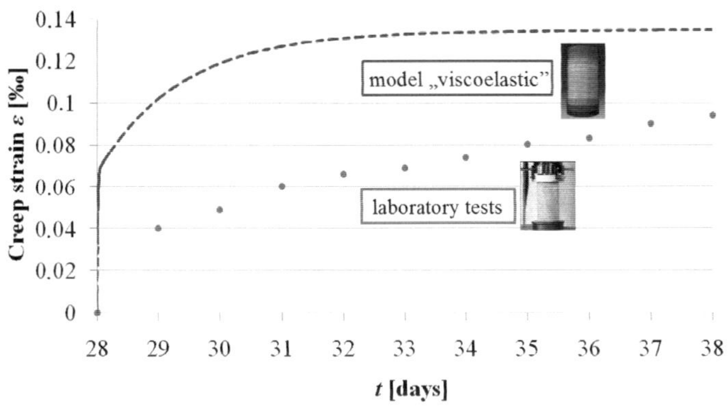

Figure 4. Comparison of the dependences of creep strain in samples of the 1-st group (t_0 = 28 days), obtained in the experiment and in numerical simulation.

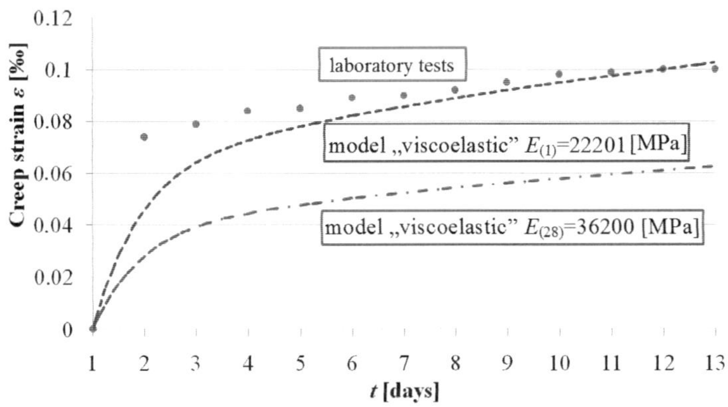

Figure 5. Comparison of the dependences of creep strain in samples of the 2-st group (t_0 = 1 day), obtained in the experiment and in numerical simulation.

Testing the performance of the "viscoelastic" model with setting the values of the dimensionless moduli of material relaxation using formulas (16*b*) and (17*b*) with an abrupt change in load value gave quite satisfactory results (figure 6).

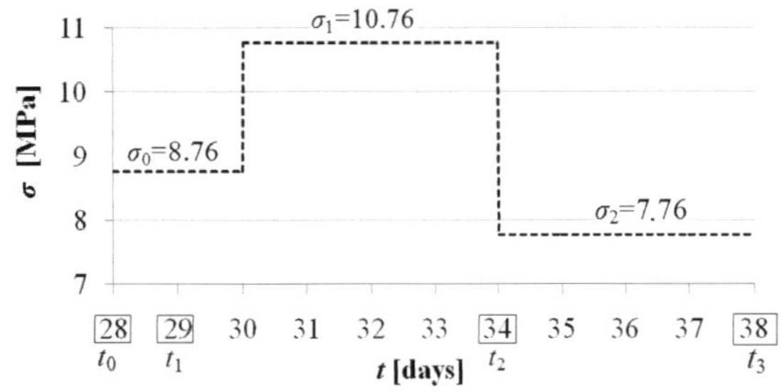

a)

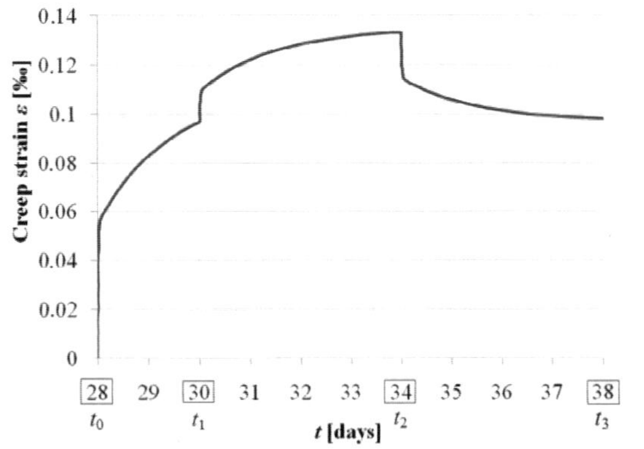

b)

Figure 6. The change in creep strain in time with an abrupt change in load value:
a) change in stress over time;
b) change in creep strain in time.

Conclusion

Application of the computational creep model for calculating the shallow reinforced concrete structure

Justified consideration of the material creep when calculating characteristic reinforced concrete structures, such as, for example, flat arches and shells, can reveal a significant change in their stress-strain state over a long time. So when the studing of the destruction causes of reinforced concrete pavement of the Transvaal-Park sports and entertainment complex (SEC) in 2004 in Moscow at the age of 2 years [11, 12], the software package SIMULIA Abaqus was also used. The coating shell had the shape of a sector with the radius of 74.87 meters (figure 7).

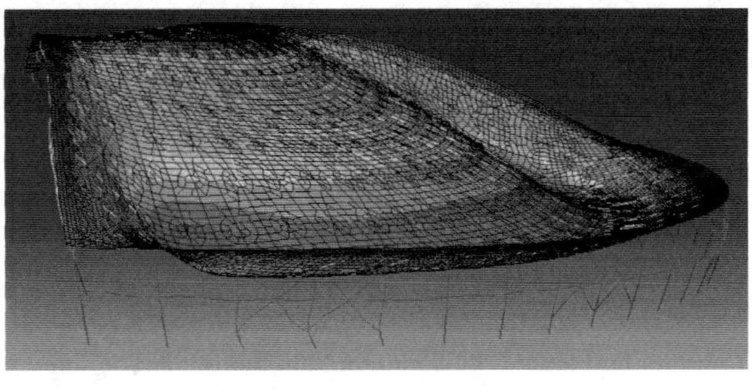

Figure 7. The deformed calculation model of the Transvaal-Park SEC coating structure.

In the present work, when calculating this construction, the "viscoelastic" concrete creep model calibrated in accordance with the requirements of [6] was implemented. Calculation taking into account the creep of concrete showed a significant change in the stress-strain state of the structure over time, caused by a decrease in the curvature of the reinforced concrete shell of the coating. So Von Mises stresses

$$\sigma_{Mises} = \sqrt{\frac{(\sigma_x - \sigma_y)^2 + (\sigma_y - \sigma_z)^2 + (\sigma_z - \sigma_x)^2}{2} + 3\cdot(\tau_{xy}^2 + \tau_{yz}^2 + \tau_{zx}^2)} \tag{19}$$

in the middle steel support column of the Transvaal-Park SEC coating construction reached the elastic limit of the used steel Re = 245[MPa] at the age of t = 2 years (figure 8).

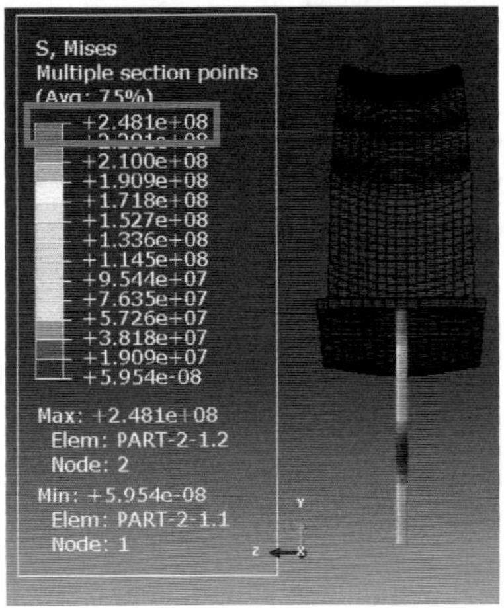

Figure 8. Stresses in the middle support of the Transvaal-Park SEC at the age of t = 2 years.

References

[1] Bondarenko V M and Bondarenko S V 1982 *Engineering Methods of the Nonlinear Theory of Reinforced Concrete* (Moscow: Stroiizdat) (in Russian)

[2] Prokopovich I Ye and Zedgenidze V A 1980 *Applied theory of creep* (Moscow: Strojizdat) (in Russian)

[3] Bazant Z P 2005. *Scaling of Structural Strength*. 2-nd updated ed., Elsevier, London.

[4] Eurocode 2: *Design of concrete structures* EN1992-1-1. 2008.

[5] PN-B-03264: Pre-norma fib Model Code 2010. *Concrete, reinforced concrete and prestressed structures. Static calculations and design.*

[6] NIIZhB 1988. *Recommendations on concrete creep and shrinkage consideration when calculating concrete and reinforced concrete structures* (Moscow: Stroiizdat) p 121 (in Russian)

[7] Construction norms and rules of the RF. SNiP 52-01-2003. *Concrete and reinforced concrete structures. Principal rules* (Moscow, in Russian)

[8] Schiessel H, Metzler R, Blumen A and Nonnenmacher T F 1995 *Generalized viscoelastic models: their fractional equations with solutions* Journal of Physics A: Mathematical and General United Kingdom, 28, pp. 6567-6584

[9] Sidorov V N and Nowak K *Substantiation of the Use of Viscoelastic Material Model in Numerical Analysis the Creep of Concrete Structures* IDT 2017 - Proceedings of the 2017 International Conference on Information and Digital Technologies. ISBN: 978-1-5090-5689-7, DOI: 10.1109/DT.2017.8024319 pp. 339-343

[10] Sidorov V N and Nowak K *Numerical investigation of the long-term work of arches under material creep conditions* (2018) IOP Conference Series: Materials Science and Engineering, 456 (1), art. no. 012007.

[11] Belostotsky A M 2007 *Analiz pritchin obrushenia konstrukcij pokrytia SOK „Transvaal-park"* Sb. trudov MGSU Voprosy prikladnoj matematiki I vytchislitelnoj mechaniki **10** (Moscow: MGSU) (in Russian)

[12] Akimov P A, Wojcicki Z and Sidorov V N *Operational and Variational Formulations of Boundary Problems of Anisotropic Plate Analysis, Adapted for Application Within Discrete-Continual Methods* MATEC Web Conferences, France, Vol. 117, 2017

[13] ABAQUS – Standard, *User's Manual, Theory Manual and Keywords Manual, 19.7.1. Time domain viscoelasticity* (Boston: Hibbitt, Karlsson and Sorensen, Inc.) MA 02111-1307 USA p 1306

Comparative study of the AISC-LRFD, Eurocode 3 & SP 16.13330.2017 steel members design

T H Gebre[1,2*], E V Lebed[3] and V V Galishnikova[1]

[1] Peoples' Friendship University of Russia (RUDN University), Moscow, Russia
[2] College of Engineering and Technology, Asmara, Eritrea
[3] Moscow State University of Civil Engineering (National Research University), Moscow, Russia

Email: tesfaldethg@gmail.com

Abstract: This paper presents the comparative design procedures for steel frame elements using three different International design codes, viz. the American Institute of Steel Construction (AISC), the European Code (EC3) and the Russian Design Code (SP). It focuses on the resistance capacity of steel members subjected to different loads, such as tension, compression and flexure. It compares the formulation of all codes that identify their similarities and differences. The comparative study is performed for a compact rolled I sections with different slenderness and lateral bracing lengths. The results are presented graphically in different ways such as strength curves, Moment capacity versus lateral bracing length and normalized graphs. The comparative study shows that the resistance capacity specified by all specifications can be different for some considered cases. For compression member, SP specifications have higher capacity and economical approach. For flexural member, AISC gives higher capacity for inelastic region and EC3 gives higher capacities for elastic region but SP approach is lower for flexural capacity.

Keywords: tension member, compression member, design curve, flexural members, lateral-torsional buckling

1. Introduction

A very important task of any designer is to size the steel frame elements to be safe for the entire life of the structure. Owners may require the use of widely accepted steel design codes regardless of the location where the structure is going to be built. Nowadays, designers are exposed to the challenge of being competent with several design specifications for a particular material type. This paper, presents a comparison of different design specifications for the design of Steel frames elements and those are AISC, EC3 and SP as per the studies by different authors [1-8]. In the United States, specification for structural steel buildings was developed by the AISC which utilizes both load and resistance factor design (LRFD) and allowable strength design (ASD) formats [1, 6]. In Europe, "Design of Steel Structures, EN 1993 (EC3)" was developed by the European Committee for Standardization [3, 5]. In Russia, "Steel Structures Code SNIP II-232-81" is continuously being used in neighboring countries, after this referred to as the SP 16.13330.2011/17 specification [4, 7, 8]. The AISC specification is a compiled document whereas the EC3 and SP specifications contain parts and subparts in addition, each part is concentrated on a particular structure type. The comparative study is performed for a rolled I sections with different slenderness and lateral bracing lengths. The results are presented graphically by different means such as strength curve, non-dimensionless analysis and moment capacity versus lateral bracing curves to simplify the understanding and to detect the difference in the resistance.

2. Methods of research

2.1 Design of Members for Tension

For tension members, all specifications consider tensile yielding with gross section and tensile rupture with net section as the two primary limit states for tension members [5, 7]. The main difference between all specifications is the calculation of shear lag factor U. An elaborate treatment is tabulated in AISC. However, a less elaborate treatment is given in EC3 and SP.

2.2 Design of Members for Compression

The strength of compression member is governed by the yielding of the material for short ones, by elastic buckling for long ones and by inelastic buckling for ones of intermediate lengths. For extremely stocky members, failure may occur by compressive yielding rather than buckling. The design of columns makes an extensive use of the column strength curves. For a compression member in the absence of imperfections and assuming a linear-elastic constitutive law (Euler column), a value of the axial force can be found to trigger element instability, called elastic critical load [9-12]. The complete AISC specification for compressive strength curve is given below.

$$\frac{KL}{r} \leq 4.71\sqrt{\frac{F_y}{E}}, \quad F_{cr} = \left(0.658^{\lambda_c^{2}}\right) * F_y, \frac{F_y}{F_e} \leq 2.25$$

$$\frac{KL}{r} > 4.71\sqrt{\frac{F_y}{E}}, \quad F_{cr} = \left(0.658^{\lambda_c^{2}}\right) * F_y, \frac{F_y}{F_e} > 2.25$$

Using EC3 specification, the reduction factor (χ) has the following forms:

$$\chi = \frac{1}{\Phi + \sqrt{\Phi^2 - \bar{\lambda}^2}} \quad \text{but } \chi \leq 1.0$$

$$\Phi = 0.5[1 + \alpha\,(\bar{\lambda} - 0.2) + \bar{\lambda}^2,$$

where $\bar{\lambda}$ is the non-dimensional slenderness; α is an imperfection factor.

As stated by EC3, There are five different buckling curves for columns analysis and the imperfection factor corresponding to the appropriate buckling curve is given in EN 1993-1-1 [10]. EC3 utilizes an imperfection coefficient (α) to distinguish between different column strength curves [13-15].

Similarly, as stated by SP the reduction factor (φ) has the following forms,

$$\varphi = \frac{\left(\delta - \sqrt{\delta^2 - 39.48\bar{\lambda}^2}\right)}{\bar{\lambda}^2},$$

where φ is the coefficient of stability under central compression, whose value at $\bar{\lambda} \geq 0.4$; the value of the coefficient δ in the above formula should be calculated as given below:

$$\delta = 9.87(1 - \alpha + \beta\bar{\lambda}) + \bar{\lambda}^2,$$

where $\bar{\lambda}^2 = \lambda\sqrt{\frac{R_y}{E}}, \ \lambda\sqrt{\frac{Kl_y}{r}}$; δ and β are the conditional flexibility coefficients of the member and determined from the code depending on the type of the section [7, 8].

2.3. Design of Flexure Members for compact section

In all specifications, yielding and lateral torsional buckling are the two limit states for flexural members and they are treated separately for clarity of comparisons. For compact section (class 1 or 2 in EC3 and class 3 or 2 in SP). All specifications allow the member to reach its plastic moment capacity and the nominal moment capacity for these types of sections is determined as follows:

$$M_n = M_{pl} = Z * F_y \qquad \text{(AISC: compact)}$$

$$M_{c,Rd} = M_{pl,,Rd} = W_{pl}\frac{f_y}{\gamma_{MO}} \qquad \text{(EC3: class 1 and 2)}$$

$$M_x = \phi_b W_{cx} R_y \gamma_c \qquad \text{(SP: class 3 and 2)}$$

where M_n is the nominal moment capacity; W_{pl} is the plastic section modulus; f_y is the yield strength; γ_{M0} is the partial safety factor; W_{cx} is the section modulus of cross section about the principal axis x; φ_b is the stability factor in bending; and R_{yx} design yield strength.

For non-compact sections in AISC, the nominal moment capacity reduces linearly with an increase in the flange slenderness and varies between the plastic moment capacity (M_p) and the yield moment considering residual stresses ($0.7 \cdot M_y$). On the contrary, the nominal moment capacity is directly equal to the yield moment in the EC3 specification for class 3 cross-sections and SP specification uses reduction factor to compute the nominal moment capacity for non-compact section as well. The nominal moment capacity for members with non-compact flanges is determined as follows:

$$M_n = \left[\frac{M_p}{M_y} - \left(\frac{M_p}{M_y} - 1\right)\left(\frac{\lambda - \lambda_{pf}}{\lambda_{rw} - \lambda_{pf}}\right)\right] M_y \qquad \text{(AISC)}$$

$$M_{c,Rd} = M_{el,,Rd} = W_{el}\frac{f_y}{\gamma_{MO}} \qquad \text{(EC3: if flange is a class 3)}$$

$$M_x = \phi_b W_{cx} R_y \gamma_c \qquad \text{(SP: if flange is a class 2)}$$

2.4. Lateral Torsional Buckling (LTB) of Compact I-shaped Members

The problem of lateral torsional buckling (LTB) of steel beams has been studied extensively by many authors, including Trahair and others [16-21]. All specifications have difference with the treatment of lateral torsional buckling. In the case of AISC specification, LTB is classified in to three different parts of buckling depending on the unbraced length of the member (L_b) and it has two threshold values for unbraced length namely L_p and L_r are defined in AISC [22-25]. The C_b factor given in design specifications for non-uniform moment diagrams can be used to estimate the increased brace requirements for other loading cases [21]. The following equations summarize the nominal moment capacity for lateral torsional buckling as per the AISC specification:

$$M_n = M_p = Z * F_y \qquad \text{when } L_b \le L_p$$

$$M_n = C_b * \left[M_p - \left(M_p - 0.7S_xF_y\right)\left(\frac{L_b - L_p}{L_r - L_p}\right)\right] \le M_p \qquad \text{when } L_p < L_b \le L_r$$

$$M_n = M_{cr} = S_x \frac{C_b * \pi^2 E}{\left(\frac{L_b}{r_{rs}}\right)^2}\sqrt{1 + 0.078\frac{J}{S_x h_o}\left(\frac{L_b}{r_{rs}}\right)^2} \qquad \text{when } L_b > L_r$$

$$L_b = 1.7 r_y \sqrt{\frac{E}{F_y}}$$

$$L_r = 1.95 r_{ts}\frac{E}{0.7F_y}\sqrt{\frac{J}{S_x h_o}}\sqrt{1 + \sqrt{1 + 6.76\left(\frac{0.7F_y}{E}\frac{S_x h_o}{J}\right)^2}}$$

$$r_{ts}^2 = \frac{\sqrt{JC_w}}{S_x}$$

where S_x is the section modulus of the compression flange about the x-axis; r_{ts} is the radius of gyration of cross-section; h_o is the distance from the centroid of the top flange to the centroid of the bottom flange; L_b is the unbraced length; and L_r and L_p are the two threshold values for unbraced length for the inelastic range and C_b is the moment gradient factor.

In EC3, a reduction factor (χ) is used to compute the capacity of a member with respect to the buckling and instability [2]. This factor is strongly dependent on the member slenderness parameter λ [26]. According to EC3, the beam should be verified against lateral-torsional buckling resistance as follows: The elastic critical moment (M_{cr}) is used as the basis for the methods given in design codes for determining the slenderness of a section [27]. The elastic critical moment (M_{cr}) is similar to the Euler (flexural) buckling of a strut in that it defines a buckling load [28]. According to Clause 6.3.2.1(1) of EN 1993-1-1, the beam should be verified against lateral-torsional buckling resistance as follows [6]:

$$M_{c,Rd} = M_{pl,,Rd} = \chi_{LT} W_y \frac{f_y}{\gamma_{MO}}$$

$$\chi_{LT} = \frac{1}{\Phi + \sqrt{\Phi^2 - \beta \lambda_{LT}^2}}, \text{with } \chi_{LT} \leq 1.0 \text{ and } \chi_{LT} \leq \left(\frac{1}{\lambda_{LT}}\right)^2.$$

And Φ_{LT} is defined,

$$\Phi_{LT} = 0.5[1 + \alpha_{LT}(\lambda_{LT} - \lambda_{LT,O}) + \beta \lambda_{LT}{}^2, \qquad \lambda_{LT} = \sqrt{\frac{W_y f_y}{M_{cr}}},$$

where α_{LT} is the imperfection factor corresponding to the appropriate buckling curve; γ_{MO} is the partial factor for member instability (the recommended value of 1.0 in EC3); $W_y f_y$ is the section moment resistance; λ_{LT} is the modified slenderness; and the values of α_{LT} and β depend on the type of beam section.

SP approach also uses a reduction factor (φ) to treat lateral torsional buckling problem [22, 26]. If the member is loaded with moment in one of the principal plane only, the design buckling resistance moment (nominal moment capacity for LTB) should be calculated as follow:

$$M_{b,RD} = \chi_{LT} W_y f_y \gamma_c$$

When $\chi_{LT} \geq 0.85$, the section is in the elasto-plastic stage, as the reuslt, the Young modulus declines and the buckling factor has to be modified. The modification of buckling factor is specified in SP code and this is done by finding the coefficients α (section SP16 G.4) and ψ (SP16 Tables G.1 and G.2). The buckling factor for For members with doubly-symmetric I-sections is calculated as follows:

$$\emptyset_1 = \psi \frac{I_z}{I_y} \left(\frac{h}{L}\right)^2 \frac{E}{f_y}$$

$$\chi_{LT} = \begin{cases} \emptyset_1, & \text{if } \emptyset_1 \leq 0.85 \\ 0.68 + 0.21\emptyset_1, & \text{if } \emptyset_1 > 0.85 \end{cases}$$

where ϕ_1 is defined in the SP code (section SP 16 G1).

In addition to the buckling factor, SP also provides equations for stable length limits of the beam (SP 16.13330.2017, Chapter 8.4.4, Table 11 [4]). For a beam with unsupported length, less than the stable length, the buckling factor's calculation gives the result $\chi_{LT} = 1$. For beams with uniform moment distribution, the stable length is calculated as follows:

$$\frac{l_{ef}}{b_f} < \left(0.35 + 0.0032\frac{b}{t} + \left(0.76 - 0.02\frac{b}{t}\right)\frac{b}{h}\right)\sqrt{\frac{R_y}{E}}, \qquad \text{but} \begin{Bmatrix} 1 \leq \frac{h}{b} \leq 6 \\ 15 \leq b/t \leq 35 \end{Bmatrix},$$

where b and t are width and thickness of the compression flange h, distance (height) between the axes of the flanges.

If the limit slenderness is more than the limit value, it is necessary to provode the intermediate stiffeners for reduction of the effective length l_{ef} [8]. The nominal moment capacity is suddenly drops from plastic moment capacity for compact section and it is limited to a small lateral bracing length.

3 Result and discussion

A single column strength curve is given in the AISC also five separate curves are presented in the EC3 and three curves for SP. In general, all specifications use a non-dimensional slenderness for flexural buckling to define the reduction factor. For comparison, the combined graph of strength curves for Compression member is presented in figure 1.

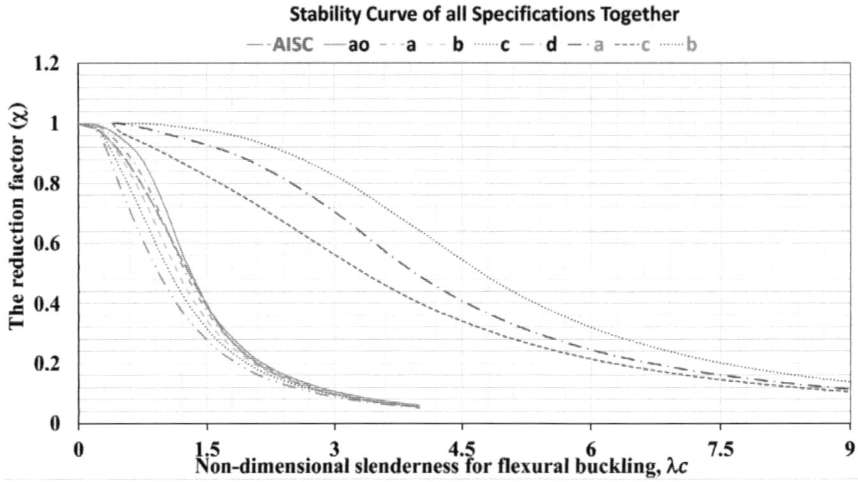

Figure 1. Combined strength curve using AISC, EC and SP.

Accordingly, all strength curves (b, c, d) for EC3 give lower capacity values as compared with AISC curve. Only a_o curve for EC3 gives higher capacities than AISC curve but the use of this curve is quite limited. In that case of SP curves (a, b, c) are not similar with AISC and EC3 somewhat different, because they have higher capacity nevertheless they are not heavy in weight, and economical approach.

Curves of nominal flexural strength of W12x30 of the steel beam sections according to AISC, EC3 and SP methods show the moment capacity of the steel sections across a wide range of lateral bracing length (L_b) and they are shown in the figures 2-4.

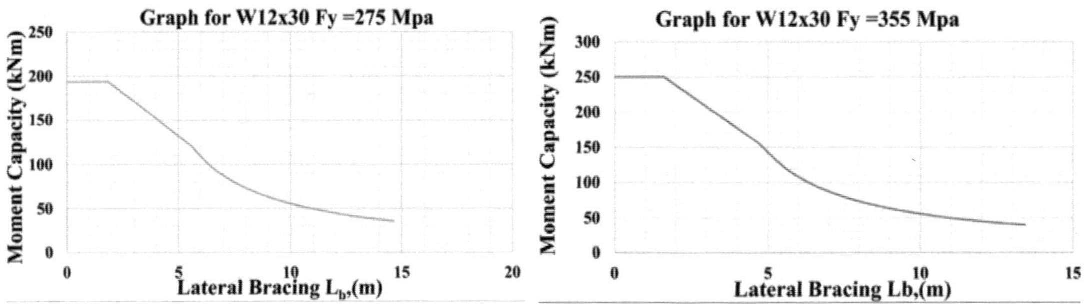

Figure 2. ΦM_nVs L_b for W12X30 using AISC.

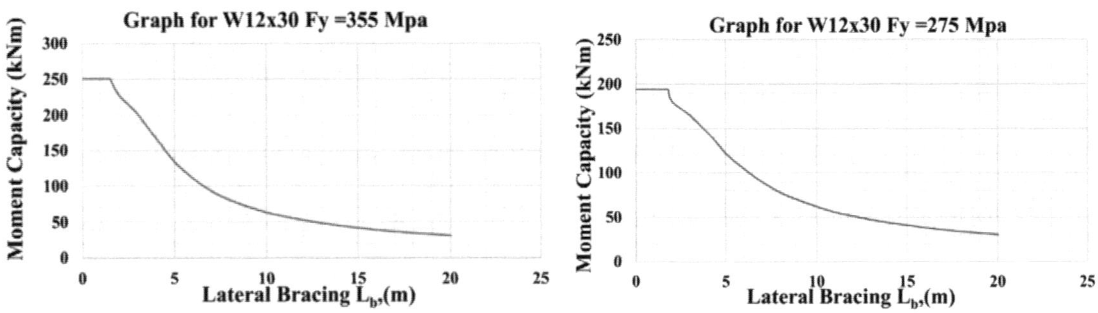

Figure 3. ΦM_nVs L_b for W12X30 using EC3.

IOP Conf. Series: Materials Science and Engineering **675** (2019) 012009 doi:10.1088/1757-899X/675/1/012009

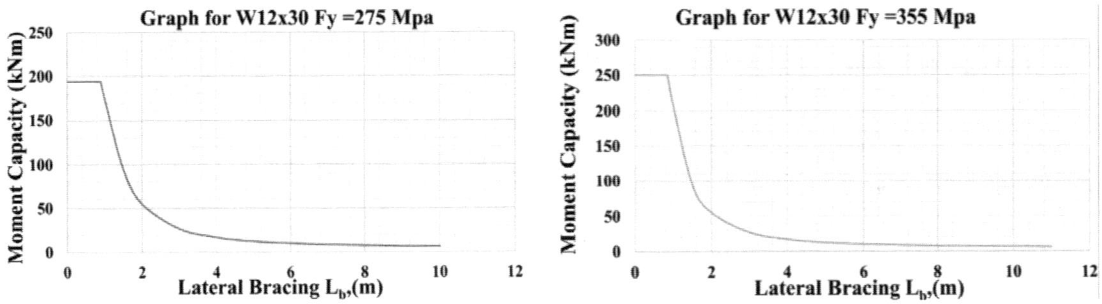

Figure 4. ΦM_nVs L_b for W12X30 using SP.

A combined curve of the nominal flexural strength of the W12x30 sections using all specifications shows the comparison of capacity across a wide range of lateral bracing distances (L_b) and they are shown in the figure 5.

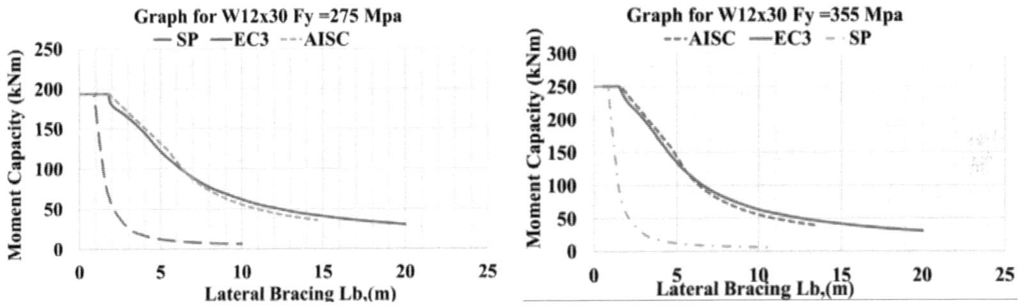

Figure 5. ΦMnVs L_b for W12X30 according to AISC, EC3 and SP.

Additionally, the same section (W12x30) is considered to compare lateral torsional buckling capacity of the members for different buckling lengths which ranges from 0 m to 20 m and normalized moment capacities curves are formulated for two different yield strengths. The normalized moment capacity for W12x30 is given in figures 6.

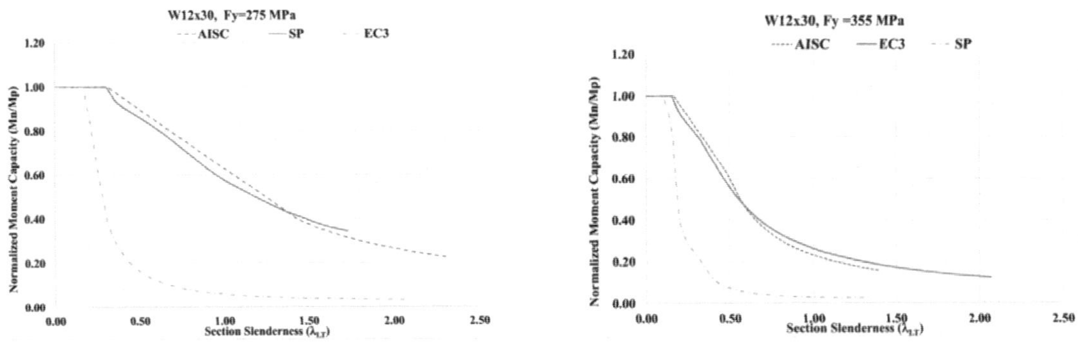

Figure 6. Normalized moment capacity vs. section slenderness for W12x30.

4 Conclusion

According to tensile capacity, all specifications are simillar but theire difference is the method used to calculate the shear lag factor. For compression membres, AISC and EC3 are almost similar however only a_o of EC3 curve gives higher capacities than AISC curve and curves (b, c, d) give lower capacity. SP curves (a, b, c) are not similar with AISC and EC3 rather different, because they have higher capacity and economical approach on the other hand SP needs a skilled engineer to follow the approach properly. For flexural members, all specifications can reach to its plastic moment capacity if the members are compact. Treatment of non-compact flanges is similar to the treatment on non-compact webs in all specifications. For slender members, the AISC specification utilizes the elastic critical buckling moment approach. In EC and SP, the post buckling reserve strength approach is utilized. According to SP, the nominal moment capacity is suddenly drops from plastic moment capacity for non-compact section and as a result, it is limited to a small lateral bracing length. Accordingly, AISC gives higher capacity in inelastic region and EC3 gives higher capacities for elastic region. SP approach is lower for flexural capacity therefore it will be uneconomical approach comparing with the other two approaches.

References

[1] American Institute of Steel Construction 2011 *Steel Construction Manual* 13th Edition p 2245
[2] EN 1993-1-1 Eurocode 3: *Design of steel structures – Part 1-1: General Rules and Rules for Buildings* CEN Brussels Belgium 2005
[3] Brown D G et al. *Handbook of Structural Steelwork* **(55)** P 440
[4] Design Code SP 16.13330.2017 *Steel Structures* (Moscow: Standartinform)
[5] Bernuzzi C and Cordova B 2016 *Structural Steel Design to Eurocode 3 and AISC Specifications* (New Jersey: John Wiley & Sons) p 536
[6] Yong D J, López A and Serna M A 2006 Beam-Column Resistance of Steel Members: A Comparative Study of AISC-LRFD and EC3 Approaches *Int. J. Struct. Stab. Dyn.* **11(2)** pp 345–361
[7] Loorits K and Talvik I 1999 A comparative study of the design basis of the Russian Steel Structures Code and Eurocode 3 *J. Constr. Steel Res.* **49(2)** pp 157–166
[8] Loorits K and Talvik I 2006 Comparative study of the buckling of steel beams in Eurocode 3 and the Russian code *J. Constr. Steel Res.* **62(12)** pp 1290–94.
[9] Lebed E V 2018 Behavior of the frames of large-span metal domes in the process of their installation *Structural Mechanics of Engineering Constructions and Buildings* **14(6)** pp 481–494
[10] Desai P R and Satish C 2018 Effect of Slenderness Ratio on Euler Critical Load for Elastic Columns with ANSYS, *Int. J. Eng. Res. Appl.* **8(5-III)** pp 40–43
[11] Galishnikova V V and Gebre T H 2018 Al-sabri S A M Saffia-Doe O Second order structural theory for the stability analysis of columns *Structural Mechanics of Engineering Constructions and Buildings* **14(3)** pp 192–197
[12] Topkaya C and Şahin S 2011 A comparative study of AISC-360 and EC3 strength limit states, *Int. J. Steel Struct.* **11(1)** pp. 13–27.
[13] Taras A and Greiner R 2010 New design curves for lateral-torsional buckling-Proposal based on a consistent derivation *J. Constr. Steel Res.* **66(5)** pp 648–663.
[14] Trahair N S 2018. Trends in the code design of steel framed structures *Adv. Steel Constr.* **14** pp 37–56
[15] Carbonell-Márquez J F, Gil-Martín L M and Hernández-Montes E 2013 Strength design optimization of structural steel members Eurocode 3 *J. Constr. Steel Res.* **80** pp 213–223
[16] Chen Z and Li J Sun L 2018 Calculation of Critical Lateral-Torsional Buckling Loads of Beams Subjected to Arbitrarily Transverse Loads *Int. J. Struct. Stab. Dyn.* **19** pp 1950-031.
[17] Trahair N S 2009 Buckling analysis design of steel frames *J. Constr. Steel Res.* **65(7)** pp 1459–63
[18] Trahair N S and Hancock G J 2003 Steel Member Strength by Inelastic Lateral Buckling *J. Struct. Eng.* **130(1)** pp 64–69.
[19] Heidari A and Galishnikova V V 2014 Direct Elastic-Plastic Limit Load and Shakedown Analysis of Steel Space Trusses With Large Displacements *Structural Mechanics of Engineering Constructions and Buildings* **3** pp 51–64
[20] Galishnikova V V and Pahl P J 2018 Analysis of frame buckling without sidesway classification *Structural Mechanics of Engineering Constructions and Buildings* **14(4)** pp 299-312
[21] Park J S and Kang Y J 2008 Flexural-torsional buckling of stepped beams subjected to pure

IOP Conf. Series: Materials Science and Engineering **675** (2019) 012009 doi:10.1088/1757-899X/675/1/012009

bending *KSCE J. Civ. Eng.* **8(1)** pp 75–82

[22] Gebre T H 2019 The development of chart based method for steel beam designs using the Russian sections *Structural Mechanics of Engineering Constructions and Buildings* **14(1)** pp 495–501

[23] Subramanian L and White D W 2016 Reassessment of the Lateral Torsional Buckling Resistance of I-Section Members: Uniform-Moment Studies *J. Struct. Eng.* **14(3)** Paper No. 04016194

[24] White D W and Duk Kim Y 2008 Unified Flexural Resistance Equations for Stability Design of Steel I-Section Members: Moment Gradient Tests *J. Struct. Eng.* **134(9)** pp 1471–1486

[25] Galishnikova V V and Gebre T H 2019 A comparative study of beam design curves against lateral torsional buckling using AISC, EC and SP *Structural Mechanics of Engineering Constructions and Buildings* **15(1)** pp 25–32

[26] Gebre T H and Negash N A 2018 The development of strength curve for compressive members using three different codes: AISC, Euro Code 3 and Russian steel construction *Proc. Engineering Systems 2018* (Moscow, Russia) pp 59–67

[27] Trahair N S, Bradford M A, Nethercot D A and Gardner L 2008 *The Behaviour and Design of Steel Structures to EC3* (Florida: CRC Press) p 513

[28] Ivan B and Jindřich M 2017 Lateral-torsional buckling of beams of mono-symmetrical cross sections loaded perpendicularly to the axis of symmetry *Theor. Anal. Euro St.* **2–3** pp 1086–1095

IOP Conf. Series: Materials Science and Engineering **675** (2019) 012010 doi:10.1088/1757-899X/675/1/012010

Window to floor ratio in the design stage in considering to visual-thermal comfort and safety in building

Nguyen Thi Khanh Phuong[1,2*], A G Tamrazyan[1], Nguyen Trung Kien[3] and Pham Van Luong[1,2]

[1] National Research Moscow State University of Civil Engineering (MGSU), Moscow, Russian Federation
[2] National University of Civil Engineering (NUCE), Hanoi, Vietnam
[3] VILANDCO Green Building Design and Consulting company, Hanoi, Vietnam

E-mail: phuongntk@nuce.edu.vn

Abstract. The building envelope with the openings play an important role in the process energy performance and ensure the well-being in a working space because of the impact of the openings to the visual and thermal comfort of the occupants. When designing buildings and structures, the standards for permissible safety should be taken into account, which are expressed by the value of critical duration of fire and the explosive load. The aim of the study is to propose the integrated approach in determining the rational Window to Floor Ratio to provide the target Daylight Factor, energy efficiency for the tropical climate, in consider to fire and the explosive load using the graphic of the relationship of the Window to Floor Ratio to the target minimum Daylight Factor. The result was shown for tropical of Vietnam, the Window to Floor Ratio was recommended of 15.2% to 18.5% corresponds to recommended target Daylight Factor of 1.35%.

1. Introduction

The Window/opening as element of the enclosing structure to ensure the target DF in the consideration with thermal comfort and safety in building. The indoor and outdoor environment are connected by the opening, design rational WFR provide the visual comfort and well-being for occupants. The research of energy consumption in Vietnam shows, the energy used for air conditioning (cooling) and lighting is the highest proportion, respectively 34% and 18% for Hanoi [1]. These two criteria are influenced by the area of the glazing. Window to Floor Ratio – the rule of thumb: Design parameters of the opening by WFR often cited as the rule of thumb [2-4]. A propose of WFR for North window façade in Malysia was estimated between 15 to 20% [5]. In European countries with a cooler climate, a window to floor ratio of 20 ÷ 25% is recommended depending on the type of building. Hopkinson argued that this ratio could provide daylight factor value of 2% at a depth of 6 m from the wall with windows [6, 5]. Perez and Capeluto (2009) [7] declare that the ratio of the window to the floor is 10 ÷ 12% for the western and eastern orientation. It is minimal for adequate natural light. In standard SS 91 42 01 [8] (Building-Daylight-Simplified method for checking the required window glass area) in Sweden simple 10% of window area to floor area is applied.

In fact, the WFR to provide good visual for a daylit space depend on the characteristic parameters of this aspect and the parameters, position of the opening. The research using the typical parameter of the aspect and the opening to define the recommendation of WFR, which afford for the target Daylight Factor (DFT). Consider the single side daylight aspect for the 3 m ÷ 3.2 m of high, dn of 6 m ÷ 9 m of depth, the standard window sill height of 0.8 m, a window head height 2.3 m ÷ 2.6 m, window high h_0 of 1.5 m ÷ 1.8 m, these parameters correspond to the common ratio d_n/h_0 of 4 ÷ 5. High WFR allows more daylight in spaces and against the effects of an internal emergency explosion by decrease pressure in volume, but it causes an increased risk of structural damage during in relate of critical duration of fire τ^{t}_{cr} by hours to the WFR, increases the heating of the room from solar radiation and visual discomfort. Accordingly, Task of the research involves the minimum of WFR to ensure the target DF and requirement of the safe level of explosive loads.

2. Methodology

2.1. The recommended target Daylight Factor for the tropical of Vietnam

The target Daylight Factor is determined when considering the following issues:

- The connectivity with the daylight climate based on cumulative diffuse horizontal illumination.
- The target illuminance value.
- The relative area of daylit space requirement.
- The relative time of usage daylight requirement in space, depends on the fixed a period of time.

Accordingly, the distribution of a daylit space is indicated by the median target Daylight Factor (DFT), which ensures that at least 50% of relative area has reached the target illuminance of 300 lux and all of the relevant floor area has reached the target minimum illuminance of 100 lux at workplane height across for half of the daylight hours per year [9-11]. This mean the ratio of DF_{TM}/DF_{T} at least should be 1/3. The relevant floor area is the entire regularly occupied floor space inside the perimeter zone of 0.5 ÷ 1 m. Based on the climate hourly typical data ASHRAE IWEC2 file of Hanoi (21.03^{0}N) [12, 13] the cumulative diffuse illuminance curve was obtained in figure 1 if consider difference fixed time period for all daylight hours from 6h to 19h (a) and for the working hour from 8h to 17h (b).

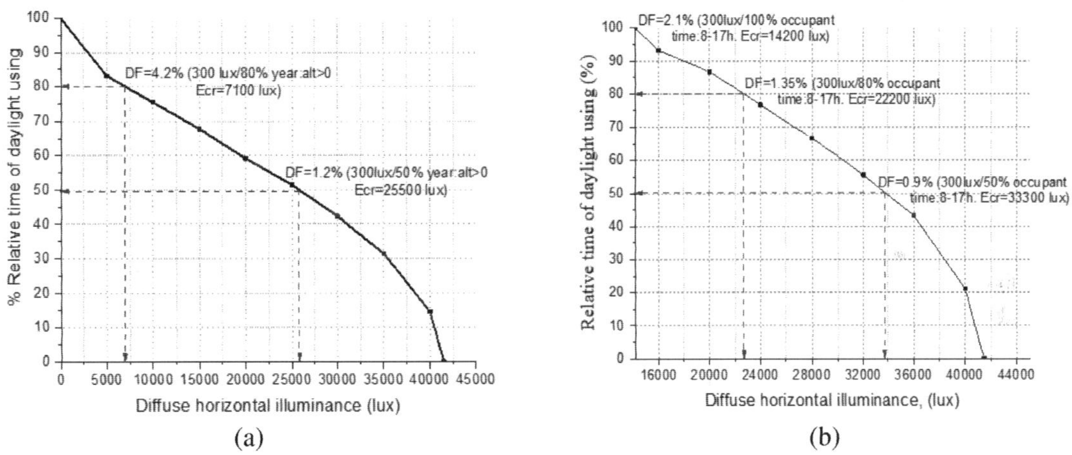

Figure 1. Cumulative diffuse illuminance curve of daylight hours from 6h to 19h (a) and working hours from 8h to 17h (b).

The relative time of usage daylight requirement in space at least 50%, the DFT values were considered at the various of percentages of daylight usage time and computed as shown in the Table 1.

Table 1. Recommendation of DF_{T} and DF_{TM} for period of working hours and daylight hours.

working hours						All daylight hour			
50%, E_{cr} = 33300 lux		80%, E_{cr} = 22200 lux		100%, E_{cr} = 14200 lux		50%, E_{cr} = 25500 lux		80%, E_{cr} = 7100 lux	
DF_{T}	DF_{TM}	DF_{T}	DF_{TM}	DF_{T}	DF_{TM}	DF_{T}	DF_{TM}	DF_{T}	DF_{TM}
0.9	0.3	1.35	0.45	2.1	0.7	1.2	0.4	4.2	1.4

2.2. Window to Floor Ratio in relation to the minimum target Daylight Factor

The more simple but approximate method for determining the size of the openings is the geometric method, in which the area of the openings is set in fractions (or %) of the floor area. This method as properly applied at the stage of study/sketch of the project to select the area of glazing. For side daylight system, according to SP 23-102-2003 and SP 52-13330-2016, the window area (by WFR criteria) can be determined by the formula (1) in terms of the minimum Daylight Factor [14]:

$$100.\frac{S_{win}}{S_{floor}}=\frac{DF_{min}.\eta_0}{\tau_0.r_0}.K_{zd}.K_z \tag{1}$$

where: S_{win}, S_{floor} - window and floor areas (m^2); DF_M - minimum Daylight Factor of the side lighting system, %, which it assigns at a point lying on the working plane at a distance of 1 m from the farthest wall; τ_0 - total light transmission coefficient of openings; r_0 - coefficient taking into account the influence of reflected light at the calculated point; K_{zd} - coefficient taking into account the shading of windows by obstructed; K_z - reserve coefficient, taking into account the pollution of the glass; η_0 - daylight characteristic windows, showing the WFR to provide the minimum Daylight Factor of 1% at an unfavorable point on the working plane, which in 1 m form the most distant wall [15].

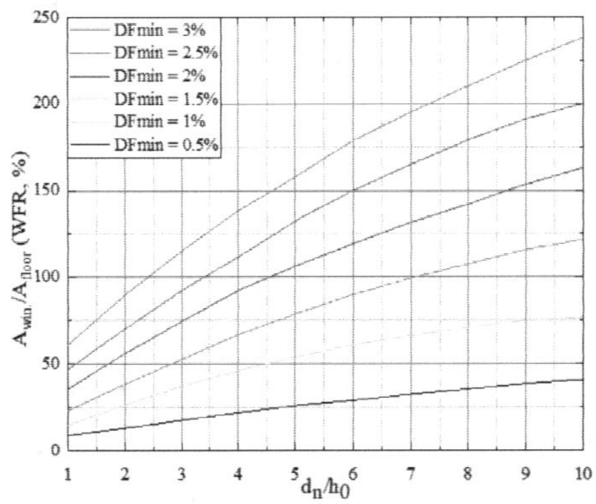

Figure 2. Graph of WFR definition with different ratios d_n/h_0.

For the most common geometrical and parameters of the side daylight system and rooms in the "Code of Practice 367.1325800.2017" [16], a graph was drawn to determine the relative area of the openings, somewhat simplifying the calculation by the formula (1). The graphics in Figure 2 can be used for preliminary determination the area of windows in buildings. The graph based on the ratio of WFR and d_n/h_0 for office buildings was developed in relation to the most common in the design practice dimensional schemes of the rooms and typical solution of translucent structures – "wooden paired opening covers". If the design adopted other types of filling openings, the values of WFR found in Figure 2 should be divided and the DFM should be multiplied by the factor k_1 given in Table 2 (SP 367.1325800.2017) [6].

Table 2. The values of coefficient k_1 (SP 367.1325800.2017)

Types of filling openings	Coefficient k_1
Single layer of window glass in steel single blind covers	1.26
Single layer of window glass, in the opening bindings	1.05
Single layer of window glass in wooden single opening binders	1.05

3. Results

3.1. Window to Floor Ratio with the target Daylight Factor

Using the graph in Figure 2, the WRF was determined from the minimum target Daylight Factor DF_{TM} shown in Table 1.

Table 3. The recommended WFR depend on DF_T and DF_{TM}

Working hours									All daylight hours					
50%, E_{cr}=33300 lux			50%, E_{cr}=22200 lux			100%, E_c =14200 lux			50%, E_{cr}=25500 lux			50%, E_{cr}=7 100 lux		
DF_T	DF_{TM}	WFR	DF_T	DF_{TM}	WFR	DF_T	DF_{TM}	WFR	DF_T	DF_{TM}	WFR	DF_T	DF_{TM}	WFR
$d_n/h_0 = 4$														

0.9	0.3	9.7	1.35	0.45	15.2	2.1	0.7	24.4	1.2	0.4	13.4	4.2	1.4	50.1
							$d_n/h_0 = 5$							
0.9	0.3	12.1	1.35	0.45	18.4	2.1	0.7	28.9	1.2	0.4	16.3	4.2	1.4	58.2

In the general, the "single layer of window glass in steel single blind covers» is often used, then k = 1.26 should be selected for the calculations. In the consideration of typical parameters $d_n/h_0 = 4 \div 5$, the results of recommendation for Window to Floor Ratio depend on the target Daylight Factor and the target minimum Daylight Factor were shown in Table 3.

3.2. Window to Floor Ratio for low energy consumption (small opening sizes)

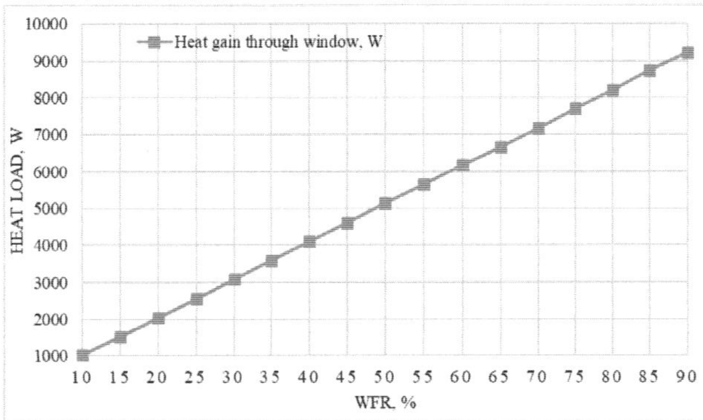

Figure 3. The influence of WFR on the heat gain.

To assess the heat transfer to the room by the heat gain through window on the influence of window sizing, a simulation with the program Heat load DACCS HkGS was conducted. The computational model is a laboratory on the 7th floor at the University of Civil Engineering in Hanoi, parameters of the room 8 x 4.2 x 4 m, the eastern window, the window to wall ratio (WWR) varies from 10.5 ÷ 90%, corresponding to the window to floor ratio (WFR) in Figure 3. The results indicate that the heat gain through window and the values of WFR are directly proportional, i.e. the greater value WFR, the greater will be heat gain through window and the more energy consumption will be.

3.3. Window to Floor Ratio, the criteria for safety in construction

Particularly acute is the issue of fire and explosion safety in high-rise office buildings due to serious consequences for people and property. In design enclosure construction with the openings, the size of the openings has an impact on the thermal hazard criteria through the determination of the critical duration of fires τ_{cr}^{t} and on the explosive loads during internal emergency explosions [17-19].

Figure 4. Gas explosion in Volgograd (Russia).

In Russia, the system of standards and regulations in construction and fire, explosion Safety include SNIP 21-01-97 «Fire safety of buildings and structures» for metal construction [20], GOST R

12.3.047–2012 “Occupational safety standards system. Fire department safety of technological processes. General requirements. Control methods” [21]. Documents work together is SNIP 2.03.01-84 “Concrete and reinforced concrete structures” [22]. The problems of the possible consequences of a fire-explosion indoors in Russia are dealt in documents [23-27]. The stability of buildings in the event of the occurrence and development of fires is ensured while maintaining the structures of the external walls, internal partitions and the flooring of load-bearing properties determined by fire resistance with the values of Critical duration of fire τ^{t}_{cr}. Reducing the risk of structural failure during fires may increase the fire resistance of reinforced concrete structures (on the basis of loss of carrying capacity), that means the smaller the opening area, the higher the fire resistance of structures [18, 28].

Experiments of the study [29] show that with a fixed height of the openings, the values of Critical duration of fire τ^{t}_{cr} depend on the ratio $\frac{S_{win}}{S_{floor}}$, which is described in the expression:

$$\tau^{t}_{cr} = \ln\left(\frac{S_{win}}{S_{floor}}.100\right) \tag{2}$$

With different height of the openings H varies from 0.1 ÷ 5 m, the critical duration of fires τ^{t}_{cr} described by formulas (3), (4) and (5):

$$\tau^{t}_{cr} = f\left(H, \frac{S_{win}}{S_{floor}}, \frac{A}{B}\right) \tag{3}$$

$$A = \frac{\pi}{3}.V_{l}^{2}.\psi_{ud}, \quad B = \frac{c_{r}.\rho_{0}.T_{0}.V}{\eta.(1\text{-}j).Q^{r}_{n}} \tag{4}$$

where: A – the value depends on the properties of the combustible material; B – the filling time of the combustion products (smoke, radiation, heat etc.) depends on the shape of the room (s); H– the openings high (m); S_{win} – the opening area (m^2); S_{floor} – the floor area (m^2); V – the room volume (m^3); ρ_0 – average gas density at the beginning of a fire ($kg.m^{-3}$); V_l – flame velocity linear ($m.s^{-1}$); ψ_{ud} – mass burnout rate ($kg.s^{-1}.m^2$); C_r – gas heat capacity at constant pressure, respectively ($J.kg^{-1}.K^{-1}$); T_0 – Room temperature at the initial time (K); η – burning ratio.

The research demonstrated that the τ^{t}_{cr} proportional dependence to the ratio $\frac{S_{win}}{S_{floor}}$, which varies from 0.4% to 65%, in other words WFR must not be exceed 65% and not depend on which level the openings are located. Internal deflagration explosions are explosions of gas-air mixtures occurring in closed and semi-closed volumes. Excessive pressure in an internal deflagration explosion in a closed volume reaches 700÷ 900 kPa. During explosions inside buildings and structures, the excess pressure should not exceed 10 ÷ 15 kPa, which is limited by the strength of building structures [18, 28]. Therefore, in buildings and premises as a result of the destruction of the enclosure construction or the operation of safety structures, the closed volume becomes open. To limit the growth of overpressure in rooms during internal deflagration explosions, glazed window openings or easily reset structures are used. The design of buildings with explosive technologies is carried out in accordance with the recommendations SNIP 31-03-2001 «Industrial buildings» [30], where it is required for each 1000 m^3 of free volume of the room to have at least 50 m^2 of released openings. Another words, the minimum values of the WFR lies within 15 ÷ 25% with a building height of 3 ÷ 5 m. It is assumed that the explosive load will not exceed 5 kPa, i.e. guaranteed the safe level [18].

4. Discussion

4.1. The proposed Window to Floor Ratio in design building for tropical of Vietnam

From the results of Table 4, it is seen that:

- In aspect, with a certain value of the target Daylight Factor, the Window to Floor Ratio is proportional with the ratio d_n/h_0 and target illuminance values.

- Reducing the openings for energy savings and increasing the fire resistance of structures along with consideration to ensure the target daylight factor and a safe level with limited explosive load will give acceptable WFR

The research results show that the rational WFR match all criteria at 80% of the working time. I.e. the minimal WFR for explosion safety should not lower than 15%. Specifically, with the target illuminance 300 lux, it corresponds to DF_T equals 1.35% and E_{cr} equals 22200 lux.

4.2. Modelling daylight in a testing space

To demonstrate compliance with the proposal, a model testing was provided with a minimum WFR value of 15% for an aspect with $d_n/h_0 = 4$, the input parameters of which are described in Table 4.

Table 4. The influence of window sizing on the heat gain.

Model (height x width x depth) (m)	3 x 9 x 6
Reflection coefficients (Floor–Walls–Ceiling)	0.2:0.5:0.7
Level of the work surface from floor (m)	0.8
parameters of window (height x width)	1.5 x 5.4
Working hours (hours)	8h00 ÷ 17h00

In that regard, the simulation results must meet the criteria for a daylit space: a target illuminance of 300 lux is achieved across at least 50% of the relevant floor area for at least 80% of the working hours (or more than 50% of all daylight hours).

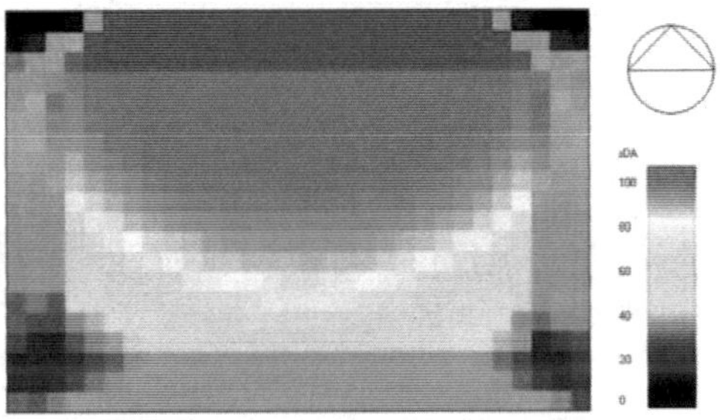

sDA (%) in accepted	UDI (%) in accepted	ASE (%) in accepted
65.48	97.6	56.85

North

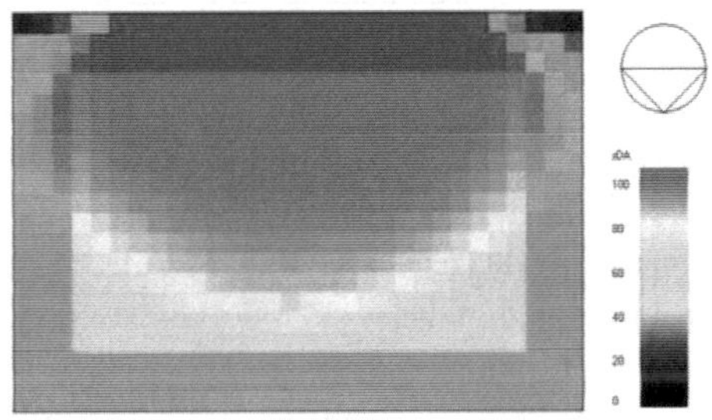

sDA (%) in accepted	UDI (%) in accepted	ASE (%) in accepted
71.43	93.2	20.24

South

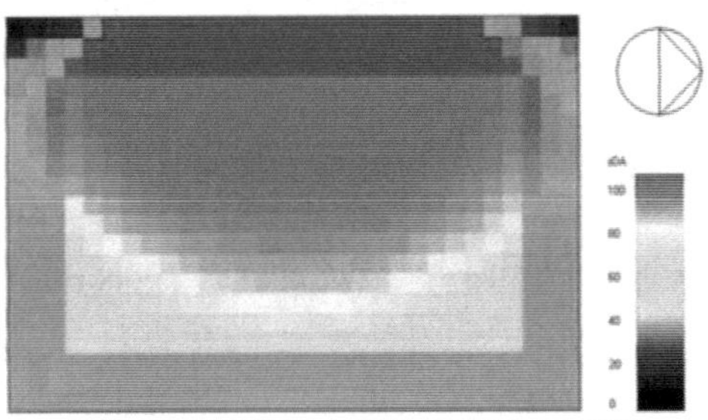

sDA (%) in accepted	UDI (%) in accepted	ASE (%) in accepted
66.67	97	22.62

East

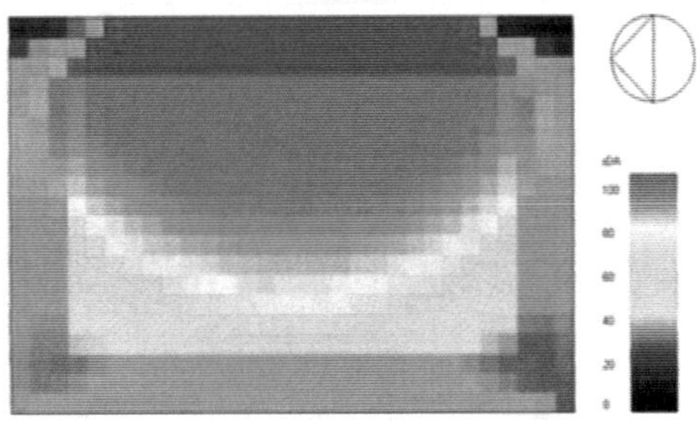

sDA (%) in accepted	UDI (%) in accepted	ASE (%) in accepted
69.94	95.8	8.04

West

Figure 5. The simulation results

The values $sDA_{300,80\%}$ for different orientations were determined by program Design Builder V5.2. Radiance illuminance simulation. The simulation using the values indicate changes of daylight in time and space are described in the Approved Method [31]:

- IES Spatial Daylight Autonomy (sDA): describes how much light is available during standard operating hours. In particular, it describes the percentage of floor area that receives at least 300 lux (for offices) within 80% of working hours per year in this case study. The sDA value between 55% and 74% indicates the space in which natural lighting is "nominally taken".

- Annual Sunlight Exposure (ASE): One metric of the probability of visual discomfort is the number of hours that direct sunlight can potentially enter a space, which is called ASE. In particular, ASE measures the percentage of floor space that receives at least 1000 lux for at least 250 busy hours per year. In the supporting research, ASE of daylit spaces were predicted to have no more than 10%, in other words, the accepted ASE should be at least 90%.
- Useful Daylight Illuminance (UDI): For UDI we specify the metric pass threshold value of 50% means that a given cell must achieve an illuminance of between 100 and 3000 lux for at least 50% of the occupied hours in the year.

The simulated results were shown in Figure 5. The highest value of the sDA is 71.43% for the southern orientation and decreases to 65.48% for the northern orientation. These sDA values are within the "nominally accepted" of 55÷74% and correspond to the recommendation sDA performance criteria in document [30]. The UDI achieves an illuminance of between 100 and 3000 lux for 93% to 97%. But the accepted values $ASE_{1000,250h}$ for the western orientation was 8.04% and increased to 56.85% for the northern orientation, which did not meet the criteria. Thus, with these spaces, in order to enhance the natural lighting condition, it is necessary to consider technical solutions for enhancing the sDA and the ASE in range by using Sunscreens and/or light shelves [32-34].

5. Conclusions and future work

This paper concludes by proposing method define the relation of WFR to the target DF. An integrated approach to determining WFR was taking into account in consideration of safety requirements. For Vietnam, the recommended WFR from 15.2% to 18.4% for the DF_T 1.35. The research gives opportunity to design building envelope in accordance with the adaptation to the climate of the region, makes the recommendations for the regulation of Daylight Factors and WFR in the early stage of design enclosure construction with openings. On this basis, rational daylight usage was ensured by the working time and the function of a space.

However, the focus of this study limits to the quantity and quality of daylighting performance in impact of the WFR of unprotected daylight systems. A dynamic daylighting assessment shows that using direct sunlight from the sky through the openings causes of visual discomfort, which demonstrated by barely meet accepted criteria $ASE_{1000,250h}$. Therefore, the technical solutions to increase the visual comfort by having good daylight distribution as by using the sunscreen, light shelf and/or shading devices should be more specific in further researches.

References:

[1] QCVN09:2013/BXD 2013 *National Technical Regulation on Energy Efficient Buildings "Guiding the application on national regulation"* (Ministry of Construction, Hanoi) p 8
[2] Aries M B, Veich J A and Newsham G R 2010 *Journal of Environmental Psychology* **30** 533-41
[3] Design Code SP 23-102-2003 *Natural lighting of residential and public buildings* (Moscow: Standartinform) p 81
[4] Design Code SP 52-13330-2011 *Daylighting and artificial lighting* (Moscow: Standartinform) p 80
[5] Mirrahimi S, Ibrahim N L N and Surat M 2013 *Research Journal of Applied Sciences, Engineering and Technology* **6(5)** 931-35.
[6] Hopkinson R G 1964 *Hospital lighting: Some basic principles* (London: Heinemann)
[7] Perez Y V and Capeluto I G 2009 *Applied Energy* **86** 533-41
[8] Swedish Standard SS 91 42 01 Building design - Daylight - Simplified method for checking the required window glass area (Swedish)
[9] Mardajevic J and Christoffersen J 2013 *Proc. CIE Centenary Conference* (Paris, France) 15-16
[10] Mardajevic J, Christoffersen J and Raynham P 2013 *Lux Europa 2013 12th European Lighting Conference* (Krakow, Polad)

[11] Mardaljevic J and Christoffersen J 2017 *Building and Environment* **113** 200-9
[12] Phuong N T K. and Solovyov A.K. 2018 *Scientific Journal Matec Web of Conferences* ***193***, Hochiminh (EDP Science Publishing) p 8 https://doi.org/10.1051/matecconf/201819304013
[13] Nguyen T K P et al. 2019 *Advances in Intelligent Systems and Computing* ***982*** (Springer Nature Switzerland AG 2020) 35-47 https://doi.org/10.1007/978-3-030-19756-8_4
[14] Design Code SP 52.13330.2016 *Daylighting and artificial lighting* (Moscow: Standartinform) p 106
[15] Solovyov A K 2008 *Physics of the environment* (Moscow: ABC Publisher) p 341
[16] Design Code SP 367.1325800.2017 *Residential and public buildings. Rules for the design of natural and combined lighting* (Moscow: Standartinform) p 121
[17] Baker W E, Cox P A, Westine P S, Kulesz J and Strehlow R A Explosion 1983 *Hazards and evaluation* (Elsevier Science) p 840
[18] Tamrazyan A and Avetisyan L 2014 *Applied Mechanics and Materials* **638-640** p 62-65
[19] Clavin P and Williams F A 2012 *Philosophical Transactions of the Royal Society A: Mathematical, Physical and Engineering Sciences* **370** 597-624
[20] Design Code SNIP 21-01-97 1997 *Fire Safety of Buildings and Structures* (Moscow: Standartinform)
[21] Russian Standard GOST P 12.3.047-2012 *Occupational safety standards system. Fire safety of technological processes. General requirements. Methods of control* (Moscow: Standartinform) p 96
[22] Design Code SNIP 2.03.01-84 1985 *Concrete and Reinforced Concrete Structures* (Moscow: Gosstroy USSR) p 81
[23] Doronin F L, Trukhanova L N and Fomina M V 2014 *Scientific and Engineering Journal for Construction and Architecture* **1** 33-40
[24] Tamrazyan A, Avetisyan L 2016 *MATEC Web of Conferences* **86** *Paper № 01029*
[25] Tamrazyan A G, Avetisyan L A 2016 *Procedia Engineering* **153** p 721-725
[26] Kabantsev O V, Tamrazyan A G 2014 *Magazine of Civil Engineering* **49(5)** p 15-26
[27] Tamrazyan A 2014 *Applied Mechanics and Materials* **475-476** p 1563-1566
[28] Mishuev A V, Kazenov V V, Komarov A A, Gromov N V, Luk'yanov A V and Prozorovsky D V 2012 *Fire and Explosion Safety* **21(3)** 49-56
[29] Rubtsov V B 1998 *The dynamics of fire hazard factors and the calculation of the critical duration of a fire in industrial premises* DSc Thesis (Moscow Institute of Fire Safety, Moscow, Russia)
[30] Design Code SNIP 31-03-2001 2001 *Factory and Industrial Occupancies*. (Moscow: Standartinform) p 17
[31] Heschong L, Wymelenberg V D, Andersen M, Digert N, Fernandes L, Keller A and Reinhart C 2012 *Approved Method: IES Spatial Daylight Autonomy (sDA) and Annual Sunlight Exposure (ASE)* (No. STANDARD) (The Illuminating Engineering Society of North Ameria, New York)
[32] Lim Y W, Ahmad M H and Ossena D R. 2013 *Indoor Built Environment* **20** 932-51
[33] Ochoa C E, Aries M B C, Van Loenen E J and Hensen J L M 2012 *Applied Energy* **95** 238-45
[34] Berardi U and Anaraki H K 2015 *Energy Procedia* **78** 1793-98

Authors's background

Name	Prefix	Research Field	Email
Nguyen Thi Khanh Phuong	PhD. student	Daylighting assessment, Building energy efficiency, Green building	phuongntk@nuce.edu.vn
Tamrazyan A.G.	Doctor of Technical Sciences, Professor	Reinforced concrete structures, Reliability assessment, Engineering safety of buildings	tamrazian@mail.ru

Nguyen Trung Kien	Msc. Architecture (UK), Lotus AP	Energy simulation, Building energy assessment	ntk.vilandco@gmail.com
Pham Van Luong	PhD. student	Building physics, construction Heat Engineering	luonganhhagiang@gmail.com

IOP Conf. Series: Materials Science and Engineering **675** (2019) 012011 doi:10.1088/1757-899X/675/1/012011

Radon risk reduction to population

S P Vysotskii* and L G Levchenko

Donbass National Academy of Civil Engineering and Architecture, Makeevka, Ukraine

Email: sp.vysotsky@gmail.com

Abstract. Ranking the risks of deaths from various factors affecting the human body shows the leading effect of radon. The main source of radon entry into the dwelling is the soil under the building. Health risks analysis was performed at various levels of radon concentrations. Based on analyses of radon exposure data made in the USA, analytical dependencies of health risks are derived depending on the radon concentration. Analysed the path of penetration of radon in the home. The cost of the work to protect the home from radon penetration has been estimated. Analysis of the appearance of oncological diseases of the respiratory organs for residents of the region according to statistical data depends on the number of floors. The approach to the choice of the level of protection from radon exposure according to the ALARA principle – to ensure the most reasonable achievable level of radon concentration is substantiated. The analysis of the protective properties of various building materials is performed.

1. Introduction

From primitive times, humanity lives in a habitat with different levels of radioactivity. Getting rid of natural levels of radioactivity is almost impossible. Radioactivity is caused by the decay of potassium-40 isotopes, rubidium-87 and atoms of three radioactive families – derivatives of uranium-238 (U-238), uranium-235 (U-235) and thorium-232 (Th-232) [1–5].

The largest share of radiation 80 % of people receive from natural sources (emission of radon gas into the atmosphere, strontium, etc.). Almost 17 % of radiation exposure a person acquires during the passage of medical examinations. And less than 0.1 % of the total number of radioactive radiation accounts for nuclear power plants and other facilities of this type [6–8].

We take into account the fact that the greatest radioactive contamination occurs as a result of natural sources, one of which is radon gas (^{222}Rn). Radon is a natural radioactive gas, it is heavy, tasteless, colorless, odorless. Radon is formed as a result of the decay of uranium and thorium. Vaporizing from the ground or standing out from building materials (concrete, granite, slate, red clay bricks, etc.), radon accumulates in mines, basements, on the first floors of houses. The radon concentration can be 8 times higher than the background values, if the room is not ventilated at a certain frequency.

Radon, entering the human body in two ways: when breathing, with the air or through the digestive system along with water from underground sources (wells, springs, wells). Penetrating into the body, radon can begin to destroy living cells. Especially great danger is the inhalation of air with a radon concentration of more than 200 Bq/m^3. Irradiation with high doses of radiation can cause lung cancer, leukemia and other types of cancer [9–13].

The disintegration of radon nuclei in human lungs causes microburns, which can lead to lung cancer, damage to the chromosomes of the bone marrow, as well as damage to the sex, hematopoietic, and immune cells.

The most dangerous are the effects of two toxins: nicotine and the decay products of radon. In figure 1 shows the United States Environmental Protection Agency (EPA) on the number of deaths due to exposure to radon.

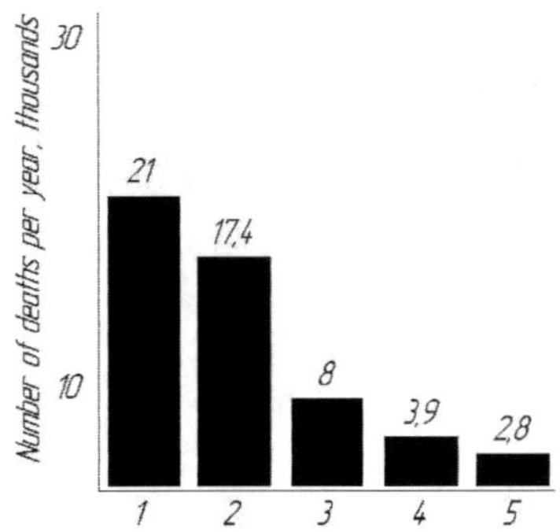

Figure 1. Ranking the number of deaths from various causes
1 – exposure to radon; 2 – drunk driving; 3 – fall; 4 – drowning; 5 – fire.

Radon enters the environment from soil, building materials, water, and the combustion of natural gas. The average concentration of radon in atmospheric air is about 5 Bq/m3, and varies significantly in different zones of the earth's surface (Figure 2) [14–16].

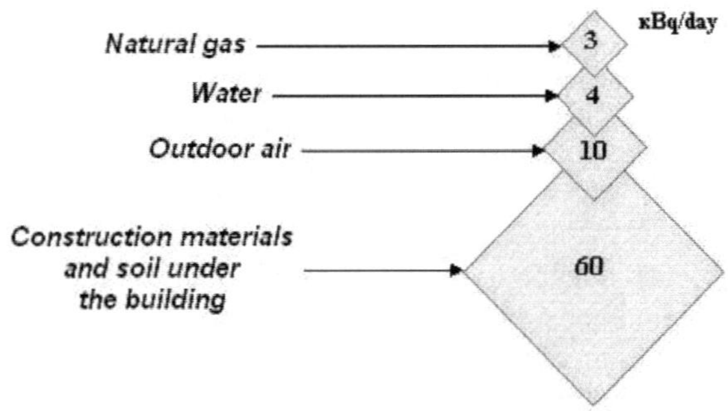

Figure 2. Release of radon into the environment [17–19].

2. Purpose of research

The purpose of research is a quantitative assessment of the effect of radon on health, sources of radon in homes and the choice of priority methods to reduce the negative impact of radon.

3. Presentation of the main research material

Let us consider the main methods of reducing the radon activity in buildings when choosing methods to reduce radon hazard. Obviously it is necessary to follow the rule to reduce the level of radon as much as possible. It is believed that the level in most homes can be reduced to 100–150 Bq/m^3. In houses and structures that are operated, the average annual radar-222 EDPA should not exceed 100 Bq/m^3. In Russia, the norm for buildings put into operation is 100 Bq/m^3, and those used are 200 Bq/m^3). It must be remembered: the urgency of action depends on the concentration of radon in the house, the sooner it is necessary to take measures to reduce its concentration or, even, to resettle the residents.

If, as a result of the research, the radon concentration is [20–22]:

- 7400 Bq/m^3 or higher, it is necessary to reduce the level to acceptable performance. It is recommended to do this in a few weeks. If possible, consider the appropriateness of temporary resettlement until the radon level in the house is lower.
- 740–7400 Bq/m^3 – this level is much higher than allowable for dwellings, it is necessary to reduce it to a lower level. It is recommended to do this for a few months.

- 200–740 Bq/m^3 – this level is also unacceptable for dwellings. Everything necessary is needed to reduce the level to 150 Bq/m^3 or lower. We recommend doing this for several years or less if the results are closer to the upper limit of the interval.
- 150 Bq/m^3 – the level acceptable for dwellings or slightly exceeds it.

It is also necessary to take into account the specific living conditions that may affect the degree of risk and necessitate additional measures. The danger of exposure to radon depends on the amount of radon penetrating the room and the time people spend in it.

In the United States, an in-depth analysis of the risks of lung cancer was performed for residents of two states, depending on the current level of exposure. At the same time, risks for residents of Minnesota and Iowa differ significantly [23, 24]. Despite the fact that the effective values of 4 pCi/dm^3 are close to the value recommended by the World Health Organization (2.7 pCi/dm^3 or 100 Bq/m^3), for residents of Minnesota with an exposure level of less than 4 pCi/dm^3 the risk will be 20 % and at the level of 2.7 pCi/dm^3 – 10 %. At the same levels for Iowa, the risk is 70 % and 60 %, respectively.

Over a 74-year period, approximately 50,000 Minnesota residents and 20,000 Iowa residents can protect themselves from death associated with lung cancer by applying measures to reduce the radon concentration in dwellings at the current level of 4 pCi/dm^3.

The authors mathematically processed the logistic dependences of the risk levels of lung cancer depending on the radon concentration (Figure 3). At the same time, the level of risk for Iowa residents is unsubscribed by dependency:

$$R = Rn^{1,536} \cdot \left(Rn^{1,536} + 4,29\right)^{-1} \quad (1)$$

and for residents of Minnesota:

$$R = Rn^{1,94} \cdot \left(Rn^{1,94} + 53,3\right)^{-1} \quad (2)$$

where R – the risk level in fractions of a unit; Rn – the radon concentration level, pCi/dm^3.

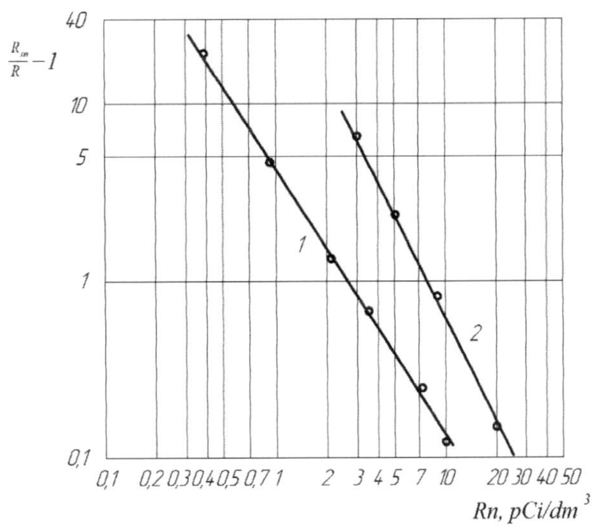

Figure 3. Dependence of the reduced risk of lung cancer from the concentration of radon.

It should be noted that in the home there is also another source of danger – carbon monoxide. However, a comparison of the risks of deaths shows that the number of deaths due to the prevention of carbon monoxide will be 30 deaths per year compared to measures to reduce the radon level of about 700 deaths per year.

The number of oncological diseases of the respiratory organs in three districts of the city of Gorlovka was analyzed. Statistical data on diseases in residential areas, depending on the number of floors, are presented in figure 4.

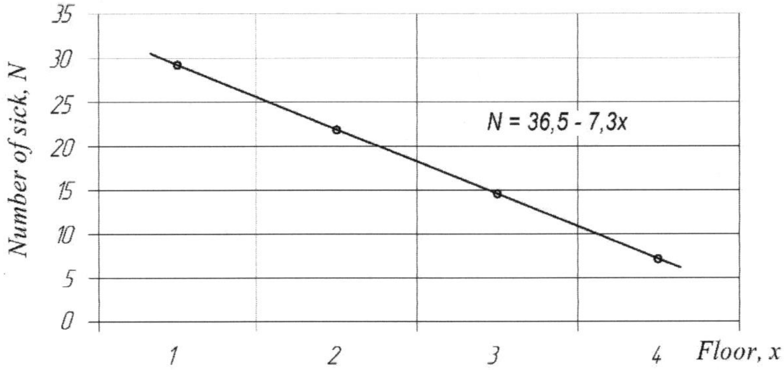

Figure 4. Correlation of the number of oncological diseases from the number of floors

The data obtained allow us to statistically predict the number of patients with upper respiratory tract (bronchi, lungs) in high-rise buildings ($x > 7$) and proves that the incidence decreases with increasing height of residence.

Sources of income are glad she is in the building shown in figure 5. Radon enters basements and basements through lightweight aggregates of concrete blocks and walls of hollow bricks that come into contact with radon-bearing soil.

When placing the foundation slab directly above the soil, radon penetrates through the gap between the walls and the slab below the floor level, permeable concrete aggregates and unconsolidated passages for water, sewage, gas and electrical wiring pipelines.

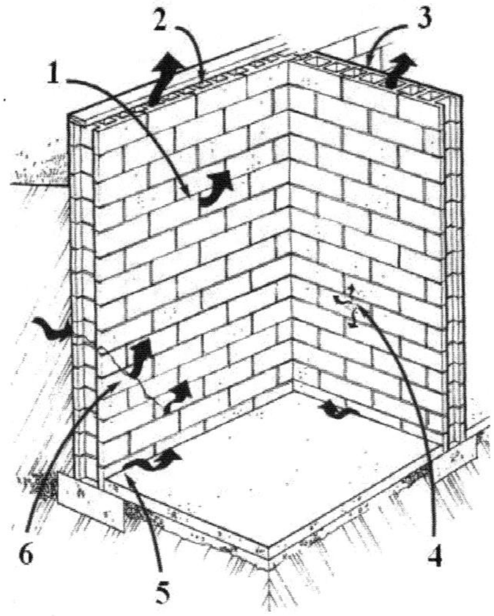

Figure 5. Typical Radon Penetration Paths
1 – omissions or lack of mortar (binder) between bricks; 2 – open cavities in the upper part of wall panels; 3 – open cavities in the upper part of the building blocks with penetration into the space between the walls and floors; 4 – pores on the surface of building blocks; 5 – joints between the walls and the floor; 6 – cracks in the masonry or in the joints filled with mortar.

The calculated data on the concentration of free radon, which is not included in the crystal lattice of minerals and is able to move along cracks and pores, is given in table 1 [29, 30].

Table 1. The content of free radon in various rocks.

Breed	The concentration of uranium, g/t	Density, g/cm^3	Porosity,%	K_{em}, %	Radon, Bq/m^3
Sandstones	2.9	2.5	20	30	133
Clay	4.0	2.0	20	40	200
Carbon shale's	15.0	2.6	20	15	500
Coal	3.5	1.3	15	35	100
Sedimentary rocks	2.5	1.8	20	55	80

The development of architectural and planning groups of protective measures in the practice of the CIS countries [25, 26] is aimed at:

- the establishment of an intermediate environment (ventilated basement and technical floors, basements) on the route of radon from the soil and the use of ground floor rooms to accommodate premises with a limited time for people to stay in them;
- maximum restriction of premises designed for the long stay of people from places of radon accumulation;
- reduction of radon intake from the soil into the air of the premises, designed for the long stay of people on the basis of engineering and design solutions;
- a decrease in the volumetric activity of radon in the indoor air created by ionizing sources of construction production on the basis of the technical group of protective measures.

To assess the effectiveness of protective measures, it is advisable to determine the insulation characteristics of the screens.

The flow of radon through the barrier decreases with increasing thickness. To take into account the thickness, as an independent parameter that determines the tightness, we use the quotient on the sample thickness and the diffusion length.

The protective properties of insulating materials are estimated by the radon diffusion length "l". The value of "l" corresponds to the thickness of the layer of material, when passing through which the intensity of the flow of radon decreases by e = 2.72 times and is determined by the formula [27, 28]:

$$l = \sqrt{\frac{D}{\lambda}} \tag{3}$$

where λ – the radon decay constant, equal to $2.10 \cdot 10^{-6}$ cm^{-1}; D – radon diffusion coefficient, m^2/s.

The diffusion coefficient of the studied building materials in the order of 10^{-9} to 10^{-6} m^2/s corresponds to a diffusion length of 6 to 100 cm. Thus, the material must be thicker, up to 300 cm, in order to obtain a certain level of protection. Only some concrete, especially with polymer impurities, can prevent diffusion. The polymer fills most of the porous system and closes the path of diffusion through the material.

Insulating materials exhibit different properties with a diffusion coefficient less than 10^{-11} m^2/s. The diffusion length is less than 2 mm. For example, the content of carbon black or graphite in black butyl rubber does not lead to an open structure and high permeability to radon.

The emanation of radon in the material of an ionizing source characterizes the emanation coefficient η, the value of which for primary materials is from 0.02 to 0.46. Higher values of η are characteristic of secondary minerals that have arisen in rocks as a result of changes in their physic-chemical conditions, and can reach up to 0.9.

The coefficient of emanation depends on the process temperature for the main groups of building materials (concrete, ceramic products). The lowest values of η were found in materials subjected to high-temperature processing in the process of technological production (Table 2).

It is established that the process of changing η in the process of heat treatment is irreversible and occurs due to the removal of physically and chemically bound water and as a result leads to the formation of a new chemical mineral.

Table 2. Effect of firing temperature on the coefficient of emanation of the produced material by firing η=f(toc).

Type of material	η	Type of material	The dependence of the coefficient of emanation from temperature t_{oc} η=f(toc)	Emission coefficient η=f(toc)
Sandstone	0.17	Ceramic wall tiles	106	0.05
Clay	0.17	Ceramic floor tile	1100	0.01
Clay (fireproof)	0.13	Ceramic brick	1250	0.003
Brick (raw)	0.12	Refractory brick	1500	0.0027

In the implementation of work to isolate residential buildings from radon intake in world practice, the rule of the need to ensure that the level of concentration is as low as permissible, ALARA (As low reasonable achievable).

In table 3 shows the analysis of the profitability of measures to protect against the effects of radon adopted in the United States (cost-effectiveness analysis – CEA). The World Health Organization (WHO) International Radon Program recommends this approach. At the same time, cost per life saved (CLS) is preferred as a measure of effectiveness. Naturally, in each individual country or even in a region, these values may differ significantly.

Table 3. The profitability of the implementation of measures to reduce the level of radon concentration [31]

Parameter	unit of measurement	Basic cost level	Lower cost level	Upper value of costs	Lower threshold
Radon Threshold	pCi/dm^3	4	4	4	2.7
Dose	pCi/m^3	150	150	150	100
Rn concentration measurement	USD	50	25	100	50
Equipment installation	USD	1400	800	2540	1400
Fan repair	USD/year	12	ten	14	12
Fan power	W	80	24	150	80
Electricity cost	USD/kWh	0.10	0.08	0.12	0.10
Additional payments for heat or cold	USD/year	190	70	400	190

4. Conclusions

1. Analysis of the effectiveness of measures affecting the radon emanation rate in residential areas shows that to achieve an acceptable level of radon concentrations, it is necessary to use insulating screens. The priorities of screen selection along the radon diffusion length in the material are given.

2. Based on the mathematical analysis of the work of foreign researchers, analytical dependencies of the risk level of diseases on radon concentration were derived.

3. The approach to choosing the level of protection from radon exposure according to the ALARA principle is justified - to ensure the most reasonable attainable level of radon concentration.

4. Analysis of the number of oncological diseases in the city of Gorlovka, Donetsk region. It shows a linear decrease in the level of diseases with an increase in the floors of residence.

References

[1] Clement C H, Tirmarche M, Harrison J D, Laurier D, Paquet F, Blanchardon E and Marsh J W 2010 *ICRP Publication 115. Annals of the ICRP* **40(1)** pp 1–64

[2] Schubert M, Peña P, Balcázar M, Meissner R, Lopez A and Flores J H 2005 *Radiation Measurements* **40(2)** pp 633–637

[3] Burghele B and Cosma C 2013 Intercomparison between radon passive and active *Rom. Journ. Phys.* **58S** pp 56–62

[4] Amin R M 2015 A study of radon emitted from building materials using solid state nuclear track detectors *J. Radiat. Res. Appl. Sci.* **8(4)** pp 516–522

[5] Hassan N M, Hosoda M, Iwaoka K, Sorimachi A, Janik M, Kranrod C, Sahoo S K, Ishikawa T, Yonehara H, Fukushi M and Tokonami S 2011 *Prog. Nucl. Sci. Technol.* **1** pp 404–407

[6] Zhang Z H, Zhu L, Zhuo X, Yi H, Zhang H and Pei Z 2011 China indoor radon levels and control of new measures *Proc. 12th International Conference on Indoor Air Quality and Climate 2011* (Austin, USA) vol 3 http://toc.proceedings.com/18032webtoc.pdf

[7] Al-Zoughool M and Krewski D 2009 Health effects of radon: A review of the literature *Int. J. Radiat. Biol.* **85** pp 57–69

[8] Field R W 2014 Radon Occurrence and Health Risk *Occup. Med. Secrets, Hanley Belfus, Philadelphia* **1999** pp 2–9

[9] Mitchell A L, Griffin W M and Casman E A 2016 Lung Cancer Risk from Radon in Marcellus Shale Gas in Northeast U.S. Homes *Risk Anal.* **36(11)** pp 2105–2119

[10] Oh S S, Koh S, Kang H and Lee J 2016 Radon exposure and lung cancer: Risk in nonsmokers among cohort studies *Ann. Occup. Environ. Med.* **28** http://dx.doi.org/10.1186/s40557-016-0099-y

[11] Chen J 2013 Canadian lung cancer relative risk from radon exposure for short periods in childhood compared to a lifetime *Int. J. Environ. Res. Public Health* **10** pp 1916–1926

[12] Du Z, Liu J and Shan S 2017 An Open-Data Analysis of Heterogeneities in Lung Cancer Premature Mortality Rate and Associated Factors among Toronto Neighborhoods https://arxiv.org/pdf/1705.08516.pdf

[13] Tomášek L 2004 Temporal patterns of lung cancer risk from radon and smoking - consequences to remediation measures *Radiat. Prot. Dosim.* **36** https://inis.iaea.org/search/search.aspx?orig_q=RN:36010945

[14] Rani B, Singh U and Maheshwari R 2011 Menace of Air Pollution Worldwide Soc. Educ. 2

[15] Perrier F, Crockett R G M and Gillmore G K 2012 Preface "Radon, health and natural hazards II" *Nat. Hazards Earth Syst. Sci.* **12** pp 799–803

[16] Al Zabadi H, Mallah K and Saffarini G 2015 Indoor exposure assessment of radon in the elementary schools, Palestine *Int. J. Radiat. Res.* **13** pp 221–228

[17] Becker K 2003 Health Effects of High Radon Environments in Central Europe: Another Test for the LNT Hypothesis? *Nonlinearity Biol. Toxicol. Med.* **1** pp 3–35

[18] Keskikuru T, Kokotti H, Lammi S and Kalliokoski P 2001 Effect of various factors on the rate of radon entry into two different types of houses *Build. Environ.* **36** pp 1091–1098

[19] Watson R J, Erichsen E, Finne T E and Ganerød G V 2015 Radon from building materials *Mineralproduksjon* **1170** pp 7–13

[20] Shoqwara F, Dwaikat N and Saffarini G 2013 Measurement of Radon Exhalation Rate from Building Materials Journal of Physics **2(1)**

[21] Karpińska M, Mnich Z, Kapała J, Barula A, Szpak A and Augustynowicz A 2008 Indoor radon concentrations in concrete slab buildings situated in green hills housing estate in Białystok,

Poland *Polish J. Environ. Stud.* **17** pp 33–38

[22] Gandolfo G, Lepore L, Pepperosa A, Remetti R and Franci D 2017 Radiation protection considerations on radon and building materials radioactivity in Near Zero Energy Buildings *Energy Procedia* **140** pp 13–22

[23] Stamat I P, Svetovidov A V, Stamat D I, Gorsky G A and Venkov V A 2009 Estimation of the contribution of radon emanation from the surface of facing products to the population exposure *Radiation Hygiene* **2(4)** pp 16–22

[24] Khaydarov R, Gapurova O, Khaydarov R, Garipov I, Cho S Y and Nikolova E 2018 Silica-based organic nanocomposite for prevention of hazardous gas seeping into buildings through construction materials *New Polymer Nanocomposites for Environmental Remediation* pp 643–647

[25] Keller G, Hoffmann B and Feigenspan T 2001 Radon permeability and radon exhalation of building materials *Sci. Total Environ.* **272** pp 85–89

[26] Arvela H 2001 Radon mitigation in blocks of flats *Science of the Total Environment* **272** p 137

[27] Keller G and Hoffman B 2000 The Radon Diffusion Length as a Criterion for the Radon Tightness (Hamburg: University of Saarland) pp 1–4

[28] Al-Sharif A and Abdelrahman Y S 2001 Factors Affecting Radon Concentration in Houses *Turkish J. Phys.* **25** pp 153–158

[29] Tuccimei P, Castelluccio M, Soligo M and Moroni M 2009 Radon exhalation rates of building materials: experimental, analytical protocol and classification criteria In: Building Materials Chapter 7

[30] Khalid N, Majid A A, Yahaya R and Yasir M S 2014 Radiological risk of building materials using homemade airtight radon chamber *AIP Conference Proceedings vol 1584* (American Institute of Physics Inc.) pp 207–210

[31] Renken K J and Konopacki S J 1993 An innovative radon mitigation-energy conservation retrofit system for residential buildings *Air Waste* **43** pp 310–315

IOP Conf. Series: Materials Science and Engineering **675** (2019) 012012 doi:10.1088/1757-899X/675/1/012012

The method of displacements calculating in the joints of timber structures

A S Markovich*, M Abu Mahadi, S A Gazizova, D A Miloserdova

Peoples' Friendship University of Russia (RUDN University), Moscow, Russia

Email: markovich-as@rudn.ru

Abstract. This article is devoted to the peculiarities of calculating compliance modifications of the timber elements on mechanical linkages. Here we analyze in detail the method of C M Kochenov using in the design standards of the Russian Federation. This method allows to determine accurately the number of links that are able to resist the shear force in the linkage, however, this method does not include analyzing of shear deformation in modifications. And it is a disadvantage. In this article the author considers another approach, based on the idea of determining the shear deformation of the mechanical linkage, installed in the connection. To calculate the shear deformations of connections in the linkage it is necessary to conduct a preliminary analysis of the construction, to establish the type of connection and thickness of the connected elements, to determine the number of shear planes, the number of established joints and the level load on the joint. After determining these values, structural analysis design is performed. This article describes in detail the theoretical aspects of this method, used assumptions and limitations. It is considered a test case for validation of the considered methods. The FEM joints model is investigated. The forces and displacements of the bonds obtained by the FEM model are compared with the result of the theoretical calculation.

1. Introduction

According to the current norms of design of timber structures [1, 2], calculation of the dowel joints of wooden elements perform according to experimental-theoretical method developed in 1953 by Professor, Doctor of technical Sciences V.M. Kochenov [3]. Russian scientists have developed other practical methods of calculation in the assumption of elastic work of the dowel pin in the joint, which are given in the special literature [4-8].

The objective of current study was analyzing and comparing the result of the theoretical calculation with the result of FEM model.

2. Materials and Methods

The method of V M Kochenov is based on the idea of the elastic-plastic work of dowel pin in the joint with the following assumptions:

1. the dowel pin is considered as a thin bar of small bending stiffness, contacting with elastic base - wood of a dowel pin gainer; thus the dowel pin itself is endowed with properties of the ideal elastic-plastic material with strongly expressed areas of elastic 1-2 and plastic 2-3 behavior (figure 1), tensions on a site of plasticity are accepted for a steel dowel pin - a yield strength of steel, and for a dowel pin gainer - the design bearing resistance of timber;
2. shearing breaking strain of dowel joint is defined as the sum of elastic and plastic deformations: $u_l + u_{pl} = 2u_l$;
3. at deformation of a connection it is assumed, that the central axis of a dowel pin before reaching of a yield strength of steel remains a rectilinear and orthogonal with plane of shear, and in view of possible formation as a result of lumber shrinkage of gaps between connecting surfaces the friction between them are neglected;

4. connection destruction is considered to be the moment of exceeding the breaking strain, therefore the breaking strain of dowel joints is limited by the norms of design and is accepted equal 1.5... 2 mm.

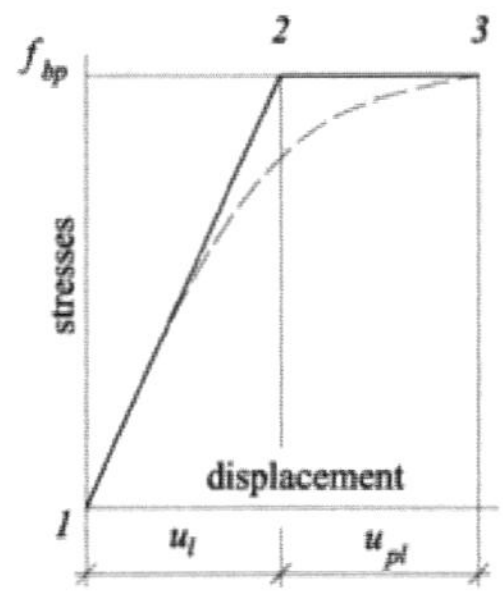

Figure 1. Work diagram of dowel pin *R-u*.

Guided by these assumptions, the calculation of the two shear wood dowel pin joints is to determine the minimum bearing capacity of the dowel pin [*T*] on one plane of shear, based on three possible variants of destruction:

- crushed of the dowel pin gainer in the edge elements

$$[T_1] = 0.8cd\,; \tag{1}$$

- crushed of the dowel pin gainer in the middle elements

$$[T_2] = 0.5cd\,; \tag{2}$$

- and excessive deformations due to bending of the dowel pin itself

$$[T_3] = 1.8d^2 + 0.02a^2 \le 2.5d^2\,; \tag{3}$$

where a and c – the thickness of the edge and middle elements, d – the diameter of the dowel pin.

After finding the minimum bearing capacity of the dowel pin [*T*], it is enough to simply determine the amount of dower pins, that is necessary for the perception of shearing force *N* by the formula:

$$n_н = \frac{N}{[T]n_s}. \tag{4}$$

The use of this method allows to reliably design the dowel pin joints in the case of known beforehand load.

However, in practice, especially in the structural survey and evaluation of the bearing capacity of existing timber structures, there is a need to determine the compliance of connections.

For these purposes, the following technique can be used.

The basis of this technique are the assumptions of elastoplastic behaviour and breaking strain of the joint, laid in the basis of the method of V.M. Kochenov.

To evaluate the stiffness of the whole joint, it is necessary to know the relative stiffness of each brace established in the joint. The numerical relative stiffness of *i*-th brace equals shear force at displacement of *i*-th brace on 1 mm:

$$K_{n,i} = \frac{N_i}{u_i}. \tag{5}$$

In view of absence of numerical description of this characteristic in Russian code, the linear stiffness should be defined according to the corresponding experimental diagrams of «*N-u*» tests of separate joints, or to be guided reference materials of foreign standards [10-19].

For example, according to the annex to EuroCode 5 [10], the first branch stiffness of the connection, corresponding to the elastic work of the joint, can be determined by the semi-empirical formula depending on the diameter of the dowel d_i and the density of the wood of the joined elements ρ:

$$K_{n,i} = \frac{d_i\sqrt{\rho^3}}{23}, \tag{6}$$

On the basis of the accepted assumption about the breaking strain of the dowel pin joint taking into account that $u = u_l + u_{pl} = 2u_l$, the average stiffness of the connection beyond the proportional elastic limit will be equal to:

$$K_{m,i} = 0.5K_{n,i}. \tag{7}$$

In this case the average linear stiffness of joint at installation of dowels of the same diameter will be:

$$K_m = K_{s,i}n_n n_s, \tag{8}$$

where n_n and n_s – the number of dowel pins and plane of shear in the joint.

Then displacement of a brace in a plane of one shear fracture at uniform redistribution of shearing force N between dowels will be equal:

$$u_i = \frac{N}{K_{m,i}n_n}. \tag{9}$$

2.1. Test example

As a test example, let's consider a simple double-shear joint (figure 2), elements of which are connected to each other by means of wooden mending plates on both sides. The joint is stretched by force N = 15.35 kN. Thickness of the connected elements c = 50 mm, overlays – a = 25 mm, the density of wood pine ρ = 500 kg/m^3.

In this case the smallest bearing capacity of the joint on one shear will be determined from the bending condition of the dowel by the formula (3):

[T] = 1.277 kN.

To perceive the connection of tensile force N requires the installation of 6 pieces of steel dowel diameter 8 mm.

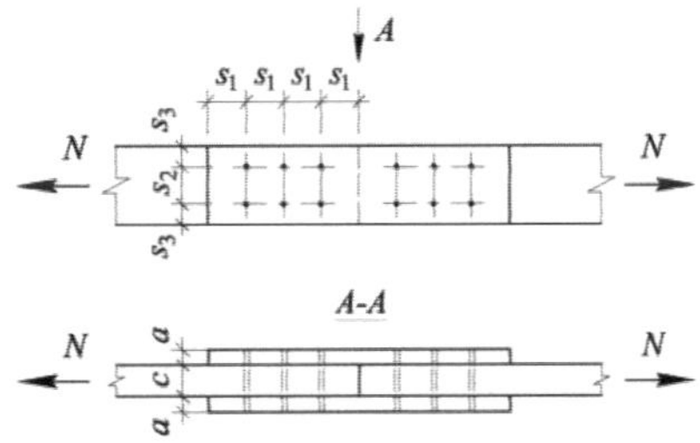

Figure 2. Double-shear stretched joint.

Now let's consider the inverse problem – we shall define a deformation capacity of the same joint on the described method.

The initial linear stiffness of the brace (dowel) according to the formula (6) is:

$K_{n,i}$ = 8·500$^{1.5}$/23·1000 = 3.888 kN/mm.

The average stiffness of the brace according to the formula (7) is equal to:

$K_{m,i}$ = 0.5·3.888 = 1.944 kN/mm.

Then displacement the brace in the plane of one shear by the formula (9) will be:

u_i = 15.35/1.944 · 6 = 1.316 mm.

As the calculation showed, the result is well correlated with the breaking strain of the dowel joint – 1.5... 2 mm, accepted as a criterion of reliability in the method of V.M. Kochenov.

It should be noted that the calculation of the deformation capacity of timber joints by the current norms is not regulated.

In view of this, it is recommended to define stiffness of joints for each characteristic case on the basis of experimental researches in accordance with recommendations [9], also at determination of stiffness it is possible to be guided by reference materials foreign standards [11-23].

IOP Conf. Series: Materials Science and Engineering **675** (2019) 012012 doi:10.1088/1757-899X/675/1/012012

2.2. FEM model
Consider the FEM model of the joint shown in Fig. 2. For modeling of wood were used the massive finite elements (spatial problem of elasticity theory). Steel dowel pins with a small bending stiffness, modeled spatial bars. The calculation was performed in the program complex Lira-Sapr.

Geometry of the joint: a = 25 mm; c = 50 mm; h = 125 mm. The distance between the dowel pins d = 8 mm) was $s_1 = s_3$ = 31.25 mm; s_2 = 62.5 mm.

Tensile force N = 15.35 kN was exerted on the joint.

Timber was considered as a orthotropic material with the following physical and mechanical properties:

E_x = 11700000 kN/m^2; E_y = 510000 kN/m^2; E_z = 430000 kN/m^2; G_{xy} = 500000 kN/m^2; G_{yz} = 710000 kN/m^2; G_{xz} = 1140000 kN/m^2; $\nu_{xy} = 0.03$; $\nu_{yx} = 0.49$; $\nu_{xz} = 0{,}41$; $\nu_{zx} = 0.037$.

Deformations in the massive finite elements were calculated by formulas:

$$\varepsilon_x = \left(\sigma_x - \nu_{yx}\sigma_y - \nu_{zx}\sigma_z\right)/E_x \,;\, \varepsilon_y = \left(\sigma_y - \nu_{xy}\sigma_x - \nu_{zy}\sigma_z\right)/E_y \,;$$
$$\varepsilon_z = \left(\sigma_z - \nu_{xz}\sigma_x - \nu_{yz}\sigma_y\right)/E_z \,;\; \varepsilon_{xy} = \tau_{xy}/G_{xy} \,; \qquad (10)$$
$$\varepsilon_{yz} = \tau_{yz}/G_{yz} \,;\; \varepsilon_{yz} = \tau_{xz}/G_{xz} \,.$$

The FEM model (figure 3) included 47629 nodes, 38592 elements. The number of nodes unknown was 139779.

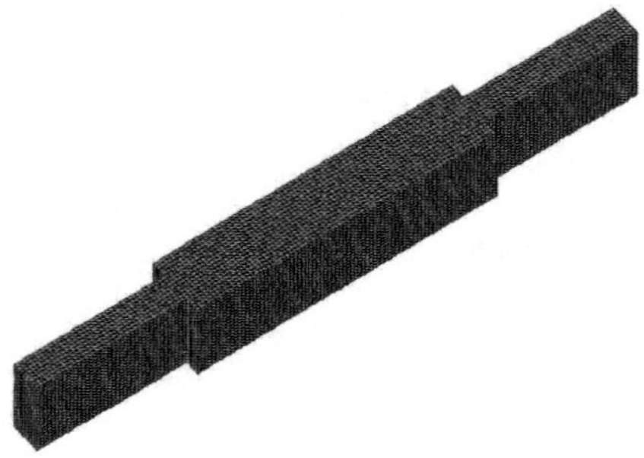

Figure 3. FEM-Model of joint

3. Results and discussion
As a result of FEM analysis the special features of the joint deformation were investigated (figure 4).

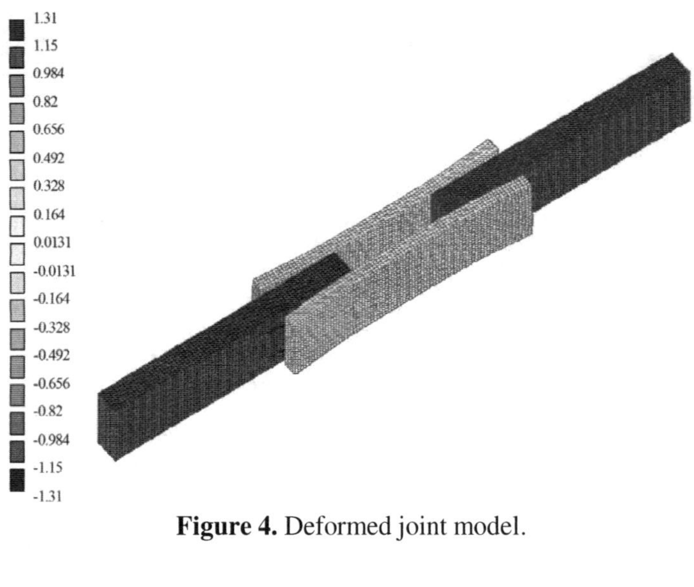

Figure 4. Deformed joint model.

The maximum value of the brace displacement in the joint (figure 5) was 1.26 mm.

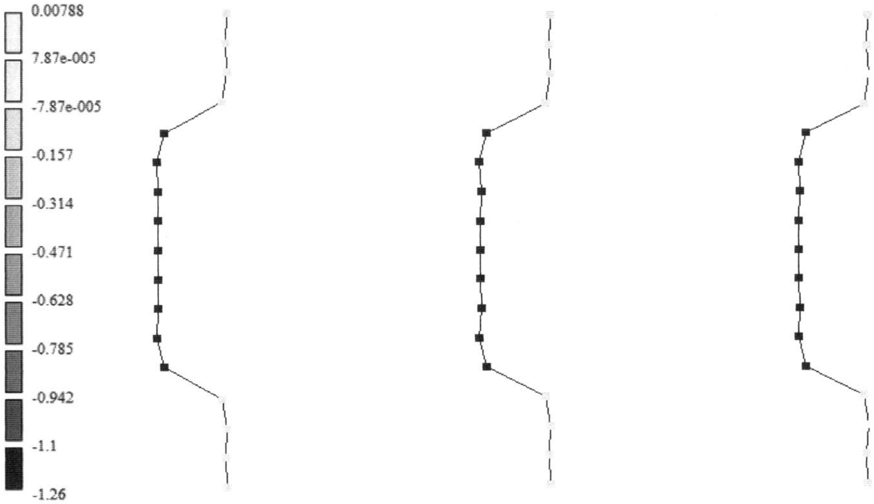

Figure 5. Brace displacement in the joint.

Into consideration, the value of the brace displacement on the considered theory was 1.316 mm (discrepancies in comparison with the FEM calculation – 4 %).

FEM analysis showed that in the deformation of elements shear forces between braces are distributed almost evenly. The maximum value of the shear force in the dowel pin on one plane of the shear fracture was 1.27 kN (figure 6), which coincides with the analytical calculation according to the formula (3).

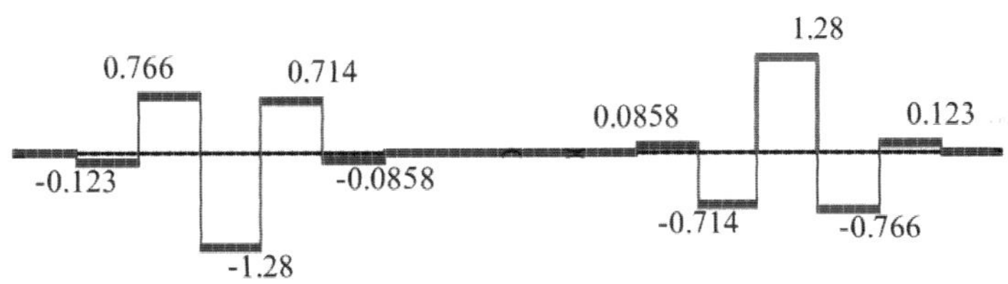

Figure 6. Shear forces distribution in dowel pin.

4. Conclusion

The forces and displacements of the bonds obtained by the FEM model have been compared with the result of the theoretical calculation. The results of the FEM analysis confirm the reliability of the proposed method of determining the deformation capacity of joints of timber elements on cylindrical dowel pins.

References

[1] Design Code SP 64.13330.2017 2017 *Timber structures* (Moscow: Standartinform)

[2] *Guidelines on Designing Timber Structures to SNiP II-25-80* CNIISK im. V A Kucherenko 1986 (Moscow: Strojizdat Publ.) p 215

[3] *Durability of Elements of Joints of Timber Constructions* 1953 (Moscow: Gosstrojizdat Publ.) p 320

[4] Timber and Plastic constructions 1986 (Moscow: Strojizdat Publ.) p 543

[5] Rasha I K, Pertseva O N, Lazareva A Yu, Martynov G V 2017 Computational modelling of random distribution of stresses for wooden structures *Magazine of Civil Engineering* **69 (1)** 23-33.
[6] Dmitriev P A 1973 Research of Long Durability of Joints of Timber Elements on Steel Nails *Construction and Architecture* **5** 28-35
[7] Dmitriev P A, Strizhakov Ju D, Shvedov V N 1999 *About Calculation Asymmetrical Nails Joints of Timber Elements with Steel Linings Construction* (Novosibirsk: News of higher educational institutions) **5** 10-15
[8] Shvedov V N, Grohotov A B 2000 Effective Method of Joints of Timber Constructions Increase of Efficiency of Rural Construction *The International Collection of Works of NGAU* 61-63
[9] *Recommendations on Test of Joints of Timber Constructions* 1980 CNIISK im. V A Kucherenko (Moscow: Strojizdat Publ.) p 40
[10] Designers' Guide to Eurocode 5 2012 *Design of Timber Buildings*. EN 1995-1-1 A.J. Porteous and P. Ross. 978-0-7277-3162-3 (London: Ice publishing) p 220
[11] British Standard BS 5268-2:2002 *Structural use of timber Code of practice for permissible stress design, materials and workmanship* (London: BSI) p 170
[12] British Standard BS EN 408:2010+A1:2012 2012 *Timber structures Structural timber and glued laminated timber Determination of some physical and mechanical properties* (London: BSI) p 42
[13] Calgaro J-A, Gulvanessian H, Holicky M 2002 *Basis of structural design. Designer's guide to EN 1990* (London: Thomas Telford) p 192
[14] British Standard BS EN 912:2011 2011 *Timber fasteners. Specifications for connectors for timbers* (London: BSI) p 52
[15] Dias A M P G, Cruz H, Lopes S M R, Van de Kuilen J-W 2010 Stiffness of dowel-type fasteners in timber–concrete joints *Proceedings of the ICE-Structures and Buildings* **163 (584)** 257-266
[16] British Standard BS EN 13271:2002 *Timber fasteners Characteristic load-carrying capacities and slip-moduli for connector joints* (London: BSI) p 18
[17] British Standard BS EN 1383:1999 *Timber Structures Test Methods Pull-Through Resistance of Timber Fasteners* (London: BSI) p 8
[18] Technical standard DIN 1052-2008 2008 *Design of timber structures - General rules and rules for buildings* (Berlin: Beuth Verlag GmbH) p 239
[19] Technical standard DIN 1052-10-2012 2012 *Design of timber structures Part 10 Additional provisions* (Berlin: Beuth Verlag GmbH) p 19
[20] Varenik A S Varenik K A 2014 Long bearing capacity of wooden structures *Structural mechanics of engineering constructions and buildings* **2** 23-30
[21] Svintsov V P, Liutov L V 2015 Experimental research on thin steel plate-to-timber joints *Structural mechanics of engineering constructions and buildings* **6** 16-20
[22] Markovich A S 2016 To the question of stress state of joints of wooden elements with the frontal cuttings *Structural mechanics of engineering constructions and buildings* **4** 21-25
[23] Razzakov S J, Juraev B G, Juraev E S 2018 Sustainability of walls of individual residential houses with a wooden frame *RUDN Journal of engineering researches* **14(5)** 427-435

Name	*Prefix*	*Research Field*	*Email*	*Personal website*
A S Markovich	Ph.D. in Engineering Science, Associate Professor	Calculation and design of buildings and structures	markovich-as@rudn.ru	http://engr.rudn.ru
M Abu Mahadi	Ph.D. in	Calculation and	moham_d@mail.ru	http://engr.rudn.ru

	Engineering Science, Associate Professor	design of buildings and structures		
S A Gazizova	Master's Degree Student	Calculation and design of buildings and structures	s.a.gazizova@gmail.com	http://engr.rudn.ru
D A Miloserdova	Ph.D. student	Calculation and design of buildings and structures	milos-dasha@yandex.ru	http://engr.rudn.ru

IOP Conf. Series: Materials Science and Engineering **675** (2019) 012013 doi:10.1088/1757-899X/675/1/012013

Analysis of the theory of calculation of fiber-reinforced concrete with non-steel fibers

A S Markovich*, D D Koroteev, M I Abu Mahadi, D A Miloserdova

Peoples' Friendship University of Russia (RUDN University), Moscow, Russia

Email: markovich-as@rudn.ru

Abstract. The fiber-reinforced concrete finds more applications in modern construction. This material has improved strength and deformation characteristics. The fiber-reinforced concrete is considered to be one of the most promising building materials in the world practice. The experience of using this material in developed countries, such as the USA, UK, Japan, Germany, Italy, France, Australia, has shown the good technical and economic indicators in building constructions. Recently, the base of standards for preparation, calculation and design of such concretes has been prepared and introduced in Russia. The method of determination of the class of the fiber-reinforced concrete by residual tensile strength, the calculation of the strength of standard sections, the diagrams of the deformation is discussed in the article based on the new Code of Practice. The obtained results of analysis of the theory of calculation suggest further studies of the properties and physical characteristics.

1. Introduction

The uniformly distributed reinforcement of the entire volume of concrete eliminates the disadvantages of unreinforced concrete. In particular, it is a low flexural strength and shrinkage cracking during hardening and using of structures. This type of reinforcement can significantly increase the service life of reinforced concrete structures.

The fiber-reinforced concrete is a type of reinforced concrete dispersed-reinforced with various high-strength modified fibers (steel and non-steel). At the beginning of the 20th century, Russian engineer V P Nekrasov one of the first in the world proposed dispersion-reinforced concrete and conducted researches its building performance [1]. As the micro armature, they used pieces of wire of small diameters. Foreign scientists also worked on this topic, for example, Gary Porter in 1910 stated that the mechanical characteristics of concrete increase approximately 8 times with the introduction of wire and nails [2]. However, the idea of dispersed reinforcement was gradually forgotten, and only in the 1960s, it was rapidly developing again, due to the urgent need to improve the physical and mechanical properties of concrete in tension and increase the fracture toughness.

Currently, the nomenclature of the used fibers is very extensive. According to the accepted classification they are divided:

- According to modulus of elasticity on high modulus (steel, carbon, glass, etc.) and low modulus (polypropylene, viscose, etc.);
- According to origin on natural (asbestos, basalt, wool, etc.) and artificial (viscose, polyamide, etc.);
- According to the main material on the metal (most often steel) and non-metal (synthetic, mineral).

The code of practice «Fiber reinforced concrete structures and precast products with non-steel fibers. Design rules» was published in 2017 [3]. The industry-specific code of practice «Design and basic principles for technology for production of fiber reinforced concrete structures» was the first published document on this theme [4].

The fiber-reinforced concrete like any other structural material, has strength and deformation characteristics. Compared with the concrete without dispersed reinforcement, the fiber-reinforced concrete is characterized by high strength under various types of impact: tension, shear, crack strength. And this material is characterized by crack toughness. The strength properties of the fiber-reinforced concrete have been studied for a long time, and they are well studied by russian and foreign experts [5-12]. After analysis of the scientific literature on the subject of fiber-reinforced concrete [13-20] had been carried out, investigation of definition of mechanical characteristics of this material was determined.

2. Method of research

According to the code of practice [3] analysis of the theory of determining the residual tensile strength of the fiber-reinforced concrete was carried out.

The main strength characteristics of fiber-reinforced concrete:

- The standard value of the resistance of the fiber-reinforced concrete to axial compression $R_{fb,n}$;
- The standard value of the resistance of the fiber-reinforced concrete to axial tension $R_{fbt,n}$;
- The standard value of the residual resistance of the fiber-reinforced concrete to axial tension $R_{fbt2,n}$;
- The standard value of the residual resistance of the fiber-reinforced concrete to axial tension $R_{fbt3,n}$.

The standard and calculation value of the resistance of the fiber-reinforced concrete to axial compression are accepted according to the code of practice [12] for a similar class of the ordinary concrete.

The standard value of the resistance of the fiber-reinforced concrete to axial tension and the residual resistance of the fiber-reinforced concrete to axial tension are determined from the results of tests of control samples for axial tension.

It is required to make a test beam with a cut for the action of a concentrated load to determine the residual tensile strength of the fiber-reinforced concrete (figure 1).

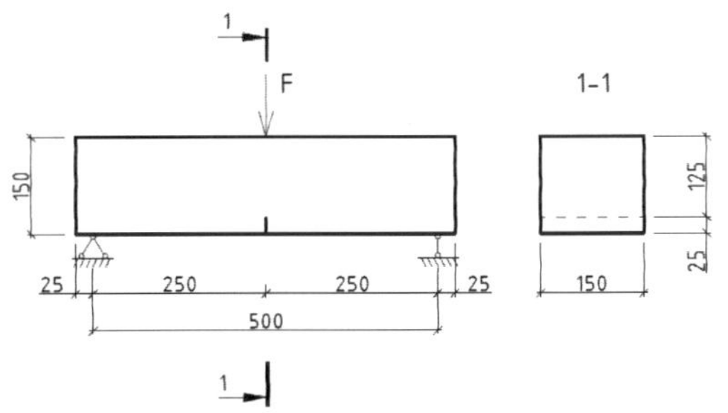

Figure 1. The scheme of the test beam (width b = 150 mm, height h = 150 mm, length L = 550 mm).

The cut width should be less than 5 mm, depth - (25 ± 1) mm, the distance between the tip of cut and the upper side of the test beam h_{sp} = (125 ± 1) mm.

As a result of testing, a diagram «F-a_F» is drawn (figure 2). Where a_F is the displacement of the external side of the cut.

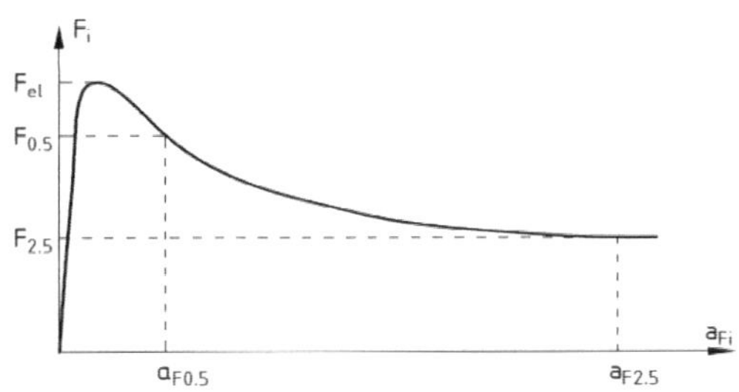

Figure 2. The diagram of "load – displacement of the external side of the cut".

For the sample, the strength values are determined taking into account the inelastic properties of the fiber-reinforced concrete according to the formulas:

$$R_{F0.5} = \frac{3F_{0.5}l}{2bh_{sp}^2}k_{F0.5}, \tag{1}$$

$$R_{F2.5} = \frac{3F_{2.5}l}{2bh_{sp}^2}k_{F2.5}, \tag{2}$$

where $F_{0.5}$ is the load value corresponding to the displacement of the external side of the cut $a_F=0.5$ mm; $F_{2.5}$ is the load value corresponding to the displacement of the external side of the cut $a_F=2.5$ mm; $k_{F0.5}=0.4$ and $k_{F2.5}=0.34$ are coefficients of accounting for inelastic deformations in the fiber-reinforced concrete of the stretched zone of the test beam.

The standard values of the residual resistance of the fiber-reinforced concrete to axial tension are determined according to the formulas:

$$R_{fbt2,n} = R_{F0.5,m}(1-1.64\upsilon_{F0.5,m}), \tag{3}$$

$$R_{fbt3,n} = R_{F2.5,m}(1-1.64\upsilon_{F2.5,m}), \tag{4}$$

where $R_{F0.5,m}$ and $R_{F2.5,m}$ are average values of the residual tensile strength of the fiber-reinforced concrete; $\upsilon_{F0.5,m}$ and $\upsilon_{F0.5,m}$ are coefficients of variation determined by the formulas:

$$\upsilon_{F0.5,m} = \frac{S_{F0.5,m}}{R_{F0.5,m}}, \tag{5}$$

$$\upsilon_{F2.5,m} = \frac{S_{F2.5,m}}{R_{F2.5,m}}, \tag{6}$$

where $S_{F0.5,m}$ and $S_{F2.5,m}$ are the quadratic means determined by the formulas:

$$S_{F0.5,m} = \sqrt{\frac{\sum_{i=1}^{n}(R_{F0.5,i} - R_{F0.5,m})^2}{(n-1)}}, \tag{7}$$

$$S_{F2.5,m} = \sqrt{\frac{\sum_{i=1}^{n}(R_{F2.5,i} - R_{F2.5,m})^2}{(n-1)}}, \tag{8}$$

where n is the number of the test beam.

According to the code of practice [3], if axial tensile tests are not carried out, the class of the fiber-reinforced concrete for axial tensile strength can be assigned according to the results of the bending tests.

In this case, the value of strength is determined by the formula:

$$R_{Fel} = \frac{3F_{el}l}{2bh_{sp}^2}k_{Fel}, \tag{9}$$

where F_{el} is maximum value of the load in the range of the displacement of the external side of the cut in the range $0 < a_F < 0.05$ mm; $k_{Fel}=0.6$ is coefficient of accounting for inelastic deformations in the fiber-reinforced concrete of the stretched zone of the test beam.

In this case the standard value of the resistance of the fiber-reinforced concrete to axial tension is determined by the formula:

$$R_{fbt,n} = R_{Fel,m}(1-1.64\upsilon_{Fel,m}), \tag{10}$$

where $R_{Fel,m}$ is the average tensile strength of the fiber-reinforced concrete; $\upsilon_{Fel,m}$ is the coefficient of variation determined by the formula:

$$\upsilon_{Fel,m} = \frac{S_{Fel,m}}{R_{Fel,m}}, \qquad (11)$$

where $S_{Fel,m}$ is the quadratic mean determined by the formula:

$$S_{Fel,m} = \sqrt{\frac{\sum_{i=1}^{n}(R_{Fel,i} - R_{Fel,m})^2}{(n-1)}}. \qquad (12)$$

The connection between deflection and displacement are determined by the formula:

$$f = 0.85a_F + 0.04, \qquad (13)$$

where a_F is a value:

- If the distance y from the axis of the displacement transducer to the lower face of the test beam is zero, it is taken equal to the value measured during the test;
- Otherwise this value is calculated by the formula:

$$a_F = a_{Fy}\frac{h}{h+y} \qquad (14)$$

3. Results and discussion

As a result of the testing and determining the standard value of the resistance of the fiber-reinforced concrete to axial tension, the class of the fiber-reinforced concrete can be chosen, taking into account the subclass "a", "b", "c", "d" and "e" according to the code of practice [3].

According to the code of practice [3], the minimum standard value of the resistance of the fiber-reinforced concrete to axial tension is 0.77 MPa (for the class of the fiber-reinforced concrete $B_{ft3}1i$ with the subclass index $i=e$), and the maximum - 16 MPa (for the class of the fiber-reinforced concrete $B_{ft3}8i$ with the subclass index $i=a$).

The analysis of the formulas for calculating the standard value of the resistance of the fiber-reinforced concrete to axial tension has shown a large using of empirical coefficients, the physical meaning of these cannot be explained.

According to the code of practice [3], the working diagram of deformation of the fiber-reinforced concrete to axial compression is the curved diagram with a falling branch, or a simplified three-line diagram with parametric points of the compressed fiber-reinforced concrete as for the ordinary concrete (figure 3). The working diagram of deformation of the fiber-reinforced concrete to axial tension is the simplified three-line diagram (figure 4). The tensile stresses of the fiber-reinforced concrete σ_{fbt} are determined depending on the relative deformations ε_{fbt} according to the formulas of the code of practice [3].

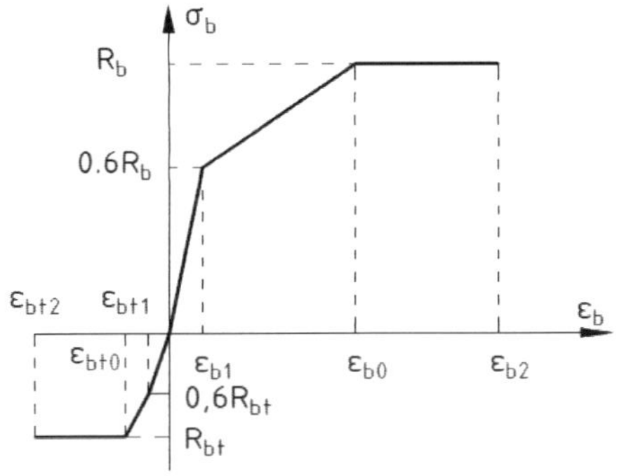

Figure 3. The diagram of the deformation of the ordinary concrete according to the code of practice [12].

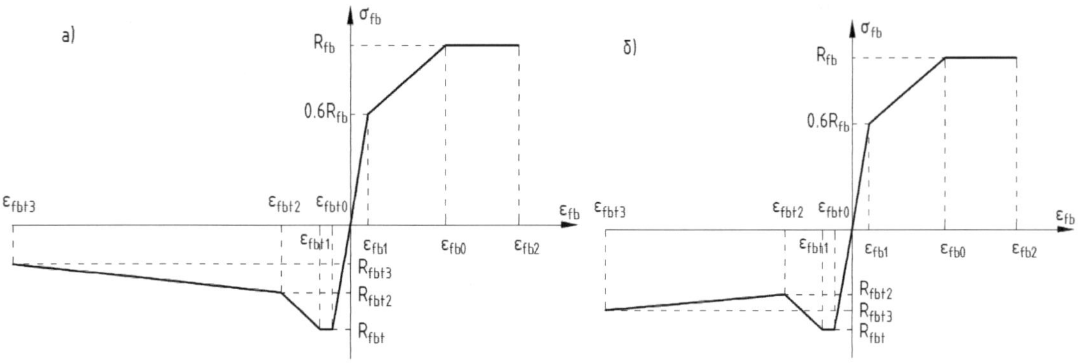

Figure 4. The diagrams of the deformation of the fiber-reinforced concrete under compression and tension: $R_{fbt3} / R_{fbt2} < 1$; b) $R_{fbt3} / R_{fbt2} > 1$.

The stress diagram in the compressed zone of the fiber-reinforced concrete and the ordinary concrete is taken a triangular shape (as for an elastic solid), and for a tensile zone of fiber-reinforced concrete - a trapezoidal shape with stresses in the tensile section (figure 5).

The calculation of the bent elements of critical force for the fiber-reinforced concrete and the ordinary concrete is produced from the condition:

$$M \leq M_{ult} \tag{15}$$

where M is the bending moment of the external load; M_{ult} is the breaking bending moment that can be perceived by the section of the element, $M_{ult} = R_{fbt}W_{pl}$ - for the fiber-reinforced concrete; $M_{ult} = R_{bt}W$ - for the ordinary concrete.

The section modulus is different for the fiber-reinforced concrete and the ordinary concrete.

For example, the elastoplastic section modulus of the fiber-reinforced concrete element of a rectangular cross-section for an extreme tensile fiber (without working reinforcement) is determined by the formula:

$$W_{pl} = \frac{bh^2}{3.6} \tag{16}$$

The elastoplastic section modulus of the ordinary concrete for an extreme tensile fiber is determined by the formula:

$$W = \frac{bh^2}{6} \tag{17}$$

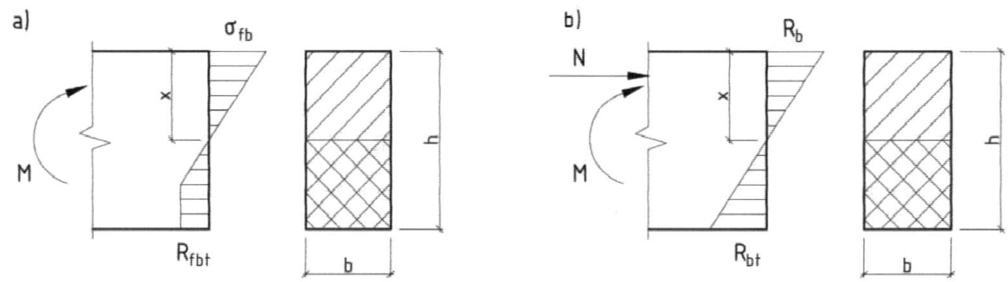

Figure 5. The design of stresses and stress diagram across the rectangular cross-section without reinforcement in its strength calculation: a) for fiber-reinforced concrete; b) for ordinary concrete.

The calculation of the strength of standard sections of the fiber-reinforced concrete and the concrete elements with working reinforcement is made depending on the ratio between the value of the relative height of the compression side of the fiber-reinforced concrete and the concrete accordingly $\xi = \frac{x}{h_0}$. The value of the boundary relative height of the compression side ξ_R is determined taking into account the relative deformation of the fiber-reinforced concrete ε_{fb2} and the ordinary concrete ε_{b2}, the calculation value of the critical relative deformations of the reinforcement ε_s.

$$\xi_R = \frac{x_R}{h_0} = \frac{\omega}{1 + \frac{\varepsilon_s}{\varepsilon_{fb2}(\varepsilon_{b2})}} \qquad (18)$$

where ω is a characteristic of the compression side of fiber-reinforced concrete (ordinary concrete).

According to the code of practice [5] the values of relative deformations of the compressed fiber-reinforced concrete and the concrete is allowed to take equal $\varepsilon_{fb2} = \varepsilon_{b2}$.

When calculation of the flexural fiber-reinforced concrete and the concrete elements of rectangular cross-section with the working reinforcement (figure 6), the value M_{ult} and the height of the compression side are determined according to the standards (table 1):

Table 1. The calculation of the flexural fiber-reinforced concrete and the concrete elements.

	The fiber-reinforced concrete element	The concrete element
The breaking bending moment, M_{ult}	$M_{ult} = R_{fb}bx(h_0 - 0.5x) - R_{fbt3}b(h-x) \times (\frac{h-x}{2} - a) + R_{sc}A'_s(h_0 - a')$	$M_{ult} = R_b bx(h_0 - 0.5x) + R_{sc}A'_s(h_0 - a')$
The height of compression side, x	$x = \frac{R_s A_s - R_{sc}A'_s + R_{fbt3}bh}{(R_{fb} + R_{fbt3})b}$	$x = \frac{R_s A_s - R_{sc}A'_s}{R_b b}$

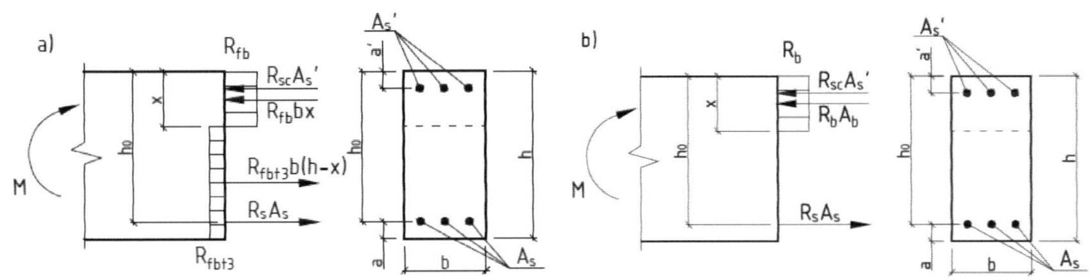

Figure 6. The design of stresses and stress diagram across the rectangular cross-section with the reinforcement in its strength calculation: a) for the fiber-reinforced concrete; b) for the ordinary concrete;

4. Conclusion

1. The conducted analysis of the theory of calculation of strength characteristics of the fiber-reinforced concrete according to the code of practice [3] has shown a large using of empirical coefficients to take into account inelastic deformations in the tensile side.

2. To determine the class of the fiber-reinforced concrete by residual tensile strength, it is an imperative to carry out the experiment of the test beams for the bending.

3. The minimum standard value of the resistance of the fiber-reinforced concrete to axial tension is 0.77 MPa (for the class of the fiber-reinforced concrete $B_{ft3}1i$ with the subclass index $i=e$), and the maximum - 16 MPa (for the class of the fiber-reinforced concrete $B_{ft3}8i$ with the subclass index $i=a$). For the ordinary concrete, according to the code of practice [6], the minimum standard value of the resistance to axial tension is 0.39 MPa (for the class of the concrete B3.5), and the maximum – 3.8 MPa (for the class of the concrete B100). The standard value of the resistance of the fiber-reinforced concrete to axial tension is analyzed to be about 4 times greater than that of the ordinary concrete.

4. The standard value of the resistance of the fiber-reinforced concrete to axial tension was happen to be about 4 times greater than that of the ordinary concrete.

5. The stress diagram in the compressed zone of the fiber-reinforced concrete and the ordinary concrete is taken a triangular shape (as for an elastic solid), and for a tensile zone of the fiber-reinforced concrete - a trapezoidal shape with stresses in the tensile section.

6. The obtained results of analysis of the theory of calculation of fiber-reinforced concrete with non-steel fibers suggest further studies of the properties and physical characteristics. Series of experiment is planned to conduct due to researching the theory data. The main purpose of experimental researches will be proof or refutation of the application of the theory of calculation of ordinary concrete structures to fiber-reinforced concrete.

References

[1] Nekrasov V P 1909 The newest techniques and tasks of reinforced concrete equipment: a system of available bond *Proc. XII Congress of Russian Cement Technicians* (St Petersburg: Pentkovsky Press) pp 294 - 348

[2] Porter H F Preparation of Concrete from selection of materials to final disposition *Proceedings of the National Association of Cement Users* (Michigan: American Concrete Institute) **6** p 191

[3] Design Code SP 297.1325800.2017 2017 *Fiber reinforced concrete structures and precast products with non-steel fibers. Design rules* (Moscow: Standartinform)

[4] Design Code VSN 56-97 1997 *Design and basic principles for technology for production of fiber reinforced concrete structures* (Moscow: Department of Construction)

[5] Rabinovich F N 2004 *Composite based on dispersed reinforced concrete. Questions of theory and design, technology, design* (Moscow: Association of Construction Universities) p 560

[6] Rybasov V P 1979 To assess the strength of fiber-reinforced concrete. Fiber-reinforced concrete and its application in construction *Collection of scientific works of Scientific Research Design and Technological Institute of Concrete and Reinforced Concrete 1979* (Moscow, USSR) pp 125-130

[7] Brandt A M 2009 *Cement-Based Composites: Materials, Mechanical Properties and Performance* (Boca Raton: CRC Press) p 544

[8] Kluyev S V 2011 Experimental researches of fine-grained concrete constructions *Structural Mechanics of Engineering Constructions and Buildings* **4** 71-75

[9] Granovsky A V, Galishnikova V V and Berestenko E I 2015 Ways of improving the competitiveness of reinforcing meshes on the basis of basalt fiber and prospects of their use in construction *Structural Mechanics of Engineering Constructions and Buildings* **3** 59-63

[10] Kharun M, Koroteev D D, Dkhar P, Zdero S and Elroba S M 2018 Physical and mechanical properties of basalt-fibered high-strength concrete *Structural Mechanics of Engineering Constructions and Buildings* **14(5)** 396-403

[11] Teryanik V, Podnebesov P 2010 New ways of strengthening of compressed elements reinforced concrete structures *RUDN Journal of Engineering Researches* **2** 36-39

[12] Design Code SP 63.13330.2012 2013 *Concrete and reinforced concrete structures. Design requirements* (Moscow: Standartinform)

[13] Klyuev S V, Khezhev T A, Pukharenko Yu V, Klyuev A V 2018 The Fiber-Reinforced Concrete Constructions Experimental Research *Materials Science Forum* **931** 598-602

[14] Lesovik R V, Klyuyev S V, Klyuyev A V, Netrebenko A V, Kalashnikov N V 2014 Fiber Concrete on Composite Knitting and Industrialsand KMA for Bent Designs *World Applied Sciences Journal* **30 (8)** 964-969

[15] Klyuev S V, Klyuev A V, Vatin N I 2018 Fine-grained concrete with combined reinforcement by different types of fibers *MATEC Web of Conferences* 245, 03006

[16] Chaohua Jiang, Ke Fan, Fei Wu, Da Chen 2014 Experimental study on the mechanical properties and microstructure of chopped basalt fiber reinforced concrete *Journal of Materials & Design* **58** 187-193

[17] Jongsung Sim, Cheolwoo Park, Do Young Moon 2005 Characteristics of basalt fiber as a strengthening material for concrete structures *Journal of Composites Part B: Engineering* V. **36 (6-7)** 504-512

[18] Bunsell A R 2005 *Fundamentals of Fibre Reinforced Composite Materials (series In Material Science And Engineering)* (Bristol: Institute Of Physics Publishing (gb)) p 398

[19] Brouwers H J H, Radix H J 2005 Self-Compacting Concrete: Theoretical and experimental study *Cement and Concrete Research* **35** 2116–2136

[20] Wang W, Liu J, Agostini F, Davy C A, Skoczylas F, Corvez D 2014 Durability of an Ultra High Performance Fiber Reinforced Concrete (UHPFRC) under progressive aging *Cement and Concrete Research* **55** 1–13

Raft foundations design charts: development and applications

S A Pechorskaya

Peoples' Friendship University of Russia (RUDN University), Moscow, Russia

Email: sophya_pechorskaya@yahoo.com

Abstract. Presented in this paper design charts for estimating effective depth of raft slabs on elastic foundation are developed in compliance with punching shear design procedures provided by Russian Design Code (SP) for reinforced concrete. Codes' formulas are generalized to include only independent parameters omitting unnecessary calculations of critical section geometry. Proposed approach considers raft slab effective depth as a function of cross section sizes of vertical members (walls, columns) only. In the conclusion reasonable range of raft slab thickness to column sizes ratios have been established for numerical tests and experiments included in fundamental studies of moderately thick plates and raft slabs resting on elastic foundations.

Keywords: Raft foundations, punching shear capacity of reinforced concrete slabs, moderately thick plates, plates on elastic foundations, design charts.

1. Introduction

Structural frames of tall buildings consist of walls and columns which transfer high gravity and lateral loads to the soil bed through the foundations. For medium-stiff and stiff soils stratum underlying the land plot of the project, it is raft slab which becomes preferable option for building foundations.

The most general principles of raft engineering are outlined by ACI Committee 336 [1] and include three major steps as follow: punching shear design, bending analysis and settlements check, the results of which depend upon large variety of the design parameters discussed in [2-4].

It is common practice to carry out raft foundation analysis with FEM-software utilizing classical thin plates FE for modeling raft slab. However this approach does not produce accurate results in the close proximity of columns and walls supports. This problem could only be solved with moderately thick plates (Midlin-Reissner plate) theory which has been widely employed to develop a number of analytical and numerical methods suitable for raft foundation analysis [5-16].

Ratio of raft slab depth to the dimensions of column footprint area h_0/h_c is a key parameter affecting the accuracy of the results produced by moderately thick plates theory. Therefore it is important to establish a reasonable range of h_0/h_c values to be used for verification tests and experiments. Such range can be obtained by simplified engineering method similar to those described in [17-19]. It is based on the assumption that effective depth of raft slab h_0 is controlled by punching shear force caused by axial column load F distributed over column footprint area $h_c \times h_c$ which sizes are resulted from the columns design. Therefore, two main parameters of punching shear calculation procedure – perimeter of punching shear critical section u and design load F transferred by column to the raft slab – are not independent. Hence, the purpose of this work is to work out simplified procedure for raft slabs depth calculations which would allow to establish practical range of h_0/h_c ratios to be used in the fundamental studies of moderately thick plates on elastic foundation.

2. Methods of research

Brittle mode of failure caused by punching shear is always a concern. Therefore it is common practice to take most conservative assumptions to punching shear design such as:

– punching shear capacity of a raft slab equals to full axial capacity of a column;

– punching shear capacity of a raft slab is provided by concrete only, while vertical reinforcement and soil pressure are neglected.

For writing punching shear expressions reference is made to figure 1a (drawing 16 [21]). Vertical member of the structural framing is represented by square column with cross-section sizes $h_c \times h_c$, which is built of normal weight concrete with compressive strength R_b^{col} and steel reinforcing bars with total area A_{sc} and design strength R_s. Column has reinforcement ratio $\mu = A_{sc}/h_c^2$ .Raft slab is built of normal weight concrete with R_{bt}^{raft} tension strength. It's effective depth equals to h_0.

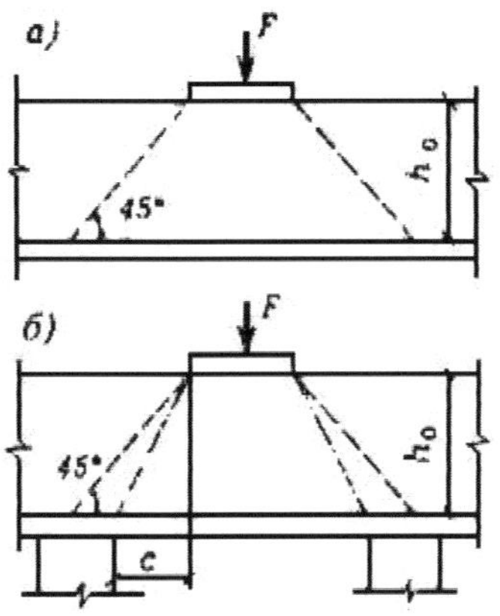

Figure 1. Punching shear design schemes a) raft slab on elastic foundation; b) pile cap

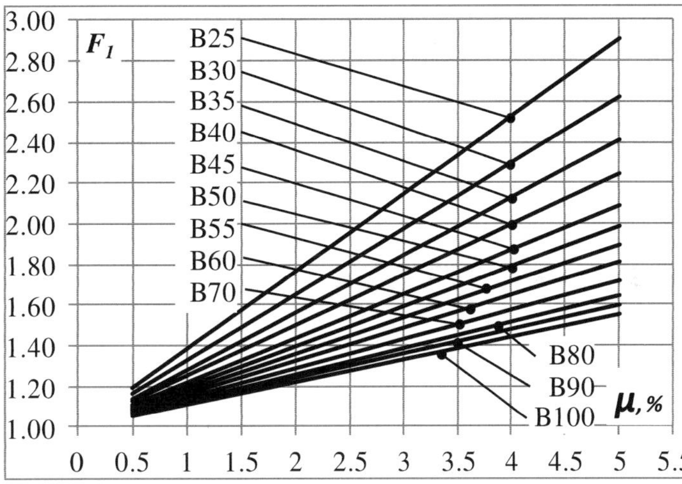

Figure 2. Columns capacity diagrams $F_1 = N_{ult}/\left(h_c^2 R_b^{col} \gamma_{b1}\gamma_{b3}\right)$

According to [21] design load F applied to a raft slab over $h_c \times h_c$ column footprint area shall not exceed punching shear capacity $F_{b,ult}$ of raft slab critical section, located at a distance $d/2$ from the face of the column, i. e. $F \le F_{b,ult}$. $F_{b,ult}$ is defined by expression (f.8.88 [20]):

$$F_{b,ult} = A_b\left(\gamma_{b1} R_{bt}^{raft}\right), \tag{1}$$

where $\gamma_{b1} = 0.9$ is concrete strength reduction factor for long-term static loads (i.6.1.12a [20]).

$A_b = u \cdot h_0$ is an area of punching shear critical section and it is calculated as follows:

$$A_b = u \cdot h_0 = 4\left(h_0 + h_c\right)h_0 = 4\left(h_0^2 + h_0 h_c\right), \tag{2}$$

where $u = 4\left(h_c + h_0\right)$ is a perimeter of punching shear critical section.

Substituting (2) in (1) we obtain:

$$F_{b,ult} = 4\left(h_0^2 + h_0 h_c\right)\left(\gamma_{b1} R_{bt}^{raft}\right). \tag{3}$$

At the same time, design load F may not exceed axial capacity of a column N_{ult}, i.e. $F \le N_{ult}$,

where $N_{ult} = \left(h_c^2 - A_{sc}\right)\left(\gamma_{b1}\gamma_{b3}R_b^{col}\right) + A_{sc}R_s = h_c^2\left(\gamma_{b1}\gamma_{b3}R_b^{col}\right) + A_{sc}\left(R_s - \gamma_{b1}\gamma_{b3}R_b^{col}\right)$

$$= \left[h_c^2 + A_{sc}\left(R_s/\left(\gamma_{b1}\gamma_{b3}R_b^{col}\right) - 1\right)\right]\left(\gamma_{b1}\gamma_{b3}R_b^{col}\right) = h_c^2\left[1 + \mu\left(R_s/\left(\gamma_{b1}\gamma_{b3}R_b^{col}\right) - 1\right)\right]\left(\gamma_{b1}\gamma_{b3}R_b^{col}\right) \tag{4}$$

$\gamma_{b3} = 0.85$ is concrete strength reduction factor for vertical elements higher than 1.5 m (i.6.1.12c [20]).

It is reasonable to conclude that punching shear capacity of raft slab critical section $F_{b,ult}$ shall be bigger than column axial capacity N_{ult}, i. e. $F_{b,ult} \geq N_{ult}$. Combining expressions (3) and (4), we get:

$$4\left(h_0^2 + h_0 h_c\right)\left(\gamma_{b1} R_{bt}^{raft}\right) \geq h_c^2 F_1 \left(\gamma_{b1}\gamma_{b3} R_b^{col}\right), \quad (5)$$

where

$$F_1 = \left[1 + \mu\left(R_s / \left(\gamma_{b1}\gamma_{b3} R_b^{col}\right) - 1\right)\right] \text{(see figure 2).} \quad (6)$$

Divided by $4h_c^2\left(\gamma_{b1} R_{bt}^{raft}\right)$ expression (5) yields to quadratic equation for unknown variable $\left(h_0/h_c\right)$:

$$\left(h_0/h_c\right)^2 + \left(h_0/h_c\right) - \left(F_1/4\right)\left(\gamma_{b3} R_b^{col} / R_{bt}^{raft}\right) \geq 0. \quad (7)$$

Discriminant of equation (7) equals to $\sqrt{F_1\left(\gamma_{b3} R_b^{col} / R_{bt}^{raft}\right) + 1}$, and one positive root is defined as:

$$h_0/h_c \geq \left(\sqrt{F_1\left(\gamma_{b3} R_b^{col} / R_{bt}^{raft}\right) + 1} - 1\right)/2. \quad (8)$$

Graphs shown on figure 2 represent unitless parameter F_1 that is a ratio of axial capacity of reinforced concrete column N_{ult} with reinforcement ratio μ to axial capacity of concrete column without reinforcement $F_1 = N_{ult} / \left(h_c^2 R_b^{col} \gamma_{b1}\gamma_{b3}\right)$. Therefore, section size h_c of a square column subjected to an axial load F can be calculated as follows:

$$h_c = \sqrt{F / \left(F_1 R_b^{col} \gamma_{b1}\gamma_{b3}\right)} \quad (9)$$

3. Results and Discussion

3.1 Raft slabs effective depth design charts

To facilitate raft slab effective depth calculations even further it seemed to be reasonable to provide design charts for reinforced concrete columns and raft slabs built of specific materials. Calculated for $R_{sc} = 435$ MPa those charts and tables are presented below on the figure 3a – 3f of this paper. Columns reinforcement ratio varies from 0.5% to 5%.

Particular range of columns concrete grades was considered for each concrete grade of raft slabs. From the practical experience it was assumed that raft slabs are built with normal weight concrete of B25 – B50 grades. Concrete grade selected for a raft slab may not be higher than concrete grade of any vertical member including columns and walls. On the other hand, column concrete grade usually does not exceed raft concrete grade by more than twice. For example, B25 – B50 concrete grades is found to be reasonable range for raft slab built of B25 concrete, and so on.

3.2 Simplified design procedure of raft slab thickness calculations

Formula (8) derived in the previous section allows to calculate directly raft slab effective depth h_0 as a function of square column section size h_c and mechanical properties of raft slab and column materials:

- compression strength of reinforced concrete column R_b^{col},
- compression strength of column reinforcement R_{sc} and it's reinforcement ratio μ,
- tension strength of concrete raft slab R_{bt}^{raft}.

When conservatively sizing raft thickness for full axial capacity of a column N_{ult} it is not necessary to know actual column load F applied to the raft slab. All calculations related to the geometry of punching shear critical section are also omitted since it's already accounted by formula (8). As soon as all of the listed above parameters $-R_{bt}^{raft}$, R_b^{col}, R_{sc}, μ, h_c- are known from the building superstructure design stage, raft slab thickness can be calculated with one single formula (8) without any iterations.

IOP Conf. Series: Materials Science and Engineering **675** (2019) 012014 doi:10.1088/1757-899X/675/1/012014

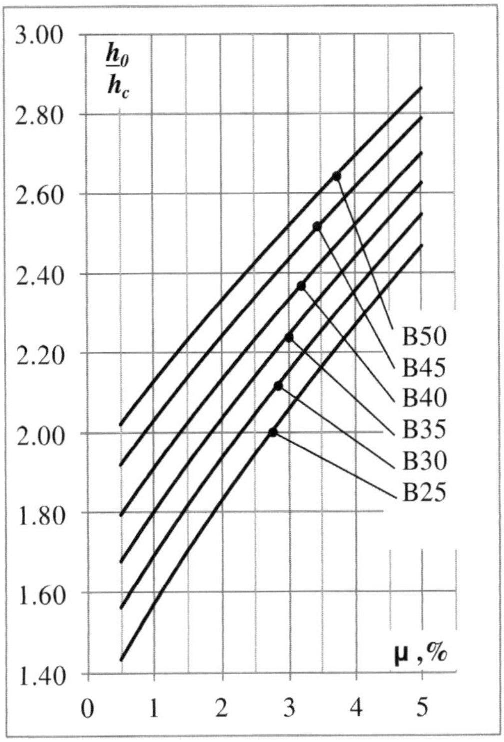

Figure 3a. Columns (B25-B50) – raft slab (B25) design chart

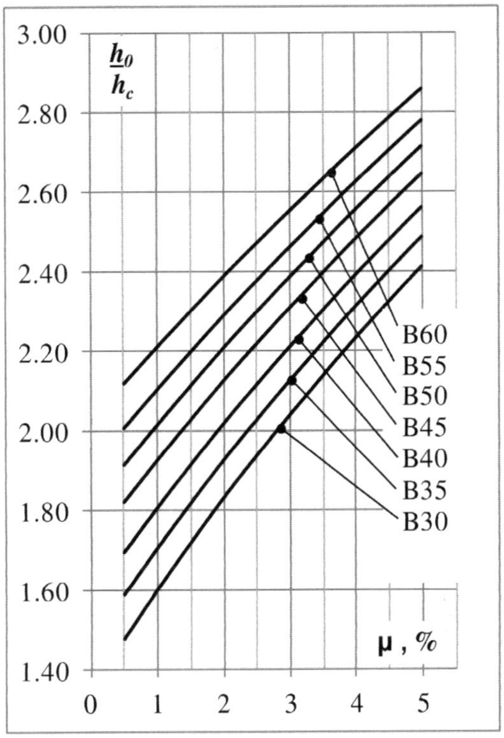

Figure 3b. Columns (B30-B60) – raft slab (B30) design chart

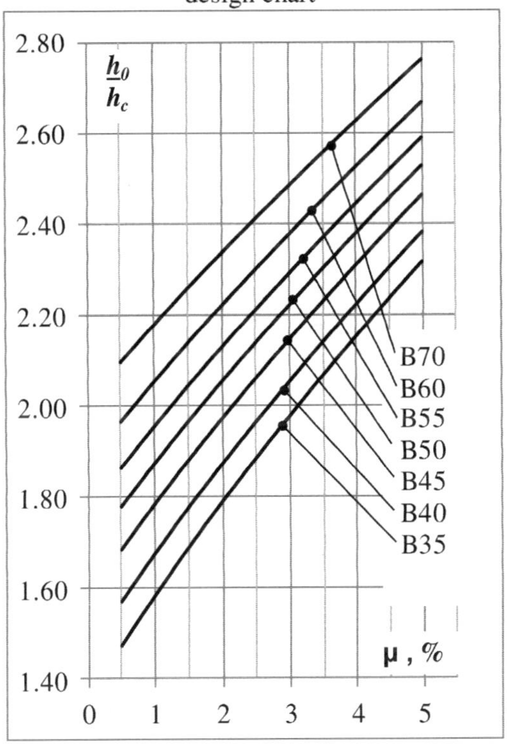

Figure 3c. Columns (B35-B70) – raft slab (B35) design chart

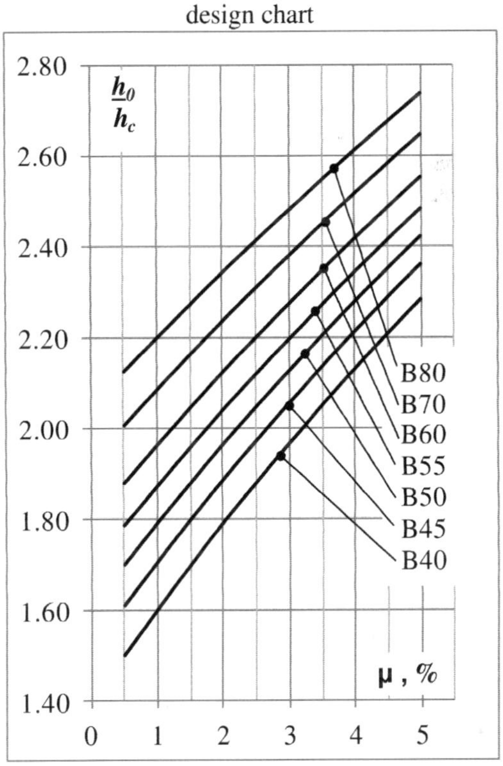

Figure 3d. Columns (B40-B80) – raft slab (B40) design chart

IOP Conf. Series: Materials Science and Engineering **675** (2019) 012014 doi:10.1088/1757-899X/675/1/012014

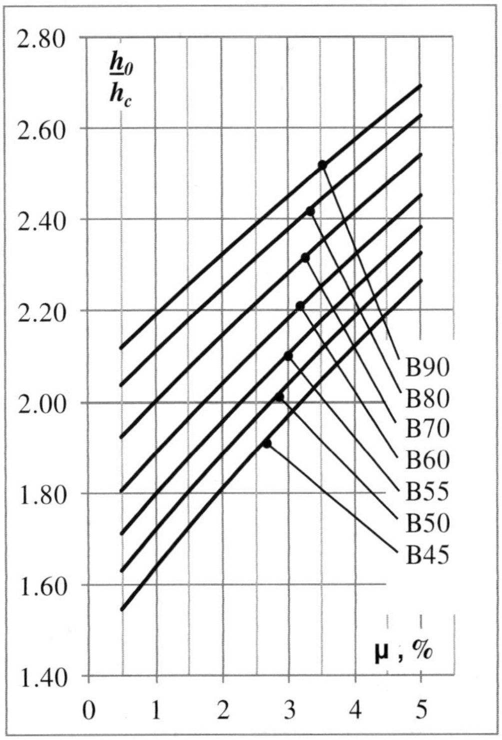

Figure 3e. Columns (B45-B90) – raft slab (B45) design chart

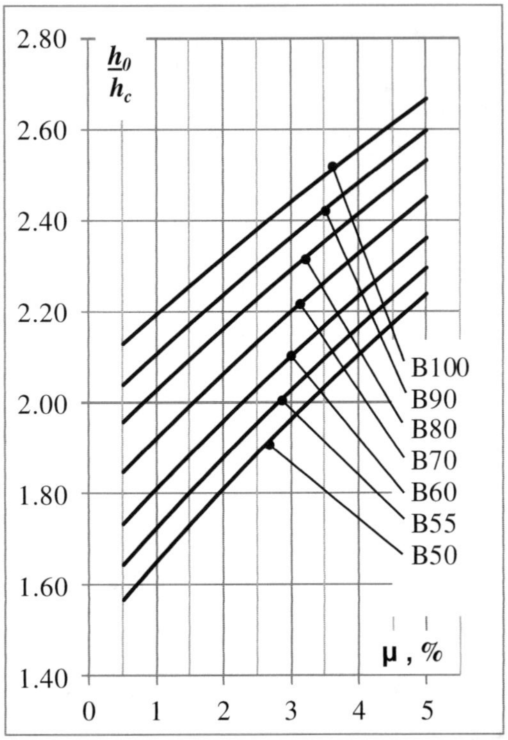

Figure 3f. Columns (B50-B100) – raft slab (B50) design chart

3.3 Analysis of design assumptions and design optimization

Design charts presented on the figure 3a – 3f are calculated for square columns with most conservative assumptions described above. Therefore, h_0/h_c values shown in Appendix A represent extreme maximums which can be further optimized on the latest design stages if needed. Appendix A is also applicable for rectangular columns since its perimeter would always be bigger than perimeter of square columns with the same area. Formula (8) can be easily modified to consider actual shape of column cross section with given area A_c and perimeter p_c, and to replace column axial capacity N_{ult} with column load F reduced by value of shear strength provided by raft slab transverse reinforcement and soil pressure:

$$h_0/h_c \geq \left(\sqrt{F_{red}/\left(A_c R_{bt}^{raft} \gamma_{b1}\right)+1}-1\right)\Big/2,\ h_c = p_c/4 \qquad (10)$$

3.4 Comparison with raft slabs design charts developed previously

Quite complex approach have been utilized by authors of [19] for developing raft slab thickness design charts based on specific building geometry, loadings and geotechnical conditions. Therefore practical application of those charts is much more limited than simple design diagrams presented herein. However, within the most common domain of generalized engineering parameters both approaches show good correlation of the results quantitatively and qualitatively.

3.5 Example problem

Using design charts presented on figures 2 and 3a estimate size h_c of a square column subjected to an axial load $F = 1500t$ and calculate effective depth h_0 of raft slab assuming column reinforcement ratio $\mu = 2\,\%$, column concrete grade B40 and raft concrete grade B25.

1. For column concrete grade B40 and $\mu = 2\%$ on figure 2 find $F_1 = 1.5$.
2. Calculate $h_c = \sqrt{F/\left(F_1 R_b^{col}\gamma_{b1}\gamma_{b3}\right)} = \sqrt{1500t/\left(1.5\cdot 2200t/m^2\cdot 0.9\cdot 0.85\right)} = 0.771m \approx 0.78m$
3. Find $h_0/h_c = 2.1$ - design chart on figure 3a
4. Calculate effective depth of raft slab: $h_0 = 2.1\cdot h_c = 2.1\cdot 0.78m = 1.638m \approx 1.64m$

Design requirement $F \le F_{b,ult}$ can be checked by substituting h_c and h_0 calculated above into (3): $F = 1500t < F_{b,ult} = 4\left(h_0^2 + h_0 h_c\right)\left(\gamma_{b1} R_{bt}^{raft}\right) = 4\left((1.64m)^2 + 1.64m\cdot 0.78m\right)\left(0.9\cdot 105t/m^2\right) = 1500.2t$.

4. Conclusion

In this work effective depth of the raft slabs is considered as a function of an axial capacity of the column. This approach is quite reasonable since such important design parameters as building geometry (number of the floors, column spacing etc) and specific loading conditions are already accounted in the column design itself. Then it allowed to develop raft slabs design charts applicable to any buildings or structures subjected to any loading conditions prescribed by design codes [20-23].

As it was demonstrated in the example problem solved above in section 3.5, proposed design method utilizing raft slabs design charts in conjunction with column capacity diagrams (figure 2) allows calculating column sizes and raft slab effective depth in 4 simple steps which greatly facilitate the design work flow of structural engineering practitioners. Since the design charts are strictly based on the structural requirements and formulas provided by SP design codes [20, 21], they can be used as a part of regular design practice followed by submitting the results to authorized institutions as a part of construction permit process.

It can also be noted that presented design charts can be easily adjusted to include concrete cylinder strength $f_c^{'}$ and safety factors φ_i used by Eurocode 2 [23] and ACI 318 [22] as design input data instead of .concrete cubic strength of concrete R_b / R_{bt} and safety factors γ_{b1}, γ_{b2} γ_{b3} utilized by Russian SP [20, 21].

Finally, analysis of the data presented on developed raft slab design charts leads to the conclusion especially important for researchers conducting fundamental study of moderately thick plates on elastic foundations. It can be observed that calculated practical values of raft slabs effective depth to column size ratio h_0/h_c vary from 1.4 to 2.9. These values define the most practical range for numerical tests and experiments carried out by researchers to verify newly proposed design methods, fundamental approaches and solutions.

References

[1] ACI Committee 318 2014 *Building Code Requirements for Structural Concrete (ACI 318-14) and Commentary* (Farmington Hills: American Concrete Institute)
[2] Hemsley J A 2000 *Design Applications of Raft Foundations* (London: Thomas Telford Ltd) p 608
[3] Gupta S C 2008 *Raft Foundations: Design and Analysis with a Practical Approach* (New Delhi: New Age International) p 148
[4] Ali S 2017 *Comparative Study of Different Analysis Techniques of Mat Foundation* (Lambert Academic Publishing) p 124
[5] Voyiadjis G Z and Kattan P I 1986 Thick rectangular plates on an elastic foundation *J. Eng. Mech.* **112(11)** pp 1218-1240
[6] Pechorskaya S A, Ikrin V A and Khomyak V P 2009 Solution of the axisymmetrical problem of the bending of a plate on a Winkler bed with allowance for displacement of normals and layer pressure *Soil Mech. Found. Eng.* **46** 186
[7] Han J B and Liew K M 1997 Analysis of moderately thick circular plate using differential quadrature method *J. Eng. Mech.* **123(12)** pp 1247-1252

[8] Korhan O, Daloglu A and Karakas A 2013 Parametric study for thick plates resting on elastic foundation with variable soil depth *Appl. Mech.* **83(4)** pp 549-558

[9] Ahmed M, Mohamed M H, Mallick J and Hasan M A 2014 3D-analysis of soil-foundation-structure interaction in layered soil *Open Journal of Civil Engineering* **4** pp 373-385

[10] Ghugal Y M and Sayyad A S 2010 A Static flexure of thick isotropic plates using trigonometric shear deformation theory *Journal of Solid Mechanics* **2(1)** pp 79-90

[11] Rouzegar J and Sharifpoor R A 2015 Flexure of thick plates resting on elastic foundation using two-variable refined plate theory *Archive of Mechanical Engineering* **LXII(2)** pp 181-203

[12] Shen H S and Yu L 2004 Nonlinear bending behavior of Reissner–Mindlin plates with free edges resting on tensionless elastic foundations *Int. J. Solids Struct.* **41** pp 4809-4825

[13] Sayyad A S and Ghugal Y M 2018 Bending of Shear Deformable Plates Resting on Winkler Foundations According to Trigonometric Plate Theory *J. Appl. Comput. Mech.* **5(3)** pp 187-201

[14] Cheng Y M and Law C W 2008 Problems with a popular thick plate element and the development of an improved thick plate element *Struct. Eng. Mech.* **29(3)** pp 327-349

[15] Abdalla J. A. and Ibrahim A. M. 2006 Development of a discrete Reissner-Mindlin element on Winkler foundation *Finite Elem. Anal. Des.* **42** pp 740-748

[16] Karasin H, Gulkan P and Aktas G 2015 A finite grid solution for circular plates on elastic foundations *KSCE Journal of Civil Engineering* **19(4)** pp 1157-1163

[17] Karasin A and Aktas G 2014 An approximate solution for plates resting on Winkler foundation *International Journal of Civil Engineering and Technology* **5(11)** pp 114-124

[18] Shukla S K, Gupta A and Kumar D 2012 Development of design charts for the deflections of a loaded circular plate resting on the Pasternak foundation *Int. J. Eng. Inv.* **1(5)** pp 56-59

[19] Al-Shayea N and Zeedan H 2012 A new approach for estimating thickness of mat foundations under certain conditions *Arab. J. Sci. Eng.* **37(2)** pp 277-290

[20] Design Code SP 63.13330.2012 2011 *Concrete and Reinforced Concrete Structure* (Moscow: Minregion of Russia)

[21] Design Code SNiP 2.03.01-84* 1985 *Concrete and Reinforced Concrete Structure* (Moscow: ZITP Gosstroy USSR)

[22] ACI Committee 336 1988 Suggested Analysis and Design Procedures for Combined Footings and Mats *ACI Struc. J.* **85(3)** pp 304-324

[23] EN 1992-1-1 Eurocode 2*: Design of concrete structures – Part 1: General rules and rules for buildings* CEN Brussels Belgium 2002

Development of methods for buildings calculation from sandwich panels for transformable low-rise buildings

A A Pleshivtsev[1*], M I Abu Mahadi[2], A C Markovich[2], P C Chiadighikaobi[2]

[1] National research Moscow State Construction University (MGSU), Moscow, Russia
[2] Peoples Friendship University of Russia (RUDN University), Moscow, Russia

Email: perspektiva-aa@mail.ru

Abstract This paper focuses on calculating the main loads of prefabricated low-rise buildings from sandwich panels of arbitrary structural elements and connections in the form of a rigid spatial system. The main objective of this research are the theoretical and experimental substantiation of rational parameters of quickly erected transformable low-rise buildings from sandwich panels for the development of optimal design solutions to ensure spatial rigidity. The research used empirical research method - observation, experiment, measurement, comparison and analysis. The quickly erected transformable low-rise buildings of sandwich panels, like a prismatic shell consisting in the transverse direction of two closed rectangular contours, can bear an arbitrary power load. Determination of the cost of materials for additional structural elements in the form of ties (embedded parts), combining the individual elements in a holistic design that provides spatial rigidity of the structure.

Key words: prefabricated low-rise buildings of sandwich panels; power load; spatial stiffness; doubly connected shell; deformed state.

1. Introduction

Sandwich panels are in active use as enclosing constructions for different types of buildings [1]. Loads and temperature changes are essential in the analysis of sandwich panels' behavior. The sandwich panel is a multi-layer structural element consisting of thermal insulation located between flat or profiled metal sheets and fixed to the buildings bearing structure by means of threaded screws. These screws often create "thermal bridges" cross the sandwich panel system [2-5].

The construction of low-rise buildings for various purposes in the mass modern construction is determined by their constructive-technological and space-planning decisions, as well as the requirements for ensuring the reliability of buildings [6].

Unsuccessful design solution (flat design scheme of the structure) causes the presence of an incorrect distribution of material and partly the work of the structure in the spatial version. This causes economic losses in the construction of structures and buildings.

Currently, the existing facilities can be divided into two main groups:

i. structures with a strongly pronounced nature of spatial work and;
ii. structures with a weakly pronounced nature of spatial work.

The first group of structures includes: domes, warehouses, arches, various shells (cylindrical, flat, double Gaussian curvature, negative Gaussian curvature, etc.), generous coverings, extended and high-rise frame-panel buildings. A clear line on the accessories in this group is difficult to hold [7].

In [8], the authors revealed the nature of the deformation of the fastening system on the basis of a series of preliminary numerical experiments performed with the application of a uniformly distributed static load to the system, as well as loads arising from the difference in temperatures outside and inside the building. Calculated results obtained by authors and compared with theoretical calculations, however, during the experiment, the joint operation of the sandwich panel system with the building frame was not taken into account.

Investigations in this direction will make it possible to more accurately determine their stress – strain state, taking into account the boundary conditions, to design, construct and make them easier and more economical [9].

The second group includes the usual buildings: industrial workshops (multi-span in cross section), garages, hangars, pavilions, etc. Such structures very often can be represented as a set of trusses, arches, frames and other elements and structures interconnected by different types of connections into the spatial framework of structures. Basically, the supporting frame corresponds to a flat design scheme of a conventional building. The load from the coating is transferred to the trusses, from the trusses to the columns and from the columns to the foundations and the basis [10-12].

In such design schemes of structures, material costs are additionally required for additional structural elements in the form of various connections (embedded parts, welding materials, etc.). These links unite the individual elements into an integral structure and provide the spatial rigidity of the structure. Thus, it turns out that many constructions made up of flat structural elements work only partially as spatial systems and are too material-intensive [13-15].

This paper considers development of methods for calculating buildings from sandwich panels of three-layer, for quickly erected transformable low-rise buildings. Derivation of numerical methods for development of sandwich panels for structures will be considered.

2. Materials and methods

When calculating, the presented building is considered as a prismatic envelope of a multiply connected section. Bearing frame of the building creates longitudinal external walls and floors of the first and second floor. In the longitudinal direction along the three nodal points of the upper belt, a closed multiple-connected cross-section, external vertical loads are applied, and at the extreme points there are additional expansions (horizontal load) from the coating.

This prismatic shell with dimensions in the plan of 6 × 9 m. along the whole contour is supported by longitudinal and transverse supports (Figure 1). The cross section of the shell is divided into separate sections of different sizes, and then set of longitudinal strips is obtained. As a result, the horizontal and vertical stripes with a length of 9 m form a doubly connected shell (Figure 2).

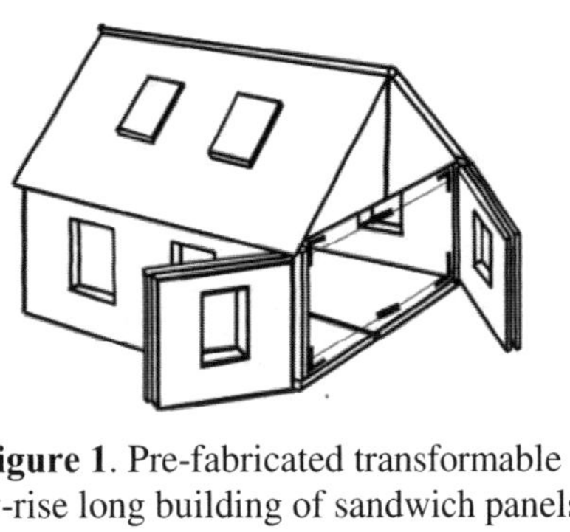

Figure 1. Pre-fabricated transformable low-rise long building of sandwich panels.

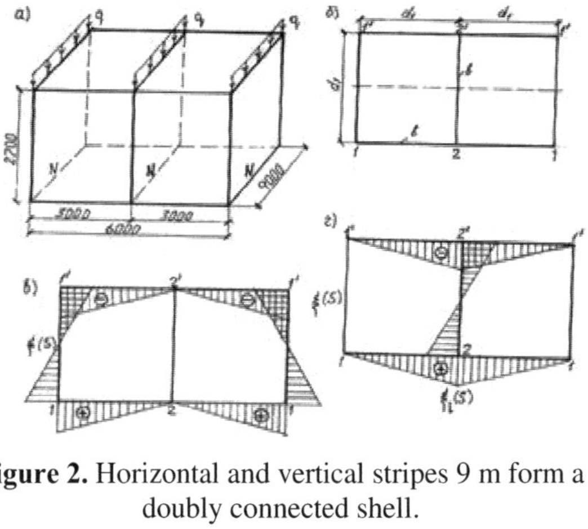

Figure 2. Horizontal and vertical stripes 9 m form a doubly connected shell.

IOP Conf. Series: Materials Science and Engineering **675** (2019) 012015 doi:10.1088/1757-899X/675/1/012015

The width of the vertical stripes d_1 = 2.7 m., The width of the horizontal stripes d_2 = 3 m. In places of pairing, bands locate stringers, working only in tension and compression. The role of stringers is played by wooden bars with a rectangular section. Geometric and mechanical characteristics of the extreme and middle stringers are given in table 1.

Table 1. Geometric and Mechanical Characteristics of Extreme and Medium Stringers.

№		$u_2(x)$	$v_1(x)$	$v_2(x)$	P_i
1	$\gamma a_{11}D^2 - a'_{11}$	$-a'_{12}$	$-b_{11}D$	$-b_{12}D$	$\frac{1}{G}P_1$
2	$-a'_{21}$	$\gamma a_{22}D^2 - a'_{22}$	$-b_{21}D$	$-b_{22}D$	$\frac{1}{G}P_2$
1	$b_{11}D$	$b_{12}D$	$C_{11}D^2 - \gamma s_{11}$	$C_{12}D^2 - \gamma s_{12}$	$\frac{1}{G}q_1$
2	$b_{21}D$	$b_{22}D$	$C_{21}D^2 - \gamma s_{21}$	$C_{22}D^2 - \gamma s_{22}$	$\frac{1}{G}q_2$

At the end, the multiple-connected shell has supporting diaphragms that are rigid in their plane and flexible from their own plane. The design of the end walls allows presentation of their work as a hinge. The end walls in the vertical plane (in their own plane) are not deformed, and in the horizontal direction (their plane) are considered absolutely flexible.

The design of the horizontal plates, i.e. the floors of the first and second floors are such that the longitudinal displacements of their cross sections cannot be obstructed, i.e. normal section forces will not be able to accept (or will accept them in a small volume).

Therefore, it is believed that these plates do not work in tension – compression, but work mainly in shear. Longitudinal normal forces are perceived only by stringers. Horizontal plates can perceive transverse normal forces and only after reinforcement with frames transverse bending moments.

In this case, the presented prismatic shell of a multiple-connected section without special reinforcement will have transverse frames in the transverse direction with a moment-less structure.

As a result, the shell reinforced in the longitudinal direction, by stringers operating exclusively in tension – compression is taken as the design scheme of the presented prismatic shell.

Thus, given the physical and structural properties of the building in question, when calculating it for a vertical load, it is sufficient to consider two states: *symmetric and anti-symmetric*.

Further, the calculation method is developed and used variation theory of V.Z. Vlasov [16] reduction of two-dimensional and three-dimensional problems to one-dimensional.

When calculating the required unknowns, the functions of longitudinal and tangential displacements are taken, which are represented as a product of two functions. One function is the desired one, and the second characterizes the pattern of its distribution in the transverse direction. It is believed that the prismatic shell has infinite in the longitudinal and finite in the transverse direction the degree of freedom. According to Vlasov, such systems are called discrete-continual.

The position of a point on the middle prismatic surface of the shell is determined by the coordinates x and s. The first coordinate indicates the point's distance from the initial, extreme cross section x = 0, and the second coordinate s determines the location of the point in the cross-sectional plane relative to the initial line (assumed as the initial) of the generator.

The longitudinal $u_i(x,s)$ and transverse $v_j(x,s)$ displacements of an arbitrary point with coordinates (x, s) on a prismatic surface are represented as the following finite expansions:

$$u(x,s) = \sum_{i=1}^{m} u_i(x)\xi_i(s) \quad (1)$$

$$v(x,s)=\sum_{j=1}^{n} v_j(x)\eta_j(s) \tag{2}$$

Here, the first equation expresses the deformed state of an elementary transverse strip of width dx = 1, carried out by longitudinal displacements $u_i(x,s)$. The required functions $u_1(x)$, $u_2(x)$, ..., $u_m(x)$ are taken as longitudinal displacements of m nodes of the shell. The functions $\xi_1(s)$, $\xi_2(s)$, ..., $\xi_m(s)$ corresponding to the longitudinal displacements must satisfy the conditions of continuity and continuity of the longitudinal displacements.

The functions $\xi_i(s)$ for the specified method of selecting longitudinal displacements $u_i(x)$ are non-zero only in straight sections of the contour of the cross section of the shell, at point i have a single value and vary in these areas according to a linear law. At the remaining points of an elementary transverse strip of width dx = 1, these functions are equal to zero. Moreover, the chosen method of constructing a function $\xi_i(s)$ is not exclusively unique and depends on the choice of the desired functions $u_i(x)$.

Equation (2) determines the deformed state of the elementary strip dx = 1 in the cross section plane x = const. The transverse strip dx = 1 is considered as a core frame of a multiply connected section with inextensible, tensile or partially tensile elements. Then, for the required functions in equation (2), functions $v_j(x)$ are taken that characterize the shape of the deformation of the core system in its plane.

The number of degrees of freedom n of the rod system under consideration is equal to the number of unknown functions $v_j(x)$ in its own plane and is determined by the formula:

$$n=2m-c \tag{3}$$

where m – number of nodes; c – number of rods.

We compose the expressions of possible work of external and internal forces acting on an elementary transverse strip of a prismatic shell of a multiple-connected section on its possible displacements in and out of the plane. The resulting system with a total number of (m + n) differential equations will represent the integral form of the equilibrium condition of an elementary transverse strip of width dx = 1.

Thus, the desired functions $u_i(x)$ and $v_j(x)$ will be determined from the specified systems of differential equations:

$$\int \frac{\partial\sigma}{\partial x}\xi_k(x,s)dF-\int \tau\xi_k'(x,s)dF+\int P_x\xi_k(x,s)ds=0;\quad (k=1,...,m) \tag{4}$$

$$\int \frac{\partial\tau}{\partial x}\eta_h(x,s)dF-\sum v_h\int\frac{M_kM_h}{EY}ds+\int P_s\eta_h(x,s)ds=0;\quad (h=1,...,n) \tag{5}$$

These equations include normal and tangential stresses, bending x = const.

In these equations, the extreme members with a plus sign express the work of external forces P_x and P_s both the work of normal and shear internal forces.

The middle term with a minus sign represents the work of the shear forces τdF on the shear deformation represented by the derivative $\xi_i'(s)$, as well as the work of the bending moment $\sum v_h(x)M_h(s)$ on the mutual angle of rotation $\frac{M_k(s)}{EY}ds$.

Normal and tangential stresses $\sigma=\sigma(x,s)$; $\tau=\tau(x,s)$ acting in the cross sections of the prismatic shell are considered uniformly distributed throughout the thickness of the section of the prismatic shell and are expressed by the known formulas:

$$\sigma = E\frac{\partial u}{\partial x}$$
$$\tau = G\left(\frac{\partial u}{\partial s} + \frac{\partial v}{\partial x}\right) \quad (6)$$

Taking into account the decomposition (1) and (2) for these stresses, we obtain the following expressions:

$$\sigma(x,s) = \sum_{i=1}^{m} u_i'(x)\xi_i(s) \quad (7)$$

$$\tau(x,s) = G\left[\sum_{i=1}^{m} u_i(x)\xi_i'(s) + \sum_{j=1}^{n} v_j'(x)\eta_j(s)\right] \quad (8)$$

where E and G are the deformation and shear moduli; $dF = \delta dS$ – differential cross-sectional area of the shell.

$M_k(s)$ and $M_h(s)$- represent the bending moments of the transverse strip dx = 1 - the frames corresponding to the elementary states of deformation of this frame with $v_k = 1$ and. $v_h = 1$.

$J(s)$ -moment of inertia of the transverse strip width dx = 1. If the shell in sections has only rectangular shapes, then the moment of inertia will be determined by the following formula:

$$J = \frac{1\cdot\delta^3}{12} \quad (9)$$

In the case when the shell is reinforced with additional cross-frames, then the moment of inertia of the cross-section of the shell is calculated taking into account the average moment of inertia of these frames; in other words, the moment of inertia per unit length of the shell is taken.

After substituting the values of normal and shear stresses into (4) and (5), σ_i, τ_i from formulas (7) and (8) we obtain the system of ordinary differential equations of the second order:

$$\gamma\sum_{i=1}^{m} a_{ij}u_i'' - \sum_{i=1}^{m} a_{ij}' u_i - \sum b_{ik}v_k' + \frac{1}{G}P_{xi} = 0$$
$$\sum_{i=1}^{n} b_{ij}u_i' + \sum_{k=1}^{h} C_{kh}v_k'' - \gamma\sum_{k=1}^{n} S_{hk}v_k + \frac{1}{G}P_{sk} = 0 \quad (10)$$

The coefficients of this system are determined by the formulas:

$$a_{ij} = \int \xi_i(s)\xi_j(s)dF;\ b_{ik} = \int \xi_i'(s)\eta_k(s)dF;\ C = \int \eta_h(s)\eta_k(s)dF$$
$$a_{ij}' = \int \xi_i'(s)\xi_j'(s)dF;\ b_{hi} = \int \xi_h(s)\eta_k'(s)dF;\ S_{hk} = \frac{1}{E}\int \frac{M_h(s)M_k(s)}{EY}ds \quad (11)$$

In these formulas, the integrals apply to all elements of the cross section of a multiple-connected prismatic shell.

Based on the Betty theorem on the reciprocity of the elastic system, here is the following equality of coefficients:

$$a_{ij} = a_{ji};\ b_{ik} = b_{ki};\ S_{hk} = S_{kh};$$
$$a_{ij}' = a_{ji}';\ C_{hi} = C_{jk} \text{ when h = k} \quad (12)$$

The free terms of the system of equations (10) in the presence of the surface forces of the shell $P_x(x,s)$; $P_s(x,s)$ are according to the formulas:

$$P_x(x) = \int P_{ix}\xi_i(s)ds \quad (13)$$

$$P_s(x) = \int P_{ks}\eta_k(s)ds$$

Free members P_x, P_s are generalized external forces. $P_x(x)$ is calculated as the work of external surface forces $P_x(x,s)$ on the longitudinal movement $\xi_i(s)$ of the elementary transverse strip in the state $u_j = 1$. The transverse force $P_s(x)$ is calculated as the work of external surface contour forces on the contour displacements of the elementary strip $\eta_h(s)$ in its plane in the state $v_h = 1$.

The system of ordinary differential equations of the second order in expanded form is given in Table 1. In this table, each element is a linear ordinary differential operator with a constant coefficient.

Operators D^2 and D show that the corresponding required function is required to take the second and first derivative with respect to the x coordinate.

By the method of academician L.N. Krylov [17], symmetric system (m + n) of ordinary linear differential equations with constant coefficients (10) can be reduced to an equivalent differential equation of order r (m + n). However, if there is a large order of the system of equations, the reduction method becomes cumbersome and its implementation becomes more complicated. Therefore, the solution of equation (10) is easier to search in rows. Depending on the boundary conditions, the desired functions $u_i(x)$ and $v_j(x)$ longitudinal and tangential displacements are decomposed into trigonometric series.

$$u_i(x) = \sum_{i=1}^{n} u_i \cos\frac{(2n-1)\pi x}{l}$$
$$v_k(x) = \sum_{k=1}^{m} v_k \sin\frac{(2n-1)\pi x}{l}, \quad (14)$$

where $u_i(x)$ and $v_k(x)$ – unknown coefficients of the n-th row member; L – shell length (distance between the end diaphragms).

The longitudinal normal voltage of the i – th state is represented as the following series:

$$\sigma_i(x,s) = E\xi_i(s)u_i{}' = -E\xi_i(s)\sum_{i=1}^{n} u_i \frac{(2n-1)\pi}{l}\sin\frac{(2n-1)\pi x}{l} \quad (15)$$

With the hinge of the prismatic shell, the boundary conditions are as follows:

$$x = 0;\ \sigma_i(0) = 0;\ v_k(0) = 0$$
$$x = l;\ \sigma_i(l) = 0;\ v_k(l) = 0 \quad (16)$$

In the presence of a uniformly distributed external vertical load with intensity q applied along the edges (nodes) of the prismatic shell, the decomposition form of the external load will have the following form:

$$q(x) = \sum_{i=1}^{n} \frac{4q}{(2n-1)\pi}\sin\frac{(2n-1)\pi x}{l} \quad (17)$$

The system of algebraic equations for the n-th term of the expansion is obtained by substituting a differential matrix $D^2 - \frac{(2n-1)^2\pi^2}{l^2}$ for the first and fourth quadrants instead of the operator (see Table 1), and in the second and third quadrant D is replaced by $\pm\frac{(2n-1)\pi}{l}$. Free terms are multiplied by a factor $\frac{4}{(2n-1)\pi}$. The final system of algebraic equations is given (see table 1).

Solving a system of algebraic equations table 1, we obtain for each term the decomposition of the longitudinal u_i and transverse (vertical) v_k displacements.

Thus, the type of functions u_i and v_k is set. After their substitution in (1) and (2) for the n – th term of the expansion, we find the values of the sought functions:

$$
\begin{aligned}
u_n(x,s) &= \sum_{i=1}^{m} u_i(x)\xi_i(s) = \sum_{i=1}^{m} u_i \cos\frac{(2n-1)\pi z}{l}\xi_i(s) \\
v_n(x,s) &= \sum_{j=1}^{k} v_j(x)\eta_j(s) = \sum_{j=1}^{k} v_j \sin\frac{(2n-1)\pi z}{l}\eta_j(s)
\end{aligned}
\tag{18}
$$

On the basis of the data obtained (meaning the displacement of characteristic points), the stress – strain field of the prismatic shell is established by the usual methods of structural mechanics.

3. Results

The prismatic shell of a quickly erected transformable low-rise buildings from sandwich panels, consisting in the transverse direction of two closed rectangular contours, is considered. The geometric dimensions of the shell elements are given. The cross section of the shell has two axes of symmetry: vertical and horizontal. Such a shell can carry an arbitrary power load. In the particular case, it is possible to consider the option when an external vertical load applied on the ribs (along the nodes) according to an arbitrary law acts on the shell. Due to symmetry, the external load can be expanded and its two states under load can be considered: the shell is operating under symmetric and anti-symmetric loads.

Each horizontal plate of the shell has a width d_2 (cm), vertical plates have a width d_1 (cm), let us assume that in the first approximation all plates have the same thickness δ_i. For each plate, the law of flat sections is assumed to be fair. Also, in the first approximation, the plates on their width are considered inextensible. Note also that formulas (11; 13) allow us to take into account the compressibility of the plates. For the generalized coordinates $\xi_1(s)$, $\xi_2(s)$ the functions that are symmetric with respect to the vertical and back symmetric with respect to the horizontal plane are selected (see Fig. 1). With this choice of functions $\xi_1(s)$, $\xi_2(s)$ the sought generalized displacements $u_1(x)$, $u_2(x)$ will be numerically equal to the longitudinal displacements at the nodal points of the cross-section contour.

The functions $\xi_1(s)$, $\xi_2(s)$ determine two independent components of the cross section deplanation related to the equilibrium state of the shell, symmetrical about the vertical plane of symmetry (figure 3).

In principle, horizontal movements or movements of vertical plates are taken as generalized movements. Type of functions $\eta_j(s)$ when $v_1 = v_2 = 1$ are given (see figure 4).

In the case of a symmetrical load, the deflections of horizontal plates in the horizontal direction are assumed to be zero. In figure 4 the arrows indicate the positive direction of deflection of the plate. Based on the above, the generalized equilibrium conditions of a strip of width dx = 1 in the form of ordinary differential equations with the unknown functions $u_1(x)$, $u_2(x)$, $v_1(x)$, $v_2(x)$.

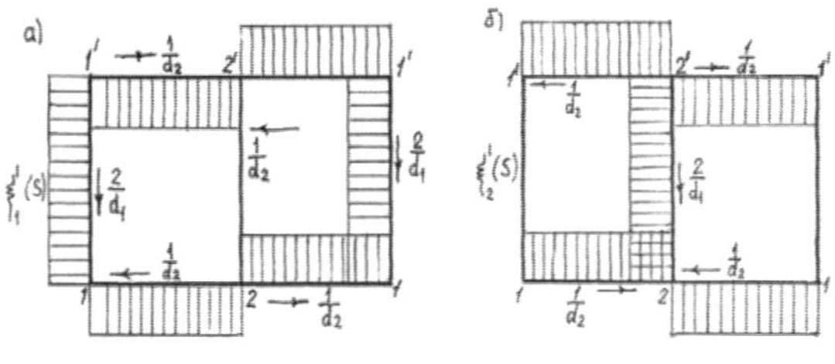

Figure 3. State of equilibrium of the shell, symmetrical with respect to the vertical plane of symmetry.

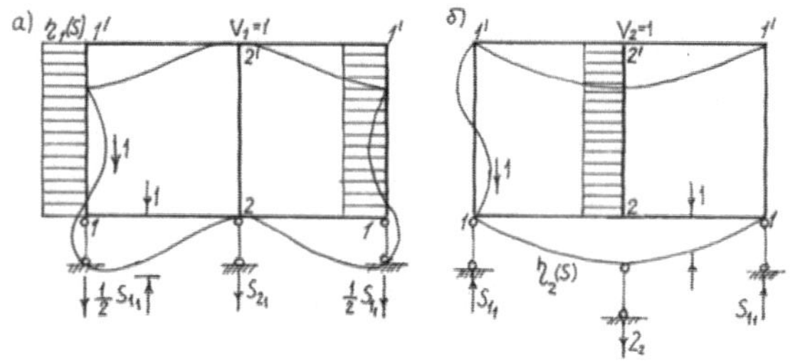

Figure 4. Positive direction of plate deflection.

4. Conclusion

Derivatives $\xi_1'(s)$, $\xi_2'(s)$ of the given functions $\xi_1(s)$, $\xi_2(s)$ in certain parts of the cross-section contour are constant values (not dependent on the s coordinate). The arrows indicate the directions in which the corresponding differentiable function increases.

Similarly, the functions of the transverse distribution of displacements are considered $\eta_j(s)$. They characterize the deformation of an elementary transverse strip in the section plane x = const. The degree of freedom of an elementary transverse strip, as a flat hinge-rod system, under the condition of the inextensibility of its elements in a symmetric state is equal to two. This means that the deformed symmetric state of the contour of the shell in the plane of its cross section is determined by two parameters. These parameters are generalized transverse displacements $v_1(x)$, $v_2(x)$ depending on the x coordinate.

References

[1] Li D, Deng Z, Chen G, Xiao H and Zhu L 2017 *Composite Structure* 1692017 pp 29–41
[2] Ali A M, Tomko M and Demjan I 2015 Experimental investigation and theoretical analysis of polystyrene panels with load bearing thin-walled cold-formed elements *Apld. Mechs. & Mats.* **769** p 145–52
[3] Sedláková A et al 2015 Optimal construction solutions of the external wall structures *Theory & Practice* **8** pp 157–63
[4] Lobanov D S, Vildeman V E, Babin A D and Grinev M A 2015 Experimental research into the effect of external actions and polluting environments on the serviceablity of fiberreinforced polymer composite materials *Mech. of Comp. Mats.* **1(51)** pp 69–76
[5] Ćetković M and Vuksanovic G 2014 Thermal analysis of laminated composite and sandwich plate using layer wise finite element *Const. of Uniq. Buldgs & Struc.* **5(20)** pp 7–14
[6] Galyamichev A V, Kirikova V A, Gerasimova G E and Sprince A 2018 *Magazine of Civil Engineering* **78(2)** pp 30-46
[7] Vedishcheva I S, Ananin M Y, Ali A M and Vatin N I 2018 *Magazine of Civil Engineering* **78(2)** pp 116-127
[8] Ananin M Yu and Vedishcheva Yu S 2015 (Yekaterinburg: Uralsk Federal University Press) p 51-5
[9] Novikov M, Polushin V and Zimin S 2016 *MATEC Web of Conference* **53** 01038
[10] Rybakov V, Dyakov S, Sovetnikov D, Azarov A and Ivanov S 2018 *MATEC Web of Conference.* **245** 08007
[11] Rybakov V, Molchanova N, Laptev V, Suslova A and Sivokhin A 2016 *MATEC Web of Conference.* **53** 01017
[12] Pavlenko A D, Rybakov V A, Pikht A V and Mikhailov E S 2016 *Magazine of Civil Engineering* **67(7)** pp 55-69
[13] Lalin V, Rybakov V and Sergey A 2014 *Applied Mechanics and Materials* **578-579** pp 858-63
[14] Tusnina O 2014 A finite element analysis of cold-formed Z-purlins supported by sandwich panels *App. Mech. & Mats.* **467** pp 398-403
[15] Danilov A and Tusnina O 2014 The joints of cold-formed purlins *J. of App. Eng. Sci.* **12(2)** pp 153-158
[16] Vlasov V Z 1962 *General theory of shells and its applications in engineering* (Moscow: Academy of Sciences USSR Press)
[17] Krylov A N 1932 *Some Differential Equations of Mathematical Physics* (Publishing House Gostekhizdat, Leningrad, USSR)

IOP Conf. Series: Materials Science and Engineering **675** (2019) 012016 doi:10.1088/1757-899X/675/1/012016

Development trends of residential buildings in Moscow city

L Savelieva[1,2*], N Kalinina[1,2] and E Ogienko[2]

[1] Moscow Institute of Architecture (State Academy), Moscow, Russia
[2] Peoples' Friendship University of Russia (RUDN University), Moscow, Russia

E-mail: savelyeva-lv@rudn.ru

Abstract. There is a large-scale construction of a new type of residential real estate – apartments in Moscow. The attractiveness of prices and the location of the complexes attract a huge number of buyers. Two terms: apartments and luxury housing become synonymous, but this is a misunderstanding and substitution of true concepts. The authors set the tasks to specify the definition of "apartments" in the understanding of specialists - architects; to identify the main features and parameters of this new for Russia concept; scientifically describe the main trends in the development of this type of real estate as a modern and relevant object of architectural design. Three types of apartments were determined as a result of the conducted analysis: apart-hotels; apartments in a multi-unit dwelling; "multi-form" apartments in a residential complex. It is revealed that the apartments are ranked by the convenience, size and quality of services provided. The authors identified the following classes: economy, comfort, business and elite. The article discusses some legislative nuances, design standards, engineering solutions for apartment buildings, as well as the features of their location in the structure of the city of Moscow.

1. Introduction

Currently, Moscow is rapidly building apartments. Attractiveness of prices and location of complexes attract a huge number of buyers of so-called "elite housing". Two terms: apartments and luxury accommodation become synonymous. Are they really same [1]?

The concept of apartments has come to Russia from the West, where the real estate is quite common. The "apartment" term (originally French appartement) refers to personal living rooms in large houses. In the XVIII century in Russia this term was used for the so-called personal accommodation in estates by analogy with the Royal apartments in the palaces of the French monarchs. Apartments as a form of a real estate have appeared relatively recently in our country, namely in the 90-ies of XX century.

The authors set the tasks to specify the definition of "apartments" in the understanding of specialists - architects; to identify the main features and parameters of this new for Russia concept; scientifically describe the main trends in the development of this type of real estate as a modern and relevant object of architectural design.

The objectives of the study are to systematize the legislative base of rental housing and technical information on the arrangement of apartments for the professional architectural community; consider the most important conditions for the functioning of apartment complexes and indicators that determine the effectiveness of the design during construction in Moscow; to determine the main types of such a living unit, their classification by comfort and ranking by other characteristics.

Studies have shown, that currently mainly articles on the problems of managing apartment buildings in terms of energy saving [2, 3 etc.], physical housing modernization [4 etc.], new housing concepts [5 et al.] etc. are being published in leading international journals. In Russian and international foreign scientific journals, the question of the widespread development of the apartments in Moscow and the problems of architectural, social, economic and legal nature connected with this fact have never been covered. Moreover, nobody has ever conducted any classification of apartments for various signs. In this article we have tried to fill this gap in international publications. The purpose of this study is the scientific description and definition of a new type of residential real estate for Russia - apartments. Especially important is the generalization and systematization of legislative and regulatory acts of the Russian Federation, that regulate the design, construction and operation of the apartments.

2. Materials and methods

The study of the stated issue was carried out on the basis of studying and analyzing the published materials. To clarify the causes of the emergence of a new type of residential real estate in Russia, as well as its further development, the authors used an analytical research method.

3. Results and discussion

One of the reasons for the appearance of apartments in Russia was the economic growth of the early years of 2000, when the building boom began across the country. The Moscow-City office complex has become the symbol of that time in Moscow. Soon after the take-off, the 2008 economic crisis ensued, as a result of which, the demand for offices fell, and the metropolitan developers were forced to look for ways to save their investments, transferring part of the Moscow-City areas to apartments. Let us list some more factors that served as prerequisites for the development of apartments, as a new format of real estate: the lack of sites for residential construction; the need for housing near the city center at affordable prices; Moscow government decree on the prohibition of point construction.

Apartments in other countries are called a real estate for rendering tourist services. The apartments are designed for temporary residence. Usually, an aparthotel or apartment complex managing company undertakes all the obligations for the operation of the apartment, its maintenance, searches for tenants and dealing with them. The owner of this property can temporarily reside in his apartment, in agreement, and actually is a passive recipient of the rental income.

In Russia there were so-called revenue houses similar to modern "apartments" already existed in the XVIII century. They became particularly widespread in Russia in the middle of the XIX century [6]. The industrial revolution, the influx of people into cities and the great demand for workers and middle class apartments - all this contributed to housing construction at an accelerated pace and in a large scale. Until 1917 about 40% of housing was the revenue houses in Moscow. With the outbreak of World War I, construction in Russia ceased, and no longer revenue houses were built. The pre-revolutionary revenue house in a certain sense was a forerunner of modern house-complexes which is nowadays significantly spread in Moscow commercial construction, and in many ways can become a prototype of future new revenue houses [7].

The idea of the revival of revenue houses, voiced by a number of senior officials of the capital in the beginning of the XXI century, unfortunately, has not been fully implemented so far [8]. The main reason is that, projects of revenue houses are of no interest to investors because of low payback. "Prospects in the existing form are not attractive for private business; the project does not have any perspectives. The goals of private business are maximum profitability, quick payback and reliability of investments. <…> It turns out that it is more profitable to build a house and sell it" [9]. According to experts of the Blackwood real estate consulting company, the payback period of the revenue house in the elite class is 7-12 years, in the economy class these terms can be more than 20-25 years [9].

In 2015 Mikhail Men', Russian Minister of Housing and Utility Infrastructure, at the plenary session of the IV Russian Investment and Construction Forum (RICF) noted: "The law on the development of rental housing and virtually all by-laws have been adopted. Our task is to achieve a 7% share of rental houses in the total of residential real estate, as we hope, in 2016. According to experts, rental housing can be a serious alternative to buying apartments, and regions need to push their developers to implement such projects using all the means" [10]. In July 2014, Federal Law No. 217-FZ "On Introducing Amendments to the Housing Code of the Russian Federation and Certain Legislative Acts of the Russian Federation Regarding Legislative Regulation of Relationships for Renting Residential Premises of the Housing Fund for Social Use" came into force, aimed at

IOP Conf. Series: Materials Science and Engineering **675** (2019) 012016 doi:10.1088/1757-899X/675/1/012016

regulating legal relations in the sphere of renting living premises. The federal law provides for the creation of legislative basis to form a market with affordable rent housing and the development of a housing fund for non-commercial use for citizens who have a low level of income. This law introduced the concept of a "rented house" which is the same as a revenue house. Since the entry into force of this law, namely from July 22, 2014, it became possible to have legal relations with a new real estate objects in the Russian Federation - a rented house. According to paragraph 1 of Art. 91.16 of the RF Housing Code, a rented house is recognized as a building which or all premises in which belong to ***one person*** right of ownership and which or all the living quarters of which are intended to be granted to individuals for possession and use for residence.

All legislative initiatives described above apply to a rental housing fund. Nowadays the apartment is a commercial real estate in Russia which belongs to a category of non-residential premises. In the last few years, the authorities have been trying to equate apartments with the housing property or recognize them as a separate type of a real estate, but so far this has not happened [11]. This is the main difference between apartments and revenue houses. A revenue house is intended for renting as a residential area to live, apartments are intended for renting as a non-residential area for living. This key moment opens a lot of opportunities for investors, developers, landlords, etc. It is this fact that contributes to the mass construction of the elite real estate in the best Moscow districts! So why are the apartments so attractive to investors and buyers? There are several reasons.

First, as already mentioned, it is the payback time of the project. It is profitable to sell and gain profit immediately or in the shortest possible time. The apartments are sold under the guise of a residential housing to individuals and, accordingly, and it brings profit, often, even at the construction stage.

Secondly, the density of housing development and, accordingly, the profit is also higher for apartments. This is worth considering in more detail. The very term "apartments" is not found in the Russian by-law acts. This concept is used in construction and tourism activities. According to the Order of the Russian Ministry of Culture dated 11.07.2014 No. 1215 [12], the apartment is "a room in a facility with an area of at least 40 m^2, consisting of two or more rooms (living / dining / bedroom) with a kitchenette." Therefore, apartments are non-residential premises with no permanent residence registration [13]. You can get only temporary registration (maximum for 5 years), and only in those complexes that have the status of an apart-hotel. Under the legislation of the Russian Federation, apartments can be compared with hotels or hostels. Accordingly, the design standards must comply with the Construction Rules 160.1325800.2014, "Buildings and multifunctional complexes. Design rules", where hotel rooms and apartments have the same name. Thus, the norms for insolation and noise insulation of housing do not apply to apartments which make it possible to densify housing development and to receive additional profits.

Thirdly, since the apartments are not intended for permanent residence, there are no problems with providing social services (kindergartens, schools, polyclinics, etc.) That is, densifying the existing buildings or building up neighborhoods of city apartment complexes, the developer is not obliged to provide the social infrastructure for the temporary residents. In 2013 Kuznetsov C.O., Moscow Chief Architect, declared about the necessity to legally specify all procedures related to rental housing and said that "apartments do not differ from an ordinary housing. The inhabitants of apartments demand for a social infrastructure" [14]. Undoubtedly, it is a considerable minus but the situation can be improved by projecting and constructing complexes with apartment premises on the ground floors allocated for shops, private clinics, fitness centers, centers for additional and pre-school education etc. One more problem should be related to the above-mentioned problems – a developer is not obliged to provide apartment complexes with the same number of parking lots as residential complexes and it leads to the difficulties with parking and storing private cars [15].

What does a buyer of an apartment acquire? Less expensive apartments (compared with the similar flat in the same district and even in the same house) without a possibility of a permanent registration with a higher coefficient of taxation (0.5% from a cadastral value, flats have 0.1%) higher utility payments for "non-residential premises". It is good if a buyer knows all the details of the purchase and apartments are not his only living place but he buys it with the aim of investing and multiplying money through leasing these premises.

There are three distinctive from flats moments to be paid attention to while buying and further exploitation of apartments:

- A physical person who bought an apartment is not subjected to the law №102-FZ "On mortgage services (mortgage security)";

The owner of non-residential premises has a more complicated process of a recovery against mortgaged premises; the interests of a pledger are also protected in a different way [16];

- The apartments are no subjected to the norms of a housing law. It is not possible to set up Association of home owners (TSZh) within the framework of this law but it is possible to set up Association of non-residential premises owners (TSN). TSZh lays stress on management and joint use of a mutual property while ownership relations are realized in TSN, i.e. they do not have the right of management. But there are exceptions: should the apartments be in a multi-unit dwelling, the management will be in accordance with TSZh norms;

- In apartments the owner can make replanning without coordination with the bodies of local government. In this case it is important not to reconstruct.

The possibility of redevelopment - in apartments, as in non-residential premises, is much wider than in classic apartments (for example, you can move a bathroom into a room, etc.). However, certain restrictions still remain - it is impossible to demolish the supporting structures, and the openings in the supporting structures also require special reinforcement and coordination.

Mobility of a modern person, labor migration, tourism is just some of the factors that require temporary housing for an individual or a group of people, such as a family. Apartments become such temporary home. Unlike hotels, they have a number of advantages: low price (the cost is fixed and does not depend on the number of residents), the availability of kitchen, household appliances, etc. Therefore, apartments are beneficial for long-term rent.

When designing apartments, changes are possible regarding the rules and regulations used in housing construction, and as a means of accommodation, these units of real estate are typologically closer to the objects of the tourist industry, i.e. they can be equated rather to hotel rooms of the "highest category".

Today, in practice, the following types of apartments can be distinguished: apartments in the multifunctional complexes, residential apartments and rental housing, condo-hotels and apartment hotels, office and service apartments. Three categories of real estate can be purchased in the property, and the rest are leased for a long-term or short-term lease, but no matter for how long the premises are rented, it provides hotel services.

Three types of apartments were determined as a result of the conducted analysis:

- apart-hotels (commercial real estate for a short-time leasing which assumes hotel services);
- apartments in a multi-unit dwelling (commercial real estate for a temporary living in a complex with office rooms, fitness centers, shops etc.);
- "multi-form" apartments in a residential complex (commercial real estate for a temporary living in a complex of a residential house).

Apart-hotels are commercial real estate properties that are rented, where private buyers purchase apartments for onward rent. Apart-hotels usually have hotel-type services. The development of tourism is in the list of priority state programs of the Russian Federation. With an appropriate level of development of the tourist infrastructure, first of all, to which the places of temporary accommodation of tourists belong - hotels, apartment hotels, motels, it will be possible to increase the reception of Russian and foreign tourists several times. Prudent and practical travelers, as a rule, prefer to be accommodated in apart-hotels, as more economical and comfortable accommodation. Therefore, increasing the pace of construction of this type of buildings is justified by the significant benefits of the socio-economic development of the capital region.

The multi-format apartment complex is a city in a city where all the needs and living conditions of temporary and permanent residents of the metropolis are provided.

Apartments in the multifunctional complex as a type of real estate have become very popular due to the opportunity to live near work. Most often, the apartments are located on the upper floors of multifunctional complexes. Their area varies from 40 square meters to 300 square meters. Apart from apartments, the multifunctional complex usually includes infrastructure facilities: a fitness club, catering establishments, parking, office sector, shopping center, co-working space, etc. The construction of multifunctional complexes with apartments in the peripheral areas of the city provides the people with inexpensive real estate, offers a diverse range of activities and increases the attractiveness of the district. To some extent, the introduction of such architectural solutions helps to reduce the daily commuting in the city. The main advantages of such real estate include the fact that residents of apartments within the framework of multifunctional complexes are able to use the hotel service and services of shopping and entertainment establishments included in the structure of the complex.

There is a managing company in all three types of apartments. It takes care of all the furnishing and maintenance of the rooms, which are equipped with high-quality finishes and are fully equipped.

The equipment under the tenant is already supervised by the management company, and although it is not so much the quantity of services provided that matters as their quality, the current state of the technical equipment is of great importance. Currently, the availability of one or other equipment is convenient not only for clients, but also for a landlord and a developer, making it more competitive. Along with technical progress, the standards for technical equipment of modern premises have increased (the availability of wireless Internet, sensor lock technology, devices, temperature and humidity indicators, multi-level adjustable lighting, etc.).

Intelligent control systems allow you to control both each component of the system individually and the entire system in the complex by pressing buttons on the control panel. Intelligent control allows the user to control such system functions as automatic or semi-automatic control of room engineering systems, such as lighting (including blinds control systems) and climate maintenance systems, and much more. In addition, modern intelligent systems can prevent the wasteful handling of electricity and reduce costs without compromising the comfort of residents.

Apartments are ranked by convenience, size and quality of services provided. In accordance with the difference in the comfort characteristics of the apartments, they are divided into the following classes: economy, comfort, business and elite [17].

To identify the trends and features of the apartments, the most outstanding and unique projects and buildings of Moscow in recent years were reviewed and studied, in accordance with the classification of comfort. For a more detailed consideration, two complexes were selected (the cheapest and the most expensive apartments) from each class and analyzed according to various parameters: from architectural and urban planning to economic and social-functional.

We considered such characteristics of the complexes as the area, number of floors, cost, location and distance from the metro, the year of commissioning of the object, and the availability of infrastructure, which was studied in more detail later on in selected complexes (Table 1).

Table 1. Comparative analysis of the apartments as part of multifunctional complexes.

Name	Number of floors	Total living area of the object in sq. m	Availability of infrastructure	Location [a]	Minimal price of the apartments (million rubles)	Accessibility to the metro, 5/10 minutes (250/500m)	Putting into operation, year, quarter	Planning solutions, number of rooms
				Elite class from 65.3 million rubles				
Klenovy DOM	4-5 fl.	11 200/5 000	+	CAD	73	5 minutes	2016	2, 3, 4, 5, 6
Cloud Nine	6 fl.	16 300 /10 100	+	CAD	65.3	5 minutes	2018	2, 3, 4, 5
				Business class from 5,2 million rubles				
Kvartal	20 fl.	95 470/30 934	+	SEAD	5.2	5 minutes	2019	1, 2, 3, 4, 5,

JAZZ Neskuchny HOME &SPA	69 fl.	228 000/190 000	+	SAD	10.9	10 minutes	2019	6 1, 2, 3, 4, 5
Clubny dom na serpukhovskom valu	8 fl.	256 700	+	CAD	16.9	5 minutes	2017.4 quarter	1, 2, 3, 4,
CITY PARK	20 fl.		+	CAD	14.8	5 minutes	2018	1, 2, 3, 4, 5
I'M	7 fl.	22 400	+	CAD	13	5 minutes	2015. 1 quarter	2, 3, 4
Wine House	7 fl.		+	CAD	35	10 minutes	2016	1, 2, 3
Vorobyov dom	4-13 fl.	41 670	+	WAD	17.4	5 minutes	2017. 3 quarter	1, 2, 3
Comfort class from 3,6 million rubles								
Rezidencyi kompozitorov	16 fl.	130 000	+	SAD	5.6	10 minutes	2020	1, 2, 3, 4, 5
Serdce stolocy	19 fl.	250 963	+	SWAD	3.8	10 minutes	2017. 4 quarter	Studios, 1, 2, 3, 4, 5-
Tsarskaya ploshad	13 fl.	290 000	+	NAD	6	5 minutes	2018	Studios, 1, 2, 3
Trikolor	58 fl.	275 620	+	NEAD	6.4	10 minutes	2015. 1 quarter	Studios, 1, 2, 3
Marshal	9 fl.	202 000	+	NWAD	9.7	5 minutes	2018. 4 quarter	1, 2, 3
Union Park	23 fl.		+	NWAD	3.9	10 minutes	2019	1, 2, 3
Nevsky	26 fl.		+	NAD	3.6	5 minutes	2018	1, 2, 3
Economic class from 2,5 million rubles								
Level Amurskaya	39 fl.	236 600	+	EAD	2.5	10 minutes	2020. 4 quarter	Studios 1, 2, -
River Park	13 fl.	273 000	+	SAD	4.3	10 minutes	2018. 3 quarter	1, 2, 3, 4, 5-
Yasny	10-13 fl.	185 000	+	SAD	5	10 minutes	2018. 3 quarter	Studios, 1, 2, 3, 4,
Savyolovsky sity	30 fl.	320 000	+	NEAD	4.7	10 minutes	2017. 4 quarter	1, 2, 3, 4,
Park Legenda	10-13 fl.	700 000	+	SAD	5.4	5 minutes	2018. 3 quarter	1, 2, 3
Letny sad	20 fl.	283 000	+	NAD	3.6	10 minutes	2018. 4 quarter	Studios, 1, 2, 3, 4,
Vodny	25 fl.	180 928	+	NAD	6.9	5 minutes	2018. 2 quarter	1, 2, 3
PerovSky	23 fl.	47 100	+	EAD	4.6	5 minutes	2019	1, 2, 3
Fili Grad	16-31 fl.	186 700	+	WAD	6	10 minutes	2018	1, 2, 3, 4

CAD – Central Administrative District; SAD – South Administrative District;
NAD – North Administrative District; WAD – West Administrative District;
EAD – East Administrative District; SEAD – South-East Administrative District;
SWAD – South- West Administrative District; NEAD – North-East Administrative District;
NWAD – North-West Administrative District

4. Conclusions

Apartments are a popular type of property for developers and owners, because at the moment the number of transactions compared to 2016 has increased by 1.5 times. Construction of apartments is attractive for developers in view of a low cadaster cost of the land (it is 4 times less that in regions aimed at a housing fund) and simplified procedure of a housing development.

North West Administrative District, Northern Administrative District, North East Administrative District, Eastern Administrative District are Moscow perspective regions to develop a given type of a real estate.

Factors contributing to this are as follows:

- Relatively cheap cadaster cost of the land and consequently a low price for apartments;
- Favorable ecological environment, availability of a big number of green fields and water resources, all the above forms recreation zones for the population rest;
- A large number of undeveloped plots.

By architectural features, the apartments in the structure of multifunctional objects are spacious apartments with an improved layout (increase in the number of bathrooms, two-level zoning), located in prestigious residential complexes with a cozy courtyard, high-quality furnishings with panoramic windows and elevated ceilings.

Trends of the development of the apartments include the increase in their comfort, and in apartment complexes - improvement and variety of infrastructure. In spite of a slight decrease in number of new projects offered in the real estate market the demand for them has a stable level and this type of property can be defined as one of the most liquid, and therefore, promising for investment.

But, in spite of the prospects and attractiveness of apartments for developers and their owners, we list the main and not yet resolved problems identified in the article once again:

1. The concept of "apartments" in Russia is not defined at the legislative level and so far this type of property has not received official classification.
2. In the large cities of Russia, "hotel complexes", which are actually housing, continue to be built, over which the threat of "social burden" hangs.
3. The control over the apartment carries the Association of home owners (TSZh), although the apartments are a segment of the hotel business.

The solution of these problems - the "legalization" of apartments, which spontaneously spread in the residential and hotel real estate sector, requires a comprehensive inter-agency approach with the participation of the Ministry of Culture of the Russian Federation, namely the Federal Agency for Tourism and the Ministry of Construction and Housing and Communal Services.

References

[1] Mironov N 2017 The apartments are out of fashion. *Newspaper "Evening Moscow"* 28 November 2017 Article 244 (27807)

[2] De Feijter F J, Van Vliet B J M and Chen Y 2019 Household inclusion in the governance of housing retrofitting: Analysing Chinese and Dutch systems of energy retrofit provision *Energy Research & Social Science* **53** pp 10-22

[3] Hamid A A, Bagge H and Johansson D 2019 Measuring the impact of MVHR on the energy efficiency and the IEQ in multifamily buildings *Energy and Buildings* **195** pp. 93-104

[4] Vesela L and Synek J 2019 Quality Control in Building and Construction *3rd World Multidisciplinary Civil Engineering, Architecture, Urban Planning Symposium* (WMCAUS 2018) **471**

[5] Borsos A, Kokas B and Bachmann B 2019 Who Owns Our Future? How to Find a Home? 3rd World Multidisciplinary Civil Engineering, Architecture, Urban Planning Symposium (WMCAUS 2018) **471**

[6] Kirichenko E I 1982 *Russian architecture of the 1830s-1910s* (Moscow: Publishing House "Iskusstvo") pp 120-124

[7] Blagovidova N G 2001 *Quarters of "profitable" houses in the planning and building of modern Moscow* PhD Thesis (Moscow Architecture Institute, Moscow)

[8] Well forgotten old. Luzhkov wants to build profitable apartment houses in Moscow (Moscow: Radio Liberty) URL: https://www.svoboda.org/a/461261.html

[9] Pogorelova Y 2011 Profitable house-2. *The newspaper Kommersant.* URL: https://www.kommersant.ru/doc/1667343

[10] Men' M 2015 The share of rental housing in the total amount of residential real estate can reach 7% (Moscow: Ministry of Construction and Housing and Communal Services of the Russian Federation) URL: http://www.minstroyrf.ru/press/dolya-arendnogo-zhilya-v-obshchem-obeme-vvoda-zhiloy-nedvizhimosti-mozhet-sostavit-7/

[11] Draft of Federal Law "On Amendments to the Housing Code of the Russian Federation and Certain Legislative Acts of the Russian Federation regarding the Regulation of Certain Legal Relations arising in Connection with Multifunctional Buildings Construction" URL: http://www.consultant.ru/law/hotdocs/51113.html/

[12] Order of the Ministry of Culture of the Russian Federation dated July 11, 2014 N 1215 "On approval of the classification for the tourist industry objects, including hotels and other accommodation facilities, ski slopes and beaches, carried out by accredited organizations"

[13] Agureeva O 2013 Apartments: the disputed format of "apartments". URL: https://www.metrinfo.ru/apartment/articles/apartamenty-spornyy-format-kvartir-privlekatelnaya.110586.html

[14] Kuznetsov S 2014 *The apartment has more minuses than pluses – Kuznetsov*. URL: https://www.m24.ru/articles/arenda-zhilya/13022014/37470

[15] Construction Rules 42.13330.2011 Urban planning. Planning and developing of urban and rural settlements (Moscow: Publishing House Kodex)

[16] Federal law N 102-Fed.Law (edited 25.11.2017) "On mortgage services (mortgage security)" (Moscow: Ministry of Law of Russia)

[17] Tretyakova D 2014 What does it cost us to build apartments? URL: http://www.miel.ru/press/6712

Authors' background

Name	*Prefix*	*Research Field*	*Email*	*Personal website*
Larisa Savelieva	Associate Professor	Architecture, digital technologies in architecture, education	savelyeva-lv@rudn.ru	
Natalia Kalinina	Associate Professor	Architecture, public building architecture, education	kalinina_arch@mail.ru	
Eugene Ogienko	Master Student	Architecture, urban planning, residential and public buildings	tengue@list.ru	

IOP Conf. Series: Materials Science and Engineering **675** (2019) 012017 doi:10.1088/1757-899X/675/1/012017

Formation of architectural objects for extreme habitat conditions in the context of innovative paradigms

N A Saprykina

Moscow Architectural Institute (State Academy), Moscow, Russia

E-mail: nas@markhi.ru

Abstract. Article is devoted to an actual problem of defining features of architectural objects and their construction for extreme habitat conditions in the context of innovative paradigms. The purpose of article is to consider approaches to dealing with the creation of the artificial habitat environment for extreme conditions, uses to ensure the safety of the existence of technological innovation of the future. It has been established that depending on natural conditions distinguish areas of very extreme conditions, requiring the need to protect the person from the surrounding aggressive environment. A leading approach to the study of the problem based on the search for new, including innovative solutions, home and life-support systems, making full use of technical innovation and creative development of advanced technologies. As a result of research defined the features of construction for each extreme region, and also identify specific requirements for the design of artificial environment. Materials can be useful to the theory and practice of forming the space habitat not only for extreme, but for normal conditions as opens completely new possibilities in architecture and construction.

1. Introduction

A distinctive feature of our times is the growing dynamism of society associated with the accelerated development of scientific and technological progress, the expansion of the geography of activity of people, increased population mobility, migration, etc. This entails changes in all spheres of human activity. When creating architectural objects and their construction in extreme conditions this poses a challenge for modern architecture completely new security concerns population.

The solution to the problem of creating an artificial habitat environment in extreme conditions has always attracted the attention of scientists, philosophers, engineers, architects, doctors, environmentalists, sociologists, psychologists and other specialists, hence the complexity study of the problem. The *solutions of this problems* has attracted many professionals involved in the formation of spatial environment in extreme conditions and proposes concrete solutions and developments performed in the framework of this concept.

Of special interest are problems formation of space habitats in the cold aggressive environment, considered in the works by Galeev S.A. [3], Saprykina N.A. и Saprykin I.A. [2], as well as in regions of hot desert in research by Esaulov G.V. [4], are notes the urgency and the need to develop techniques of creating architectural objects to reduce aggressive influence factors of natural environment on the human body.

The establishment of underground space habitats in studies Rogozhnikova M.A. [6] and Saprykina N.A. [5] is notes the desirability of its application in various fields - from Space Mountain abandoned mines and before the creation of the in-depth underground homes. In conditions of high mountains and seismic areas in the work of Saprykina N.A. [7] in the context of environmental approaches to creating residential environment suggested buildings and structures with the Organization of systems of autonomous power supply.

Formation of habitats during construction environment on water and underwater in research Kizilova S A [8] noted the usefulness of alternative architecture principles, and in the organization of the artificial habitable environment in cosmos in research Gogolkina O.V. [12] and Saprykina N.A. [13] notes the need to develop innovative new approaches to design and construction. In addition, the formation environment of habitation under extreme conditions in the study Saprykina N.A. [15] notes the relevance of the development of innovative energy technologies based on information design.

Stringent requirements when designing architectural in extreme environment of habitats require to the architect to think comprehensively and require extensive research in connection with exceptional complexity and relevance of this issue.

The relevance of the issue. The solution to the problem of creating an artificial habitat environment for extreme conditions as an integrated ecological system will provide an opportunity to increase the level of compliance of architectural sites to the needs of society, reduce material costs their construction and increase the comfort and safety of their stay in areas with aggressive environment. In this regard, it is important to note also the need to review traditional tools architecture and use of other fields of science and technology, made possible thanks to recent developments in the field of electronic technology.

The *purpose* of this article is to review approaches to dealing with the creation of the artificial habitat environment for extreme conditions, uses to ensure the safety of the existence of technological innovation of the future. It is warranted to identify ways and methods for creating an artificial habitat Environment that practically are not used in modern architecture and construction practice.

2. Materials and method of research

The complex technological and economic requirements for creating such architectural objects cause the search for new approaches and use such *analytical methods* of research as: integrated design object as a system, a method of structural analysis and the informational methods. The main and common characteristics of extreme areas are: remoteness and isolation from developed areas, their inaccessibility, complex and adverse and impossible without the appropriate technical equipment conditions the existence of the people. This also includes the temporariness of human habitation and the autonomy of the existence of settlements, a specific demographic composition of the population, etc.

Depending on natural conditions distinguishes areas of very extreme conditions, requiring the need to protect the person from the surrounding aggressive environment: cold (North, Arctic, and Antarctica), hot (desert), sharply continental (highlands). Depending on the extreme factors and conditions of construction environment include the creation of architectural sites in the northern regions, on the water, underwater and underground, construction in seismic and other hazardous areas with aggressive environment. In connection with the expansion of the range of those areas it influences both on Earth and beyond. If we consider the particular construction separately for each extreme, you can highlight a number of specific requirements to design out there to artificial environment.

2.1. The organization of space habitats in the cold aggressive environment - architecture in high latitudes

During the planning and construction of inhabited places in the conditions of the North of the main town-planning challenge is to protect the person from the adverse effects of the external conditions of the bioclimatic improvement environment, as well as the operating conditions of the residential settlements. Specifics are climatic conditions in these areas, weak industrial base construction and lack of reliable vehicles require a special approach to the design and construction of residential homes and buildings cultural and community purpose.

Specific extreme natural and climatic conditions of a large part of the construction areas with increased duration of the winter period, and a number of other regional factors hamper the development of agriculture and many industries industry. The remoteness of the areas from cities and other human settlements entails a lack or shortage of local workforce and the need to provide shelter

IOP Conf. Series: Materials Science and Engineering **675** (2019) 012017 doi:10.1088/1757-899X/675/1/012017

and various types of social services. The absence in most cases of local building materials necessitate transportation here designs and elements of buildings or transport them by air.

All of this necessitates the use, particularly in the early stages of development, the mobile settlements with smaller lifetimes, that with the brevity of their lifetime reduces construction costs. Residents of mobile settlements a number of specifics because of settlements object (building power lines, oil and gas pipelines, roads, railways, etc.) to move from one place to another, and as a result of conduct relatively long period under unfavorable conditions [1]. Here, effectively making transformable spatial structures that you can use to quickly create objects for different purposes. When establishing such settlements, it is advisable to use the principles of *transformation* and *mobility*. The prefabricated modular buildings and complexes have not only high performance characteristics. They survive as low temperatures in Northern and southern high temperature thermal insulation thanks to modern environmentally friendly materials [2].

The harshness of the climate and low temperature necessitated insulation interior space from the surrounding building environment, as well as place the highest demands on thermal insulation of buildings. The presence of permafrost, accompanied by thawing in violation of topsoil, leads to the loss of bearing capacity building, which requires the use of pile foundations. If there is not enough sunlight for short day construction of the façade should ensure maximum transmission of sunlight during the winter and protection from overheating in the summer.

Addressing issues of design and construction of these types of settlements fairly considered in detail in the literature. Therefore, we can mention some basic design requirements in the conditions of the North. These include: the formation of a compact planning structure of residential objects, to effectively use its territory and reduce communication; introduction of construction planning techniques regulating microclimate and reduce aggressive impact factors of natural environment on the human body. Incorporation of these requirements is particularly important to develop sound development proposals with extreme northern conditions, so as to determine to what extent and how buildings and structures should isolate a person from natural environment [3].

2.2. The formation of architectural objects in the hot aggressive environment habitats

In extreme hot *desert* regions there are some similar requirements as for the northern regions. There is a general problem of protection from extreme environmental conditions of environment. Despite unfavorable natural conditions (excessive solar radiation, dust, the mobility of the sands etc.), there is a deeper exploration of the industrial potential of deserts where there are deposits of valuable mineral fossil. This in turn raises the issue of organization of production and residential complexes. In the same way as in northern areas, there is a shortage of local building materials, as well as materials with low thermal conductivity and temperature resistance [4].

One of the main challenges when designing of buildings and structures in hot desert is the reduced heat load on them. For this purpose, it is necessary to take into account the close connection orientation, insolation and ventilation buildings, as well as the height and density of building. An important requirement is the arrangement of the buildings into a single closed space with internal green and water yards, with the device in the walls of small light apertures [5]. In addition, it is necessary to use different devices to protect against sunlight and hot winds, to capture favorable air currents and its cooling. Here can be used by the dynamic alternative architecture techniques: from the use of the simplest transformable Sun shields and devices before the introduction of the solar- wind and bioenergy.

2.3. The features create underground space habitats

Experience in designing and creating an artificial habitat environment in extreme conditions would be useful in the creation of *underground* construction projects. In the conditions of *city building* they include engineering and transport facilities, trading enterprises and public nutrition, entertainment, administrative and sporting facilities. In addition, objects of utilities and communications, warehouse objects, industrial objects, objects of engineering equipment, etc.

However, ways to of underground urban development applies not only to address urban challenges. Great value they acquire to use *mountain abandoned mines*, where it is possible to have objects in the most various applications: warehouses, cold storage, garages, some industrial companies, archives of securities documents, depositories, mushroom plantation, greenhouses, museums, churches, resorts, etc. [6].

The emergence of problems of underground construction in *cities* due to such factors as lack of town of vacant land, the need for a loosening of the historical building, speed and continuous motion public, special and individual transport, preservation of historical and architectural ensembles, development systems of cultural and community welfare and public services, etc.

To resolve conflicts related to particular cultural or historical value of the plot, as well as for energy savings and conservation ecology construction sites, due to a slight disruption of natural environment and visual compositions, there is the trend of creating *in-depth underground dwellings*. In this regard, one of the most interesting destinations in modern architecture is the use of natural materials, borrowed directly from the clay (for example, earthen architecture) [7].

For hundreds of years, this experience was used by the peoples of Africa, Asia, South America and other regions of southern latitudes due to the fact that the clay is a natural and universal battery heat and provides a balanced internal microclimate due to the good ventilation qualities. In addition, clay is a material that allows fast enough to erect buildings of this kind. All this made contemporary architects interested in the possibilities of earthen construction.

Thus, the main positive qualities using earthen construction are its cheapness, environmental friendliness and brief terms of erection of the object. Underground and earthy construction with certain advantages, is gaining increasing recognition, and modern the level of technology promotes to its broad development. The experience of such construction gives examples of its versatile use in various fields of human activity that has great prospects.

2.4. The peculiarities of formation of high mountains habitats environment and seismic areas

Under extreme conditions of *high mountains and seismic areas* there is a need for construction engineering objects here, as well as residential and public buildings. Development in these mining enterprises, agricultural production, research and other objects puts the problem of adapting buildings and structures to the specifics of their operation in such conditions for comfortable and stay safe people [8].

In these areas, features of the construction are defined by meteorological and geophysical conditions. The ways to confront the earthquakes, snow avalanches, to occasional minor rockslides and the others in practice and research quite known. They are taken into account when drafting the construction techniques, choosing designs and create architectural and planning solutions construction sites [9]. It is obvious that the production objects in such areas it is advisable to carry out expeditionary or the urgent-methods. Application of highly mobile buildings and structures, with using of systems of autonomous power supply is indispensable for the permanent residence of the population in these areas in case of natural disasters.

2.5. The peculiarities of formation of habitats environment during the construction on water and under water

Due to the fact that almost all the listed scope extreme on Earth comparatively widely understood, it is advisable to pay particular attention to the organization of the space in the most aggressive areas - on the water and under the water. Use of riches of the oceans to expand territory on land, as well as the development of the economic potential of the oceans as a source of food, mineral and organic resources attracted the attention of construction in this environment. Experience of the ocean opens new frontiers and gives the precedents of organization an artificial habitat precedent in aggressive environment.

When construction on the water poses a number of problems with regard to the need to overcome the immense water spaces (construction of super bridges), due to the scarcity of land (floating cities and airfields), and mining (underwater drilling oil wells). Known is projects of construction of super huge bridges (architect I. Friedman) and floating cities (P. Maymon). In connection with the shortage of space on the islands, this problem tries to solve the Japanese architects and specialists (K.Tange, K. Kikutake, V. Jonas, W. Katavolos, etc.). Such artificial islands already exist, for example, in the form of enormous platforms, with which drilling of undersea oil wells (in Russia in the construction of the town of oil rocks). The entire water complex includes residential, public and industrial premises, as well as a stand-alone subsidiary of electro-energy devices [9].

Artificial islands projects-airfields located at some distance from the coast, developed in the United States, England, and Japan that can solve the problem of coastal towns, where there are no plots for the construction of airfields. This also solves the problem of harmful noise exposure created powerful and ultra-fast planes in addition; this solves the problem of obstacles related to terrain that prevented the aircrafts during takeoff and landing. Of course, in economic and technical relations such island airfields still disadvantageous, but environmental benefits obvious.

As in any extreme area, this is a problem for creation of comfortable living and working conditions of people forced on duty sometime is in conditions of exclusion and isolation from the rest of the world. It is clear that here will find widespread use of alternative architecture principles. When this is solved a number of economic and environmental challenges: organization of autonomous energy supply using natural renewable energy sources. It is possible that the design and construction of artificial islands-recreation centers will lead to the emergence of a new original direction in architecture because mastering the coastal shelf allows you to embark on a new stage of organization artificial human habitat environment [9].

No less attention attracts and underwater construction. Here, as in space exploration, is a choice between automatic or human. Because the current level of science and technology does not solve all the problems just by using machines, creating a comfortable and reliable artificial human habitat is one of the important tasks. This is due to the additional costs in connection with the operation of the underwater research facilities and the adaptation of the human body.

It is obvious that the main difficulty in the organization of underwater sites are the civil and erection works. Therefore, all underwater homes mostly created on the land, and then descend into the water, fully ready to operate as mobile. In addition to using the principles of mobility when you create small confined spaces of houses underwater, it is also important to use the principles of transformation of inhabited space in connection with the necessity of its multifunctional use [9].

The high requirements for the creation of underwater inhabited objects other than purely technical, include a large architectural complex mental tasks. In small confined spaces underwater homes apart from the principles of mobility are important principles for the transformation of inhabited space in connection with the necessity of its multifunctional use. In addition to all necessary amenities they provide a favorable psychological atmosphere, which in turn determines the efficiency of manufacturing operations.

Because underwater and space environment aggressive to humans, then here is the most acute problem of creating volumetric-spatial solutions industrial-residential complexes with the sustenance of a closed system (when using natural renewable energy sources). So much so that studies in this area to develop recommendations for the use of scientific and technological achievements in development and advance their use.

2.6. The creating artificial habitable environment in the cosmos - weightlessness architecture

Extreme conditions in cosmos affect the Organization and creation of artificial habitat Environment depending on the distance from ground: orbital objects in near-Earth space and interplanetary objects, as well as run in outer space. Here the specific factors are determining environment for her specific requirements and evaluation criteria. This is primarily due to weightlessness as a factor that explains of our view of the usual environment and creates certain difficulties for human existence in space [10].

Zero gravity defines the specific conditions of existence and requires modifying the properties of the construction materials used. This allows you to use the construction techniques that cannot be applied in the face of the Earth. So, expanding transformable structures, previously soaked in special structures under the influence of vacuum and solar radiation provide great advantages over conventional methods of construction. Unlimited opportunities arise in connection with the creation of space composite materials competing with conventional materials.

Architecture of weightlessness is an area of unlimited possibilities in shaping of form. But since the construction of space objects on a variety of technical reasons will be done on Earth and transported in a compact form, there is a need to create transformable kinematic structures. Transformation of structural systems requires not only at the stage of transportation, but also at the stage of exploitation. Organization of the normal flow of functional processes in closed, confined space would require its multipurpose use, which is possible only through transformation [11].

In the face of the problem of the touch hunger in an enclosed space can be solved through the use of controlled dynamic light-color forms, sound and heat of the climate, for example, simulating natural rhythms [12]. Thus, when create a programmed environment in extreme conditions can be applied the principles of kinematic formation. Development and production of these designs requires new approaches to design and construction organization that promotes their application on the Earth conditions.

2.7. Innovative energy technologies in the formation environment of habitation under extreme conditions

The problem of energy saving, resource saving and money spent in the construction and operation of buildings requires the development of architectural and construction techniques to enhance their energy efficiency through the use of alternative modes renewable sources of energy and new clean technologies. In many countries, this trend is reflected in the architecture of its further development in the so-called "autonomous house", which is an example of self-organization of life of people. Therefore, the establishment of autonomous small power plants based on renewable energy and supplying local consumers, has obvious advantages. Providing power and heat supply systems here is a huge social problem.

Implementation of environmental principles in the notion of electronic offline home based on progressive science and construction equipment holds great promise for use in the architecture of future scientific and technological developments, generating new ideas. This enables you to unleash the potential of architecture and best meet the needs of society [13]. Therefore, the establishment of autonomous small power plants based on renewable energy and supplying local consumers, has obvious advantages. Providing power and heat supply systems here is a huge social problem.

Using the principles of information technologies can provide successful urban development outside the direct spatial dependence on infrastructure, i.e. do without creating major highways and power grids. This will be possible when using alternative energy sources and eco-technology, satellite communications and the Internet. In this case, typology of architecture restocked with new kinds of architectural objects [14].

In resettlement system extreme areas, a new element of infrastructure represents are information-dispatching and technological centers. In regard they are new architectural element of the infrastructure of the future architectural complexes and managerial and technological centers. They do not require a lot of maintenance personnel and can be located at significant distances from each other, that it is important for extreme conditions [15].

Within this issue of great importance also and is conducting research and experimental work to create architecture planning and volumetric-spatial solutions industrial-residential complexes non-waste technology conditions for permanent and temporary residence. Such complexes needed for industry construction, agriculture (vegetable, livestock and poultry), science, recreation and tourism (recreational complexes and protection of environment), etc.

3. Results and discussion

Innovative approaches to formation of architectural objects of extreme habitats identified in this review in the context of the following fields marked.

3.1. The organization of space habitats in the cold aggressive environment. The main design requirements in the North are: forming a compact planning structure of residential education, to effectively use its territory and reduce of communications; introduction of construction planning techniques regulating microclimate and reduce aggressive impact factors of natural environment on the human body. When establishing such settlements to use the principles of transformation and mobility.

3.2. The formation of architectural objects in the hot aggressive environment habitats. An important requirement is the arrangement of the buildings into a single closed space with internal green and water yards, with the device in the walls of small light apertures, as well as the use of various devices to protect against sunlight and hot winds, to capture favorable air currents and it's cooling.

3.3. The features create underground space habitats. The emergence of underground construction is evident in various spheres:

- To use the mountain abandoned mines, where it is possible to have objects in the most various purposes (warehouses, refrigerators, parking garages, some industrial companies, archives and documents, securities depositories, mushroom plantation, greenhouses, museums, churches, resorts, etc.). The text should be set to single line spacing.
- In the cities due to the disadvantage of free territories the necessity of disintegration of the historical building, the development of high-speed and continuous traffic, preserving historical and architectural ensembles, etc.
- To create the in-depth underground dwellings using natural materials, borrowed directly from the Earth, which is inexpensive, environmentally friendly and concise terms of erection of the object.

3.4. The peculiarities of formation of high mountains habitats environment and seismic areas. Construction works are determined by meteorological and geophysical conditions. In the face of disasters such as earthquakes, avalanches, rock falls, etc. require the use of highly mobile buildings and structures, with the organization of systems of autonomous energy supply. It is indispensable when the permanent residence of the population in these areas in case of natural disasters.

3.5. The peculiarities of formation of habitats environment during the construction on water and under water:

- Organization of artificial water habitats environment linked to the expansion of the territories on land and development of the economic potential of the oceans. Here will find widespread use of alternative architecture principles - organization of autonomous energy supply using natural renewable energy sources, because it solved a number of economic and environmental objectives.
- Creating a comfortable and reliable artificial human habitat environment under of water supply-residential complexes associated with autonomous closed life support system. The high requirements for the creation of underwater inhabited objects in addition to purely technical include large complex architectural and psychological tasks.

3.6. The creating artificial habitable environment in the cosmos - weightlessness architecture. Weightlessness is determines the specific conditions and requires the use of the properties of the building materials and techniques that are impossible to implement in the face of the Earth. There is a need to create transformable kinematic structures, which requires new approaches to design and construction.

3.7. Innovative energy technologies in the formation environment of habitation under extreme conditions. Developing progressive direction in the architecture, applying new technologies when creating architectural objects in extreme conditions, associated with the new attitude to environmental values, environment and save energy conservation. This will be possible when using alternative sources of energy and environmental technology, satellite communications and the Internet.

4. Conclusion.

Construction in the extreme conditions requires of the use of innovative methods and approaches in architectural and technological formation of objects. As a result of the research in this article describes new approaches to creating artificially environment of habitat for extreme conditions which uses to ensure the safety of the existence the technological innovation of the future. In this regard, identified approaches and techniques for creating an artificial environment in such conditions:

- The establishment of the compact planning structure of inhabited space for effective use of the territory and reduce of communication, as well as for the regulation of microclimates and reduce aggressive impact of factors of natural environment on the body for human in cold conditions.
- The creating of the space habitats with the use of various devices for protect against sunlight and the hot winds, for capture favourable air currents and its cooling for architectural objects in hot aggressive environment.
- The establishment of underground space habitats in various spheres: to use the mountain abandoned mines, establishing systems of high-speed and continuous traffic of transport, creating the in-depth underground dwellings using natural materials.
- The using of highly mobile buildings and structures, with the organization of systems of autonomous energy supply in conditions of high mountains habitats environment and seismic areas, as well as in the event of disasters.
- The using of the principles of alternative architecture with uses natural renewable sources of energy to create a comfortable and reliable artificial environment human habitation on the water and under the water.
- The creating of the transformable kinematic structures, manufacturing of which requires the new approaches to design and construction, in the artificial habitable Wednesday under space conditions.
- The creating of the space in extreme conditions of habitats like integrated ecological system using innovative energy technologies.

The main objective of the considered problem is to create a programmed environment in extreme conditions Arctic, desert, underground, underwater, Alpine, space and other habitats. Here is created an unusual environment and erased the line between technology and architecture. This extends the creative possibilities and generates a new direction in architecture - *technology architecture*. The problem of creating an artificial habitat environment in extreme conditions will require adjustments to the existing method of design, since it cannot be solved purely by architectural means. The term "architecture" in this case, loses his usual concept. It means not only the building but also the totality of phenomena occurring around the person.

The practical significance. Development of principles of designing and creating project proposals for the organization of pre-fabricated settlements in areas with extreme natural conditions is of great importance in different fields of human activity. Experience in creating such objects will be useful when organizing seasonal recreation and tourism, festivals and other events and shows, as well as socio-political and religious events.

Research in this area will develop recommendations on the application of scientific and technological achievements in the development of extreme areas and for normal conditions, as well as on long-term use of the achievements, the emergence of which are very probably. Here the technological architecture becomes a customer who requires certain new technical achievements and even scientific discoveries.

References

[1] Sugars A N 1994 Houses for rural development in the North (Leningrad: Strojizdat) p 256

[2] Saprykina N A, Saprykin I A 2018 Ecology approaches to creating architectural objects as the basis for the formation of living Environment IOP Conf. Series: Materials Science and Engineering **451** 012131 https://doi.org/10.1088/1757-899X/451/1/012131

[3] Galeev S A 2015 Adaptation of architectural systems to extreme conditions environment / Electronic scientific journal "APRIORI" / Series: *Natural and technical science* **4** http://www.apriori-journal.ru/index.php/journal-estesvennie-nauki/id/769

[4] Architecture and urbanism in extreme natural and techno-genic influences 2012 *Edited by GV Esaulov* (St. Petersburg: Nestor-History) 166 p

[5] Saprykina N A 2019 Innovative Conceptions of Natural Light Using as an Essential Component of the Formation of Architectural Space *Light & Engineering* **27(2)** pp 23-31

[6] Rogozhnikova M A 2013 Typology of the underground dwelling house *Architecture and Modern Information Technologies* **1(22)** pp 1-15 https://marhi.ru/AMIT/2013/1kvart13/rogozhnikova/abstract.php

[7] Saprykina N A 2018 Sustainable development of spatial habitat environment as a challenge to civilization *IOP Conference Series: Materials Science and Engineering* **463** 022052 http://iopscience.iop.org/article/10.1088/1757-899X/463/2/022052/meta

[8] Kizilova S A 2019 Form and functional features of modular floating structures *E3S Web of Conferences* **91** pp 1-6 05013 https://doi.org/10.1051/e3sconf/20199105013

[9] Saprykina N A 2018 The basics of dynamic morphology in architecture: *textbook for high schools* (Moscow: Arhitektura-S) p 372

[10] Kholodova LP, Zhuikov S S 2012 Background architecture for the future "Astro-architecture" *Proc. International Science Conference FAD TOGA* (Khabarovsk: Pacific National University) pp 458-462 http://elibrary.ru/item.asp?id=18963967

[11] Saprykina N A 2018 Forecasting technology as a method of modeling and building Smart City concept *IOP Conference Series: Materials Science and Engineering* **365** 022068 https://doi.org/10.1088/1757-899X/365/2/022068

[12] Gogolkina O V 2019 Parametric Architecture in the Formation of Recreational Complexes *IOP Conference Series: Materials Science and Engineering* **463** 022066 https://doi:10.1088/1757-899X/463/2/022066

[13] Saprykina N A 2016 Synergetic approaches to formation of architectural space *Architecture and Modern Information Technologies* **4(37)** pp 171-178 http://marhi.ru/AMIT/2016/4kvart16/PDF/AMIT_2016-4(37)_Saprykina_PDF.pdf

[14] Ryabov AV 2010 Alternative energy objects in modern architectural environment: *Proc. Scientific-Practical conference "Science, education and experimental design"* **1** pp 203-206 (Moscow: MARCHI)

[15] Saprykina N A 2018 New approaches to formation of info-space future as a separate category of architectural environment *Architecture and Modern Information Technologies* **1(42)** pp 317-340 http://marhi.ru/AMIT/2018/1kvart18/23_saprykina/index.php

Authors' background

Name	Prefix	Research Field	Email	Personal website
Saprykina Natalia Alexseevna	Full Professor, Doctor of architecture	1. Architectural objects in extreme conditions. 2. Sustainable development of architectural Environment.	nas@markhi.ru	-

IOP Conf. Series: Materials Science and Engineering **675** (2019) 012017 doi:10.1088/1757-899X/675/1/012017

		3. Dynamic shaping in architecture. 4. Architectural futurology and Bionics. 5. Innovative energy technologies in architecture.		

The interaction between mathematics and architecture

E V Sergeeva*, E A Moskvina and O A Torshina

Nosov Magnitogorsk State Technical University, Magnitogorsk, Russia

E-mail: sew1108@yandex.ru

Abstract. Mathematics and architecture have always been closely interrelated, the main methods of design in architecture has always belonged to mathematics. At the present stage in connection with the development of computer technology and the Internet began a new round in the relationship between mathematics and architecture. A competent architect should have excellent skills in painting and sculpture, as well as a good knowledge of mathematics. We came to the conclusion that the purpose of mathematical training in the University of students of construction specialties is the development of their mathematical competence. This requires: 1) students should be able to solve applied mathematical problems; 2) students should carry out projects related to their future professional activities; 3) students should be able to create mathematical models. We have developed criteria, indicators and levels of development of mathematical competence of students of the University, which allows you to organize control and measuring materials to assess the level of mathematical training of students.

1. Introduction

Since Ancient times, mathematics and architecture have been closely interrelated, the main methods of design have always belonged to mathematics. Even in Ancient Greece, mathematics, with its geometry of Euclid, was one of the sections of architecture. This idyll was broken in the 18th century, when the first Engineering school was established in Paris in 1747. Since then, mathematics and architecture have evolved each on their own. A new stage in the relationship between mathematics and architecture began with the development of computer technology and the Internet. For successful work architects should be able to solve applied mathematical problems. These applied mathematical problems include, for example, the problem of determining the strength of structures, structural optimization, stability and control modes of their operation. A competent architect should have excellent skills in painting and sculpture, but also good knowledge of mathematics. The ideas of the architect will remain a dream on paper, if he cannot calculate the design of the designed structure. Therefore, while studying at the University, the future architect should study well many fundamental sections of higher mathematics, including the theory of probability and mathematical statistics, and get the skills of mathematical modeling.

We studied various sources of information concerning the application of mathematics in architecture and construction. Some of these works are devoted to mathematical methods in architecture, others-to architectural creativity and form formation, there are works devoted to the methodology of design, as well as the history of architecture.

Publications concerning mathematical methods in architecture can be divided into two groups. The first such group includes works that consider the interaction of mathematics and architecture in a certain historical era. To this group can be attributed, for example, "Treatise on architecture" Antonio Filaret [1], which is a guide of his time on the design of civil architecture. This may also include V.S. Lesovik [2], S. Lysov, M. Balzannikov, V. Evstropov, M. Lysov [3], A. V. Voloshinov [4] and others. Another group includes works that study the variety of applications of modern methods of mathematics in architectural design, considering the modern view of the proportions of the "Golden section" in the architecture of ancestors. Such works are the works of V. Erofeev [5], T.V. Nazmeeva, N. I. Vatin [6], V. I. Sazonov [7], Yu. V. Nemirovsky, N. A. Fedorova [8] and others. The authors have the latest modern research on this issue: W. Wu, A. Tesei, S. Ayer [9], S. I. Rifaat [10], E. I. Saqan, A. K. Ashmawy [11], Choudhary A, Dogne N [12]. We studied these and other works and drew conclusions about the great importance of the study of mathematics for future architects, and identified the problem of mathematical training of students of construction specialties of the University. The analysis of methodical literature showed that the question of development of mathematical competence of students of construction specialties insufficiently studied. It is necessary to develop the conditions for the development of mathematical competence of students of construction specialties.

For the successful implementation of these tasks it is important to develop a culture of logical thinking, the ability to creatively non-standard approach to solving professional issues, the ability to find the necessary information in books, reference books, the Internet. That is, to be a competent architect, a student must have an average, or better, a high level of development of mathematical competence. Which, in turn, is an integral part of the professional competence of the architect.

We consider the mathematical competence of students as an integral quality of a specialist, manifested in the free possession of a system of professionally significant mathematical knowledge and skills, in the ability to independently carry out meaningful multidirectional cognitive activity and creatively solve professional problems of various levels of complexity [13].

We have developed criteria, indicators and levels of development of mathematical competence of University students, which makes it possible to organize control and measuring materials to assess the level of mathematical training of students [13].

Students-architects having studied the course of mathematics should be able:

1) formulate the problem in mathematical language,
2) build a mathematical model from the available data,
3) choose the necessary mathematical apparatus for the solution,
4) develop an algorithm to solve the problem,
5) if necessary, use numerical methods to solve the problem using a computer,
6) using the methods of mathematical statistics to draw conclusions.

Thus, the aim of mathematical training of students of architects is to develop their mathematical competence.

2. Materials and methods

Mathematics should not be separated from the future professional activity of students, from their daily lives. If students feel the importance of mathematics for themselves, they will study it with enthusiasm. We try to use this in our work, so when studying each topic, we focus on solving applied problems, performing applied tasks and projects. Let us consider some of these problems on the topic "Surfaces".

1) What is the amount of material required for covering three columns of cylindrical shape with a height of 5 m and a diameter of *50 cm*? Take into account the consumption of material for waste and seams – 15%.

2) A wooden ball with a mass of 630 g and a diameter of 20 cm. Find the density of the wood used.

3) It is necessary to make two pipes with a length of 7 m and a diameter of *30 cm*. How many square meters of sheet metal will be needed if the seams need to add 3% of the area of its side surface?

4) Reinforced Concrete panel has dimensions 600x120x22cm. Along its entire length there are six cylindrical holes whose diameter is 14 cm. Find the mass of the panel if the density of the material is 2500 kg/m^3

Studying the topic "Integration of functions" we consider the following tasks:

1) calculate the coordinates of the center of mass of the figure, bounded by the arc of the ellipse $x = 2\cos t$, $y = 3\sin t$, and coordinate axes (Consider 1 quarter).

2) calculate the volume of the surface of the ellipse at $a=2$, $b = 3$, $c = 1$.

In mathematics classes, we also use the technique of self-preparation of tasks by students according to known formulas. Students receive a table with formulas and for a certain allotted time, for example,

10-15 minutes should come up with the condition of the problems that will be solved by these formulas.

For example, given table 1, it is necessary to create a problem in the study of the topic "A certain integral".

Table 1. Relationship between values

Values	Derivative calculation	Integral calculation
m – thin rod weight ρ – linear density	$\rho(x) = m'(x)$	$m = \int_{x1}^{x2} \rho(x)dx$
Q – amount of heat C – heat capacity	$\rho(t) = Q'(t)$	$Q = \int_{t1}^{t2} c(t)dt$
S – movement v – speed	$v(t) = S'(t)$	$S = \int_{t1}^{t2} v(t)dt$

Another technique that we use in mathematics, is the technique of " its support." Students make their own reference notes on the topics studied. Abstracts can be made both in handwriting and on the computer. To teach to make a basic abstract we teach students directly, in the study of the very first topics in high school, as a rule, the "Matrix". The teacher introduces students to the algorithm of drawing up a reference abstract on a given topic and work with him. Students learn to identify the basic concepts on this topic, to find their essential properties, to make algorithms for solving typical tasks. Be sure to reference the syllabus must reflect the practical application of the topic. The algorithm of drawing up a reference abstract includes the following stages: the study of theoretical material on a given topic; drawing up a plan; the development of symbols; determining the logic of the presentation of the material and its layout in blocks; color design of the reference abstract. After mastering the algorithm, students at home make their reference notes, they can use them in further practical classes. Making the basic abstract on the studied subject, the student fixes the received knowledge, systematizes them, learns to isolate the main points in a subject, to represent the studied material visually. And most importantly, he is using a variety of information sources, looking for the use of this topic in other subjects and their future professional activities.

Any future architect should be able to build a mathematical model from the available source data. Since in the future it may face the tasks of organization, planning and management of activities, and they are characterized by a plurality of possible solutions, often great uncertainty and dynamic processes. In the process of working on a plan for the construction of any construction object, it is necessary to compare different options and choose the most suitable one. To solve all these problems, the student must learn modeling.

Under the model, we understand the generalized, idealized, visual-logical image of the investigated object (process), more accessible to study than the process itself [13].

Modeling allows us to identify the essential features of the real object for the researcher, helps to make experiments. The results of calculations and experiments obtained on the model are analyzed and conclusions about the real object are made.

The process of mathematical modeling includes three stages: 1) formalization, 2) solving the problem within the model, 3) interpretation. Teaching students mathematical modeling, we pay special attention to the work on the stages of formalization and interpretation, because these two stages cause the greatest difficulties for students [13].

In mathematics classes there is a comparison of real objects and their mathematical models. For example, when studying a linear function, an analogy is drawn between the linear function $y = kx + b$ and

a) the length of the circle and its radius $C = 2\pi r$;

b) the dependence between the length of the rod and the heating temperature $l = (1 + at)$;

c) the speed of sound in the air depending on the temperature $v = 331 + 0{,}6t$.

For students, there are several problems to solve which it is necessary to make a linear function and explore it. Consider such tasks.

1) A Resident of our city put in the Bank 20 thousand rubles, which are charged monthly 7%. The Deposit amount is calculated using the simple interest formula. It is necessary to Express the dependence of the Deposit amount on the number of months that this Deposit will lie in the Bank and answer the questions:

a) what is the Deposit amount in 3 years?

b) in how many years the Deposit amount can be increased by 2 times?

According to the data of the problem with students the function $y = 20000 + 1400 \cdot n$, then answer the questions of the problem.

2) Two mines *A* and *B* are located at a distance of *150 km* on the highway. Mine a produces *200 tons* of ore per day, and mine $B - 100$ tons of ore per day. Where do you need to build an ore processing plant, so that the number of ton-kilometers for its transportation is the least?

Make a function and explore it: $y = 200x + 100(150 - x) = 100x + 15000$.

Homework for students give examples of linear function related to future profession or life.

In order to teach students mathematical modeling, we consider the subject problems in the classroom, translating them into mathematical language.

Under the topic "Sequence. The ranks of" students are offered problem about two shooters (masters and novice) on the shooting range. The sequence of hits for each arrow can be represented as a series of numbers, but with the proviso that the competition will continue indefinitely (since the sequence is an infinite set of numbered elements). Watching the master, we will see that no matter how small a circle we take, starting with a certain shot, all subsequent hits will be inside this circle. Hence, the sequence of hits is committed (converge) to the center of the target, thus the target is the limit of a sequence of hits. Watching the newcomer on the shooting range, you can outline around the center of the target, a circle of a certain radius, such that whatever number we make come true, there is a hit with a large number that falls outside this circle. This means that the sequence of hits does not tend (does not converge) to the center of the circle. Then, in this example, we consider the Cauchy criterion (convergence criterion of the sequence). Think of any distance. Then find a number so that the distance between any two hits with large numbers would be less than the intended distance. If, no matter how small a distance we plan, we always succeed, then the sequence satisfies the Cauchy criterion. From this it follows that it converges.

As another example, you can consider the problem of candy, as in the song: "You half – and I half." Candy is divided between the girls in half, the resulting share one of them again, bisects, etc. From a share of the candy you can get to the numbers. A decreasing sequence of numbers is formed. Further, the convergence of this sequence can be solved.

The illustration of the theorem " every monotonically decreasing and bounded from below numerical sequence has a limit "We build on the identification of the answer to the question:" Is there a limit of sports capabilities of a person?". The question of records of freestyle swimming for 100 meters is considered. Not so long ago, the record was to overcome a hundred meters in 50 seconds, and now he is beaten. There are no eternal records, there is always someone who improves the previous record, reducing the time of the race. On the scale of the results of swimming freestyle for a hundred meters there is a point to which the sequence of record achievements tends. No matter how small a neighborhood of this point we take, starting with some member of the sequence, all the other members will lie in this neighborhood. A similar example of the theorem that every monotonically increasing and bounded from above numerical sequence has a limit is the example with the records of weightlifters.

We give another theorem as an example: "A Series is called convergent if the sequence of its partial sums has a limit; this limit is called the sum of the series. If the sequence of partial sums has no limit, the series is called divergent". Students are offered the subject problem connected with family life: "In some family every year on birthday of the daughter on a door jamb a notch growth of the birthday girl is noted. Over the years, on the jamb there is a whole ladder marks. In this example, we can observe two sequences: the first-a sequence of values of sprouts from year to year, the second – a sequence of values of growth. These two sequences are related to each other. The second sequence is obtained from the first addition. Growth is the sum of all previous years ' gains. The members of the second sequence are supposed to be summarized, they are called members of the series. The sum of several n first terms of the series is called the n-th sum of the series (partial sum of the series). Over time, the marks on the door jamb converge, touching, more closely to each other, that is, the sequence of growth values displayed on the door jamb, has a limit (the series converges).

The presented examples illustrate the process of studying the topic: "Sequences; Ranges». Similarly, you can study all the topics of mathematics, as well as to consider the concepts and theorems in the study of the "Function", "Differential and integral calculus", "Functions of several variables", "Functional series".

Using the considered methods of actualization, we change the students 'understanding of mathematics, creativity in it, increase interest in its study, increase the importance of new knowledge in the field of mathematics, which ultimately stimulates the development of students' motivation to study mathematics, as evidenced by the results of a survey conducted at the end of the experimental work [13].

Mandatory sections of mathematics for students of construction specialties is the theory of probability and mathematical statistics. In these areas of mathematics, many problems are of applied importance. Here are examples of tasks performed by students in mathematical statistics in Microsoft Excel.

Target 1. Find the average value and the standard error percentage of compressive strength (random value X) 100 test grades of concrete (or the same – for the percentage of resistance of the same grades of concrete to rupture (random value Y)). To estimate the degree of deviation of the empirical distribution of this measurable trait from the normal distribution. (Data are taken from laboratory work)

Target 2. According to task 1, check the statistical hypothesis of coincidence with the normal distribution of one measurable feature of the population.

To get acquainted with the main tasks of statistical testing of hypotheses, with frequently used methods of testing the hypothesis of normal distribution. To study the solution of the problem of consistency of theoretical and statistical distributions.

In order to make students aware of the importance of mathematics for themselves, they perform creative project tasks in large sections of mathematics, which study inter-subject projects. All creative project tasks performed by students meet the following requirements:

1) existence of an open problem;
2) in the process of the assignment students go through all stages of the project;
3) creative project task is aimed at achieving personal and socially significant results;
4) the task reflects the applied importance of mathematics [14].

Before the first topic of the research is presented to the students, a methodological seminar is held, which explains the theoretical foundations of the design, demonstrates the completed projects on various topics, discusses the pros and cons of these projects. When carrying out the first projects on the basis of the principle of cooperation by the teacher together with students the main directions of work are considered, important characteristics of future projects are noted. In the future, in the process of working on subsequent projects gradually increases the level of independence of students, and the teacher is enough only to voice the topic. After each voiced research problem, students are given at least three weeks to develop the project.

We give examples of projects performed by students related to the solution of professional problems in which mathematics performs an applied function. Thus, in the study of the topic "Curves of the second order and spatial bodies" students architects carried out projects on the theme "the Existence of unusual figures in our world." They made models of figures, geometric bodies, prepared multimedia presentations, reports. The most interesting topics of the projects: "Impossible figures",

"Penrose Triangle", "Impossible Reutersvard's triangle", "Mobius Strip". These projects are carried out with a creative approach on the part of students, a huge share of imagination and a fairly close connection with their profile orientation [15].

A mandatory theme of group projects is "the Golden section in architecture and life". With this mathematical formula, the architect calculates the proportions of his creations.

Great success among the students produce projects that show the relationship of mathematics with the future profession, with everyday life. For example, carried out projects of the future giving or future house, connected mathematics with the specialty and others. We have considered examples of such projects in our previous articles [16].

According to the results of work on the projects, student conferences were held. Students made presentations, demonstrated and defended their projects. The best papers were then published in the journal of Nosov Magnitogorsk State Technical University "Youth. Science. Future".

The main research methods in our work were: analysis, comparison, synthesis, generalization, specification, abstraction and modeling.

3. Results

The aim of our work, as we noted above, was the development of mathematical competence of students. The experimental work was carried out, which was attended by 72 students studying in the field of "Architecture" in the Nosov Magnitogorsk State Technical University. Two groups of students-architects, one experimental and the other control, were chosen for the experiment. The control group of students studied the fundamentals of higher mathematics, classes in it were standard. In the experimental group, in addition to standard training, students solved applied tasks and tasks on each topic, carried out projects, including interdisciplinary ones, and spoke at conferences. That is, we have implemented a methodology developed by us, different from the methods of other authors.

To justify the reliability of the results obtained in the experiment, two problems were solved. Firstly, the methods of mathematical processing of the experimental results were selected, and secondly, the representativeness of the sample was proved.

To process the results of the experiment, the statistical Pearson criterion χ^2 was used to justify the reasons for the changes in the results. The null statistical hypothesis reads as follows: "The levels of mathematical competence of students of the experimental and control groups will not differ as a result of the experiment."

At the beginning of the experimental work, criteria and indicators of the level of development of mathematical competence of students were developed [16]. Processing the experimental data, each indicator of the level of development of mathematical competence of students we evaluated on a three-point scale.

At the beginning of the first semester, primary data were determined, which showed a fairly low level of development of mathematical competence of students.

According to the selected criteria and indicators, the level of development of mathematical competence in each of the selected groups was checked at the beginning of the study of mathematics at the University (the beginning of the first semester) and at the end of the study of mathematics at the University (the end of the third semester). The results are shown in table 2.

At the end of the course of mathematics at the University was repeated diagnosis.

4. Discussion

The experiment showed an increase in the number of students at the average and high level of development of mathematical competence. As the survey of students showed at the end of the experiment, many students realized the importance of mathematics for themselves, their future professional activity.

Table 2. Results of development of mathematical competence of students in the process of experimental work.

Group	Number of persons	Stage	Levels						χ^2 obser.
			low		medium		high		
			Number	%	Number	%	Number	%	
EG	38.0	Initial	23.0	60.5	11.0	29.0	4.0	10.5	0.1
		End	3.0	8.0	24.0	63.0	11.0	29.0	10.5
KG	34.0	Initial	20.0	58.8	11.0	32.4	3.0	8.8	-
		End	13.0	38.0	17.0	50.0	4.0	12.0	-

In determining the dynamics of the level of development of mathematical competence of students will use the following statistical indicators of time series:

1. The average (*C*), we measured according to the formula $C = \frac{a + 2b + 3c}{100}$, where *a, b, c* – expressed as a percentage of the number of students who are respectively at low, medium and high levels of mathematical competence.

2. The efficiency coefficient (K), which reflects the effectiveness of the experimental study, is determined by the formula $K = \frac{C(EG)}{C(KG)}$, where *C (EG)* – the average value of the experimental group, *C (KG)* – the average value of the control group.

3. The index of absolute growth (*G*), shows the difference between the final and initial values of any studied indicator, it was calculated by the formula $G = \Pi(final) - \Pi(initial)$, where *Π (final)* – final value of the indicator, *Π (initial)* – initial value of the indicator [16].

The data obtained during the experiment are shown in table 3.

Table 3. Table of results of development of mathematical competence of students during the forming experiment.

Group	*Stage*	*C*	*K*	*G*
EG	Initial	1.50	1.27	0.71
	End	2.21		
KG	Initial	1.51	–	0.23
	End	1.74		

According to the results of table 2, we see that at the level of significance $\alpha = 0.05$ in the experimental group χ^2 obser. > χ^2critical (χ^2critical= 5.99), that is, the condition of application of applied problems, tasks and implementation of projects in the process of mathematical training is statistically significant for the development of mathematical competence of students. Based on the results of the work, we conclude that the successful development of mathematical competence of students of architects of the experimental group is not accidental, but is the result of the application of applied problems, tasks and creative projects in mathematics classes.

Conclusions

All of these design decisions in architecture are mathematically justified, proved. According to the results of our study, we can draw the following conclusions, a student, who studying in the field of "Architecture", will not take place in the future as an architect, if he do not know the theoretical foundations of mathematics, will not be able to build a mathematical model on the condition of the problem, will not be able to apply the methods of mathematical statistics for the processing of experimental data. That is, a student at the end of the University should be well developed mathematical competence.

This requires:

1) students should be able to solve applied mathematical problems;
2) students should carry out projects related to their future professional activities;
3) students should be able to create mathematical models.

We also made sure that the best results in the experimental group suggest that the application of applied problems, tasks and projects in the classroom of mathematics allows more qualitative development of mathematical competence of students of architects.

The purpose of our research is carried out, problems have been solved.

References

[1] Filarete A 1999 *Treatise on Architecture* (Moscow: Russian University) 448 p
[2] Lesovik V S 2009 Science at the University: directions and prospects of development *Higher education in Russia* **12** p 103-108 https://elibrary.ru/item.asp?id=12985128
[3] Lysov S, Balzannikov M, Evstropov V and Lysov M 2015 Ways of modernization of education system of construction specialists taking into account requirements of professional standards *Procedia Engineering* **117** p 218-224
[4] Voloshinov A V 1999 The old Testament Trinity of Andrey Rublyov: geometry and philosophy *Leonardo* **32(2)** p 103-112 https://elibrary.ru/item.asp?id=3613724
[5] Erofeev V 2016 Frame construction composites for buildings and structures in aggressive environments *Procedia Engineering* **165** p 1444-1447
[6] Nazmeeva T V and Vatin N I 2016 Numerical studies of incised C-shaped compressed elements with initial imperfections Nazmeeva *Engineering and construction journal* **2(62)** p 92-101 https://elibrary.ru/item.asp?id=26723936
[7] Sazonov V I 2014 The Unity of Views on Geometric Harmony in Architecture and Arts Srednevekovoi Central Asia and Egypt, Greece, Medieval Russia *Bulletin of Kemerovo state University of culture and arts* **28** p 138-145 https://elibrary.ru/item.asp?id=21989295
[8] Nemirovsky Yu V and Fedorova N A 2019 Simulation of limit States of a curvilinear reinforced rotating disk *Mathematica Montisnigri* **44** p 84-99 http://www.montis.pmf.ac.me/vol44/7.pdf
[9] Wu W, Tesei A and Ayer S 2019 Closing the Skills Gap: Construction and Engineering Education Using Mixed Reality - A Case Study *Proceedings - Frontiers in Education Conference FIE 2018-October* 8658992
[10] Rifaat S I 2019 The Multidisciplinary Approach to Architectural Education: Bridging the Gap between Academic Education and the Complexities of Professional Practice *IOP Conference Series: Materials Science and Engineering* **471** 082067
[11] Saqan E I and Ashmawy A K 2018 An alternative approach for practical experience in undergraduate engineering curricula *World Engineering Education Forum - Global Engineering Deans Council WEEF-GEDC 2018* 8629774
[12] Choudhary A, Dogne N and Maheshwari S 2014 Mathematics and Architecture: Importance of Geometry *NCAICT: National Conference on Advances in Information and Communication Technology TEQIP-II/EE/AICMT-5* https://www.researchgate.net/publication/297167812
[13] Sergeeva E V 2017 *Development of mathematical competence of students in the process of professional training in technical profiles*: autoab. kand. pedagogical sciences p 179 https://dlib.rsl.ru/viewer/01006654914#?page=1
[14] Sergeeva E V 2018 Project activity of students of construction specialties in universities *IOP Conference Series: Materials Science and Engineering* **451** 012119
[15] Sergeeva E V 2018 Role of projects in the study of mathematics *SHS Web of Conferences* **50** 01157
[16] Sergeeva E V 2016 Criteria determining the level of development of mathematical competence of students *World of science: online journal* **4(1)** http://mir-nauki.com/PDF/37PDMN116.pdf

IOP Conf. Series: Materials Science and Engineering **675** (2019) 012019 doi:10.1088/1757-899X/675/1/012019

Aerodynamic features of buildings with runways unit

A Kazurov[1*] and O Suslova[2]

[1] Architectural Workshop "Popov & architects", Moscow, Russia
[2] Peoples' Friendship University of Russia (RUDN University), Moscow, Russia

E-mail: kazurov-aleksandr@yandex.ru

Abstract. Currently, opportunities are being developed and are being considered for using flying taxis in large cities due to promising types of air transport (Flying cars, passenger drones). In the second half of the twentieth century, projects for the development of helicopter transport were developed in the UK and the USA. Within the framework of this project, projects for heliports and buildings that interact with helicopter transport were developed in the UK. This experience is very important in the design of buildings with a runway unit, interacting with advanced air transport. Based on an analysis of projects in the UK and the USA, the authors developed experimental aerodynamic schemes of buildings with a landing unit. These projects were tested in the Virtual Wind Tunnel and implemented with the recommendations received by the authors at TsAGI (Central Institute of Aero hydrodynamics). In architectural design of buildings interacting with air transport, aerodynamics is the most important tool for shaping the spatial and compositional solutions of architectural objects.

1. Introduction

XX century gave the world a lot of amazing discoveries and developments, including a flying car. Understanding of the need for decisive action in the organization of the transport network of the city turned the eyes of architects and engineers in the direction of urban air transport since the middle of the XX century. Ideas of flights in the city are reflected in the works of architects and artists of science fiction, such as "Einem Partnership", Klaus Brugle, Arthur Badebo and others. The futuristic projects of the future depict an active symbiosis of the city with all modes of transport including air transport which is emphasized. By the middle of the XX century, these ideas began to take shape thanks to the appearance of the first civilian serial helicopters for air traffic in London and projects of buildings with large helipads [1]. Large-scale helicopter routes between cities and the nearest countries were planned. However, these projects did not allow to continuein this directionactive development of air transport because of technical shortcomings of helicopters of that time [2, 3, 4].

The development of technology and the increase in the number of different vehicles has led to a significant complication of the transport infrastructure such as major highways, interchanges, tunnels and parking lots, as well as some changes in the design of buildings, which often include underground parking and helipads. However, the existing design methods of designing objects in large cities no longer meet the requirements of modern life in the city with a rapid increase in transport per capita. The current problems force the developers of vehicles to create projects of new modes of transport, allowing to increase transport mobility and efficiency in the city, as well as to solve environmental problems through the use of alternative energy sources [5, 6]. Thus, it generates a new round in the development of urban air transport, which has a serious chance of success due to major breakthroughs in engineering, automation and electronics. Currently, helicopter transport in the megacities of the world is developing rapidly, and passenger traffic is increasing every year. It is very likely that helicopters, and later passenger drones, will occupy a large segment in the urban transport system [7].

The promising air transport includes passenger drones of various designs, for example, a passenger drone from Airbus. It should be noted that this group of transport has a similar mechanics of flight with helicopters 'figure 1' [8, 9]. Therefore, it is necessary to study the main features and developments in the design of buildings with helipads. It will help to make the most accurate forecast of how the architecture of buildings can change, interacting with urban advanced air transport. Usually in modern studies on the aerodynamics of buildings with a helipad, the authors consider narrow technological and engineering problems [10]. However, to create form-building methods for buildings with a helipad, research in architectural aerodynamics is necessary. This study offers architects some methods that allow the design of buildings with helipads not to lose the architectural expressiveness of the author's intention [11].

Figure 1. Prototypes of modern flying cars and passenger drones.

In addition to urban ideas and concepts, appropriate principles should be developed for the formation of architectural objects with a runway unit. These principles will allow to contact with a new type of transport. The existing helipads are not ready to receive such transport, they often do not have a control tower and space for temporary storage of aircraft. Also, they are not adapted for use without prior preparation [12]. New types of buildings can be created according to the "aircraft carrier" principle."Aircraft carrier" is a kind of floating city with an airfield, which ideally combines these two functions. Buildings should have special landing platforms: with markings, landing lights and anti-icing elements.

2. Materials and methods

Airflow modeling was computed using Autodesk's Flow Design computer program, which simulates airflow around architectural concepts. This is necessary to test ideas early in the development cycle. This virtual wind tunnel software is intended for designers and architects. The study presents forms for buildings with a helipad were obtained experimentally. The main purpose of experiments with the shape of the building, to achieve a steady air flow in the area of the helipad. The main purpose of experiments with the shape of the building is to achieve a stationary air flow in the area of the helipad for buildings with a runway unit. The forms obtained by the authors were analyzed for changes in the internal layout of the building with the runway unit. For the analysis, the authors created sketchy 3D models with functional zoning of buildings. Images of the air flow around the buildings were obtained by the authors in the Autodesk "Flow Design" computer program.

Air transport has fairly high safety requirements, not only for operation, but also for the infrastructure with which it interacts. One of the most important conditions for the safe use of a new type of air transport will be the requirements for the aerodynamics of buildings. There are several types of cities: "plains", "hills" and "hollows", which corresponds to different heights of urban development and the terrain [13, 14, 15]. These types of building directly affect the height of the landing pads (consoles, roofs, courtyard spaces) in urban areas and their location on the master plan of the city in terms of the wind rose. These parameters are necessary to create favorable aerodynamic conditions above the landing pads in the city.

For the experiments were selected the main types of aerodynamic schemes of buildings on which helicopter pads can be located. In the course of the experiment, 9 models of buildings were blown through the virtual wind tunnel: a cylindrical, rectangular tower type, a cavity (rectangular with a courtyard), an open courtyard cavity, T-shaped in plan, L-shaped in plan, C-shaped in plan, trefoil, H-shaped in plan, cruciform in plan 'figure 3', 'figure 4'. The dimensions of the models were given as close as possible to the overall dimensions.

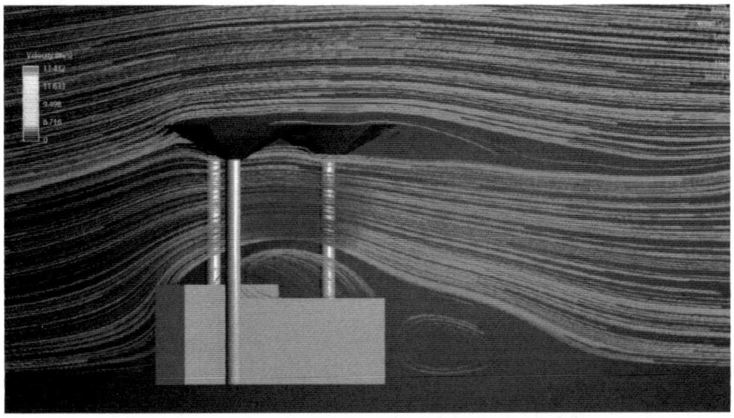

Figure 2. Visualization of the air flow lines demonstrates a very successful aerodynamic design of the building due to the absence of turbulent vortices over the landing pad. Pilkington's Glass Age Development Committee Visualization is made in the program Flow design. The air flow velocity is 10 m/s.

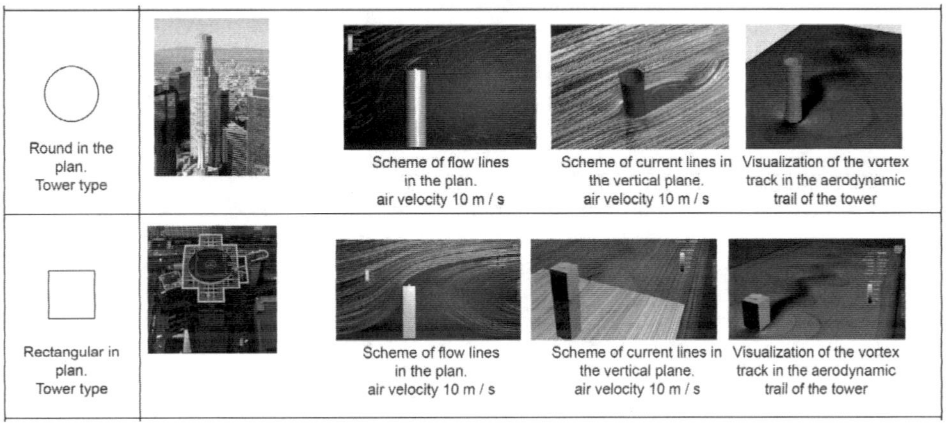

Figure 3. Airflow around tower-type buildings.

All models of buildings have a height of 30 meters and dimensions in the plan in the range from 30-50 meters to 60-80 meters, depending on the configuration of the floor plan. The experiment revealed relatively similar nature of the wind flow over all the buildings except the cavities. When the air flow around the buildings forms several characteristic swirls on the building and the same aerodynamic footprint. The air escaping from the roof of the building is called a vortex layer. Under this layer formed outgoing turbulence. The aerodynamic footprint in the plan is a vortex track in the form of whirlwinds, diverging in a checkerboard pattern. On buildings with such aerodynamic characteristics, landing sites are usually installed on the roof in a stationary air flow.

The cavity has favorable conditions for the landing of an aircraft due to the formation of a steady vortex inside the cavity (inner courtyard) and low air velocity. Experiments have shown that the landing site should be placed in the center of the vortex, where the air flow is minimal. The experiments also revealed the movement of the vortex in the direction of the wind to the opposite wall of the cavern and the appearance of the second vortex, provided that the size of the cavern courtyard exceeds 90 meters. When the air flow rate is from 7 to 10 m / s, the speed of the air flow in the yard is about 3-4 m / s. In square in plan yard spaces with a size of more than 70-90 meters, a vortex moves in the direction of the air flow 'figure 5'.

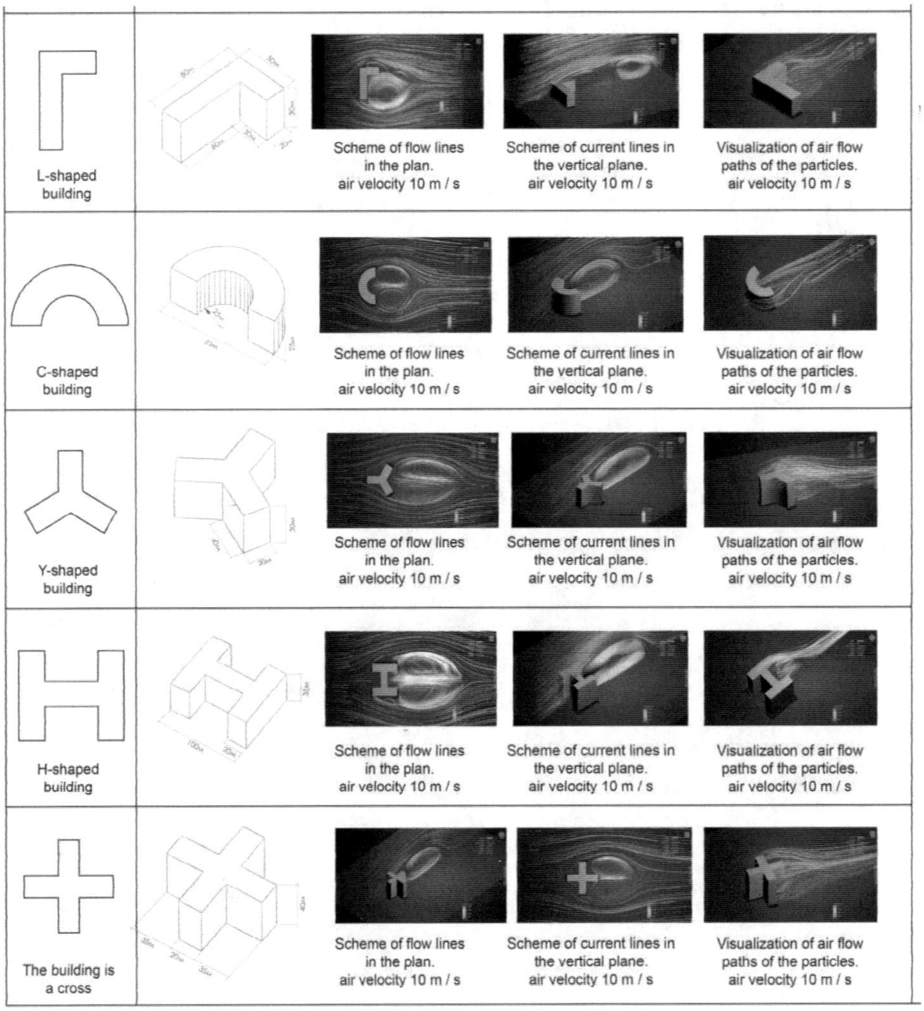

Figure 4. Airflow around the most common forms of buildings.

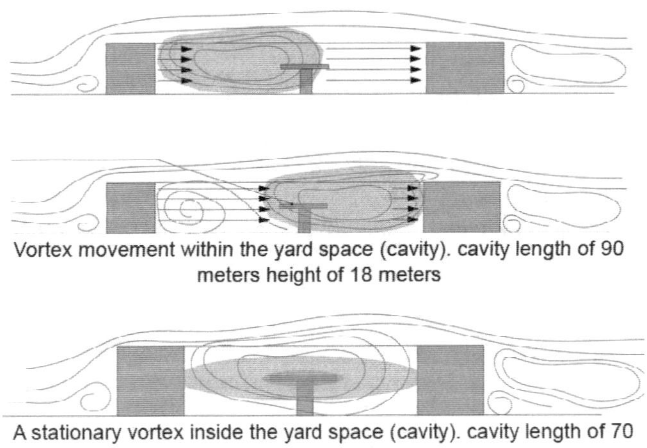

Figure 5. The movement of the vortex in the direction of the air flow inside the cavity of more than 90 meters. In the caverns with a courtyard up to 70 meters there is a stationary whirlwind.

3. Results

Experiments were carried out with typical forms of buildings of prismatic configuration and the American heliport "Port Autority". Was modeled several aerodynamic designs of buildings with favorable aerodynamic regime. Various forms of the building were applied.

The building with the "inverted wing". The most common version of the location of landing pads - the roof. This is the highest point for landing among the surrounding buildings, which can be relatively easy to land.

The air flow of such buildings is not stationary, because the air flows on the leeward side of the building are alternately separated. The angular value between the landing pad and the flow disruption depends on the air flow rate and the time from the beginning of the blowing, as well as the height of the building.

To ensure that the landing pad is always in a stationary air flow, it can be placed on a streamlined platform and raised above the main body of the building. The air becomes freely circulated to the platform with two sides. The platform has a profile resembling the profile of an inverted wing.

The building step shape allows landing on the roof due to the stationary air flow over the roof. Air flow failure is observed only at the edge of the roof from the incoming air flow. This effect is provided by the stepped structure of the facade, as it slows the upward flow from the windward side of the building. This prevents the flow from the edge of the roof from the incoming air flow. This solution provides a steady wind over the landing pads. However, it has its limitations for landing aircraft due to the maximum possible wind speeds for landing on the building.

Caverns or yard spaces. Wind flows move over the roofs of buildings in cities on the plains, where the buildings are about the same height. They do not create disturbances in the airspace along the streets, as they coincide with the direction of the wind. The cavern has the most favorable conditions for landing due to the formation of a stationary air flow inside it and a low speed of air movement in it. Experiments have shown that the landing pad should be placed in the center of the vortex, where the movement of the air flow is minimal. Thus, the motion of particles forms a stationary vortex flow. The speed here is much lower than above the roof of the building. Experiments were carried out on computer models at different air flow rates. At a wind speed of 10 m/s, the movement of particles inside the cavity is 3-4 m/s, and at a wind speed of 15 m/s, the movement of particles inside the cavity is 5-6 m/s. The experimental results show that the air velocity inside the cavity is usually 3 times lower than above the roofs.

The main features of buildings with a landing pad are as follows: it has three options for the location of the landing pads on the building: at the level of the roof, on the console, in the cavern. There is a need to create an optimal aerodynamic regime over the runways to ensure safety during take-off and landing. The runway unit has a set of main and auxiliary premises. It has special damping structures on the runways. It has lighting, meteorological and navigation equipment 'figure 6'.

4. Discussion

The principles of formation of buildings with the runway unit were developed. They showed a great potential of architectural and planning diversity of buildings of a new type. Three schematic diagrams of buildings were made in the Autodesk Flow Design program. They differ in the location of the runway unit. This affects the space-planning decisions of the project. However, it should be noted that the dimensions of the buildings of these concepts may vary depending on the specifics of the project and its location in the building. The illustrations show the methods of forming buildings with a runway unit, which will require more detailed study.

Buildings with a console location of the runway unit can have three options for the location of consoles. First, the fan location, which is most suitable for tower-type buildings. In this case, the

IOP Conf. Series: Materials Science and Engineering **675** (2019) 012019 doi:10.1088/1757-899X/675/1/012019

consoles can be in the same level around the perimeter of the building. Secondly, the consoles can be located at different heights. This option is suitable for multifunctional high-rise buildings, which will allow quick access to the necessary functional area of the tower, and landing pads will be in the aerodynamic shadow of the tower. In these two variants, the runway unit is located in the body of the building, and the landing pads are consoles. In the third variant, the runway unit is a large console that extends beyond the building. This option is suitable for buildings of any configuration.

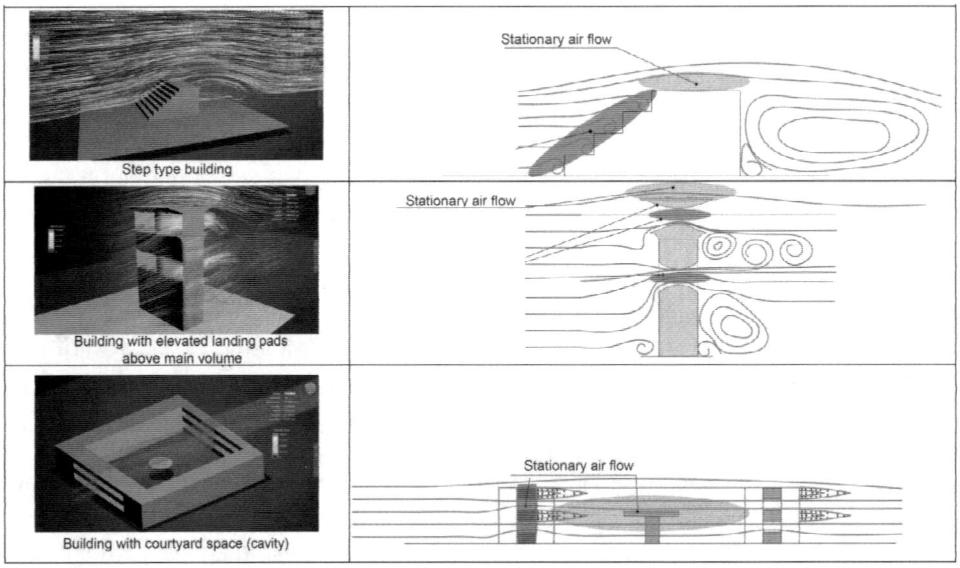

Figure 6. Experimental projects of building shapes to create a favorable aerodynamic regime over the helipad.

Buildings with a runway unit on the roof also have three schemes. The building with a terraced location of the runway units is suitable for long buildings with different functional purpose. In the building "aircraft carrier" runway unit is located on two parts of buildings. The roof of such a building is divided into two parts for takeoff and landing. On the roof there are Islands with entrance groups and elevator halls, a control tower and lifts for air transport. This building is most suitable for emergency services, fire protection, medical centers, which are located in a dense urban area. This type of building will allow in emergency situations to send and receive a large number of aircraft per minute. The building with a raised runway unit above the main building is designed for cities with one high-rise buildings. This is optimal for aerodynamic qualities and the most convenient for the location of the runway unit for landing.

Two principles have been developed for buildings with a runway unit in the cavern: the first - the runway unit is located in the courtyard of the building. The second - the landing pads of the unit are deepened in the caverns on the roof of the building. This location is suitable for low-rise buildings, such as transport hubs, shopping centers, technology parks 'figure 7'.

The basic principles of planning decisions were developed. A set of rooms and functional areas for the runway unit has been given in these principles. The runway unit consists of several functional zones: landing pad, pad to lift-off area for pre-training, entry gateway for transportation, parking, vertical transportation, aircraft. In these schemes, some variations may be used depending on the shape and type of building, such as departure and arrival areas. They can be separated or combined to carry the character. The runway unit is separated by technical floors and fire compartments. The unit has its own entrance groups, which include security posts, checkpoints, reception desks, waiting areas, etc., which are located in the departure and arrival areas.

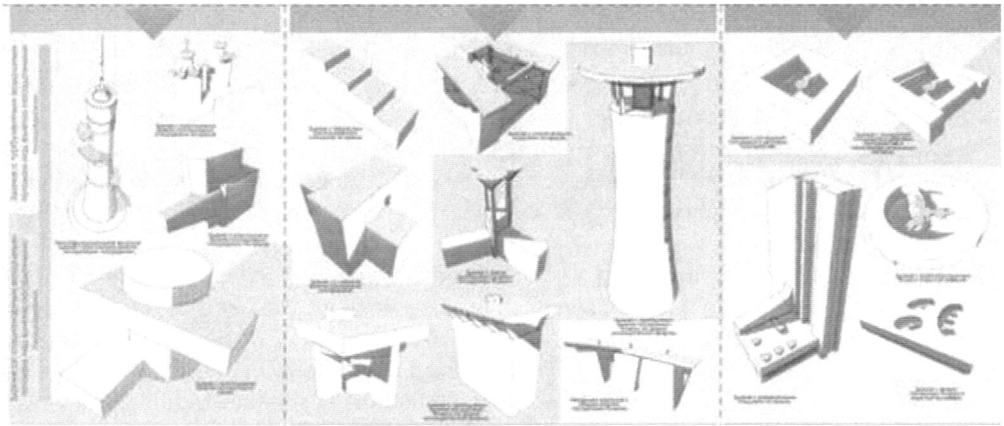

Figure 7. Schemes of architectural forms of buildings with optimal aerodynamic properties (experimental building designs with a runway unit).

The buildings are interacting with air transport will change in comparison with buildings with a helipad. First, the shape of the building will be dictated by aerodynamic calculations based on aerodynamic analysis. It will be taking into account the individual characteristics of the surrounding buildings and terrain. Secondly, the internal structure of the building will change, as it will become important for regular flights of promising air transport. The building has a number of facilities and engineering solutions that will allow you to efficiently and safely take and send advanced air transport and provide high capacity for passengers. The Airbus transport system takes into account the principle of hybridization of the aircraft with the car in the building interacting with the perspective air transport. In this building there are vertical communications that can provide communication with the street and road network of the city.

Thirdly, it is necessary to organize the storage and maintenance of aircraft. The above features of buildings interacting with advanced air transport can be combined into one "unit", which is built into the structure of the building. Such a "unit" in the building makes the building look like a heliport or aircraft carrier, where the flight deck is the roof, console or cavern, and auxiliary functions are located inside the building. Based on this, you can define a new type of building – *a building with a runway unit*. The runway unit in the building is a complex of premises and engineering measures providing reception, dispatch, basing of aircraft with a system of vertical take-off and landing and air transportation services [16].

5. Conclusion

The experiments were considered: building with an elevated helipad on the roof, the building is a stepped form and the building with a helipad in the courtyard space. This is necessary in order to create the best aerodynamic conditions over the runways. In these experiments, form-generating methods were used, which allow to organize the optimal aerodynamic characteristics above the helipad. Also, these methods allow architects and designers to create the expressive shape of a building. Consequently, when designing buildings with a *runway unit*, an architect can use the results of these experiments to create the most expressive architectural appearance of the building. Also experiments with aerodynamic schemes of buildings affect the internal structure of the building. Therefore, the concept of runway unit in a building as a universal modular system was considered. It will allow to fully provide the functions of the 'airport on the roof' for any aerodynamic configuration. Further research is needed to study the stages of development of a new type of air transport for urban planning, the economy, and the safety of using passenger drone above the city.

References

[1] Dodge M and Brook R 2014 Dreams of helicopter travel in the 1950s and Liverpool's undeveloped plans for a city centre heliport *Transactions of the Historic Society of Lancashire* 163 pp 111-125

[2] Beall A 2016 London's scrapped 1931 designs reveal a wheel-shaped runway at King's Cross *Daily Mail* Available at: https://www.dailymail.co.uk/sciencetech/article-3678907/London-s-crazy-plan-elevated-airport-Scrapped-1931-designs-reveal-dangerous-wheel-shaped-runway-King-s-Cross.html

[3] Dodge M 2013 Vertical urbanism and the forgotten plans for heliports *Liverpool Seminar* Available at: https://personalpages.manchester.ac.uk/staff/m.dodge/Forgotten-Heliports-Liverpool

[4] Novak M 2013 Skyport One: The Airport of the Future from 1957 *Gizmodo* Available at: https://paleofuture.gizmodo.com/skyport-one-the-airport-of-the-future-from-1957-1474861998

[5] Garrett J T 2012 Terrafugia flying car completes first phase of flight testing *New Atlas* Available at: https://newatlas.com/terrafugia-flying-car-test-flight/23141/

[6] Lisboa M R, Ottieri J C and González A E 2015 Noise Annoyance Due to Wind Flow Interaction with a Building's Facade *Open Journal of Acoustics* **5(1)** pp 1-10

[7] CityAirbus demonstrator passes major propulsion testing milestone 2017 Available at: https://www.airbus.com/newsroom/press-releases/en/2017/10/cityairbus-demonstrator-passes-major-propulsion-testing-mileston.html

[8] Pradeep P and Wei P 2018 Energy Efficient Arrival with RTA Constraint for Urban eVTOL Operations *AIAA Journal* doi:10.2514/6.2018-2008

[9] Pring-Mill D 2018 Dubai drone taxi service could impact urban infrastructure *Engineering News-Record* Available at: https://www.enr.com/articles/44267-dubai-drone-taxi-service-could-impact-urban-infrastructure?v=preview

[10] Dziubiński A 2016 CFD analysis of wind direction influence on rooftop helipad operations safety *Transactions of the Institute of Aviation* **1(242)** pp 7-22 doi: 10.5604/05096669.1202199

[11] De Voogt A J 2007 *Helidrome Architecture* (010 Publisher, Rotterdam) Available at: https://www.worldcat.org/title/helidrome-architecture/oclc/150255416

[12] Advisory Circular 2012 *Heliport Design* (US Department of Transportation) p 196 Available at: https://www.faa.gov/documentLibrary/media/Advisory_Circular/150_5390_2c.pdf

[13] Myagkov M and Alekseeva L 2014 Wind pattern features in typical city models *AMIT* **1(26)** pp 1-15 Available at: https://marhi.ru/AMIT/2014/1kvart14/myagkov/myagkov.pdf

[15] Elshaer A et al 2016 Aerodynamic shape optimization of tall buildings using twisting and corner modifications *Proc. 8th International Colloquium on Bluff Body Aerodynamics and Applications* (Boston, USA)

[16] Kazurov A and Suslova O 2018 Features city buildings with a strong take-off and landing module *Architecture and Modern Information Technologies* **3(44)** pp 167-182 Available at: https://marhi.ru/eng/AMIT/2018/3kvart18/09_suslova_kazurov/index.php

Estimation of the runoff of the hills of the city of Portoviejo-Ecuador to assess the degree of flooding in the region

A F Campos Cedeno[1], P A Salas Guillen[1], J L Macias Ramos[1], E K Sinichenko[2*] and I I Gritsuk[2,3]

[1] Universidad Tecnica de Manabi, Portoviejo, Ecuador
[2] Peoples' Friendship University of Russia (RUDN University), Moscow, Russia
[3] Russian Academy of Sciences Water Problems Institute, Moscow, Russia

Email: sinichenko_ek@pfur.ru

Abstract. Portoviejo, capital of the province of Manabi is one of the oldest cities on the Ecuadorian coast. It is characterized by a stable temperate climate with an average annual temperature of around 24ºC, with annual rainfall patterns of up to 500 mm. It is a city located in a valley, with average height of 40 m. It is surrounded by hills that have heights that reach up to 280 meters above sea level, which potentiates that the continuous runoffs that occur in the rainy season causes problems related to landslides and floods in the lower part. In the present work an estimation of the runoff coming from the hills located in the northeastern areas of the city where the parishes and citadels Progreso, Cevallos, San Pablo and Briones are located. The capacity of conduction of existing rainwater collectors in comparison with the runoff values obtained is also analyzed. The results obtained constitute an approximation of runoff values for return periods of 1, 2, 5, 10, 20, 50 and 100 years, which are an input for the repowering of existing works and for the design of infrastructure works complementary to the control of floods and landslides in the area.

Key words: hydrology, rainfall, return periods, climate change, runoff control

1. Introduction

Portoviejo, capital of the province of Manabí, founded on March 12th, 1535, is one of the oldest cities on the Ecuadorian coast. It is characterized by a stable temperate climate with an annual average temperature of around 24^{0}C. The average rainfall for the years 2000 to 2009 is 596.20 mm per year, the wettest years are 2000 with 733 mm and 2008 with 823 mm; while the driest are 2001, 2003 and 2009 with rainfall below 500 mm. Maximum precipitation values were obtained in the years 1997 and 1998 during the El Niño phenomenon [1].

The subject of floods has been the subject of study by professionals and institutions at a regional and international level, however, for the city of Portoviejo-Ecuador, it is a very specific problem from which little or nothing has been done until now.

As an example of studies related to flooding and excess of runoff lamination in urban areas in different parts of the world, among others, we can mention that in the Atemajac river basin, in the State of Jalisco-Mexico, Rubén Ernesto Hernández et al. They analyze the problem of overflowing the river and its possible damages while presenting a methodology where they try to standardize the factors that intervene in environmental vulnerability from a social and economic approach [2].

On the other hand, in Buenos Aires-Argentina, the Economic Commission for Latin America and the Caribbean, analyzes the problem of flooding and overflow of the Salado River in the province of Santa Fe with the purpose of proposing projects for the rehabilitation of affected homes by floods and to meet needs related to water disasters that have occurred [3].

ZW Kundzewicz et al, members of the Collaborative Innovation Center on Forecast and Evaluation of Meteorological Disasters / School of Geographical Science, Nanjing University of Information Science and Technology (NUIST), Nanjing, China, in the article "Flood risk and its reduction in China" analyze that the growth of flood risks in China is due to anthropogenic and climatic factors [4].

Related to the floods, Mohammed Sarfaraz Gani Adnan, in the journal Science of The Total Environment, in his article entitled "Have coastal embankments reduced flooding in Bangladesh?", analyzes the effect of polders on the increase of the flooded pluvial area [5].

A very interesting investigation is carried out by Leila Goodarzi et al, where a flood warning system based on forecasts of atmospheric assemblies is developed [6].

Among the investigations related to the topic of floods that have taken place in Japan, we have the one of Tomahiro Tanaka et al, where it is highlighted that the frequency of an extreme flood happening with greater probability is linked to the treatment that happens upstream of the hydrographic basin, so if in the city of Kyoto, specifically in the Yodo River, control is achieved through a digital model of flood - inundation floods that happen upstream are regulated by hydraulic regulation works, it will be less likely to happen an extreme flood within the big cities [7].

Worldwide, many researchers deploy their work with topics related to hydrological studies aimed at determining the risks and solutions for surface runoff that cause flooding in inhabited territories[8-13].

For the city of Portoviejo-Ecuador, for the first time, the problem of floods is approached as a result of the rolling of the runoff from the hills. In these moments in which the situation of the floods has worsened, the Technical University of Manabí has assigned us the task of technically substantiating the problem and proposing guidelines to solve the problem, which is the reason for the present investigative work.

The city is located in a valley, whose average elevation has been estimated at 40.00 meters above sea level, and in the northeast and southwest areas it is limited by hills whose maximum levels oscillate between 250 ~ 300 meters above sea level.

The parishes of Progreso, Pacheco, San Pablo and Briones are located in the lower part of the hills of the northeast sector. During the rainy season they experience landslide problems, with consequent flooding due to surface runoff and water erosion. zone.

For the control of runoff there is a sewer system, which is insufficient, since the lower area is flooded, causing year-to-year urban problems of vehicular and human traffic, in addition to material losses.

With the purpose of contributing with design elements that allow to diagnose the current situation of the runoff and its control works, in the present work, based on the methodology proposed by the Soil Conservation Service of the United States (Department of Agriculture United States. USDA Natural Resources Conservation Service Soils. 2015) (Soil Conservation Service – SCS), complemented with the techniques of Geographic Information Systems (GIS), runoff has been estimated by sectors for return periods 1, 2, 5, 10, 20, 50 and 100 years.

2. Materials of research

The basis for this work is the 24 hour maximum rainfall records provided by the Ecuadorian Institute of Hydrology and Meteorology of Ecuador (INAMHI), meteorological station M005 located in the Botanical Garden of the Technical University of Manabí, latitude 01° 02' 26" S, length 80° 27' 54" W, period 1982-2013. Figure 1 shows the geographical location of the M005 station within the territory of the Manabí Hydrographic Demarcation (DHM) (Secretary of Water of Ecuador) [14].

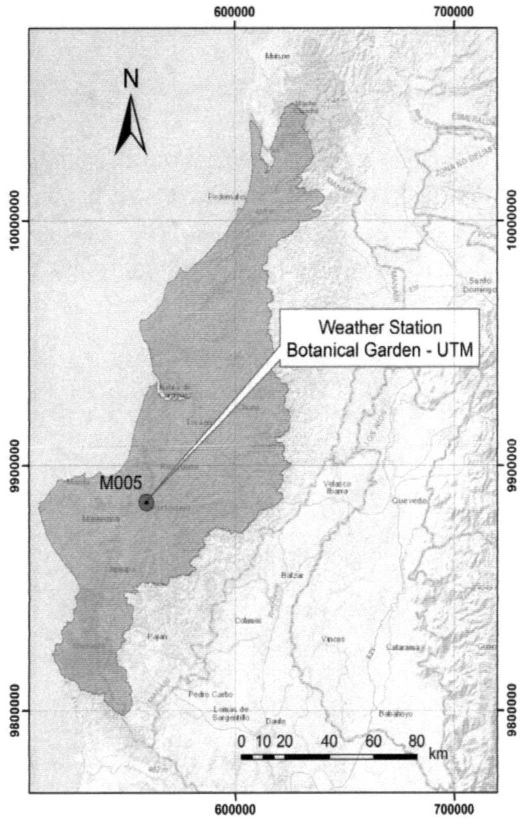

Figure 1. Location of the weather station M005 - Portoviejo – UTM.

The total records of maximum rainfall in 24 hours used in the investigation amount to 384. Table 1 presents the respective records, period 1982-2013 (National Institute of Hydrology and Meteorology. Meteorological Yearbooks) [15].

Table 1. Registration of maximum precipitations in 24 hours, mm, Portoviejo Station – UTM (M005) Period 1982 – 2013

Year	P max 24, [mm]	Year	P max 24, [mm]	Year	P max 24, [mm]	Year	P max 24, [mm]
1982	48.90	1990	37.80	1998	86.70	2006	56.70
1983	120.50	1991	20.90	1999	69.40	2007	46.80
1984	47.30	1992	69.70	2000	41.30	2008	53.30
1985	40.00	1993	25.30	2001	147.90	2009	26.50
1986	54.70	1994	66.50	2002	50.80	2010	52.30
1987	65.60	1995	60.10	2003	38.40	2011	37.00
1988	18.30	1996	40.80	2004	57.90	2012	59.30
1989	131.40	1997	131.10	2005	79.70	2013	107.60

IOP Conf. Series: Materials Science and Engineering **675** (2019) 012020 doi:10.1088/1757-899X/675/1/012020

For the determination of the morphometry of the micro watersheds, digital elevation models with a resolution of 3 meters were used, facilitated by the GAD of Portoviejo. Likewise, the aforementioned institution provided basic information regarding the vegetation cover and infrastructure works for the management of surface runoff in the zone.

Six zones of analysis have been considered (figure 2), each of them being the micro basin or micro basins in which each parish is located: Progreso, Pacheco, San Pablo and Briones. The physiographic parameters of the micro watersheds were determined with the help of specialized software for the management of geographic information. The total area of study has been estimated at 327.03 ha. Through the S.C.S. method, runoff is determined with the formula (1)

$$Q = \frac{(P - 0.2S)^2}{P + 0.8S} \tag{1}$$

where Q – surface runoff height, mm; P – rainfall, mm; S – maximum potential difference between P and Q, mm.

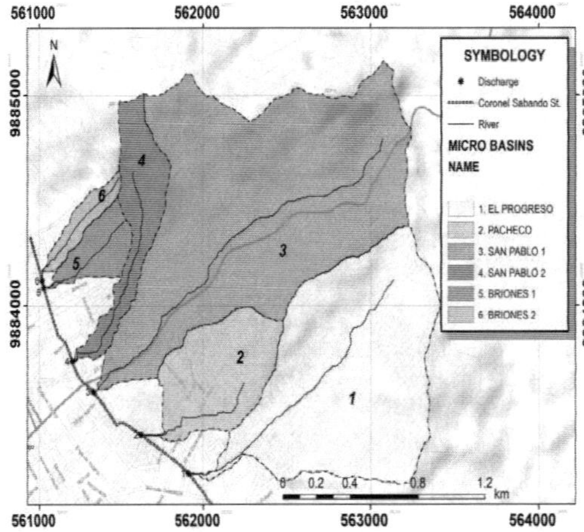

Figure 2. Study zones.

The parameter S, the maximum potential difference between P and Q, can be determined through the so-called *CN curve number*, which characterizes the soil types and vegetation cover of the area. The calculation formula in metric units is as follows:

$$CN = \frac{25400}{254 + S} \tag{2}$$

$$S = \frac{25400}{CN} - 254 \tag{3}$$

For the valuation of the curve numbers of the micro basins, was used the map of coverage and land use of the continental Ecuador 2013-2014, scale 1: 100000, developed by the Ministry of Agriculture, Livestock, Aquaculture and Fisheries (MAGAP) with the Ministry of the Environment (MAE) (Applications in Land Management and Integral Management of Watersheds. 2015).

Surface runoff has been estimated at the discharge points where the steep slopes of the hills end, this is at Coronel Sabando and Avenida Seminario streets. Table 3 shows the coordinates of the projected discharge points in the UTM system WGS84, ZONA 17 S, M.

Based on the shapes of the storm sewer network, facilitated by the GAD-Portoviejo and with the digital elevation model of the area, the slopes of the main rainwater collectors were estimated, which were used to determine the maximum capacity of driving them. Then, the values obtained were compared with the runoff of the micro basins of the study areas in order to validate the existing urban drainage network.

3. Results

The weighted results of the curve numbers for each of the micro-watersheds are contained in table 2.

The peak flows of surface runoff, for various return periods, were estimated for discharge points, whose coordinates are contained in table 3 and figure 3.

Table 4 shows the peak flows for return periods of 1, 2, 5, 10, 20, 50 and 100 years, expressed in m^3/s.

Table 5 compares the maximum conduction capacity of the rainwater collectors with the runoff flows determined in this study, which should be received by the collectors located in Coronel Sabando and Av. Seminario streets.

Table 2. Data of micro basins with weighted values of CN curve numbers.

Micro basins	Name	Area basin **(ha)**	CN weighted
1	Progreso	100.74	74
2	Pacheco	32.63	81
3	San Pablo	155.15	74
4	San Pablo 2	22.00	81
5	Briones 1	10.35	81
6	Briones 2	6.16	81

Table 3. Sites of runoff discharge.

Discharge	**Zone**	**UTM coordinates WGS84**		
		X (m)	**Y (m)**	**Z (m)**
1	Progreso	561907.62	9883202.53	45.75
2	Pacheco	561614.76	9883388.06	45.04
3	San Pablo 1	561332.10	9883592.02	40.81
4	San Pablo 2	561207.42	9883736.05	43.19
5	Briones 1	561027.46	9884086.86	43.53
6	Briones 2	561017.29	9884119.28	43.59

Table 4. Peak runoff estimated at Coronel Sabando and Av. Seminario streets expressed in m^3/s

Micro basins	**Name**	**Return Periods**						
		1 year	2 years	5 years	10 years	20 years	50 years	100 years
1	Progreso	0.0540	0.3205	0.8076	1.3456	1.9388	2.7195	3.4499
2	Pacheco	0.0176	0.1799	0.4761	0.7148	0.9579	1.2657	1.5451
3	San Pablo 1	0.0828	0.4919	1.2393	2.0651	2.9753	4.1734	5.2944
4	San Pablo 2	0.0118	0.1200	0.3174	0.4765	0.6386	0.8438	1.0301
5	Briones 1	0.0059	0.0300	0.0600	0.2383	0.3193	0.4219	0.5150
6	Briones 2	0.0033	0.0338	0.0894	0.1343	0.1800	0.2378	0.2903

Table 5. Confrontation of capacity of collector of rain water with the runoff of the hills for several return periods

Zone	Collector		Runoff [l/s]						
	Diameter [mm]	Flow [l/s]	Return Periods						
			1 year	2 years	5 years	10 years	20 years	50 years	100 years
Progreso	800	2800	54	321	808	1346	1939	2720	3450
Pacheco	400	140	18	180	476	715	958	1266	1545
San Pablo 1	500	700	83	492	1239	2065	2975	4173	5294
San Pablo 2	200	40	12	120	317	477	639	844	1030
Briones 1	200	20	6	30	60	238	319	422	515
Briones 2	250	80	3	34	89	134	180	238	290

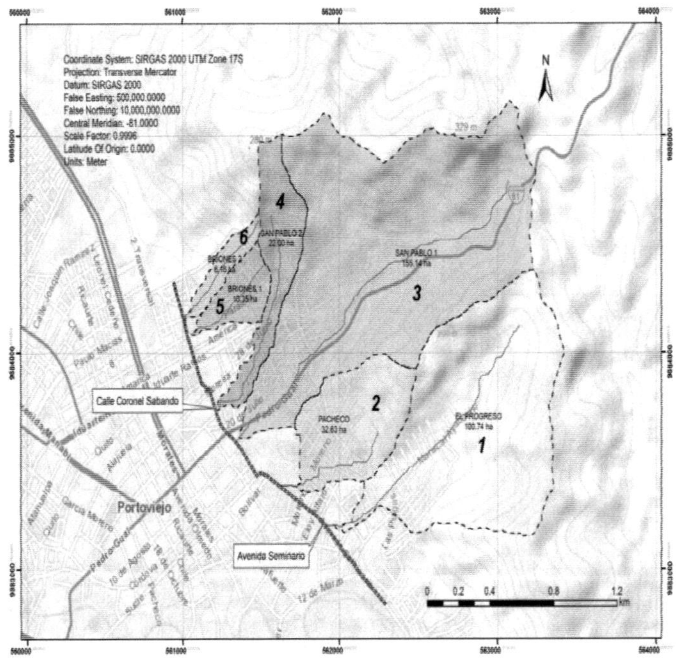

Figure 3. Micro basins of hills of Portoviejo city.

4. Conclusions

1. The total area of the micro basins in which the parishes Progreso, Pacheco, San Pablo and Briones are located amounts to 327.03 ha, corresponding to them, respectively, 100.74, 32.63, 177.15, 16.51 ha.
2. For return periods of 1, 2, 5, 10, 20, 50 and 100 years, the basin with the highest runoff is the one where the San Pablo parish is located (micro-basins 3 and 4), values ranging from 0.05 to 3.45 m^3/s.
3. The lowest runoff values were obtained for the Briones basins (micro-basins 5 and 6), which are between 0.009 and 0.805 m^3/s.
4. Minor values of surface runoff threw the micro basins where the Progreso and Pacheco parishes are located. For Progress zone were obtained values between 0.05 to 3.45 m^3/s, and 0.018 to 1.55 m^3/s, for the Pacheco parish.

5. The estimated values constitute an input for the institutions that are dedicated to the management of runoff from the hills, with the purpose of controlling landslides and floods in the zone.
6. It was determined that the collectors that receive the runoff from the micro-basins have sections that allow flow rates for periods of return of less than 5 years, with the exception of the collector of the Progreso citadel of 800 mm in diameter, which has a maximum driving capacity of 2800 l/s, close to the flow of 50 years of return period. This situation explains that, very often, in the rainy season, the lower parts of the study areas experience problems related to serious floods.
7. In terms of current regulations, contained in the Ecuadorian Construction Code, Part IX, Sanitary Works (Section 5.1.5 Rainwater Design Flows)(Water Secretariat, Ecuadorian Code for the Construction of Sanitary Works. 2012, Quito), for watershed micro drains, design flows of rainwater collectors must correspond to 10-year return periods, which is not true for existing conditions.
8. Based on the analysis carried out, the corresponding instances of urban development should consider this situation in order to deepen these studies in order to reinforce the existing works and / or propose complementary projects that tend to mitigate the environmental impacts caused by the insufficiency of the capacity of driving of the existing storm sewer.

5. References

[1] Campos A F, Sinichenko E K 2017 *Características de Sistemas Fluviales Pequeños y Recursos Hídricos de la Demarcación Hidrográfica de Manabí, Perspectivas de Desarrollo* (Moscow: RUDN University Press) p 241
[2] Hernandez R E, Pina H B, Ramirez A I 2017 Flood risk analysis: Methodology and application to the Atemajac basin *Tecnologia y Ciencias del Agua* **8(3)** pp 5-25
[3] *Evaluación del impacto de las inundaciones y el desbordamiento del Río Salado en la provincia de Santa Fe, República Argentina, 2003* (Santiago: Economic Commission for Latin America and the Caribbean) https://repositorio.cepal.org/handle/11362/28461
[4] Kundzewicz Z W, et al. 2019 Flood risk and its reduction in China *Advances in Water Resources* https://doi.org/10.1016/j.advwatres.2019.05.020
[5] Adnan M S G, Haque A and Hall J W 2019 Have coastal embankments reduced flooding in Bangladesh? *Science of the Total Environment* **682** pp 405-416
[6] Goodarzi L, Banihabib M E and Roozbahani A 2019 A decision-making model for flood warning system based on ensemble forecasts *Journal of Hydrology* **573** pp 207-219
[7] Tanaka T, et al. 2017 Impact assessment of upstream flooding on extreme flood frequency analysis by incorporating a flood-inundation model for flood risk assessment *Journal of Hydrology* **554** pp 370-382
[8] Bin Ghomash S K, Caviedes-Voullieme D and Hinz C 2019 Effects of erosion-induced changes to topography on runoff dynamics *Journal of Hydrology* **573** pp 811-828
[9] Meng X, et al. 2019 Flood control operation coupled with risk assessment for cascade reservoirs *Journal of Hydrology* **572** pp 543-555
[10] Wasko C and Nathan R 2019 Influence of changes in rainfall and soil moisture on trends in flooding *Journal of Hydrology* **575** pp 432-441
[11] Chang C-M, Yeh H-D, Chuang M-H 2019 Spectral analysis of temporal variability of nonlinear and nonstationary rainfall-runoff processes *Journal of Hydrology* https://www.sciencedirect.com/science/article/pii/S0022169419305311
[12] Sungkaew S and Teerawatananon A 2017 Survey of flood-tolerant bamboos in 2011 flooding in Thailand *Agriculture and Natural Resources* **51(6)** pp 499-503
[13] Reitz M and Sanford W E 2019 Estimating quick-flow runoff at the monthly timescale for the conterminous United States *Journal of Hydrology* **573** pp 841-854
[14] Campos A F, Sinichenko E K and Gritsuk I I 2014 Features of water mode of the rivers of the province of Manabi (Ecuador) *RUDN Journal of Engineering researches* **4** pp 41-46
[15] Campos A F, Sinichenko E K and Gritsuk I I 2015 Analysis of data of meteorological stations of the province of Manabi *RUDN Journal of Engineering researches* **3** pp 41-49

IOP Conf. Series: Materials Science and Engineering **675** (2019) 012020 doi:10.1088/1757-899X/675/1/012020

Authors' background

Name	*Prefix*	*Research Field*	*Email*	*Personal website*
A F Campos Cedeno	Ph.D. in Technical Sciences, Associate professor	hydraulics, engineering hydrology, hydrotechnical construction, geographic information systems	acampos@utm.edu.ec	https://www.utm.edu.ec/
P A Salas Guillen	Graduated Student, Project assistant	hydraulics, engineering hydrology, hydrotechnical construction, topography	psalas@utm.edu.ec	https://www.utm.edu.ec/
J L Macias Ramos	Civil and Electrical Engineer, Associate professor	hydraulics, engineering hydrology, hydrotechnical construction, electrification	jlmacias@utm.edu.ec	https://www.utm.edu.ec/
E K Sinichenko	Ph.D. in Technical Sciences, Associate professor	hydraulics, engineering hydrology, hydrotechnical construction.	sinichenko_ek@pfur.ru	http://www.rudn.ru/
I I Gritsuk	Ph.D. in Technical Sciences, Associate professor	hydraulics, engineering hydrology, hydrotechnical construction	gritsuk_ii@pfur.ru	http://www.rudn.ru/

Water supply and demand of the hydrographic demarcation of Manabi, Ecuador

A F Campos Cedeno[1], X H Valencia Zambrano[1], C A Cevallos Castro[1], E K Sinichenko[2*] and I I Gritsuk[2,3]

[1]Universidad Tecnica de Manabi, Portoviejo, Ecuador
[2]Peoples' Friendship University of Russia (RUDN University), Moscow, Russia
[3]Russian Academy of Sciences Water Problems Institute, Moscow, Russia

Email: sinichenko_ek@pfur.ru

Abstract. Manabi is the only province in Ecuador that does not receive contribution of the runoff coming from the thaw of the snow-covered of the Andes mountain. This situation impacts severely on the agriculture activities. This study presents an estimate of water supply and demand of the Manabi hydrographic demarcation based on the hydro meteorological data and the water needs of the population by 2050. The research topic is very topical since it contributes to providing the necessary inputs for the planning and use of water resources in order to generate projects that tend to cover the current water deficit. The methodology used to estimate the water supply is based on the precipitation-runoff model proposed by the Soil Conservation Service of the United States of America. In fact, in spite of the annual volume of runoff has been estimated at 9658.83 hm^3, against 2388.93 hm^3, which is the average annual volume of water to meet the demands, from July to November it can be seen that the deficit reach at 330.27 hm^3. The results obtained constitute a valuable tool for specialists and governmental and non-governmental institutions responsible for the control, planning and development of water resources in the region.

Key words: hydrology, rainfall, runoff, water balance, planning and development

1. Introduction

At a global level, the subject of the study of water balances has been subject to various purposes and needs, which are related to the planning of water resources to meet consumptive and non-consumptive demands, for the control of water dynamics in the water. environment and for different purposes.

Among the many existing investigations, we have the work of Moreira et al "Assessment of terrestrial water balance using remote sensing data in South America". This study investigated the potential of assessing the water balance in South America, at multiple scales, using satellite remote sensing precipitation and evapotranspiration datasets, terrestrial water storage changes from Gravity Recovery and Climate Experiment, and discharge measurements [1].

In the field of agriculture we have the work of Kisekka et al "Uncertainties in leaching assessment in micro-irrigated fields using water balance approach", where water balance is a common approach used to estimate leaching in agricultural fields and was applied in this study to assess field-scale leaching and the associated uncertainties for an almond orchard under drip and micro-sprinkler irrigation systems [2].

In the work "Impacts of projected change in climate on water balance in basins of East Africa" was observed and large-scale climate variables (predictors) are obtained from the national meteorological agency of Ethiopia and international databases. BROOK90, a physical-based hydrological model, is used to assess the impacts of the projected change in precipitation and maximum and minimum temperature (T-max, and T-min) on the water balance [3].

Manabi is one of the largest provinces of Ecuador, which has an area of 18940 km^2 (6.70% of the national territory) and is considered "specialized" in agriculture compared to the other provinces of the country. The large size of the agricultural sector is the key to the development of the economy of Manabi. In addition, the agricultural sector is one of the most diverse in Ecuador, due to the size of the province, the stable climate and the topography of the lands suitable for cultivation. In Manabi an important series of agricultural products is cultivated, such as coffee, cocoa, bananas, citrus fruits (lemon, grapefruit, orange, tangerine and passion fruit), corn and rice. More than 14000 hectares of land for irrigation produce pepper (2000 ha), tomatoes (3000 ha), watermelon (2000 ha), cassava (1000 ha), fine lemon (1000 ha), pearl onion (3000 ha) and fruit of the passion (2000 ha) [4-5].

To guarantee agriculture in the central area of the province there are 2 irrigation systems. The first of 15000 ha has as source of supply the Poza Honda dam, and the second of 14250 ha, which is supplied by the dam "La Esperanza". The two systems are part of the so-called Chone-Portoviejo Integrated System. All the rivers of the hydrographic demarcations of Ecuador, excepting those of the demarcation of Manabi, receive contribution of the runoff coming from the thaw of the snow-covered ones of the mountain range of Andes. This situation has a negative impact on the development of the province of Manabi, especially in the field of agriculture, because rainfall is distributed non-uniformly in the dry and rainy periods throughout the year. The rainy period, from December to May, is accompanied by excess runoff with serious flooding, while, in the dry period, from June to November, due to the absence of rains, when the flow in rivers is less than 1 m^3/s, there is a large water deficit, since the rivers practically dry up. Modular minor flows vary between 1.99 and 15.87 l/s/km^2, compared with the other provinces, having input thaws Andes, where these flow rates above 80 l/s/km^2 [6].

The main rivers of Manabi, from north to south are: Coaque, Don Juan, Jama, Briceno, Chone, Grande, Carrizal, Chico, Portoviejo, Salado, Sancan, Jipijapa, Buenavista and Ayampe. Since 2011, the development of Ecuador's water resources is planned, not from its territorial-administrative base, but from its division by hydrographic demarcations (Figure 1). In accordance with this, 9 state entities have been created, called "demarcations", which work under the general direction of the Water Secretariat (SENAGUA). The total area of the Manabi hydrographic demarcation is 11484 km^2.

This study presents an estimate of the water supply and demand of the Manabí hydrographic demarcation on the basis of hydro meteorological data and the water needs of the population by 2050 [7-8].

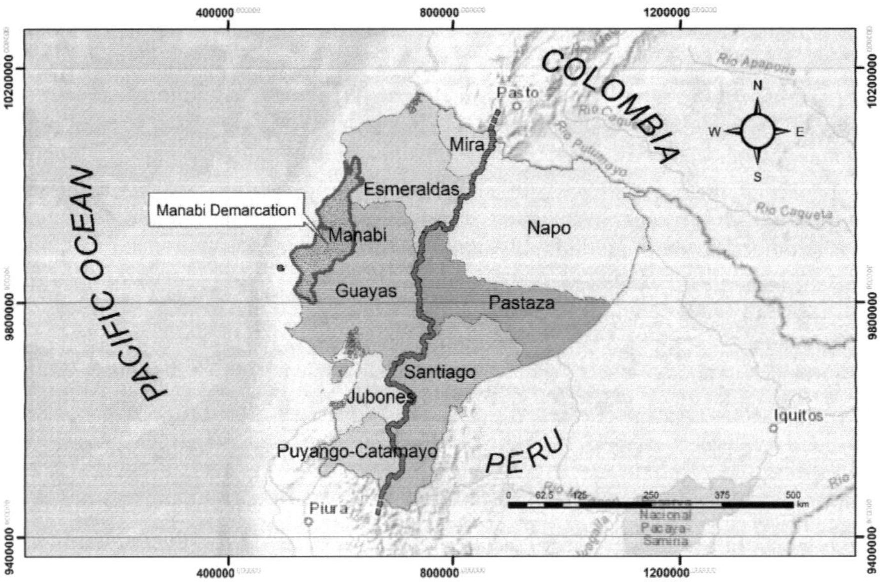

Figure 1. Hydrographic demarcations of Ecuador.

2. Materials and methods

For the execution of this work the following material was used: 1) Monthly rainfall of 34 meteorological stations located in the territory of the province of Manabi for the period 1963-2013 taken from the yearbooks and public institutional documents of the Institute of Hydrology and Meteorology (INAMHI); 2) Thematic digital GIS information of the Manabi hydrographic demarcation taken from the geoportals of the Water Secretariat (SENAGUA), the Military Geographical Institute (IGM) and the Ministry of Agriculture, Aquaculture and Fisheries (MAGAP) of Ecuador, 3) Water demands contained in the Plan for the Development of Water Resources of the Province of Manabi (PHIMA 1990).

In sequential order, the tasks that were carried out are: 1) Filling of missing data in the monthly rainfall records; 2) Estimation of multi-year average rainfall; 3) Distribution of monthly multi-year average precipitation; 4) Development of multi-year medium precipitation isohyet map; 5) Temporal distribution of annual rainfall; 6) Analysis of the variability of interannual precipitation; 7) Estimation of runoff and evapotranspiration by hydrological units of the Manabi hydrographic demarcation (DHM); 8) Estimation of the multi-year volume of water by precipitation (supply); 9) Estimation of annual average water demand; and, 10) Analysis of water balance: excess / deficit.

The main methods of approach and methodologies used to estimate the water supply and demand of the DHM, are sustained below.

For the filling of missing data in the monthly precipitation records for the period of analysis, the orthogonal correlation method was used, due to the fact that by applying this method, higher correlation coefficient values are obtained, giving greater reliability to the synthetic data generated.

On the basis of the digital files type shapes that were obtained through the geoportals of development institutions, such as: SENAGUA, INAMHI, IGM and MAGAP, with the help of specialized GIS software, digital elevation models with resolution of 3.00x3.00m, which were input, for the generation of important information for the achievement of the proposed objectives.

For the determination of the morphometry of the micro watersheds, digital elevation models with a resolution of 3,00 meters were used, facilitated by the MAGAP. Likewise, the aforementioned institution provided basic information regarding the vegetation cover and infrastructure works for the management of surface runoff in the zone.

Fifty-six sub-basins of analysis have been considered. The physiographic parameters of the micro watersheds were determined with the help of specialized software for the management of geographic information. The total area of study has been estimated at 11484 km^2.Through the Soil Conservation Service (S.C.S.) method, runoff is determined with the formula:

$$Q = \frac{(P - 0.2S)^2}{P + 0.8S} \tag{1}$$

where Q — Surface runoff height, mm; P — Rainfall, mm; S — Maximum potential difference between P and Q, mm.

The parameter S, the maximum potential difference between P and Q, can be determined through the so-called CN curve number, which characterizes the soil types and vegetation cover of the area. The calculation formula in metric units is as follows [9]:

$$CN = \frac{25400}{254 + S} \tag{2}$$

$$S = \frac{25400}{CN} - 254 \tag{3}$$

For the valuation of the curve numbers of the micro basins, was used the map of coverage and land use of the continental Ecuador 2013-2014, scale 1:100000, developed by the Ministry of Agriculture, Livestock, Aquaculture and Fisheries (MAGAP) with the Ministry of the Environment (MAE) [10].

Total evaporation, also called evapotranspiration, is a magnitude that includes transpiration, evaporation of soil and evaporation of water, contained in the stems and leaves of plants during rain. It can be estimated by the water balance method, with which the annual evapotranspiration standard E_0 equals rainfall x_0 minus runoff y_0 [11-12]:

$$E_0 = x_0 - y_0 \tag{4}$$

The multiannual volume of runoff by micro basin, which is the value corresponding to the annual average value of the water supply, is determined as the product of the runoff layer by the area of each of the hydrological units of the DHM, being the offer of the demarcation, the sum total of the partial values obtained.

The water demand was estimated on the basis of the Water Resources Development Studies of the Province of Manabi carried out in 1990 (PHIMA) by national institutions with cooperation of international organizations such as: Manabi Rehabilitation Center (CRM), National Institute of Water Resources of Ecuador (INERHI), National Development Council (CONADE) and the Organization of American States (OAS) [13].

The estimation of water demands was made on the basis of a projection of population growth until the year 2050. With this, the approximate population to the year 2020 was estimated at 2130000 inhabitants and total irrigated area at 51900.00 ha. Table 1 shows the demands considered in the PHIMA projects.

3. Results and discussion

With the complete series of the 34 rainfall stations, the average multiannual monthly rainfall for the period 1963 - 2013 was determined. After analyzing the results obtained, it is determined that the average monthly rainfall range between 0.96 - 229.75 mm. Of all the rainfall, for the dry season (June - November) corresponds 10.12%, and the remaining 89.88%, to the rainy season (December - May). This is illustrated in figure 2.

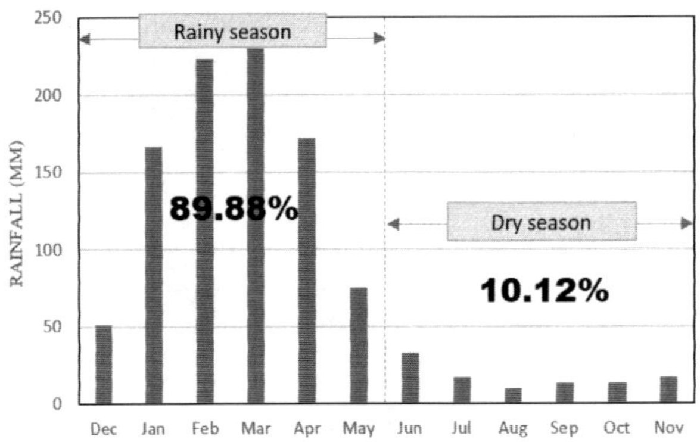

Figure 2. Distribution of average monthly rainfall.

The analysis shows that the average multi-year values of rainfall in Manabi vary from 200 mm in the west to 2800 mm in the east.

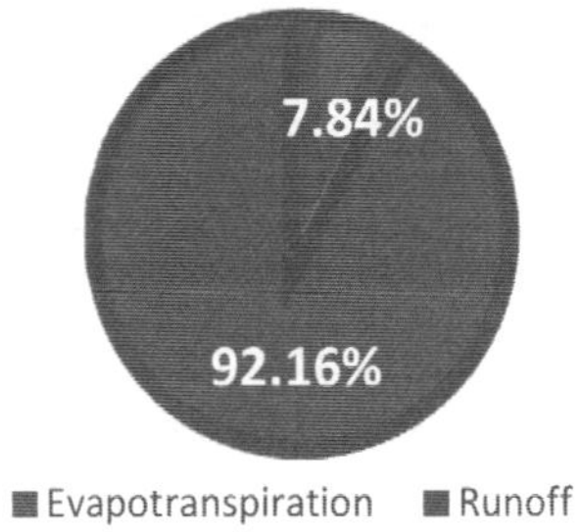

Figure 3. Distribution of average annual rainfall.

The total volume by evapotranspiration has been estimated at 822.12 hm^3 / year, which constitutes 7.84% of the total rainfall.

The annual average global volume by rainfall in the Manabi hydrographic demarcation (DHM) was estimated at 10480.95 hm^3, of which 9658.83 hm^3 correspond to runoff (92.16% of rainfall). The results of the average annual distribution of rainfall are shown in figure 3.

As a graphic result of the spatial analysis of the multi-year average rainfall, the isohyet map of the area was obtained by means of triangulations (figure 4).

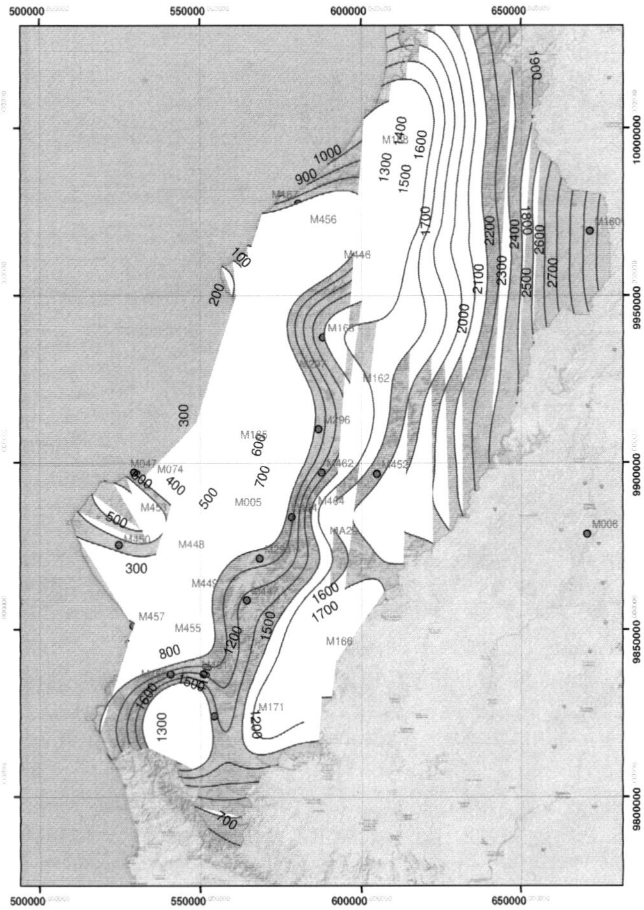

Figure 4. Isohyet map of province Manabi

4. Conclusions

1. The present study constitutes a diagnosis of the current situation of the hydrographic demarcation of Manabí in terms of supply and demand of water. The offer, based mainly on surface runoff, has been determined on the basis of consistent rainfall data of 51 years, while the demand includes needs to cover the consumption of potable water, industrial use, irrigation and hydroenergy generation.
2. According to the estimates made in 1989, the population growth of Manabi by 2020 amounted to 2130000 inhabitants, on the other hand, the projections of INEC 2010 to the year 2050, show a growth of 2085438 inhabitants, that is to say that there is a decrease, which is explained due to the global migratory processes. With these considerations, the average annual water demands of the province of Manabi for potable water and irrigation until the year 2050 amount to 2388.93 hm^3.
3. Considering a uniform monthly distribution of water demands for irrigation and AA.PP. and the percentage distribution of the average monthly rainfall, previously estimated, the general values of

excess and water deficit for the DHM were determined. With these results, on the one hand, the average annual volume of water to meet the demands is 2388.93 hm^3, on the other hand, the annual volume of runoff has been estimated at 9658.83 hm^3, meaning that the latter exceeds demand by 304%. At first glance, one might think that there is no deficit, but in fact there is, and it amounts to 330.27 hm^3. This is explained by the fact that there is a marked nouniformity in the distribution of annual runoff, especially in the dry season (June-November), when the percentage of monthly distribution drops to 0.96% in the month of August. The results are shown in the table 1.

Table 1. Water excess/deficit.

Month	Water demand	Runoff	Excess	Deficit
	(hm^3/year)			
Dec	199.08	481.92	282.84	-
Jan	199.08	1575.83	1376.75	-
Feb	199.08	2113.66	1914.58	-
Mar	199.08	2174.28	1975.20	-
Apr	199.08	1623.09	1424.01	-
May	199.08	713.05	513.97	-
Jun	199.08	311.88	112.80	-
Jul	199.08	160.07	-	39.01
Aug	199.08	92.50	-	106.58
Sep	199.08	126.50	-	72.58
Oct	199.08	124.18	-	74.90
Nov	199.08	161.87	-	37.21
Total	**2388.93**	**9658.83**	**7600.17**	**330.27**

4. Many times the decision making to allocate resources for the construction of hydraulic works must be based on solid diagnoses on the problem, and the results of this research lead to establish the need to project and build dams and reservoirs in the territory of the Hydrographic Demarcation of Manabí that allow to store in the rainy season the water and to use it in the dry periods, with which it would be giving solution to this oppressive and historical problem water shortage.
5. The results obtained constitute a valuable tool for specialists and governmental and non-governmental institutions responsible for the control, planning and development of water resources in the region.

5. References

[1] Moreira A A, et al. 2019 Assessment of terrestrial water balance using remote sensing data in South America *Journal of Hydrology* **575** pp 131-147
[2] Kisekka I, et al. 2019 Uncertainties in leaching assessment in micro-irrigated fields using water balance approach *Agricultural Water Management* **213** pp 107-115
[3] Gebrechorkos S H, Bernhofer C and Hülsmann S 2019 Impacts of projected change in climate on water balance in basins of East Africa *Science of The Total Environment* **682** pp 160-170
[4] Instituto Nacional de Estadisticas y Censos [Electronic Resource] *Como crecera la poblacion del Ecuador? Retrieved* 2015 Access mode: http://www.inec.gob.ec/proyecciones_poblacionales/presentacion.pdf
[5] Instituto Nacional de Estadisticas y Censos [Electronic Resource] *Poblacion y Demografía del Ecuador. Retrieved* 2015 Access mode: http://www.ecuadorencifras.gob.ec/
[6] Secretaria Nacional del Agua [Electronic Resource] *El Agua en el Ecuador* 2012 Access mode: http://agua-ecuador.blogspot.ru/2012/03/iniciamos-este-blog-para-compartir.html

[7] Campos A F, Sinichenko E K and Gritsuk I I 2014 Features of water mode of the rivers of the Province of Manabi (Ecuador) *RUDN Journal of Engineering Researches* **4** pp 41-46

[8] Campos A F, Sinichenko E K and Gritsuk I I 2015 Analysis of data of meteorological stations of the province of Manabi *RUDN Journal of Engineering Researches* **3** pp 41-49

[9] U.S. Army Corps of Engineers [Electronic Resource] *Hydrologic Modeling System* 2015 Access mode: http://www.hec.usace.army.mil/software/hec-hms/.

[10] Ministerio de Agricultura y Ganaderia, IICA, and CLIRSEN [Electronic Resource] *Sistemas de Informacion Geografica para Aplicaciones Agropecuarias en el Ordenamiento de Territorio y Manejo Integral de Cuencas* 2015 Access mode: http://www.uazuay.edu.ec/promsa/ecuador.htm.

[11] Aybaca Casar E J 1984 *Runoff from the rivers of Ecuador* PhD Thesis (Moscow State University, Moscow, Russia)

[12] Zheleznyakov L V, Legovskaya T A, Ovcharov E E 1984 *Hydrology, Hydrometry and Regulation of Runoff* (Moscow: Editorial KOLOS) 432 p

[13] Centro de Rehabilitacion de Manabi. Resumen Ejecutivo [Electronic Resource] *Plan Integral de Desarrollo de los Recursos Hídricos de la Provincia de Manabí (PHIMA)* 1989 Access mode: http://www.oas.org/usde/publications/Unit/oea41s/begin.htm

Exchanging traditional pumping unit for a "Geron" hydraulic drive

A N Drozdov*, S A Blokhin, D X Pak, G S Zemlianskii, M P Yurtov and A Y Antonenkova

Peoples' Friendship University of Russia (RUDN University), Moscow, Russia

Email: drozdov_an@pfur.ru

Abstract. This work was devoted to analysis of the efficiency of the application of a "Geron" hydraulic drive as an alternative to a beam pumping unit. The issue of replacement old drives for new one was considered in the context of economic profitableness taking into account field conditions of Russia and neighbor countries. To prepare this scientific article, different data about field experience of usage the machines in different countries and fields were asked for and obtained. Thanks to information got general characteristics of both machines with their advantages and disadvantages based generally on technological, energetical and economic data were composed. After this the drives were compared with each other. As a result, hydraulic drive "Geron" was taken to be a good alternative of traditional pumping unit.

1. Introduction

Oil companies were always interested in high level of fluid extraction. Out of the fact that modern extraction is mostly mechanical, pumps and drives play very important roles in production. Consequently, any offers of modernizations in this field are always hotly debated.

One of that offers is exchanging traditional pumping units for hydraulic drives. Some experts has already mentioned about prospects of this operation in own articles, for example, Karchagin V. [1] Molchanov A. [3,4], Blokhin S. [5], Kiselev A. V.[6], Grigoriev S. [7], Seduch A.[8].

The main purpose for this research was to determine if such replacements do interesting from the practical point of view.

Currently, one of the main trends in the development of pumping units is an increase in the stroke length of the suspension point of the rod string, however, in the implementation of "classical" mechanical schemes using a lever converting mechanism, such machines become enormous and inconvenient to maintain. Besides that they require creating large foundations. As practice shows, balancing pumping units with a stroke of more than 4.2 m are not used on an industrial scale, and the actual maximum stroke is about 3.5 m.

Hydraulic drives were originally developed in the United States as devices that provide stroke lengths of more than 6 m. Operation of such hydraulic actuators was successful, however, hydraulic actuators did not receive wide industrial use due to maintenance complexity, bulkiness, and low reliability. Attempts to repeat these structures using the domestic element base of the hydraulic drive also were not successful.

In the 70s-80s of the last century, reliable and effective hydraulic drives for oil production using rod pumps appeared. Simultaneously domestic production of parts of the hydraulic drive base began to be established: pumps, valves, seals, distributors, etc. Development G.V. Molchanov's hydraulically operated rocking chair, known under the code "AGN" was the first design that was mass-produced and began to be widely used in the fields.

The use of a tubing string as a balancing load was a distinctive feature of this installation, which, in combination with a volumetric hydraulic drive, made it possible to install it directly at the wellhead without a special foundation. Specially designed sealing devices of the hydraulic drive ensured reliability and compactness of installations, as well as the use of elements of the volumetric hydraulic drive developed and used in road machine-building engineering and construction.

In addition to large stroke lengths and reduced metal consumption, the labor intensity of installation maintenance has decreased; additional possibilities have appeared for adjusting the drive in order to strictly regulate the conditions in the rod string, maintain plant performance, etc. These advantages were made possible through the use of elements of hydraulic automation, and in some cases, microprocessors that control the apparatus of hydraulic actuators.

In addition to the mentioned drive, from the 60s of the last century to the present, many hydraulic drives have been developed. Studying their designs, it is easy to trace the path of search and development of fundamentally new for that time schemes, many of which are still the basis for modern equipment. Currently, about 100 units of hydraulic installations are in operation. Positive indicators of their work in combination with the continuous modification of the elemental base of the volumetric hydraulic drive indicate the prospects of this direction in the development of drives.

Lower at the picture 1 a principal scheme of modern hydraulic drive "Geron" can be seen.

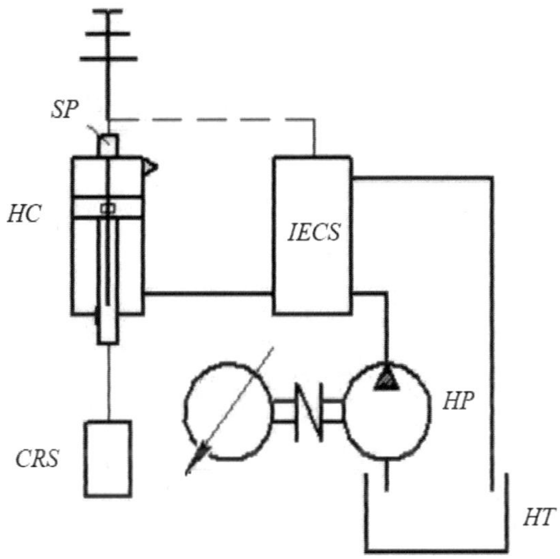

Figure 1. Principal scheme of modern hydraulic drive "Geron". HC - hydraulic cylinder; CSR - column of sucker rods with deep well pump; HP - hydraulic pump; HT - hydraulic tank; IECS - an intelligent electronic control system; SP - linear system for feedback position of the piston HC.

Hydraulically driven units can be particularly effectively used for trial well operation, well operation in remote and hard-to-reach areas located in marshy areas or ramps. But, as practice shows, fundamentally new equipment does not completely replace the old one.

Moreover, today a significant part of the oil well stock of the Russian Federation operates at the expense of pumping units. The tendency is unlikely to change in close future. Some specialists analyzed this problem, for instance technical director of the "Chernushkaneft" Vladimir Karchagin and leader of the "PSM-Impacs" Sergey Grigoriev, confirm prospects of hydraulic drives but at the same time say that such machines won't soon replace traditional equipment [1].

The beam-pumping unit is a machine whose design has been improved throughout the history of the oil industry. In the existing constructions of changes is not expected. Energy intensity and specific power of beam pumping unit are determined by strength properties of materials, changes of which are also not expected. At the same time, the resultativeness of modern machines of this type are rather low [2]. Previously successful technology fairly outdated is not able to provide high performance. For this reason, some specialists say that the installation of the pumping units on the wells can no longer be considered cost-effective. This work offers an objective look at things.

The thing is that the work of pumping units, as a mechanical drive of sucker-rod pump is characterized by the following main disadvantages:

- The pumping unit operation requires a special foundation that causes the inconvenience of mounting it;
- Changing the operating mode of the Hydraulic Drive for Deep Well Rod Pump "Geron" (HDDWRP) driven by the pumping unit is possible only when the well is completely stopped, since it is necessary to adjust the number of swings and the stroke length of the rod, which in turn requires changing the pulleys, shifting the crank fingers, and adjusting the counterweights;
- The need to hire maintenance personnel and operators.

The solution of these problems is proposed by the company "Konkov's Petro-Hydraulic drives". It engaged in manufacturing "Geron" - a hydraulic drive that allows, due to minor innovations, to significantly reduce power consumption, save on installation and operating personnel, change the mode without stopping the operation of the well. Its image is presented in figure 2 [5].

Figure 2. The Hydraulic Drive for Deep Well Rod Pump "Geron".

Hydraulic drives "Geron" have already found application in the fields of many countries, including Russia, Turkmenistan, Kazakhstan, Colombia. On the territory of the Russian Federation, in the period from 2015 to 2017, companies such as "Orenburgneft", "RN-Nyaganneftegaz", "TATOYLGAZ", "Kara Altyn", "Turkmennebit Group of Companies" have tested and adopted. In a number of foreign countries, trial operation is planned in the near future. Figure 3 shows the geography of the use of hydraulic drives "Geron".

2. Methods of research

The main method of research was comparison. Results of hydraulic drive field test were analyzed and comparised with data from design documentation. According to it technical and economic efficiency of «Geron» was estimated and some conclusions made.

To form in readers' minds the real image of «Geron» here, in this chapter, full information about hydraulic drive testing sample was given.

Not so long ago in 2018 pilot tests were performed for "RN-Nyaganneftegaz". A comparative analysis of "Geron" and pumping unit will be conducted according to this recent project.

First of all, it should be said that for this company, HDDWRP -140-6.0 was chosen as a sample for testing with electricity. According to the project, this drive should provide:

- The operation of one or several submersible borehole pumps with a wellhead loading up to 14 tons;
- Uninterrupted operation of the facility represented by the well stock of the electric centrifugal pump unit with a liquid flow rate of up to 30 m^3 / day, characterized by a periodic mode of operation.

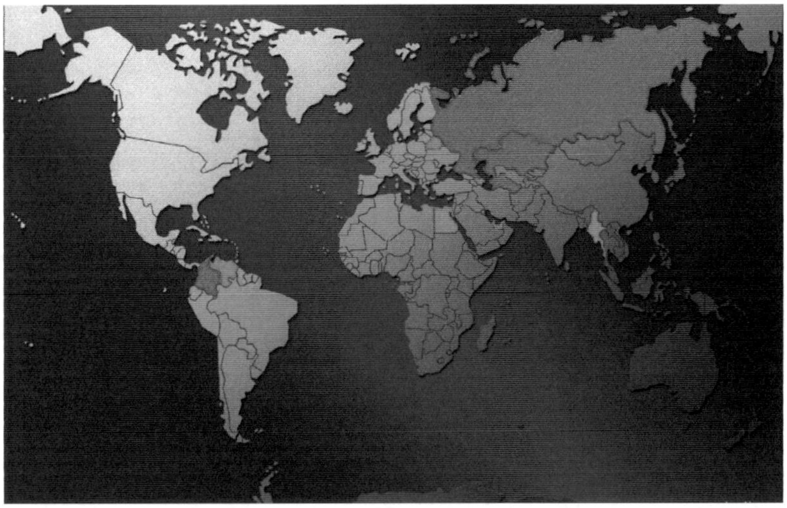

Figure 3. The geography of the use of the "Geron" by countries (red color is already used; blue - trial operation is underway / trial operation is planned).

The need for the implementation of the "Geron" in the company arose due to a decrease in oil recovery at the stock of wells equipped with electric centrifugal pump unit and the desire to transfer wells from the periodic mode (34.9%) to the permanent mode of operation.

In this regard, several main objectives of pilot field operation were identified:

- Confirmation of the operational reliability of HDDWRP 140x6-0A during operation of the drive-in field conditions of the "RN-Nyaganneftegaz";
- Qualitative verification of the stated reduction in electricity consumption;
- Identification of the intellectual capabilities of the "Geron";
- Achievement of key performance indicators (no downtime due to equipment failure);
- Reduction prevention of fluid production of candidate wells;
- Reduction of specific electricity consumption (kW) per 1 m3 of produced fluid in comparison with the previous method of operating a candidate well by 10%;
- Simplicity and ease of maintenance.

The main technological advantages provided by the "Geron" over the base variant / alternative technical solution are: the possibility of operating a low flow well stock in a constant mode of the sucker-rod pumping unit at depths of over 2000 meters; the ability of this drive to convert the energy of the weight of the pump rods into electrical energy and return it to the supply network (when the drive rod moves down) due to the use of a frequency converter with a recovery function and a motor-generator providing operation at nominal load on the wellhead up to 140 kN (14 mc).

The disadvantage compared with the basic version / alternative technical solution was only the lack of confirmation of the resource reliability of the tested equipment in the field conditions of the "RN-Nyaganneftegaz".

Tests of the hydraulic drive "Geron" were conducted from November 2017 to May 2018. HDDWRP -14O-6.0A "Geron" with a standard weight changer of the type RRS-18 which was mounted on a dedicated well No. 5515, hive No. 15 of the Sever-Talinka Strengthened oilfield assets and put into operation with the following parameters:

- Depth of the submersible pump: 2450 m;
- The number of double moves was set to: 2.5 per minute;
- The stroke length of the hydraulic cylinder rod: 4 m;
- Load on the hydraulic cylinder rod during lifting: 9–9.5 t;
- Load when lowering on the hydraulic cylinder rod: 4 - 5 tons.

3. Results and discussion

During the operation of the hydraulic drive, it has been established that the drive complies with the design documentation and the declared technical parameters, namely:

1. The noise level does not exceed 65 dB;
2. The number of double strokes can vary widely depending on the stroke of the hydraulic cylinder rod in the range from 0.2 to 4 strokes per minute;
3. The smoothness of the course is regulated continuously by an electronic control system;
4. Automatic switching on - turning off the heating and cooling systems of the working fluid according to the temperature sensor readings;
5. Continuous full-flow filtration of working fluid with control of clogging of filter elements;
6. Automatic start-up after shutdown - power on with saving settings;
7. Display on the display of the current operating parameters of the hydraulic drive: the temperature of the working fluid, the load on the rod, the operating pressure in the system, the current time, the number of double strokes, the total operating time of the drive in hours, the reason for stopping the equipment, the passage of position sensors (upper and lower);
8. Display of the dynamogram at the request of the operator with a memory of up to 20 dynamograms.

During the testing period, the following happened:

- 01.02.2018 – a technical service was carried out by the representatives of the "Konkov’s Petro-Hydraulic drives";
- 03.07.2018 - equipment shutdown due to metal rupture through the body of the axle hanger HDDWRP -140-6.0A, which led to the bending of the polished rod and the hydraulic cylinder shaft HDDWRP -140-6.0A (total simple well amounted to 325 h. 12 min., because of the delay in admission of workers of the "Konkov’s Petro-Hydraulic drives" of a simple well due to the fault of the "Konkov’s Petro-Hydraulic drives" 19 hours 45 minutes).

On 08.05.18 pilot field tests HDDWRP -140-6,0A produced by "Konkov’s Petro-Hydraulic drives" from the moment of the beginning amounted to 181 days (the actual hours worked including stops were 167 days; operating ratio - 0.93).

During the reporting period, there was one failure of controllable equipment which led to well shutdown because of the RRS-18 produced by "Neftekh" failed.

There was no significant reduction in the specific electricity consumption (kW) per 1 m3 of the produced fluid during operation of the hydraulic drive of the submersible sump pump HDDWRP -140-6.0 compared with the previous method of operating the candidate well.

Table 1. The test results.

Number of test	Period of time	Type of equipment	$Q_{liq.}$ m^3/day	Energy used kW* day	Energy used, kWh/day	Specific energy used kW/1m^3
1	23.09-25.09.16	ECP 50/2400, SEM 45-117 (CES)	7	118,2	4,9	16,9

2	28.01-01.02.17	HDDWRP -140-6,0	9	547	21,4	60,8
3	28.01-01.02.17	HDDWRP -140-6,0 (recuperation)	9	537,5	21	59,7
4	01.05-01.06.17	HDDWRP - 140-6,0 (recuperation)	8	483	20,1	60,4
5	01.06-06.06.17	HDDWRP -140-6,0 (recuperation)	7	459	19,2	65,6
6	06.09-09.09.17	HDDWRP -140-6,0 (recuperation)	8	491	20,4	61,2
7	09.11.17 08.05.18	HDDWRP -140-6,0 (modernized with recuperation	8	285	11,9	35,6

Thanks to data shown technical and economic efficiency was estimated:

1. There was 1 stop 07.03.2018 (after 118 days of continuous operation of the drive) due to the failure of the weight tool, the test goal is 0, the proportion in the estimate is 40%, if 1 stop is considered, then the estimated value will decrease to 0.9.

2. During the specified period of testing, the volume of produced fluid was 8 m3 (based on acts for the period of operation), the goal was achieved, the proportion was -30%, the estimated value was 1.

3. Reduction of specific electricity consumption to 14 kW per 1m3 of produced fluid, when operating a sucker-rod well pump for hydraulic HDDWRP -140-6.0 compared with the previous method of operating a candidate well: Target value - 10%, estimated value - 0, specific weight -10%.

4. Simplicity and ease of maintenance of ground equipment (drive of a sucker rod pump of a hydraulic pump HDDWRP -140-6,0 manufactured by "Konkov's Petro-Hydraulic drives"): estimated value - 1, specific gravity - 20%.

The percentage of efficiency is defined as the sum of the values of the product of the estimated coefficient by the specific weight, for each indicator.

Total: 0.9 * 40 + 1 * 30 + 0 * 10 + 1 * 20 = 86%

4. Conclusion

The obtained results indicate that the equipment HDDWRP -140-6.0 produced by "Konkov's Petro-Hydraulic drives" can be considered a worthy alternative to the pumping unit.

At the same time, it is worth noting the prospect of reducing the cost of electricity, since during the pilot operation at other facilities, "Geron" proved to be very worthy in this regard. Thus, for example, at the pilot operations in the "Orenburgneft", the measurements of consumed and delivered electricity were confirmed by the manufacturer's stated functions of returning electricity to the network, thereby ensuring a reduction in the cost of electricity consumption in the amount of 13.4 thousand rubles / month.

According to the obtained results, the specialists of the "Orenburgneft" developed a program for the implementation of equipment, according to which in 2015 ten such installations were installed [8].

5. References

[1] *Karchagin V and Grigoriev S 2014 Gidravlicheskiy privod vs stanok-kachalka i tsentrobezhnyy nasos [Hydraulic drive vs machine-swing and centrifugal pump]* (Moscow: Neftegaz №10) pp 74-75

[2] Molchanov A G 2007 *Stanki-kachalki: problemy i perspektivy sovershenstvovaniya [Rocking machines: problems and prospects for improvement]* (Moscow: Industrial lists №10) / URL: http://www.promved.ru/articles/article.phtml?id=1298&nomer=47

[3] Molchanov A G 2014 Puti dal'neyshego sovershenstvovaniya shtangovykh skvazhinnykhnasosnykh ustanovok [Ways of further improvement of sucker-rod pumping units] (Moscow: Drilling and oil № 2) p 3-10

[4] Molchanov A G, Pevnev V G and Tarasov K V 2013 *Gidravlicheskiy privod shtangovogo skvazhinnogo nasosa s inertsionnym uravnoveshivaniyem [Hydraulic drive of a sucker-rod well pump with inertial balance]* (Moscow: Territory of oil and gas № 4) pp 52-55

[5] Blokhin S A 2015 *Neft, zoloto, demantoidi [Oil, gold, demantoids]* (St. Petersburg: Innovative Russia №12) p 31-33

[6] Kiselev A V, Starostenko E V and Sukhinov A V 2009 *Efficiency of applying a hydraulic drive for a well sucker-rod pump installation [Effektivnost'primeneniya gidravlicheskogo privoda ustanovki skvazhinnogo shtangovogo nasosa]* (Moscow: Oil Industry №11) pp 60-61

[7] Grigoriev S L and Demidov O V 2010 *Gidroprivod shtangovogo skvazhinnogo nasosa - novaya tekhnologiya dobychi nefti [Hydraulic drive of a sucker-rod well pump - a new technology of oil production]* (Moscow: Territory of oil and gas №10) p16

[8] Sedikh A A, Stepanov Yu G and Abdulin A F 2016 *Rezul'taty provedeniya opytno promyslovykh ispytaniy gidroprivoda shgn s funktsiyey rekuperatsii elektroenergii na fonde pao «orenburgneft'» [The results of the experimental field test SRP hydraulic drive with electric power recovery function on the foundation of PJSC "Orenburgneft"]* (Moscow: Engineering Practice №3) p 44

IOP Conf. Series: Materials Science and Engineering **675** (2019) 012023 doi:10.1088/1757-899X/675/1/012023

Water and gas mixture characteristics calculation and its possible injection variants

A N Drozdov*, I M Narozhnyy and A M Mereutsa

Peoples' Friendship University of Russia (RUDN University), Moscow, Russia

Email: drozdov_an@pfur.ru

Abstract. The aim of this article is the practical application of calculation methodology of water and gas volume injection into the reservoir and optimal parameters of water and gas mixture in Samodurovskoe field in Russia. Calculations were performed using two technologies: water alternating gas (WAG) and simultaneous water and gas injection (SWAG). For each case the optimum volume of water and gas mixture and its components which are injected into the reservoir was calculated. The variant of alternative water and gas injection cannot be applied in oilfield because of existing restriction to maximum flowrate of associated gas in standard conditions. In Samodurovskoe field can be applied water-alternated-gas (WAG) technology. Proper array of gas and water ration phases in reservoir condition with account of associated gas usage was calculated. Optimization of water-gas mixture volume ratio components showed that the biggest effectiveness of oil recovery cannot be reached resulting from abovementioned restriction. This calculation methodology can be used in job execution, it is simple and practical.

1. Introduction

The methods of enhanced oil recovery, practical usage of associated gas, ecological safety are essential problems of present-day oil industry in Russia and abroad. For today there is a tendency to deterioration of oil field development conditions, it relates to depletion of reservoir with easily recoverable oil reserves. In such conditions conventional flooding won't be effective. For this reason, the interest grows to use technologies which increase effectiveness of field development such as alternated-gas injection (WAG and SWAG) [1-12]. To estimate effectiveness of this technology and practical application the appropriate calculations should be made. There are three variants of alternated-gas injection implementation:

- Variant 1. Long water alternating gas (WAG).
- Variant 2. Alternate water alternating gas (WAG).
- Variant 3. Simultaneous water and gas injection (SWAG).

2. Materials and methods

Calculations of optimal parameters for the implementation of WAG, the optimal volume of water-gas mixture pumped into the reservoir and its constituents (water and gas phases) were carried out according to the methods described in the monograph [1] with the change of symbols of some variables. The calculated data on the Samodurovskoye field were taken from the work [2], also by the procedure from this work the bottom-hole pressure of injection well was calculated. In techniques several generally accepted formulas for more closer definition of calculations were added.

3. Results

The data for calculation for the Samodurovskoye field is presented in table 1.

Table 1 The data for calculation for the Samodurovskoye field

Parameter	Designation	Measure	Value

A number of injection wells	N_{skv}	w.	11
A well depth	H_{skv}	m.	1796.6
An inner diameter of production string	D_{vn}	m.	0.06
A water flow in the injection well during the implementation of the reservoir pressure maintenance system using conventional technology	Q_{vV}	m³/d	130
A volume of water injected into the reservoir during the implementation of the reservoir pressure maintenance system using conventional technology	$Q_{vV\ VGV\ pl}$	m³/d	1535
A reservoir pressure	P_{pl}	MPa	16
An injection wellhead temperature	T_y	K	268
An injection bottom-hole temperature	T_{zab}	K	304
A reservoir temperature	T_{pl}	K	306
A reservoir density of oil	$\rho_{n\ pl}$	kg/m³	854
A density of oil in standard conditions	ρ_n	kg/m³	863
A water-mass density in reservoir conditions	$\rho_{v\ pl}$	kg/m³	1158
A water-mass density in standard conditions	ρ_v	kg/m³	1161
A gas density in standard conditions	ρ_g	kg/m³	1.15
A formation volume factor	B_o	unit fraction	1.030
A gas oil ratio	G	m³/t	18.9
A maximum value of the flow of associated gas in standard conditions	$Q_{g\ max}$	m³/d	60274

3.1. Calculation of optimal parameters for the implementation of WAG

In the implementation of WAG on the current site to compensate for the selection, is adopted by the existing reservoir pressure maintenance system, the volume of pumped water, which, in terms of gas, in reservoir conditions, will be

$$Q_{gV\ VGV_{pl}} = b_{V\ pl}\, Q_{vV\ VGV}\ , \text{m}^3/\text{d}.,$$

where $b_{v\ pl}$ – the coefficient of cubic expansion of water, in reservoir condition $b_{v\ pl} \approx 1$,

$$Q_{gV\ VGV_{pl}} = Q_{vV\ VGV}\ .$$

The volume of the injected gas in standard conditions is determined from the formula

$$Q_{gV\ VGV} = \frac{Q_{gV\ VGV_{pl}}\, P_{pl} T}{P T_{pl}\, Z_{pl}}\ , \text{m}^3/\text{d}., \tag{1}$$

where $Q_{vV\ VGV}$ – the volume of water injected into the reservoir during the implementation of the reservoir pressure maintenance system using conventional technology in standard conditions, m³/d.;

P_{pl} – reservoir pressure, MPa; T_{pl} – reservoir temperature, K; P and T – pressure and temperature in standard conditions, P = 0.1 MPa, T = 293 K; Z_{pl} – z-factor in reservoir conditions.

By placing the T1 reservoir basic data we will get

$$Q_{gV\ VGV} = \frac{1535 \times 16 \times 293}{0.1 \times 306 \times 0.8179} = 287\ 524\ \text{m}^3\text{/d}.$$

To determine Z_{pl} can be used approximation formulas of P.D. Lyapkov:

$$P_{pr} = \frac{P}{P_{kr}} = \frac{16}{4.5} = 3.5\ \text{, MPa},$$

$$T_{pr} = \frac{T}{T_{kr}} = \frac{306}{190.7} = 1.6\ \text{, K},$$

$$\text{for } 0 < P_{pr} \leq 3.8 \text{ and } 1.17 \leq T_{pr} < 20$$

$$Z_{pr} = 1 - P_{pr}\left[\frac{0.18}{T_{pl}-0.73} - 0.135\right] + \frac{0.016 {P_{pr}}^{3.45}}{{T_{pr}}^{6.1}},$$

$$Z_{pr} = 1 - 3.5 \times \left[\frac{0.18}{1.6-0.73} - 0.135\right] + 0.016 \times \frac{3.5^{3.45}}{1.6^{6.1}} = 0.8179.$$

The resulting volume of gas does not correspond to the maximum allowable value of the volume injected into the reservoir $Q_{gV\ VGV} > Q_{g\ max}$, therefore this embodiment of the WAG is not suitable.

In implementing WAG technology essential injection well head pressure is:

$$P_y = \sqrt{\frac{{P_{zab}}^2 - \theta {Q_{gV\ VGV}}^2}{e^{2S}}}\text{, MPa,} \tag{2}$$

where P_{zab} – bottomhole pressure of the injection well, MPa; $Q_{gV\ VGV}$ – gas flow in standard conditions, m³/d., determine from (1); θ – parameter, from formula:

$$\theta = 1.19 \times 10^{-6} \lambda\, \frac{{Z_{sr}}^2 {T_{sr}}^2 (e^{2S}-1)}{{D_{vn}}^5}, \tag{3}$$

where D_{tv} – inner diameter of production string, m; λ – hydraulic resistance coefficient of gas current in production string, accepted λ = 0.025; e – base of the natural logarithm, e = 2.7; S – dimensionless parameter, by formula:

$$S = \frac{0.0314 \bar{\rho}_g H_{skv}}{Z_{pl} T_{sr}}, \tag{4}$$

where $\bar{\rho}_g$ – gas relative density:

$$\bar{\rho}_g = \frac{\rho_{n\ pl} B_o - \rho_n}{G},$$

where H_{skv} – well depth, m; T_{sr} – average temperature, K, determined as:

$$T_{sr} = \frac{(T_y + T_{zab})}{2}\text{, K.}$$

Offered calculation methodology allows to determine the optimal volume of gas and its pressure on wellhead, providing its injection into reservoir in this volume and use this data to assess the optimal volume and time limits of the technology implementation.

3.2. The method of calculating the optimal volume of injected into the reservoir water-gas mixture and its components (water and gas phases) during the implementation of SWAG [1]

Effective oil displacement will be done in case of gas and water ration phases in reservoir under the condition:

$$k_{VGV\ pl} = \frac{Q_{g\ VGV\ pl}}{Q_{v\ VGV\ pl}} = 0.2 \div 0.5.$$

A calculation is made by next formulas:

$$Q_{vV\ VGV\ pl} = \frac{Q_{vV\ comp}}{1+k_{VGV\ pl}}\ , \text{m}^3/\text{d.}, \tag{5}$$

$$Q_{gV\ VGV\ pl} = Q_{comp} - Q_{v\ VGV}\ , \text{m}^3/\text{d.}, \tag{6}$$

$$Q_{gV\ VGV} = Q_{gV\ VGV\ pl}\frac{P_{pl}T}{PT_{pl}\,Z_{pl}}\ , \text{m}^3/\text{d.} \tag{7}$$

In implementing SWAG at the site of the Samodurovsky field under this condition $k_{VGV\ pl} = 0.2$:
- injected water consumption by (5)

$$Q_{vV\ VGV\ pl} = \frac{1535}{1+0.2} = 1279.2\ \text{m}^3/\text{d.},$$

- consumption of injected gas under SWAG in reservoir conditions by (6)

$$Q_{gV\ VGV\ pl} = 1535 - 1279.2 = 255.8\ \text{m}^3/\text{d.},$$

- consumption of injected gas under SWAG in standard conditions by (7)

$$Q_{gV\ VGV} = 255.8 \times \frac{16\times293}{0.1\times306\times0.8179} = 47\,914.5\ \text{m}^3/\text{d.}$$

Under $k_{VGV\ pl} = 0.3$:
- injected water consumption by (5)

$$Q_{vV\ VGV\ pl} = \frac{1535}{1+0.3} = 1180.8\ \text{m}^3/\text{d.},$$

- consumption of injected gas under SWAG in reservoir conditions by (6)

$$Q_{gV\ VGV\ pl} = 1535 - 1180.8 = 354.2\ \text{m}^3/\text{d.},$$

- consumption of injected gas under SWAG in standard conditions by (7)

$$Q_{gV\ VGV} = 354.2 \times \frac{16\times293}{0.1\times306\times0.8179} = 66\,346\ \text{m}^3/\text{d.}$$

Under $k_{VGV\ pl} = 0.4$:
- injected water consumption by (5)

$$Q_{vV\ VGV\ pl} = \frac{1535}{1+0.4} = 1096.4\ \text{m}^3/\text{d.},$$

- consumption of injected gas under SWAG in reservoir conditions by (6)

$$Q_{gV\ VGV\ pl} = 1535 - 1096.4 = 438.6\ \text{m}^3/\text{d.},$$

- consumption of injected gas under SWAG in standard conditions by (7)

$$Q_{gV\ VGV} = 438.6 \times \frac{16\times293}{0.1\times306\times0.8179} = 82\,155.1\ \text{m}^3/\text{d.}$$

Under $k_{VGV\ pl} = 0.5$:
- injected water consumption by (5)

$$Q_{vV\ VGV\ pl} = \frac{1535}{1+0.5} = 1023.3\ \text{m}^3/\text{d.},$$

- consumption of injected gas under SWAG in reservoir conditions by (6)

$$Q_{gV\ VGV\ pl} = 1535 - 1023.3 = 511.7\ \text{m}^3/\text{d.},$$

- consumption of injected gas under SWAG in standard conditions by (7)

$$Q_{gV\ VGV} = 511.7 \times \frac{16\times293}{0.1\times306\times0.8179} = 95\,847.6\ \text{m}^3/\text{d.}$$

Summary data of calculations is represented in the table 2.

Table 2. Summary data of calculations while applying SWAG.

Parameters	Value when $k_{VGV} = Q_{gV\ pl} / Q_{vV}$

	0	0.2	0.3	0.4	0.5
The volumes of components of the water-gas mixture injected into the reservoir:					
- a water in reservoir conditions, m³/d.	1535	1279.2	1180.8	1096.4	1023.3
- a gas in standard conditions, m³/d.	0	47914.5	66346	82155.1	95847.6
million. m³/ year	0	17	24	30	35

Taking into account maximum value of the flow of associated gas in standard conditions $Q_{g\ max}$ = 60274 m³/d., it can be concluded that Effective oil displacement will be done in case of gas and water ration phases in reservoir condition in a range of $0.2 \leq k_{VGV\ pl} \leq 0.265$.

3.3. The calculation of the pressure at the bottom of the injection well.

$$P_{zab} = \left(P_y + \rho_v g H_{skv} - \Delta P_{tr}\right) \times 10^{-6} \text{, MPa,} \tag{8}$$

$$\Delta P_{tr} = \lambda \frac{H_{skv} V^2}{2\times10^6 D_{vn}} \rho_в \text{, MPa,} \tag{9}$$

where λ – hydraulic resistance coefficient, determined by Altshul formula

$$\lambda = 0.11 \left(\frac{68}{Re} + \frac{\Delta}{D_{vn}}\right)^{0.25}, \tag{10}$$

Re – Reynolds number

$$Re = \frac{V\, D_{vn}\, \rho_v}{\mu_v}, \tag{11}$$

V – water velocity in production string

$$V = \frac{4\, Q_{skv\ v}}{86400\, \pi\, D_{vn}{}^2} \text{, m/s,} \tag{12}$$

$\mu_в$ – dynamic viscosity of stratum water, Pa*s. It can be determined by P.D. Lyapkov's formula [2]

$$\mu_v = \frac{0.0014+38\times10^{-7}(\rho_v-1000)}{10^{0.0065\times t}} \text{, Pa*s.} \tag{13}$$

Calculations of friction pressure drop will be made. At first, μ_v will be found by formula (13), by taking temperature equal to t = 20 °C.

$$\mu_v = \frac{0.0014+38\times10^{-7}(1158-1000)}{10^{0.0065\times20}} = 1.48 \times 10^{-3} \text{ Pa*s.}$$

Calculate *V* by formula (12)

$$V = \frac{4\times130}{86400\times3.14\times 0.06^2} = 0.5324 \text{ m/s.}$$

Reynolds number by formula (11)

$$Re = \frac{0.5324\times0.06\times1158}{1.48\times10^{-3}} = 24994\,.$$

Hydraulic resistance coefficient by formula (10), taking the value of Δ=0.003 m

$$\lambda = 0.11 \left(\frac{68}{24994} + \frac{0.003}{0.06}\right)^{0.25} = 0.05271\,.$$

Substitute the value in formula (9), friction pressure loss will be found ΔP_{tr}

$$\Delta P_{tr} = 0.05271 \frac{1796.6\times0.5324^2}{2\times10^6\times0.06} \times 1158 = 0.2590 \text{, MPa.}$$

Bottom hole pressure P_{zab} in injection well will be found by (8)

$$P_{zab} = (9.5 + 1158 \times 9.81 \times 1796.6 - 0.2590) \times 10^{-6} = 20.4 \text{, MPa.}$$

To determine pressure on wellhead in implementation of SWAG, calculations of generalized hydrodynamic pressure gradient in production string.

$$P_y = P_{zab} - \sum_1^{m-1} \left(\frac{dp}{dh}\right)_m h_{sh} \text{, MPa,} \tag{14}$$

where $(dp/dh)_m$ – generalized hydrodynamic pressure gradient in the 1st calculated section of tubing, MPa/m; h_{sh} – calculation step, m; m – any calculated section or number of calculated sections in production string from 1 to m; P_{zab} – bottom hole pressure in injection well, MPa.

3.4. Optimization of water-gas mixture volume ratio components, which is injected in a reservoir in implementation of WAG or SWAG.

Density of water-gas mixture, which is injected in volume of injection to withdrawal ratio in reservoir conditions, is determined from formula:

$$\rho_{GVS\,pl} = (\rho_{g\,pl}\, Q_{gV\,VGV\,pl} + \rho_{v\,pl}\, Q_{vV\,VGV\,pl})/Q_{sm\,V\,pl} \text{, kg/m}^3\text{,} \tag{15}$$

where $\rho_{v\,pl}$ – water density in reservoir conditions, kg/m^3 , is allowed to equal water density in standard conditions, that is $\rho_{v\,pl} = \rho_v$; $\rho_{z\,pl}$ – density of hydrocarbon gas in reservoir conditions will be found next:

$$\rho_{g\,pl} = \rho_g \frac{P_{pl}T}{PT_{pl}Z_{pl}} \text{, kg/m}^3\text{,} \tag{16}$$

$$\rho_{g\,pl} = 1.15 \frac{16 \times 293}{0.1 \times 306 \times 0.8179} = 215.4 \text{ kg/m}^3\text{,}$$

substitute the value $\rho_{g\,pl}$ in formula (15), the density of water-gas mixture will be found

$$\rho_{GVS\,pl} = \frac{215.4 \times 255.8 + 1161 \times 1279.2}{1535} = 1003.4 \text{ kg/m}^3\text{,}$$

$$\rho_{GVS\,pl} > \rho_{n\,pl}\,.$$

Water-alternated-gas injection effectiveness increase can be achieved by means of usage of this technology in the quality of water-gas mixture displacement working agent in density equal to the oil which is displaced in reservoir conditions

Density of water-gas mixture in reservoir conditions, which is equal to the oil which is displaced in reservoir conditions, is achieved by the means of providing in the flow of consumption:

— gas in reservoir conditions, determined by formula which is received by the way of solution relative to $Q_{gV\,VGV\,pl}$, trying to find $\rho_{GVS\,pl} = \rho_{n\,pl}$

$$Q_{gV\,VGV\,pl} = \frac{Q_{sm\,V\,pl}(\rho_{v\,pl} - \rho_{n\,pl})}{\rho_{v\,pl} - \rho_{g\,pl}} \text{, m}^3\text{/d.,} \tag{17}$$

$$Q_{gV\,VGV\,pl} = \frac{130 \times (1161 - 854)}{1161 - 215.4} = 42.2 \text{ m}^3\text{/d.,}$$

— water in reservoir conditions, determined by formula:

$$Q_{vV\,VGV\,pl} = Q_{sm\,V\,pl} - Q_{gV\,VGV\,pl} \text{, m}^3\text{/d.,} \tag{18}$$

$$Q_{vV\,VGV\,pl} = 130 - 42.2 = 87.8 \text{, m}^3\text{/d.,}$$

where $Q_{sm\,V\,pl}$ – consumption of water-gas mixture, which is injected in a reservoir in the process of WAG implementation, in reservoir conditions, m^3/d., accepted $Q_{sm\,V\,pl} \approx Q_{vV\,comp}$; $\rho_{v\,pl}$ – density of water in reservoir conditions, kg/m^3, assume that $\rho_{v\,pl} \approx \rho_v$; ρ_v – density of water in standard conditions, kg/m^3,

$$k_{VGV\,pl} = \frac{Q_{gV\,VGV\,pl}}{Q_{vV\,VGV\,pl}} = \frac{42.2}{87.8} = 0.48\,.$$

That is as $k_{VGV\,pl} = 0.48$ maximum effectiveness of water-alternated-gas injection will be achieved.

4. Discussion

Usage of alternated-gas injection has remarkable advantages over other advanced recovery methods:

— increasing of conformance factor and oil recovery factor;

— rational utilization of associated gas.

With the purpose of successful practical usage of this technology, it is necessary to calculate volumes of water and gas injected into the reservoir, determine the appropriate injection option. That's the reason of necessity in selecting a convenient method of such parameter's calculation. The data which was obtained in the present article shows the ability of described methodology usage and the possibility of water-alternated-gas (WAG) technology realization in Samodurovskoe field.

Conclusions

For conditions of different reservoirs is necessary to select optimum parameters while injecting water and gas. Proceeding from Samodurovsky field calculations, next conclusions can be made:

— volume of gas consumption in alternative injection of water and gas in reservoir falls short of maximum flowrate condition of associated gas in standard conditions;

— taking into account that $Q_{g\ max}$ = 60274 m^3/d., effective oil displacement will occur in correlation of gas and water phases in reservoir conditions ranging from 0.2 to 0.265.

References

[1] Mullaev B T 2018 Proyektirovaniye i optimizatsiya tekhnologicheskikh protsessov v dobyche nefti [Design and optimization of technological processes in oil production] (Almaty: Andronum Publishing House) p 534

[2] Drozdov N A 2011 Issledovaniye fil'tratsionnykh kharakteristik pri vytesnenii nefti vodogazovymi smesyami i razrabotka tekhnologicheskikh skhem nasosnoezhektornykh sistem dlya vodogazovogo vozdeystviya na plast [Investigation of filtration characteristics in the oil displacement of water-gas mixtures and development process flow of pump-ejecting systems for SWAG injection] PhD dissertation (Moscow: Gubkin University Press) p 168

[3] Drozdov A N, Egorov Y A and Telkov V P 2007 Vodogazovoye vozdeystviye: issledovaniye protsessa vytesneniya neftey razlichnoy vyazkosti primenitel'no k Shumovskomu mestorozhdeniyu [Water-gas impact: a study of the process of displacement of oils of different viscosity with reference to the Shumovsky field] *Territoriya neftegaz [Territory of Oil-Gas]* **4** p 56-61

[4] Drozdov A N 2014 Utilizatsiya neftyanogo gaza s ispol'zovaniyem sushchestvuyushchey infrastruktury promyslov [Utilization of associated petroleum gas with using of existing field infrastructure] *Neftyanoye khozyaystvo [Oil industry]* **4** p 74-77

[5] Lisin Yu V, Korshak A A and Golofast S L 2016 Universal'nyye kharakteristiki zhidkostno-gazovykh ezhektorov [Universal characteristics of liquid-gas ejectors] *Neftyanoye khozyaystvo [Oil industry]* **10** p 102-104

[6] Tarasov M Yu, Zobnin A A, Zyryanov A B, Panov V E and Magomedsherifov N I 2009 Razrabotka i promyslovyye ispytaniya tekhnologii utilizatsii nizkonapornogo neftyanogo gaza s pomoshch'yu struynykh kompressorov [Development and field tests of technology of low-pressure oil gas utilization with the help of jet compressors] *Neftyanoye khozyaystvo [Oil industry]* **2** p 43-45

[7] Bunkin N F, Drozdov A N, Drozdov N A, Kozlov V A, Tuan V M and Fouilhe V L 2017 Suppression of the Coalescence of Gas Bubbles in Aqueous Electrolyte Solutions: Dependence on the External Pressure and Velocity of Gas Flow through a Column with Liquid *Physics of Wave Phenomena* **5(3)** p 219-224

IOP Conf. Series: Materials Science and Engineering **675** (2019) 012023 doi:10.1088/1757-899X/675/1/012023

[8] Drozdov A N and Drozdov N A 2012 Laboratory Researches of the Heavy Oil Displacement from the Russkoye Field's Core Models at the SWAG Injection and Development of Technological Schemes of Pump-Ejecting Systems for the Water-Gas Mixtures Delivering *Proceeding of the Society of Petroleum Engineers Oil Conference* (Calgary, Canada) p 872-878

[9] Drozdov A N, Malyavko E A, Alekseev Y L and Shashel O V 2011 Stand Research and Analysis of Liquid-Gas Jet-Pump's Operation Characteristics for Oil and Gas Production *Proceeding of the Society of Petroleum Engineers Annual Technical Conference and Exhibition 2011* (Denver, USA) p 2122-2130

IOP Conf. Series: Materials Science and Engineering **675** (2019) 012024 doi:10.1088/1757-899X/675/1/012024

Electrohydraulic effect as an example of electrophysical technologies application in the oil industry

A N Drozdov*, I M Narozhnyy, D X Pak, V B Ludupov and G S Zemlianskii

Peoples' Friendship University of Russia (RUDN University), Moscow, Russia

Email: drozdov_an@pfur.ru

Abstract. This work was devoted to studies of electrohydraulic effect applicability in the oil industry. The major aim was to determine if the effect would be really useful in cleaning field pipelines from different precipitations such as calcium and barium sulphates, calcium and magnesium carbonates, sodium chloride salts and also tar and paraffin. In addition to it, another fields of possible applicability of the electrohydraulic effect were supposed to be find. Realisation of purposes put was achieved though reconstituting Lev Yutkin's layout for electrohydraulic impact and conducting several special experiments. Their thoughtful description with necessary accompanying illustrations can be find at the pages of this article. Results of experiments carried out, have proved the electrohydraulic impact effectiveness as a mean for cleaning pipes. Moreover, some results gave a reason to think that the effect worth applying to for increasing of the wellbore zone permeability as well as for extracting of solid minerals.

1. Introduction

Electrohydraulic technologies occupy a separate niche in the series of electrophysical technologies. Despite the relative simplicity of the execution, due to their specific features, they allow to effectively perform tasks that are not solved or difficult to solve by traditional methods. According to scientists involved in this issue (Mayer V.V. [5], Meriin B.V [1], Blazin B.S. [2], Kurets V.I., Eremin V.Ya. [3]., Perevyazkina E.N. [4] and others) such technologies are used in mechanical engineering, agriculture, construction, hydrometallurgical production, in medicine, and even in oil industry [12].

For example, electrohydraulic effect (EHE) processing is applied to: reduce stress in welds (EHE produced in the casting cleaning bath, allows to reduce the residual stresses in the weld by 50-70%) [1], to destroy large boulders (in cases when explosives cannot be used due to the risk of clogging up the area with debris) [2, 3], to create organic fertilizer (thanks to EHE effect, it can be easily produced from peat) and food (especially meat) [4,5], to disinfect water (the effect is based on rendering pathogenic bacteria, even the most unpleasant, like E-coli.) [6], for the destruction of stones in the bladder and the removal of kidney stones with the help of an external source of electrohydraulic exposure.

The main reason for carrying out new experiments, during which it became possible to identify the strengths of the electrohydraulic impact (EHI), was the desire to find a new efficient and inexpensive way to clean the field pipelines from various types of precipitations. In addition to this, the research tasks included the search for other areas of electrohydraulic effect applicability in the oil industry.

The operation of oil wells and field pipelines is often complicated by precipitations of various composition and genesis. Clogging conglomerations in most cases are represented by salt sediments: calcium and barium sulphates, calcium and magnesium carbonates, sodium chloride, as well as asphalt-resin and paraffin sediments.

Precipitating on the walls of the producing strings and field pipelines, such sediments effect on the equipment operation process extremely negatively. There is a risk of the development of metal corrosion, energy losses significantly increase either. Besides that a need to increase the pressure arise (some cases when the pressure in the pipeline was increased by 3.5 times are already known), which is certainly a negative factor too.

Popular Methods of combating undesirable sediments are divided into chemical (the use of inhibitors of salt sedimentation), physical (usually presented by the effects of a magnetic field and ultrasonic vibrations) and technological.

In this scientific article, the effectiveness of another physical method based on the use of electrohydraulic effect was analyzed. The discoverer of the mechanism of this phenomenon allowing transform electric energy into mechanical energy was the Soviet physicist Lev Aleksandrovich Yutkin. Thanks to the EHE model that was recreated according to his schemes, it became possible to carry out a number of tests, the results of which can be found on the pages of this work. The results of these tests are an interesting material which can be highly valued in the oil industry.

2. Methods

To conduct research, a special replica of the EHI was decided to be recreated according to the scheme proposed by L.A. Yutkin [7].

The principle of the installation is based on the interaction of high-voltage electrical pulses with a liquid [8]. Between the electrodes lowered into a vessel with water (ordinary running water was used as a liquid), a spark jumps, and results in instant excitement of the interelectrodes water. This happens due to the fact that "cavitation seeds" (gas bubbles) generated by a spark expand and collapse with the release of large energy (many times greater than the energy expended to start the process) [9,10,11].

The process itself can be described as follows. At first, the transformer generates a low voltage of 24V and transmits it to the ignition coils, which, in turn, rise this voltage. Then the high voltage is transmitted to a diode rectifier, designed to "cut off" the negative half-cycles of the signal. After that the signal goes to a capacitor. To regulate both the required values of the voltage and the capacitance accumulated by the capacitor, a formative gap, acting as a simple spark gap, was introduced into the circuit. It transmits the charge from the capacitor to one of the electrodes through the air, while the rest of the charge from the capacitor goes straight to the other electrode. As a result of the interaction of the electrical impulses (which are emitted by the electrodes) with the liquid, a cavitation process occurs, it being accompanied by high and super high excess hydraulic pressures capable of performing useful mechanical work. After termination of the overpressure action, a condensation (relaxation) of previously formed liquid vapors occurs [9].

After the layout of the EGW was assembled, researches were carried on. The experimental part of the work consisted of two stages. So, at first, experiments on cleaning pipes with the help of EHE were carried out, then experiments on the destruction of solid rock - diorite took place.

To carry out the experiment on the cleaning of pipes from contamination, a small metal tube 150 mm long and 26.8 mm in diameter with a wall thickness of 3.2 mm was used. In order to simulate sediments of salts, asphaltenes and paraffins, this tube was filled with plaster. After solidification, a hole with approximate diameter of 14.5 mm was drilled (figure 1a) in the plaster. The tube was placed in a plastic dish and filled with water, so it was possible to lower the holder of electrodes into the hole drilled and realize the EHE.

When carrying out this experiment, some measurements of the voltage values at the output from the capacitor were made thanks to using of Hantek DSO5202P oscilloscope. As a result the character of relationship between voltages, intershock period and the distances in the forming gap (in this case, 3 mm) was determined. The measurement results can be seen at the figure 5a.

Due to getting good results in the experiment, it was decided to continue this studies, but already with more powerful precipitation model. For this purpose a tube of larger diameter and all the same plaster should have been used. But, in order to follow the development of cracks, a transparent cylindrical vessel with a diameter of 100 mm was used (figure 2a) was used instead of standard iron tube. Since such a replacement does not affect the quality of the experiment, it can be considered to be absolutely reasonable.

As previously, the vessel was filled with plaster, the height of the cylinder of plaster being equal to the same 150 mm. The only difference was that, this time, not all the cross-sectional area of the transparent vessel was filled in with plaster (alabaster), but only the space between its wall and a special tube (diameter 20 mm), installed in advance in the center of the vessel. Then, the tube was removed after the plaster got hardened. This decision made it possible not to drill a hole in the plaster.

As a result, all conditions for accurate repeat of the previous experiment (but already with a more powerful clogging imitation) were created.

The next stage was devoted to experiments on the destruction of the diorite core sample. These experiments were carried out in order to find out how significant the EHE effect can be while destroying more or less hard rocks (diorite takes the 6th line in the Mohs scale).

Initially, several whole core samples (diameter 62.9 mm, length 110 mm) and several semi-cylindrical samples were available. In order to assess the physical characteristics of the material (diorite), it was decided to place one of the whole core samples under a press (figure 3). The results gave an opportunity to judge the hardness of the rock.

To make the realization of the experiment more convenient, two semi-cylindrical core samples were selected for further working at. Both halves were placed in a special plastic form filled with water. Electrodes were placed directly between them. The forming spark gap was adjusted to the meaning of 4mm.

It should be noted that the voltage values were measured using an oscilloscope like in the case of the previous experiment, but this time the forming gap was 4 mm (figure 5b).

3. Results

During the first experiment (cleaning of small-diameter tube from plaster clogs), electrohydraulic impact was being applied for 1 minute. Forming gap at the beginning of the experiment was approximately equal to 3 mm. After each electrohydraulic shock, the electrode holder was moved along the drilled hole (on average, at 5 mm distance). This was done to have the entire contamination area under the effect. As a result, the tube was cleaned from gypsum (figure 1).

a) b)

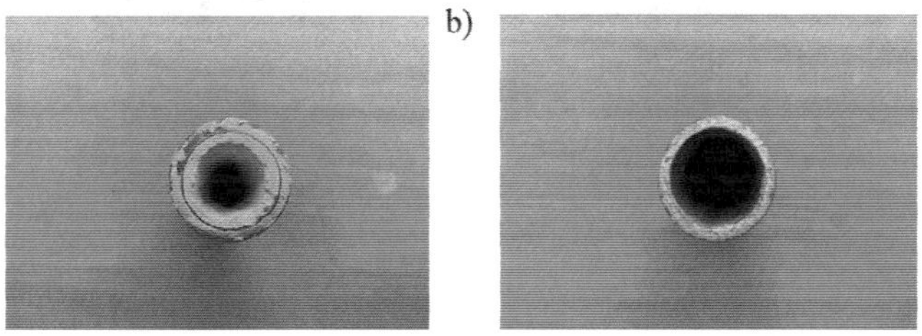

Figure 1. The result of usage of electrohydraulic effect in the tube (a - before EHE realization; b - after EHE realization).

Keeping on carrying out the same experiment in a cylindrical vessel, a decision of leaving the distance in the forming gap unchanged (3 mm) while changing longitude of the experiment in time was taken. Since the thickness of the gypsum wall this time was significantly greater, EHI was being used for 7 minutes. During this time, a whole network of cracks in the workpiece was successfully formed (figure 2b).

a) b)

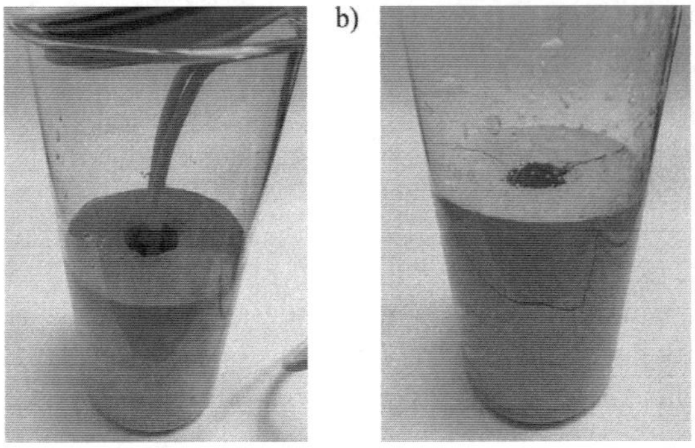

Figure 2. Results of the EHE application in a transparent cylinder (a - before EHI realization; b - after EHI realization).

At the second stage of the laboratory tests, while preparing for the experiments with diorite, a CYBER-PLYS EVOLUTION press was used to determine the physical characteristics of the core sample: the maximum load of power that the sample withstood was 706.285 kN, while the maximum pressure was 212.630 MPa. Basing on the data obtained, it was concluded that the compressive strength of the material is quite high, comparable to the strength of granite, which everybody knows.

a) b)

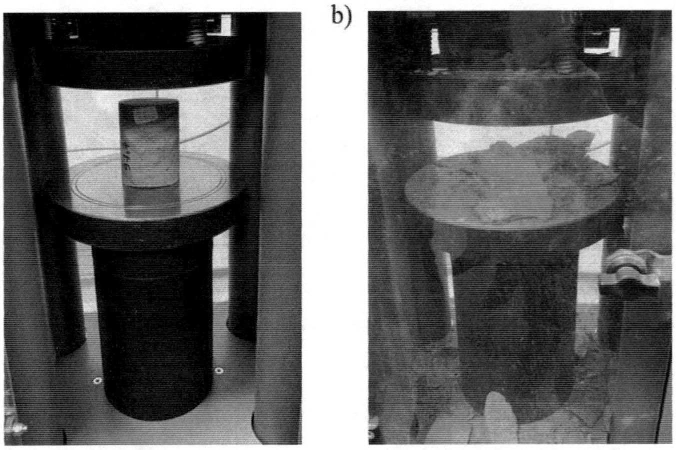

Figure 3. Diorite core sample in compression tests (a - before testing; b - after testing).

The experiment with diorite lasted for 3 minutes (results can be seen at the figure 4). During this time, about 30 electro-hydraulic shocks were carried out. As a result, semi-cylindrical samples split in halves.

Figure 4. Diorite semicylinders split as a result of EHE application.

As mentioned above, during the experiments, an oscilloscope was used. It measured the voltage at the output from the capacitor in dependence on the time between electro-hydraulic shocks. From figure 6a it becomes clear that with a three-millimeter formating gap, a peak voltage of 8.8 kV was recorded; the interstrike period was 2.12 sec. In the experiment with a 4 mm gap, the oscilloscope gave the following data: the peak voltage slightly increased to 10.2 kV, and time between shocks increased to 3.08 seconds (Figure 5b).

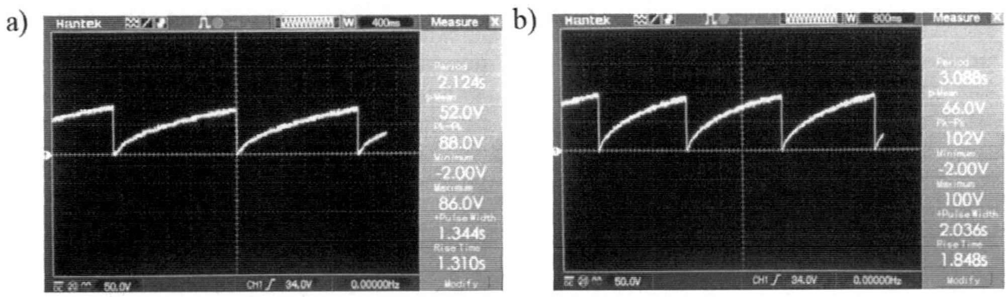

Figure 5. Dependence of the voltage at the outlet of the capacitor on the time between electro-hydraulic shocks (a - with a forming gap of 3 mm, b - with a forming gap of 4 mm).

Summing up the results, it should be noted that the EHE happened to be rather effective in all experiments. Electrohydraulic impact succeed in: clearing of the tube from mechanical clogs; creating a network of cracks in massive plaster billet and in splitting the diorite core sample.

4. Discussion

Studies have confirmed the assumption that the Yutkin electrohydraulic effect can be successfully applied in the oil industry.

It should be emphasized that all the experimental studies were successfully completed: local tube resistances represented by gypsum precipitations were removed, and diorite was split. No occurrence of any negative consequences of EHE usage was noticed during the tests.

Data on the dependence of the voltage and the period between shocks on the length of the forming gap were also obtained.

Based on the results of the experiments, alternative ways of using the electro-hydraulic effect in the oil industry were proposed (in addition to cleaning pipes from salt, resins and paraffins).

The results of tests done have become a reason to think that such effect would be very effective for increasing of the wellbore zone permeability (this could be achieved through increasing amount of fractures and clearing from clogging particles). Perhaps it will be possible to perform a number of studies in this field in the future.

In addition to this, the fact that Yutkin effect can be used while extracting solid minerals has become obvious. Its ability to chop rather hard rocks likely to be very useful during extraction (According to the Mohs scale diorite has a hardness of 6).

5. References

[1] Meriin B V 1985 *Elektrogidravlicheskaya obrabotka mashinostroitel'nykh izdelii [Electrohydraulic Treatment of Machine Engineering Products]* (Leningrad: Publishing House Mashinostroenie) p 118

[2] Blazin B S 1983 *Obrabotka prirodnogo kamnya elektricheskimi razryadami [Natural stone processing by electrical discharges]* (Moscow: Publishing House Nauka) p 257

[3] Eremin V Ya 2002 Razryadno-impul'snyye tekhnologii na stroykakh Rossii [Discharge-pulse technology settings in Russia] *Stroyklub* **2** pp 11-15

IOP Conf. Series: Materials Science and Engineering **675** (2019) 012024 doi:10.1088/1757-899X/675/1/012024

[4] Nagdalian A A and Oboturova N P 2013 Development of electrohydraulic technology of meat salting *Proc. 5th International scientific conference. European Science and Technology* (Munich, Germany) p 514

[5] Oboturova N P, Evdokimov I A, Nagdalian A A, Kulikov Y I and Gusevskaya O A 2015 The study on the influence of the electrohydraulic effect on the diffusion coefficient and the penetration depth of salt into muscle tissues during salting *Foods and Raw Materials* **2** pp 74-81

[6] Perevyazkina E N 1982 Priroda obezzarazhivayushchego deystviyaimpul'snogo elektricheskogo razryada v vode [Nature of the disinfecting action of a pulse electric discharge in water] *Proc. All Scientific Conference* (Tomsk, Russia) pp 10-12

[7] Yutkin L A 1986 *Elektro-gidravlicheskiy effekt i yego primeneniye v promyshlennosti [Electrohydraulic effect and its application in industry]* (St. Petersburg: Publishing House Mashinostroenie) p 330

[8] Charred K A and Roy N A 1971 *Electric discharges in water* (Moscow: Publishing House Nauka) p 155

[9] Mayer V V 1989 *Kumulyativnyy effekt v prostykh opytakh [Cumulative effect in simple experiments]* (Moscow: Publishing House Nauka) p 173

[10] Shalnev K K, Varga J J and Sebestyen G 1967 Scale-effect investigation of cavitation erosion using the energy parameter (Erosion by Cavitation or Impingement ASTM STP 408) p 220

[11] Knapp P, Daily J and Hammit F 1974 *Cavitation* (Moscow: Publishing House Mir) p 687

[12] Demakhin A G 2008 *Effekt elektrogidravlicheskogo vozdeystviya na neft' [The effect of the electrohydraulic action on oil] Proc. Modern Problems of APK and Natural Use* (Saratov, Russia) 66-79.

Protection of pipeline systems from water hammer using oscillatory circuits

F V Rekach[1], S L Shambina[1*] and Yu V Belousov[1,2]

[1] Peoples' Friendship University of Russia (RUDN University), Moscow, Russia
[2] Bauman Moscow State Technical University (National Research University), Moscow, Russia

E-mail: shambina_sl@mail.ru

Abstract. The issues related to reduction of the amplitude of high pressure waves in long pipes are relevant in connection with the danger of failure of some hydraulic system's sections. The article deals with the reduction of high pressure by simultaneous separation of fluid flow for long hydraulic systems. Numerical experiments are carried out in order to identify the most cost-effective dimensions of the equipment, as well as the most profitable in terms of energy damping locations of this equipment.

Keywords: hydraulic schemes, pressure fluctuations, water hammer, long pipelines.

1. Introduction

A sharp increase or decrease of pressure in the hydraulic system can lead to serious accidents with great economic consequences, as well as possible human victims. The situation in which there is an extended destruction of pipes with accompanying equipment or the occurrence of plastic deformations [1], which negatively affect the functioning of the pipeline scheme is especially dangerous. To prevent such catastrophes, devices capable of damping excessive (or reduced) pressure are developed tested. Theoretical, numerical and practical research in this area is carried out.

This article focuses on problems of harmful pressure fluctuations' reduction by means of energy dampers separating the flow. The idea of this device's functioning is that with a significant increase (or decrease) of the pressure in the hydraulic system, the flow is directed along the path where it meets the hydraulic resistance. The principle of operation of these energy dampers is as follows: when the pressure increases in the pipeline sections where there is a constant or fluctuating flow discharge (and hence the corresponding flow velocity), the flow of liquid is blocked. An oscillating circuit is created. If you block the flow of fluid at the same time at two points of the segment, then there are two waves: high pressure and vacuum. They interact and partially extinguish each other in the process of mutual movement.

The computational part of the numerical program was implemented in C++, the graphical representation of information was made in Marple.

2. Analysis of the stated issue

The founders of hydraulics as a science in Russia were full members of the Russian Academy of Sciences: M.V. Lomonosov (1711-1732), Leonard Euler (1707-1783) and Daniel Bernoulli (1700-1782). Hydraulics received a theoretical basis as a result of their theoretical and practical studies. Russian scientist N.E. Zhukovsky (1847-1907) has made a significant contribution to the development of hydraulics [2, 3]. He was the first one who developed the theory of water hammers in pipes and gave a classical solution to a wide range of technical issues in the field of hydraulic supply and hydraulic engineering.

The theory of unsteady fluid motion was created and developed by I.A. Charny [4] and D.A. Fox [5]. In reference book [6] the values of hydraulic resistances are given. The theory and practical modeling of the equipment stabilizing pressure in short pipelines is considered in the works of R.F. Ganiev, H.N. Nizamov and E.I. Derbukov [7]. Pressure stabilizers can be used in heat pipes, water supply lines, sewage systems, as well as in any schemes in which the movement of liquid is realized. The equipment serving for this purpose is also considered in [8, 9]. The resonance phenomena were studied in the works of Glikman [10]. The fluctuations in the feed systems of liquid fuel in the rockets are considered in the works [11, 12].

The equations of fluid motion and continuity of fluid flow are shown below:

$$\frac{\partial}{\partial x}\left(\rho g z+p+\alpha\rho\frac{v^2}{2}\right)+\alpha'\rho\frac{\partial v}{\partial t}+\frac{\rho\lambda}{2D}v\,|\,v\,|=0,\ \left[N/м^3\right], \tag{1}$$

$$v\frac{\partial p}{\partial x}+\frac{\partial p}{\partial t}+\rho c^2\frac{\partial v}{\partial x}=0,\quad \left[N/(м^2\cdot\sec)\right], \tag{2}$$

where x - coordinate along pipe's axe; ρ -mass density of liquid (const); g - acceleration of free fall; z - the height of the considered point; p - pressure in the pipeline; α - the coefficient of Coriolis (adopted α =1); v - average flow velocity; α' - the ratio of Businesc (adopted α'=1); t - time; λ - hydraulic resistance coefficient; D - the pipe diameter; c - propagation velocity of waves of increased pressure.

The system (1) - (2) consists of two hyperbolic partial differential equations. All variable parameters are changed according to the marching scheme, that is, they are developing with the change of time. These equations require the dependence *H=H (Q)* or one constant condition (*Q* or *H*) at the boundaries.

The method of characteristics is used as a numerical method, which is discussed in details in [13]. The absolute hydrodynamic head *H* expressed in *m* and mass flow rate *Q* [m^3/sec] are taken as the main characteristics. The computational part of the numerical program is implemented in C++, the graphical representation of information is made in Marple.

The paper deals with kinetic energy of the liquid flow [14, 15]. The problem is how to reduce it with the use dampers. The idea is as follows: when the pressure at point *A* rises to a certain value, the signal of the long flow closing by the valves is transmitted almost instantly (the velocity of the electromagnetic field propagation reaches 300 000 km/sec). The best location of the points of the flow stop and graphics of the obtained values of pressure are investigated. The question of further elimination of the threat of a significant increase in pressure is studied. The problems associated with transient fluid motion are discusses in [16, 17]. Some ideas have been gathered from these works.

The investigated hydraulic scheme is shown in figure 1. It consists of 18 nodes and 21 pipes.

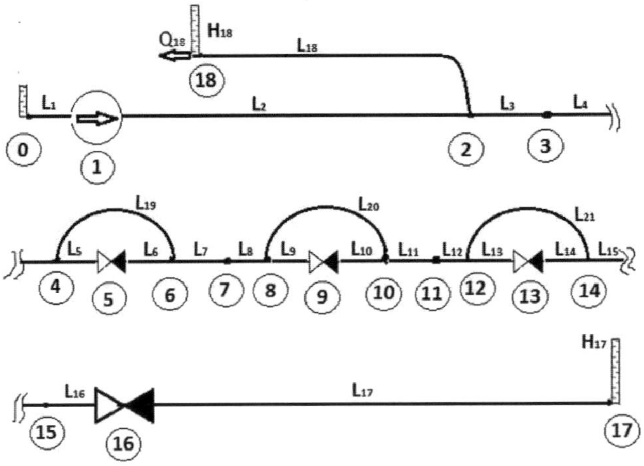

Figure 1. The hydraulic scheme.

The pumping station is installed in node 1, the dependence $H=H(Q)$ is given. The boundary condition in nodes 18 and 17 is a constant head. All the numerical experiments had one goal: to create braking areas in elements 19, 20 and 21 for ensuring that the pressure in the system was significantly less than the pressure without braking areas. The valves are installed in nodes 5, 9 and 15 (points *D, C* and *B*), these valves receive a signal of operation when the pressure in node 15 (point *F*) rises above the pressure specified in the task condition.

3. Numerical calculations and the main results

Now we'll consider the hydraulic scheme shown in figure 2 in details.

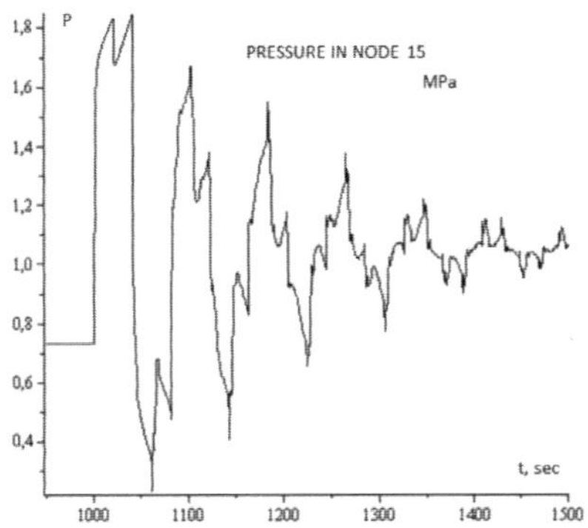

Figure 2. Pressure at node 15 without energy dissipation with the water head $H_{18}=60$ m.

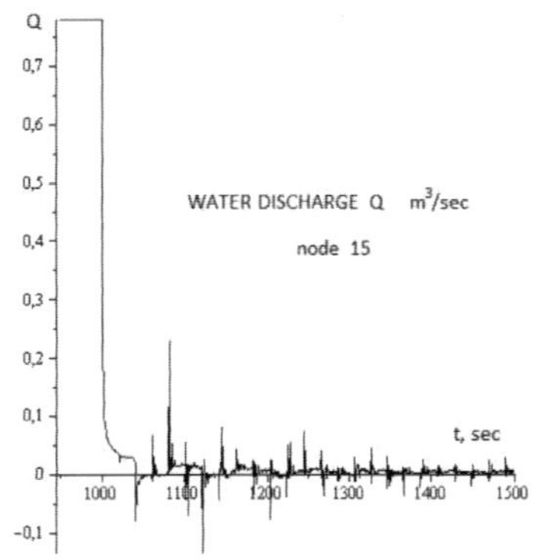

Figure 3. Water discharge at node 15 without energy dissipation with the water head $H_{18}=60$ m.

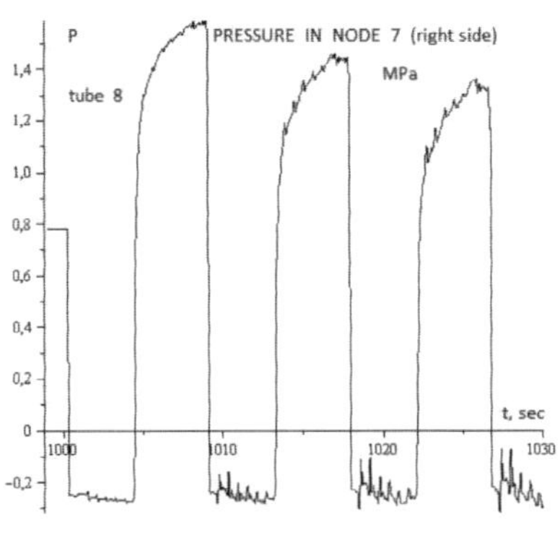

Figure 4. Pressure to the right from the valve ($t_{cl}=0$) in node 7 when the signal from node 15.

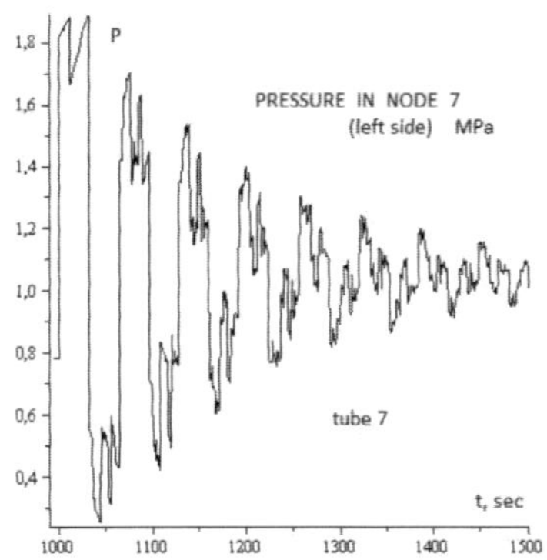

Figure 5. Pressure to the left from the valve ($t_{cl}=0$) in node 7 when the signal from node 15.

Test problem 1:

Initial data: $L_1=200\ m$, $L_2=10000\ m$, $L_3=2700\ m$, $L_4=300\ m$, $L_5=L_6=200\ m$, $L_7=2700\ m$, $L_8=300\ m$, $L_9=L_{10}=200\ m$, $L_{11}=2700\ m$, $L_{12}=300\ m$, $L_{13}=L_{14}=200\ m$, $L_{15}=200\ m$, $L_{16}=200\ m$, $L_{17}=20000\ m$, $L_{18}=12000\ m$, $L_{19}=300\ m$, $L_{20}=300\ m$, $L_{21}=300\ m$. The diameter of elements 1 – 18 are $D_{1\text{-}17}=1.0\ m$, $D_{18}=0.5\ m$, the diameters of tubes 19 - 21 are variable. Water heads in nodes 0, 17 and 18 are $H_0=10$, $H_{17}=50$, $H_{18}=60$ respectively. The speed of steady motion of the liquid in the elements 3 - 17 is 0.93 m/sec, the coefficient of hydraulic resistance of pipes (except pipes 19 - 21) is equal to $\lambda=0.023$. The velocity of high pressure waves $c=1020\ m/sec$. The parameters of the pump unit $H=H(Q)$ are given by three points: $H_1=96$, $H_2=80$, $H_3=42\ [m]$, $Q_1=0.08$, $Q_2=2.24$, $Q_3=4.0\ [m^3/sec]$.

The scheme with a constant water intake in the node 18 $Q=0.19\ m^3/sec$ was tested. However, this boundary condition (which is the only one in node 18) is too strict to obtain good results. After the introduction of a constant head in the node 18 (H_{18}=60 m), the pressure and flow in node 15 (point *F*) have the following form as shown in figure 2 and 3. The maximum pressure of *1.85* MPa is in good agreement with the theory.

Let's consider the case when the electrical signal is instantly transmitted to node 7 from node 15. Figure 4 shows the pressure to the right of node 7 when the valve ($t_{cl}=0$) in node 7 is actuated after the head exceeds 85 m in node 15. The formula $t_{cl}=0$ means that the valve closes instantly. The electrical signal (from node 15) spreads with a speed of about 300,000 km/sec. The vacuum wave goes from node 7 to node 16. A wave of increased pressure simultaneously moves from node 16 to node 7. In the Fig. 4 it is seen that the period of pressure fluctuation between nodes 7 and 16 after the addition of these waves is about 9 sec. This is much less than the oscillation period of the entire system (about 64 sec.) from nodes 1, 18 to node 7 (Figure 5).

Problem 2. The initial data are changed (compared to the test problem 1): $\lambda_{20}=0.04$, $d_{20}=0.1$ m. The valve in node 9 closes instantly when the head in node 15 reaches 85 m. In this problem, gate valves in nodes16 and 9 close almost simultaneously. As a result of the addition of waves, the oscillation period in node 15 (and throughout from node 9 to node 16) is two times less than if the boundary condition of the constant head $H_9=const$ (Figure 6). The second positive result is that the peaks of the maximum pressures decreased. An interesting fact is that with a decrease in the friction coefficient in the pipe 20 (or an increase in the diameter of the pipe 20) the pressure even rises with time. This is probably due to the influence of flow to the left of node 8.

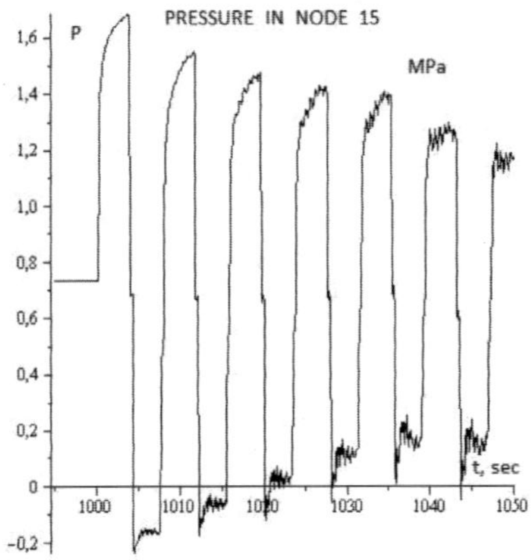

Figure 6. Pressure in node 15 when the gate in node 9 ($t_{cl}=0$) closes from a signal of node 15.

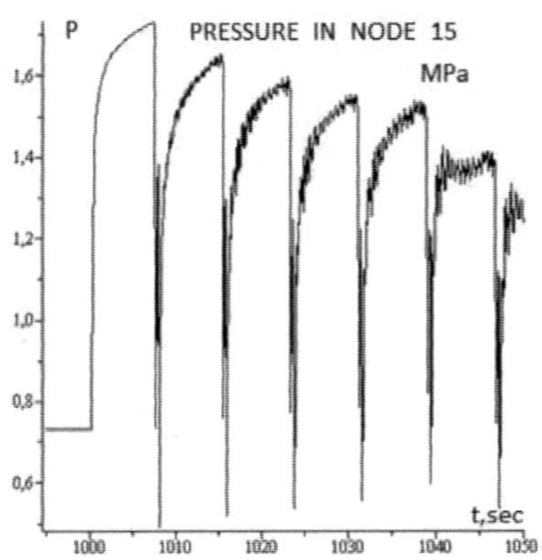

Figure 7. Pressure in node 15 when the valve in node 9 is actuated ($t_{cl}=0$) from the head increase in node 9 up to 85 m.

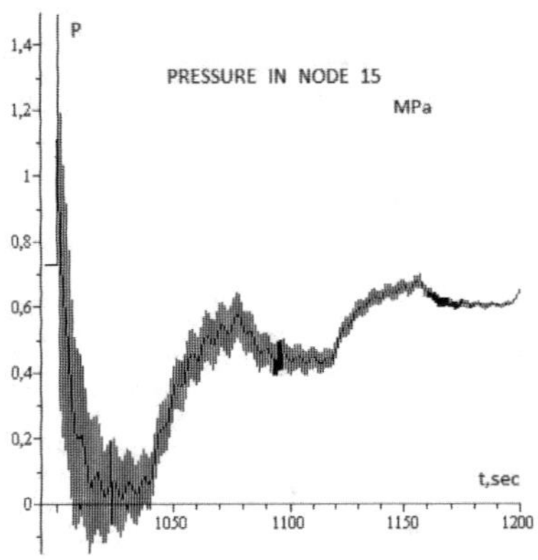

Figure 8. Pressure in node 15 when the valves in nodes 5, 9,13 close ($t_{cl}=0$) from pressure increase in node 15 over 85 m.

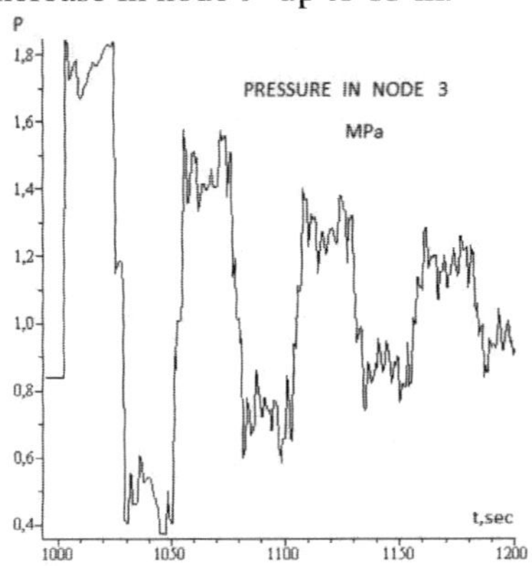

Figure 9. Pressure in node 3 when the valves in nodes 5, 9,13 close ($t_{cl}=0$) from pressure increase in node 15 over 85 m.

Problem 3. It differs from problem 2 that the valve in node 9 closes when the pressure in node 9 exceeds 85 m (Figure 7). It is interesting that there is no vacuum wave from node 9 and therefore the average pressure is much higher than in problem 2. In addition, high pressure peaks are wider. All this suggests that it is more difficult to extinguish the overpressure in this case than in the example of problem 2.

Problem 4. It differs from the problem 2 is that the valves in the nodes 5, 9, 13 close when the pressure in the node 15 exceeds 85 m (Figure 8 and 9). In the problem the oscillation frequency is high in node 15 (Figure 8), because the distance between node 13 and 16 is not large (600 m). This creates good conditions for breaking the flow. It should be noted that the pipeline from node 1 to node 5 is poorly protected, but the period of pressure fluctuations in node 3 is less than without electrical devices.

4. Conclusions

1) Introduction of valves, which close almost simultaneously create a wave of vacuum. It adds up with a wave of high pressure. As a result, the pressure at the gate valve in node 15 is significantly less than without them (Figure 2, 4). The period of pressure fluctuations is two times less, so it is much easier to extinguish such oscillations than the oscillations of all systems (Figure 7).
2) Valves, closing simultaneously divide the flow into shorter sections. Here the mileage of high pressure and vacuum waves is correspondingly less. This greatly facilitates the fight against the increased pressure. Comparing figure 5 and figure 2 we see that the periods of pressure fluctuations are different (64 sec. and 82 sec.)
3) Introduction of self-actuating valves when pressure increase (Figure 7) have a higher average pressure and wider pulses of increased pressure, which makes it difficult to stabilize (equalize) the pressure.
4) Figure 9 shows that the introduction of three breaking circuits effectively dissipates high pressure.
5) Creating of short circuits well dissipates energy, due to frequent fluctuations of water discharge.
6) Pressure waves with a short period can be well extinguished by means of pressure stabilizers.

References

[1] Rynkovskaya M 2018 Plastic deformations occurring in shells with developable middle surfaces during bending *IOP Conf. Series: Materials Science and Engineering* **371** 012054
[2] Zhukovsky N E 1949 *Collected works in 7 volumes Volume 2: Hydrodynamics Volume 3: Hydraulics. Applied mechanics* (Publishing House Gostekhizdat, Moscow) p 487
[3] Zhukovsky N E 1949 *On Water Hammer in Water Pipes* (Publishing House Gostehizdat, Moscow) p 106
[4] Charny I A 1951 *Unsteady movement of a real liquid in pipes* (Publishing House Nedra, Moscow) p 224
[5] Fox D A 1981 *The hydraulic analysis of the unsteady current in pipelines* (Publishing House Energoizdat, Moscow) p 248
[6] Idelchik I E 1992 *Reference book on hydraulic resistance* (Publishing House Mashinostroenie, Moscow) p 672
[7] Ganiev R F, Nizamov H N and Derbukov E I 1996 *Wave Stabilization and Prevention of Accidents in Pipe-Lines* (Publishing House MGTU, Moscow) p 260
[8] Rekach F V, Shambina S L and Sinichenko E K 2017 Influence of pressure stabilizer perforation area on character of unsteady fluid motion in hydraulic systems. *Int. J. of Mech. Eng. and Robotics Res.* **6(4)** pp 268-71
[9] Svintsov A P, Kharun M I and Mukarzel S A 2015 Valve head for water fittings with high regulatory capacity *Mag. of Civ. Engin.* **58(6)** pp 8-18
[10] Glikman B F 1986 *Mathematical models of pneumohydraulic systems* (Moscow: Nauka) p 368
[11] Kolesnikov K S 1971 *Longitudinal oscillations of a liquid-propellant rocket engine* (Publishing House Mashinostroenie, Moscow) p 259

[12] Natanzon M S 1977 *Longitudinal self-oscillations of a liquid rocket* (Publishing House Mashinostroenie, Moscow) p 208
[13] Rekach F V 2010 Calculation of fluctuations in circular cylindrical shells with pressure stabilizer by method of characteristics *J. Struct. Mechan. of Engin. Constr. and Build.* **1** pp 60-65
[14] Samuel S Pegler 2016 The dynamics of confined extensional flows *J. Fluid Mech.* **804** pp 24-57
[15] Pietro Scandura, Carla Faraci, Enrico Foti 2016 A numerical investigation of acceleration-skewed oscillatory flows *J. Fluid Mech.* **808** pp 576-613
[16] Young W R and Wolfe C L 2014 Generation of surface waves by shear-flow instability *J. Fluid Mechanics* **739** pp 276-307
[17] Kerswell R R 2016 Energy dissipation rate limits for flow through rough channels and tidal flow across topography *J. Fluid Mechanics* **808** pp 562-75

Acknowledgements
This paper was financially supported by the Ministry of Education and Science of the Russian Federation on the program to improve the competitiveness of Peoples' Friendship University of Russia (RUDN University) among the world's leading research and education centres in the 2016-2020. This publication was prepared with the support of the "RUDN University Program 5-100".

Modeling of the conjugation of vortex flows with downstream in ANSYS

L E Schesnyak

Peoples' Friendship University of Russia (RUDN University), Moscow, Russia

Email: shchesnyak-le@rudn.ru

Abstract. Using the new technology of longitudinal circulation currents which allows dissipating the excess kinetic energy, thus preventing riverbed erosion and increasing the reliability and environmental safety of the structure. In modern practice of the design of water discharge structures of high-pressure dams, those special properties of flows are widely used. This allows to solve a number of important engineering problems arising in the practice of operating the structures by using the new innovative approaches to. In the paper, the author offers an interesting solution to the engineering problem – A new mathematical modelling has been created. For this study, the standard software package Fluent in ANSYS R 19.0 is used and the required outcome is compared with experimental results. Consequently, the necessary parameters of the swirling flow which are required when evaluating the conditions for the coupling of the water discharged with the downstream of the dam structure are depend on the attained results.

Keywords: vortex flow, downstream, water discharge, spillway, drifting flow, modelling, ANSYS.

1. Introduction

The conjugation of vortex flows with downstream is the component where the flow existing the runner is decelerated, thereby converting to the excess of kinetic energy in to the static pressure, as a consequence increase the global power of the turbine. In practice of hydraulic engineering, recently, the new technologies for dissipating the excess kinetic energy of the stream are introduced, particularly in spillway structures of hydroelectric power stations and rotary-conjugating structures on canals. The principle of these technologies lies in the use of special properties of longitudinal circulation currents that allow solving a number of important engineering problems arising in the actual operation of structures, namely:

- Dissipating the excess kinetic energy of the flow within the culvert;
- Prevention of dangerous riverbed erosion and the collapse of coastal slopes in the downstream;
- Protection of streamlined surfaces from cavitation damages allowing to increase the reliability and environmental safety of hydraulic engineering objects.

As an example of using these technologies, where the features of longitudinal circulation currents are implemented, can be mentioned the vortex shaft spillway of the Tehri Dam in India [1]. In recent years, extensive scientific research has been carried out on longitudinal circulation currents and methods of their hydraulic calculation [2-8]. However, high estimation of the work of the vortex spillways of the Tehri Dam is given by different authors, who are engaged in the design of spillway structures. It provides a basis for the continuation and more detailed study of longitudinal circulating currents, especially in the propagation of swirling turbulent jets which are discharged from the spillways into the downstream. Those scientific and engineering problem requires new innovative approaches for its solution. To solve these problems, it is reasonable to use inexpensive and affordable models describing the conjugation of vortex flow with downstream. The same physical phenomenon admits a variety of different models, which are divided into physical and mathematical [9-11]. Physical models are extremely reliable and convenient for solving engineering problems, but they are very expensive, particularly when modeling the fundamentally non-linear once its scale cannot be changed. The creation of physical models of real hydroelectric power stations in the laboratory presents difficulties.

On the contrary, mathematical models offer an extremely economical solution to engineering tasks but always should be remembered that every mathematical model takes into account only some phenomena with approximate accuracy. According to Samarsky et al., mathematical modeling includes three stages [10]:

- Designing on the basis of general laws of nature, and one or another empirical material of a correct mathematical task, called the mathematical model of the phenomenon;
- Creation of an algorithm for solving this task
- Verification of the model by comparison with full-scale experiment and with other simpler models.

In hydrodynamic tasks the boundary-value problems for partial differential equations are used as a model. Such problems are rarely solved analytically, therefore, they usually apply numerical and grid methods for solving differential equations. Theoretically, the flow of water during discharge is most accurately described by the Navier-Stokes equation [12–17].

The application of grid method implies that the scale order of the stack step of the solution can be well approximated by linear functions but in turbulent problems this assumption is not fulfilled. However the Navier-Stokes equation appears to describe well acceptable structure of the flows therefore, it is poorly computable by the finite difference method. So, models based on the direct solution of the Navier-Stokes equations are not used in engineering calculations. From the mathematical point of view, the Navier-Stokes equations are extremely complicated in addition at this time the existence and smoothness of the solution for Navier-Stokes equations in R3 have not been proved [18]. At first, the Reynolds decomposition between mean and fluctuating (e.g., turbulent) components is introduced and applied to the Navier-Stokes equations, resulting in the so-called Reynolds-Averaged Navier-Stokes equations (RANS, in the Favre formulation valid for compressible flows) [20–24]. An unclosed system of differential equations is obtained, Because of the nonlinearity behavior of the equations. Its closure system is performed in various ways, which leads to different models of turbulent flow. Known closure methods are based on certain assumptions where the substantiation of which presents significant difficulties and contains parameters that are determined empirically. At the same time, these parameters were selected for several decades on flows whose parameters varied in very large ranges. Therefore, it can be expected that these models will create an adequate model of conjugation of vortex flow with downstream for the hydroelectric power stations on the right scale.

In the present paper, the $k - \varepsilon$ model is used to create a complete mathematical model of a hydroelectric power station with a vortex discharge, and this is verified by comparing with a reduced physical model of the hydroelectric power station. For calculations, the standard tool package Fluent in ANSYS R 19.0 is used for working with empirical models of turbulent flows. This system provides an access to many packages which focused on modeling of phenomena from various areas of physics but combined with the general idea – discretization of partial differential equations by the method of finite elements analysis (FEA), and by the modern classification refers to FEA software. The choice of this system is primarily due to the fact that the ANSYS Inc. was established in 1970 specifically for the development of FEA software and now has accumulated a lot of experience in this matter. Currently, Fluent ANSYS is the most famous and universal software for modeling the turbulent flows, new software packages are always tested to coincide with the results of Fluent [24]. It also described in detailed documentation and furthermore included with ANSYS, incl. the User Guide and the training literature.

It should be noted that we have at our disposal a mathematical model of the vortex flow, which does not take viscosity into account and cannot describe the conjugation with stream, but it is solved analytically. In this study, a comparison will be made with this simple mathematical model, specifically, as suggested by Samarsky et al. [9], a hierarchy of vortex flow models will be designed. In Summary, the aim of the present study can be formulated through creation and testing of a model for the conjugation of vortex flow with downstream along with its verification by carrying out a complete experiment on a small scale.

The objectives of this study is to develop a mathematical model of conjugation of vortex spillways with the downstream by ANSYS modelling and proving the outcome with the experimental results.

2. Methods

2.1. Mathematical Model

The vortex flow is described as a pipe of constant circular cross-section and in the physical model its diameter is 50 mm. An impeller was placed at the entrance of the flow to create swirl (Figure 1). The downstream for simplicity is described as a rectangular parallelepiped and its dimensions are taken smaller than the physical model in order to save the calculating resources.

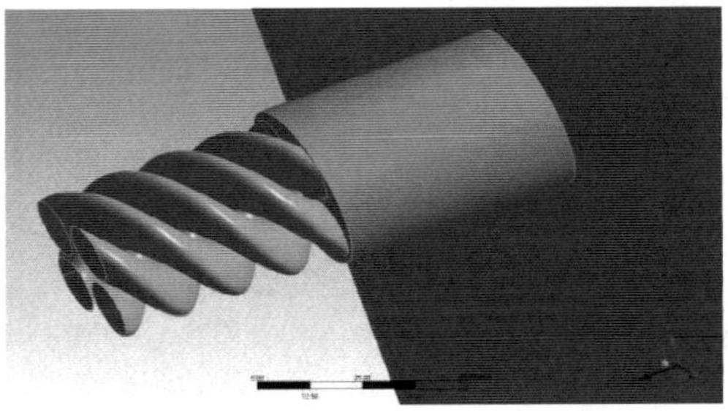

Figure 1. Discharge with the 4-channel impeller included in the downstream.

The modeling is based on the continuity equation and the averaged Navier-Stokes equation (Reynolds-averaged Navier-Stokes equation, RANS)

$$\frac{\partial U_i}{\partial x_i} = 0$$

$$\rho\frac{\partial U_i}{\delta t} + \rho U_j\frac{\partial U_i}{\partial x_j} = -\frac{\delta P}{\partial x_i} + \frac{\delta}{\partial x_j}(2\mu S_{ij} + \rho\tau_{ij}$$

where U and P – averaged velocity and pressure, respectively, and

$$S_{ij} = \frac{1}{2}\left(\frac{\partial U_i}{\partial x_j} + \frac{\partial U_j}{\partial x_i}\right)$$

For the density and viscosity of water, the standard values are taken. In order to close this system, it is also necessary to determine the components of the Reynolds stress tensor T_{ij}. We will use the standard model proposed by Launder and Sharma in the early 1970s [9]. This is the most popular model for describing turbulent flows [25]. In the Fluent module, other popular models of turbulent flow are implemented, computational experiments have shown that the $k - \varepsilon$ and $k - \omega$ models give approximately the same results, so we are considering the k and ε model [11, 25]. To describe this model, it is necessary to introduce the kinetic energy of turbulent pulsations

$$K = \frac{1}{2}\overline{u_i' * u_i'}$$

And the rate of dissipation

$$\varepsilon = v\overline{\frac{\delta u_i'}{\partial x_k}\frac{\delta u_i'}{\partial x_k}}$$

Within the framework of the model, it is assumed that the Reynolds stress tensor is expressed through the functions as follows:

$$\tau_{ij} = 2v_T S_{ij} - \frac{2}{3}k\delta_{ij}$$

where the kinematic viscosity v_T is determined from considerations of dimensionality as $v_T = C_\mu k^2/\varepsilon v_T$.

The coefficient C_μ is assumed to be constant, and was determined by the empirical data as $C_\mu = 0.009$. The dynamics of these values, characterizing turbulence, are described by two equations which are similar to the Navier-Stokes equation. The dynamics of the energy k is described by the equation below:

$$\frac{\delta k}{\delta t} + U_j \frac{\delta k}{\partial x_j} = \tau_{ij} \frac{\partial U_i}{\partial x_j} - \varepsilon + \frac{\delta}{\partial x_j}\left[\left(v + \frac{v_T}{\delta_k}\right)\frac{\delta k}{\partial x_j}\right]$$

$$\frac{\delta \varepsilon}{\delta t} + U_j \frac{\delta \varepsilon}{\partial x_j} = C_{\varepsilon 1} \frac{\varepsilon}{k} \tau_{ij} \frac{\partial U_i}{\partial x_j} - C_{\varepsilon 2} \frac{\varepsilon^2}{k} + \frac{\delta}{\partial x_j}\left[\left(v + \frac{v_T}{\delta_\varepsilon}\right)\frac{\delta \varepsilon}{\partial x_j}\right]$$

where $\delta_k = 1.0$. The dynamics of the rate of dissipation is described by equation and the coefficients are taken empirically as given below:

$$C_{\varepsilon 1} = 1.44, C_{\varepsilon 2} = 1.92, \ \sigma_\varepsilon = 1.3$$

Adding these two equations to RANS gives 6 equations with 6 unknowns, four of which P, U_1, U_2 *and* U_3 are characterized as the flow on average, and the remaining two unknowns considered with the turbulence of the flow. We consider only the steady flow, i.e. substitute $\delta t = 0$. We can add boundary conditions of the impermeability of the walls of discharge and stream for these six equations then set the rate at the discharge entrance. In all calculated experiments, a velocity of 10 cm/sec was taken at the inlet of the discharge, which ensured a noticeable swirling of the flow in the discharge. The conditions at the ends of the stream were selected in such a way that to provide a drifting flow of 4 cm/sec.

Within the framework of the described geometry and model, the ANSYS R 19.0 system was used for calculations. For the numerical solution of the system of partial differential equations with the resulting system of partial differential equations, the discretization by the finite element method was performed. A more detailed grid was used for the discharge than for the stream. The system of algebraic equations obtained after such a discretization is solved by means of an iterative process. The convergence is estimated from the difference of speeds, k and ε, found at neighboring steps. In all experiments, for 100 iterations, we were able to achieve accuracies for velocities about 10^{-3} m/sec (Figure 2). This is more than enough for engineering calculations.

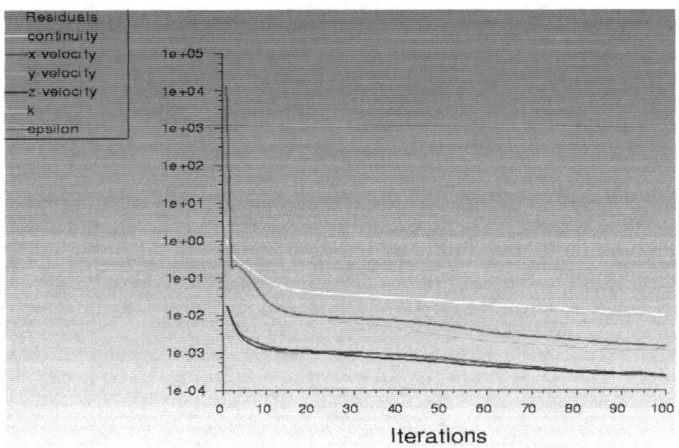

Figure 2. Deductions for estimating the convergence of the iterative method, the velocity values are within 10^{-3} m/sec.

In this Method, conjugation devices were used for the water outlet below the downstream level. Different angles of the drifting flow (0 - 90°) were used for the study, which are 30°, 45° and 90°. As a downstream, a hydraulic chute was used, which allows to create a drifting flow and also a dimensional grid was applied in the place of the outlet of the swirling flow.

The control of velocity in the downstream (behind the swirling device) was carried out using FlowTrecker2 (FT2). FT2 uses a reliable Acoustic Doppler Velocimetry (ADV) technology, tested by experts from around the world for decades in hydraulic laboratories and in a wide range of operating conditions. The modified and improved acoustic ADV sensor for the FT2 provides unprecedented accuracy, especially with minimal water flow and in shallow water. 2D data in the horizontal plane (2D and 3D options available) allows to perform the most comprehensive quality control and get an idea of the flow parameters.

Determination of the actual parameters of the swirling jet in a fixed space filled with the same fluid, as well as in a drift flow for different modes with different residual swirl including the initial section (L ≤ 10d). Choose the modes with the most suitable for economic efficiency and safety of operation of hydraulic structures. Confirm the relationship between model data and actual values.

3. Results and Discussion

A series of computational experiments was performed in which the swirling flow entered the hydraulic chute at various angles, and the obtained results were compared with real observations. The following three cases were studied:

1. The vortex flow is perpendicular to the flow in the downstream (Figure 3);
2. The vortex flow enters the downstream at an angle of 30° (Figure 4);
3. the vortex flow enters the downstream at an angle of 60° (Figure 5).

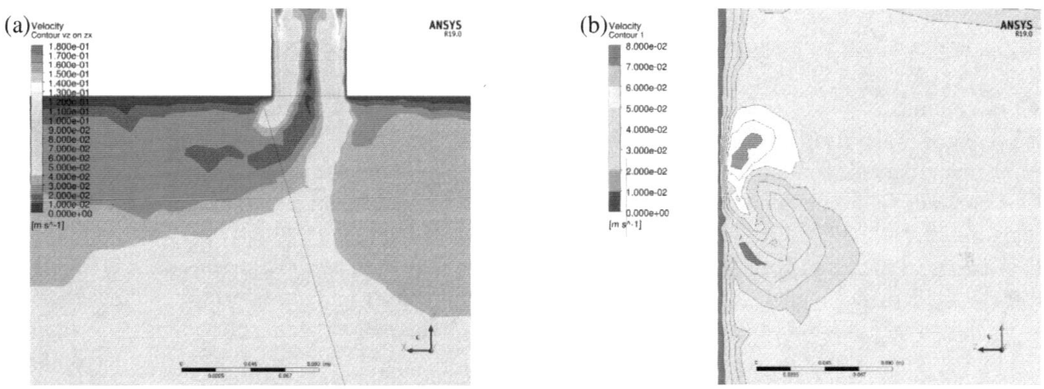

Figure 3. Velocity module for the 1st case, horizontal section (a) and vertical section (b); vertical section was made along the plane, marked on the horizontal section by the black line.

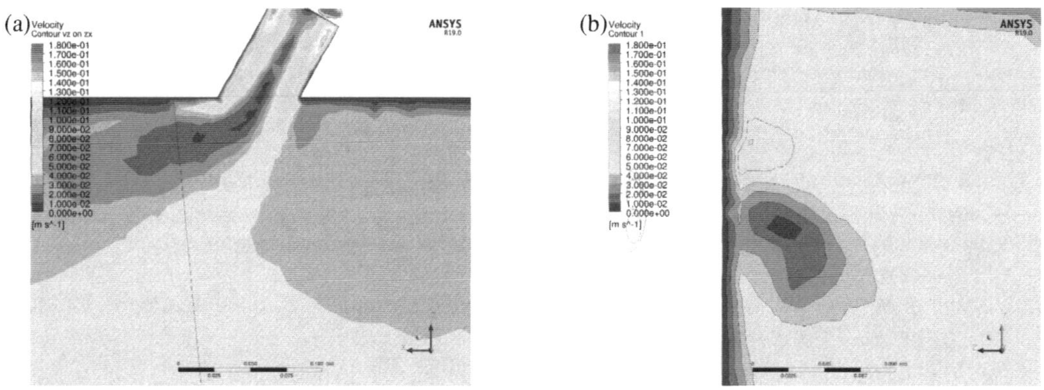

Figure 4. Velocity module for the 2nd case, horizontal section (a) and vertical section (b); vertical section was made along the plane, marked on the horizontal section by the black line.

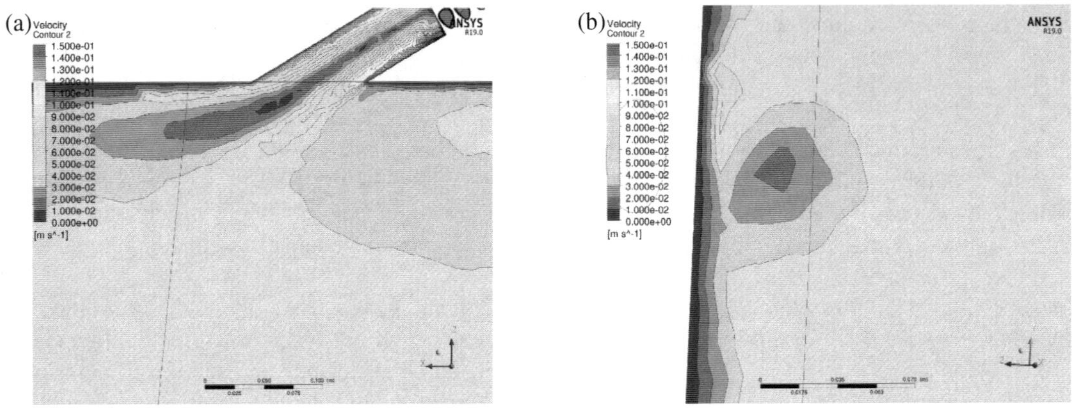

Figure 5. Velocity module for the 3rd case, horizontal section (a) and vertical section (b); vertical section was made along the plane, marked on the horizontal section by the black line.

A velocity of 10 cm/sec on the course to the discharge and 4-channel impeller provides a noticeable vortex flow in the discharge. As predicted by the analytical model of Zhivotovsky [5], the maximum velocities were reached near the walls, and a barely noticeable reverse flow appeared on the axis. Since our model, in contrast to the analytical model, takes viscosity into account, the vortex flow is straightened after a certain distance from the swirler. In the stream near the entrance of the vortex flow, a pocket emerges at very high velocities. These velocities are directed against the drifting flow, but not in the direction of the discharge axis, i.e. the velocity is extinguished in the direction of flow. The smaller the angle between the discharge and the drifting flow, the more noticeable this phenomenon.

4. Conclusion

On the basis of the experimental study, the following results were achieved:

1. A mathematical model of conjugation of vortex spillways with the downstream has been created;
2. A model was implemented in the ANSYS Fluid system and a series of numerical experiments was performed;
3. A physical model of conjugation of vortex spillways with the downstream has been created;
4. A series of experiments was performed using modern measuring equipment FlowTrecker2;
5. A mathematical model is verified by comparison with the real experiment.

The obtained results allow to determine the necessary parameters of the swirling flow required in evaluating the conjugation conditions of the spillway structures with the downstream.

References

[1] Bhosekar V V C, Deolalikar P B and Sridevi M I 2012 Hydraulic design aspects for swirling flow at vertical drop shaft spillways *ISH Journal of Hydraulic Engineering* **13(3)** pp 78–89

[2] Zhivotovsky B A 1995 *Spillway and interfacing structures with swirling flow* (Moscow: RUDN University Press)

[3] Galant M A, Zhivotovsky B A, Novikova I S, Rodionov V B and Rozanova N N 1995 Features of vortex tunnel spillways and hydraulic conditions of their operation *Hydrotechnical Construction* **9** pp 16–22

[4] Volshanik V V, Zuykov A L and Mordasov A P 1990 *Swirling flows in hydraulic structures* (Moscow: Publishing House Energoatomizdat) p 280

[5] Volshanik V V, Orekhov G V, Zuykov A L and Karelin V Y 2000 Engineering hydraulics of swirling fluid flows *Hydrotechnical Construction* **11** pp 23–26

[6] Mitrofanova O V, Podzorov G D and Pozdeeva I G 2000 Vortex structure of swirl flows *International Journal of Heat and Mass Transfer* **65** pp 225–234

[7] Gesheva E S, Shtork S I and Alekseenko S V 2017 Investigation a single-spiral vortex in a swirl flow *MATEC Web of Conferences* **115** 02025

[8] Akhmetov D G and Akhmetov T D 2015 Swirl flow in vortex chamber *Physical and mathematical Sciences* **4(6)** pp 109-120
[9] Samarsky A A and Mikhailov A P 2001 *Primery Mathematical modeling: Ideas, Methods* (Moscow: Publishing House Fizmatlit] p 320
[10] Samarskii A A and Mikhailov A P 2018 *Principles of Mathematical Modeling; Ideas, Methods, Examples* (Frorida: CRC Press) p 360
[11] Kartashev A L and Krivonogov A A 2016 Mathematical Modelling of Vortex Generation Process in the Flowing Part of the Vortex Flowmeter and Selection of an Optimal Turbulence Model *Bull. South Ural State Univ. Ser. Mathematical Model and Program* **9(4)** pp 105–116
[12] Sangwha Y 2019 Exact Solution of Navier-Stokes Equations *International Journal of Advanced Research in Physical Science* **6** pp 39–43
[13] Vellando P, Puertas J, Bonillo J and Fe J 2000 Finite element solution of the Navier-Stokes equations using a SUPG formulation *Advances in Computational Engineering and Sciences* pp 856–861
[14] Alam M and Saha S 2017 Normal stress differences and beyond-Navier-Stokes hydrodynamics *EPJ Web of Conf* **140** 11014
[15] Ramm A G 2019 Solution of the Navier–Stokes problem *Appl. Math. Lett.* **8** pp 160–164
[16] Ramm A G 2019 On the Navier-Stokes problem *J. Adv. Math.* **16** pp 8262–8266
[17] Willis A P 2017 The Open pipe flow Navier–Stokes solver *Software X* **6** pp 124–127
[18] Launder B 2003 Application of the energy-dissipation model of turbulence to the calculation of flow near a spinning disc *Int. Commun. Heat Mass Transf.* **1(2)** pp 131–137
[19] Sun S, Jenkins E W, Zhangxing C and Geiser J 2011 Mathematical and numerical modeling of flow and transport *Jornal of Applied Mathematics* Article ID 901380.
[20] Parra T, Castro F, Rodriguez M A, Szasz R Z, Robert Z, Gutkowski A and Perez R 2015 Numerical Simulation of Swirling Flows - Heat Transfer Enhancement *Journal of Fluid Flow, Heat and Mass Transfer* **2** DOI:10.11159/jffhmt.2015.001
[21] Westerweel J 2009 Advanced Experimental Methods For Turbulent Shear Flows *Proc. 6th Int. Symp. on Turbulence and Shear Flow Phenomena* pp 779–784
[22] Teresa P, Perez J R, Szasz R Z, Rodriguez M A and Castro F 2015 Numerical Modelling of flow pattern for high swirling flow *EPJ Web of Conferences* **92** 02059
[23] Belov I A and Isaev S A *Modeling of turbulent flows* (St. Petersburg: Publishing House)
[24] *ANSYS Fluent Theory Guide. Release 15* 2013 (Canonsburg: ANSYS Inc.)
[25] Georgiev D J, Roberts A J and Strunin D V 2007 Nonlinear Dynamics on Centre Manifolds Describing Turbulent Floods: K-W Model *Discrete and Continuous Dynamical Systems* **2007** pp 1-10

The penetration modeling of flat obstacles in Ansys Autodyn program

O K Kroytor

Peoples' Friendship University of Russia (RUDN University), Moscow, Russia

Email: olegkroytorge@gmail.com

Abstract. The article is devoted to mathematical modeling of flat obstacles punching in Ansys Autodyn software package with the use of finite element method. The tasks of punching flat, composite, two-layer barriers with a drummer at a right angle and 45 degrees angle are considered. With the help of modeling in Ansys Autodyn, the behavior of obstacles is investigated in time and after they are punched by a drummer. In the work, numerical and semi-analytical solutions of the problems were obtained, the results of the solution were presented, comparisons were made and corresponding conclusions were made. A numerical simulation of high-speed collision of barriers with a drummer was carried out, and the rate of through penetration of barriers consisting of several layers was obtained. The layers in the barriers are made up of various materials. The results of numerical simulation of the barrier and the impactor with a conical head are given.

Keywords: Ansys Autodyn, mathematical modeling, simulation of high-speed collision, semi-analytical solution, punching flat obstacles.

1. Introduction

Modern aircrafts, including spacecraft, artificial satellites, orbital stations, are continuously subject to collisions with various particles (large cosmic dust, fragments of meteorites, and so on) during operation. Constant collisions, especially high-speed collisions, lead to an accumulation of damage in the material of the apparatus construction and in the future to failure [15]. In this regard, the above devices are supplied with special protective coatings. As a rule, the protective coating consists of a material other than the outer layer of the structure, since the protective layer has a different purpose. For example, if the main material of the outer layer of the aircraft is metal (alloy), then ceramics can act as a protective layer. The advantages of using ceramics for protective layers are, firstly, that ceramics have greater strength, at least along one of their directions, due to the anisotropy of physical and mechanical properties, in contrast to metals. Secondly, ceramics better absorb energy during collisions.

The existence of high-performance computing, the availability of the necessary mathematical and software allow us to consider mathematical modeling as one of the most effective ways to assess the strength, predict durability and optimize structures and manufacturing processes, which are based on the processes of impact and explosion. One of the most effective automated engineering analysis systems is the Ansys Autodyn software package [1]. This work also uses the Ansys Autodyn package. This package for numerical calculations of shock processes is also used in [2, 14].

A number of works are devoted to the problems of modeling of obstacles punching, in which obstacles consisting only of ceramics, only of metals and composite are considered. However, the most interesting are the works devoted to modeling of multilayer barriers [5, 6].

A review of papers in which analytical dependencies were proposed for calculating the limiting velocities of penetration of two-layer barriers is given in [3]. A common drawback of these works is the fact that the most important components of the process of breaking through the barrier are not taken into account [12]. First, the destruction of the impactor and the associated decrease in the kinetic energy of the impactor are not taken into account. Secondly, the destruction of ceramics and the cost of this kinetic energy of the impactor are not taken into account consistently or not at all.

This article proposes a model that takes into account the processes of deformation and destruction of the striker, the destruction of the ceramic cone and the deformation of the substrate.

The aim of this work is to calculate the rate of through punching of flat, composite obstacles by the striker, by methods of numerical and semi-analytical solutions, as well as to compare the results.

2. Materials and method of research

2.1. Mechanical statement of tasks

Elastic steel drummer with conical head of structural steel (Structural Steel) hits at an angle of 90° and 45° degrees on the combined obstacles, rigidly fixed at the edges (figure 1). Barriers consist of a set of plates, which consist of different materials (table 1) and are rigidly fixed at the edges.

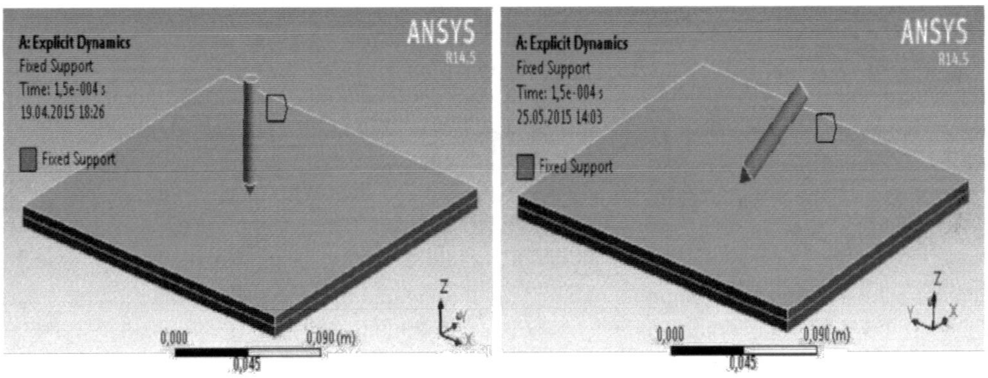

Figure 1. The results of the decision in the program.

Plate size - 20×20 cm;
The thickness of each of the plates - 0.5 cm;
Diameter of the drummer - 0.9 cm;
The height of the drummer - 5.5 cm.

Table 1. Barrier properties.

Barrier number	Barriers consist of the following materials:	Weight, kg	Barrier thickness, cm
1	1 layer is made of B_4C (boron carbide); 2 layer is made of S7 grade steel	2.052	1
2	1 layer is made of 1600 grade steel; 2 layer is made of S7 grade steel	3.1292	1

Boron carbide B_4C has a low density and other useful parameters that make this substance indispensable in the manufacture of protective products armored plates. The characteristics of the materials that make up the barriers were taken from the works [7, 8].

2.2. Mathematical formulation of the problem.

The equations underlying the problem include the equation of motion, the equation of energy, the equations describing the property of the material, the law of conservation of mass, contact, initial and boundary conditions.

The boundary conditions the plate along the contour is rigidly fixed, mathematically it can be represented as $u|_{x=\mp a}=0$, $u|_{z=\mp a}=0$ where a is the length and width of the plate.

Initial conditions: $u|_{t=0}=0$, $\dot{u}|_{t=0}=v_i$,

where: v_i is the speed given at the initial moment of time.

For the material of the drummer, we use an elastic material model. For her, Hooke's law is fulfilled:

$$\sigma_{ij}=\lambda J_1(\varepsilon)\delta_{ij}+ 2\mu\varepsilon_{ij}, \tag{1}$$

where ε_{ij} is the strain tensor, $J_1(\varepsilon)$ is the first invariant of the strain tensor, λ , μ are Lame constants. $\mu=\frac{E}{2(1+v)}$, $\lambda=\frac{Ev}{(1\text{-}2v)(1+v)}$, E is Young's E-module, ν is Poisson's coefficient.

For the plate material, an elastic-plastic material model was used. The condition of the beginning of plasticity: $(\sigma_1+\sigma_2)^2+(\sigma_2+\sigma_3)^2+(\sigma_3+\sigma_1)^2=\left(\sqrt{2}\sigma_m\right)^2$,

where: $\sigma_1,\sigma_2,\sigma_3$ are the main stresses, σ_m is the yield strength.

Contact interaction of two surfaces is described using the following formulas: $z_1+z_2=\alpha\text{-}(w_1\text{-}w_2)$ for points of contact area, and $z_1+z_2>\alpha\text{-}(w_1\text{-}w_2)$ for points outside the contact area.

Moving points M_1, belonging to one contact surface and M_2 belonging to another contact surface, occurring due to deformation of parts of bodies near the point of tangency, we denote by w_1 and w_2, and the movement of points at the origin is w_1^0 and w_2^0. Obviously, the quantity $\alpha = w_1^0 + w_2^0$ represents the convergence of contiguous bodies as a result of their compression, at which the distance between points M_1 and M_2 decreases by the value of α - $w_1^0 + w_2^0$ [9].

2.3. Methods for solving the problem

2.3.1. Semi-analytical method for solving the problem

In the work of Ilyushin [10] considered the analytical method of solving the problem of penetration of barriers, and obtained the formula of the speed of the projectile after penetration of the projectile solid and multilayer barriers

2.3.2. The method of solving the problem in Ansys Autodyn

To calculate tasks in ANSYS, the finite element method is used. The essence of this method is that the continuous medium is replaced by a finite number of structural elements of finite size, connected with each other only at the nodal points. Thus, we superimpose a grid on our object and consider the movement of particles in time.

The process of implementation of FEM in the problem was divided into several stages:
– the choice of a mathematical model;
– the creation of a geometric model of the object;
– setting the properties and models of materials;
– the creation of a finite element grid;
– setting the boundary conditions;
– modeling loads;
– setting the solver;
– the solution of equations and the calculation of stresses in the nodes;
– analysis of the results[13].

In this study, we used a spatial grid of tetrahedral or hexahedral. The computational grid in the area of collision of the barrier with the striker was crushed to improve the accuracy of the calculation. Poor grid quality, even in the case of accurate modeling of other stages can lead to incorrect results. On the one hand, the smaller the size of the final element, the higher the accuracy of the calculation, on the other hand, too thick mesh weights the model and increases the calculation time [11].

2.4. Numerical and semi-analytical solution of problems

Consider the interaction of obstacles and the drummer at right angles. The thickness of the barriers 1 cm, cross size: square 20 cm on the edges of the plate clamped obstacles. From the list of obstacles (table 1), you want to determine the minimum speed of the striker for through punching obstacles.

The numerical results of the tasks (table 2) were obtained by a series of excrement in the Ansys Autodyn software package.

For semi-analytical solution of the problems, the equations obtained in the works of Ilyushin [1] were used. And the following values were used:

Cone height - 0.6975 cm;

The radius of the cone - 0.45 cm;

$R = 0.45$ cm (the radius of the base of the drummer);

$H = 5.5$ cm (the height of the drummer);

$m = 0.025145$ kg (the mass of the projectile);

$\varphi = 30°$ (the angle between the axis of the drummer and the perpendicular to the contact plane of the drummer and the obstacle);

$\mu = 0.8$ (friction coefficient);

$k = 3.5 \cdot 10^8$ Pa (kg / m · s^2) (shear yield strength).

Table 2. Results of numerical solution of the problem at an angle of 90°.

Barrier	V_0 at an angle of 90 °, m / s	φ_0, radians	e , cm
1	795	2π/5	0.9231
2	720	2π/5	0.9549

where V_0 is the average minimum speed of the drummer for the through penetration of the barrier, obtained in Ansys Autodyn; e is the maximum deflection of the obstacle, taken from experiments; φ_0 is the bend angle of the destructive part of the barrier, taken from experiments.

Semi-analytical solution of the problem is calculated using Matlab. Writing down the necessary equations and variables in Matlab, and calculating the average velocity of the striker (V) for punching obstacles, the following results were obtained (table 3).

Table 3. Results of semi-analytical solution of the problem at an angle of 90°.

Barrier	V at an angle of 90 °, m / s
1	-
2	733.95

3. Results

3.1. The results of the process of breaking down barriers in Ansys Autodyn

The process of impact proceeds as follows. First, an object moving at a certain speed enters into contact with the system, and the speed of its movement decreases sharply. The system is in motion. However, due to the inertia of the mass of the system, its particles do not begin to move simultaneously. It takes some time for the deformation wave to spread from the point of impact throughout the system. The leading edge of the wave moves through the system at the speed of sound propagation in a given medium. After contact, the object moves together with the shock-sensing system, and the speed of their movement as the deformation and elastic forces of the system gradually decreases.

At the obtained speeds in Ansys Autodyn the drummer breaks through the obstacles and causes serious deformation of the barrier in the impact area. When the striker penetrates into the barrier, the elements of the plates under the striker begin to be removed (figure 2), due to the impact of stresses on the Mises in the barrier sharply increase, after which there is a sharp drop in stresses. The results are partially presented.

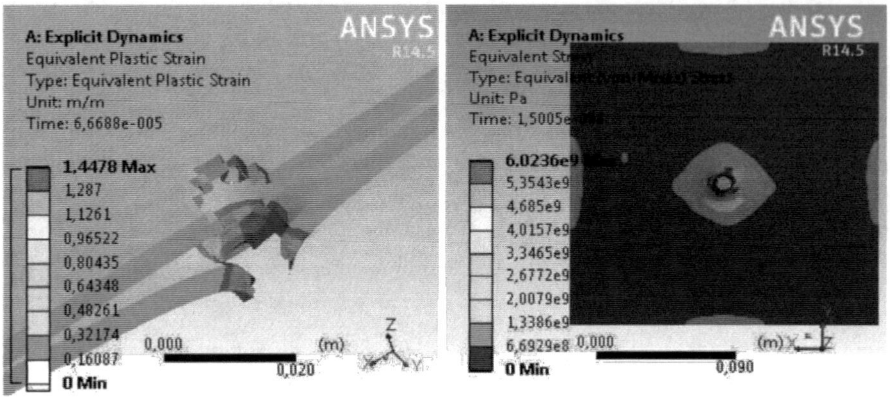

Figure 2. The results of the decision in the program.

3.2. Comparison and verification of results

Table 4 shows the results of solving problems obtained by numerical and semi-analytical methods.

Table 4. Comparison of results.

Barrier	Semi-analytical *V* at an angle of 90°, m / s	V_0 at an angle of 90°, m / s	V_0 at an angle of 45°, m / s
1	-	795	1057
2	733.95	720	1030

where V_0 is the average minimum velocity of the striker for through penetration of the barrier obtained in Ansys Autodyn (numerical solution); *V* is the average minimum speed of the striker for through penetration of the barrier obtained by the semi-analytical method.

First of all, we note the high efficiency of using a metal substrate in terms of resistance to penetration of the composite barrier - an increase in the thickness of the all-ceramic barrier leads only to a slight decrease in the residual velocity, while the use of substrates gives a greater effect.

From the list of obstacles, on the basis of the results of the numerical and semi-analytical solution, the strongest barrier consisting of a ceramic layer (B_4C boron carbide) and a metal layer (Steel S7) was determined.

Calculations made with the help of Ansys Autodyn software package show how significantly different the rate of penetration of the barrier at right angles and at an angle of 45°, as well as the results obtained by numerical and semi-analytical solution of the problem differ by about 5-10%. This indicates a fairly accurate result obtained when solving problems of this type in the program Ansys Autodyn.

Comparing the numerical results obtained in this work with the experimental and numerical calculations presented in [12] in tables 1 and 5, which also considers the punching of two-layer barriers with a substrate, the difference reaches about 7-13%. Given that the paper [12] considers barriers consisting of other materials and the thickness of the ceramic layer varies from 5 to 11 mm, and the thickness of the substrate from 5-8 mm, it can be concluded that the results obtained in this paper have a good match with the results presented in [12].

Conclusions

1. The perforation penetration rates of two-layer ceramic-metal and metal barriers at angles of 90 and 45 degrees were obtained, taking into account the structural characteristics and physicomechanical properties of the materials of the impactor and obstacles.

2. Verification of the velocities obtained by a numerical method in the Ansys Autodyn package and experimental ones was carried out. The speeds obtained by a numerical method with the correct setting of the solver in the Ansys Autodyn program differ from the experimental ones by 5-10%.
3. When calculating the speed by the analytical method, the detected deficiencies in the calculations of Ilyushin and introduced corrections.

Acknowledgments
The author is grateful to professor Zhidkov Alexander Vasilievich for problem statement and discussion of presented above results.

References

[1] ANSYS Advantage: Russian Edition 2012 *Engineering and Technical Journal* **17** p 68 URL: http://www.ansysadvantage.ru/fileadmin/archive/ANSYS-ADVANTAGE-Rus-17.pdf
[2] Kobylkin I F and Gorbatenko A A 2016 Phenomenological model of perforation ceramic plates *Vestn. Mosk. Gos. Tekh. Univ. im. N.E. Baumana: Mashinostr* [*Herald of the Bauman Moscow State Technical University: Mechanical Engineering*] **6(111)** pp 62–73 URL: https://elibrary.ru/item.asp?id=27669644
[3] Ben-Dor G, Dubinsky A and Elperin T 2013 High-Speed penetration dynamics: Engineering models and methods (Singapore: World Scientific Publishing) p 696
[4] Kobylkin I F 2017 Mechanics of penetration of ceramic targets *Combustion, Explosion, and Shock Waves* **53(1)** pp 110–115
[5] Kobylkin I F 2017 Penetration of double-layer targets with an outer ceramic layer and optimization of their structure *Combustion, Explosion, and Shock Waves* **53(4)** pp 483–489
[6] Mossakovsky P A, et al 2015 Investigation dissipative factors in puncturing multilayered woven obstacles *Problems of strength and plasticity* **77(4)** pp 385–392 URL: https://elibrary.ru/item.asp?id=25281092
[7] Johnson G R and Holmquist T J 1999 Response of boron carbide subjected to large strains, high strain rates, and high pressure *J. Appl. Phys.* **85(12)** pp 8060–8073
[8] Johnson G R and Cook W H 1983 A constitutive model and data for metals subjected to large strains, high strain rates and high temperatures *Proc. 7th International Symposium on Ballistics 1983* (Hague, Netherlands) pp 541–547
[9] Demidov S P 1979 *Theory of elasticity* (Moscow: Publishing House Vysshaya Shkola) p 432
[10] Ilyushin E A, Korotkina N R and Ilyushin A A 1945 *Scientific works* vol. 1 (Moscow: Publishing House Fizmatlit) p 352
[11] Ivanov D V and Dol A V 2016 *Introduction to Ansys Workbench: Textbook. method. manual for students of natural sciences* (Saratov: Amirit Press) p 56
[12] Kobylkin I F and Gorbatenko A A 2018 Analytical model of trough-penetration of ceramic-metal protective barriers and optimization of such barriers *Vestn. Mosk. Gos. Tekh. Univ. im. N.E. Baumana: Mashinostr.* [*Herald of the Bauman Moscow State Technical University: Mechanical Engineering*] **2(119)** pp 17–30 URL: https://elibrary.ru/item.asp?id=32808948
[13] Singiresu S. Rao 2018 *The Finite Element Method in Engineering* vol. 6 (Elsevier) p 782 URL: https://doi.org/10.1016/C2016-0-01493-6
[14] Aitor Arriaga, Rikardo Pagaldai, Ane Miren Zaldua, Alicia Chrysostomou, and Michael O'Brien 2009 Impact testing and simulation of a polypropylene component. Correlation with strain rate sensitive constitutive models in ANSYS and LS-DYNA *Elsevier* **29(2)** pp 170-180 URL: https://doi.org/10.1016/j.polymertesting.2009.10.007
[15] Erkai Watson, Max Gulde and Stefan Hiermaier 2017 Fragment Tracking in Hypervelocity Impact Experiments *Combustion, Explosion, and Shock Waves* **vol. 204** pp 170–177 URL: https://doi.org/10.1016/j.proeng.2017.09.770

IOP Conf. Series: Materials Science and Engineering **675** (2019) 012028 doi:10.1088/1757-899X/675/1/012028

Methodology for calculating the stress-strain state of robotic system made of composite materials

K Z Khayrnasov

Moscow Aviation Institute (National Research University), Moscow, Russia

Email: kamilh@mail.ru

Abstract. The methodology for calculating the stress-strain state of robotic systems made of composite material under dynamic load is considered. The problem is solved by the finite element method. Formulas for determining the reduced stiffness of a multilayer composite material are given depending on the location of the base of the composite in layers, as well as in three-layer structures consisting of external bearing layers and a layer of aggregate between them. The method of placing the base layers of a composite material in the structure under study along the lines of maximum stress is considered. Particular attention is paid to the accounting of gearboxes, gears and support bearings in robotic systems. The results obtained in relation to the multi-stage dynamic stand, made of composite material.

Keywords: composite materials, robotic systems, multi-stage stands, finite element method, stress-strain state.

1. Introduction

Composite materials are widely used in recent years due to the ability to change the characteristics of the material depending on the location of the base of the composite material. High strength characteristics and a relatively small specific gravity of composite materials made it possible to use them in various areas of technology: aircraft manufacturing, shipbuilding, automotive, etc [1]. At the same time, the use of composite materials in robotics is not enough. This is primarily due to the presence in these designs of gear rims, gearboxes, bearing supports made of homogeneous materials and therefore not compatible with the main composite structure. In connection with the use of composite materials in various fields of technology, various methods are being developed for the calculation of multilayer composite structures. Modern numerical methods are used, mainly the finite element method [2, 3]. Currently, there are a sufficient number of computer-aided design systems for calculating structures made of composite materials [4], but given the many factors that affect the bearing capacity of composite structures, the method for calculating composite materials is being developed and improved [5-7].

Consideration of the stress-strain state of a robotic structure containing gearboxes, bearings and gears is a complex task. But at the same time, taking into account their interaction allows to take into account the work of all parts of the structure during operation. Therefore, the solution to this problem is an important and relevant topic.

Consider the definition of the stress-strain state of structures made of composite materials on the basis of a multi-stage stand of semi-natural modeling (hereinafter referred to as a stand). The use of composite material in the stands has not yet been applied. At the same time, the main characteristic of the stand performance is small inertial characteristics and maximum rigidity, which largely depend on the specific strength of the material. Composite material has a higher specific strength compared to homogeneous metals. For example, the specific strength of magnesium alloy, the most fully used material in the manufacture of the stand is 12-15 kilometers, titanium - 26-40 kilometers, while this indicator for composite material is 50-100 kilometers.

The main disadvantage of the composite material is its low rigidity in the direction that does not coincide with the base of the composite. At the same time, the composite material is multi-layered; therefore, varying the layer-by-layer direction of the base of the composite can achieve a sufficient rigidity of the structure [8]. Consideration of the stress-strain state of a robotic structure containing gearboxes, bearings and gears is a complex task. But at the same time, taking into account their interaction allows to take into account the work of all parts of the structure during operation.

Methods for counting bearings, gears and especially gearboxes in robotics are not yet fully developed. Therefore, the methods of accounting for such elements in robotic systems are important and relevant.

2. Method and solution building

In this paper, the stand made from composite material is considered. The stand includes both shell elements of different curvature, as well as parts that are not amenable to direct approximation by finite elements, such as, for example, bearing supports, gearboxes, and toothed rims. Therefore, when approaching the bearing supports, a rod structure was applied with rigidity equal to that of the bearing support. The gearbox was replaced with a core system with stiffness that matched the rigidity of the gearbox; the same procedure was used for gears [9].

The basic design of the stand was approximated by multilayer shell finite elements of different curvature, including three-layer finite elements with external carrier layers and a filler between them. The filler prevented the approach of the bearing layers and perceived the shear stresses

Figure 1 shows a stand approximated by finite elements.

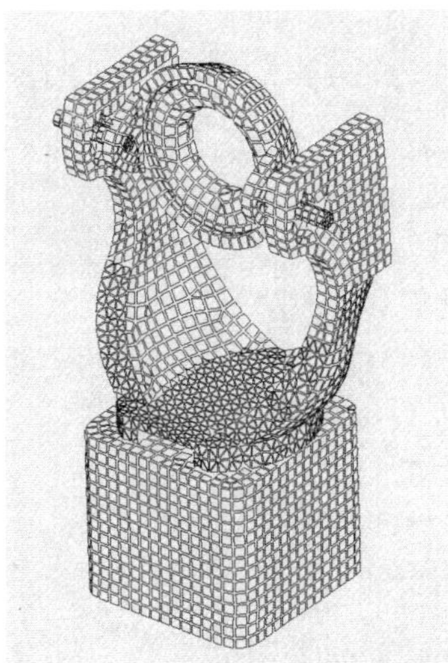

Figure 1. The finite element approximation of the stand.

2.1. Stress-strain equations

The characteristics of a composite material having layers with a different arrangement of the base, in the method of reduced stiffness, can be obtained as follows.

The connection matrix of stresses and strains for an orthotropic material, the orthotropic axes of which coincide with the axes of coordinates, in the plane-stressed state have the form

$$\boldsymbol{\sigma} = (E)\boldsymbol{\varepsilon}, \tag{1}$$

where $$(E) = \begin{pmatrix} Q_{11} & Q_{12} & 0 \\ Q_{21} & Q_{22} & 0 \\ 0 & 0 & Q_{66} \end{pmatrix}, \boldsymbol{\varepsilon}^T = (\varepsilon_s,\ \varepsilon_\theta,\ \varepsilon_{s\theta}),\ \boldsymbol{\sigma}^T = (\sigma_s,\ \sigma_\theta,\ \sigma_{s\theta}),$$

$$Q_{11} = E_s/(1 - \nu_{s\theta}\nu_{\theta s}),\ \ Q_{12} = \nu_{s\theta}E_s/(1 - \nu_{s\theta}\nu_{\theta s}),$$

$$Q_{21} = \nu_{\theta s}E_s/(1- \nu_{s\theta}\nu_{\theta s}),\ Q_{22} = E_\theta/(1- \nu_{s\theta}\nu_{\theta s}),\ Q_{66} = G_{66}.$$

Rotation of the axes of coordinates at an angle θ converts the matrix of elastic coefficients to the form

$$(\bar{E}) = \begin{pmatrix} \bar{Q}_{11} & \bar{Q}_{12} & \bar{Q}_{16} \\ \bar{Q}_{21} & \bar{Q}_{22} & \bar{Q}_{26} \\ \bar{Q}_{61} & \bar{Q}_{62} & \bar{Q}_{66} \end{pmatrix},$$

where $\bar{Q}_{11} = c^4Q_{11} - s^4Q_{22} + 2(Q_{12} + 2Q_{66})s^2c^2,$

$$\bar{Q}_{12} = (Q_{11} + Q_{22} - 4Q_{66})s^2c^2 + (s^2+c^2)Q_{22}, \bar{Q}_{16} = (c^2Q_{11} - s^2Q_{12} + (Q_{12} + 2Q_{66})(s^2-c^2))sc,$$

$$\bar{Q}_{22} = s^4Q_{11} - c^4Q_{22} + 2(Q_{12} + 2Q_{66})s^2c^2,\ \bar{Q}_{26} = (s^2Q_{11} - c^2Q_{12} - (Q_{12} + 2Q_{66})(s^2-c^2))sc,$$

$$\bar{Q}_{66} = (Q_{11} - 2Q_{12} + Q_{22})s^2c^2 + (s^2-c^2)Q_{66},\ s = \sin\theta\ ,\ c = \cos\theta.$$

Layer deformations at a distance z from the middle surface take the form

$$\boldsymbol{\varepsilon} = \boldsymbol{\varepsilon}^o + z\boldsymbol{\chi}^o,$$

where $\boldsymbol{\varepsilon}^o$ are the deformations of the middle surface, $\boldsymbol{\chi}^o$ are the curvature changes.
Substituting these relations into equation (1) we get

$$\boldsymbol{\sigma} = (\bar{Q})\boldsymbol{\varepsilon}^o + z(\bar{Q})\boldsymbol{\chi}^o.$$

If the stresses are expressed through normal efforts and moments

$$\boldsymbol{N} = \int_{-h/2}^{h/2}\boldsymbol{\sigma}dz,\ \boldsymbol{N}^T = (N_s,\ N_\theta,\ N_{s\theta}), \boldsymbol{M} = \int_{-h/2}^{h/2}\boldsymbol{\sigma}zdz,\ \boldsymbol{M}^T = (M_s,\ M_\theta,\ M_{s\theta}). \quad (2)$$

Here is designated $\boldsymbol{N}$ – are the membrane forces, $\boldsymbol{M}$ – are the bending moments.
Conducting the integration of expressions (2) we have

$$\begin{pmatrix} \boldsymbol{N} \\ \boldsymbol{M} \end{pmatrix} = \begin{pmatrix} A & B \\ B & D \end{pmatrix}\begin{pmatrix} \boldsymbol{\varepsilon}^o \\ \boldsymbol{\chi}^o \end{pmatrix},$$

$$(A_{ij},\ B_{ij},\ D_{ij}) = \int_{-h/2}^{h/2} Q_{ij}(1, z, z^2)\, dz\, , (i,j = 1,2,3).$$

Denoted here $A_{ij} = \sum_{k=1}^{n}\bar{Q}_{ij}(h_k - h_{k-1}),\ i,j = 1,2,6,$

$$B_{ij} = \sum_{k=1}^{n}\bar{Q}_{ij}({h^2}_k - {h^2}_{k-1}),\ i,j = 1,2,6,\ \ D_{ij} = \sum_{k=1}^{n}\bar{Q}_{ij}({h^3}_k - {h^3}_{k-1}),\ i,j = 1,2,6. \quad (3)$$

The designations in the formula (3) are shown in figure 2.

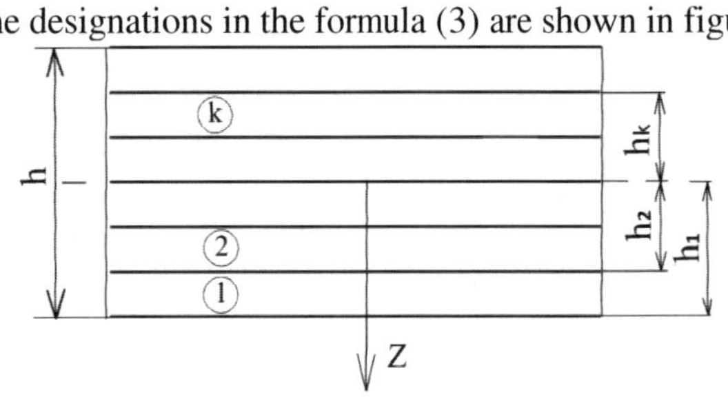

Figure 2. Multi-layer composite material.

Consider the three-layer structure of a composite material, which consists of two external carrier layers and a layer of aggregate between them. For carrier layers apply the equations discussed above. The displacements of the filler are defined as the arithmetic average of the displacements of the carrier layers. Angle of rotation as the difference in the displacements of the bearing layers divided by the thickness of the composite material.

IOP Conf. Series: Materials Science and Engineering **675** (2019) 012028 doi:10.1088/1757-899X/675/1/012028

Figure 3 shows a three-layer composite structure with two outer carrier layers and a layer of aggregate between them. The displacements of the bearing layers are denoted by u_1 and u_2, the angles of rotation through φ_1 and φ_2, the thicknesses are indicated as h_1 and h_2. For the filler, the thickness is equal to h_o and the displacement of the neutral line u_0.

The displacements of the neutral line of the aggregate and the angle of rotation of the normal during deformation are determined from the relations:

$$u_o = (\bar{u}_1 + \bar{u}_2)/2,\ \varphi_o = (\bar{u}_1 - \bar{u}_2)/h_3,\ \bar{u}_1 = u_1 - h_1 e_{13}/2,\ \bar{u}_2 = u_2 + h_2 e_{23}/2$$

where u_o is the displacement and φ_o is the rotation angle of the neutral filler line, r are the shell radius, e_{13}, e_{23}- are the rotation angles around the coordinate lines, $\varphi_s, \varphi_\theta$ are the angles of inclination of the meridian to the axis of the shell and the angle in the circumferential direction

$$e_{13} = \left(\frac{\partial w}{\partial s} + u\frac{\partial \varphi}{\partial s}\right);\ e_{23} = -\frac{1}{r}\left(\frac{\partial w}{\partial \theta} - v\sin\varphi\right).$$

Here denoted u is the indicated by displacement in the meridional direction, w is the displacement in the normal direction, v is the displacement in the circumferential direction of the shell.

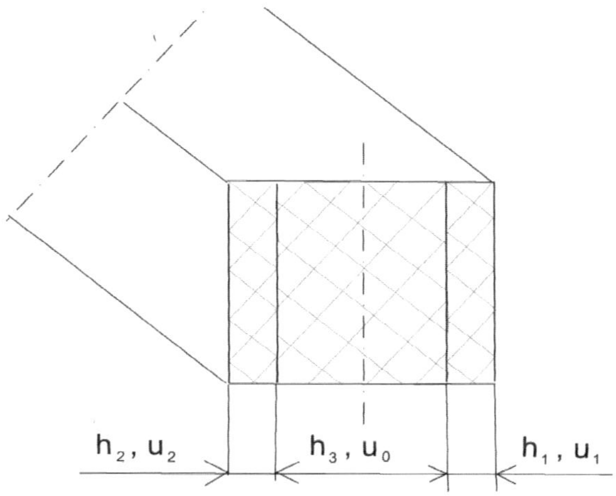

Figure 3. Three-layer composite structures with two outer carrier layers and a layer of aggregate between them.

Displacements in the placeholder at a distance z from the neutral axis can be written as

$$v_0(z) = v_0 + z\varphi_0 = 0.5(v_1 - h_1e_{13}/2 + v_2 + h_2e_{23}/2) + z(v_1 - h_1e_{13}/2 - v_2 - h_2e_{23}/2)/h_3,$$

$$u_0(z) = u_0 + z\varphi_0 = 0.5(u_1 - h_1e_{13}/2 + u_2 + h_2e_{23}/2) + z(u_1 - h_1e_{13}/2 - u_2 - h_2e_{23}/2)/h_3,$$

$$w_0(z) = w_1 + z(w_1 - w_2)/h_3.$$

Thus, one can consider a model of a shell structure consisting of several layers.

The stand design was considered as a three-layer structure with two external carrier layers and filler between them. The filler perceived only shear stresses. To determine the location of the base of the composite material along the lines of maximum stress in the first approximation, a stand of a homogeneous material was calculated. As a result of the calculation, the location of the maximum voltage lines was determined. At the second stage, a stand made of composite material was considered. The basis of the composite material was located mainly along the lines of maximum stress obtained by calculating the stress-strain state of the stand from a homogeneous material. Subsequently, the arrangement of the composite material was varied to achieve its best position. As a result of the iterations, the optimal location of the composite material was determined from the point of view of the minimum stress-strain state of the stand under the load. Thus, it is possible to determine the arrangement of the layers of the composite material for their best arrangement in terms of the stress-strain state.

IOP Conf. Series: Materials Science and Engineering **675** (2019) 012028 doi:10.1088/1757-899X/675/1/012028

2.2. *Matrix of the resolving equation*

To determine the matrix of the resolving equation, it is necessary to determine the potential energy of the structure deformation. The potential strain energy for a multi-layer composite construction is separated from the equation:

$$U=0.5\iint[\boldsymbol{\varepsilon}^{oT}(A)\boldsymbol{\varepsilon}^{o}+\boldsymbol{\chi}^{oT}(B)\boldsymbol{\varepsilon}^{o}+\boldsymbol{\varepsilon}^{oT}(B)\boldsymbol{\chi}^{o}+\boldsymbol{\chi}^{oT}(D)\boldsymbol{\chi}^{o}]\,dA.$$

Here notation: (A), (B), (D) are membrane, bending-membrane, and bending rigidity matrices of the layers, $\boldsymbol{\varepsilon}^o = (\varepsilon^o{}_s,\ \varepsilon^o{}_\theta,\ \varepsilon^o{}_{s\theta})$ are strains in the middle surface plane, $\boldsymbol{\chi}^o = (\chi^o{}_s,\ \chi^o{}_\theta,\ \chi^o{}_{s\theta})$ are curvatures, and $dA = Rdsd\theta$ is the area of integration.

The matrix of the resolving equation or the stiffness matrix for a multilayer composite construction is obtained by differentiating the potential deformation energy by generalized displacements

$$\partial U/\partial \boldsymbol{q}=(K)\boldsymbol{q},$$

where U is the strain energy of the structure, and q is the displacements at the nodes.

2.3. *External force vector*

To determine the vector of external forces, it is necessary to find the work of external forces on possible displacements. The work of external forces can be determined from the ratio

$$W = \int \boldsymbol{P}\,\boldsymbol{u}\,dA,$$

where $\boldsymbol{P}$, $\boldsymbol{u}$ are the vectors of external forces and the vector of displacements.

The vector of external forces is determined by differentiating the work of external forces by generalized displacements

$$\boldsymbol{Q} = -\,\partial W/\partial \boldsymbol{q}.$$

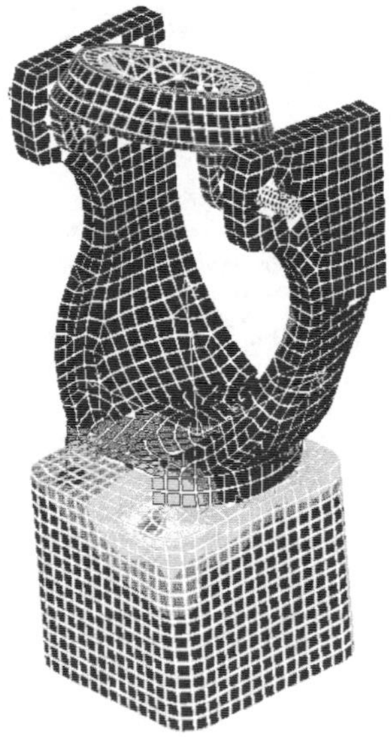

Figure 4. The stress-strain state of the stand when it hits the locking device

2.4. *System of resolving equations*

The definition of the stress-strain state of a structure is reduced to solving a system of linear equations derived from the Lagrange equations [9]

$$(K)\boldsymbol{q} = \boldsymbol{Q}. \tag{4}$$

IOP Conf. Series: Materials Science and Engineering **675** (2019) 012028 doi:10.1088/1757-899X/675/1/012028

Here (K) is the stiffness matrix, $\boldsymbol{Q}$ is the vector of external forces, and $\boldsymbol{q}$ is a generalized displacement.

As a result of the solution of equation (4), a picture of the stress-strain state of the stand of the semi-natural modeling was obtained upon striking the locking (Figure 4). The applied force was determined based on the D'Alembert's principle, and was applied to the locking device of the stand.

3. Results

The stress-strain state of a robotic structure under the action of an external load has been studied. A method for determining the characteristics of a multilayer composite material is presented. Dependencies are obtained for determining displacements in a three-layer composite shell consisting of external carrier layers and a filler layer between them. The method of determining and positioning the base of the composite material along the lines of maximum stress is considered. The results are shown in the figures. Gears, bearings and gearboxes approximated by the method proposed by the author. The results obtained are in good agreement with the experimental results.

4. Conclusion

The study showed that the use of composite material for robotic systems is important and relevant. One of the main characteristics of robotic systems are inertial characteristics, which are the best in composite materials. The lack of rigidity of products made of composite materials is compensated for by the correct positioning of the base of the composite material along the lines of maximum stresses. To find the direction of the maximum stresses of the structure, it is necessary to calculate the stress-strain state of the structure from a homogeneous material. Determine the lines of maximum stress and then place the basis of the composite material in the composite structure along these lines. A new approach to the approximation of such elements as support bearings, gearboxes, gears is proposed. The proposed method compares favorably with the methods used in [10-12], where the bearings are directly approximated by finite elements, which leads to an unjustified increase in the number of finite elements. Errors in the calculation of large systems of equations negate the direct approximation of bearings. Moreover, gearboxes that have a direct approximation do not lead to satisfactory results.

In this paper, the rigidity of the gearbox is calculated separately and the gearbox is replaced by a system of connecting kernels with similar rigidity. Also gears, bearings are approximated by the structure of the kernels, the rigidity of which coincides with these elements. The convergence of the research results was checked by thickening the finite element mesh. If the results did not differ by more than 2%, the results of the study were considered reliable. Comparison of the results of the study of the stress-strain state with the available experimental results showed that the use of composite material in robotic systems is a promising direction.

Calculations of a robotic system using a bench as an example showed that the mass of the composite structure as compared to the magnesium alloy traditionally used in the production of robotic systems decreases by at least 5-7%, and, accordingly, the inertia characteristics decrease. Thus, the robotic system reacts faster to the signal and improves the positioning accuracy by reducing the inertial forces of the channels of the robotic system.

The calculations of the robotic system on the example of the stand showed that the proposed method of calculating systems containing gearboxes, gears and support bearings, gives good results that reliably reflect the behavior of structures in working conditions. The proposed method of calculating the stress-strain state of robotic structures can be used for structures made of composite and homogeneous materials.

References

[1] Jones R M 1999 *Mechanics of Composite Materials* (London: Taylor & Francis) p 538

[2] Zienkiewicz O C, Taylor R L and Zhu J Z 2013 *Finite element method: its basis and fundamental* (Oxford: Butterworth-Heinemann) p 756

[3] Segerlind L J 1976 *Applied Finite Element Analysis* (New York: John Wiley & Sons) p 422

IOP Conf. Series: Materials Science and Engineering **675** (2019) 012028 doi:10.1088/1757-899X/675/1/012028

[4] Imgrund M C 1992 *ANSYS® Verification Manual* (Canonsburg: Swanson Analysis Systems)
[5] Burton W S and Noor A K 1994 Three-dimensional solutions for therymomechanical stresses in sandwich panels and shells *J. Eng. Mech. ASCI* **120(10)** pp 2044-2071
[6] Kamanyu P and Lambert L 2006 Methodology for the design of mechanically bonded joints in laminated composite materials *Comp. Sci. Technol.* **66** pp 3004-3020
[7] Alfano G and Crisfield M 2001 A Finite Element Interface Models for the Delamination Analysis of Laminated Composites: Mechanical and Computational Issues *Int. J. Num. Meth. Eng.* **50** pp 1701-1736
[8] Volmir A S and Khayrnasov K Z 1982 Stability of toroidal composite shells *Mechanics of composite materials* **3** pp 454-459
[9] Khayrnasov K Z 2002 Methodology for optimizing the layout of multi-step dynamic stands for semi-natural modeling *RUDN Journal of Engineering Researches* **1** pp 37-41
[10] Roos R, Kress G and Ermanni P 2007 A post-processing method for interlaminar normal stresses in doubly curved laminates *J. Composite Structures* **81** pp 463-470
[11] Houpert I 2014 An enhanced analytical approach for ball and roller bearings calculation *J. of Tribology* **136** pp11105-11116
[12] Hirani H, Athre K and Biswas S 1999 Dynamically Loaded Finite Length Journal Bearings: Analytical Method of Solution *ASME J. of Tribology* **121** pp 844-852

Comparison of n-LSB embedding approach in different color spaces

Ozcan CATALTAS* and Kemal TUTUNCU

Selcuk University, Konya, TURKEY

* Corresponding author: ozcancataltas@selcuk.edu.tr

Abstract. Today, secure communication is one of the biggest goals that people want to achieve. There are many different methods to ensure security in data communication. Steganography is one of the methods that can be used in secure data communication. In this study, n-LSB embedding method based on reducing the amount of distortion generated by the classical Least Significant Bit substitution method that we introduced in another study has been applied in both RGB color space and HSI, XYZ and YCbCr color spaces. In the embedding process, 3 message bits were embedded in each color channel. 7.94 kb data is embedded on 3 different cover images. The obtained results were compared with the PSNR criterion, which is the criterion of image quality evaluation, and it is seen that proposed n-LSB approach gives better results in HSI color space for 3 cover images. The highest PSNR value obtained was 83.907 dB.

Keywords: Image steganography, n-LSB embedding, Color spaces, Image quality.

1. Introduction

Secret message transmission technique, which contains hidden messages and no one can notice these messages except the sender and the receiver, is called "Steganography". Even the person who is sent to him through Steganography will receive secret message only with his key. The purpose of steganography is to hide the existence of the message [1].

Steganography, which is the process of conveying information in the information or storing a confidential message in another message, is quite different from cryptography, which is a code of science. Cryptography is a method that can be accessed by the recipient only after they are subjected to a certain decryption method. Steganography prevents information from being accessed by other people by storing the information to be transmitted in other information. In Steganography, there must be a communication channel and a type of cover media where the message will be moved. However, cryptography is a communication method that uses a password to prevent the message from being exposed and does not care whether the communication is monitored and does not need to keep the privacy of the message. The greatest advantage of steganography by cryptography is that it is never known whether a person who sees the information has a secret message within the transmitted message. In other words, if today's advanced technologies are considered, there is a possibility of breaking the password no matter how strong it is, while steganography is not known and therefore does not attract attention. Steganography takes its power from "confidentiality", while cryptography does this through the "encryption algorithm" [2-9].

Four types of media are generally used as cover media in steganography. These are text, images, audio and video. Since images are used more frequently in communication among people, studies on steganography have focused on the image.

In Steganography, text, image, audio and video types can be used as a message file, and also any digital file can be used.

Image steganography methods are divided into two main categories. These are spatial domain methods and frequency domain methods. In the spatial domain, the message file is embedded directly in the pixels of the image. In the frequency domain, the image is first converted to frequency coefficients. The embedding operation is then done to these frequency coefficients. The LSB substitution method is one of the most basic methods in both domains.

This paper organized as follows: In Section 2, similar studies have been mentioned. In Section 3, LSB substitution method, n-LSB substitution method and color spaces have been mentioned. In Section 4, details and results of the study are given. In Section 5, the obtained results are evaluated and interpreted.

2. Related works

In our study [3], we proposed a new method to reduce the deformation resulting from the application of the classical LSB substitution method. We applied this method called n-LSB substitution in the 2 and 3-LSB substitution method. As a result of the tests performed, an increase of 6.6% was achieved according to the classical LSB substitution method in PSNR.

In another study we have done [4], the classical LSB substitution method has been tested in different color spaces. The cover image was transformed into RGB, HSI, YCbCr, XYZ and YIQ color spaces, and the stego images were compared according to the image quality evaluation criteria. The results showed that the highest PSNR values were obtained with HSI color space.

In [5], the most suitable color space for image steganography was investigated. They found that HSV color space from RGB, YCbCr, HSV and Lab color spaces is the most suitable color space for time complexity and image quality. They also proposed a new method of steganography using the HSV color space.

In [6], data hiding was performed by using LSB substitution method on RGB, YCbCr, XYZ, CMYK, YIQ and HSV color spaces. The resulting stego images were compared according to the image quality evaluation criteria and they claimed that the CMYK color space was the most suitable color space for embedding.

3. Materials and methods

3.1. Least significant bit substitution method

The LSB (Least Significant Bit) means the bit that, if changed, makes the least change in an image pixel. In other words, converting the LSB from 0 to 1 or from 1 to 0 will cause the least distortion on the cover image relative to other bits. Therefore, the basis of data hiding algorithms is mostly based on LSB substitution [1-9].

Just as the LSB of a pixel can be changed, the least significant 2 bits or the least significant 3 bits can be changed. During embedding for this, the next 3 bits of message file is changed to the last 3 bits of the cover image pixel. As a result, the amount of data that can be embedded on the image is increasing, as the distortion on the image is relatively increased.

3.2. n-LSB substitution method

The n-LSB method, proposed in [3], is based mainly on reducing the distortion that occurs in the classical LSB method (Figure 1). This process is done as follows:

First, the message file is embedded on the cover image using the classic 3-LSB (or 2-LSB) substitution method. After embedding, the cover and stego image pixels are compared one by one. The difference values must be between -7 and 7 in decimal. If the difference is between -4 and 4, no action is taken on the related pixel. If the difference is 5, 6, or 7, the first 1 encountered from the 4th bit is converted to 0. That is, decimal number 8 is subtracted from the related pixel. If the difference is -5, -6, or -7, the first 0 encountered from the 4th bit is converted to 1. That is, the decimal number 8 is added to the related pixel. As a result of these operations, there will be no loss of information since the first 3 bits of the pixels have not changed [3].

IOP Conf. Series: Materials Science and Engineering **675** (2019) 012029 doi:10.1088/1757-899X/675/1/012029

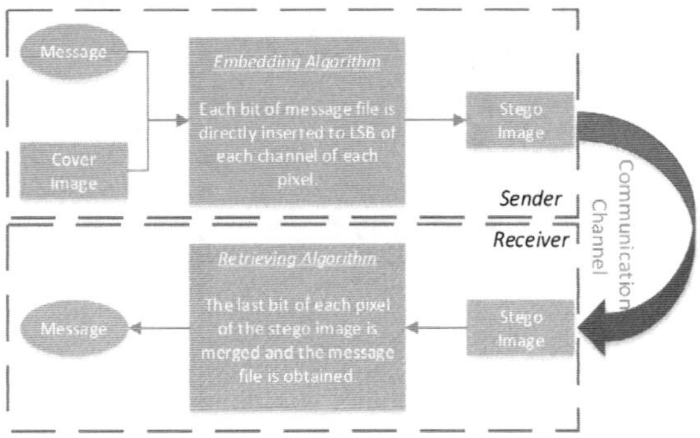

Figure 1. LSB substitution method.

On the receiver side, since there is no operation to change the first 3 bits on the stego image, the message information will be extracted according to the classical 3-LSB substitution method.
The flowchart of proposed n-LSB substitution method was given in figure 2.

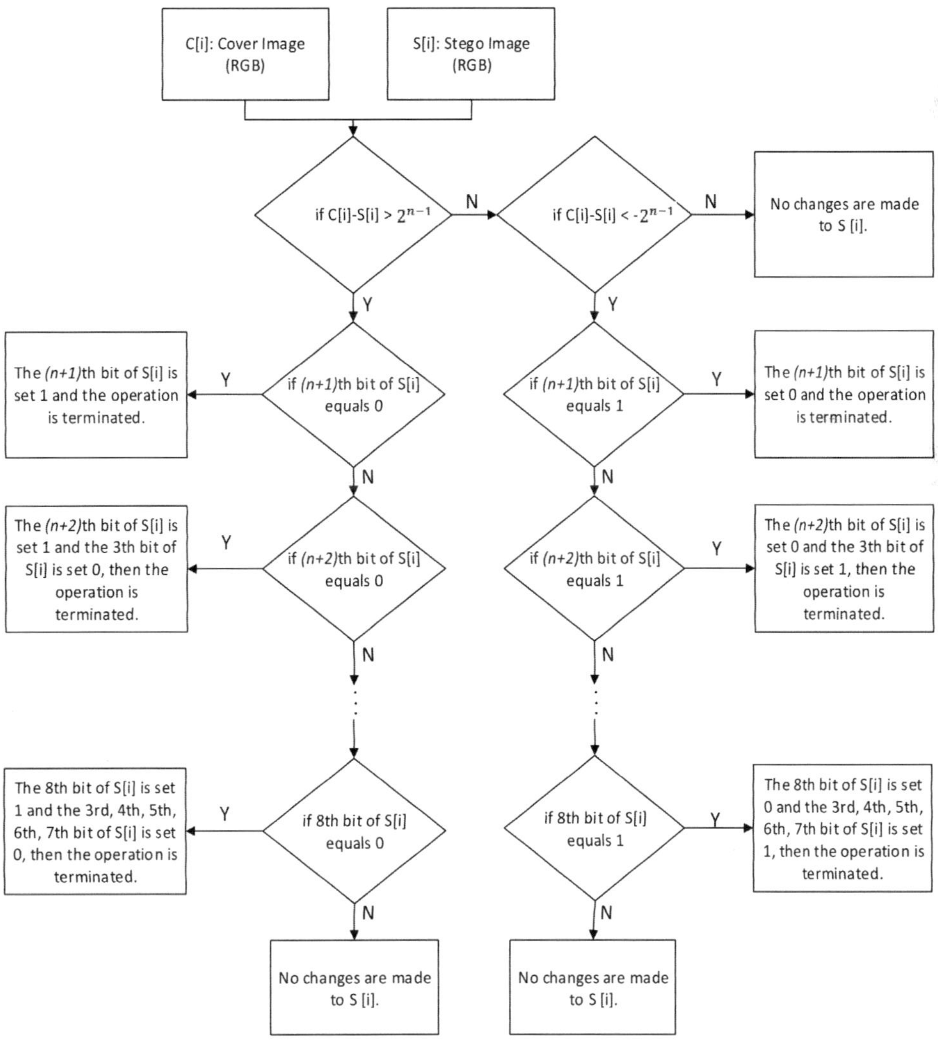

Figure 2. n-LSB embedding algorithm.

3.3. Color spaces

Due to the high variety of colors, the need for grouping these colors is emerged. In order to group and standardize these colors, the concept of color space has emerged. Each color space has its own structure to describe the color set [7].

In this study, the classical and n-LSB method is applied to HSI, XYZ and YCbCr color space as well as RGB color space. These color spaces to RGB color space conversion formulas are given in (1), (2), (3), (4), (5) and (6).

$$I = \frac{1}{3}(R+G+B) \tag{1}$$

$$\theta = \cos^{-1}\left[\frac{((R-G)+(R-B))/2}{\sqrt{(R-B)^2+(R-B)(G-B)}}\right] \tag{2}$$

$$H = \begin{cases} \theta & B \le G \\ 360-\theta & B > G \end{cases} \tag{3}$$

$$S = 1 - \frac{3}{R+G+B}\left[\min(R,G,B)\right] \tag{4}$$

$$\begin{bmatrix} X \\ Y \\ Z \end{bmatrix} = \begin{bmatrix} 0.412453 & 0.357580 & 0.180423 \\ 0.212671 & 0.711600 & 0.072169 \\ 0.019334 & 0.119119 & 0.950227 \end{bmatrix} * \begin{bmatrix} R \\ G \\ B \end{bmatrix} \tag{5}$$

$$\begin{bmatrix} Y \\ Cb \\ Cr \end{bmatrix} = \begin{bmatrix} 77/256 & 150/256 & 29/256 \\ -44/256 & -87/256 & 131/256 \\ 161/256 & -110/256 & -21/256 \end{bmatrix} * \begin{bmatrix} R \\ G \\ B \end{bmatrix} \tag{6}$$

3.4. PSNR

Image quality can be divided into two main classes: objective and subjective quality. Subjective quality methods are based on qualitative human perception. Objective quality methods are based on mathematical algorithms. Objective quality methods are designed with the human vision system in mind. Objective methods need to have a high connection with subjective quality methods in order to be useful [8].

One of the most used objective quality method is Peak Signal to Noise Ratio (PSNR). The PSNR formula is given in (7). In (7), MSE is the average of the square of the difference between the original image and the stego image. The difference between the images decreases as the PSNR value increases.

$$PSNR = 20 * \log_{10}\frac{255}{\sqrt{MSE}} \tag{7}$$

4. Results and discussion

In this study, we aimed to investigate the effect of the n-LSB method on different color spaces, which we have proposed in a previous study. Experiments have been done on RGB, HSI, XYZ and YCbCr color spaces. 3 different images were used as cover image. The resolutions of the images used are; 225x225, 229x220 and 350x348. All images are in ".bmp" format. The message file is 7.94 kB plain text. No compression or encryption is done before the message file is embedded. The embedding process is done in pixel-by-pixel format so that each pixel has the last 3 bits.

Initially, the message file was hidden using the n-LSB method on the image named "Lena" at a resolution of 225x225 pixels. Then, this cover image was transformed into HSI, XYZ and YCbCr color spaces, and embedding was performed using n-LSB method. The obtained cover and stego images are given in figure 3.

The message file was hidden using the n-LSB method on the image named "Baboon" at a resolution of 229x220 pixels. Then, this cover image was transformed into HSI, XYZ and YCbCr color spaces, and embedding was performed using n-LSB method. The obtained cover and stego images are given in figure 4.

Finally, the message file was hidden using the n-LSB method on the image named "Tiger" at a resolution of 350x348 pixels. Then, this cover image was transformed into HSI, XYZ and YCbCr color spaces, and embedding was performed using n-LSB method. The obtained cover and stego images are given in figure 5.

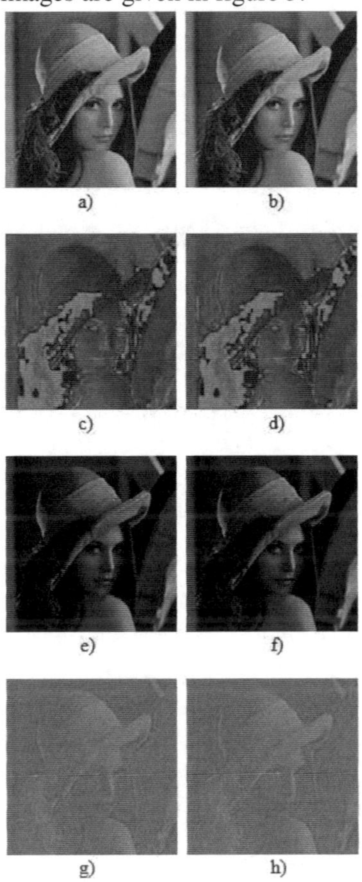

Figure 3. a,c,e and g are cover images, in RGB, HSI, XYZ and YCbCr color spaces, respectively. b,d,f and h are stego images, in RGB, HSI, XYZ and YCbCr color spaces.

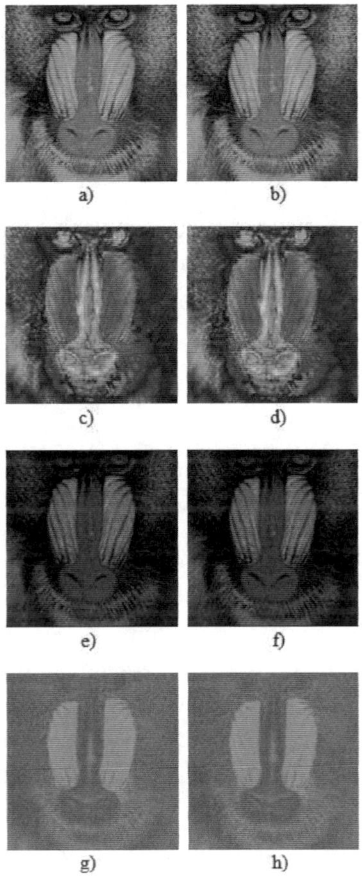

Figure 4. a,c,e and g are cover images, in RGB, HSI, XYZ and YCbCr color spaces, respectively. b,d,f and h are stego images, in RGB, HSI, XYZ and YCbCr color spaces.

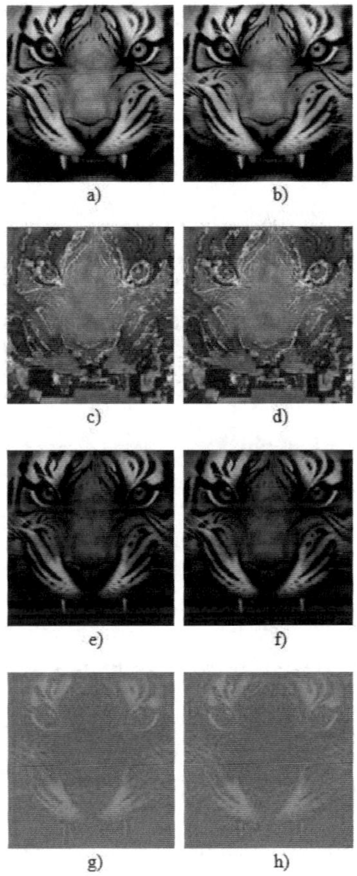

Figure 5. a,c,e and g are cover images, in RGB, HSI, XYZ and YCbCr color spaces, respectively. b,d,f and h are stego images, in RGB, HSI, XYZ and YCbCr color spaces.

After embedding, all stego and carrier images obtained were subjected to PSNR analysis. The PSNR values obtained are given in Table 1.

IOP Conf. Series: Materials Science and Engineering **675** (2019) 012029 doi:10.1088/1757-899X/675/1/012029

Table 1. Obtained results.

Image	Color Space	PSNR (dB)
Image 1 "Lena"	RGB	49,7006
	HSI	80,1997
	XYZ	78,7413
	YCbCr	49,7126
Image 2 "Baboon"	RGB	49,4791
	HIS	80,0773
	XYZ	78,7411
	YCbCr	49,7260
Image 3 "Tiger"	RGB	53,3519
	HSI	83,9074
	XYZ	82,5740
	YCbCr	53,6346

When we examine Table 1, the highest PSNR value in 3 cover images was obtained with HSI color space. With the XYZ color space, a PSNR value close to this value was obtained. In the 3 cover images, the worst PSNR value was obtained with RGB color space. The highest PSNR obtained was 83.9074 dB.

5. Conclusion

In this study, the success of the n-LSB substitution method we introduced in a previous study was tested in different color spaces. The message file was hidden on 3 different cover images using the n-LSB substitution method. PSNR values of obtained stego images and cover images, which are one of the image quality evaluation criteria, were calculated. According to the results, the HSI color space gave the best result in the whole cover images. The lowest PSNR value was obtained by RGB color space. When we compare the results with the results of our another study, which we tested the classical LSB substitution method on different color spaces, it was observed that the HSI color space gave the best results, similarly. In future studies, the proposed n-lsb method can be applied in the frequency domain techniques and the results can be examined.

6. References

[1] Farhan H and Alwan Z 2018 Improved method using a two Exclusive-OR to binary image in RGB color image steganography *International Journal of Engineering & Technology* **7(4)** pp 4295-99

[2] Yadav R, Patidar K, Kushwah R and Saxena G 2018 A review and analysis based on image encryption techniques *ACCENTS Transactions on Information Security* **3(10)**

[3] Cataltas O and Tutuncu K 2017 Improvement of LSB based image steganography *Proc. Research World International Conference* (Rome, Italy) pp 36-40

[4] Cataltas O and Tutuncu K 2017 Comparison of LSB Image Steganography Technique in Different Color Spaces *Proc. International Conference on Artificial Intelligence and Data Processing* (Malatya, Turkey)

[5] Muhammed K, Ahmad M, Sajjad M and Rho S 2016 Evaluating the Suitability of Color Spaces for Image Steganography and its Application in Wireless Capsule Endoscopy *Proc. International Conference on Platform Technology and Service* (Jeju, South Korea)

[6] Meenakshi R and Kuppusmay K 2014 The Use of Least Significant Bit Technique at Various Color Spaces for secure Steganography with Performance Evaluation *Proc. International Journal of Advanced Research in Computer Science and Software Engineering* **4(6)** pp 1047-51
[7] Gonzalez R C and Woods R E 2008 Digital Image Processing 3rd ed., Pearson Education, Inc.
[8] Wang Z, Bovik A C, Sheikh H R and Simoncelli E P 2004 Image quality assessment: from error visibility to structural similarity *IEEE Transactions on Image Processing* **13(4)** pp 600-612
[9] Manisha S and Sharmila T S 2019 A two-level secure data hiding algorithm for video steganography *Multidim Syst Sign Process* **30(2)** pp 529-42

Authors' background

Your Name	Title*	Research Field	Personal website
Özcan ÇATALTAŞ	Phd. Candidate	Information Security, Steganography, Artificial Intelligent	
Kemal TÜTÜNCÜ	Asistant Prof.	Steganography, Data Mining, Artificial Intellegent	

Tree-seed algorithm in solving real-life optimization problems

M A Sahman[1*], A C Cinar[2], I Saritas[1] and A Yasar[3]

[1] University of Selcuk, Konya, Turkey
[2] Turkish State Meteorological Service, Konya Division, Konya, Turkey
[3] University of Selcuk, Konya, Turkey

Corresponding author: asahman@selcuk.edu.tr

Abstract. Tree-seed algorithm (TSA) is a nature-inspired and population-based algorithm for solving continuous optimization problems. The tree-seed relationship is the main motivation of this algorithm. TSA has only two peculiar parameters which are the total number of trees in the stand (pop) and the controller of the seed production (search tendency). Although many problems have been solved in the literature by TSA which is the successful optimizer for low dimensional unconstrained continuous problems, real-life problems have not been addressed yet. In this study, six continuous unconstrained real-life optimization problems (gas transmission compressor design, optimal capacity of gas production facility, gear train design, frequency modulation sounds parameter identification, the spread spectrum radar polyphase code design with 10 decision variables, the spread spectrum radar polyphase code design with 20 decision variables) have been solved. It is seen that choosing the number of the population as 50 and the value of search tendency as 0.1 is appropriate according to experimental results for these problems.

Keywords: Tree-Seed Algorithm, Real-Life Problems, Optimization, Unconstrained Continuous Problems

1. Introduction

The purpose of optimization is to provide maximum benefit with minimum cost. Optimization problems are solved with exact methods such as Linear Programming and also heuristic methods such as TSA. Due to some optimization problem, NP-Hard problems, cannot be solved in polynomial time, heuristic methods (not always guarantee the optimum but satisfies the near optimum) are proposed so as to solve these kinds of problems. Since 1975, more than 300 metaheuristic algorithms have been proposed for solving optimization problems. One of these methods is TSA proposed by Kıran [1] in 2015. TSA models the relationship between trees and their seeds to solve optimization problems. In TSA, the ground is search space, trees and seeds are potential solutions for the problem. Due to the fact that TSA is a population-based a heuristic algorithm, the information sharing and communication with each other in the population (swarm intelligence) is used in order to create new candidate solutions. Diversification and intensification phases are controlled by ST in the TSA. A random tree is used for seed creation in diversification phase and the best tree for seed creation in intensification phase. Unconstrained continuous optimization problems have been solved by TSA effectively. In literature, various TSA versions can be found for example constrained continuous [2], discrete binary [3] and parallel [4-6]. In this study, real-life optimization problems (ROPs) have been solved via TSA. This is the first study for solving ROPs through TSA. The original TSA is introduced briefly in section 2, the details of the experimental setup are given in section 3, the results and discussion are presented in section 4 and concludes of this paper are given in section 5.

2. Tree-seed algorithm

TSA is a nature-inspired, population-based, stochastic, metaheuristic optimization algorithm in swarm intelligence category. TSA is proposed by Kiran [1] based on the relationship between trees and seeds for the solving of continuous optimization problems. Seeds are used as reproduction and propagation mechanisms. Seeds are transported in accordance with their structure. Some seeds are carried from tree to different places by holding on to birds 'feathers, animals' furs and clothes. Thus, in the suitable environment seeds grows and becomes a tree. Ground is considered to be the search space of optimization problem. By the way, both of the trees and seeds are considered the solutions of optimization problem. Exploration and exploitation process of the algorithm is so important for the metaheuristic methods to obtain qualified solutions. At the stage of exploration, trees are scattered in the search space randomly, and the seeds of the same characteristics with the trees are used in the exploitation stage. Seed number of each tree is determined randomly, but not less than one. It is recommended that the number of random seeds is between 10% and 25% of the total number of trees. Seeds are derived according to ST used to choose either the best tree or the randomly selected tree during seed creation and using two seed production equation given as follows:

$$S_{k,j} = T_{i,j} + \alpha_{i,j} \times (B_j - T_{r,j}) \tag{1}$$

$$S_{k,j} = T_{i,j} + \alpha_{i,j} \times (T_{i,j} - T_{r,j}) \tag{2}$$

where, $S_{k,j}$ is *jth* dimension of the kth seed derived from *ith* tree , $T_{i,j}$ is the *jth* dimension of *ith* tree, B_j is *jth* dimension of achieved best tree location so far, $T_{r,j}$ is *jth* dimension of the *rth* tree which is selected randomly in population, $\alpha_{i,j}$ is scale factor that is randomly generated in the range of [-1,1]. A greedy selection is carried out between the seeds of each tree in the search process and the best seed is determined. If the solution quality of best seed is better than its own tree, the tree dries and the best seed substitutes that tree. TSA algorithm is executed until the termination criteria are met. For detailed information, pseudocode, and flowchart, readers can look at these [1-8] works.

3. Experimental setup

In this study, we used six real world engineering optimization problems (gas transmission compressor design, optimal capacity of gas production facility, gear train design, frequency modulation sounds parameter identification, the spread spectrum radar polyphase code design with 10 decision variables, the spread spectrum radar polyphase code design with 20 decision variables). These problems are unconstrained and highly nonlinear. The maximum function evaluations numbers (maxFEs), dimensions and bounds are shown in Table 1. The Pop is taken as 50 and 100. The ST is taken as 0.1, 0.5 and 0.9.

Table 1. The properties of benchmark suite

No	Problem Name	D	maxFEs (Dx10000)	Bounds
F1	Gas transmission compressor design	3	30000	$10 \le x_1 \le 55$ $1.1 \le x_2 \le 2$ $10 \le x_3 \le 40$
F2	Optimal capacity of gas production facility	2	20000	$17.5 \le x_1 \le 40$ $300 \le x_2 \le 600$
F3	Gear train design	4	40000	$12 \le x_i \le 60$ $i = 1, 2, 3, 4$
F4	Frequency modulation sounds parameter identification	6	60000	$-6.4 \le x_i \le 6.35$ $i = 1, 2, 3, 4, 5, 6$
F5	The spread spectrum radar polyphase code design	10	100000	$0 \le x_i \le 360$ $i = 1, \ldots, 10$
F6	The spread spectrum radar polyphase code design	20	200000	$0 \le x_i \le 360$ $i = 1, \ldots, 20$

3.1. Gas transmission compressor design problem

The aim of this design problem which has 3 decision variables (x1, x2 and x3) to find the optimal design parameters when 100 million cu. Ft. of gas per day are delivered with minimum cost for a gas pipeline transmission system [9]. x1 is the length between compressor stations (in miles), x2 is the compressor ratio and x3 is the pipe inside diameter (in inches). Mathematical model of Gas Transmission Compressor Design Problem is given as follow (Eq. 3).

$$\begin{aligned}\text{Min F1(x)} &= 8.61 \text{ x } 10^5 x_1^{1/2} x_2 x_3^{-2/3} (x_2^2 - 1)^{-1/2} + 3.69 \text{ x } 10^4 x^3 \\ &+ 7.72 \text{ x } 10^8 x_1^{-1} x_2^{0.219} - 765.43 \text{ x } 10^6 x_1^{-1}\end{aligned} \tag{3}$$

3.2. Optimal capacity of gas production facility

The optimum combination of capacities of the production facilities of an oxygen production and inventory system is found in this optimization problem. There are 2 decision variable that need to be optimized. First decision variable x1 is oxygen production rate and second decision variable x2 is oxygen storage pressure. Mathematical model of Optimal Capacity of Gas Production Facility problem is given as follow (Eq. 4).

$$\begin{aligned}\text{Min F2(x)} &= 61.8 + 5.72 x_1 + 0.2623 \left[(40 - x_1) \ln \left(\frac{x_2}{200}\right) \right]^{-0.85} \\ &+ 0.087 (40 - x_1) \ln \left(\frac{x_2}{200}\right) + 700.23 x_2^{-0.75}\end{aligned} \tag{4}$$

3.3. Gear train design

In this problem, there are four decision variables (x1 to x4). These decision variables represent the number of teeth in gear which must be an integer and in the range of [12,58]. Mathematical model of Gear Train Design problem is given as follow (Eq. 5).

$$\text{Min F3(x)} = \left\{ \frac{1}{6.931} - \frac{x_1 x_2}{x_3 x_4} \right\}^2 \tag{5}$$

3.4. Frequency modulation sounds parameter identification

This problem is highly complex multimodal problem, having strong epitasis and have 6 decision variables which is in the bounds of [-6.4 , 6,35] . Mathematical model of Frequency Modulation Sounds Parameter Identification problem is given as follow (Eq. 6).

$$\begin{aligned}\text{Min F4(x)} = \sum_{t=0}^{100} &((x_1 \text{ x } \sin(x_2 \text{ x } t \text{ x } 3.6 + x_3 \text{ x } \sin(x_4 \text{ x } t \text{ x } 3.6 + x_5 \text{ x } \sin(x_6 \text{ x } t \text{ x } 3.6)))) \\ &- (1.0 \text{ x } \sin(5.0 \text{ x } t \text{ x } 3.6 + 1.5 \text{ x } \sin(4.8 \text{ x } t \text{ x } 3.6 + 2.0 \text{ x } \sin(4.9 \text{ x } t \text{ x } 3.6)))))^2\end{aligned} \tag{6}$$

3.5. The spread spectrum radar poly phase code design

This problem is min-max continuous optimization problem and it have a lot of local optimum. The spread spectrum radar polyphase code design problem has 10 (for F5) and 20 (for F6) decision variables. Mathematical model of The Spread Spectrum Radar Poly Phase Code Design problem is given as follow (Eq. 7-8).

$$\text{Min F6(x)} = \text{Min F5(x)} \tag{7}$$

$$\begin{aligned}
&\text{Min F5(x)} = \max\left\{f_1(X),\ldots,f_{2m}(X)\right\}\\
&X=\left\{(x_1,\ldots,x_n)\in \mathrm{R}^n \mid 0\le x_j\le 2\pi,\ j=1,2,\ldots,n \ and\ m=2n-1\right\}\\
&f_{2i-1}(x)=\sum_{j=1}^{n}\cos\left(\sum\nolimits_{k=|2i-j-1|+1}^{j} x_k\right),\ i=1,2,\ldots,n\\
&f_{2i-1}(x)=0.5+\sum_{j=i+1}^{n}\cos\left(\sum\nolimits_{k=|2i-j-1|+1}^{j} x_k\right),\ i=1,2,\ldots,n-1\\
&f_{m+i}(x)=-\,f_i(x),\ i=1,2,\ldots,m
\end{aligned} \tag{8}$$

The mathematical models of these problems are presented in Eq.3-8. The gas transmission compressor design problem has 3 decision variables. The optimal capacity of gas production facility problem has 2 decision variables. The gear train design problem has 4 decision variables. The frequency modulation sounds parameter identification problem has 6 decision variables.

4. Results and discussion

Six unrestricted real-world optimization problems were solved by TSA and the results were shown in the following tables. For each optimization problem, population value was changed from 50 to 100 and ST value was changed to 0.1, 0.5, 0.9 respectively and its effect on the quality of the solution was investigated.

Table 2. The peculiar parameters analysis for F1 problem

F1	Mean	Std.Dev.	Median	Min	Max
Pop=50 ST=0.1	**2.9644E+06**	1.4209E-09	2.9644E+06	2.9644E+06	2.9644E+06
Pop=50 ST=0.5	**2.9644E+06**	7.6369E-10	2.9644E+06	2.9644E+06	2.9644E+06
Pop=50 ST=0.9	**2.9644E+06**	2.3964E-09	2.9644E+06	2.9644E+06	2.9644E+06
Pop=100 ST=0.1	**2.9644E+06**	3.2281E-06	2.9644E+06	2.9644E+06	2.9644E+06
Pop=100 ST=0.5	**2.9644E+06**	9.0824E-05	2.9644E+06	2.9644E+06	2.9644E+06
Pop=100 ST=0.9	**2.9644E+06**	1.5505E-03	2.9644E+06	2.9644E+06	2.9644E+06

Table 3. The peculiar parameters analysis for F2 problem

F2	Mean	Std.Dev.	Median	Min	Max
Pop=50 ST=0.1	**1.6984E+02**	1.4454E-13	1.6984E+02	1.6984E+02	1.6984E+02
Pop=50 ST=0.5	**1.6984E+02**	1.4454E-13	1.6984E+02	1.6984E+02	1.6984E+02
Pop=50 ST=0.9	**1.6984E+02**	1.4454E-13	1.6984E+02	1.6984E+02	1.6984E+02
Pop=100 ST=0.1	**1.6984E+02**	1.4454E-13	1.6984E+02	1.6984E+02	1.6984E+02
Pop=100 ST=0.5	**1.6984E+02**	1.4454E-13	1.6984E+02	1.6984E+02	1.6984E+02
Pop=100 ST=0.9	**1.6984E+02**	1.4454E-13	1.6984E+02	1.6984E+02	1.6984E+02

Table 4. The peculiar parameters analysis for F3 problem

F3	Mean	Std.Dev.	Median	Min	Max
Pop=50 ST=0.1	**3.0389E-12**	4.6722E-12	6.1269E-13	3.5396E-16	2.0617E-11
Pop=50 ST=0.5	7.0400E-12	1.4887E-11	4.2169E-13	1.7825E-14	6.6906E-11
Pop=50 ST=0.9	9.2863E-12	3.2745E-11	7.3145E-13	1.7528E-15	1.7885E-10
Pop=100 ST=0.1	5.8605E-12	1.5436E-11	5.7964E-13	1.0574E-14	8.0754E-11
Pop=100 ST=0.5	3.3289E-12	4.5034E-12	1.1487E-12	1.6826E-15	1.6498E-11
Pop=100 ST=0.9	3.3492E-12	6.1042E-12	8.2565E-13	8.6298E-16	2.6227E-11

Table 5. The peculiar parameters analysis for F4 problem

F4	Mean	Std.Dev.	Median	Min	Max
Pop=50 ST=0.1	**2.7113E+01**	1.3265E+00	2.7390E+01	2.3293E+01	2.8799E+01
Pop=50 ST=0.5	2.7342E+01	1.3747E+00	2.7418E+01	2.3010E+01	2.9213E+01
Pop=50 ST=0.9	2.7697E+01	1.9379E+00	2.8347E+01	1.9066E+01	2.9897E+01
Pop=100 ST=0.1	2.8310E+01	1.0380E+00	2.8700E+01	2.5596E+01	2.9785E+01
Pop=100 ST=0.5	2.8556E+01	9.3447E-01	2.8662E+01	2.5926E+01	3.0133E+01
Pop=100 ST=0.9	2.8491E+01	1.4119E+00	2.9136E+01	2.3818E+01	2.9842E+01

Table 6. The peculiar parameters analysis for F5 problem

F5	Mean	Std.Dev.	Median	Min	Max
Pop=50 ST=0.1	**9.1763E-03**	5.4369E-03	7.1746E-03	1.1593E-03	2.1297E-02
Pop=50 ST=0.5	1.0510E-02	4.8403E-03	9.3799E-03	2.9600E-03	2.1945E-02
Pop=50 ST=0.9	9.8215E-03	5.4371E-03	1.0676E-02	5.7712E-04	2.0180E-02
Pop=100 ST=0.1	1.0620E-02	6.2614E-03	9.2922E-03	2.2842E-03	2.5794E-02
Pop=100 ST=0.5	9.7471E-03	5.5547E-03	9.1413E-03	8.5971E-04	2.1881E-02
Pop=100 ST=0.9	9.7416E-03	5.4833E-03	8.9009E-03	3.1135E-03	3.0710E-02

Table 7. The peculiar parameters analysis for F6 problem

F6	Mean	Std.Dev.	Median	Min	Max
Pop=50 ST=0.1	1.3623E-02	7.2883E-03	1.3174E-02	2.1955E-03	3.0215E-02
Pop=50 ST=0.5	1.2773E-02	6.0711E-03	1.1923E-02	1.9110E-03	2.7479E-02
Pop=50 ST=0.9	**8.9200E-03**	4.8231E-03	8.6281E-03	2.8247E-04	2.1422E-02
Pop=100 ST=0.1	1.3744E-02	6.8534E-03	1.3688E-02	2.6005E-03	2.6798E-02
Pop=100 ST=0.5	1.2564E-02	6.4110E-03	1.1238E-02	2.3099E-03	2.9027E-02
Pop=100 ST=0.9	1.0205E-02	4.9953E-03	9.8464E-03	2.0434E-03	2.3014E-02

It has been seen in results from the tables above for F1 and F2 problems, Pop and ST parameters do not affect the final solution. The same results are obtained with all different conditions. When Pop is taken 50 and ST is taken 0.1 the best results are obtained for F3, F4 and F5 problems. For F6 problem, the best condition is obtained when Pop is taken as 50 and ST is taken as 0.9.

Also, the convergence graphs (Fig.1) of the TSA in six real-life optimization problems prove that the ideal population size is 50 and ST is 0.1.

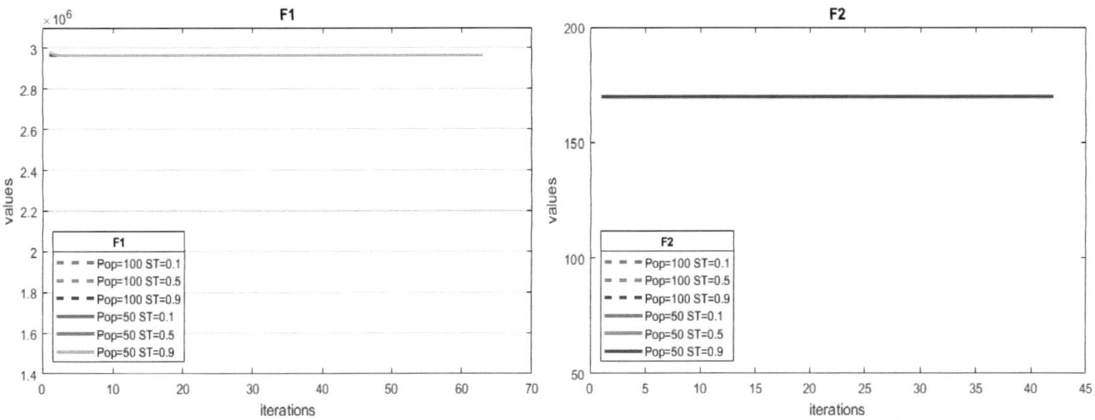

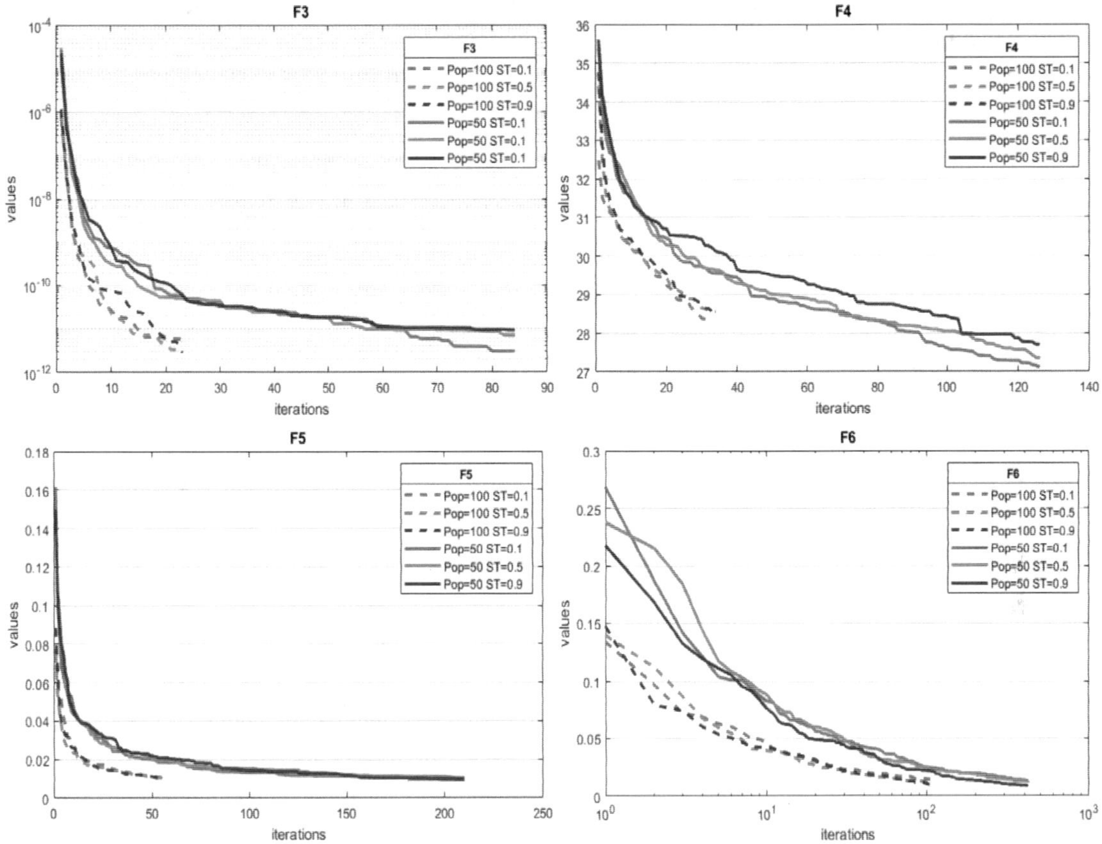

Figure 1. Convergence graphs for the benchmark suites' (F1 to F6)

5. Conclusion

In this study, six continuous unconstrained real-life optimization problems (gas transmission compressor design, optimal capacity of gas production facility, gear train design, frequency modulation sounds parameter identification, the spread spectrum radar polyphase code design with 10 decision variables, the spread spectrum radar polyphase code design with 20 decision variables) were solved with TSA. The experimental results showed that TSA is an alternative solver for these type problems.

References

[1]. Kiran M S 2015 TSA: Tree-seed algorithm for continuous optimization *Expert Systems with Applications* **42(19)** p 6686-6698

[2] Babalik A, Cinar A C and Kiran M S 2018 A modification of tree-seed algorithm using Deb's rules for constrained optimization *Applied Soft Computing* **63(C)** pp 289-305

[3] Cinar A C and Kiran M S 2018 Similarity and logic gate-based tree-seed algorithms for binary optimization *Computers & Industrial Engineering* **115** pp 631-646

[4] Cinar A and Kiran M 2016 A Parallel Version of Tree-Seed Algorithm (TSA) within CUDA Platform *Proc. International Scientific Conference on Applied Sciences* (Konya, Turkey)

[5] Cinar A C 2016 Ağaç-tohum aalgoritması için cuda tabanlı bir paralel programlama yaklaşımı *Proc. Fen Bilimleri Enstitüsü* (Konya, Turkey)

[6] Cinar A C and Kiran M S 2018 A parallel implementation of Tree-Seed Algorithm on CUDA-supported graphical processing unit *Journal of the Faculty of Engineering and Architecture of Gazi University* **33(4)** pp 1397-1409
[7] Cinar A C, Iscan H and Kiran M S 2018 Tree-Seed algorithm for large-scale binary optimization *KnE Social Sciences* **3(1)** pp 48-64
[8] Cinar A C and Kiran M S 2017 Boundary conditions in Tree-Seed Algorithm: Analysis of the success of search space limitation techniques in Tree-Seed Algorithm *Proc. International Conference on Computer Science and Engineering 2017 (UBMK)* IEEE
[9] Cui L, et al. 2018 A smart artificial bee colony algorithm with distance-fitness-based neighbor search and its application *Future Generation Computer Systems* **89** pp 478-493

Authors' background

Name	*Prefix*	*Research Field*	*Email*
Mehmet Akif Sahman	Assistant Professor Dr.	Artificial Intelligence, Optimization	asahman@selcuk.edu.tr
Ahmet Cevahir Cinar	PhD Candidate	Artificial Intelligence, Optimization	ahmetcevahircinar@gmail.com
Ismail Saritas	Professor Dr.	Artificial Intelligence, Optimization	isaritas@selcuk.edu.tr
Ali Yasar	Assistant Professor Dr.	Image Processing	aliyasar@selcuk.edu.tr

Classification of Parkinson disease data with artificial neural networks

A Yasar[1], I Saritas[2*], M A Sahman[2], A C Cinar[3]

[1] Department of Mechatronics Engineering, Selcuk University, Konya, Turkey
[2] Department of Electrical and Electronics Engg, Selcuk University, Konya, Turkey
[3] Turkish State Meteorological Service, Konya Division, Konya, Turkey

Corresponding author: isaritas@selcuk.edu.tr

Abstract. An artificial neural network system has been developed to detect Parkinson's Disease (PD). Three samples were taken from each patient and included in the system. The importance of the study is based on the development and use of a new subject-based ANN approach that takes into account the dependent nature of the data in a replicated measure-based design. In order to evaluate the performance of the proposed system, an audio replication-based experiment was performed to differentiate healthy people from PD patients. The UCI Experiment consisted of 80 subjects, half of whom were affected by PD. Although the proposed system has a reduced number of subjects, the system is able to distinguish people with PD from an acceptable degree of healthy people with an accuracy rate of 94.93% in an artificial neural network.

1. Introduction

Parkinson's disease is the second most common neurodegenerative disease affecting 2-3% of the human population at age 65 and older [1]. According to the Parkinson's Disease Foundation, 7-10 million people worldwide are infected with this disease [2]. As a result of the depletion of dopaminergic nigrostriatal neurons, deterioration occurs in movement (tremor, stiffness, slow movements, unstable posture…). Depending on the movement of the articulators, voice and speech may change. Non-dopaminergic changes may also affect language, cognition, and mood, which may affect communication [3]. Parkinson's Disease has symptoms such as tremors, loss of gestures while talking, slowing movements, not waving arms while walking, stiffness in one or more extremities, slow, small step or foot-dragging walking, soft and low voice or monotonous speech, oozing from mouth, difficulty swallowing, constipation, excessive sweating, drop in blood pressure, pain, and muscle spasms. The causes of Parkinson's disease (PD), the second most common neurodegenerative disease, are still unknown. The current idea is that parent gene mutations cause only a small proportion of all cases, and in most cases, non-genetic factors probably play a role in interacting with susceptibility genes. Numerous epidemiological studies have been conducted to identify such non-genetic risk factors [4]. For people with Parkinson's disease, voice and speech are also affected because voice and speech are dependent on laryngeal, respiratory and joint functions. Vocal disruption is assumed to be one of the earliest symptoms of the disease [5]. From the early stages of Parkinson's disease, there are abnormalities that prevent speech perception by listeners, but acoustic analysis of recorded speech signals can provide objective evaluation [6]. Our study was performed using Parkinson's Dataset with replicated acoustic features in the UCI Machine Learning database [7]. System was trained with artificial neural network and an estimation process was realized which decides whether the data belong to Parkinson's patient or not by taking data such as Credentials received from individuals, Registration number, Gender, Pitch local perturbation measures data: relative jitter (Jitter_rel), absolute jitter (Jitter_abs), relative average degradation (Jitter_RAP) and step perturbation coefficient (Jitter_PPQ), amplitude perturbation measure data: local Shimmer (Shim_loc),

Shimmer in dB (Shim_dB), 3-point amplitude perturbation coefficient (Shim_APQ3), 5-point amplitude perturbation coefficient (Shim_APQ5) and 11-point amplitude perturbation coefficient (Shim_APQ5) and 11-point amplitude perturb , Harmonic-to-noise ratio criteria: 0-500 Hz (HNR05) in the frequency band 0-1500 Hz (HNR15), 0-2500 Hz (HNR25), 0-3500 in the range of harmonic-noise ratio Hz (HNR35) and 0-3800 Spectral measurements in Hz (HNR38) based on the Mel frequency cepstral coefficient of the range 0 to 12 (MFCC0, MFCC1, ..., MFCC12) and their derivatives (Delta0, Delta1, ..., Delta12), Density of recurrence period entropy RP (RPDE), degraded fluctuation analysis (DFA), Pitch period entropy (PPE), Glottal-to-noise excitation ratio (GNE). Recently, artificial neural networks have been used as an important model in the classification of disease data.

2. Materials and Methods

Our study was performed using Parkinson's Dataset with replicated acoustic features in the UCI Machine Learning database [7].

2.1. Data set

A total of 80 people over the age of 50 participated in the creation of the data set. A total of 40 participants (22 males and 18 females) were healthy and 27 males and 13 females that form the remaining 40 participants were having Parkinson's disease. The mean (± standard deviation) age of the participants was 66.38 ± 8.38 for the control group and 69.58 ± 7.82 for people with Parkinson's disease.

2.2. Speech recordings

At this stage, the vocal task of the participants was to make as constant a voice as loud and as loud as possible. This phonation was held for at least 5 seconds in one breath. This task was repeated three times per person, all of which were regarded as copies, creating 240 data sets for a total of 80 participants. Speech data collected from the participants was recorded using a portable computer with an external sound card (TASCAM US322) and a headband microphone (AKG 520) with a cardiogram pattern. The resulting digital audio recordings were performed at a frequency of 44.1 KHz and 16 bit / sample resolution using Audacity software (version 2.0.5). Different data fields were created according to the different characteristics of the obtained voice recordings [2].

2.3. Artificial neural network

Artificial neural network is a model inspired by biological neural networks used to predict a function connected to multiple inputs. Basically; ANN was inspired by a biological neural network consisting of a large network of interconnected neurons. These links are weighted based on past experience; Thus; it is an adaptive network and is capable of learning [8]. The general structure of artificial neural networks is shown in Fig.1. Artificial neural networks are widely used in pattern recognition and classification problems. During the training of ANN, the most appropriate algorithm should be chosen for the correct conclusion of the problem. The most widely used approaches used in the training phase of multilayer networks are batch back propagation (BBP), uintessential propagation (QP), conjugate gradient descent (CGD) and Levenberg-Marquardt (LM) training algorithms. In our study, classification process with LM algorithm was preferred. The Levenberg - Marquardt (LM) algorithm, developed by Kenneth Levenberg and Donald Marquardt, is a combination of gradient descent characteristics and Newton's method in back propagation [9]. It is a nonlinear optimization algorithm based on the use of second order derivatives. The weight calculation equation used in the training algorithms is shown in eq. (1) [10].

$$W=(J^TJ+\mu U)^{-1}J^Te, \tag{1}$$

where J is Jacobian matrix, which is composed of derivatives of errors at each weight, μ is a scalar; U unit matrix; and e represents the error vector of the network.

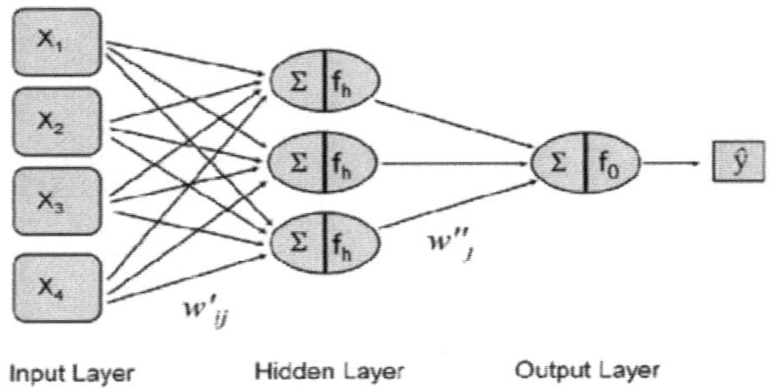

Figure 1. General Structure of Artificial Neural Network.

3. Result and discussion

The study is based on gender and 44 acoustic characteristics that can be divided into five categories. These include: noise characteristics, local perturbation measurements, amplitude local perturbation measurements, nonlinear measurements and spectral envelope measurements. Neural Network Pattern Recognition tool (nprtool) was used in MATLAB to perform the classification process [11]. Using this tool, input values of 45 properties whose entries are defined and classified in 1 output category, ie Parkinson's Disease or Not. The data were randomly divided into 3 sample groups of artificial neural networks. These are:

- Training: 80% random data (192) from the data set is given to the network during the training and the network is trained accordingly.
- Validation: Random 5% data from the data set (12) was used to terminate training when generalization stopped the development of the network.
- Test: 15% random data (36 pieces) from the data set were used to measure the performance parameters of the network during training and verification procedures. Figure 2 shows the customized view of ANN.

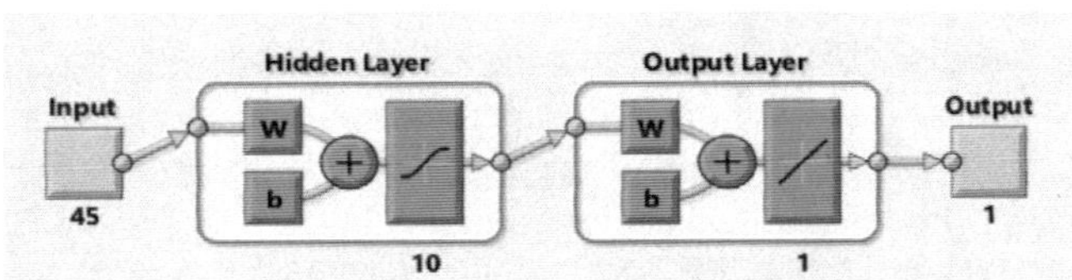

Figure 2. Customization of ANN

When the ANN is run according to Levenberg - Marquardt algorithm, the regression graph of the training data set is shown in figure 3. Figure 4 shows the regression graph of the validation dataset

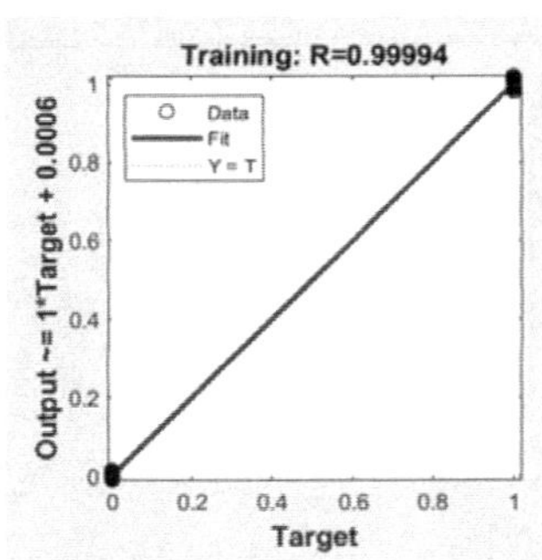

Figure 3. Training Regression.

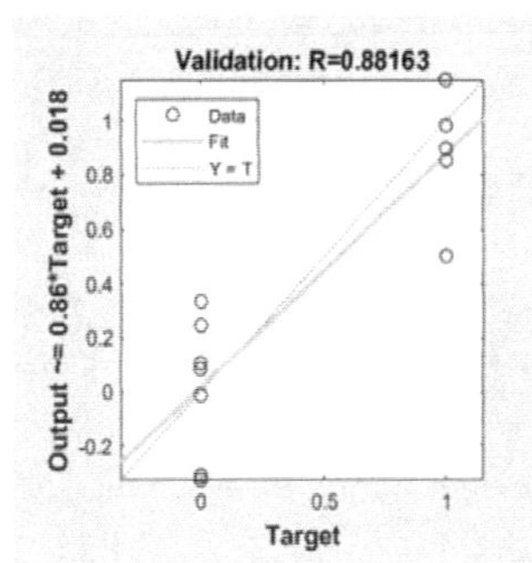

Figure 4. Validation Regression.

Figure 5 shows the regression graph of the test data set. The regression graph of our entire data set is shown in figure 6.

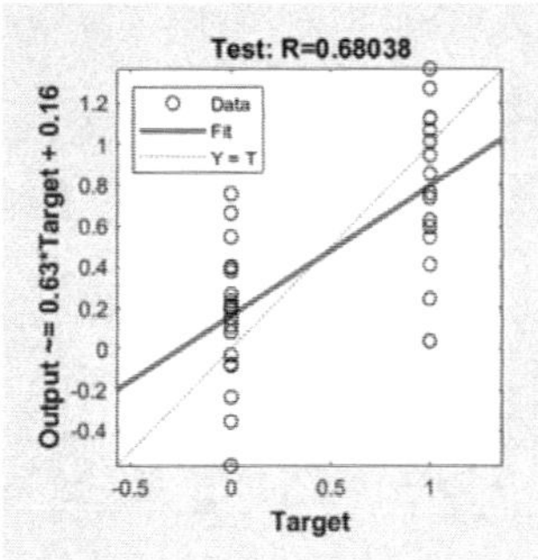

Figure 5. Test Regression.

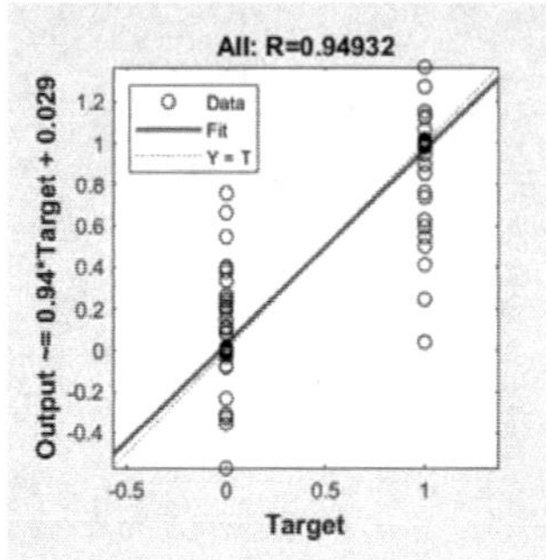

Figure 6. All Dataset Regression.

After the completion of ANN operation, the output and the graph of the target output are shown in figure 7.

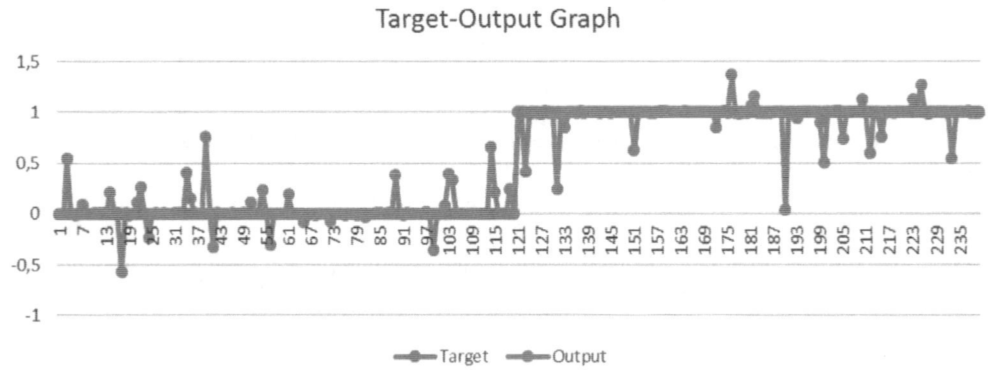

Figure 7. Target-Output Graph

Figure 8 shows the error histogram graph of the Training, Validation and Test data set. In figure 9, the error histogram graph of the whole data set is given.

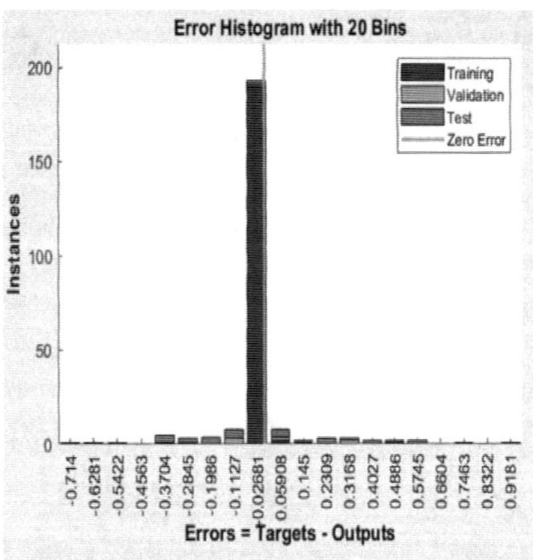

Figure 8. The error histogram graph of the Training, Validation and Test data set.

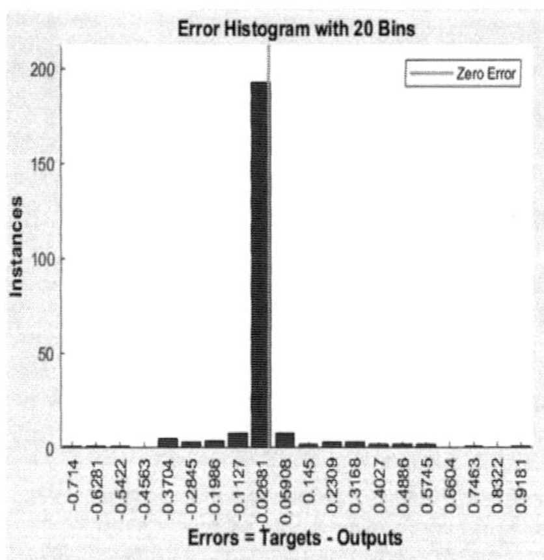

Figure 9. The error histogram graph of the whole data set.

The MATLAB program is used for executing the process. The model uses the trainlm by Levenberg – Marquardt function for training. The MATLAB program randomly divides input variables and target variables into three sets. 80% of the samples are assigned to the training set, 5% to the validation set, and 15% to the test set. It is predicted that it can produce better results with machine learning methods that can be done on classification methods and data used and it is thought that these studies will be done in the future.

4. Conclusion

Artificial neural networks have been shown to exhibit good classification performance of the dataset [7]. In our study, an accuracy of 94.93% was obtained for the whole data set. MATLAB is one of the most widely used software for this purpose. Nowadays, research studies in this area are developing and various classification techniques are applied in order to obtain more accuracy.

Acknowledgments
This study has been supported by Selcuk University's Scientific Research Unit (BAP).

References

[1]. Poewe W, et al 2017 Parkinson disease *Nature reviews Disease primers* **3** 17013
[2]. Naranjo L, et al 2016 Addressing voice recording replications for Parkinson's disease detection *Expert Systems with Applications* **46** pp 286-292
[3]. Miller N 2009 Communication changes in Parkinson's disease *Revista de Logopedia, Foniatria y Audiologia* **29(1)** pp 37-46
[4]. De Lau L M and Breteler M M 2006 Epidemiology of Parkinson's disease *The Lancet Neurology* **5(6)** pp 525-535
[5]. Duffy J R 2013 *Motor Speech Disorders-E-Book: Substrates, differential diagnosis, and management* Elsevier Health Sciences
[6]. Harel B T, et al 2004 Acoustic characteristics of Parkinsonian speech: a potential biomarker of early disease progression and treatment *Journal of Neurolinguistics* **17(6)** pp 439-453
[7]. Carlos J 2019 Parkinson Dataset with replicated acoustic features Data Set URL: https://archive.ics.uci.edu/ml/datasets/Parkinson+Dataset+with+replicated+acoustic+features+
[8]. Manik S, Saini L M and Vadera N 2016 Counting and classification of white blood cell using artificial neural network (ANN) *Proc. IEEE 1st International Conference on Power Electronics, Intelligent Control and Energy Systems 2016*
[9]. Hagan M T and Menhaj M B 1994 Training feedforward networks with the Marquardt algorithm *IEEE transactions on Neural Networks* **5(6)** pp 989-993
[10]. Can A, Dagdelenler G, Ercanoglu M and Sonmez H 2019 Landslide susceptibility mapping at Ovacık-Karabük (Turkey) using different artificial neural network models: comparison of training algorithms *Bulletin of Engineering Geology and the Environment* **78(1)** pp 89-102
[11]. Manik S, Saini L M and Vadera N 2016 Counting and classification of white blood cell using artificial neural network (ANN) *Proc. IEEE 1st International Conference on Power Electronics, Intelligent Control and Energy Systems 2016*

Authors' background

Your Name	Title*	Research Field	Personal website
Ismail Saritas	Prof. Dr.	ANN, Fuzzy, Control, Smart Control	https://www.selcuk.edu.tr/teknoloji/elektrik_elektronik/akademik_personel/bilgi/5879/en
Ali Yasar	Assist. Prof. Dr.	Image Processing, AI, ANN	https://www.selcuk.edu.tr/teknoloji/elektrik_elektronik/akademik_personel/bilgi/9499/en

IOP Conf. Series: Materials Science and Engineering **675** (2019) 012031 doi:10.1088/1757-899X/675/1/012031

Mehmet Akif Sahman	Assistant Professor Dr.	Artificial Intelligence, Optimization	https://www.selcuk.edu.tr/teknoloji/elektrik_elektronik/akademik_personel/bilgi/8680/tr
Ahmet Cevahir Cinar	PhD Candidate	Artificial Intelligence, Optimization	-

Tree based classification methods for occupancy detection

M Koklu*, K Tutuncu

Selcuk University, Konya, TURKEY

* Corresponding author: mkoklu@selcuk.edu.tr

Abstract. Latest smart buildings are not only be intelligent to allow occupant to control the light, heating, cooling, gas and other systems but also focuses on occupancy detection since accurate occupancy detection can result in saving energy up to 42% as can be seen in literature. For this aim, different autonomous systems including sensors, actuators, microcontroller and etc. are at the development phase for smart buildings. At this point, determination of classification methods to detect the occupancy together with hardware plays crucial role. Having done in this study 3 different classification methods that is based on Machine Learning Methods were applied on benchmark dataset named Occupancy by UCI Machine Learning Repository, 2016. The classifiers are Random Forest, Decision Tree and Bagging. They were chosen by following two principals. First one is to have classifier methods that were not use in literature for benchmark dataset and the second one almost never usage of tree based classifiers in the literature. The Occupancy dataset consists of light, temperature, humidity, CO_2 and occupancy. It has been seen that the highest accuracy or prediction ratio was obtained as 99,368% by Decision Tree method namely; Random Forest. This result was compared with the results of the studies on the same benchmark dataset. It has been seen that it is the second best accuracy ratio after Fuzzy Granulation (Fgf) method among 16 different Machine Learning Based classification methods. Additionally, Decision Tree and Bagging had the accuracy ratio of 99.222% and 99.207, respectively. These ratios are also higher than other methods used in literature but Fgf. Thus this study showed how decision tree can be promising for occupancy detection.

Keywords: Classification methods, occupancy detection, random forest, decision tree, bagging

1. Introduction

The growth in population of the world, the climate change, unconscious usage of energy resources, global warming and other factors are threatening current energy resources of the world. As all other sectors the building sector responses to this problem and tries to build SMART buildings. The basic aim of SMART building is to provide an environment that is comfortable and healthy with less energy consumption. At this point, it is very important to control light, air-conditioning and other ambient related conditions with respect to number of people or occupant and ambient parameters. Mostly sensing, actuating and control technologies are used with Information Technologies to get environmental information and provide desired comfort and health [1]. Since motion sensors are not sensitive enough to detect the occupant who is hold still, video and some other sensors such as humidity, CO_2, temperature, light and etc. are employed for detection of occupant. Since privacy and also requirement of large data storage are essential factors for everyone the video is not preferred for these applications. Thus, determining whether or not the occupant is detected needs different classification methods to evaluate the information coming from sensors. Once the occupant is detected related actuators will be triggered and then desired ambient conditions will be provided. So, the goal of occupancy detection will be provided as reducing inefficient energy consumption. The studies [2-4] showed that when appropriate and accurate occupancy detection system is employed energy saving ratio can be achieved between 30% and 42%. The authors of the study [5] claimed that they reduced energy consumption by saving 37% ratio of energy in commercial building by occupancy detection. It is claimed in the study [6] that between 29% and 80% energy consumption reduction was achieved when the occupancy related insights were used as an input for Heating Ventilation Air Conditioning (HVAC) control system. These amounts of energy saving show how important to study in the field of occupancy detection is. The contribution of this paper to this field is to investigate the successes of the different classification methods for occupancy detection on specific benchmark dataset that is obtained from UCI Machine Learning Repository, 2016 [7]. The results of this study were only compared with the previous studies whose methods were applied on the same benchmark dataset to be consisted.

The rest of the paper is organized as follows. In the next (Section 2), a literature survey that only includes the previous studies that applied on the same benchmark Occupancy dataset is presented in the table to be more readable and comparable. In Section 3, the Occupancy dataset is introduced and then information about 3 Machine Learning Based Classification Methods, namely Random Forest (RF), Decision Tree (DT) and Begging is presented. In section 4, the results of the applications of applied methods are presented, and also suggestions for future study are presented.

2. Theoretical background

In order to be consistent and comply with rationality only the studies whose methods were applied on the same benchmark dataset is listed in this section as table. Since Occupancy by UCI Machine Learning Repository was delivered in 2016 [6,13], the studies belongs to year 2016 and later were listed in Table 1 to be more clear and compact. As can be seen from the studies in table 1, the highest accuracy ratio is obtained by Yanyong Huang et al. [10] as 99.74% using Fuzzy Granulation method.

Table 1. Literature survey on the benchmark dataset

Method	Accuracy Ratio (%)	References
Fuzzy Granulation (Fgf)	99,740	Yanyong Huang et al., 2018 [10]
Artificial Neural Network (ANN)	99,061	Ozcan et al., 2017 [12]
SVM with Standart Particle Swarm Optimisation (SVM with SPSO)	98,980	Shinichi Yamada et al., 2017 [9]
FS-KC	98,840	Shinichi Yamada et al., 2017 [11]
Ultra-fast Optimization Multiple Kernel Learning (MKL 2)	98,670	Shinichi Yamada et al., 2017 [11]
Kernel Construction (KC)	98,580	Shinichi Yamada et al., 2017 [11]
Lp-Norm Multiple Kernel Learning (MKL 1)	98,080	Shinichi Yamada et al., 2017 [11]
SVM	98,050	Shinichi Yamada et al., 2017 [11]
Feature Selection (FS)	97,960	Shinichi Yamada et al., 2017 [11]
K-Nearest Neighbour (KNN)	97,655	Sachin Kumar et al., 2018 [8]
Logistic Regression (LR)	97,029	Sachin Kumar et al., 2018 [8]
Gradient boosting machine	96,807	Sachin Kumar et al., 2018 [8]
Naive Bayes (NB)	96,402	Sachin Kumar et al., 2018 [8]
Linear Discriminate Analysis (LDA)	96,156	Sachin Kumar et al., 2018 [8]
Classification and regression tree	92,495	Sachin Kumar et al., 2018 [8]
Support vector machine (SVM)	78,770	Sachin Kumar et al., 2018 [8]

3. Materials and methods

3.1. Dataset description

Occupancy data set is received from UCI machine learning repository [7, 13]. The dataset contains input feature space with attributes such as temperature, humidity, light and CO_2, and it consists of 20.560 samples.

3.2. Classification methods

The classifier methods used in this study were chosen such a way that they haven't been used in literature for Occupancy benchmark dataset. Followings are definition and information about the classifier techniques of this study.

Random Forest (RF): Breiman (2001) A random forest is a classifier consisting of a collection of tree structured classifiers {h(x,Θk), k=1, ...} where the {Θk} are independent identically distributed random vectors and each tree casts a unit vote for the most popular class at input x [14].

Decision Tree (DT): The decision tree classifier is one of the possible approaches to multistage decision making; table look-up rules, decision table conversion to optimal decision trees and sequential approaches are others. The basic idea involved in any multistage approach is to break up a complex decision into a union of several simpler decisions, hoping the final solution obtained this way would resemble the intended desired solution [15].

Begging: Bagging predictors is a method for generating multiple versions of a predictor and using these to get an aggregated predictor. The aggregation averages over the versions when predicting a numerical outcome and does a plurality vote when predicting a class. The multiple versions are formed by making bootstrap replicates of the learning set and using these as new learning sets [16].

3.3. Commonly-accepted performance evaluation measures

The correctness of a classification can be evaluated by computing the number of correctly recognized class examples (true positives), the number of correctly recognized examples that do not belong to the class (true negatives), and examples that either were incorrectly assigned to the class (false positives) or that were not recognized as class examples (false negatives). These four counts constitute a confusion matrix shown in figure 1 for the case of the binary classification. [17].

TP, FP, TN and FN are first calculated depending on the obtained result. Then, Classification Accuracy, Sensitivity, Specificity, Positive Predictive Value, Negative Predictive Value and ROC Area are found by using TP, FP, TN and FN values. Followings are the definitions of related terms and Table 2 shows how to calculate these terms.

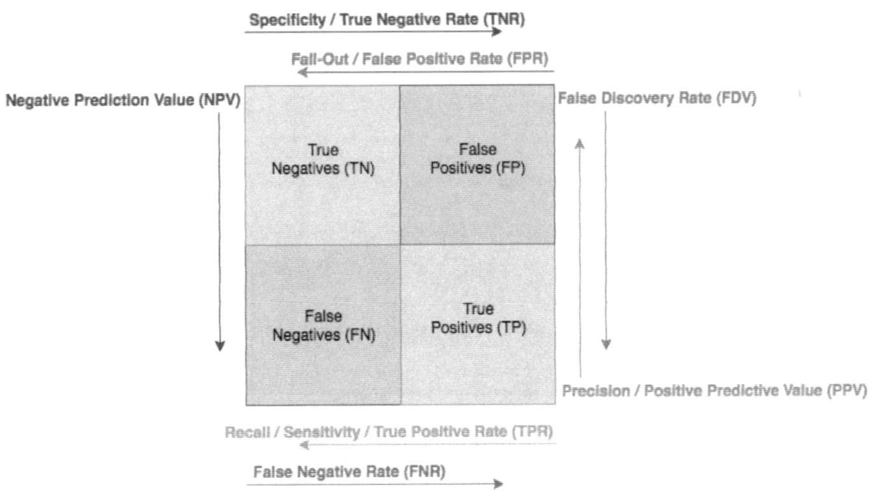

Figure 1. Confusion matrix for binary classification.

Classification Accuracy: It refers to the total number of records that are correctly classified by the classifier. Accuracy of a classifier is defined as the percentage of test set tuples that are correctly classified by the model.

Sensitivity: Refers to the true positive rate that means the proportion of positive tuples that were correctly identified.

Specificity: Refers the rate at which a test or diagnostic method sets a correct (ie negative) diagnosis for a patient who is not ill.

Positive Predictive Value: The fraction of retrieved instances that are relevant.

Negative Predictive Value: The result that the modeling makes a negative prediction.
Area Under Curve (AUC): Classifier's ability to avoid false classification.

Table 2. Performance values measures for classifiers

Performance Evaluation Measures	Formula
Classification Accuracy (%)	$\frac{TP+TN}{TP+FP+TN+FN} x100$
Sensitivity (%)	$\frac{TP}{TP+FN} x100$
Specificity (%)	$\frac{TN}{FP+TN} x100$
Positive Predictive Value (%)	$\frac{TP}{TP+FP} x100$
Negative Predictive Value (%)	$\frac{TP}{TN+FN} x100$
Area Under Curve (AUC)	$\frac{1}{2} x \left(\frac{TP}{TP+FN} + \frac{TN}{TN+FP} \right) x100$

4. Results and discussions

In order to apply RF, DT and Begging classification techniques; first Occupation dataset was normalized, then 10-layer cross folder was applied on normalized dataset and finally related classification algorithms were applied on the dataset. To evaluate the performance measures of classifiers the formula presented in table 2 were used.

The "normalization" of the data plays very important role with respect to not corrupting the relationship between the variables, the accuracy of the analysis and the network performance. Normalization is a squeezing process between the upper and lower bounds of the activation function used for each data in the dataset. The activation function used in the analysis is "sigmoid" and the data are squeezed to the range [0,1]. Before application of classifiers Occupancy dataset is normalized by using equation (1).

$$x_i = \frac{x_i - x_{min}}{x_{max} - x_{min}} \quad (1)$$

For 10-layer cross-validation method the given dataset is divided into 10 parts as shown in figure 2 [18]. Any method that needs to be trained and tested will use k different training and test sets (where k=number of data/10), and still achieves k success measures accordingly. For this reason, to determine the success of cross-validation, the arithmetic mean of the obtained k performance measures is taken [19, 20]. For occupancy dataset that consists of 20560 samples 10-fold cross-validation is applied by dividing it into the groups; where each group has 2056 elements. Since there are 10 groups, 10 different test results are formed. The average success of these tests will be the overall success, that is, the success of the method to be applied.

After applying 10-fold-cross validation and obtaining the results for RF, DT and Bagging methods the confusion matrix for three classifiers shown in Table 3 is obtained.

Table 3. Confusion matrix obtained from Random Forest, Decision Tree and Bagging classifiers

Actual	**RF Predicted**		**DT Predicted**		**Bagging Predicted**	
	P	**N**	**P**	**N**	**P**	**N**
P	4703 (TP)	47 (FN)	4696 (TP)	54 (FN)	4709 (TP)	41 (FN)
N	83 (FP)	15727 (TN)	106 (FP)	15704 (TN)	122 (FP)	15688 (TN)

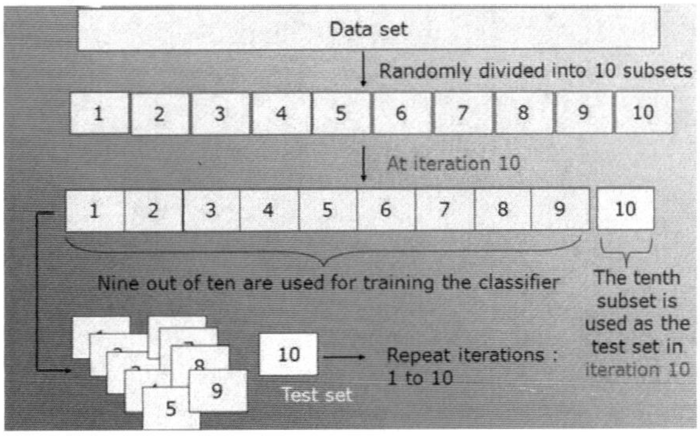

Figure 2. 10-Fold cross-validation method.

By using confusion matrix presented in table 3 and the formulas presented in table 2 performance evaluation measures of RF, DT and Bagging were calculated and presented in table 4.

Table 4. Performance evaluation measures for RF, DT and Bagging classifiers

Performance Evaluation Measures	Random Forest	Decision Tree	Bagging
Classification Accuracy (%)	99,368	99,222	99,207
Sensitivity (%)	99,011	98,863	99,137
Specificity (%)	99,475	99,330	99,228
Positive Predictive Value (%)	98,266	97,793	97,475
Negative Predictive Value (%)	99,702	99,657	99,739
Area Under Curve (AUC)	0,992	0,991	0,992

5. Conclusion

As it's mentioned before Occupancy dataset by UCI was used to be benchmark dataset. It should be noted that this dataset has allows user to have binary classification since the output variables called "occupancy" and has a value of either 1 or 0. Additionally this dataset was published in 2016 by UCI and only the studies from 2016 and then were listed in literature survey and compared with this study. When the RF, DT and Bagging were chosen for Machine Learning Based classification methods two principals were carefully followed. First one is to have classifier methods that were not use in literature for benchmark dataset and the second one almost never usage of tree based classifiers in the literature. Thus, the author of this paper would like see the success and/or failure of tree based classifiers. It has been seen that the highest accuracy or prediction ratio was obtained as 99,368% by Random Forest. So, it is the second best accuracy ratio after Fuzzy Granulation (Fgf) method among other 16 methods in the literature that we can reach. Additionally. DT and Bagging classifier showed higher accuracy than other 15 methods in the studies listed in literature section. Thus this study showed how decision tree can be promising for occupancy detection.

6. References

[1.] Lam K P, et al 2009 occupancy detection through an extensive environmental sensor network in an open-plan office building *Proc. 11th Intl. IBPSA Conf.* (Glasgow, Scotland)

[2.] Erickson V L, et al 2011 OBSERVE: Occupancy-based system for efficient reduction of HVAC energy *Proc. 10th Intl. Conf. on Information Processing in Sensor Networks* (Chicago, USA) pp 258–269

[3.] Erickson V L, et al 2014 Occupancy modeling and prediction for building energy management *ACM Trans. Sensor Networks* **10(3)**

[4.] Dong B and Andrews B 2009 Sensor-based occupancy behavioral pattern recognition for energy and comfort management in intelligent buildings *Proc. 11th Intl. Building Performance Simulation Association Conference*

[5.] Brooks J, et al 2014 An experimental investigation of occupancy-based energy-efficient control of commercial building indoor climate *Proc. IEEE 53rd Annual Conf. on Decision and Control* (Los Angeles, USA) pp 5680–5685
[6.] Brooks J, et al 2015 Energy-efficient control of under-actuated HVAC zones in commercial buildings *Energy Build* **93** pp 160–168
[7.] Lichman M and Bache K 1998 *UCI machine learning repository* (University of California, USA) URL: https://archive.ics.uci.edu/ml/datasets/Occupancy+Detection+
[8.] Kumar S et al 2018 Machine Learning-Based Method and Its Performance Analysis for Occupancy Detection in Indoor Environment *Proc. Intl. Symp. on Signal Processing and Intelligent Recognition Systems* (Bangalore, India) pp 240-251
[9.] Yamada S and Neshatian K 2017 Hyper-parameter Search in Support Vector Machines using PSO with Cellular Fitness Approximation *Proc. 2017 IEEE Symposium Series on Computational Intelligence* (Honolulu, USA) DOI: 10.1109/SSCI.2017.8285406 (978-1-5386-2726-6/17/$31.00
[10.] Huang Y, et al 2018 Dynamic Fusion of Multi-source Interval-valued Data by Fuzzy Granulation *IEEE Transactions on Fuzzy Systems* DOI 10.1109/TFUZZ.2018.2832608
[11.] Yamada S and Neshatian K 2017 Kernel Construction and Feature Subset Selection in Support Vector Machines *Proc. Asia-Pacific Conf. on Simulated Evolution and Learning* (Shenzhen, China) pp 605–616 https://doi.org/10.1007/978-3-319-68759-9_49
[12.] Cataltas O, Tutuncu K and Koklu M 2017 Occupancy detection through light, temperature, humidity and CO2 sensors using ANN *Intl. Journal of Industrial Electronics and Electrical Engineering* **5(2)** pp 63-67
[13.] Candanedo L M and Feldheim V 2016 Accurate occupancy detection of an office room from light, temperature, humidity and CO_2 measurements using statistical learning models *Energy and Buildings* **112** pp 28-39
[14.] Breiman L 2001 Random Forests *Machine Learning* **45(1)** pp 5-32
[15.] Safavian S R and Landgrebe D 1991 A survey of decision tree classifier methodology *IEEE transactions on systems, man, and cybernetics* **21(3)** pp 660-674
[16.] Breiman L 1996 Bagging predictors *Machine learning* **24(2)** pp 123-140
[17.] Sokolova M and Lapalme G 2009 A systematic analysis of performance measures for classification tasks *Information Processing & Management* **45(4)** pp 427-437
[18.] Randell T 2018 Classification and Prediction (Data Mining: Concepts and Techniques) URL: http://slideplayer.com/slide/8148400/
[19.] Golub G H, Heath M and Wahba G 1979 Generalized cross-validation as a method for choosing a good ridge parameter *Technometrics* **21(2)** pp 215-223
[20.] Shao J 1993 Linear model selection by cross-validation *Journal of the American statistical Association* **88(422)** pp 486-494

Acknowledgments
This study has been supported by Scientific Research Project of Selcuk University.

Authors' background

Name	*Prefix*	*Research Field*	*Email*
Murat KOKLU	Assist. Prof. Dr.	Artificial intelligence and applications	mkoklu@selcuk.edu.tr
Kemal TUTUNCU	Assist. Prof. Dr.	Artificial intelligence and applications	ktutuncu@selcuk.edu.tr

IOP Conf. Series: Materials Science and Engineering **675** (2019) 012033 doi:10.1088/1757-899X/675/1/012033

Comparison of typical access networks with AMOOFDM based access networks

H A Yazbeck[1*] and V V Belyaev[1,2]

[1] Peoples' Friendship University of Russia (RUDN university), Moscow, Russia
[2] Moscow Region State University, Moscow, Russia

E-mail: 1042175153@pfur.ru

Abstract. Access networks are under constant development looking for customer's satisfaction by enhancing the internet service speed and the cost of the service. Moreover, trying to serve as many residential units as possible by extending the distance reach of the access network to cover a part of a big city or to reach a rural area in a range up to 60 km. Nowaday's networks still experience limitations in services, coverage area, and still pricey for average customers. However, a significant percentage of these networks still using the terrestrial radio wave telecommunications to distribute the service inside or between city blocks, while others use the fiber to the home (FTTH). In this article, we are comparing the performance of the currently operated access networks to the adaptively modulated optical orthogonal frequency multiplexing (AMOOFDM) based access networks, according to the bit-rate, and the distance reach of these networks; highlighting the advantages of the AMOOFDM system using some types of the semiconductor optical amplifier (SOA) as an intensity modulator.

Keywords: wireless communication; digital subscriber line (DSL); passive optical network (PON); adaptively modulated optical orthogonal frequency multiplexing (AMOOFDM); semiconductor optical amplifier (SOA).

1. Introduction

Internet infrastructure is vital for civilisation as same as water and electricity. International Telecommunication Union (ITU) declared that in 2018, the world reached the 50/50 milestone of internet usage. Where it is the base of e-health, education and catastrophe aid. Notably, optical networks spread with the developing internet of things (IoT). So, the number of globe's copper networks shrank by 6%, while optical fiber connections boosted by 28%, and that just in one year (2016-2017). The plan for the fiber networks is to satisfy services up to 10 million residences and enterprises by 2020 [1].

Fiber technology is easy to upgrade, it presents symmetrical bandwidth, driving to a unique performance with high speed, low cost and low energy consumption resulting in a chance to create new sorts of services and applications [2]. However, in Europe, as mentioned in [3] expanding broadband networks by 10% results in an increment of 1% in GDP per capita per year and a 1.5% extra work productivity for the subsequent 5 years. Furthermore, investments in broadband play a functional role in education quality and promote developing countries and help rural regions.

In developing countries, there are still around 4 billion people don't have access to the internet. One of the causes of the shortage of internet is inhabiting rural regions and in this case fiber networks are the best solution, with a shared infrastructure, established with other utilities such as power cables [4] and using AMOOFDM system for such cost-sensitive networks, not only because it is more cost-effective, but also, it has a long-distance reach.

Access networks not only for providing the internet but also, are used to transmit and receive all kinds of data we generate and demand from the international under-ocean optical network, that connects the continents (figure 1) [5].

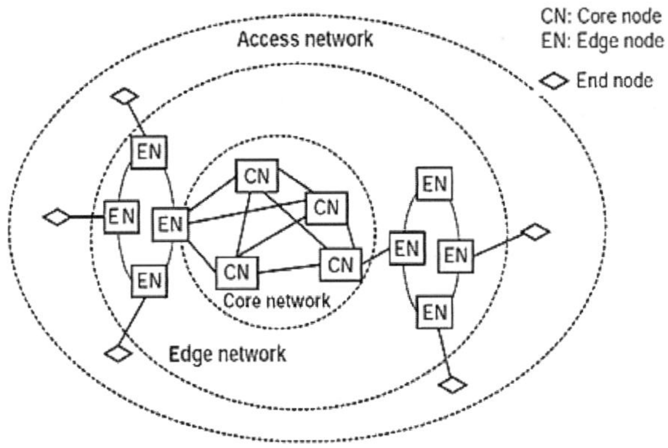

Figure 1. global network architecture [5].

This network is denominated the core network or Wide Area Network (WAN); it uses the mesh network as a physical topology. Moreover, inside continents, there are vast networks that connect the countries, where each country contains edge networks or Metropolitan Area Networks (MANs) for the cities, it utilizes the ring network as a physical topology. Inside each city, several access networks or Local Area Networks (LANs) to connect the city blocks, buildings and residential units, it uses the star network as a physical topology. Access networks can be wireless (mobile networks) or wireline (copper networks, fiber optics). However, for the core and edge networks, they are always optical networks.

1.1 Mobile networks

The mobile networks have a backbone of an optical fiber system to transmit and receive the data faster and to be connected to the global fiber ocean network. However, the distribution of the service is through microwave telecommunications that divide the entire country into hexagonal cells, and at the center of each cell a base transceiver station (BTS) (see figure 2) in touch with each customer in its territory to serve his calls, internet and data [6]. BTS covers a city block if the buildings are not too high and there is not too much steel in the concrete structure of buildings. Wireless Fidelity Wi-Fi is also a wireless media for telecommunications, and it is the easiest way to distribute the service indoor. Each time the user wants to use the internet he doesn't have to connect a cable into his computer, he can use laptops, mobile phones or any other gadget to access the internet via Wi-Fi router, that ends the access networks, receiving the service from wireless access points or wireline (copper, fiber optics).

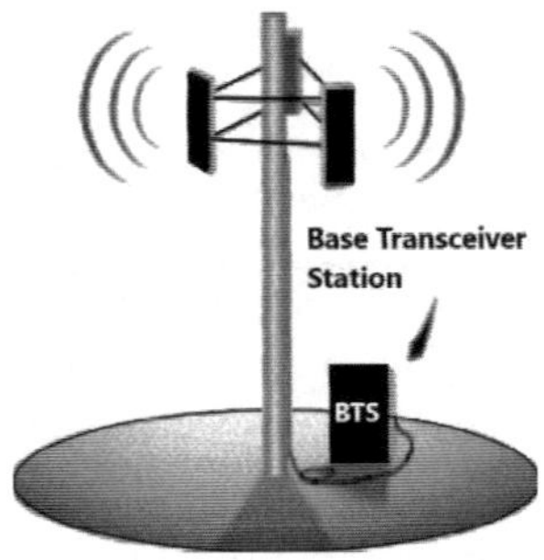

Figure 2. Base Transceiver Station BTS [6].

1.2 Copper networks

Digital subscriber line DSL (Figure 3) still used worldwide particularly in Europe and North America [7], to distribute data and video simultaneously with the voice services over the ordinary telephone lines, without affecting the analogue transmission of the telephone services. There are many types in the digital subscriber line technology known as xDSL. The first generation is the asymmetric digital subscriber line ADSL, used as a small network. ADSL followed by an enhanced version like ADSL2 and ADSL2+, noting that ADSL2+ was a significant development in the distance reach wise. The second generation of the DSL technology is the very-high-bit-rate VDSL; it took the DSL service to another level in terms of network speed compared to the first generation. Also, it was updated to a better version (VDSL2) improving the distance reach and the bandwidth. Both generations of DSL saved too many investments because of the already existing copper telephone network, and this is the reason made it useful until the current time. Although it is very beneficial to exploit the existing copper network, it has many drawbacks. The main disadvantage is the oxidation of the copper wires. It affects the transmission signal by highly increasing the resistance of the transmission line that leads to signal loss. Not only oxidation decreases the performance of the copper network, but also the current leakage to other lines or ground, not tight wire contacts, inappropriate wire cut, moist mediums and many others [8].

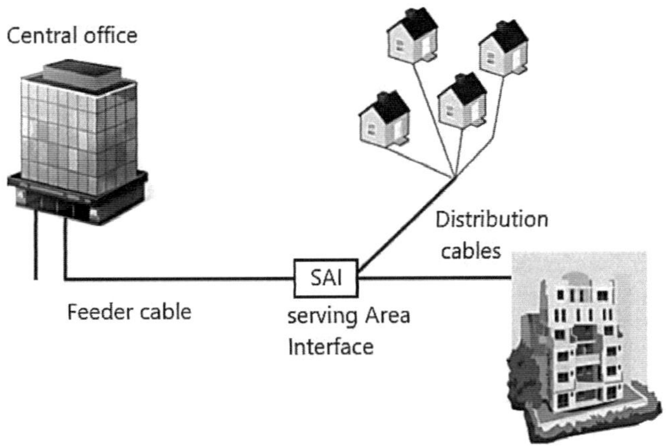

Figure 3. Digital subscriber line DSL architecture [8].

1.3 Optical fiber networks

The optical fibers are a glass pipes consists of a core and cladding; each has a different density. Therefore, the laser propagates inside the core carrying the information in the form of light modulated signals. At the transmitter, electrical data is converted to the optical domain to be transmitted through the fiber cable, and at the receiver, it is converted back to an electrical domain.

Fiber networks have too many advantages over any other type of networks. The main advantage is that they have a broad bandwidth allow them to send a considerable amount of data simultaneously in both upstream and downstream without affecting each other. Moreover, they are thin, and light compared to copper networks. Although copper networks physically are more robust, fiber optics networks have a too much longer life-span, too low signal loss, and they have no signal interference, this means good quality signals are received very fast, with less cost.

It is the fastest media of communication now, wherever it can send in one single fiber upstream data and downstream data each on a different wavelength (λ). Nowadays used optical access networks are called E-PONs (Ethernet passive optical networks) and G-PONs (Gigabit passive optical networks). The PON contain an optical line terminal (OLT), which is the service provider that provides the data over the optical fiber cable and passively distribute it to the optical network units (ONUs) or residential units using an optical splitter that split each wavelength λ to the relevant destination (figure 4) [9].

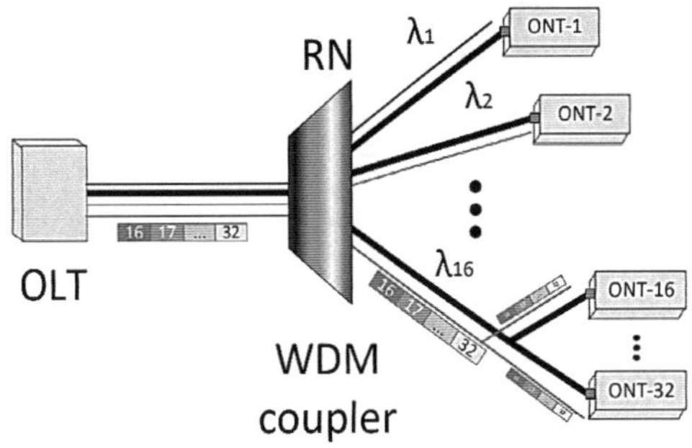

Figure 4. Passive optical network [9].

1.3.1 AMOOFDM system

Although the GPON technology is successful, it has some drawbacks. That's why science endures finding a solution for each disadvantage or substitute the system by another more successful technology. The adaptively modulated optical orthogonal frequency division multiplexing (AMOOFDM) proved that it's way better than others. It consists of an electrical OFDM transmitter, generating an electrical OFDM signal. This signal is converted to an optical signal using an electrical to optical converter (E/O), and then the optical OFDM signal is being sent to the ONU side via single mode fiber. The transmission procedure involves a negotiation between the transmitter and the receiver to determine the modulation format, to send as much data as possible. That's why it's designated an adaptive modulation. At the receiver side, the optical OFDM signal is converted back to electrical using a positive-intrinsic-negative (PIN) photodiode and processed by the electrical OFDM receiver. So just like most of the existing optical networks AMOOFDM system based on the intensity modulation and direct detection (IMDD) concept. The simplicity and the low cost of the IMDD systems render it profitable for cost-sensitive optical access networks since coherent optical systems are more expensive and more involved in the design. The AMOOFDM system setup illustrated in figure 5 [10].

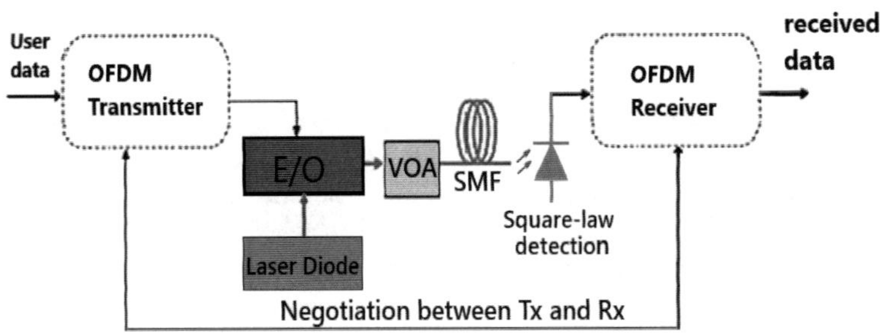

Figure 5. AMOOFDM system [10].

In this system (E/O) converter can be employed as a semiconductor optical amplifier (SOA). SOA is an optoelectronic device, that can under appropriate operating conditions amplify an incident light signal. A basic structure of an SOA is shown in figure 6 its central element called the active region that provides optical gain when the amplifier is electrically pumped. Pumping is necessary to achieve the population inversion and the stimulated emission. An integrated waveguide is used to contain the wave signal in the active region. This active region has many shapes that determine its type and operation; for example, when the active region designed as dots, it is called quantum dot SOA (QD-SOA). When the SOA have mirrors on the input and output facets, it is named reflective SOA (RSOA). Moreover, SOA can be used as in a cascaded way to improve its operation, like the two-cascaded SOA (2C-SOA) it is a two Bulk SOAs connected in series (the output of the first device used as an input to the second device).

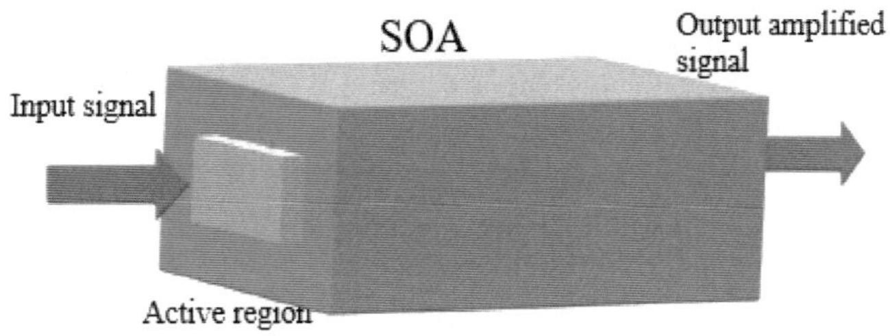

Figure 6. The SOA structure.

In the available to us literature, we didn't face any comparison of the AMOOFDM optical access networks with other varieties of access networks. Therefore, this study aims to compare the AMOOFDM access networks in terms of bit-rate, distance reach and splitting ratio with wireless, copper and other optical fiber networks.

2. Materials and methods

The publications in the field of access networks (wireless, copper, fiber optics) for the latest 10 years were gathered. All the information about the speed and distance reach were collected in tables. The data for each class of access networks were analyzed and compared, to find the best solution for the cost-sensitive networks. The proper splitting ratios were calculated for each. Then, only the bests of each sort of access networks were compared to each other. Eventually, a conclusion was derived.

3. Results and discussion

As shown in table 1, the wireless telecommunication technologies are listed starting from the 1st generation (1G) cellular network till the current 4th generation (4G) network, and the near future 5th generation (5G) network. Comparing them to the Wi-Fi (IEEE 802.11a) and Wi-Max technologies (IEEE 802.16e). The 1st generation cellular network offered 2.4 Kbps bit-rate for the area of the cell coverage, and this was used only for calls. While it increased to 150 Kbps in the 2nd generation with the addition of data exchange. In the 3rd generation, it reached the 42 Mbps which was a good improvement over the 2nd generation for that time, but it still not good enough to support the bandwidth demanding applications like live streaming of high definition videos, online gaming and the voice over IP (VOIP) different applications. This problem was solved in the 4th generation, where the bit-rate reached the 100Mbps, and it is noticed that it is similar in bit-rate to the Wi-Max, but the coverage area is way less than Wi-Max. By 2020 [11], the 5th generation (5G) will be widely applied with a bit-rate up to 1 Gbps that gives the possibility of a variety of new applications to be set on the service. When talking about wireless technology, we must mention a viral way of using internet data. It is the Wi-Fi, where it's bit-rate up to 54 Mbps with a limited coverage area to 9 meters only. However, it's good to know that it's a straightforward way to use the data signals inside buildings and living units, notably the ones who don't contain too much steel. Because steel shields are the greatest enemies for wireless communication. As a result, the 5th generation is the best choice to use wireless communications, but it still not in practice in all capitals. While the Wi-Fi can't be avoided for indoor usage.

Table 1. Wireless telecommunication technologies distance reach and bit-rates [7, 12].

Technology	Mobile					Wi-Fi	Wi-Max
	1G	2G	3G	4G	5G		
Coverage area (km)	Cell area	Cell area	Cell area	Cell area	Cell area	0.009	64
Bit-rate (Mbps)	0.0024	0.15	42	100	1000	6-54	100

Table 2 depicts the wireline telecommunication technologies; some of them are still in use and others extinct. Starting from the coaxial cables, ending with the ITU-T G.9701 series fast access to subscriber terminals (G. Fast) technology, through all the sorts of digital subscriber line (xDSL) from ITU-T

G.992.1/2 Asymmetric digital subscriber line (ADSL), to ITU-T G.993.2 very high bit-rate digital subscriber line 2 (VDSL2). Coaxial cable networks are not in use anymore because it sends the data for less than 100 meters, although it allows 100 Mbps bit-rate. While, the ADSL transferred the data for 2 km with a rate of only 4 Mbps, but the next generation ITU-T G.992.3 (ADSL2) increased the rate to 12 Mbps. The ITU-T G.992.5 (ADSL2+) doubled the bit-rate to 25 Mbps, and the tripled the distance to 6 km, that was an exciting result distance wise, wherever it was better than the ITU-T G.993.1 (VDSL) that had only 1.5 km range, although the rate of data transmission was 52 Mbps. The ITU-T G.993.2 (VDSL2) doubled the distance to 3 km and the bit-rate to 100 Mbps, and that was a good trade-off between the distance-reach and the data-rate for that time. A big boost in the bit-rate to 1 Gbps applied by the (G. fast) but unfortunately, it is very limited in the distance to less than 0.5 km. We can select network technology according to the situation. If we need to serve more speed or more residential units, we can select G. fast. Else, if we want to serve far city blocks, we can select ADSL2+. Nevertheless, the best compromise between the speed and the distance is by selecting the VDSL2.

Table 2. Wireline telecommunication technologies distance reach and bit-rates [7].

Technology	xDSL						Coaxial
	ADSL	ADSL2	ADSL2+	VDSL	VDSL2	G. Fast	
Distance reach (km)	2	2	6	1.5	3	0.5	<0.1
Bit-rate (Mbps)	4	12	25	52	100	1000	100

In table 3 (a & b), there are two types of optical fiber access networks. The typical used passive optical networks (PONs) and the Adaptively modulated optical orthogonal frequency digital multiplexing (AMOOFDM) based PONs, employing a variety of intensity modulator (IM), optimizing the bandwidth to get the maximum possible bit-rates. In the conventional PONs (table 3a), we have the IEEE 802.3ah Ethernet-PON (E-PON) with a distance 20 km and a speed of 1.25 Gbps. With a splitting ratio of 1:64 it can serve approximately 20 Mbps per optical network unit (ONU) or residential unit while the ITU-T G.984.x Gigabit-PON (G-PON) doubles the bit-rate to 2.5 Gbps with a splitting ratio of 1:64. Hence, approximately 40 Mbps per residential unit, further for a distance between 20 and 60 km. In the AMOOFDM based PONs (table 3b) when the reflective and the quantum-dot semiconductor optical amplifier (RSOA-IM) and (QD-SOA-IM) were applied as an intensity modulator an extraordinary bit-rate up to 30 Gbps achieved to a distance reach of 60 km. While using the semiconductor optical amplifier intensity modulator (SOA-IM), the distance stretched to 80 km, which is 4 folds longer than the usual PONs. However, the best was applying the two-cascaded semiconductor optical amplifier as an intensity modulator (2C-SOA-IM), it prolonged the distance reach up to 120 km. That made it a robust full-service access network. As noticed that all the AMOOFDM based PONs have an equivalent bit-rate about 30 Gbps, and this proposes that we can control the splitting ratio according to our aim. To serve less residential units with a very high bit-rate, we can use 1:64 splitting ratio that offers roughly 470 Mbps per one residential unit, or owning a fast-enough speed but serving too many customers using a splitting ratio of 1:256, that offers approximately 120 Mbps per a residential unit. Moreover, that substitutes the expense of the network, presenting it more economical than the other currently utilized networks, comparing 256 to only 64 customers, that means 4 multiples of the profit.

Table 3. Fiber optics telecommunication technologies (a) typical, (b) AMOOFDM based, distance reach and bit-rates [7, 10, 13, 14, 15]

	Technology	Typical PON	
		E-PON [7]	G-PON [7]
(a)	Distance reach (km)	20	20-60
	Bit-rate (Gbps)	1.25	2.5

		AMOOFDM			
	Technology	(SOA-IM) [10]	(RSOA-IM) [13]	(QD-SOA-IM) [14]	(2C-SOA-IM) [15]
(b)	Distance reach (km)	80	60	60	120
	Bit-rate (Gbps)	30	30	30	30

4. Conclusion

This work is a comparison of different types of access networks in the cities judging each by its bit-rate, splitting ratio and distance reach. Both wireless and wireline access networks have been reviewed. In the wireless networks, it is evident that the 5th generation is the best solution for the mobile networks where it supports 1 Gbps in the cell region, and any other kind of access networks can't replace it, but it still not available enough. In the wireline networks, the best in the copper networks was the VDSL2. It can serve 64 residential unit with a speed of approximately 1.5 Mbps. In the optical fiber access networks, the best was the AMOOFDM based access network using the two-cascaded SOA (2C-SOA-IM), using a single fiber it can serve 256 residential unit with a speed of 120 Mbps for each, instead of 64 residential unit with a speed of 40 Mbps for G-PON. As a summing-up, 2C-SOA-IM in AMOOFDM offers service for 4 times more residential units than the best existing access networks (G-PON), with 3 multiples more speed per each unit, with a distance reach up to 6 folds longer. It is way too much economical and beneficial.

Acknowledgement

This research was supported by the Russian Foundation for Basic Researches RFBR. Grant № (19-07-00602_a).

References

[1] Phillippa B, Ivan V, Ahone N, et al. 2018 *The State of Broadband 2018: Broadband Catalyzing Sustainable Development* (Switzerland, Geneva: International Telecommunication Union) chapter 2 pp 10-33

[2] Ivette O and Filippo M 2015 *Socio-Economic Benefits of High Speed Broadband* (Brussels: DG connect) chapter 2 pp 6-12

[3] Angelika Z, Niels-Erik B, Romauld K, et al. 2018 *Broadband in the EU Member States: Despite Progress, Not All the Europe 2020 Targets Will Be Met* (Luxembourg: Publication office of the EU) chapter 1 pp 1-12

[4] Gallegos D, Narimatsu J, Batori O, et al. 2018 *Innovative Business Models for Expanding Fiber-Optic Networks and Closing the Access Gap* (Washington, D.C.: World bank group) chapter 1 pp 20-26

[5] William S and Ivan D 2010 *Orthogonal Frequency Division Multiplexing for Optical Communications* (Elsevier) chapter 3 pp 53-118

[6] Larry D and Lars D 2011 *Digital Forensics for Legal Professionals, Understanding Digital Evidence from the Warrant to the Courtroom* (USA: Syngress) chapter 33 pp 225-37

[7] Yiran M and Zhensheng J 2017 *Evolution and Trends of Broadband Access Technologies and Fiber-Wireless Systems* (Switzerland: Springer) chapter 2 pp 43-75

[8] Fredrik L 2009 *Estimation of Line Properties in the Copper Access Network* (Lund, Sweden: Tryckeriet i E-Huset) pp 1-17

[9] Samael H, Jose A and Jose L 2018 *«Passive optical networks: Introduction» Wiley Encyclopedia of Electrical and Electronics Engineering* (Catalonia, Spain: Wiley Online Library) pp 1-20

[10] Jinlong W, Ali H, Roger G and Jianming T 2009 Semiconductor optical amplifier-enabled intensity modulation of adaptively modulated optical OFDM signals in SMF-based IMDD systems *Lightwave Tech.* **27(16)** pp 3678-88

[11] Oluwakayaode O, Muhammad Ali I, Junaid Q and Arjuna S 2016 Will 5G see its blind side? evolving 5G for universal internet access *e-prints* **1** p 1603

[12] SagarKumar P, Vatsal S and Maharshi K 2018 Comparative study of 2G, 3G and 4G *Int. J. of scientific Reseach in comp. sci. Engineering* **3(3)** pp 1962-4

[13] Jinlong W, Ali H, Robert G, Emilio H, Xi Z, Saad M and Jianming T 2010 Adaptively modulated optical OFDM modems utilizing RSOAs as intensity modulators in IMDD SMF transmission systems *Optics express* **18(8)** pp 8556-73

[14] Ali H, Mohamad H, Jinlong W, Ammar S and Jianming T 2011 Theoretical investigations of quantum-dot semiconductor optical amplifier-enabled intensity modulation of adaptively modulated optical OFDM signals in IMDD PON system *Optics express* **19(25)** pp 25696-711

[15] Ali H, Mohamad H, Haidar T, Layaly M, Ammar S, Ali A, Ali H, Elias G and Jianming T 2014 Two cascaded SOAs used as intensity modulators for adaptively modulated optical OFDM signals in optical access networks *Optics express* **22(13)** p 15763

New data on mineral complexes of the Turgoyak limestone deposit (South Ural)

A F Georgievskiy[1*], V M Bugina[1], O Yu Rogova[2], and A A Georgievskiy[3]

[1] Peoples' Friendship University of Russia (RUDN University), Moscow, Russia
[2] Geological-prospecting college of Miass, Miass, Russia
[3] Multi-specialty middle school №1577, Moscow, Russia

Email: georgievskiy-af@rudn.ru

Abstract. The main purpose of the research was to ascertain additional development potential of the oldest mining plant in the Southern Urals - the Turgoyak limestone deposit (Miass District, Chelyabinsk region, South Ural). An additional site dike survey work and mineral veins enclosed in carbonate rock strata and overlapping Low Devonian jasperoid rock was carried out. The main purpose of that exploration was a detailed study of the chemical and mineral composition of both dikes and host rock and their secondary alterations. In addition to photographic documentation and mapping, an interrelation between dikes and veins and theirs contacts was determined. Also the connection with them of newly formed sulfide and other secondary minerals was detected (recorded). Subsequently rock and mineral samples were analyzed by microscopic methods, X-ray diffraction and X-ray phase analysis, X-ray fluorescence analysis by RLP-3-01 device, a DRON-7 diffractometer, and a DELTA Premium portable spectrometer. Those investigations resulted in establishment of silicate and calcareous skarns, spatially associated with diabase dikes at the Turgoyak flux limestone deposit. In addition to the skarns, vein polymineral bodies were recorded there. The most interesting of those are carbonate-quartz-axinitic and axinitic ones. The veins of this composition contain superimposed post-skarn sulfide mineralization, with signs of increased gold content. Given the development at the deposit of the dyke complex and the skarns forming on it, it is possible to forecast the existence of a granitoid intrusion in deep horizons that can generate industrial ore mineralization.

Keywords: skarns, dikes, axinite, gold ore mineralization, veins, X-ray phase analysis

1. Introduction

The main purpose of the research is to show additional reserves of the oldest mining enterprise in the Southern Urals related to the prospects of detecting skarn type ores. Conducted researches show the possibility of detecting a hidden skarn mineralisation, potentially gold-bearing in the area of the Turgoyak flux limestone deposit.

Figure 1. Miass district of the Chelyabinsk region

IOP Conf. Series: Materials Science and Engineering **675** (2019) 012034 doi:10.1088/1757-899X/675/1/012034

Miass district of the Chelyabinsk region (Figure 1) is a major centre of mechanical engineering and mining with a well-developed energy and transport infrastructure. The Turgoyak deposit is among its oldest enterprises. It has been developed by quarrying since the end of the forties of the last century, and the limestone mined here goes to the metallurgical plants of the Urals, Central and Western Russia.

From a geological viewpoint, the deposit belongs to Turgoyak strata of the Lower Devonian age [1, 2] and is part of a giant tectonic block, which is enclosed in a serpentinite melange of the Main Ural fault [3-5]. Within that block, the strata is deformed into a syncline structure, stretching 25 km in the northeast direction, with a width of 140 to 700 m. Over the sixty years of operation of the quarry, the deposit has been thoroughly studied. In the central part, it is divided by a lateral faulting into the southern and northern blocks, where limestones steeply dip to the east, as well as to the south-east with angles of 65 - 85° (Figure 2a). The limestone are gray and bluish-gray, lenticularly layered (Figure 2b), of high quality. Taking into account their average chemical composition (CaO — 55.19%; MgO — 0.33%; SiO_2 — 0.33%; S — 0.02%; P_2O_5 — 0.04%) [2,6] they consist of calcite (97.0%) and as a minor impurity include dolomite (1.5%), siliceous and clay minerals (1.5%).

Figure 2. Limestones of the Turgoyakskoye stratum (a - rock exposure, southern flank of the deposit; b - ore limestone with a lenticular-layered structure; c - microscopic structure of limestone; at the top, large remnants of recrystallized echinoderms. Thin section, polars crossed.

The rocks are marbled, but sometimes contain relics of recrystallized echinoderm fossils (Figure 2c). The recrystallization of limestone is enhanced in the contact areas of diabase dykes and calcite veins, which cross both limestone and overlying jasperoid. The dikes either vertical or acclivous, can be traced up to 300 m and have a thickness of 0.5 - 5 m.

Figure 3. Jasperoids of the stratum overlying the limestones (a - Jasperoid ore with indistinct fine horizontal stratification; b - ore with elements of gradational layering; c - the microscopic structure of jasperoid; in the center - inclusion of phosphate recrystallized to the state of apatite. Thin section, plane polarized light.

The transition from limestone to jasperoid rock is gradual, through the interlayering in the range of 2 - 3 m. The jasperoid is dirty green in color with gray, brown and yellowish hues. Macroscopically, it is solid monolithic rock with hardly visible horizontal and lenticular stratification (Figure 3a). In some cases, clastic structures with an unclear gradation constitution are observed (Figure 3b). In thin sections, they are composed of weakly crystallized chalcedony-like quartz with fine fragments of carbonate, mica flakes, rare quartz sand grains, and also disseminated pyrite. In addition, there are grains and inclusions of phosphate, recrystallized to the state of microcrystalline apatite (2.5% P_2O_5) (Figure 3c).

2. Materials and methods

As part the RUDN University students' geological practice, an additional site exploration of the southern flank of the Turgoyak cast mine was carried out. The exploration was accompanied by detailed photographic documentation of the rock material and sampling for various types of analyses. The analyses were carried out at RUDN University and RGGRU University's labs. In addition to optical microscopy phase, structural, and fluorescent X-ray methods were used to determine the mineral and material composition of the samples using an RLP-3-01 device, a DRON-7 diffractometer, and a DELTA Premium portable spectrometer.

3. Results and discussion

The field and cameral analytical studies made it possible to establish the development of skarn rock at the Turgoyak deposit (Figure 4) and closely related compound rock veins with axinite mineralization.

Figure 4. Skarns (4a - calcareous skarn in limestone; 4b - skarn ore; 4c - microscopic structure of the apodiabase silicate skarn), Thin section, polars crossed.

Skarns are represented by silicate and carbonate varieties. The first are formed upon diabase dykes, often replacing them entirely. The latter make up slim (0.2–1.5 m) margins around the first ones and appear only in cases when diabase dikes intersect limestone (Figure 4a). Macroscopically diabase and skarn look very similar to each other. The presence of the latter is obvious only because of the typical dirty green color hues (Figure 4b). In thin sections, skarns are composed of slightly crystallized mineral aggregates, the identification of which causes considerable difficulties (Figure 4c). Calcite (by twinkling and abnormal interference colors under the microscope), amphibole (by cleavage with an angle of 57°), garnet (by isotropism), epidote (by bright multicolored "lights") and quartz are identified with certainty. X-ray phase analysis supplements information about minerals (Figure 5).

According to its results, skarns consist of epidote (2.90; 2.81; 2.53; 2.40 Å), calcite (3.85; 3.03; 1.91; 1.87 Å), quartz (3.34; 2.28; 1.98 Å), amphibole (8.44; 3.13 Å), chlorite (14.2; 7.08; 2.01 Å), vesuvian (2.59; 2.45 Å), wollastonite (2.16; 1.82 Å) and garnet. Potassium spar (4.25; 4.02 Å) and scapolite (marialite) are also found in silicate skarns (3.20; 2.68; 2.12 Å). Identified chemical components are: Fe - 3.27%; Mn -0.43%; Ca — 6.19%; K -2.24%; Ti - 0.28%; further in ppm: V - 30; Ga - 11; Cr - 450; Ni - 145; Sr - 59; Zr — 6; Zn - 52; W - 8; Ta - 100. Veins bodies, along with the well-known calcite veins, are also represented by polymineral buildups. In addition to the usual minerals (calcite and quartz), axinite is involved in their formation (Figure 6).

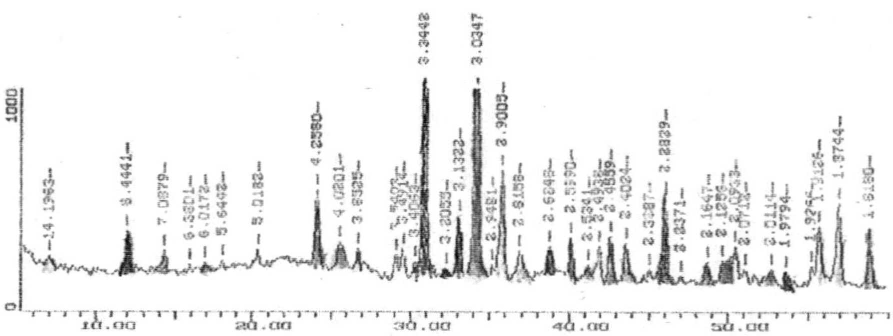

Figure 5. X-ray diffraction (XRD) patterns of main skarn minerals.

The latter either fills the central parts of the veins or composes independent bodies up to 1 m width (Figure 7).

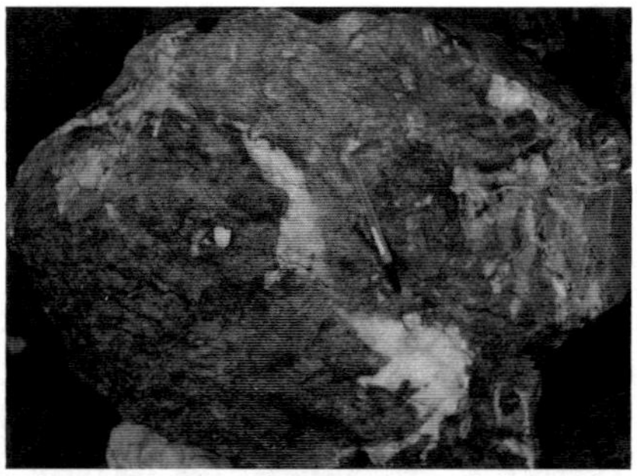

Figure 6. Massive quartz-axinite vein ore in skarn.

Figure 7. Axinite from monomineral vein.

Axinit has a vitreous lustre, light brown in color with shades of pink; under the microscope it forms aggregates of translucent tabular wedge-shaped crystals up to 1.5 cm in size. The hardness of the axinite on the Mohs scale is 6.5 - 7. In ultraviolet rays, it emits a faint glow. In thin sections, axinite varies from colorless to brownish shades and has a high relief and positive refractive index (Figure 8a). It has low double refraction, with gray interference colors. When the analyzer is on, there is a noticeable dispersion of the optical axes (the mineral does not go completely dark) (Figure 8b).

a b

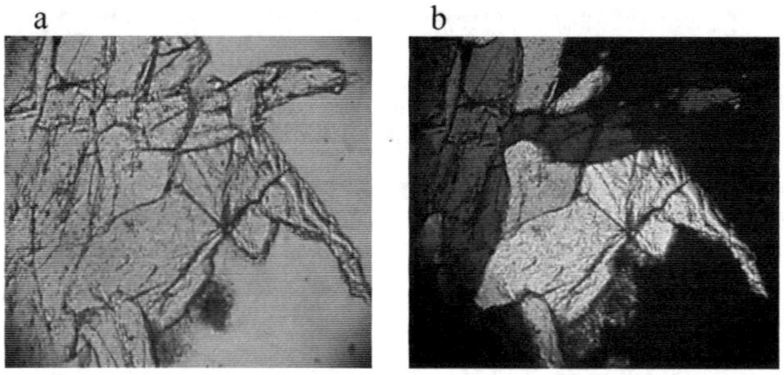

Figure 8. Axinite under the microscope, thin section (a – plane polarized light, b – polars crossed).

According to XRD analysis, the separated fraction relates to magnesioaxinite ($Ca_2MgAl_2BSi_4O_{15}OH$) with cell parameters: a = 7.138 Å; b = 9.161 Å; c = 8.943 Å; α = 91.9; β = 98.1; γ = 77.5 (Figure 9).

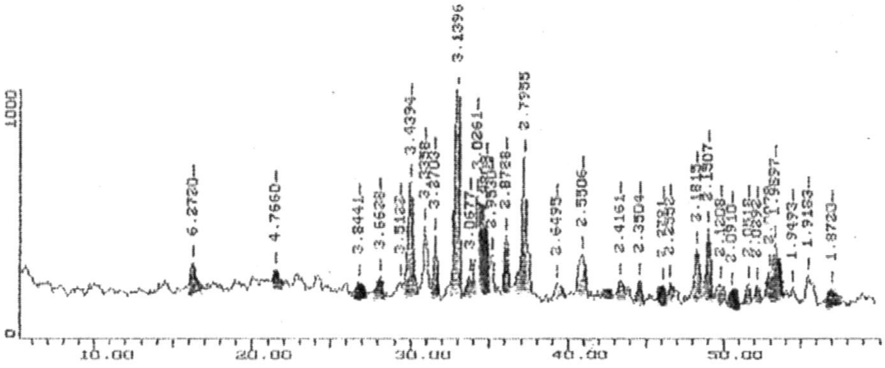

Figure 9. X-ray diffraction (XRD) patterns of fraction rich in axinite.

The increased sulfur content is noteworthy, which is most likely due to sulphide minerals' ingress in the sample. The latter accompany axinite veins, and associate with their selvages (Figure 10).

Figure 10. Sulphide minerals in veins of axinite.

Sulfides are represented mainly by the veinlets of pyrite and, partly, by chalcopyrite. It is important to emphasize that, in addition to selvages, the presence of pyrite is also noted in the near-contact zones in limestone, not in veinlet form, but in the form of pockets and solitary crystals up to 1 cm in size (Figure 11). The formation of axinite veins occurred after the skarn formation. The veins associate with fissures in skarns (Figure 12). Taking into account field observations and microscopy data, it is possible to outline a few stages in skarn-vein formations. At the first stages, the skarnification, most likely, had an alkaline nature of metasomatism. This is indicated by the development of alkaline amphibole halos in the host skarn rock (Figure 13).

Figure 11. Pyrite pockets in limestone.

Figure 12. Vein bodies in skarns.

Figure 13. Alkali-amphibole, in thin section, polars crossed.

The latter in thin sections, in addition to cleavage at an angle of 57°, is distinguished by low (gray) interference colors, the dispersion of optical axes, and the development of twins. According to X-ray phase analysis, it relates to amphibole close to crossite of Na_2 (Mg, $(Fe^{+2})_3$), (Al, $Fe^{+3})_2$ Si_8 $O_{22}(OH)_2$ [7] (Figure 14 and 15).

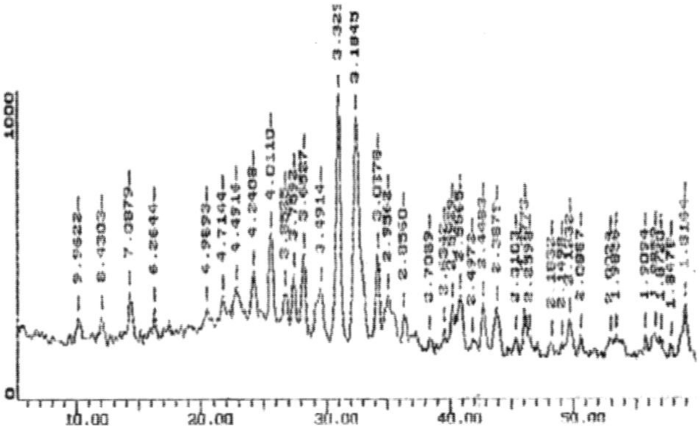

2 200470 N (Na,Ca)2(Fe,Mg,Al)5(Si,Al)8O22(OH)2
Sodium Magnesium Iron Aluminum Silicate Hydroxide / Crossite
Кроссит

1. 8.4454	100	11. 2.6039	14	21. 1.8703	2
2. 4.8878	2	12. 2.5457	8	22. 1.8108	4
3. 4.7906	10	13. 2.3967	8	23. 1.6493	6
4. 4.5210	12	14. 2.3326	4	24. 1.5988	8
5. 3.4105	12	15. 2.3037	2	25. 1.5913	6
6. 3.2900	14	16. 2.2832	4	26. 1.4458	12
7. 3.1508	90	17. 2.1787	12	27. 1.3672	8
8. 2.9703	6	18. 2.0368	8	28. 1.3491	6
9. 2.8248	20	19. 2.0167	4		
10. 2.7288	25	20. 1.8850	4		

Figure 14. X-ray diffraction (XRD) patterns of alkali-amphibole and its X-ray parameters.

Signs of the alkaline nature of the process is the presence of skarn minerals of the scapolite group – marialite ($Na_4Al_3Si_9O_{24}Cl$) and potassium feldspars ($K[AlSi_3O_8]$. Subsequently, the geochemical environment probably changed to a less alkaline one, which caused amphibole corrosion and mass development of fine-grained mineral aggregate of calcite ($CaCO_3$) and calcium aluminosilicates (epidote Ca_2Al (Fe^{3+}, $Al)_2$ $[Si_2O_7][SiO_4]O[OH]$; wollastonite Ca_3 (Si_3O_9); vesuvian $(Ca, Na)_{19}$ $(Al,Mg,Fe)_{13}$ $(SiO_4)_{10}$ $(Si_2O_7)_4$ $(OH,F,O)_{10}$ and garnet $Ca_3Al_2[SiO_4]_3$) in diabases.

As a result, skarnified rocks almost without original volcanics features appear on the flanks of the diabase dikes. Finally, at the final stage of skarn formation, low alkalinity conditions appear, favorable for the mass development of quartz, for recrystallization of epidote (Figure 16) and for calcite veins' formation. It can be assumed that this stage proceeded on the background of high tectonic activity of the region, accompanied by fragmentation and rock fracturing. In thin sections, such a scenario of events is indicated by a common presence of twin gliding with distinct deformation structures in calcite (Figure 17).

Figure 15. Crossite, thin section, polars crossed.

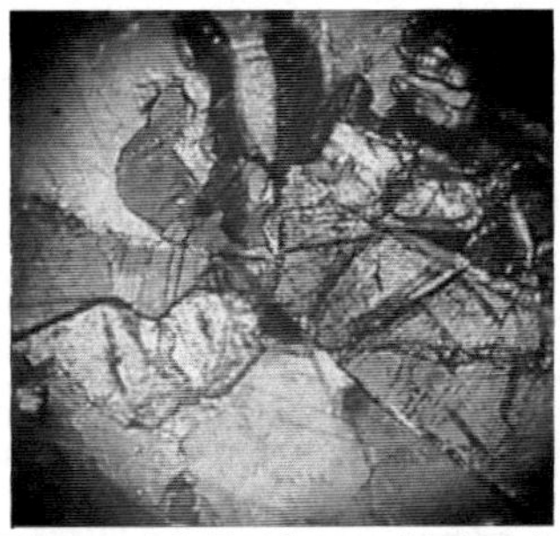

Figure 16. Epidote with bright interference colours, thin section, polars crossed.

Figure 17. Twins gliding in calcite, thin section, polars crossed.

The final stage of mineral formation is associated with calcite-quartz-axinite veins' formation and the chalcopyrite-pyrite and pyrite precipitates accompanying them. This rock assemblage requires special attention, since according to the data obtained by the RLP-3-01 device (x-ray fluorescent spectrometer), its gold content is recorded at up to several dozens of ppm (Figure 18).

The consanguinity of skarns and granitoid intrusions is widely known and proven by many researchers [8-17]. Therefore, taking into account the above, it can be assumed that at the Turgoyak deposit at deep level under the limestone bank there is a granitoid massif which is associated with skarns formation. If this is true, then as we approach the intrusion, we should expect an increase in the activity of the skarnification processes and the growth of their ore potential. Moreover, this concerns not only gold, but also other skarn ore mineralization: Cu, Zn, Pb, W, B and others [8-13,15-16]. Of particular interest is boron. As is known, it can accumulate in limestone skarns, forming industrial agglomerations of datolite minerals (CaB [SiO_4] OH) and danburite (Ca [$B_2Si_2O_8$]) [8-10,18]. Often in such cases they are accompanied by axinite [9,10,18].

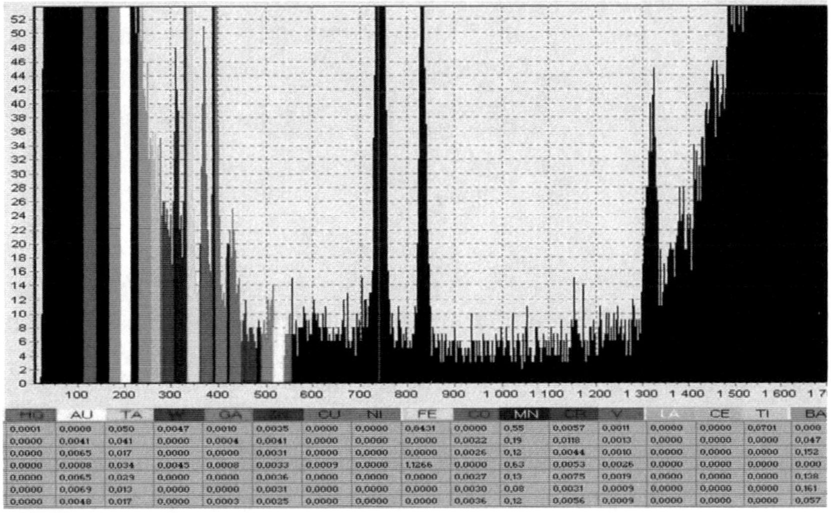

Figure 18. X-ray fluorescence analysis patterns of fraction with sulphides.

At the Turgoyak deposit, the presence of this mineral is established in the form of veins, spatially associated with identified skarns. Thus, as can be seen from the above, there is every reason to expect at the deep horizons of the deposit a concentration of not only gold, but also boron industrial mineralization. To confirm this conclusion, it is necessary to carry out a full range of prospecting and exploration operations with the use of drilling technologies.

4. Conclusion

Taking into account the research results, it can be assumed that at deep level under the limestone bank at the Turgoyak deposit there is a granitoid massif which is associated with skarns formation. Therefore, an increase in the activity of the skarnification processes and the growth of their ore potential should be expected, and this concerns the skarn ore mineralization, such as Au, Cu, Zn, Pb, W, B and others.

5. References

[1] Puzhakov B A, Savel`ev V P., Kuzneczov N S., Shox V D et all 2013 Ob"yasnitel'naya zapiska k Gosudarstvennoj geologicheskoj karte Rossijskoj Federacii. Masshtab 1: 100 000. Seriya Ural'skaya. 1 List N-41-CHelyabinsk. Sankt-Peterburg).. [Explanatory note to Geological State Map of Russian Federation] Saint- Petersburg 415 p https://vsegei.ru/ru/info/pub_ggk1000-3/Uralskaya/n-41.php

[2] Rapoport M 2006 Geologicheskoe stroenie i razvitie Yuzhnogo Urala (V predelah CHelyabinskoj oblasti). [Geology and formation of South Ural, Chelyabinsk region] *Geological journal of Ural* **3(51)** pp 3–20 https://elibrary.ru/item.asp?id=9510159

[3] Rusin A I, Rusin I A and Krasnobaev A A 2007 Mafit-ul'tramafitovye kompleksy Urala: geodinamicheskie aspekty. Geodinamika formirovaniya podvizhnyh poyasov Zemli. Materialy mezhdunarodnoj nauchnoj konferencii. Ekaterinburg: Institut geologii i geohimii UrO RAN. [Mafit-ultramafit complexes of Ural: aspects of geodynamic. Geodynamic of world mobile belts] *Proc. Int. Conf. (Ekaterinburg: Publishing House of Institute of Geology and Geochemistry, Urals Branch of Russian Academy of Science)* pp 260–4 https://elibrary.ru/item.asp?id=19478505

[4] Ryl'kov S A, Rybalka A V and Ivanov K S 2013 Glubinnoe stroenie i metallogeniya Urala: sopostavlenie glubinnoj struktury YUzhnogo, Srednego i Polyarnogo Urala [Abyssal

geology and metallogeny of Ural: Correlation between geology of South, Middle and Polar Ural] *Lithosphere* **1** pp 3–16 https://elibrary.ru/item.asp?id=19096331

[5] Seravkin I B, Znamenskij S E and Kosarev A M 2003 Glavnyj Ural'skij razlom na YUzhnom Urale: struktura i osnovnye etapy formirovaniya [The Main Ural belt on South Ural: structure and general formation stages] *Geotectonics* 3 pp 42–64 https://www.researchgate.net/publication/263673308

[6] Bochkarev K I 2008 Chelyabinskaya oblast'. Ensiklopediya. [The region of Chelyabinsk. Encyclopedia] Kamennyj poyas **6** p 943 https://search.rsl.ru/ru/record/01004377643

[7] Leake B E, Woolley A et al. 2004 Nomenclature of amphiboles: Additions and revisions to the International Mineralogical Association's amphibole nomenclature *American Mineralogist* **89** pp 883–7 https://www.researchgate.net/publication/299271929

[8] Eremin N I 2007 Nemetallicheskie poleznye iskopaemye [Non-metal mineral resources] (Moscow: Publishing House Academkniga of Moscow State University) p 461 https://search.rsl.ru/ru/record/01003082952

[9] Karas' O A, Ratkin V V 2014 Dal'negorskoe skarnovoe mestorozhdenie: etapnost' i istochniki veshchestva borosilikatnyh rud (Sihote-Alin') [Dal'negorsk Skarn Deposit: stages and original materia sources] *Bulletin of Russian Academy of Science* **455(4)** pp 444–6 https://elibrary.ru/item.asp?id=21299067

[10] Starostin V I, Ignatov P A 2004 Geologiya poleznyh iskopaemyh. [Geology of mineral resources] (Moscow: Publishing House Academichesky prospect) p 512 http://www.geokniga.org/books/11604

[11] Eui-Jun K, et al. 2018 New discoveries, skarn zonation, and skarn textures at the Geodo Mine in the Taebaeksan Basin, South Korea *Geosciences Journal* **22(6)** pp 881–9 https://link.springer.com/article/10.1007/s12303-018-0048-6

[12] Fedoseev D, Pakhomova V, Gvozdev V et al. 2014 Kordonnoe scheelite-sulfide skarn deposit: Criteria of relations to igneous rocks differing in age *Geology of Ore Deposits* **56** pp 431–9. https://www.researchgate.net/publication/278326700

[13] Soloviev S G and Kryazhev S G 2019 Geology, mineralization, and fluid inclusion characteristics of the Koitash redox-intermediate W–Mo skarn and W–Au stockwork deposit, western Uzbekistan, Tien Shan *Miner Deposita* https://doi.org/10.1007/s00126-019-00869-5

[14] Soloviev S G and Krivoshchekov N N 2011 Vostok-2 gold-base-metal-tungsten skarn deposit, Central Sikhote-Alin, Russia *Geol. Ore Deposits* **53** 478 https://doi.org/10.1134/S1075701511060080

[15] Salemink J and De Jong A F 1988 Petrochemical and $^{18}O/^{16}O$ Characteristics of 'W-Skarn Associated' and 'W-Barren' Granitoids in the (E-) Pyrenees and NW Portugal. In: Boissonnas J., Omenetto P. (eds) Mineral Deposits within the European Community *Special Publication of the Society for Geology Applied to Mineral Deposits* **6** Springer, Berlin, Heidelberg https://link.springer.com/chapter/10.1007/978-3-642-51858-4_4

[16] Yu M 2019 Skarn Zonation and Mineral Geochemistry of the Galinge Skarn Deposit. In: Metallogenic Mechanism of the Galinge Polymetallic Iron Skarn Deposit, Qiman Tagh Mountains, Qinghai Province. Springer Theses (Recognizing Outstanding Ph.D. Research). Springer, Singapore https://link.springer.com/chapter/10.1007/978-981-10-7907-8_2

[17] Misra K C 2000 Skarn Deposits In: Understanding Mineral Deposits. Springer, Dordrecht

[18] Karas O A, Pakhomova V A and Ekimova N I 2014 Granitoid massif of the Dal'negorsk borosilicate deposit in the Primorye region of Russia: The role of the intrusion in the formation of the boron mineralization, *Russian Journal of Pacific Geology* **8(3)** 213–23 10.1134/S181971401403004X https://www.researchgate.net/publication/271949609

IOP Conf. Series: Materials Science and Engineering **675** (2019) 012035 doi:10.1088/1757-899X/675/1/012035

Genesis and evolution of the Hamama volcanic massive sulphide deposits, Egypt

A S Mahmoud[1*], V V Dyakonov[1] and A E Kotelnikov [2]

[1] Russian State Geological Prospecting University, Moscow, Russia
[2] Peoples' Friendship University of Russia (RUDN University), Moscow, Russia

E-mail: halim.geologist@mail.ru

Abstract. Hamama Zn-Pb-Cu VMS deposits are hosted along the contact of bimodal metavolcanic sequence and their volcaniclastics in association with quartz-carbonate matrix. Ore forming process include multi-stage steps. 3D interpolation of drill hole database shows that the ore body was affected by multidirectional faults and fracture zones, that later were occupied by carbonates. The origin of carbonates is approached using a C and O isotopic compositions in whole-rock samples. The isotopic data revealed signs of both hydrothermal-sedimentary and hydrothermal-metasomatic processes. The primary gold and sulphide minerals were deposited from exhalations, forming massive sulphide ore then the tectonic processes due to the emergence of two successive phases of granitoids followed by the of post-tectonic dykes provided war to the introduction of carbonates that changed the configuration of the ore deposit and affected its grade. Later diagenetic and metasomatic processes resulted in enrichment of gold in the oxidation zone.

1. Introduction

Volcanic massive sulphide (VMS) deposits are an interesting type of deposits for both by the miners and the geologists. They are not only a source of precious metals like gold and silver, but also base metals like Cu, Zn, Pb, and represent an important source of Co, Sn, Cd, Te, Bi, Mn, Gr, Ga and Ba. Some of them also contain trace amounts of Hg, Sb, Se, In, and As [1, 2], which save companies from the price fluctuation of the gold and silver. The Hamama VMS deposit, central Eastern Desert, Egypt, is one of several mineralized metavolcanic provinces in the Eastern Desert (ED). The prominent feature of this deposit is its association with carbonates. The Western Zone of Hamama is estimated to contain an indicated source of 137000 ounces and thus is considered one of Egyptian's promising gold resources [3]. However, the ore grade is lower than normal VMS deposits. Few data are available about the genesis ED VMS and their relation to associated carbonates. The obtained results significantly expand the current understanding of the genesis of the Hamama deposit and explain the causes of its low grade, which can be used in a comparative analysis with similar genetic type.

2. Materials and Methods

The field relationship of the host-rock metavolcanics, ore and carbonate were documented during two field trips. More than 100 thin polished sections were prepared for petrographic and mineralogical studies using transmitted-reflected light microscope. The drilling database of the company "Aton resources" representing analyses of 8440 drill hole samples and 4475 trench sample were used for 3D modelling and ore type classification. Seven samples picked manually from quartz-carbonate veins, nests and lenses from the Hamama deposit. The isotope analyses were carried out in the Institute of Geology of Ore Deposits, Petrography, Mineralogy and Biochemistry (IGEM- RAS), on a Deltaplus mass spectrometer (Thermo-Finnigan). Calibration in international scales V-PDB and V-SMOW was carried out using external (NBS-18, NBS-19, MCA-8) and laboratory (ATC-1) standards.

3. Results

3.1. Geology of the ore deposit

The Hamama area is occupied mainly by basic and intermediate metavolcanic rocks (meta-basalt, basaltic meta-andesite and meta-andesite) with pillow lava occupy the lower section (Figure 1). To the south, the metavolcanics are in contact with different types of volcaniclastic rocks (massive, lapilli and banded tuffs and breccia) intercalated with alternative bands of iron. Quartz carbonate veins occupy the contact of the metavolcanics and tuffs with different degrees of sulphide mineralizations forming a long channel of gossan on the surface. These carbonates can be divided into two types, a) non-mineralized, pure carbonate thin bands (1-15 cm thick); and b) mineralized thick layers of carbonates associated with sulphide mineralization (or oxides in the oxidation zone). The meta-basaltic sheet contains multiple felsic units, mainly meta-dacite and meta-rhyolite (Figure 1). The area was intruded by post-tectonic small quartz diorite dyke-like intrusion in the central part. A feldspar-quartz phyric rhyolite dyke extends in NE-SW direction, cutting the ore body, metavolcanics and volcaniclastics (Figure 1). Both acidic and basic metavolcanic rocks host extensive alteration zones around felsic metavolcanics and later granitoid bodies. The area was subjected to extensional tectonics resulted in the formation of a set of NW-SE strike-slip faults. Nubian sandstone (Cretaceous) covers the western and southern parts of the area with small outcrops.

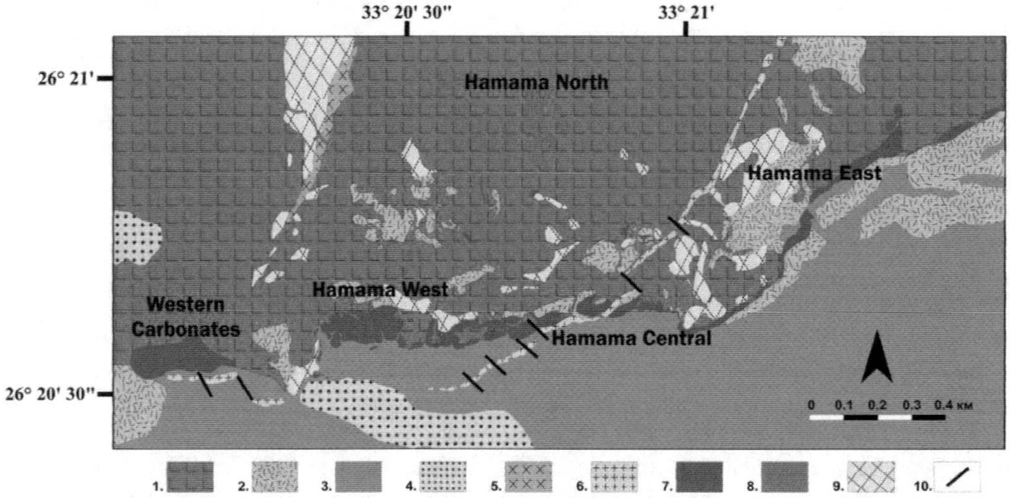

Figure 1. Geological map of the Hamama ore deposit. Legend: 1. Basic metavolcanics with pillow lava; 2. Acidic and intermediate metavolcanics; 3. Tuffs with banded iron formation; 4. Nubian sandstone; 5. Quartz diorite; 6. Rhyolite dykes; 7. Quartz-carbonate veins; 8. Quartz veins; 9. Alteration Zones; 10. Faults.

The metavolcanics, volcaniclastics and quartz-carbonate veins were exposed to long periods of tectonic deformation expressed by their large dipping angles. The drilling programs performed on the ore zones showed that the mineralized horizon is steeply dipping [4]. The stratigraphic sequence is overturned with dip ranges from sub-vertical to moderately overturned to the northwest, so the stratigraphic top is to the southeast [5].

3.2. Mineralogical characteristics and paragenesis

The ore samples are characterized by low to moderate sulphide mineralization (5-30%). Some drill cores revealed massive ore continuing up to 70 %. Mineralogically, it is composed in an average of about 30 % modal quartz with amorphous silica, 25% dolomite, 15% calcite, 10% pyrite, 10% other sulphides, 5% feldspar and clay minerals, and 5% chlorite with other constitutes. The overall identified ore minerals of the fresh ore are pyrite, sphalerite, chalcopyrite, galena, and covellite, with varying amounts of cinnabar, arsenopyrite, stibnite, enargite, iodargyrite, greenockite, acanthite, and tetrahedrite. The oxide equivalents include hematite, goethite, limonite, zincite, and litharge. Sampled fresh ore from drill cores can be distinguished into two groups: a. Pyrite-rich (Figure 2a-f), and b. Polymetallic sulphides (Figure 2g, h). The pyrite-rich ore is the most common type. In terms of massiveness, the ore can be divided into two types: a) vein and disseminated ores, containing low-thickness veins of ores, as well as interlayers of poorly mineralized disseminations of pyrites, with other sulphides in small quantities. b) Massive ore with more than 30 % of ore minerals (Figure 2g, h). Sphalerite is the main ore mineral in this type, followed by pyrite, chalcopyrite, and galena.

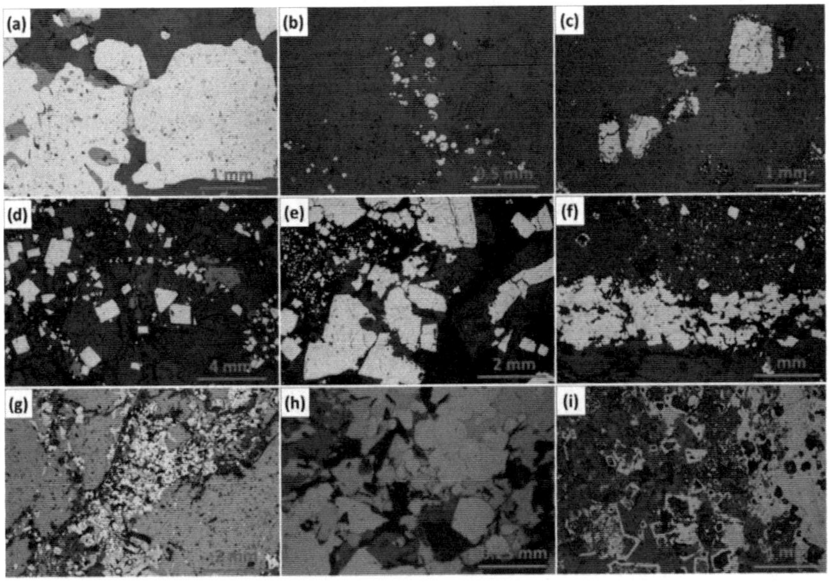

Figure 2. a) chalcopyrite filler between Py-1 grains, which contains inclusions of sphalerite; b) Fine pyrite framboids (Py-2), c) Reserved nests of framboid bacteria built on (Py-1) pyrite cubes; d) fine grained pyrite (Py-3) and sphalerite deposited along carbonate boundaries; e) coarse, fractured grains of py-1 and fine-grained py-3 in the highly altered parts; f) vein of pyrite (Py-1) and fine scattered (Py-3); g) inclusions of galena and pyrite in sphalerite; h) inclusion of pyrite, and chalcopyrite (replaced partially to covellite) in tetrahedrite and sphalerite; i) alteration products of sulphide mineralizations preserving cube form of pyrite.

Table 1. Paragenetic sequence of hydrothermal and supergene stages of the Hamama deposit.

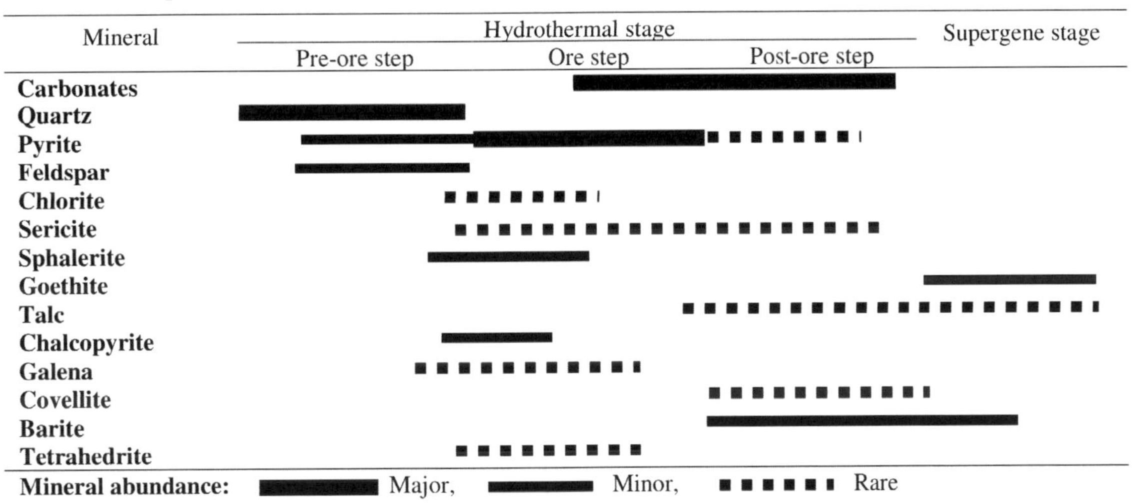

The process of ore formation was rather complex, expressing signs of both hydrothermal-sedimentary and hydrothermal-metasomatic processes. The hydrothermal and supergene stages of the ore formation are represented in (table 1) discussed below:

3.2.1. The pre-ore step is represented by dissemination of the ore minerals with the hosting rock-forming minerals and their alteration products. The mineral assemblage of this stage is represented by plagioclase, amorphous silica, quartz, calcite, barite, chlorite, sericite, kaolin, pyrite (Py-0), few galena, sphalerite and chalcopyrite. The mineral assemblage of this stage is preserved mainly in the stringer zone in the hanging wall tuffs and carbonate lenses.

3.2.2. The ore step includes the intensive deposition of ore minerals include, in descending order, pyrite (Py-1), sphalerite, galena and chalcopyrite (Cpy-1), secondary - pyrrhotite, less common - arsenopyrite, enargite, greenockite, cinnabar, acanthite; nonmetallic minerals - quartz, calcite, dolomite, barite, chlorite, sericite, kaolinite. The mineral assemblage of this stage can be observed mainly in the massive ore from deep drill cores of the Hamama western zone.

3.2.3. The post-ore step expresses the formation of most carbonate mass and reworking of ore minerals producing some secondary sulphide minerals in the fresh ore chiefly of copper as covellite, bornite, chalcopyrite (Cpy-2), and tetrahedrite. It also includes the formation of new generations of pyrite as framboids (Py-2) and fine skeletal grains (Py-3) related to alteration of host rock.

3.2.4. The supergene stage includes the formation of secondary minerals of oxides and carbonates as hematite, goethite, limonite, acanthite, zincite, litharge, malachite, siderite, smithonite, otavit, anglesite, and rare native gold etc.

Four generations of pyrite are distinguished. Pyrite-0 represents the primary well-crystallized pyrite as inclusions in early sphalerite and galena (Figure 2h) and encountered in the massive ore. Pyrite-I composes the main mass of pyrite. Grains of pyrite-1 contain inclusions of non-metallic minerals, cracked, often coarse (up to 1.5 mm), anhedral crystals, crushed to fine-grained aggregates (Figure 2a, c, d, e). Chalcopyrite, galena, sphalerite, are partially replaced grains of pyrite-1. Framboidal pyrite represents the pyrite-2 that formed by bacterial action in reducing environment. There are also clusters of individual, framboids, cemented by non-metallic minerals (Figure 2b, c). They subjected to many reworking processes producing new forms include atoll-like pyrite. Pyrite-3 occurs as idiomorphic crystals without inclusions of other minerals and cracks. Pyrite-III grains are very fine; from thousandths to hundredths parts of a mm; form chains, or uniform disseminations in ores (Figure 2d, e, f). Sphalerite is, generally, found in a subordinate amount, except in massive ore may exceed 70 vol. %. It is found in the form of complex intergrowths or as "independent" xenomorphic grains (aggregates). Independent grains contain almost no inclusions of sulphides and non-metallic minerals and has curly boundaries corroded by carbonates. Sphalerite formed during the first stage of ore formation. A characteristic feature of the earlier sphalerite is emulsion-like inclusions of chalcopyrite rarely exceed 0.1 mm in size. Chalcopyrite occurs in the form of xenomorphic grains in accretion with ore-forming sulphides and in the form of emulsifying impregnation in sphalerite (Cpy-1), and forms late veins in the sphalerite-pyrite association. Chalcopyrite-1 (Figure 3h) most often occurs with sphalerite in the composition of xenomorphic-grained clusters, and confined to the intergranular spaces of pyrite and fills cracks in them. Chalcopyrite-2 (Figure 3a) is emulsion-like inclusions in sphalerite but probably later xenomorphic clusters and intergranular fillings, indicating that after the deposition of the main mass of sphalerite, the flow of copper-bearing ore-forming solutions did not stop. Galena occurs in the form of inclusions and deformed veins within large sphalerite masses. The spatial relationships of galena with pyrite, sphalerite indicates that the deposition of galena began later, but without a temporary interruption in ore formation.

3.3. Ore classification

Different diagrams for classification of VMS were used to define the type of the ore deposit using the database of "Aton resources" of the fresh ore sample. In terms of gold, silver and base metal contents (Cu+Zn+Pb), the majority of the ore samples (>70 %) fall in the field of base metal deposits (Figure 3a), rather than the gold deposits [6-8]. However, some ore samples fall in the field of auriferous deposits due to its relatively high silver contents rather than gold contents. The average point falls in the field of base metal deposits.

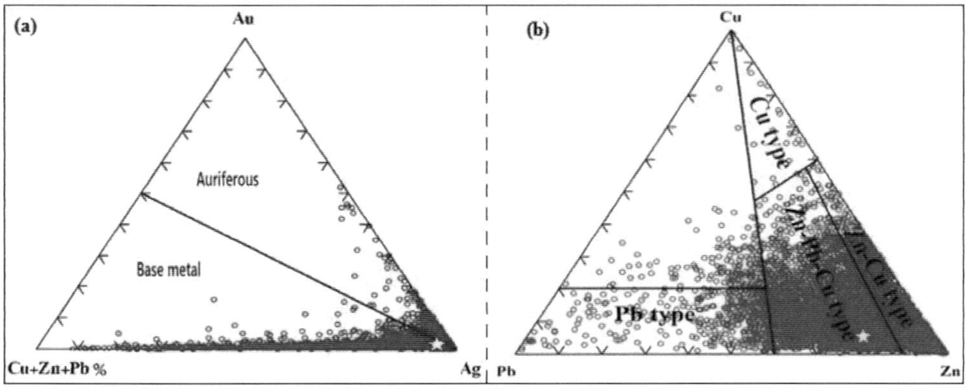

Figure 3. a) Ternary representation of the relative abundances of Au (ppm), Ag (ppm), and base metal (%) in the Hamama deposit [7] ; b) Ternary representation of the relative abundances of Cu, Zn and Pb in the Hamama deposit [9]; the average values in both plots are represented by a star.

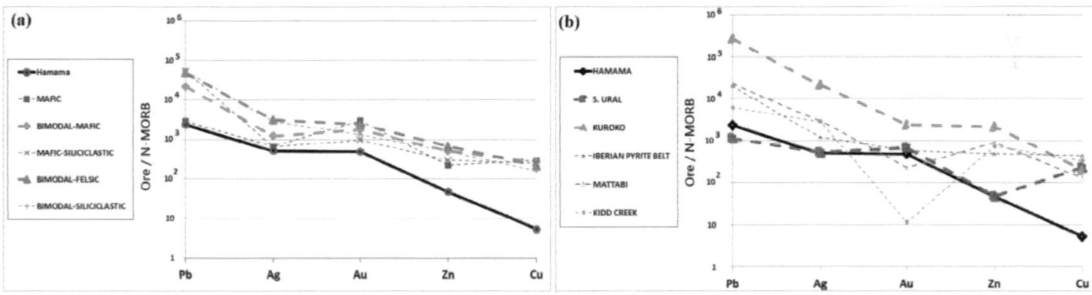

Figure 4. N-MORB normalized diagrams for average Pb, Ag, Au, Zn and Cu contents for the Hamama deposits (solid line) compared pattern with type of host rock (a) and common VMS types (b); values for N-MORB normalization and VMS host-rock types from [10, 11] ; for Sibai of South Ural from [12]; for Kuroko-type, Uchnotai, Kosaka mine, Japan from [13] ; for Kid Creek, Timmins District, Canada from [14] ; for Rio Tinto, Iberian Pyrite Belt, Spain from [15] .

The Hamama deposit belongs to Zn-Pb-Cu type (> 70 % of samples) as revealed from the ternary plot of [9] , where about 20 % fill in the field of Zn-Cu type, the remaining part fills the field Pb type with few samples fall in the Cu type (Fig. 3b). Correlation of N-MORB normalized metal contents of Hamama deposits based on host rock shows a strong correlation with bimodal felsic and bimodal mafic host rock VMS (Fig. 5a). In fact, the Hamama deposit is hosted in bimodal mafic sequence, but silicification of the deposit resulted in its higher correlation with bimodal felsic host for VMS. Correlation of N-MORB-normalized metal content of Hamama with common VMS types (Fig. 4b) shows that Hamama deposit correlates strongly with Kuroko-type and South Ural pyrite deposits (Sibai) and share some similarities with Kid Creek, Mattabi and Iberian Pyrite Belt (Fig. 5b). Hamama agrees with Kuroko-type in that both of them are Zn-Pb-Cu type. However, metal contents of all metals of Hamama are lower. Also, Hamama is more comparable to Sibai Cu-Zn pyrite [16] deposit of South Ural in metal contents except for copper where Sibai has higher copper concentrations.

3.4. C and O isotopic compositions of carbonates

Carbonation is a very common feature associated with green-schist facies metamorphosed metavolcanics of the Eastern Desert. Stern and Geinn [17], stated that large amounts of carbonate having been added to the basement during deformation and metamorphism. The origin, age of formation and the relation of these carbonates to mineralization is not clear. Seven samples picked manually from quartz-carbonate rock, manifest a diverse mineralogy. Mineralogical investigation indicates that these samples consist of calcite (H-33, AHA-050a, AHA-066b, AHA-071b), dolomite + calcite (AHA-004b, AHA-044), and calcite + smithsonite + dolomite (AHA-001). The C, and O isotopes of the carbonate samples cluster between -7.68‰ and -3.55‰ for $\delta^{13}C$ (delta-C-18) (average= - 6.26), and +11.45‰ and +22.37‰ (average= 16.94) for $\delta^{18}O$ (delta-O-18) (Table 2). The $\delta^{13}C$ isotopic signature of the carbonates from Hamama area are consistent with the value of the mantle derived rocks, ($\delta^{13}C$ ~ -5 to -6 ‰; [18-20] . Such a low value is usually could be taken to signify an absence of life, since photosynthesis usually acts to raise the value. Biological materials on the one hand are strongly depleted in $\delta^{13}C$ (- 20 to - 30 ‰; the mean value of the terrestrial biomass is - 26 ± 7 ‰ according to [21].

Table 2. The carbon and oxygen isotope contents of the separated carbonates, Hamama deposits. V-PDB: Vienna Pee Dee Belemnite standard and V-SMOW: Vienna Standard Mean Ocean Water.

Sample	Carbonate content in the sample, wt. %	$\delta^{13}C$ (V-PDB)	^{13}C (Stand. Dev.)	$\delta^{18}O$ (V-SMOW)	^{18}O (Stand. Dev.)
H-33	73	-5.54	0.04	12.59	0.06
AHA-001	99	-3.55	0.04	22.37	0.06
AHA-004b	101	-7.61	0.04	21.42	0.06
AHA-044	48	-6.38	0.04	14.54	0.06
AHA-050a	121	-7.68	0.04	19.18	0.06
AHA-066b	105	-5.91	0.04	17.03	0.06
AHA-071b	65	-6.94	0.04	11.45	0.06

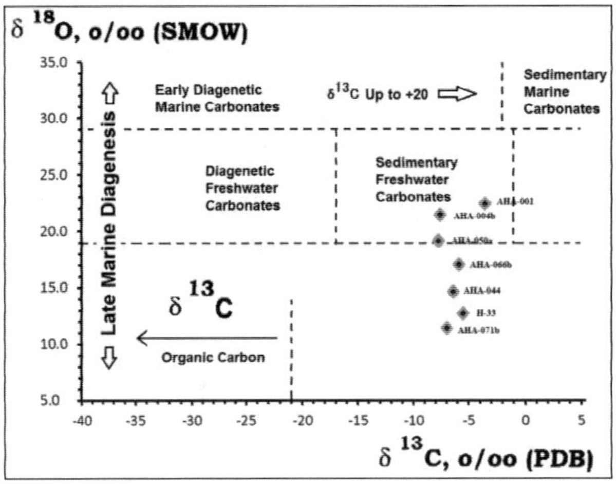

Figure 5. O and C isotopic composition of the Hamama carbonates on the Kuleshov diagram [22].

Our $\delta^{18}O$ values (from 11.45‰ to +22.37) are in agreement with the previous work of Stern and Gwinn [17] ($\delta^{18}O$ from +9.5 to + 27.4‰), carried on samples from the Central Eastern Desert of Egypt. They explained the origin of the "intrusive" carbonates of the Central Eastern Desert as mixing between remobilized sedimentary carbonates and mantle fluids of low temperatures (< 300 ° C). This inference is also consistent with the lack of any evidence for chilled margins on any of the carbonate veins. In the diagram of Kuleshov [22] (Fig. 5), three samples fill in the field of sedimentary fresh water carbonates which have $\delta^{18}O$ (V-SMOW) > 19. It is noticed that these three samples are reworked. Thus, their higher values of ^{18}O are due to further addition from meteoric water upon alteration. These results suggest occurring of some isotopic re-equilibration with a lower temperature fluid, and probably the involvement of metasomatic fluids.

3.5. 3D interpolation of drill holes database
The results of 3D interpolation of the database (Fig. 6) of Hamama deposits shows that the ore body is about 3 km length, up to 110 m thick, more than 250 m deep, and steeply-dipping with an average angle of 55 ° to the south.

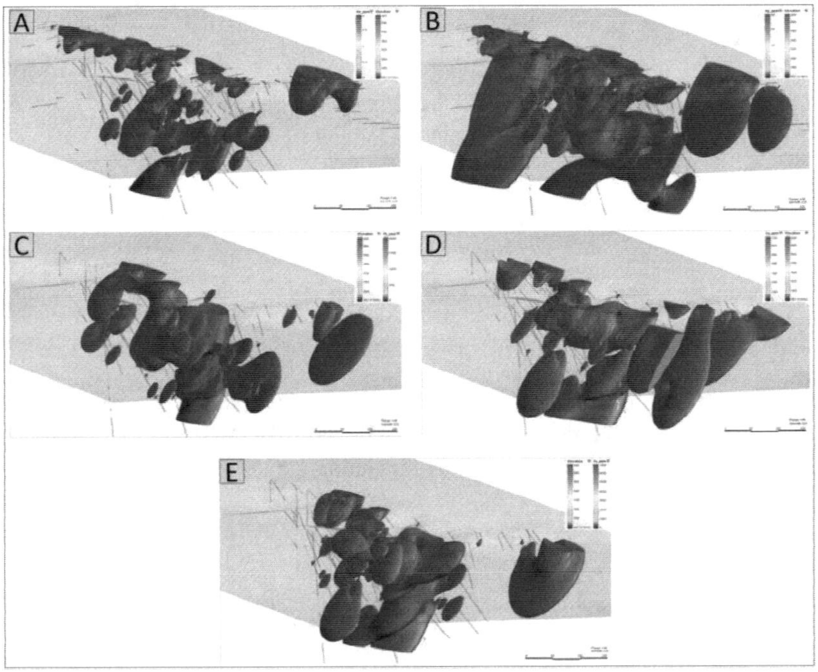

Figure 6. 3D model for distribution of Au, Ag, Pb, Cu, and Zn respectively in Western Hamama.

The high-grade ore of the Hamama West is separated into small pieces and not grouped into a continuous ore body. It is controlled by multidirectional faults and fracture zones. These faults formed after the complete formation of the ore body and strongly changed its configuration. The Hamama East appears on the surface as isolated masses dissected by many strike-slip faults due to its closest from centre of tectonic activity. In the central Hamama, the ore body is crossed by a large rhyolitic dyke, and then this dyke is dissected into small masses by a set of strike-slip faults (Figure 1). In many places, these strike-slip faults are filled with quartz-carbonate rock. The carbonates have apophasis in the surrounding metavolcanics, which increases the indication of their hydrothermal origin.

4. Discussion
Geochemically, the Hamama ore are classified as Zn-Pb-Cu (polymetallic) VMS type. However, its metal content is lower than normal VMS provinces. The field observations indicate that the Hamama VMS were exposed to long-period of tectonics. The 3D interpolation of drilling database supports this observation, where the cut-off grade is bothered and separated into small masses. The ore mineralization is hosted in carbonate matrix. The crosscutting relationship suggests that carbonatization post-dated silicification. The C and O isotope compositions support the hydrothermal origin of these carbonates. The restriction of carbonate in areas of tectonic disturbance refers to their late formation. The introduction of the carbonate that lowered the ore grade was probably contemporary to the emergence of granitoids.

Conclusions
The Hamama deposits are steeply dipping ore body classified as bimodal Zn-Pb-Cu VMS deposit. Paragenetic sequence of ore and gangue minerals is given. Mineralogical investigations indicate that

carbonate formed in later stages after ore formation. Isotopic compositions of carbonate gangue indicate hydrothermal origin and signs of reworking by later low temperature fluids. 3D model of metal contents shows that the ore body is dissected into small parts. The introduction of two phases of granitoids in the eastern and northern parts of the Hamama area, in addition to the emergence of post-tectonic dikes, played an important role in changing the configuration of the ore deposit. The formation of the Hamama deposit included three steps: a. Formation of Kuroko type VMS ores similar to modern black smokers; b. The ore body was dissected by a series of faults, and fractures. Ore became brecciated and cemented by silicates and c. The ore body then was subjected to a new tectonic stage with the introduction of a high-alkaline hydrothermal solution, filling cavities and newly formed cracks by carbonates.

References

[1] Allen R L and Weihed P 2002 *Geological Society*, London, Special Publications **204(1)** pp 13-37

[2] Hannington M, Barrie C and Bleeker W 1999 *The gaint Kidd Creek volcanogenic massive sulfide deposit, western Abitibi Subprovince, Canada* (Canada: Economic Geology Publishing Company) pp 616-72

[3] Aton Resources 2017 Independent technical report, Hamama West deposit, Abu Marawat concession, Arab Republic of Egypt 24 January 2017 *Cube Consulting* NI 43-101 p 121

[4] Alexander Nubia 2014 Summary note 7th january 2014 *Vicarage capital limited* **2** URL: https://www.js-research.de/fileadmin/user_upload/aan_vcl.pdf

[5] Alexander Nubia 2012 Technical report on the Abu Marawat concession, Egypt April 5, 2012 *Roscoe Postle Associates Inc.* NI 43-101 Report 112 URL: https://www.atonresources.com/site/assets/files/1201/abu_marawat_43-101.pdf

[6] Hannington M, Kjarsgaard I and Bleeker W 1999 *Economic geology Monograph* **10** p 61

[7] Poulsen H and Hannington M 1995 *Geology of Canadian mineral deposit types, geology of canada* **8** pp 183-96

[8] Poulsen K H 2000 *Bulletin of the geological survey of canada* **540** p 106

[9] Large R R 1992 *Economic geology* **87(3)** pp 471-510

[10] Barrie C and Hannington M 1999 *Reviews in economic geology* **8** pp 2-10

[11] Galley A G, and Koski R A, 1999, Setting and characteristics of ophiolite-hosted volcanogenic massive sulfide deposits, in Barrie C T, and Hannington M D 1999 *Reviews in Economic Geology* **8** pp 215-36

[12] Galley A G, Hannington M D and Jonasson I 2007 Volcanogenic massive sulphide deposits. Mineral deposits of canada: a synthesis of major deposit-types, district metallogeny, the evolution of geological provinces, and exploration methods: geological association of canada, mineral deposits division, special publication **5** pp 141-61

[13] Oshima T 1974 *Mining geol spec issue* **6** pp 89-100

[14] Franklin J and Thorpe R 1982 *Spec pap geol assoc can.* **25** pp 3-90

[15] Leistel J, et al. 1997 *Mineralium deposita* **33(1-2)** pp 2-30

[16] Prokin V A 1977 *Patterns of distribution of pyrite deposits in the South Urals* (Moscow: Publishing House Nedra)

[17] Stern R J and Gwinn C J 1990 *Precambrian research* **46(3)** pp 259-72

[18] Deines P 1980 *Handbook of environmental isotope geochemistry* (Chapter 9 - The isotopic composition of reduced organic carbon) Elsevier **1** pp 329-406

[19] Exley R, Mattey D, Clague D and Pillinger C 1986 *Earth and planetary science letters* **78(2-3)** pp 189-99

[20] Pineau F, Javoy M and Bottinga Y 1976 *Earth and planetary science letters* **29(2)** pp 413-21

[21] Schidlowski M 1987 *Annual review of earth and planetary sciences* **15(1)** pp 47-72

[22] Kuleshov V 2001 *Lithology and Mineral Resources* **36(5)** pp 429-44

Combined development of oil and salt layers in an oil field

V P Malyukov and K A Vorobyev*

Peoples' Friendship University of Russia (RUDN University), Moscow, Russia

E-mail: k.vorobyev98@mail.ru

Abstract. The problem of complex development and simultaneous use of hydrocarbon mining raw materials in the oil field is relevant for a number of domestic and foreign fields. Combined development of oil and salt layers in an oil field allows obtain a saline solution to intensify oil recovery from productive layer and create an underground reservoir in salt layer for underground storage of hydrocarbons, including creating an underground gas storage (UGS) in rock salt. Joint development of oil and salt reservoirs in the oil field allows to obtain a salt solution of a certain concentration for subsequent injection into the oil reservoir to intensify the capacity (underground reservoir) in the salt reservoir for underground storage of hydrocarbons, including the creation of an underground gas storage (UGS) of associated petroleum gas in rock salt. In this article, the authors used the method of theoretical, practical and experimental studies of complex development of oil and salt reservoirs in the oil field. For the first time in the world practice, joint development of oil and salt reservoirs was carried out at the Talakan oil field.

Keywords: crank punching machines, durability, high-cycle fatigue, stress concentrator, structural rigidity.

1. Introduction

The integrated development of deposits and the maximum use of georesource potential in explored deposits, as well as the multifunctional use of the waste space of mineral deposits, represent one of the main tasks of mining science [1-7].

In some oil fields, the screen is located above the productive oil layer of the salt layer of various capacities. This can be used to produce saline water with subsequent injection of saline water into the productive oil layer, and the generation capacity formed in the salt bed (underground reservoir) can serve as a depository of hydrocarbons extracted from the productive layer (natural gas, associated petroleum gas, oil). The classification of underground storage facilities in rock salt [8] by purpose and types of stored products are showed in figure 1.

The construction of underground workings-tanks for hydrocarbon reserves in rock salt is carried out by supplying a solvent and dispensing brine through a borehole [9-14].

The construction of underground workings-tanks in rock salt is largely determined by mass transfer processes in the marginal zone. When the solvent is supplied through a borehole with hydrodynamic effects at the "rock salt-fluid" boundary, mass transfer processes take place, which characterize the speed of construction of the underground production capacity and its shaping.

Furthermore, studies of the hydrodynamics of gas bubbles in a liquid are of great practical importance not only for the processes of drilling and cementing boreholes, but also in the construction of underground workings-tanks in rock salt by dissolving through boreholes, as well as for use in various mining technologies oil, for example, in the operation of boreholes jet pumps and the water-gas impact on the reservoir. Experimental studies of the coalescence of gas bubbles in aqueous electrolyte solutions [15-18] showed, in particular, that there are areas of rational salt composition and concentrations in which the coalescence of gas bubbles is suppressed, which contributes to the creation of stable gas-liquid mixtures with foamy structure.

IOP Conf. Series: Materials Science and Engineering **675** (2019) 012036 doi:10.1088/1757-899X/675/1/012036

The value of the mass transfer coefficient during the dissolution of rock salt with a high gas content increases, compared with the dissolution of the traditional salt.

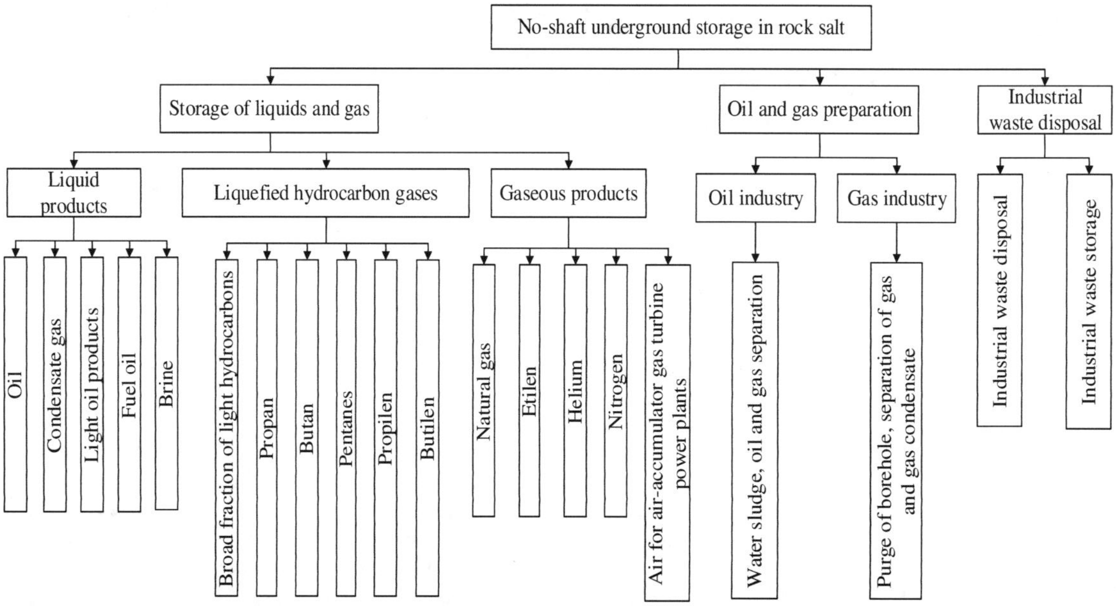

Figure 1. Classification of underground storage in rock salt by purpose and types of stored products.

The mass transfer coefficient of the sample with rebounding of salt plates and gas evolution is 30% more than that of rock salt samples without rebounding of salt plates and gas evolution.

The method of underground dissolution of rock salt in the Leykovskoye field, taking into account the significant value of gas, built in 10 volumes of mineral resources in the amount of 50 or 75 thousand m^3). At the Leikovsky salt stock with a significant gas content in the intercrystalline space of rock salt and impurities, a factor was found to increase the mass transfer coefficient of rock salt by dissolving by about 30% and faster con-struction of workings-tanks (about 10%) compared with the calculated parameters.

2. Materials and methods

The use of water flooding to intensify the extraction of oil from the productive layer is associated with the salinity control of injected water and the compatibility of the injected water with the layer of mineralized water and features of the mineral composition of the manifold [19].

The use of saline water in development of oil fields can increase by 5-40% recoverable oil reserves.

Some principal schemes of the construction and operation of underground hydrocarbon storage can be used in the combined development of oil and salt reservoirs of an oil field together with the corresponding schemes for the development and operation of productive oil reservoirs. The principal technological scheme of operation of the underground gas storage in rock salt is presented in figure. 2.

To obtain saline water, it is advisable to drill wells in the salt layer and conduct underground dissolution of rock salt, to obtain a solution of rock salt with certain concentration (combined with the actual concentration of the reservoir saline water in the oil field) for download through injection wells into the productive layer and the extraction of oil from the productive layer through production wells.

Clay minerals play a significant role in the development of oil deposits. A change in the salinity of water layer in the process of displacing oil from the reservoir, can lead to a change in the state of clay cement in the reservoir and a change in the structure of the porous medium and a change in the main filtration characteristics of the medium. The increase in clay is due to the swelling of clays. Clay dispersion sometimes occurs [20, 21].

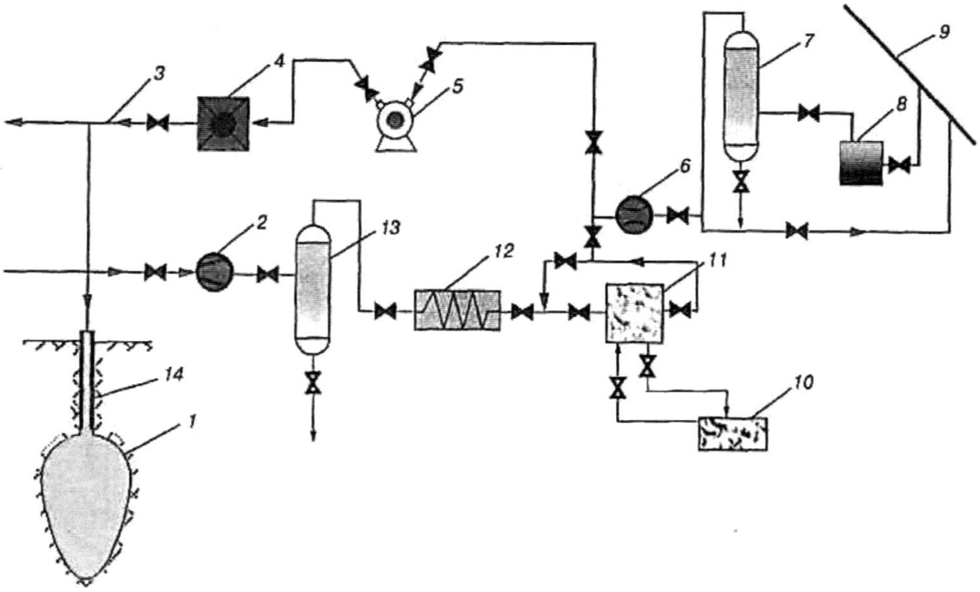

Figure 2. The principal technological scheme of operation of the underground gas storage in rock salt: 1 – underground tank; 2 – gas throttling knot; 3 - gas loop; 4 - gas refrigerator; 5 - compressor station; 6 - knot of measurement of a consumption of gas; 7 - filter separator; 8 – dust collector; 9 - main gas pipeline; 10 - absorbent regeneration installation; 11 - absorber; 12 - gas heating; 13 - separator; 14 - operational well.

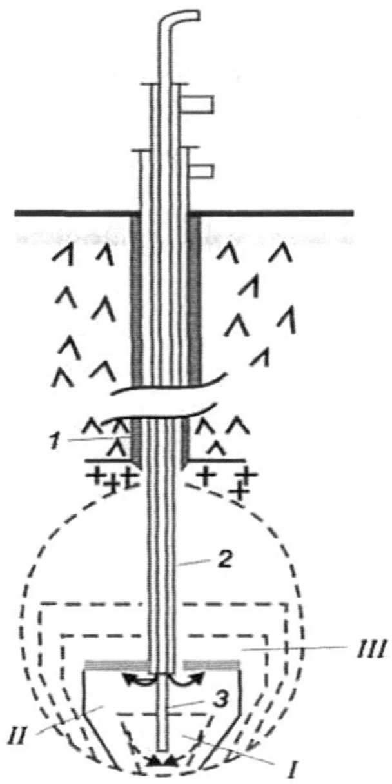

Figure 3. Scheme of a Construction of Vertical Underground Development Capacity steps (*I* – *III*) from below up (layer-by-layer technology): 1 – upsetting column; 2 – external suspended column; 3 – central suspended column.

Pumping mineralized water in a clay-containing collector, with excellent mineralization from sheeted, can reduce permeability of the collector and does it by a low-permeability collector. Control of mineralization of the pumped water and properties of clays in productive layers can significantly increase oil recovery coefficient [22].

Use of mineralization waters allows reduce hydration of clay layers, but it is desirable to use such composition of water, which is most compatible to bedded components of productive layer. Use of the mineralized water received at dissolution of salt layer of the oil field allows solve substantially a problem of compatibility of the pumped mineralized water with reservoir water and the mineral list of the collector.

With development of salt layer, power in several tens of meters with supply of solvent through a borehole on production schedules creates underground development capacity in rock salt (vertical or horizontal) for storage of hydrocarbons. In figure. 3 scheme of a construction is submitted to vertical underground development capacity by steps.

3. Results

On the Talakan oil-gas, condensate field (Sakha Republic) realization of the combined development of oil and salt layers of oil-gas condensate field with creation of an underground gas storage in rock salt and use of the mineralized solution for an intensification of oil recovery was begun [23].

According to the intensity of gas evolution when the salt of the Talakan deposit is dissolved, the mass transfer process is characterized mainly by gas evolution: from rock inclusions and from the intercrystalline space. During the dissolution of core samples from the Talakan site, gas bubbles with a diameter of 2 to 15 mm were noted.

A non-solvent (diesel fuel, gas) is used to control the underground dissolution process. Diesel fuel was used as a non-solvent in the Talakan field.

In the process of dissolving rock salt from the Tuz Galu deposit, when the rocky partition is thinning in a salt crystal, in which gas is under pressure, plates of rock salt rebound and gas is released (microdischarge).

The micro-discharge of gas from the salt crystal with the rebounding of salt plates in the process of dissolution is a continuous process of the separation of salt particles (destruction of the contact zone of the rock).

In the process of mass transfer, gas, particles of insoluble inclusions and particles of salt are released from rock salt (when gas is included in salt crystals). Depending on the stratification of the solution, insoluble heavy particles sink to the bottom of the mine, and light particles together with gas microbubbles rise up, resulting in a significant amount of particles entering the non-solvent (diesel), significantly changing its properties.

It has been found that at the roof of the production capacity, which is constructed in rock salt with a significant gas content and with the use of diesel fuel as a non-solvent, a foam layer is formed [24, 25]. The foam layer at the roof of the mine at the Talakan field is represented by a cluster of gas bubbles evenly distributed between thin layers of solid mineral particles and rocks of small size with diesel fuel and mortar. Suspended particles of rocks of different densities and sizes, like gas bubbles, are in the entire volume of the solution. On the right side of the diagram, processes of gas evolution from the intercrystalline space and particles of insoluble rocks are shown when salt is dissolved with a high gas content (typical of salt dissolution at the Leukovo field). On the left side of the diagram, the process of gas extraction from salt crystals (typical for the dissolution of salt in the Tuz-Galu field) and particles of insoluble rocks during the dissolution of salt with a high gas content is presented. Gas bubbles stand out from the rock, and gas bubbles stand out from the water supplied to the production capacity [26, 27].

4. Discussion

By results of large-scale modeling, dependences are received and the technique of a construction of horizontal developments capacities in rock salt is developed (one horizontal well constructed with use or with use of horizontal and vertical wells). The technology of construction to horizontal development capacity in rock salt is realized in the Volgograd UGS.

In the process of dissolving rock salt from the Tuz-Galu deposit, when the rocky partition is thinning in a salt crystal, in which gas is under pressure, plates of rock salt rebound and gas is released (micro-discharge).

A similar significant gas emission with the release of drilling equipment to the surface occurred during the drilling of a borehole in the Tuz-Galu field.

The definitions on the core material of the Talakan stock from the intervals of underground workings-capacities of the mass transfer coefficient of the vertical surface of the rock salt in water under conditions of natural convection revealed gassing containing mainly methane (~ 99%).

When foaming and changing the properties of the non-solvent, there is a problem with determining the location of the non-solvent in the under-ground working-capacity. To determine the location of the non-solvent, repeated gamma-gamma logging was performed after injection of an additional amount of non-solvent.

Conclusions

The heads of a number of wells on the Talakan salt stock, to reduce the external negative impact, were placed with a depth of 3 m into the ground in a protective hemispherical concrete dome with an earthen top on the surface. In the upper part of the dome there was a hatch through which the necessary operations were performed on the wellhead. Because of the leakage of threaded connections on the tip of the well, methane was released from the brine. Fire situations have arisen in a confined unventilated space at the top of a well located in a protective dome. It should be noted:

1. During the construction of workings-tanks at rock salt deposits with a significant gas content, drilling conditions are complicated. Perhaps the effect of gas released from brine and rock salt on the quality of cementing works and the tightness of the well.
2. As a result of research, the influence of gas shows during the construction of underground workings-tanks in rock salt with a significant gas content with the formation of three-phase foam in diesel fuel (nonsolvent) at the roof of the mine has been established.
3. The process of building workings-tanks is accelerated by increasing the mass transfer coefficient when salt is dissolved with a significant gas content.
4. Fire hazard increases when methane is released from rock salt in the process of dissolution on closed ends of wells.
5. When dissolving rock salt with a significant gas content, the gas goes into brine. In the production capacity, when operating an underground reservoir in rock salt with a significant gas content, in addition to the hydrocarbon product, there is a brine with a significant gas content. Gas can transfer to the stored product and affect its properties.

References

[1] Zhang Y, et al 2017 *Chemistry of Materials* **29(5)** DOI: 10.1021/acs.chemmater.6b05143
[2] Buonsanti R, et al 2011 *Journal of the American Chemical Society* **133(47)** pp 19216-39 DOI: 10.1021/ja208418z
[3] Xia S-M, et al 2018 *Environmental Technology* **40(1)** DOI: 10.1080/09593330.2017.1422555
[4] Rempel A A, et al 2017 *Scientific Reports* **7** 10769 DOI: 10.1038/s41598-017-11164-6
[5] Keraudy J, et al 2017 *Applied Surface Science* **409** DOI: 10.1016/j.apsusc.2017.02.229
[6] Peng W-C, et al 2017 *Materials Science in Semiconductor Processing* **61** pp 85-92 DOI: 10.1016/j.mssp.2017.01.005
[7] Li J, Zhang M, Li Q and Yang J 2017 *Applied Surface Science* **391** pp 184-193 DOI: 10.1016/j.apsusc.2016.06.145
[8] Bouclé J and Herlin-Boime N 2016 *Synthetic Metals* **222** DOI: 10.1016/j.synthmet.2016.03.030
[9] Wang J, et al 2016 *Nanomaterials* **6(3)** E35 DOI: 10.3390/nano6030035
[10] Wang H and Wang W 2016 *Journal of Cluster Science* **27(1)** DOI: 10.1007/s10876-015-0938-y
[11] Benkoula S, et al 2015 Water adsorption on TiO_2 surfaces probed by soft X-ray spectroscopies: bulk materials vs. isolated nanoparticles *Scientific Reports* **5(1)** DOI: 10.1038/srep15088.
[12] Bouhadoun S, et al 2015 *Applied Catalysis B: Environmental* **174-175** pp 367-375 DOI: 10.1016/j.apcatb.2015.03.022

[13] Periyat P, Saeed P A and Ullattil S G 2015 *Materials Science in Semiconductor Processing* **31** pp 658-665 DOI: 10.1016/j.mssp.2014.12.040
[14] Valeeva A A, Schröttner H and Rempel A A 2014 *Russian Chemical Bulletin* **63(12)** pp 2729-32 DOI: 10.1007/s11172-014-0806-1
[15] Nguyen T-T-N, Chen Y-H and He J-L 2014 *Thin Solid Films* **572** pp 8-14 DOI: 10.1016/j.tsf.2014.09.054
[16] Duarte D A, Massi M and Da Silva Sobrinho A S 2014 *International Journal of Photoenergy* 839757 DOI: 10.1155/2014/839757
[17] Wu C-H, Chen S-Y and Shen P 2014 *CrystEngComm* **16** 2220-29 DOI: 10.1039/c3ce42184h
[18] Chen Z, et al 2014 *Journal of Nanomaterials* DOI: 10.1155/2014/298619
[19] Melhem H, et al 2013 *Solar Energy Materials and Solar Cells* **117** pp 624-631 DOI: 10.1016/j.solmat.2012.08.017
[20] Simon P, et al 2013 *Chemical Physics* DOI: 10.1016/j.chemphys.2013.04.008.
[21] Blazevska-Gilev J, et al 2013 *Journal of Solid State Chemistry* **197** pp 337-344 DOI: 10.1016/j.jssc.2012.09.023
[22] Seo S G, et al 2013 *Journal of Materials Chemistry* **1(11)** DOI: 10.1039/c3ta00936j
[23] Cao Y, et al 2012 *Applied Surface Science* **263** pp 95-99 DOI: 10.1016/j.apsusc.2012.09.005
[24] Kostenko M G, et al 2012 *JETP Letters* **96(8)** pp 507-510 DOI: 10.1134/S0021364012200106
[25] Kostenko M G, et al 2012 *JETP Letters* **95(12)** DOI: 10.1134/S0021364012120089
[26] Mesnage A, et al 2011 *Journal of Materials Science* **46(19)** DOI: 10.1007/s10853-011-5709-z
[27] Liu P, et al 2011 *Journal of Materials Research* **26(12)** DOI: 10.1557/jmr.2011.124

The influence of gas appearance on the construction of underground tanks in rock salt

V P Malyukov

Peoples' Friendship University of Russia (RUDN University), Moscow, Russia

E-mail: v.malyukov@mail.ru

Abstract. The article describes the influence of gas bubbles released during the underground dis-solution of rock salt during the construction of underground tanks for storage of petroleum products on drilling, oil borehole drilling, operation of tanks, and acceleration of the construction of underground tanks in rock salt with a significant gas content. It is revealed that complex use of oil and salt layers of the oil field allows to obtain a mineralized solution to intensify the extraction of oil from the reservoir and to create an underground reservoir in the salt reservoir for underground storage of associated petroleum gas (APG), i.e. to create an underground gas storage (UGS) in rock salt.

Keywords: rock salt, underground production-capacity, salt dissolution, solvent, borehole, non-solvent, gas, insoluble inclusions, flotation processes.

1. Introduction

The construction of underground workings-tanks for hydrocarbon reserves in rock salt is carried out by supplying a solvent and dispensing brine through a borehole [1-7].

The construction of underground workings-tanks in rock salt is largely determined by mass transfer processes in the marginal zone. When the solvent is supplied through a borehole with hydrodynamic effects at the “rock salt-fluid” boundary, mass transfer processes take place, which characterize the speed of construction of the underground production capacity and its shaping.

The mass transfer coefficient for dissolving rock salt determines the total amount of movement of the interphase boundary into the bulk of the mass when dissolving rock salt per unit time.

The mass transfer coefficient of rock salt (K) is determined by the formula:

$$K = \frac{\Delta G}{S_{cp} \cdot t \cdot (C_n - C_0)}, \qquad (1)$$

where K – is the mass transfer coefficient of rock salt, m/s; G – weight loss of the sample, g; S_{cp} – average surface dissolution; t – duration of dissolution, s; C_n – maximum concentration of a solution of NaCl in water.

The mass transfer coefficient of the sample with rebounding of salt plates and gas evolution is 30% more than that of rock salt samples without rebounding of salt plates and gas evolution.

The method of underground dissolution of rock salt in the Leykovskoye field, taking into account the significant value of gas, built in 10 volumes of mineral resources in the amount of 50 or 75 thousand m³). At the Leikovsky salt stock with a significant gas content in the intercrystalline space of rock salt and impurities, a factor was found to increase the mass transfer coefficient of rock salt by dissolving by about 30% and faster con-struction of workings-tanks (about 10%) compared with the calculated parameters.

2. Materials and methods

At the Tuz Galu field in each crystal of rock salt is a pore filled with gas.

In the process of dissolving rock salt from the Tuz Galu deposit, when the rocky partition is thinning in a salt crystal, in which gas is under pressure, plates of rock salt rebound and gas is released (micro-discharge).

The micro-discharge of gas from the salt crystal with the rebounding of salt plates in the process of dissolution is a continuous process of the separation of salt particles (destruction of the contact zone of the rock).

The value of the mass transfer coefficient during the dissolution of rock salt with a high gas content increases, compared with the dissolution of the traditional salt.

The mass transfer coefficient of the sample with rebounding of salt plates and gas evolution is 30% more than that of rock salt samples without rebounding of salt plates and gas evolution.

The method of underground dissolution of rock salt in the Leykovskoye field, taking into account the significant value of gas, built in 10 volumes of mineral resources in the amount of 50 or 75 thousand m^3). At the Leikovsky salt stock with a significant gas content in the intercrystalline space of rock salt and impurities, a factor was found to increase the mass transfer coefficient of rock salt by dissolving by about 30% and faster construction of workings-tanks (about 10%) compared with the calculated parameters.

In Figure.1 the vertical sections of the 4T production capacity, built on the Leykovskoye field, are presented (based on sound-location materials).

According to the intensity of gas evolution when the salt of the Leukovsky deposit is dissolved, the mass transfer process is characterized mainly by gas evolution: from rock inclusions and from the intercrystalline space. During the dissolution of core samples from the Leykovsky site, gas bubbles with a diameter of 2 to 15 mm were noted.

A non-solvent (diesel fuel, gas) is used to control the underground dissolution process. Diesel fuel was used as a non-solvent in the Leykovskoye field.

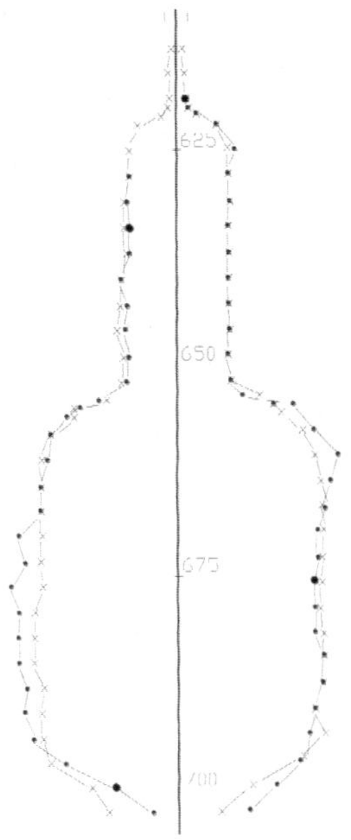

Figure 1. Vertical cross sections of underground excavation 4T according to sound colocation materials: sections: · ------- · south-north; x ------ x west-east.

On the right side of the figure, processes of gas evolution from the intercrystalline space and particles of insoluble rocks are shown when salt is dissolved with a high gas content (typical of salt dissolution at the Leukovo field). On the left side of the figure, the process of gas extraction from salt crystals (typical for the dissolution of salt in the Tuz-Galu field) and particles of insoluble rocks during the dissolution of salt with a high gas content is presented. Gas bubbles stand out from the rock, and gas bubbles stand out from the water supplied to the production capacity.

3. Results

Tables 1 and 2 present the values of the mass transfer coefficient of the vertical surface of rock salt in water under conditions of natural convec-tion at a temperature of 200°C, identified in core samples of SLE. UGS 1 Tuz Galu and UGS 2 Tuz Galu (d – core diameter, h – core height, ρ – density of the salt sample, G_1– is the initial weight of the sample, G_2 – is the weight of the sample after dissolution).

Table 1. Value of the mass transfer coefficient in vertical surface of rock salt in water under natural convection at a temperature of 20°C, identified in core samples of SLE. UGS 1 Tuz Galu

Sample number	Sample's depth, m	*d*, cm	*h*, cm	*ρ*, g/cm3	G_1, g	G_2, g	*K*, m/h
10P	1048,0-1048,2	9,44	10,01	2,20	1541,40	1273,54	0,0596
15P	1077,0-1077,3	9,40	10,01	2,17	1510,0	1223,36	0,0643
25P	1083,6-1084,0	9,45	9,95	2,23	1558,6	1290,31	0,0598
36P	1105,4-1105,6	9,41	9,96	2,126	1472,8	1297,79	0,0387
42P	1108,0-1109,0	9,38	10,08	2,19	1524,7	1262,70	0,0583
45P	1151,0-1151,35	9,41	10,0	2,12	1471,15	1167,69	0,0685
52P	1154,2-1155,0	9,44	9,87	2,11	1458,7	1126,41	0,0761
64P	1203,0-1204,0	9,47	9,91	2,18	1520,6	1234,51	0,0644
74P	1250,0-1251,0	9,44	9,96	2,24	1559,35	1298,09	0,0584
82P	1275,5-1275,8	9,51	9,96	2,15	1524,45	1186,93	0,0759
86P	1301,0-1302,0	9,47	9,97	2,20	1524,90	1223,35	0,0682
93P	1308,0-1309,0	9,50	9,90	2,18	1533,2	1247,39	0,0640

Table 2. Value of the mass transfer coefficient in vertical surface of rock salt in water under natural convection at a temperature of 200C, identified in core samples of SLE. UGS 2 Tuz Galu

Sample number	Sample's depth, m	*d*, cm	*h*, cm	*ρ*, g/cm3	G_1, g	G_2, g	*K*, m/h
27P	962,6-963,0	9,46	9,99	2,17	1523,45	1249,60	0,0610
33P	956,6-966,0	9,50	9,97	2,10	1486,72	1206,60	0,0624
44P	972,0-972,7	9,52	9,98	2,24	1590,15	1362,98	0,0499
52P	978,0-929,0	10,16	10,29	2,22	1854,2	1568,78	0,0489
55P	1033,0-1034,0	9,52	9,92	2,18	1540,9	1262,71	0,0620
68P	1044,0-1045,0	10,15	9,92	2,24	1800,7	1538,84	0,0540
68P	1044,8-1045,0	10,16	10,05	2,25	1828,75	1568,7	0,0532
74P	1049,8-1050,0	10,17	9,92	2,21	1783,3	1512,98	0,0553
80P	1105,0-1105,5	10,08	11,1	2,25	1990,21	1707,2	0,528
84P	1107,4-1108,0	10,16	10,05	2,23	1816,9	1544,2	0,0558
90P	1112,0-1112,5	10,15	9,90	2,23	1782,36	1508,0	0,0572
96P	1116,0-1117,0	10,16	10,05	2,25	1837,87	1590,2	0,0502
98P	1118,0-1118,4	10,15	10,05	2,26	1838,5	1574,59	0,0537
99P	1118,4-1119,0	10,17	10,02	2,25	1831,9	1561,0	0,0551
101P	1181,0-1181,5	10,14	9,74	2,20	1731,7	1474,4	0,0533
107P	1185,5-1186,0	9,46	9,93	2,18	1519,85	1263,44	0,0574
110P	1188,0-1189,0	9,46	9,99	2,10	1530,17	1261,82	0,0586
112P	1251,0-1251,25	9,53	9,98	2,21	1573,71	1318,17	0,0563
120P	1255,0-1256,0	9,49	9,98	2,21	1564,9	1315,85	0,0550
127P	1259,0-1260,0	9,48	9,83	2,22	1539,3	1284,74	0,0573
131P	1323,0-1323,4	9,52	10,02	2,25	1607,75	1417,28	0,0413
139P	1327,0-1327,4	10,02	8,65	2,11	1436,19	1180,07	0,0612

Average values for the mass transfer coefficient of vertical surface of rock salt, under conditions of natural convection at temperature 20°C for core of the field Tuz Galu compose for SLE. UGS 1 Tuz Galu – 0,063 m/h; SLE. UGS 2 Tuz Galu – 0,0582 m/h.

In the process of dissolving salt samples with rebounding of salt plates and gas evolution, thin plates of rock salt ~ 1-2 mm thick and ~ 2-3 mm in diameter fall out (plates of ellipsoidal shape, sometimes larger ~ 4 mm). During the entire time of sample dissolution, salt plates "bounce off" the core surface to a distance of 5 cm. When the core samples are dissolved from the Tuz-Galu field, gas bubbles from salt crystals, ~ 2 mm in diameter are released.

4. Discussion

The study of gas in a mass transfer and its influence on the process of building workings-tanks at the creation of underground storage in salt Lachowska stock was developed. Gas occurrences from rock salt are considered for various technological operations throughout the entire period of construction of workings-tanks, while gas occurrences of different intensity are recorded on a number of boreholes.

During the drilling of the 9T borehole from a depth of 502 m, there was a significant gassing of mainly methane with the release of drilling equipment to the surface. Gas composition: CH_4 – 94,38%; C_2H_6 – 2,5%; C_3H_8 – 0,27%; C_4H_{10} – 0,11%.

A similar significant gas emission with the release of drilling equip-ment to the surface occurred during the drilling of a borehole in the Tuz-Galu field.

Laboratory setup to determinate the coefficient of mass recovery of rock salt during dissolution is shown in Figure 2.

With the determination of mass transfer coefficient of rock salt core samples from the laying-excavation-tanks interval, a high content of gas in the rock was found on the Leikovsky stock (Ukraine) and the Tuz-Galu de-posit (Turkey).

The studies were performed for rock salt containing gas in crystals (Tuz-Galu field) and for rock salt with a high gas content, mainly in the intercrystalline space and impurities (Leikovsky salt rod).

Furthermore, studies of the hydrodynamics of gas bubbles in a liquid are of great practical importance not only for the processes of drilling and cementing boreholes, but also in the construction of underground workings-tanks in rock salt by dissolving through boreholes, as well as for use in various mining technologies oil, for example, in the operation of boreholes jet pumps and the water-gas impact on the reservoir. Experimental studies of the coalescence of gas bubbles in aqueous electrolyte solutions [8, 9] showed, in particular, that there are areas of rational salt composition and concentrations in which the coalescence of gas bubbles is suppressed, which contributes to the creation of stable gas-liquid mixtures with foamy structure.

The definitions on the core material of the Leykovsky stock from the intervals of underground workings-capacities of the mass transfer coefficient of the vertical surface of the rock salt in water under conditions of natural convection revealed gassing containing mainly methane (~ 99%).

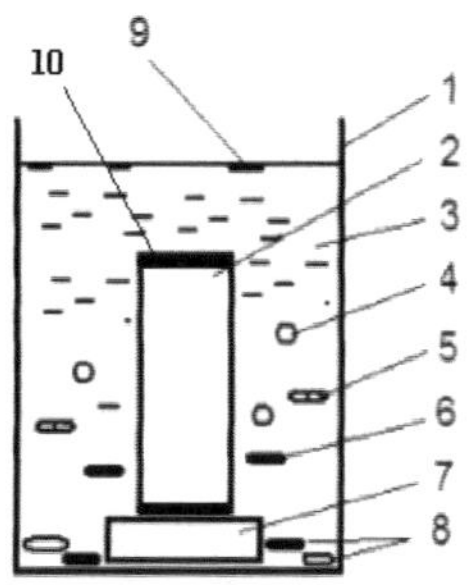

Figure 2. Diagram of the laboratory setup to determinate the mass transfer coefficient of rock salt during dissolution: 1– tank; 2 - core; 3 - solvent; 4 - gas bubbles; 5 - salt plates; 6 - particles of insoluble inclusions; 7 - support; 8 - precipitated insoluble particles and salt plates; 9 - floating light particles of impurities; 10 - insulating coating of core ends.

In the process of mass transfer, gas, particles of insoluble inclusions and particles of salt are released from rock salt (when gas is included in salt crystals). Depending on the stratification of the solution, insoluble heavy particles sink to the bottom of the mine, and light particles together with gas microbubbles rise up, resulting in a significant amount of particles entering the non-solvent (diesel), significantly changing its properties.

It has been found that at the roof of the production capacity, which is constructed in rock salt with a significant gas content and with the use of diesel fuel as a non-solvent, a foam layer is formed [10]. The foam layer at the roof of the mine at the Leikovsky field is represented by a cluster of gas bubbles evenly distributed between thin layers of solid mineral particles and rocks of small size with diesel fuel and mortar. Suspended particles of rocks of different densities and sizes, like gas bubbles, are in the entire volume of the solution.

Conclusions

The use of mineralized water can reduce the hydration of formation clays, but it is desirable to select the composition of water that is most compatible with the formation components of the productive formation.

The use of mineralized water obtained by dissolving the salt layer of the oil field can largely solve the problem of compatibility of pumped mineralized water with reservoir water and mineral composition of the reservoir.

When developing a salt formation with a capacity of several tens of meters with the supply of solvent through the drilling well, according to the technological regulations, an underground production is created-a container in rock salt (vertical or horizontal) for storage of petroleum gas along the way.

Complex use of oil and salt layers of the oil field allows to obtain a mineralized solution to intensify the extraction of oil from the reservoir and to create an underground reservoir in the salt reservoir for underground storage of associated petroleum gas (APG), i.e. to create an underground gas storage (UGS) in rock salt.

At Talakan oil and gas condensate field (Republic of Sakha) the implementation of combined development of oil and salt layers of oil and gas condensate field with the creation of an underground gas storage in rock salt and the use of mineralized solution for the intensification of oil extraction was started.

References

[1] Sousa R L 2010 *Risk analysis for tunneling projects* PhD Thesis (Massachusetts Institute of Technology, USA)

[2] Yufin S, Lamonina E and Postolskaya O 2007 Estimation of strength and deformation parameters of jointed rock masses *Proc. 5th Int. Workshop on Applications Computational Mechanics in Geotechnical Engineering* (Guimaraes, Portugal) pp 3-15

[3] Menezes A T, et al 2007 Geomechanical studies for a road tunnel in volcanic formations *Proc. World Tunnel Congress* (Prague, Czech Republic).

[4] Goodman R and Shi G 1985 *Block theory and its application to Rock Mechanics* (Prentice-Hall, USA)

[5] Cafofo P and Sousa L R 2007 Innovative underground works at Madeira Island, Portugal *Proc. 11th ACUUS Conference on Underground Space: Expanding the Frontiers* (Lavrios, Portugal) pp 137-143

[6] Eclaircy-Caudron S, et al 2007 Inverse analysis of two geotechnical works: a tunnel and a cavern *Proc. 5th Int. Workshop on Applications Computational Mechanics in Geotechnical Engineering* (Guimarães, Portugal) pp 125-142

[7] Miranda T, Sousa L R and Correia A G 2008 Bayesian framework for the deformability modulus updating in an underground structure *Proc. 42nd US Rock Mechanics Symposium* (San Francisco, USA)

[8] Lemos J V 2010 Modeling rock masses in large under-ground works *Proc. Int. Conference on Hydroelectric Schemes in Portugal* (Porto, Portugal)
[9] Elegbede C 2005 Structural reliability assessment based on Particles Swarm Optimization *Structural Safety* **27** pp 171–186
[10] Castro A T, et al 2002 *ISRM News* **7(2)** pp 24–32

Petrographic investigation and geochemical survey of rocks in Inteet area (Nile River State, Sudan)

M A E Ibrahim[1], A E Kotelnikov[1*], S A Ibrahim[2] and E M Kotelnikova[1]

[1] Peoples' Friendship University of Russia, Moscow, Russia
[2] University of Khartoum, Khartoum, Sudan

E-mail: kotelnikov-ae@rudn.ru

Abstract. The aim of the work is to estimate the Inteet area for search for gold mineralization. This area is located between Umm Trambishi and Alsalam area in Nile River State, Sudan. Surrounding territory is a gold-mining region, in which there are known deposits with estimated resources, as well as promising places where local prospecting is carried out. For a preliminary assessment of the prospects of the site is necessary to conduct geological and geochemical studies. Evaluation of the geological environment was carried out both during field work and with a petrographic microscope. 11 thin sections of rock samples have been prepared and petrographically investigated under polarized microscope the mineral compositions, textures, microstructures, degree of alteration and metamorphism have been studied. Geochemical studies were conducted on 32 samples from 25 sampling points. The main results of petrographic investigation and geochemical survey show the presence of hydrothermal alteration zones, which are associated with a high content of elements-indicators of gold mineralization. In general, the territory can be assessed as promising, but requiring further exploration and assessment work.

1. Introduction

The object of study is the Inteet area. It's a part of Alibediyya area Nile River state, Sudan (figure 1). The distance from Khartoum to the study area is about 400 km. The Inteet area is located between Umm Trambishi and Alsalam area. The economic importance of the territory is the presence of a number of gold mineralization zones [1-3], of which Alsalam and Umm Trambishi ore conceder as the main deposit. We have focused on study area about (800x450) to study the geology and gold mineralization in it, since the area already prospected by artisanal mining by locals [4].

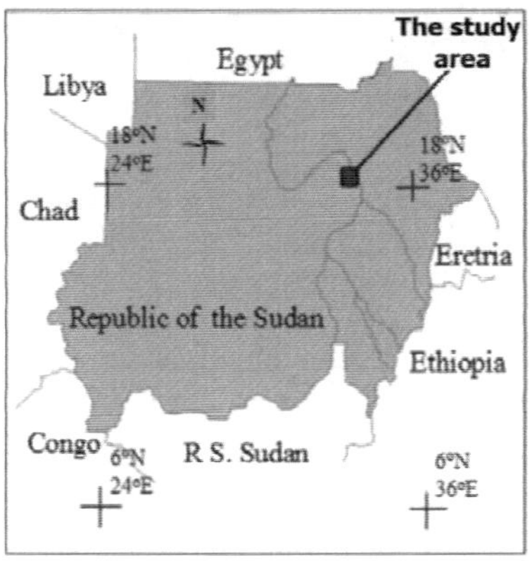

Figure 1. Location map of the study area.

The main goal of research is the initial assessment of the mineral prospecting in the Inteet area [5, 6]. The main objectives of the study are petrographic investigation of rocks and geochemical survey.

2. Methods

2.1. Petrographic investigation

During the field work samples of rocks were collected. 11 thin sections have been prepared from them at Khartoum University Laboratory of cutting and preparation of thin sections. Petrographic microscope allows investigate the mineral compositions, textures, microstructures, degree of alteration and metamorphism of rocks [7]. The main results of petrographic investigation shown in the article.

2.2. Geochemical survey

32 samples were collecting after remove the soil horizon this is was done by drilling to about 1-2 m approximately to reach the bed rocks plate. The total of pits that drilled are 25 pits, and from each pit we collected chip samples for geochemical analyses, but in certain pits we mainly need two samples (in case of presence of Quartz vein cutting across the bed rock) one sample from quartz vein and the others from bedrock. The collected samples had been prepared in form of powder for geochemical analyses and analyzed by Atomic Absorption Spectroscopy (AAS). The results of laboratory research were studied by statistical and graphical analysis [8-11], that shown in the article.

3. Results

3.1. Geology of the area

The geology of Inteet area is the same of Alibediyya area and it mostly covered by super facial deposits (1-2 m - from drilling), that covered the volcano-sedimentary sequence and low grade schistosed rocks which are not exposed in the surface the description of these volcano sedimentary sequence [12, 13] will be shown below from petrographic study. Some outcrops of quartz and Chert have been found in the area. But there is no existing of ophiolite mafic-ultra mafic. Granitic out crops are also distributed in the area. The area also influenced by the shear zones and also affected by the four phases of deformation, these very clearly in volcano sedimentary units.

3.2. Petrographic studies

Elven chip samples and out crops were collected from the area and thin sections are prepared and studded under polarized microscope [7] (figure 2-9). The description is shown below.

Figure 2. Photos of petrographic thin sections A1 (aX2.5, bX10): Hydrothermally altered metasediment (quartz-sericite-chlorite schist.) a: shows quartz, sericite chlorite assemblage & b: shows quartz, chlorite and epidote.

Thin section A1 (figure 2). The specimen shows medium-grained texture. Composed mainly of Quartz and feldspar (alkali-feldspar and plagioclase) Quartz is seemed to be of two generations (primary quartz and secondary quartz which is coarser than primary one due to silicification). Sericite is found after Feldspars. Muscovite is also detected. Considerable amount of chlorite is detected together with epidote aggregates. The secondary quartz Sericite, chloriteand epidote is due to hydrothermal activities (silicification, sericitization. chloritization and epidotization respectively). Iron oxides, zircon, and ilmenite are accessories. The rock is deformed and highly affected by hydrothermal alterations. The specimen is meta-sediment (quartz-sericite-chlorite schist).

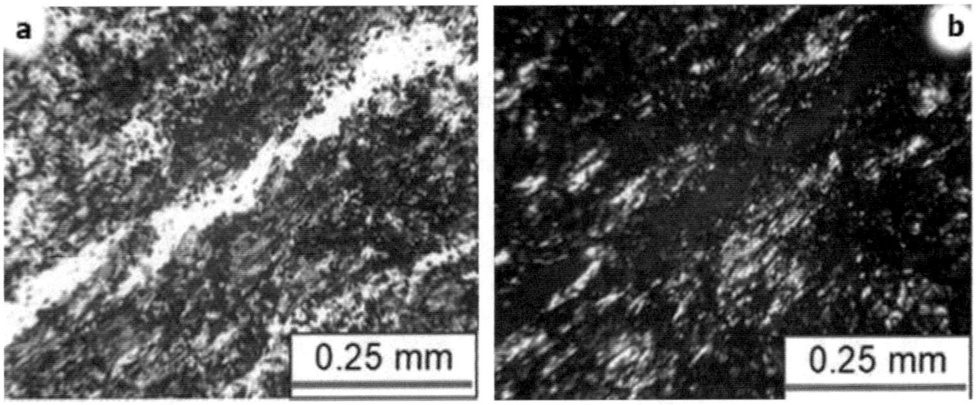

Figure 3. Photos of petrographic thin sections A2 (aX10&bX10): Hydrothermally altered Metasediment (quartz- chlorite talc schist) show quartz veinlet across quartz, chlorite and talc assemblages.

Thin section A2 (figure 3). The specimen is intensively weathered, hydrothermally altered and deformed. Quartz is detected together with some feldspar and also as vein and veinlets across the section. Chlorite and talc are found in considerable amount. Accessories are iron oxides. The rock is Metasediment highly chloritized talc schist.

Thin section A3 (figure 4). The rock consists mainly of quartz and feldspars (alkali feldspar and some plagioclase). The feldsparsare highly sericited and kaolintized Muscovite and chloriteare recognized that intensively affected by deformation and forming what is called pressure shadow texture (brittle minerals bentaround the harder one like quartz and pyrite). Pyrite is found in considerable amount due hydrothermal activities. The rock is hydrothermally altered metasediment (quartz -chlorite pyritic schist).

Thin section A4 (figure 5). The specimen is coarse-grained leucocratic containing mainly quartz and feldspar (alkali feldspar) constituting more than 85% of the total feldspars the remaining is plagioclase. Muscovite and biotite are also recognized together with epidote. Hornblende is detected as minor constituent. Iron oxides (mainly magnetite), zircon and apatite are accessories. The feldspar is highly kaolinitized. The specimen is weathered and metamorphosed and it has composition of granitic gneiss.

Thin section A5 (figure 6). In the specimen feldspars are dominant mainly plagioclase and some alkali feldspar. The Feldspars and some quartz are forming the fine groundmass. The quartz is recognized as primary (phenocrest and groundmass) and secondary constituent (amygdaloidal and veinlets). Some of the plagioclases are altered to Sericite. Chlorite and some epidote are recognized as secondary components due to hydrothermal alteration. Sulphides and iron oxides are accessories. The rock is hydrothermally altered meta-volcanic (quartz chlorite schist).

Thin section A6 (figure7). The slide is mainly consist of quartz (coarse to medium grained) about more than 95% of the section with minor iron oxides and sulphide. The specimen is ferruginous meta-chert.

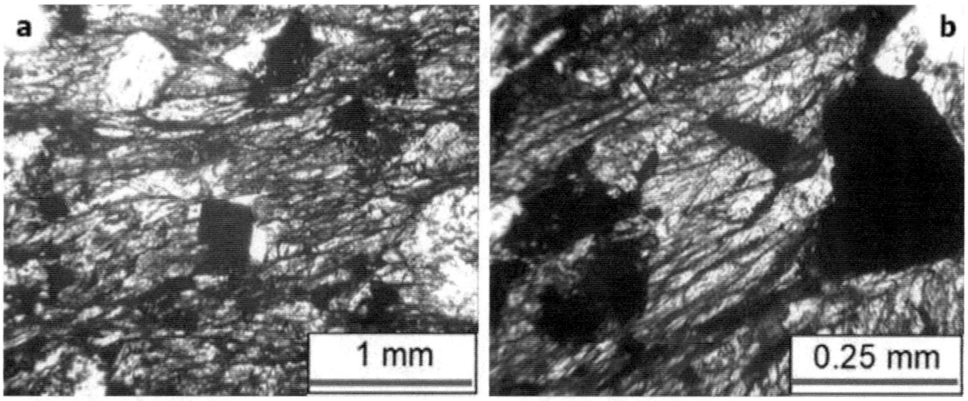

Figure 4. Photos of petrographic thin sections A3 (aX2.5 &bX10): the quartz -chlorite pyritic schist shows chlorite and muscovite shaded the quartz and pyrite grains.

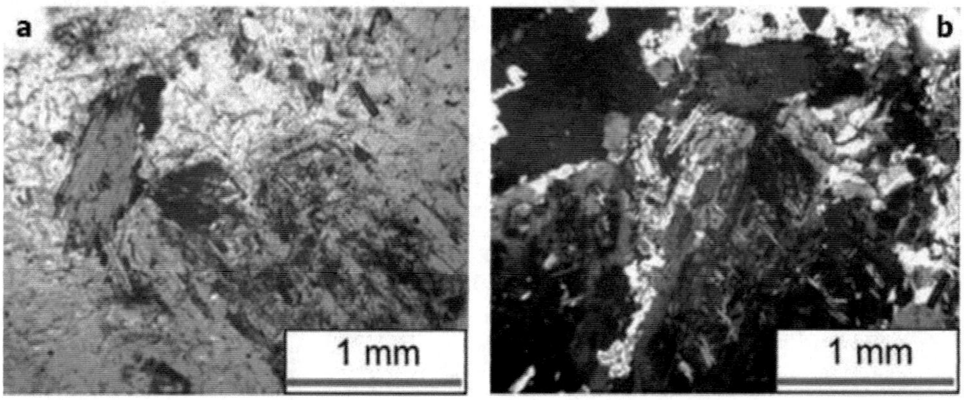

Figure 5. Photos of petrographic thin sections A4 (a X2.5 & bX2.5) a: coarse-grained rock containing mainly quartz and feldspar (alkali feldspar) and hornblende, biotite and iron oxides; b – show coarse.

Thin section A7 (figure 8). The rock is granular in texture that mainly composed of quartz, alkali feldspar (mainly microcline) and plagioclase that highly sericitized and kaolinitized in addition to considerable amount of muscovite and biotite. Iron oxides and apatite are accessories. The rock has composition of granite.

Thin section A8 (figure 9). The specimen is composed mainly of quartz more than 98% in addition to iron oxides. The specimen is quartz vein.

The description shows that the host rocks specimens are meta-sediment to meta-volcanic rocks in composition, that highly deformed weathered and affected by hydrothermal alterations. The hydrothermal processes affected the area are silicification, sericitization, chloritization, epidotization and pyritization. These host rocks are intruded by granitic to granitic gneiss and in some part granodiorite. Quartz vein is also found cutting across the host rocks. Chert and meta-chert are also detected in the area.

And therefore is a region hydro thermal solution, this means the presence of mineralization and economic ores.

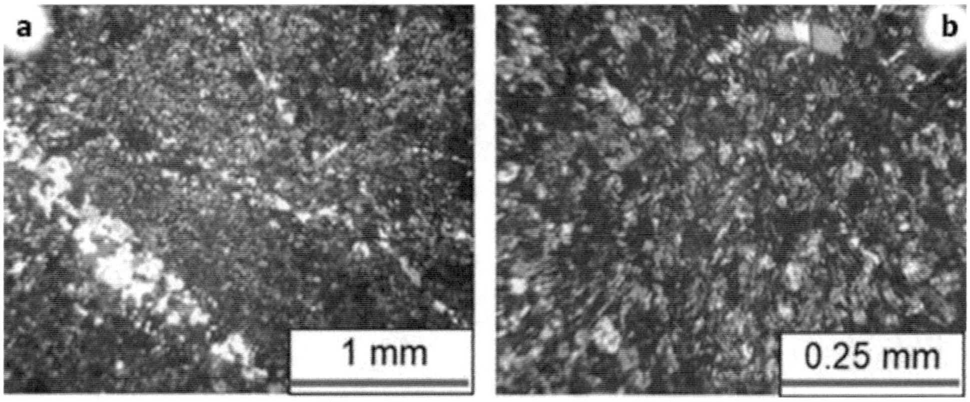

Figure 6. Photos of petrographic thin sections A5 (a X2.5 & X10): quartz chlorite schist with some grains of epidote. The hydrothermal activities cause silicification and chloritization.

Figure 7. Photos of petrographic thin sections A6 (aX2.5 &bX10): Meta-Chert shows quartz with minor iron oxides and sulphides.

Figure 8. Photos of petrographic thin sections A7 (aX2.5 & bX10): shows granitic rock with quartz and feldspars and micas.

Figure 9. Photos of petrographic thin sections A8 (ax2.5 &bX10): quartz vein.

3.3. Geochemical survey

Chip samples have been taken during the field work. The results of laboratory research (AAS) were studied by statistical and graphical analysis [14], that shown in the article. Statistical analysis shown that Au, Ag, Cu and Pb are more informative for area assessment. The result of (Au, Ag, Cu and Pb) was used in ArcGIS to construct monoelemental maps (isoconcentration maps) to locate the zones of geochemical anomalies (figure 10). Analysis of the contrast of anomalies and the spatial combination of haloes of dispersion of elements with geological elements showed a high perspective of the area. [15-16].

4. Discussion

The cut-off grade is the level of mineral in an ore below which it is not economically feasible to mine it which is 0.25ppm. In the study area the geochemical analysis gives the values of gold in ppm these values are plot as isoconcentration map made to explain the location of high gold anomaly. From Au isoconcentration map the high anomaly appeared to be concentred in southern west part of the area in point (89611) the anomaly is 0.45 ppm found in the chert cutting the meta-sediments.

There are other samples near the range and also in the southern west part (points 89601, 89610, 89612, 89615 and 89616) the meta sediments and have an anomaly of (0.21 ppm) which may give a higher values if there have been from higher depth.

The lowest anomaly concentration in the north and northern east of the area the rest of the points found in Meta sediments.

From Ag isoconcentration map the high anomaly appeared to be concentred in south, centre and the northern-east in points (89604, 89605, 89618, 89623, 89631and 89632) the anomalies are range between (1.5 – 2.2) ppm.

From Cu isoconcentration map the high anomaly appeared to be concentred in west and north in points (89608, 89623, 89630 and 89623) the anomalies are range between (95 – 140) ppm.

From Pb isoconcentration map the high anomaly appeared to be concentred in most of the area except part of southern west and northern east in points (89604, 89613, 89615, 89618, 89621, 89622, 89623, 89624, 89625, 89627, 89630 and 89631) the anomalies are range between (11.5 – 16.5) ppm.

5. Conclusions

Petrographic study showed that the host rocks specimens are meta-sediment to meta-volcanic rocks in composition, that highly deformed weathered and affected by hydrothermal alterations. The hydrothermal processes affected the area are silicification, sericitization, chloritization, epidotization and pyritization.

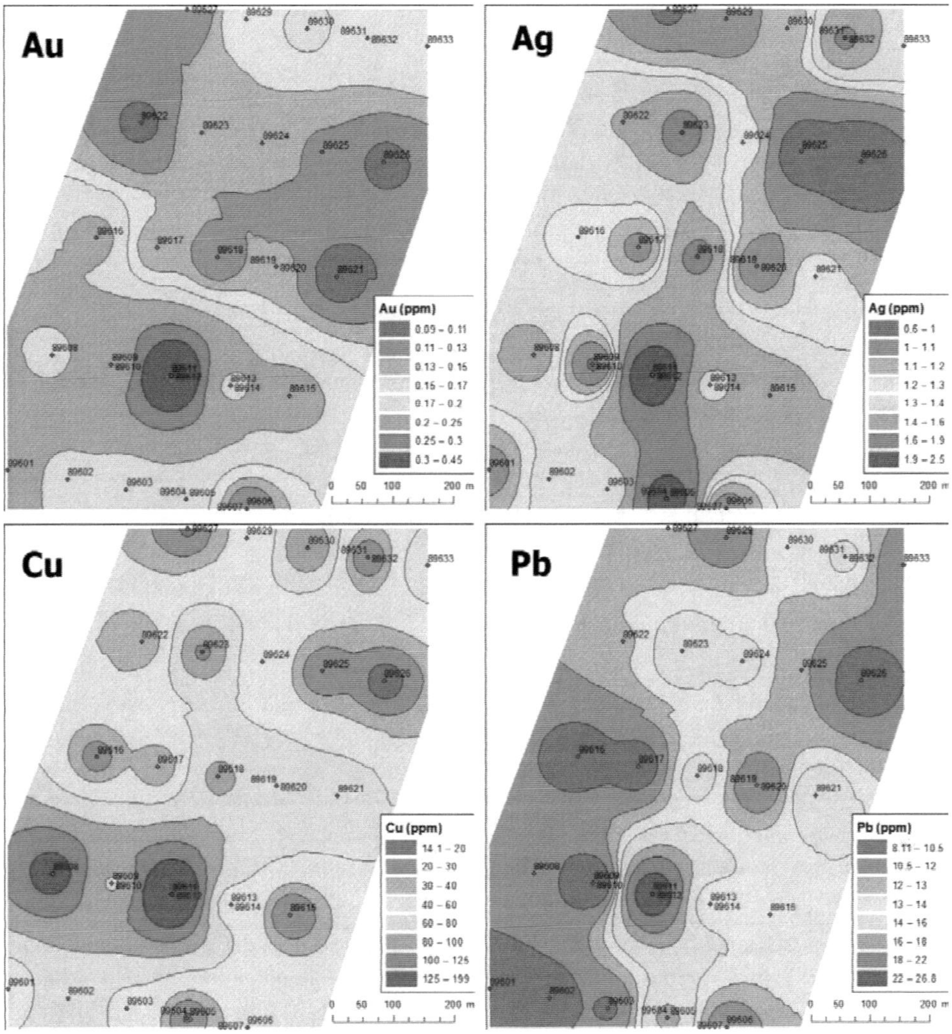

Figure 10. Isoconcentration maps of Au, Ag, Cu, Pb.

The geochemical analysis showed that the (Ag, Cu and Pb) can be considered as associated elements with the gold. The southern west part of the area consists of high anomalies (0.45-0.21) ppm and the meta-sediments cutting by the chert.

Therefore, the detected rock changes indicates the presence of hydrothermal systems that are of potential interest for the search for gold mineralization. The parameters of geochemical anomalies and their spatial location indicates the presence of gold mineralization in the southern west part of the study area.

Recommendation. The Inteet area is one of the rich areas by gold mineral, but in spite of some difficulties like lake of power sources, water sources and telecommunication.

Therefor this study recommends the following: digging trenches in the southern west part of the study area in high-geochemical anomalies to identify the direction of gold depositional; encouragement the mining in the area; enhancement the mining process for better result; preserve the environment during mining; development of methods of analysis for pathfinder elements (As, Hg, etc.).

References

[1] Lissan N H and Bakheit A K 2010 *Research Journal of Applied Sciences* **5(4)** pp 285-302 DOI: 10.3923/rjasci.2010.285.302

[2] Botros N S 2004 A new classification of the gold deposits of Egypt *Ore Geol. Rev* **25** 1-37

[3] Elsamani Y, Almuslem A A and El Tokhi M 2001 *International Geology Review* **43(12)** pp 1117-1128 DOI:10.1080/00206810109465064

[4] Bakr A 2018 Smart Artisanal Gold Mining from a Sudanese Perspective. *Biomedical Journal of Scientific & Technical Research* **8(5)** DOI:10.26717/BJSTR.2018.08.001704

[5] Altigani M A H, Elzien S M and Abdel Rahman E M (2016) *International Journal of Current Research* **8(3)** 27863-27876.

[6] Bouabdellah M, Slack J F (Editors) 2016 *Mineral Deposits of North Africa (Mineral Resource Reviews)* (Berlin: Springer) p 598

[7] Raymond L A 2007 *Petrology: the study of igneous, sedimentary, and metamorphic rocks* (Boston: McGraw-Hill) p 736

[8] Beus A A and Grigorian S V 1977 *Geochemical Exploration Methods for Mineral Deposits* (Illinois: Applied Publishing Ltd) pp 31-270

[9] Grigoryan S V 1974 Primary geochemical halos in prospecting and exploration of hydrothermal deposits *International Geology Review* **16(1)** pp 12-25 DOI:10.1080/00206817409471901

[10] Ovchinnikov L N and Grigoryan S V 1978 Geochemical prospecting for ore deposits *International Geology Review* **20(12)** pp 1413-1425 DOI:10.1080/00206817809471518

[11] Lindagato P, Li Y, Yang G, Duan F and Wang Z 2018 *Geologos* **24(2)** pp 95-109 DOI:10.2478/logos-2018-0010

[12] Schluter T 2006 *Geological Atlas of Africa, with Notes on Stratigraphy, Economic Geology, Geohazards and Geosites of Each Country* (Berlin, Heidelberg, New York: Springer-Verlag) pp 218-221 DOI:10.1017/S0016756807233321

[13] Hassan M, Amlas M, Al Zein A Z and Sheng H 2019 *Journal of Geography, Environment and Earth Science International* **21(1)** pp 1-23 DOI: 10.9734/jgeesi/2019/v21i130118

[14] Reis A P, Sousa A J, Cardoso Fonseca E 2003 Application of geostatistical methods in gold geochemical anomalies identification (Montemor-O-Novo, Portugal) *Journal of Geochemical Exploration* **77(1)** pp 45-63 DOI:10.1016/S0375-6742(02)00269-8

[15] Voroshilov V G 2009 *Geology of Ore Deposits* **51(1)** pp 1-16 DOI:10.1134/S1075701509010012

[16] Nikolaeva L A, Nekrasova A N, Milyaev S A, Yablokova S V, Gavrilov A M 2013 *Geology of Ore Deposits* **55(3)** pp 176-184 DOI: 10.1134/S1075701513030045

IOP Conf. Series: Materials Science and Engineering **675** (2019) 012039 doi:10.1088/1757-899X/675/1/012039

Prospects for development of high-temperature evaporative cooling systems of internal combustion engines with increased temperatures of the cooling body

V Konoplev*, V Bogdanov, Z Melnikov and G Belitsky

Peoples' Friendship University of Russia (RUDN University), Moscow, Russia

E-mail: konoplev_vn@pfur.ru

Abstract. The article deals with issues related to the development of the high-temperature evaporative cooling (HTEC) systems of internal combustion engines (ICE). These systems can transfer the proportion of heat that flows to the ICE into the cooling system. HTEC systems are usually part of the integrated heat recovery (IHR) systems based on large-piston ICEs. It is proposed to improve the efficiency of the operation of the HTEC systems, as well the IHR systems on their basis, by a significant increase in the temperature of the cooling body in the cooling jacket up to 200°C and even 350°C. The increased coolant temperature causes a decrease in the power of the heat fluxes from the heated surfaces thus causing the engine overheat. This reduction can be compensated by increasing the area of the cooled surfaces of the sleeves and cylinder heads. To test the theoretical proposals, a working ICE model equipped with the HTEC-based IHR system was developed. The results of its bed tests are provided. During the experiments, the temperature of the coolant in the cooling jacket reached 204°C, with the engine performance remaining stable.

Keywords: The high-temperature evaporative cooling systems of internal combustion engines

1. Introduction

This paper considers the issues related to the prospects for the development of integrated heat recovery (IHR) systems of the main engines, based on high-temperature evaporative cooling (HTEC) with significantly increased operating parameters (temperature and pressure) as compared to the current operating parameters. Use of secondary heat is an effective way to increase the efficiency of power plants of almost all types, including those based on piston internal combustion engines.

Nevertheless, the efficient heat recovery on power plants based on piston ICEs faces a number of obstacles. For example, in gas turbine installations (GTIs), all the heat released into the environment is the residual enthalpy of exhaust gases, and the major part of the secondary heat of the ICEs, almost in equal proportions, is lost both with exhaust gases and through the cooling system. Therefore, additional heat exchangers are required to recover the secondary heat released by the gas turbine, but to do the same with ICEs it certain steps should be taken to recover both the heat lost with the exhaust gases and the heat diverted to the cooling system.

2. High-temperature evaporative cooling systems of internal combustion engine

The method of high-temperature cooling (HTC) of ICEs can be implemented in classical closed-type liquid cooling systems with pressures in the cooling circuit exceeding the ambient pressure. The increased pressure in the cooling circuit makes it possible for the coolant to overheat to higher temperatures without boiling. The coolant temperatures in such systems are maintained above 115–120°C. The temperatures of coolant overheating to 115–120°C can be obtained in conventional cooling systems. The upper limit of temperatures in the HTC systems rarely exceeds 135°C, although there is information about ICE installations with HTC systems in which the water temperature reaches 180°C. High-temperature cooling with a phase of overheated coolant transiting to vapor is commonly referred to as high-temperature evaporative cooling (HTEC). Both high-temperature and high-temperature evaporative systems are widely distributed (large vessels, ICE diesel and fixed installations). In a pure form, the HTC and HTEC being used for ICE cooling, they have a rather limited, but important advantages, if compared to conventional cooling systems. First, they provide increased mechanical efficiency of the HTC-based ICE due to the lower viscosity of the oil at a higher temperature. More intensive heat exchange with the environment and a higher temperature of the cooling body allow, for a given heat flux, to somewhat reduce the surface of the heat exchangers. The high temperature of the walls of the cylinders does not allow moisture to condense on them and ensures their higher corrosion resistance, more stable thermal state of the engine parts, etc.

However, various HTEC-based schemes are mostly used in the integrated heat recovery systems of power plants based on piston ICEs. The use of a piston ICE as a basic one in a combined power plant causes a number of specific difficulties associated with the issues of secondary heat recovery released into the environment. Two ways to divert the secondary heat of a piston engine into the environment make it necessary to take a number of measures to recover both the heat released with the exhaust gases and the part of the heat lost through the cooling system.

While the recovery of heat of ICE exhaust gases is possible, as at GTIs, through the use of additional heat exchangers, effective recovery of the heat of the cooling system requires significant modification of the existing facilities. To ensure effective recovery of heat diverted to the cooling system of the engine is a rather complicated task. The key problem is that in conventional liquid cooling systems, the coolant (water, air) has, from the point of view of thermodynamics, very low parameters and cannot be used as a working body in the recovery cycle. The temperature of the coolant (water, antifreeze) in the cooling circuit does not usually exceed 85–95°C which is below the boiling point for water at zero level, i.e. the cooling body does not undergo phase transformations which provide the opportunity to use vapors of the cooling agent as the working body for the recovery cycle. The use of hot water is limited to heat supply and technological purposes. From the point of view of the possibilities of efficient conversion of heat into process, the working medium should feature a gaseous form and have the highest possible temperature at the expansion stage.The use of high-temperature evaporative cooling simultaneously solves two problems: to provide effective cooling of the engine and to recover heat that is lost at the same time. The possibility to increase the pressure in the cooling jacket of the engine allows the coolant overheating to higher temperatures. The temperatures of the beginning of vaporization are determined by the degree of pressure increase, and the saturated vapor obtained can already be used as a working fluid. Thus, for the water temperature of 135°C, the corresponding saturated water vapor pressure is 0.3 MPa, and for the water temperature of 180°C, it is 1.0 MPa. It is also worth striving to further increase the temperature of the coolant and the vapor pressure in the HTEC systems. Increase of the coolant temperature reduces the proportion of heat leaving the cylinders in the cooling system and increases the ICE thermal efficiency. Higher parameters of steam produced in HTEC systems with increased operating temperature make it possible to reduce the weight and dimensions and increase the efficiency of various heat engineering and steam power equipment. This is due to an increased temperature difference in the heat exchangers, as well increased initial temperature and pressure of the working processes in steam expansion machines. The relatively low temperature of the cooling body in the ICE cooling systems is not an accidental factor. Regardless of the type and kind, the cooling system primarily serves to remove heat, from the main, most heat-loaded engine parts to the environment. Under all conditions, this system must ensure a stable temperature state of the cylinder-piston group (CPG) parts which corresponds to the most advantageous combination of energy characteristics with minimal wear of the parts. The temperature of the cooling body is selected taking into account the provision of both the specified optimum thermal regime of the engine and the prevention of the boiling and evaporation of the cooling body in the circulation circuits in order to avoid the development of steam congestion and pump surging. When converting conventional cooling systems to the HTEC ones, these conditions are seen as the key factor limiting the maximum coolant temperature. Without taking a number of special measures, increase of the coolant temperature in conventional cooling systems, even without taking into account the pressure increase would cause thermal overstrain of the engine components, primarily the CPG, and premature engine failure. Rather accurate representation of the stress-strain state of the main, most heat-stressed parts of the engine, and a forecast of their performance, can be made only with temperature fields of these parts available. Preliminary assessment of the reliability of heat-stressed parts is usually performed using the temperatures at individual points or zones of parts.

The temperature limits at the CPG characteristic points are determined taking into account the material of the parts, the conditions of their operation, the grade of fuel and lubricants, the requirements for the engine, etc. At the same time, the possibility of deviations of the temperatures measured from standstill temperatures, from the temperatures that occur during the engine performance under operating conditions, is taken into account. Considering Russian and Western experience in the operation of engines of various types, the following temperature values can be accepted as limit ones. Maximum temperatures of aluminum alloy parts should not exceed 250–350°C, of cast iron of different types – 400–450 ° C, of steel – 450–500°C, and of heat-resistant steels – 600–650°C. Excess of these temperatures causes a change in the structure of the material, worsens its physical and mechanical properties, and leads to a sharp decrease in the strength and reliability of the parts.

Interesting variants of the further development of HTEC systems for ICE were proposed in the following works [1-3]. We would attempt to evaluate the main proposals outlined in above mentioned works. Having analyzed the operating conditions of the existing cooling systems and assessed the prospects for the development of high-temperature systems, the authors conclude that it is possible to achieve the temperatures of the cooling body in the cooling jacket of the cooling body up to 300°C and higher accompanied by reaching the corresponding pressures (up to 16.5 MPa for saturated water). However, the achievement of new parameters of high-temperature evaporative cooling systems necessitates a different degree of modification of the ICE existing designs and the maintenance of certain technological conditions relating to the cooling process.

Considering the data cited by the authors, we would summarize them:

1. Cooling each part of the cooling surface of the cylinder and head with a limited volume of the cooling body (liquid or vapor), the temperature of which is less or equal to the most optimal temperature for the corresponding cooled section.
2. The limitation of heat transfer, mainly convective, between the layers of the cooling body and the direction of displacement of the coolant mass, caused by evaporation of its part and the supply of an appropriate compensating quantity, from cooling zones with a lower temperature, to areas with a higher temperature.
3. The engine is positioned relative to the ground so that the head of the cylinder lies higher or on a level with the cylinder base
4. Increase of the surface area of the cooling surfaces of the cylinder and the cylinder head to the values that provide the necessary heat transfer to the cooling medium with a preset temperature. The specific degree of increase is determined based on a special calculation.
5. Additional measures to ensure the operation of the HTEC system with increased parameters: thermal insulation of surfaces, installation of control and safety equipment, etc.

The engine design shown in Pic. 1a, b was developed taking into account the above recommendations. The initial value for the calculation of the elements of the cooling system is the amount of heat Q kJ / h, which must be diverted from the engine to the cold-producing medium. The value of Q can be determined both through the specific amount of heat of cooling q_{cool} J / kWh and the engine output N_e kWh and through the total (heat transfer from the wall by convection) and special (heat transfer through the wall) cases of writing the Newton-Richman equation:

$$Q = q_{col} \cdot N_e = \alpha_2 \cdot F_{cool} \cdot (t_2 - t_{cool}),$$

where α_1 and α_2 – average for the time of heat exchange, heat transfer coefficients, from gases to the heated wall and from the wall to the cooling body, W / m^2·K; F_h and F_{cool} – the average area of the heated and cooled surfaces, respectively, m^2; t_h and t_{cool} – average temperatures of gases and a cooling body, respectively, °C; λ – the average coefficient of thermal conductivity of the material of the wall,

W / m·K; δ – wall thickness, m.

It can be seen from the equation that the amount of heat Q transmitted through the walls of the cylinder and the engine head depends on their size and material (influence of F_h, F_{cool}, λ and δ), engine workflow (influence of α_1 and t_h) and on cooling parameters - t_{cool} , α_2 .

These equations can be jointly solved as relative to the temperature of the cooling body, i.e. calculate it with the other known parameters, and relative to the area of the cooled surfaces:

$$F_{cool} = \frac{q_{cool} \cdot N_e \cdot \lambda \cdot \alpha_1}{\alpha_2 \cdot (\alpha_1 \cdot \lambda \cdot F_h \cdot (t_h - t_{cool}) - \lambda \cdot N_e \cdot q - \alpha_1 \cdot \delta \cdot N_e \cdot q)}$$

i.e. to obtain the required value of the area of cooled surfaces with the other known or specified parameters.

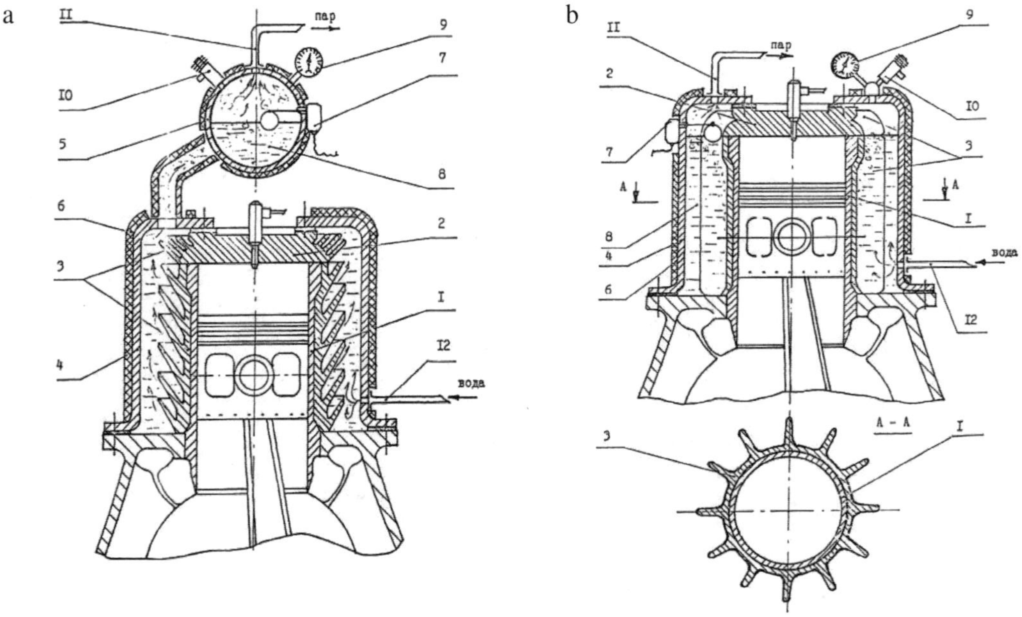

Figure 1. Schemes of the construction of cavities of the cooling jacket of a piston engine operating with increased parameters of the cooling body: a) with divided cooling zones (jacket-casing 4) and evaporation (steam separator 5) and inclined finning; b) with a combined cooling and evaporation zone (jacket-casing 4) and vertical finning;
1 - cylinder; 2 - cylinder head; 3 - cooling fins; 4 - jacket-casing; 5 - steam separator; 6 - heat-insulating and shielding layers; 7 - the automatic power supply; 8 - level of a cooling liquid; 9 - manometer; 10-safety valve; 11 - steam main; 12 - feed water supply line.

3. HTEC system experimental analysis

In addition to the theoretical aspects of the HTEC system development, the results of test of the HTEC-based working model are presented in [1, 2]: "for practical evaluation of the possibility of a significant increase in the temperature of the coolant in the ICE cooling systems and the degree of its impact on the efficiency of secondary heat recovery systems, technological methods and constructive improvements, the IHR working model was developed which ensures both the use of heat from exhaust gases and the cooling system." The works considered being little known, the experimental model of the operating installation is described below. The steam generating part is made according to the classical scheme and is represented by three corresponding devices: economizer (water heater), evaporator boiler and superheater. The model was developed based on a low-power, two-stroke ICE transferred to the HTEC with increased parameters. The basic engine is a single-cylinder, carburetor type with air-cooling and a cylinder capacity of 50 cm^3. The rated effective power is 0.75 kW at 4500 rpm. The actual compression ratio is 6.4. The transfer to the HTEC system implied that, with considerable preservation of the edges of the air cooling surfaces, the engine cylinder and the head were enclosed in a steel, sealed, heat-insulated casing. In the engine heat-insulated exhaust system, a steam superheater and a feed water heater (economizer) wounded from copper tubes were installed. Together with the casing they make up a steam generating part of the recovery system. The feedwater supply to the system is carried out by an autonomous electrically-driven pump. One of the first versions of the working model, and its final design are shown in figures 2 and 3.

Figure 2. The initial version of the engine design for operation with increased parameters of the coolant.

Figure 3. The main version of the HTEC-based working model.

Integrated experimental studies of this ICE model with secondary heat recovery were conducted on a small-sized motor test bed equipped with a balancing DC machine. The test bed is designed for testing low-power ICEs (up to 2 kW) and is equipped with the main units and equipment necessary for conducting standard engine tests of ICEs, including the indicating system and gas analyzer, and for testing individual elements and the entire heat recovery system of the compound engine as a whole. In addition to the already mentioned water pump, it has a steam condenser, steam meter, stand-alone

IOP Conf. Series: Materials Science and Engineering **675** (2019) 012039 doi:10.1088/1757-899X/675/1/012039

water boiler-evaporator and superheater with gas heating, thermometry system, various control and shut-off valves. The purpose of the research is the experimental verification of theoretical assumptions about the possibility to achieve higher temperatures of the cooling body in HTC systems and an evaluation of the engine performance under such cooling conditions. The second part is comprehensive assessment of the impact of elevated temperatures in the HTC and HTEC systems on the efficiency of integrated heat recovery systems developed on their basis. The main tasks are to record the standard speed characteristics of the engine, and to make up a heat balance both of the recovery system and of the engine as a whole, at various coolant temperatures (in the experiments, the coolant is water).

4. Results of HTEC system experimental analysis

Summarizing up, we can assume that the results of the experiments provided a positive result. Compared to the air-cooling system, the HTEC-based engine was capable to operate at coolant temperatures up to 204°C in the cooling jacket. However, it should be noted that the HTEC-based engine completely lost its efficiency and died under a sharp narrowing of the stable area, compared to a temperature of 180°C for the basic air-cooled engine. When working with the temperatures of the cooling surfaces above 150-160°C, the serial air-cooled engine generates only 50% of the rated power, while the HTEC-based engine loses only 10-15% of the power when heated, even to the maximum temperatures, but with a decrease in the zone of stable regimes, which nevertheless indicates a partial engine overheat. When compared with the water cooling system, the HTEC-based heat recovery system allows obtaining water vapor, using for this purpose about 65-80% of the heat diverted to the cooling system and lost with the exhaust gases. Further use of the steam received should ensure an overall increase of the IHR system efficiency. Graphs that demonstrate the effect of coolant temperature on the external speed characteristics of the experimental engine are shown in figure 4.

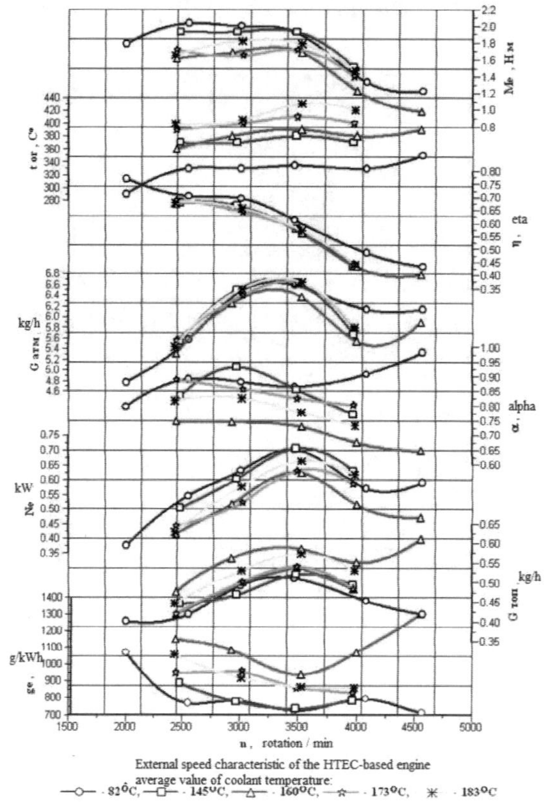

Figure 4. External speed characteristic of the engine.

Also presented the approximation equations of the experimental graphs of the external speed characteristic of the engine in the coolant temperature range 82–183°C (Table 1). The accuracy of the approximation (polynomial of the second degree) is characterized by the multiple correlation coefficient $R^2 = 0.901\text{-}0.999$.

Table 1. Approximation equations of the experimental characteristics of the engine.

$t_{cool\ av}$ °C	$M_e = f(n)$ $N \cdot m$	$N_e = f(n)$ kw	$g_e = f(n)$ g/kWh	$\alpha = f(n)$
82 (min)	M_e = -3E-07n^2 + 0,0015n - 0,0526 R^2 = 0,901	N_e = -1E-07·n^2+ 0,0008·n - 0,7789 R^2 = 0,881	g_e = 9E-05n^2 - 0,711n + 2069,9 R^2 = 0,775	α = 5E-09n^2 + 2E-05n + 0,7646 R^2 = 0,756
183 (max)	M_e = -5E-07n^2 + 0,0029n - 2,6708 R^2 = 0,982	N_e = -2E-07n^2 + 0,0012n - 1,5733 R^2 = 0,986	g_e = 0,0001n^2 - 0,8584n + 2472,1 R^2 = 0,999	α = -6E-08n^2 + 0,0003n + 0,4047 R^2 = 0,977

n – crankshaft rotation speed r / min; α – excess air ratio

5. Conclusion

Despite a significant amount of experimental data, testing only one engine equipped with the proposed HTEC-based IHR system will not give a complete answer to all questions related to the development, test and operation of HTEC systems with increased parameters. However, the fundamental problem of the possibility of development and wide implementation of this system type was solved. Considering the studies conducted, it is already possible to draw conclusions about the main dependencies and characteristic values of various parameters inherent in these cooling systems, both separately and in combination with recovery systems. Depending on the proposed type or mode of operation of the HTEC-based power plant, it is possible to select modes and system adjustments that would ensure increased efficiency of the entire power plant. Experimental analysis of the engine working with the increased parameters of the working fluid in the cooling system fully confirmed the possibility of achieving the declared parameters, and the proposed HTEC systems can be used both independently and in combination with the recovery systems. Further research is needed to conduct more in-depth analysis, to obtain more accurate dependencies, to use more rational design schemes and solutions and, ultimately, to create more efficient cooling systems than those used in ICEs now. One article is enough to provide a detailed description of neither the calculations of the HTEC system, nor the design of the current model, nor the results of the tests. It can be briefly said that the theoretical background of the development of the HTEC systems with increased parameters has been fully confirmed and the possibility of developing and operating such systems has been proved. In the experiments, the coolant temperature in the engine cooling jacket up to 200°C was reached.

References

[1] Ovsyannikov E, Gaitova T, Klyukin P and Poliakova V 2016 Device for producing and addition of hydrogen into fuel-air mixture for internal combustion engine *Alternative Fuel Transportation* **6(54)** pp 85-92

[2] Belyi Yu I and Teregulov T R 2016 Hydrogen energy: Advantages and disadvantages *Journal of Science and Education* **12** pp 8-10

[3] Fomin V.M. 2015 Hydrogen Storage Systems on Board of a Motor Vehicle, *Alternative Fuel Transportation* **1** pp 45-53

[4] Konoplev V N, Korzin A S, Melnikov Z G and Belitsky G V 2018 Hydrogen Energy in the strategy of international energy consumption and its relationship with the perspective of the development of transport technologies *Proc. International Academy of Astronautics SciTech Forum 2018* (Moscow, Russia) pp 58-63

[5] Konoplev V, Sarbaev V and Melnikov Z 2018 The main aspects of safety of hydrogen energy in relation to the motor transport process on gas engine fuels *Transportation Research Procedia* **36** pp 295-302

[6] Bogdanov V N 2013 Increasing the efficiency of heat recovery systems of internal combustion engines PhD Thesis (Moscow: Central Scientific Research Automobile and Automotive Engines Institute Press)

[7] McDowall W 2014 Exploring possible transition pathways for hydrogen energy: A hybrid approach using socio-technical scenarios and energy system modelling *Futures* **63** pp 1-14

[8] Jafari S, Dunne J F, Langari M, Yang Z, Pirault J-P, Long C A and Thalackottore J J 2016 A review of evaporative cooling system concepts for engine thermal management in motor vehicles *Journal of Automobile Engineering* **231(8)** pp 1126-1143

[9] Konoplev V N 2008 *Scientific basis for the design of motor vehicles operating on gas-engine fuels,* DSc Thesis (Moscow: Central Scientific Research Automobile and Automotive Engines Institute)

The effect of cyclic heat treatment on the structure of 5140 steel after cold plastic deformation

M Yu Malkova*, A N Zadiranov, M I Abu-Mahadi and P Dkhar

Peoples' Friendship University of Russia (RUDN University), Moscow, Russia

E-mail marianna300@yandex.ru

Abstract. The objective of this work are experimental research of steel 5140 damages appearance under cold plastic deformation and healing continuity defects characteristic study by following periodic heat treatment. The experimental research of deformation defects continuity appearance in steel 5140 with the granular and lamellar perlit under cold plastic deformation in dependence to original structure was conducted. The deformation defects appearance pattern in steel, their growth in dependence to degree of deformation, microstructure and capacity for healing in the process of cold forging are seemed to be pertinent are explored. It was concluded, that at early deformation stages independently of perlit morphology in steel 5140 the damage is related to micropore appearance in ferrite. The possibility of defects healing by periodic heat treatment is found out. For damage changes studying in the process of cold plastic deformation and healing heat treatment the following experimental design was used: the preliminary periodic heat treatment – deformation with varying degrees – healing periodic heat treatment – deformation up to failure. The periodic heat treatment with phase recrystallization is shown to bring about more full healing damages in comparison with recrystallization annealing. Herewith the microdefects critical size increases up to 2 micrometers, and invertible damage stage extend to 0.4.

Keywords: medium carbon steel, deformation defects, periodic heat treatment, perlit, phase recrystallization.

1. Introduction

The cold forging progressive methods development for high-level express operations demands nomenclature expansion of constructional steel using marks and development of multistage technical processes that consist of several cycles «deformation – heat treatment». In this regard, traditional technologies of hardening still by heat treatment with providing high-degree hardwearing and other mechanical properties of different steel marks stay still pertinent [1-7]. One of this kind of research perspective objects is medium-carbon steel 5140, that is ever-growing used in machine industry due to high mechanical and performance characteristics.

Currently, there is an active search for fundamentally new non-traditional ways to strengthen steels [8-11]. At the same time, the possibilities of traditional methods of heat treatment to improve the physical and mechanical properties of steel have not yet been exhausted.

In the process of cold forging, the metal suffers measurable plastic deformations. For this reason deformation defects appearance in steel, their growth in dependence to degree of deformation, microstructures and capacity for healing in the process of cold forging are seemed to be pertinent.

The research efficiency is proved true by theoretical underpinning good correlation and experimental results. Phenomenological failure theory allows by using results of mechanical tests for plasticity to calculate the level of metal damage in any deformation process, for which the plasticity boundary value problem [12-14] is figured out. However, periodic heat treatment influence on steel 5140 plasticity under combined stress state conditions and capacity for damages healing are not studied.

The purpose of the work is experimental research of steel 5140 damages appearance under cold plastic deformation in dependence to original structure and healing continuity defects characteristic study by following periodic heat treatment.

2. Materials and methods

For damage changes studying in the process of cold plastic deformation and healing heat treatment the following experimental design was used: the preliminary periodic heat treatment – deformation with varying degrees – healing periodic heat treatment – deformation up to failure. Periodic heat treatment was conducted under following conditions:

Conditions 1.

The preliminary periodic heat treatment included following operations:

1) heat up the sample in the furnace to 960°C,
2) soaking the sample in the furnace of 6 hours,
3) oil hardening + heat up the sample to 760°C,
4) soaking the sample in the furnace of 1 hour,
5) cooling the sample with furnace up to 680-700°C,
6) soaking the sample in the furnace of 1 hour,
7) heat up the sample to 770-780°C,
8) soaking the sample in the furnace of 1 hour,
9) cooling the sample with furnace up to 680-700°C,
10) soaking the sample in the furnace of 1 hour,
11) cooling the sample with furnace up to 300°C,
12) further air cooling the sample.

Healing periodic heat treatment included following operations:

1) heat up the sample in the furnace to 760°C,
2) soaking the sample in the furnace of half an hour,
3) cooling the sample with furnace up to 680-700°C,
4) soaking the sample in the furnace of half an hour,
5) heat up the sample to 770-780°C,
6) soaking the sample in the furnace of half an hour,
7) cooling the sample with furnace up to 680-700°C,
8) soaking the sample in the furnace of half an hour,
9) cooling the sample with furnace up to 300°C,
10) further air cooling the sample.

Conditions 2.

The preliminary periodic heat treatment hosted in following sequence:

1) heat up the sample in the furnace to 760°C,
2) soaking the sample in the furnace of 1 hour,
3) cooling the sample with furnace up to 680°C,
4) soaking the sample in the furnace of 1 hour (5 cycles),
5) cooling the sample with furnace up to 300°C,
6) further air cooling the sample.

Healing periodic heat treatment hosted in following sequence:

1) heat up the sample in the furnace to 760°C,
2) soaking the sample in the furnace of half an hour,
3) cooling the sample with furnace up to 680°C,
4) soaking the sample in the furnace of half an hour (2 cycles),
5) cooling the sample with furnace up to 300°C,

6) further air cooling the sample.

3. Results and discussion

Photographs at the figure 1 show the steel 5140 microstructure to vary widely in dependence of periodic heat treatment conditions.

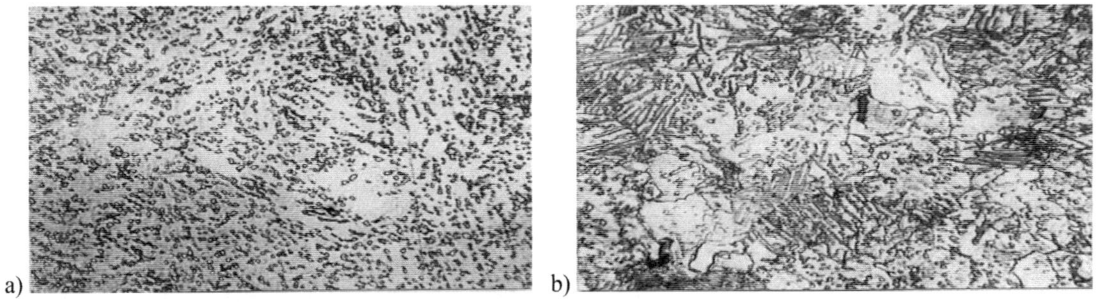

Figure 1. Steel 5140 microstructure after periodic heat treatment: 1000 magnification. a – under conditions 1; b – under conditions 2.

After the periodic heat treatment in accordance with conditions 1 the structure is 100% granular perlite with equally distribute in ferrite mold big (>1.2 micrometers) cementite granules (Figure 1a). Steel firmness after this treatment is 143 – 145 HB. In steel structure after periodic heat treatment under conditions 2 appreciable quantity of lamellar perlite (~30%) and small cementite granules (>1 micrometer), that are closely-pitched to each other, are presented (Figure 1b). The treated under conditions 2 steel has the firmness 158-164 HB.

The mentioned distinction in microstructure and steel 5140 firmness significantly effect on its plasticity in the process of cold plastic deformation. The figure 2 shows the steel plasticity diagram after periodic heat treatment under conditions 1 to be situated above the k stress state index within the studied value range.

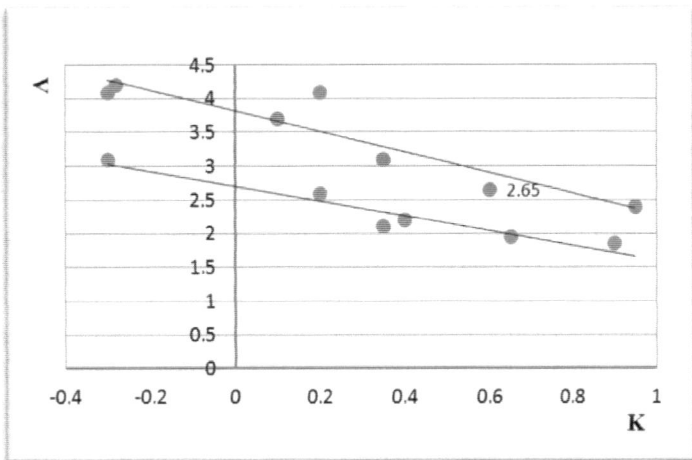

Figure 2. Plasticity steel 5140 diagram after periodic heat treatment. Conditions:
- - conditions 1;
- - conditions 2.

The structure changes and steel 5140 firmness abating while using periodic heat treatment in the capacity of preliminary heat treatment under conditions 1 drive the plasticity increase up of 30%.

Electron microscope analysis of deformed samples fractures allows to observe variant appearance and development of micro effects devises in dependence to degree of deformation **Λ** and original steel structure. Of deformation degree **Λ** < 1.35 for steel after periodic heat treatment under conditions 1 and **Λ** < 1.10 after periodic heat treatment under conditions 2, when the damage did not exceed the value of $\omega = 0.4$, certain defects in the form of spherical shape micropores, that were situated inside ferrite grains (inside cleavage facets) and along grain boundaries (figure 3a).

For one and the same ω_l value the microdefects number and size after periodic heat treatment under conditions 1 and 2 are aproximately the same. The carbide phase in steel 5140 after periodic heat treatment under presented conditions does not take part in shearing processes, that carry on the beginning deformation stages, for this reason micropores here evidently appear in accumulation dislocation points along grain and cells boundaries.

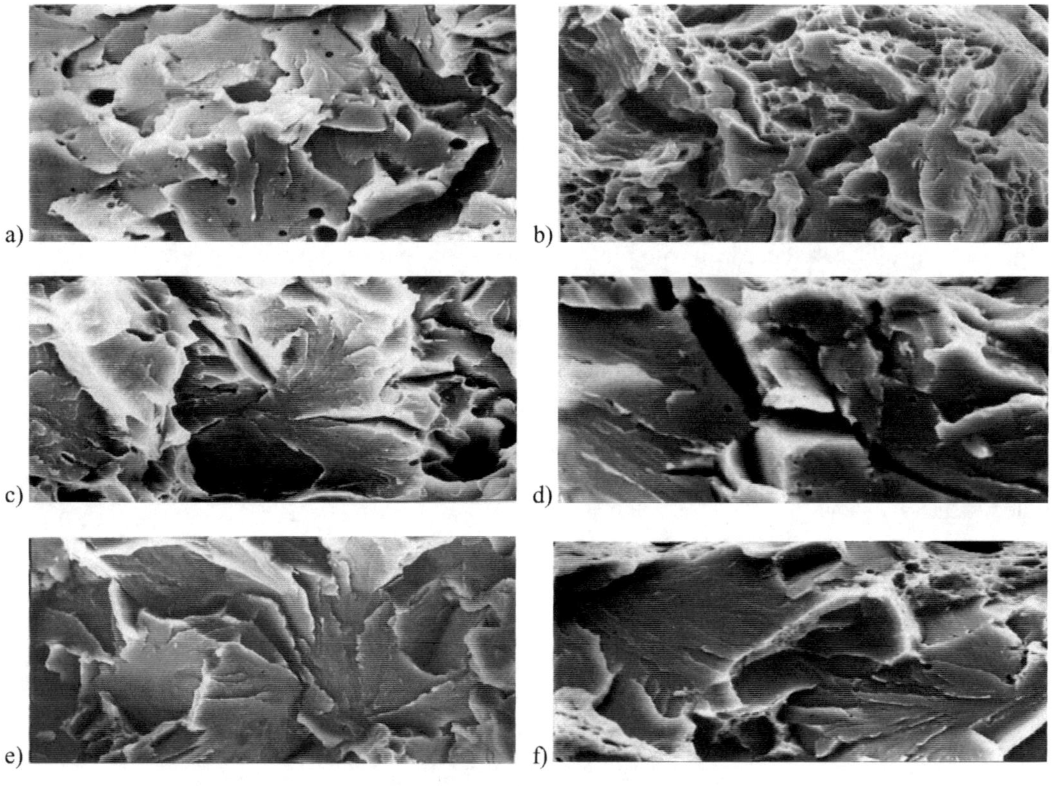

Figure 3. Steel 5140 fractures micro factors; 2000 magnification:
a) – cyclic heat treatment under conditions 1 + deformation up to value $\omega_l = 0.25$;
b) – cyclic heat treatment under conditions 1 + deformation up to value $\omega_l = 0.77$;
c) – cyclic heat treatment under conditions 2 + deformation up to value $\omega_l = 0.51$;
d) – cyclic heat treatment under conditions 2 + deformation up to value $\omega_l = 0.80$;
e) – cyclic heat treatment under conditions 1 + deformation up to value $\omega_l = 0.30$ + healing cyclic heat treatment under conditions 1;
f) – cyclic heat treatment under conditions 1 + deformation up to value $\omega_l = 0.77$+ healing cyclic heat treatment under conditions 1.

Of higher deformation degrees two microdefects development mechanisms can be observed in dependence to original steel structure. For steel with granular perlite (periodic heat treatment under conditions 1) the ω_l increase up to 0.77 sequentially leads to appreciable quantity of micropores appearance, their growth up to sizes ~ 3 micrometers due to dislocation sink, distance decrease between micropores and finally their confluence. On micro fracture patterns this mechanism in the form of ductile failure areas, that are mainly situated nearby grain boundaries (figure 3b).

The used in this research microfracturing method allows, on the one hand, to correctly estimate only collected in the form of micropore damages. On the other hand, steel fracture studying with lamellar perlite after periodic heat treatment under conditions 2 (figure 3c, 3d) shows the micropore number in samples to change a little with deformation degree raising and obviously disagree with high calculation damage values ω_1 for samples, which plasticity resource is considerably depleted yet.

The relative narrowing of the samples δ reached 52 and 63%, whereas the value δ = 68% corresponds to the moment of macroscopic destruction. This suggests that the microcracks observed in the fracture (Figure 3c, d) arise at the stage of preliminary plastic deformation, and are not secondary, formed by branching cracks in the brittle fracture samples. If ω1 = 0.51, the microcracks have size less than 3.0 microns and are located mainly along the grain boundaries and along the interface boundaries (Figure 3c). With the increase of ω1 to 0.7 – 0.8 there is a crack growth along the borders with the formation of micro-voids, the size of which is commensurate with the grain size (~ 15 microns), as well as the development of in some cases, microcracks from the oncoming dislocation clusters merge with micropores or become blunt and cease to grow as a result of the exit of their top in the previously formed pore.

The difference in the mechanism of deformation defects determines the different level of plasticity of steel 5140 with granular and lamellar perlite.individual cracks directed from the boundaries to the center of the grain body (Figure 3g). Micro cracks, unlike pores, are more effective stress concentrators, so the appearance of the main crack and the destruction of steel after cyclic heat treatment in mode 2 occurs at lower degrees of deformation.

Depending on the extent of damage to the steel at the stage of preliminary deformation during heat treatment, microdefects are healed partially or completely. At ω1 < 0.4, microdefects of the continuity of steel completely disappear (Figure 3d); at 0.4 < ω1 < 0.8, micropores and cracks up to 2.0 microns in size are completely healed. Defects of larger dimensions are only partially healed, and residual damage is observed in the steel (Figure 3e).

When heated above the temperature 727 °C, recrystallization processes develop, which is accompanied by the annihilation of dislocations, the release of accumulated deformation energy, the nucleation and growth of new grains in the deformed matrix. Migration of the boundaries of new grains through deformation defects leads to their healing [12-15]. In the subcritical temperature range (30-50 °C lower than the temperature 727 °C) microdefects with a size of no more than 1 micron are fully healed.

Further heating above the temperature 727 °C causes a polymorphic α/γ - transformation, in which there is an intense migration of interphase boundaries, disruption of interatomic bonds and acceleration of diffusion processes. The thermal cycle in the temperature range 727 °C, causing multiple repetition of α/γ - transformation, enhances healing of microdefects by the mechanism: time on the drain [16,17]. Double thermal cycle in the temperature area 727 °C under the above conditions, ensures complete healing of micro-defects with the size less than 2 microns.

When comparing the damage healing diagram in steel 5140 with similar diagrams constructed after recrystallization annealing of carbon, austenitic steel, aluminum, titanium, nickel alloys (figure 4) it can be seen that the section 0 < ω 1 < 0.4 has reversible damage, which is characterized by the healing of deformation defects in the process of heat treatment.

In the range of 0.4 < ω1 < 0.8 there is a monotonic increase in residual damage. At 0.4 < ω1 < 1, the ability to heal the damage decreases markedly. The value of ω = 0.4 are related to mass education in wrought steel of micro-defects with a size of ~2 μm. The value ω = 0.8 characterizes the completion of intense coalescence of micropores, growth of microcracks and the formation of microfibrils having dimensions commensurate with the size of the grain.

Compared with a single recrystallization annealing at subcritical temperatures, thermal Cycling with phase recrystallization extends the stage of reversible damage on average from 0.25 to 0.4, which, apparently, should be attributed to the critical size of treatable microdefects. The intensity of damage healing on the second part of the graph also increases, and the size and number of microdefects is significantly less than after recrystallization annealing.

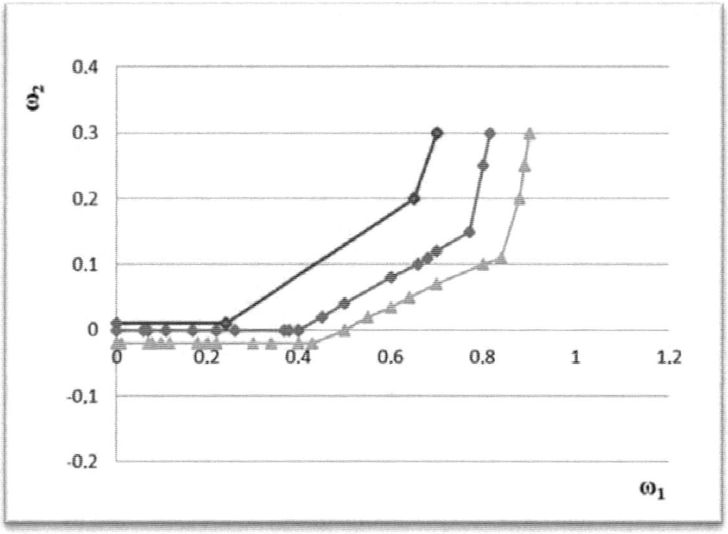

Figure 4. The generalized residual damage diagram (ω2):
◆ – after a single recrystallization annealing for a number of steels and alloys [2, 3];
for steel 5140 after cyclic heat treatment:
◆ – under conditions 1;
▲ – under conditions 2.

4. Conclusions

The steel 5140 damage on the beginning stages is determined to be attributed independently of perlit morphology to micropore appearance in ferrite. The steel 5140 periodic heat treatment with phase recrystallization leads to more full damage healing in comparison with recrystallization annealing. Herewith the microdefects critical size increases up to 2 micrometers, and invertible damage stage extend to $\omega = 0.4$.

Acknowledgment

The publication has been prepared with the support of the «RUDN University Program 5-100».

References

[1] Kleiner L M et al. 2011 *Perspective materials* **1** pp 59-67
[2] Si-Yuan L, Ke-Fu Y, Yun-Bo C, Miao-Hui W, Xue L and Xueyun G 2015 *Electrochimica Acta* **165** pp 45-55 https://doi.org/10.1016/j.electacta.2015.02.038
[3] K Amini, A Akhbarizadeh and S Javadpour 2013 *Materials and Design* **45** pp 316-322 https://doi.org/10.1016/j.matdes.2012.08.006
[4] Kandrotaitė-Janutienė R 2013 *Mechanika* **19(2)** pp 236-242
[5] Kim H, Lee J, Barlat F, Kim D and Lee M 2015 *Acta Materialia* **97** pp 435-444 https://doi.org/10.1016/j.actamat.2015.06.023
[6] Warner D H and Curtin W A 2009 *Acta Materialia* **57** pp 4267-4277 https://doi.org/10.1016/j.actamat.2009.05.024
[7] Tonysheva O A, Voznesenskaya N M, Shalkevich A B and Petrakov A F 2012 *Aviation Materials and Technologies* **3** pp 31-36.
[8] Bilenko E G and Vegera I I J. Frict 2008 *Wear* **29** pp 419-425 https://doi.org/10.3103/S1068366608060020
[9] Ryabchikov A I et al 2018 J. *Phys.: Conf. Ser.* 1115 032019

[10] Suchanek J and Kuklik V 2009 *Wear* **267** pp 2100-2108 https://doi.org/10.1016/j.wear.2009.08.003
[11] NicolaScholl, FlorianMinuth-Hadi and KlausThiele 2018 *Journal of Constructional Steel Research* 145 pp 414-424 https://doi.org/10.1016/j.jcsr.2018.02.013
[12] Hambli R and Reszka M 2002 *Int. J. Mech. Sci.* **44** pp 1349-1361
[13] Ogawa N, Shiomi M and K Osakada 2002 *Int. J. Mach. Tools Manuf.* **42.** pp. 607-614 https://doi.org/10.1016/S0890-6955(01)00149-3
[14] Botkin A V et al. 2013 *Deformation and Destruction of Materials* **7** pp 17-22
[15] Boyko Yu I, Volosyuk M A and Kononenko V G 2013 *Functional Materials* **20** pp 44-51 http://dspace.nbuv.gov.ua/handle/123456789/119800
[16] Tschopp M A and McDowell D L 2007 *Applied physics letters* **90** pp 121916-121919
[17] Zhu T, Li J, Van Vliet K J, Ogata S, Yip S and Suresh S 2004 *Journal of the Mechanics and Physics of Solids.* **52.** pp. 691-724 https://doi.org/10.1016/j.jmps.2003.07.006

The effects of operational thermal cycling on mechanical and magnetic properties of structural steels

Kyaw Zaya[1]*, Thet Paing[1], A V Kornilova[2]

[1]Moscow State Teachnical University "Stankin", Moscow, Russia
[2]Peoples' Friendship University of Russia (RUDN University), Moscow, Russia

E-mail: k.kyawzaya@yandex.ru

Abstract. This article shows the experimental results of the operational thermal cycling modeling of structural carbon steels with temperature range from the point of phase transition of type II cementite to the beginning of phase transformations of steel. In this study the influence of temperature that the number of thermal cycles and the histogram of temperature loading on the magnetic and mechanical properties of the steels were investigated. The possibilities of controlling the residual life of the elements, which are subjected to thermal cycling on changing the magnetic properties of metal, were determined. The functional dependence of the coercive force and hardness on the number of thermal cycles at a given temperature is obtained.

Keywords: coercive force, hardness, carbon steels, operational thermal cycling, phase transformations.

1. Introduction

The investigated temperature range is located near the Curie point of cementite, where cementite undergoes magnetic transformation (phase transition of II kind), in which it is considered as an Invar system [1]. Near the Curie point of cementite, there is an abnormal behavior of its elastic modulus and an increase in diffusion mobility [2]. In the Fe-Fe3C system, cementite is a metastable phase accordingly cementite near its Curie point which at 210°C must be highly sensitive to external influences [3]. The study of the properties of cementite is devoted to the work [4-9]. It is shown that within the framework of the orthorhombic lattice, variants of the cementite structure with different arrangement of iron and carbon atoms relative to each other can arise, so the cementite will be dependent on the initial heat treatment in two or more structural states. On this subject, experiments proved that high-carbon steels can be graphitized by thermal cycling near the Curie point of cementite [10]. However, there is low data in the literature review about the dependence of the magnetic and mechanical characteristics of low-carbon and medium-carbon steels in thermal cycling, which is especially important in the technical diagnosis of basic parts of machines designed for the system "specified life". Therefore, the topic of the study is relevant.

Thermal cycling above the line PSK in diagram Fe-Fe3C, the temperature of the eutectoid transformation in a metastable system – **727**°C, has several applications in the mechanical engineering such as; grinding grain, thermal insulation and concentration of harmful impurities in isolated areas of ferrite relief. Because it causes improvements in the mechanical characteristics of steels due to the partial $\alpha \rightarrow \gamma$ transformation. In the manufacture of devices, thermo cycling is used as quick test life in order to identify hidden defects manufacturing process.

However, a number of equipment parts when proceeding technological operations are subjected to temperature insolation in the temperature that ranges significantly below the eutectoid transformation temperature, for example, the frame and shafts of crank hot forging presses are subjected to cyclic temperature loads in the range from the workshop temperature to 150°C – 550°C. That thermal loading is proposed in order to introduce the term operational thermal cycling. So that the impact assessment of such action is possible, accurate prediction of large basic parts resource is a difficult task because it requires multi-factor tests for its solution. This task is particularly relevant in the technical diagnostics of machines designed for a given durability. This paper aimed to investigate the effect of thermal cycling on the magnetic and mechanical properties of structural carbon steels and to obtain studies basis of the functional dependence of structural carbon steels characteristics on the number of thermal cycles at a given temperature.

2. Materials, equipment and methods

In the experiment, certified and verified devices were used as part of the control and measuring system – Portable pulse coercimeter KIM-2M and hardness taster TKM-459M. Furnace PKE-25.1was used for heating samples. For the experiment, special samples were made of structural quality steel **25** according to GOST 1050-88 (carbon content 0.22-0.30%). The procedure of the experiments was as follows: the samples were placed in a furnace with a predetermined temperature, kept in the furnace for **3-5** minutes (the temperature of the sample was monitored by a pyrometer Testo 835-T2) and cooling in calm air. This work-cycle is repeated. During the experiment, the coercive force H_c, the value of the magnetic field strength necessary for the complete demagnetization of a ferromagnetic or ferrimagnetic substance. A/m, and the hardness HRB on a scale in Rockwell in conventional units were measured. For monitoring the microstructure, an inverted microscope CKX41 was used and for obtaining the fractograms, 3D scanner Range Vision Spectrum was used.

3. Results and Discussions

Figure 1 shows the diffractogram of destroyed sample as a result of operational thermal cycling (temperature from 20˚C to 250˚C). Obviously, the fracture occurred along the grain boundaries as a result of embrittlement. Therefore, with this type of loading, it is possible to control the value of the residual resource by the degradation of the structure. However, for operational control and technical diagnostics of critical parts, the use of structure-dependent material characteristics is more effective. It is demonstrated that a characteristic is the coercive force, which has a correlation with mechanical characteristics and structure.

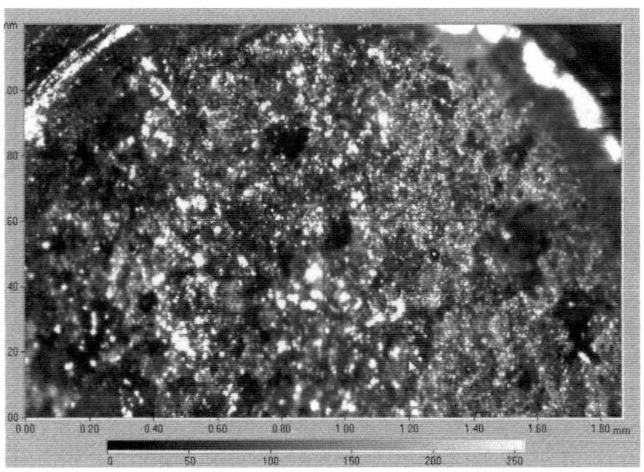

Figure 1. Diffractogram of the sample destroyed as a result of thermal cycling in the temperature from 20° to 250°C.

The results of the experiment are shown in table 1, and illustrated by the graphs in figures 2 and 3. The number **1-5** refer to the numbers of the samples under research. It is obvious hardness and coercive force of the samples increases up to 2 times in that thermal cycling near the Curie point of cementite on the third cycle. Even more with an increase in the temperature load cycles, a monotonic drop in the research values occurs. However, thermal cycling of these steels at temperatures above 300˚C (but below the PSK line of the Fe-Fe3C diagram, where austenitic transformation occurs during heating and eutectoid transformation during cooling) did not lead to the same result.

Table 1. Experimental results values of coercive force

Samples numbers	Initial state 0	The number of thermal cycles (300°C) 2	3	6	12	20
1	614	633	797	760	737	681
2	758	753	914	790	745	710
3	902	820	1105	1000	1015	753
4	835	844	903	857	805	715
5	909	894	1012	919	920	723
$\overline{H_c}$	804	789	946	865	844	716

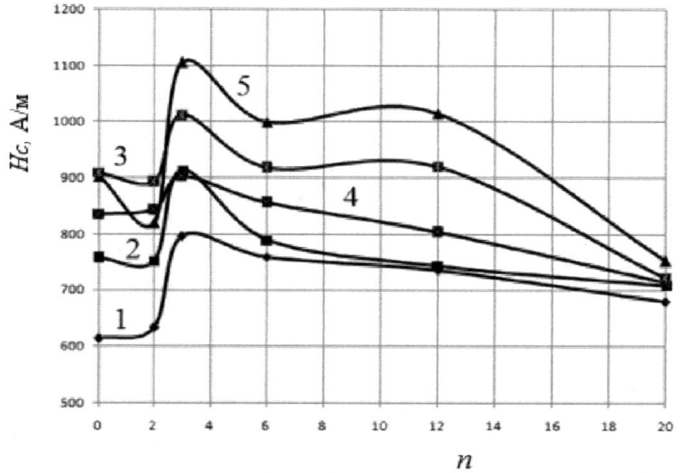

Figure 2. Influence of the coercive force of the 5 samples on the number of thermal cycles.

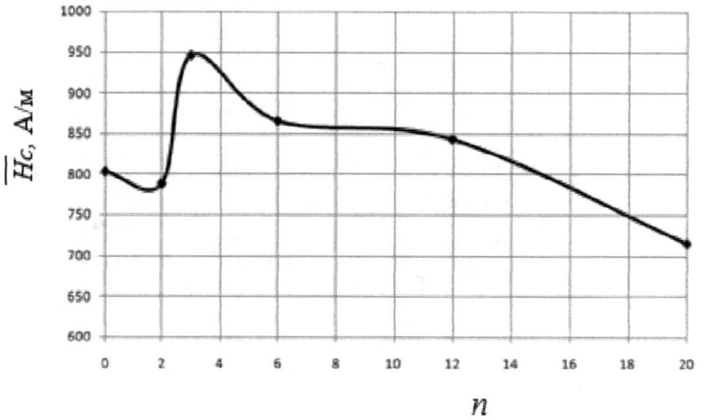

Figure 3. Effect of the average value of the coercive force on the number of thermal cycles.

In addition to the coercive force, hardness of samples was measured. Hardness measurements have a smaller spread of the obtained values due to the fact that the hardness is less structure dependent characteristic than the coercive force. But it has the same kind of dependence of the measured characteristics on the number of thermal cycles as the coercive force.

The graphic depend on the average value of hardness (HRB) with number of thermal cycles as shown in figure 4. Figures 3 and 4 were demonstrated by the following functional equations; the coercive force of the number of loading cycles $H_c = -1.136\ n^2 + 17.58\ n + 814.2$ (the coefficient of determination is 0.615), hardness $HRB = -0.271\ n^2 + 5.7085\ n + 35.818$ (the coefficient of determination is 0.76). These factors can be used for analyzing technical diagnostics. The measured values (hardness and/or coercive force) are determined the number of thermal cycles and the residual life of the structure.

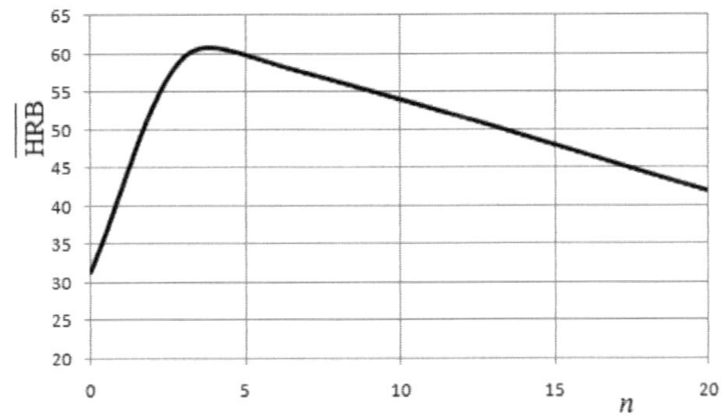

Figure 4. Effect of the average hardness of the sample on the number of thermal cycles.

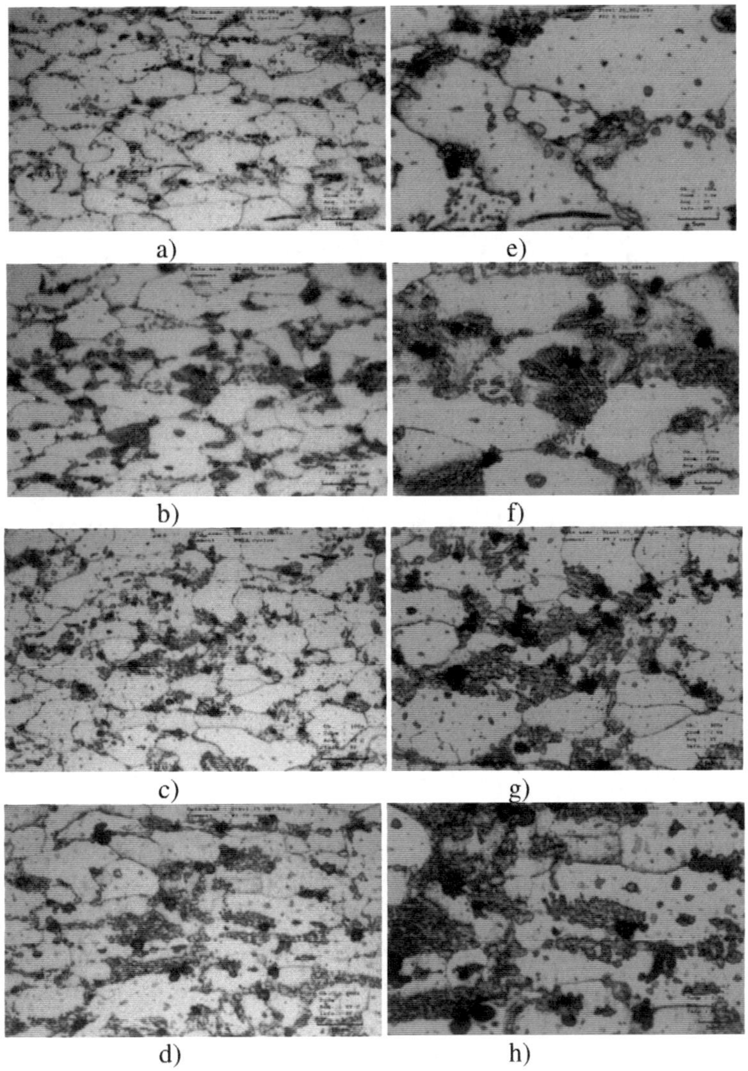

Figure 5. Structure: in the initial state, a – in the initial state; b – after 3 cycles; c – after 7 cycles; d – after 20 cycles; (e, f, g, h the same as a, b, c, d, respectively but with a larger scale).

To investigate the structure of the samples, micro-sections were made in the initial state, and then micro-sections were made from the same sample after thermal Cycling at a temperature of 300°C after 3, 7 and 20 cycles. Photos of structures are shown in figure 5 (a-h). The scale is indicated on each photo. Analysis of micro-sections showed that an increase in hardness and magnetic characteristics, and that can only be partially explained by the transformation of granular perlite into lamellar perlite. In other words, the appearance of temperature stresses in the structure can be an explanation. The relevant micro-sections showed the propagation of inter-crystalline corrosion with increasing of the number of thermal loading cycles. Corrosion continued to increase after the completion of the experiment in a room temperature. The increase in magnetic properties (coercive force) and hardness at the third cycle by temperatures near the Curie point of cementite is confirmed by thermal Cycling of flat samples from steels 20 and 45.

Furthermore, thermal Cycling by loading with 2 temperature blocks (the total number of cycles in each of the histograms 15) was carried out in order to study the magnetic properties and hardness of structural carbon steels. The results are shown in figures 6 and 7 (material of samples is steel 25). Number 1 and 2 — refer to the samples of the loading histogram.

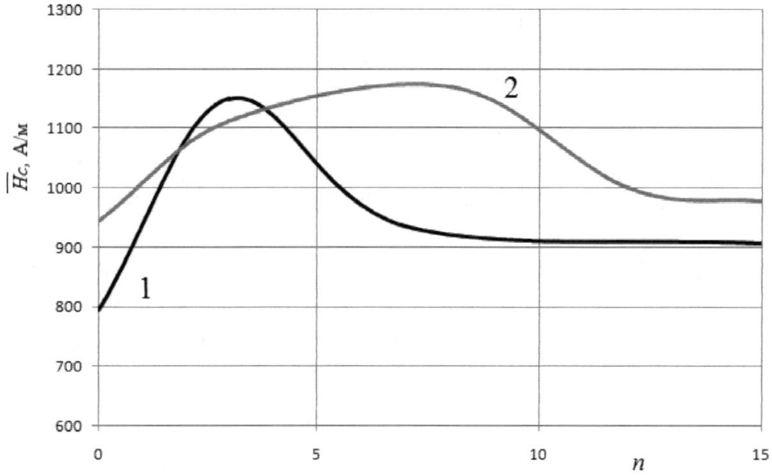

Figure 6. The influence of average value of coercive force on the number of loading cycles by two histograms (1 – 7 cycles with 550°C, then 8 cycles with 300°C & 2 – 8 cycles with 550°C, then 7 cycles with 300°C).

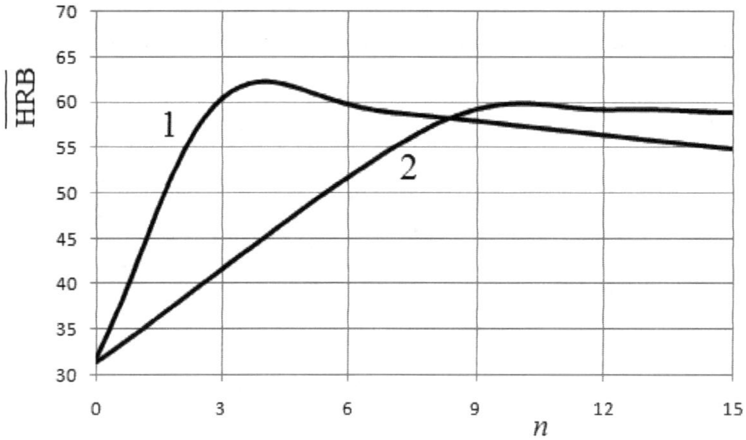

Figure 7. The influence of hardness average value on the number of loading cycles of two histograms (1 and 2 is the same action as the figure 6).

4. Conclusions

1. The possibility of controlling the residual life of elements subjected to thermal Cycling by changing the magnetic properties of the metal. In order to achieve that task it is required to create a new system which is unclouded control and measuring methods by software system, control and measuring devices and adaptive block for express-analysis of the state of the metal.

2. During the thermal cycling from 20°C to the maximum cycle temperature, it is regarded to the coercive force and hardness behavior on the number of loading cycles and cycle temperature is revealed. It is shown, that thermal Cycling near the Curie point of cementite on the third cycle increases up to twice the hardness and coercive force of structural carbon steels.

3. The analysis of micro-glyphs showed that the sharp peak increase in the hardness and magnetic characteristics can only be partially explained by the transformation of granular perlite into lamellar (figure 5, e and f). This point requires further research.

4. After the third cycle, there is a monotonic drop in the magnetic properties and hardness of the material with an avalanche-like increase in intergranular corrosion, which must be considered when carrying out technical diagnostics of parts subjected to thermal Cycling. For machines designed and manufactured according to the "specified life" system, which will be subjected to operational thermal Cycling during operation, we recommend the use of chromium and titanium alloying to eliminate the phenomenon of intergranular corrosion. However, thermal Cycling with 3 cycles from 20 ° C to 300 ° C can be used to strengthen small and medium - carbon steels in structures that will not be subjected to thermal cycles during operation.

5. References

[1] Medvedev N I, Karkina L E and Ivanovskaya A L 2006 Electronic structure and magnetic properties of α-, γ-phases of iron, their solid solutions with carbon and cementite *Physics of Metals and Metallography* **101(5)** pp 479-484

[2] Drapkin B M and Fokin B V 1980 On the young module of cementite *Physics of metals and materials science* **49(3)** pp. 649-651

[3] Titorov D B 2007 Simulation of various possible cementite structures *Physics of Metals and Metallography* **103(4)** pp 413-419

[4] Schastlivtsev V M, Yakovleva I L, Mirzaev D A and Okishev K Yu 2003 On the Possible Positions of Carbon Atoms in the Cementite Lattice *Physics of Metals and Metallography* **96(3)** pp 75-82

[5] Medvedeva N I, Ivanovskii A L and Kar'kina L E 2003 Effects of Atomic Disordering and Nonstoichiometry in the Carbon Sublattice on the Energy-Band Structure of Cementite *Physics of Metals and Metallography* **96(5)** pp 16-20

[6] Bunin K P and Baranov A A 1970 *Metallography* (Moscow: Metallurgy) p 256

[7] Tikhonova I V, Kuzovlev O V, Starikov N E and Gvozdev A E 2008 Dissolution of cementite in carbon steels by thermal Cycling *Production of Rolled Products* **8** pp 36-37

[8] Tikhonova I V, Malyarov A V, Kuzovlev O V, Gvozdev A E and Starikov N E 2009 The influence of carbon content on the dissolution of cementite in carbon steels under cyclic thermal processing *Production of Rolled Products* **5** pp 29-31

[9] Lomov S B, Sokolova T V, Mal'kova M S and Balakirev E V 2015 The dependence structure of the 40X steel cold deformation from the original structure and cyclic heat treatment *Metallurgy Engineering* **2** pp 21-24

[10] Malyarov A V 2009 Graphitization of carbon steels under cyclic thermal treatment near the Curie point of cementite *Proc. X Int. Sc. and Tech. Ural School-Seminar of Metallurgists-Young Scientists* pp 79-81

[11] Collins D A, Barkley E L, Lach T G and Byun T S 2019 Effects of thermal aging on the fracture toughness of cast stainless steel CF8 *International Journal of Pressure Vessels and Piping* **173** pp 45-54

IOP Conf. Series: Materials Science and Engineering **675** (2019) 012041 doi:10.1088/1757-899X/675/1/012041

[12] Yeddu H K, Shaw B A and Somersb M A J 2017 Effect of thermal cycling on martensitic transformation and mechanical strengthening of stainless steels – A phase-field study *Materials Science and Engineering: A* **690** pp 1-5
[13] Shao Y, et al 2019 Thermal simulation on double-pass welding of a high Cr ferritic steel *Journal of Manufacturing Processes* **43(A)** pp 9-16
[14] Baxter D J and Fordnam R J 2000 The Oxidation and Corrosion Behavior of Nonoxide Ceramic Matrix Composites *Comprehensive Composite Materials* **4** pp 221-264
[15] Li S, Ren X, Hou H 2015 The effect of thermal cycling in superplastic diffusion bonding of 2205 duplex stainless steel *Materials & Design* **86** pp 582-586
[16] Qu H, Hou H, Li P, Li S and Ren X 2016 The effect of thermal cycling in superplastic diffusion bonding of heterogeneous duplex stainless steel *Materials & Design* **96** pp 499-505
[17] Park S-J and Seo M-K 2011 Chapter 8 - Composite Characterization *Interface Science and Technology* **18** pp 631-738
[18] Elaieb M, Shuaeib F M, Saied R O and Almalki M G 2013 The Influence of Thermal Cycling on the Corrosion and Hardness of the Low Carbon Steel *Libya for Applied and Technical Science* **2** pp 24-28
[19] Ilyin A M, Tazhibaeva I L and Borisov B A 2002 Effect of thermal cycling on impurity grain boundary segregation in maraging steel *Journal of Nuclear Materials* **307(311)** pp 475-478
[20] Tataurova E V 2002 Effect of thermocycling on structure and properties of carbon steels *Russian Metallurgy (Metally)* **1** pp 67-71

Creation of a design methodology for crank punching machines of specified durability

Paing Thet[1*], **Zaya Kyaw**[1], **A V Kornilova**[2]

[1] Moscow State Teachnical University "Stankin", Moscow, Russia.
[2] Peoples' Friendship University of Russia, Moscow, Russia.

E-mail: thetpaing520@gmail.com

Abstract. The article discusses the main aspects of designing large-forging equipment of given durability on the example of crank punching machines. A scientifically based algorithm is proposed to ensure reliable operation of the structure operating in the field of high-cycle fatigue. An example of calculations is given when designing a bed for a crank hot forging press (CHFP) and a crank plate forging press defined durability on the basis of existing structures.

Keywords: crank punching machines, durability, high-cycle fatigue, stress concentrator, structural rigidity.

1. Introduction

Under market conditions, economic feasibility strictly dictates the production of the need to reduce production costs. One of the ways to reduce costs is the design and manufacture of equipment and tools by a strictly limited program of durability ordering (specified life), for example [1-8]. Figure 1 shows the dynamics of publications on the topic specified life over the past 10 years. Figure 2 shows the dynamics of patenting over the past 10 years on this topic in the countries of the world. Obviously, it is the industrialized countries that are most active in this direction, which indirectly indicates the cost-effectiveness of this approach.

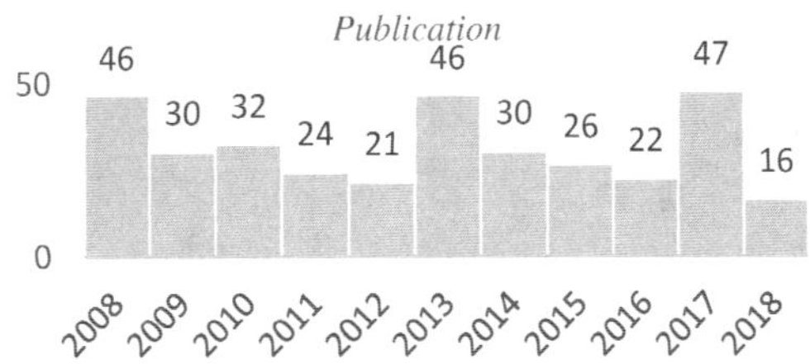

Figure 1. Dynamics of publications on the topic "specified life" over the past 10 years.

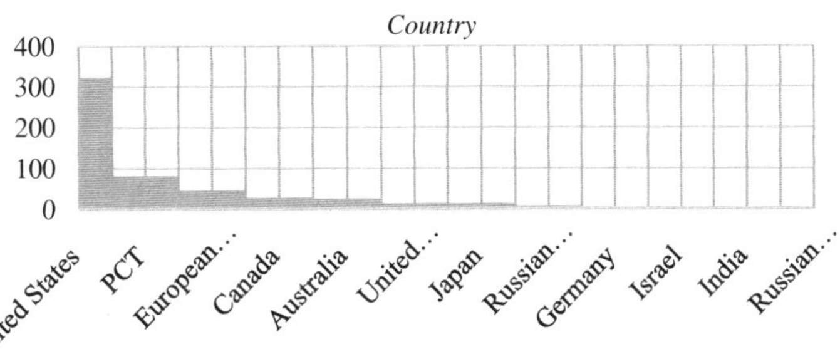

Figure 2. Dynamics of patenting on the topic "specified life" over the past 10 years in the world.

There are a number of works on designing die tools with limited durability [1-5], metallurgical equipment and aircraft structures [9]. However, such approaches are practically not applied to large metal-consuming forge-pressing equipment. But it is in this area that they can provide greater economic efficiency by reducing metal consumption, manufacturing time for large parts, reducing the load on the foundations, reducing transport costs. This is especially interesting for the industry of the Republic of the Union of Myanmar, since it does not yet have the capacity to create large parts, and the need for domestic technological machines of large forces is increasing. Therefore, the way to create technological machines of limited durability with low metal consumption is one of the options for the development of this industry in Myanmar. The same approaches can be implemented when modernizing equipment in Russian conditions, when existing machines increase the nominal force while maintaining metal-intensive basic parts (the main direction of modernizing press-forging equipment in Russia) [8].

2. Materials and Methods

As part of this work, the press frame CHFP (AJAX-CECO) with a nominal force of 20 MN and an open-type frame KD2130 with a force of 1 MN were simulated to increase the level of cyclic strength (durability) with constant metal intensity and rigidity (figure 3).

For calculation using numerical methods, the Solid-Works system was used to automate mechanical design and create solid-state parametric models of parts, assemblies and drawings. The system works in the MS Windows operating environment. Solid-Works software, that includes finite element analysis, which corresponds to a high accuracy class, proved by the International Association for Engineering Modeling, Analysis and Simulation NAFEMS.

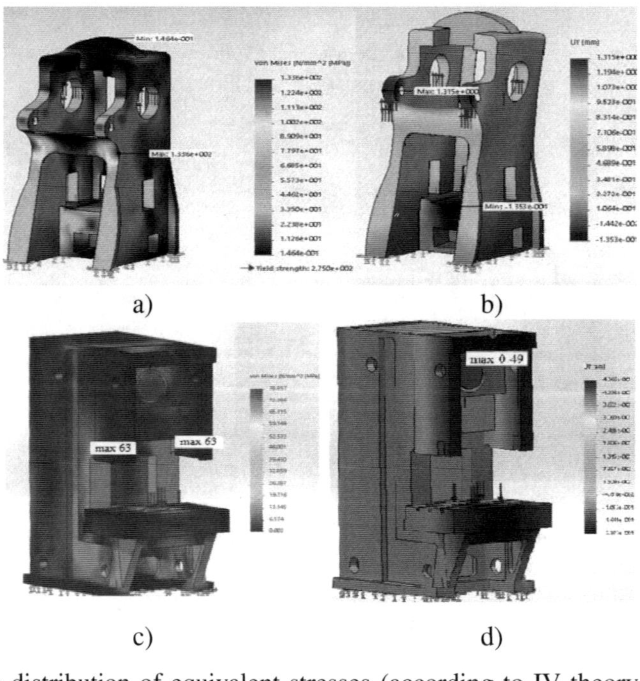

Figure 3. a, c — the distribution of equivalent stresses (according to IV theory of strength), b, d — displacements along the vertical axis; a, b — relate to the CHFP (AJAX-CECO) press frame with a nominal force of 20 MN; c, d — to the open-type KD 2130 press frame with a force of 1 MN.

3. Results and Discussions

The calculation by the numerical method (finite element method) allowed us to reveal the optimal parameters of the design concentrators for the structures under consideration according to the criterion of cyclic strength and zones, the most dangerous according to the criterion of possible origin and growth of macro-cracks. It is assumed that the limiting states regulated in [10], PLS3 (*primary limit state*) can be realized in the supports of the main shaft of the press and the guides of the slide. However, it is removable by replacements and registers. Therefore, in the future, only 3 limit states are taken into account: PLS2 — failure to operate due to the formation of fatigue cracks in metal parts at the moment of cyclic loading, ALS1 (*additional limit state*) — occurrence of fragile, quasi-brittle or viscous type with process initiation from a technological defect. In this paper, for the crank hot forging press (CHFP) we propose to take account ALS4 - the occurrence of specific degradation processes in the material of the frame during operational thermal cycling due to the influence of thermal load from the workpiece when performing technological operations on the die zone [11-13]. When designing power parts of forge-and-press machines for a given durability, the main tasks are the problems of fatigue damage accumulation under loads (PLS2), whose maximum values do not reach σ_t (yield strength) values, but there is a significant amount of loads with amplitudes greater than fatigue limit with cyclical loading [14-16].

The calculations were carried out only for the optimal according to the criterion of the cyclic strength of the concentrator with a given number of cycles to failure. The allowable stresses are calculated using the data analysis function "goal seek" in Microsoft Excel software. Accepted in the calculation of the primary data and the calculated values of some parameters are summarized in table 1. Stresses are found corresponding to the known values of the endurance limit of the frame from the table 1. The obtained values are presented in column 3 of the table 2.

Table 1. Primary data of two frames of presses

Primary data	CHFP	KD 2130
Tensile strength σ_B (reference value)	491 MPa	300 MPa
The value of the sensitivity of the material to the cycle asymmetry by dependencies [17]	0.12	0.08
The endurance limit of the frame σ_{-1c} c taking into account all downward factors according to [17]	61 MPa	72 MPa

Table 2. Results of stress calculation by a given number of cycles to failure.

No.	1	2	3	4	5	6	7
1	The number of cycles to failure (specified)	N	10^7 basic number of cycles according to [17]	$6\cdot10^6$	$6\cdot10^5$	$6\cdot10^4$	$5\cdot10^4$
2	The maximum stress allowed by the condition of ensuring a specified number of loading cycles, MPa	σ_{max}	109.11[a]	128.59[a]	216.41[a]	304.23[a]	311.18[a]
			133.33[b]	145.66[b]	201.21[b]	256.77[b]	261.17[b]
3	Stress resulting in a symmetrical loading cycle, MPa	σ_{res}	61[a]	71.9[a]	120.99[a]	170.1[a]	173.98[a]
			72[b]	78.66[b]	108.66[b]	138.66[b]	141.03[b]

[a] Values for press frame CHFP.
[b] Values for press frame KD2130.

Then the dependences of equivalent static stresses (MPa) on the number of loading cycles were constructed (figure 4). The graph was plotted only for the region of multi-cycle fatigue in semi-logarithmic coordinates. There is a linear relationship between stress and the number of cycles before failure. With this dependence, it is possible to determine which stress level corresponds to a given number of loading cycles, and then, based on the dependence connecting the stresses and the force causing them (with the optimal parameters of the concentrator defined above), designate the declared when designing for a given durability. That is, the declared force in the design (modernization) of the "specified life" system proposes to call the force that provides the specified value of durability.

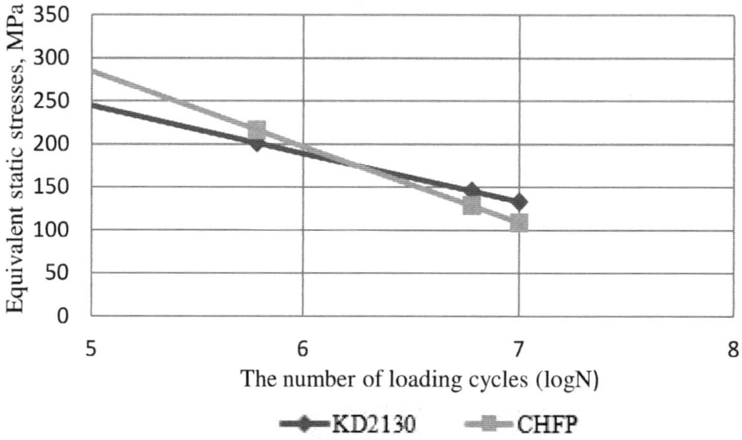

Figure 4. The influence of stresses (MPa) on the number of loading cycles (high-cycle area of fatigue) at optimal concentrators.

Table 3. Calculated performance with an increase in nominal forces.

Press	k	1	1.5	2	2.5	3	3.5	4	4.5	5
CHFP	$k \times 10^5/M$	1.42	2.14	2.85	3.56	4.27	4.98	5.7	6.41	7.12
	$N \times 10^3$	5300	920	160	27.7	4.81	0.84	0.15	0.025	0.004
KD2130	$k \times 10^5/M$	12.82	19.23	25.64	32.05	38.46	44.87	51.28	57.69	64.1
	$N \times 10^3$	∞	∞	∞	3650	990	268	73	20	5.3

A further calculation was the number of cycles to failure with an increase in technological forces with the existing version of the bed considered above CHFP and KD2130. The values of stresses in each variant were determined by calculation by numerical methods. The data are summarized in table. 3. The following notation k – the coefficient of increase of nominal power (P_n), M – the mass of the frame (70237 kg 7800 kg CHFP and KD2130 respectively).

The ratio of the applied force to the mass of the structure, as a target function, should tend to the maximum when imposing restrictions on rigidity and the required (specified) durability. Next, you need to find out the nature of the influence of technological forces on the durability of the forging and pressing machine on which the operation is performed. Low-cycle fatigue is limited to $N \leq 5 \cdot 10^5$, presses designed according to "specified life" and crank punching machines that have been modernized with an increase in the declared technological forces should not work in the area of low-cycle fatigue. On universal forming presses, the main operation is cutting / punching, the technological strength of which depends on the shear resistance and the area of the cut [18]. A graphic representation of the dependence of force on the parameters of separation operations is given in figure 5.

To ensure a specified number of cycles to failure during a technological operation on the press KD2130, it is accepted by approximating the calculated data using the $P \rightarrow \sigma_{eq}$ (FEM) $\rightarrow \sigma_{res}$ (according to [17]) $\rightarrow$ N. We obtain the following:

$$N = 4 \cdot 10^6 \cdot P^2 - 16.315 \cdot P + 2 \cdot 10^7 \quad (1)$$

The main operation at the CHFP accept hot upsetting [18]. The dependence of the force of hot upsetting of a billet of steel 45 on the height of the billet H_{ups} and diameter of the billet D_{ups} after hot upsetting is shown in figure 6.

Similarly, we obtain the following expression for the number of cycles to the destruction of the CHFP bed from process operations of hot sludge:

$$N = 1.39 \cdot 10^{-6} D_{oc}^{4} \cdot \left(1 + 0.17\frac{D_{oc}}{H_{oc}} - 0{,}33\frac{H_{oc}^{2}}{D_{oc}^{2}}\right)^{2} - 9.8 D_{oc}^{2} \cdot \left(1 + 0.17\frac{D_{oc}}{H_{oc}} - 0.33\frac{H_{oc}^{2}}{D_{oc}^{2}}\right) + 2 \cdot 10^{7} \quad (2)$$

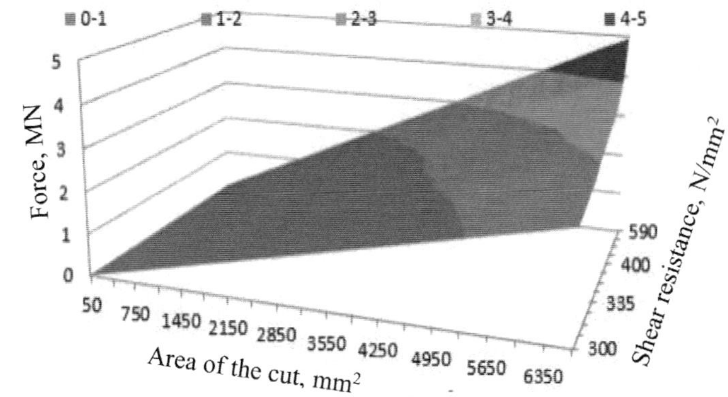

Figure 5. Graphical representation of the calculated data for the separation operations.

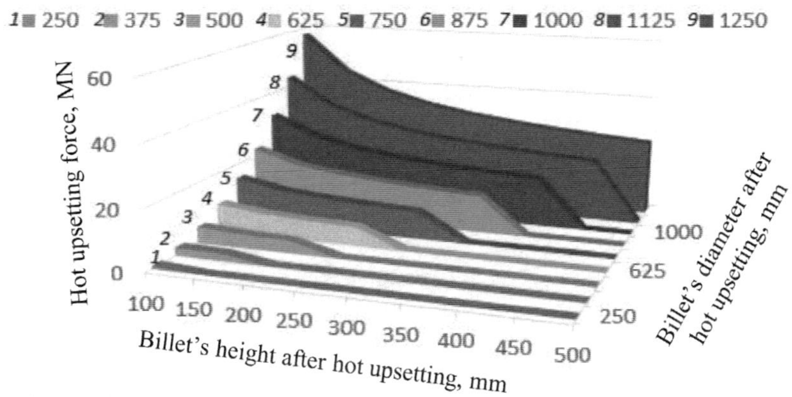

Figure 6. Graphic representation of the ratio of hot upsetting force, height and diameter of the work-piece after hot upsetting for steel 45.

Graphic representation (2) is shown in figure 7.

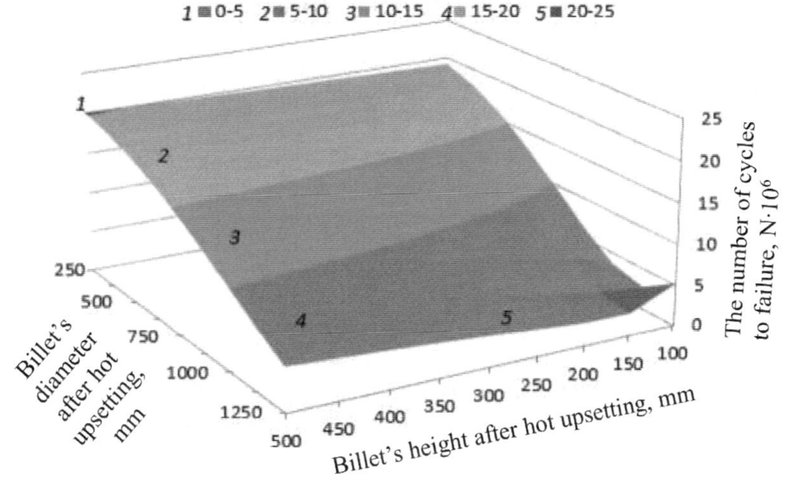

Figure 7. Graphic representation of the expression (2).

Thus, a calculation procedure has been developed, which makes it possible to ensure the durability set by the technical specification (the number of cycles to failure) of the basic parts of the crank punching machines. The main stages of calculations:

- Determination of the loads acting on the base parts of the press in the process of loading with a nominal force.
- Calculation of the finite-element model of the objects under study by numeral methods (finite element method) under the action of nominal force.
- Optimization of all structural stress concentrators of the base part by the criterion of maximum cyclic strength with a constant metal intensity: determination of the dependence of stresses and displacements on the magnitude of the applied force at various parameters of concentrators by the finite element method, calculation of loading cycles to failure at various parameters of the concentrator, according to calculations plotting the number of loading cycles to failure from the geometry/dimensions of the concentrator, optimization of the obtained dependence of n on the criterion for obtaining the maximum loading cycles by the simplex method with the imposition of restrictions on constructive considerations.
- Construction of the calculated dependence of the voltages, reduced to a symmetrical cycle, on the applied technological forces, other than the nominal, adopted in the original calculation.
- Construction of the dependence of the stresses reduced to a symmetric cycle on the number of loading cycles (inverse problem).
- Determination of the corresponding values of stresses to symmetric cycles according to the graph of the dependence of the stresses to symmetric cycles from the number of cycles to failure, according to a given number of cycles to failure.
- Determination of the dependence of the stresses, reduced to symmetrical cycles, on the applied forces, the force, allowing to ensure the specified durability. We assign this force nominal.
- Determination of stiffness and comparison of the obtained stiffness with empirical stiffness according to recommendations substantiated in the works of Lansky E.N., Banketov A.N. and Kruka A.T. Comparison of obtained stiffness with recommended values for nominal force. If the resulting value is less than the recommended, then change the design to obtain the desired values (by iterative approximation or optimization with the imposition of restrictions on stiffness / displacement).
- Determination of the order of loading (loading history) according to the hypotheses of Henry, Guts, Serensen S.V. and Manson, giving the maximum value of the residual life of the base part at the end of the last load block. This will provide a minimum scientifically based margin on fatigue life, which will allow parts to work out all the load cycles specified in its design without cracking with the minimum ratio of metal intensity / technological force.
- Determination of the number of cycles to failure with increasing technological forces at a constant mass of the structure and its rigidity.
- Determination of the parameters of technological operations according to the obtained dependencies (1) and (2), allowing to ensure the specified durability and not get into the low-cycle fatigue zone.

4. Conclusions

1. It has been proposed to introduce for a CHFP designed for a given durability, another type of limit state ALS4 - the occurrence of specific degradation processes in the material of the frame during operational thermal cycling.
2. A design methodology was developed to provide the durability of the base parts of the crank punching machines specified by the technical specification, while preserving the standardized technical documentation rigidity of the machine, based on the parameters of the main technological operations performed on the crank punching machine. For hot upsetting of billet on CHFP with a force of 20 MN, the following parameter range was considered - diameter after hot upsetting from 250 mm to 1250 mm, height after hot upsetting from 100 to 500 mm, σ_s - deformation resistance from 1 to 25.3 MPa. It is shown that it is only possible to increase the technological force on this press up to 1.5 times without

entering the low-cycle fatigue zone. For a crank punching machines KD2130 as the variable parameters of the technological cutting/punching operations were taken — the area of the cut (varied in the interval from 50 to 6700 mm^2) and the shear resistance (from 300 to 590 MPa). It is shown that the maximum increase in force is 3 times.

3. The optimal constructive concentrators of the frames of the KD2130 press and the CHFP with a nominal force of 20 MN were determined according to the fatigue strength criterion. It is shown that under cyclic loading even a slight decrease in static stresses (by 7 MPa relative to the existing version of the frame) allows to increase the number of cycles to failure by 20%. The dependence of the ratio of the technological force of a crank punching machine to the weight of the frame for presses KD2130 and CHFP with a nominal force of 20 MN is compared. In the existing version for CHFP the ratio of the nominal force to mass is 0.28 kN/kg and for press KD2130 -0.13 kN/kg.

5. References

[1] Karpukhin I I, Selishchev A I, Kornilova A V and Idarmachev I M 2018 Investigation of the Change in the Magnetic Properties of Die Steels in the Process of Manufacture and Operation of Die Tools *IOP Conference Series: Materials Science and Engineering* **287** 012006

[2] Ivanov V I, Kornilova A V and Musatov V V 2016 Technical diagnostics in accident risk assessment *Inorganic Materials* **52(15)** pp 1509-1514

[3] Kornilova AV, Selishchev A I and Idarmachev I M 2016 Application of magnetic kinds of nondestructive inspection to parts from die tool steels *Metal Science and Heat Treatment* **57(9-10)** pp 632-637

[4] Kornilova A V, Idarmachev I M, Paing T and Zayar C 2014 A method of determination of the service life of a die tool with application of magnetic methods of nondestructive control and diagnostics *Journal of Machinery Manufacture and Reliability* **43(5)** pp 439-444

[5] Kornilova AV 2005 Designing technique of limited durability tool *Journal of Machinery Manufacture and Reliability* **3** pp 76-81

[6] Sivak B A and Kornilova A V 2005 General durability and residual service life of machine parts *Tyazheloe Mashinostroenie* **3** pp 11-12

[7] Kornilova A V 2005 The technique of estimation of durability and remaining life of press-forging machines' basic parts *Forging and Stamping Production. Material Working by Pressure* **5** pp 11-16

[8] Kornilova A V and Paing T 2017 To the problem of design of forming machines with a preset durability *Forging and Stamping Production. Material Working by Pressure* **11** pp 23-29

[9] Bespalov V A 2012 Design of reinforced specified fatigue life capability airframe panels containing notches *Siberian Aeronautical Research Institute named after S.A. Chaplygin* **5(36)** pp 48-52

[10] Moskvichev V V 2018 Applied problems of fracture mechanics, resource and safety of technical systems *Procedia Structural Integrity* **13** pp 2114-2119

[11] Zdenek C 2014 Effect of heat load on a forging press *Procedia Engineering* **69** pp 897-901

[12] Zdenek C and Karel R 2016 Effect of heat load on a mechanical forging press *Annals of DAAAM & Proceedings* **27** pp 344-348

[13] Zdenek C, Karel R and Tomas K 2018 Thermal analysis of the forging press LZK 4000 *Annals of DAAAM & Proceedings* **29** pp 740-745

[14] Makhutov N A 2017 Generalized regularities of deformation and fracture processes *Herald of the Russian Academy of Sciences* **87(3)** pp 217-228

[15] Makhutov N A and Reznikov D O 2015 Application of scenario analysis in the assessment of structural reliability of complex technical systems *Journal of Machinery Manufacture and Reliability* **44(8)** pp 675-686

[16] Burdukovsky V G, Kolmogorov V L and Migachev B A 2015 Prediction of resources of materials of machine and construction elements in the process of manufacture and exploitation *Journal of Materials Processing Technology* **55(3-4)** pp 292-295

[17] GOST 25.504-82 1982 *Calculations and strength tests. Methods for calculating the characteristics of fatigue resistance* (Moscow: Standartinform) p 55
[18] Altan T, Ngaile G and Shen G 2005 *Cold and hot forging: fundamentals and applications* (Ohio: ASM International) p 341

Performance characteristics of portable air conditioner with condensate-water spray

J I Yoon*, C H Son, C G Moon, J Y Kong, J H Yoon

Department of Refrigeration and Air-Conditioning Engineering, Pukyong National University, 45 Yongso-ro, Busan, South Korea

Email: yoonji@pknu.ac.kr

Abstract. In this research, a device for injecting condensate-water generated from the evaporator into the distributor on the top of the condenser was fabricated as a method for improving the efficiency of the portable air conditioner. To compare the efficiency, the condensation temperature, blowout air temperature, power consumption, cooling capacity, and COP of the system were measured. As a result of the experiment, the condensation temperature dropped by about 6.2 ℃. Also, the power consumption of the compressor was reduced by about 0.16 kW and the cooling capacity increased by 0.2 kW. As a result, it was confirmed that the COP increased by about 16 % when the condensate-water was sprayed on the condenser.

Keywords: Portable air conditioner, Condensate-water, COP, Cooling capacity.

Introduction

The demand for air conditioners is increasing every year due to rising temperatures in summer in South Korea. In particular, the demand for the portable air conditioner is increasing in industrial plants or various fields requiring local cooling. Portable air conditioners use a method of absorbing ambient air and cooling the surroundings when the refrigerant evaporates. In addition, since the outdoor unit and the indoor unit are built in the main body, there is no need to install the outdoor unit separately, so the burden on the cost or installation space is less, but there is a disadvantage that the cooling capacity is lower than the existing system air conditioner. [1] A. Khalil et al. If the evaporator coil surface temperature is lower than the dew point temperature of the inlet air, condensation will occur on the coil surface. This is called surface cooling, which is why condensate-water is produced and humidity in the atmosphere is low. [2] A. H. Mahvi et al. Air conditioners produce an average of 36L of condensate-water per day. This amount of production is thought to be available in many ways. [3] S. Algarni et al. The study of air conditioning condensate-water recovery and application has addressed the need for building energy recovery and freshwater sustainability. [4] W. Akram et al. When using other water on earth, the scale is generated as impurities contained in water. Therefore, the application of condensate-water from the evaporator side of the refrigerator was studied. [5] John A. Bryant et al. The amount of condensate-water produced in the industrial building was quite high and the potential impact on the resulting condensate-water utilization and reuse technique was studied. [6] N. Nethaji et al. To utilize the condensate-water, an experiment was conducted to reduce the indoor circulation air temperature by installing a condensate-water drain pipe inside the wall of the building. As a result, experimental results showed that average indoor air temperature is get lowered to about 0.3~0.5 ℃ and achieved energy-saving about 7%. [7] I. N. Ardita et al. Codensate-water is reviewed as an additional cooling medium. As a result of the experiment, we found that the system air conditioner increased the cooling capacity by 4%, the COP by 7%, and decreased the power consumption by 3%.

When the air conditioner is running, condensate-water is generated as moisture in the air condenses on the evaporator. In the case of a system air conditioner, this condensate-water is removed through a drain pipe using a pump, but a portable air conditioner uses a method of manually removing condensate-water by placing a separate reservoir under the evaporator for convenience of movement. The purpose of this study is to investigate the characteristics of system performance improvement as the condensate-water evaporates and the condensation temperature is lowered by spraying the condensate-water on the condenser.

2. Experiment apparatus and methods

The compressor of the portable air conditioner was Samsung Rotary Compressor, which was selected as the 48D199IT model. At rated conditions, the cooling capacity is 19,300 Btu/h, and power consumption is 1.8 kW. The condenser consists of 4 rows of 24 stages of 9.52 mm copper tubes. Aluminum fins have a 2 mm pitch. When water is sprayed on the fins, aluminum is corrosive, so the surface of the fins is coated with urethane, which has excellent corrosion resistance to prevent corrosion. The evaporator has a 9.52 mm copper tube arranged in 2 rows of 20 stages. The pin pitch is 2 mm, and the aluminum fin of the evaporator is coated with a urethane hydrophilic coating to prevent corrosion of the aluminum fin of the evaporator due to the condensation of moisture in the evaporator. The blower motor inside the portable air conditioner was selected as 220V, 1Ph, 60Hz with 0.5 kW.

Figure 1 shows a schematic and photograph of a portable air conditioner. A spray pump is installed in a reservoir of a portable air conditioner, and a distributor is installed at the top of the condenser.

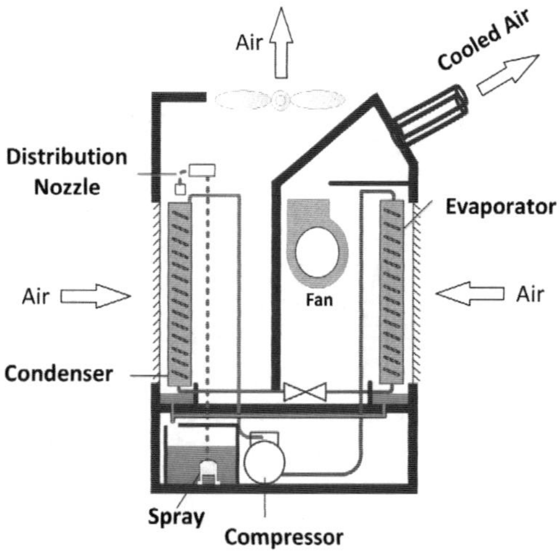

(a) Schematic diagram of portable air conditioner

(b) Photo of portable air conditioner

Figure 1. Device schematic diagram and photo.

Figure 2 is a photograph of the data acquisition and measurement equipment. The data were divided into a refrigeration cycle, and airside. The measurement data are as follows. In the refrigeration cycle, compressor suction and discharge temperature, inlet and outlet temperature of condenser and evaporator were measured using a T-type thermocouple.

The power meter was measured using Yokogawa's WT200 model, which measured the amount of power consumption. The data logger was also used Yokogawa's device, which collected all the necessary data from the experiment. Condenser side inlet/outlet air temperature and humidity are required to confirm cooling capacity and heat of condensation. For this purpose, the above data was measured from the airside. At this time, the thermocouple was fixed in the air using T-type thermocouple and humidity sensor using GHP-100T.To confirm the effect of the presence of the spray of condensate-water at a constant ambient condition, the experiment was conducted under the conditions of table 1.

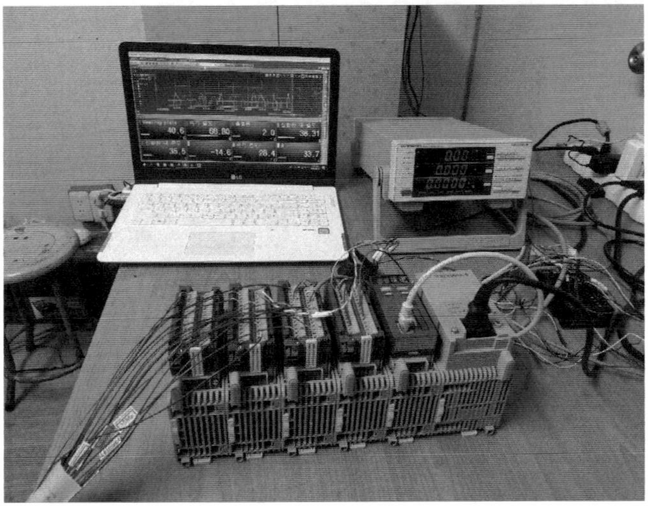

Figure 2. Data acquisition device.

Table 1. Experimental conditions

	Values	Units
Refrigerant	R-22	-
Evaporator side air flow rate	0.21	kg/s
Condensate-water temperature	27±1	°C
Atmosphere temperature	26~38	°C

The same temperature and humidity conditions were applied to confirm the improvement of portable air conditioning performance through the injection of the condensate-water upper side of the condenser. Temperature and humidity conditions were maintained using a steady temperature and humidity room. Under these conditions, power consumption, atmosphere temperature, humidity, evaporator inlet and outlet temperature, and condensation temperature were measured. To compare the cooling capacity, through temperature and humidity of evaporator side inlet and outlet, enthalpy was measured via the Psy-Chart-SAREK [8] program proposed by the Society of Air conditioning and Refrigerating Engineers of Korea. The cooling capacity was calculated through the following equation (1). The power consumption was measured by a power meter, and using the cooling capacity calculated and the power consumption measured by the measuring device are written in equation (2) to compare the device coefficients of performance.

$$Q = G * (h_1 - h_2)[kW], \tag{1}$$

where Q = Cooling capacity [kW]; G = Evaporator side air flow rate [kg/s]; h_1 = Evaporator side inlet air enthalpy [kJ/kg]; h_2 = Evaporator side outlet air enthalpy [kJ/kg].

$$COP = \frac{Q}{AW}, \tag{2}$$

where Q = Cooling capacity [kW]; AW = Power consumption [kW].

3. Experimental results and discussion

Figures 3 and 4 are experimental data to confirm the influence of atmospheric temperature. The power consumption and COP of the portable air conditioner were analyzed according to the atmospheric temperature increased. Figure 3 shows that power consumption tends to increase by 0.6 kW on average as the figure 3 atmosphere air temperature increases. It seems to be the result of decreased refrigerant circulation amount as the condensation temperature in the cycle increases with increasing atmospheric temperature.

Under the same experimental conditions, as shown in figure 4, the maximum COP decreases by 0.8 as the atmospheric temperature increases. It is thought that the cooling capacity is decreased due to the increase of power consumption and a decrease of refrigerant circulation amount caused by the rise of condensation temperature and decrease of refrigerant circulation.

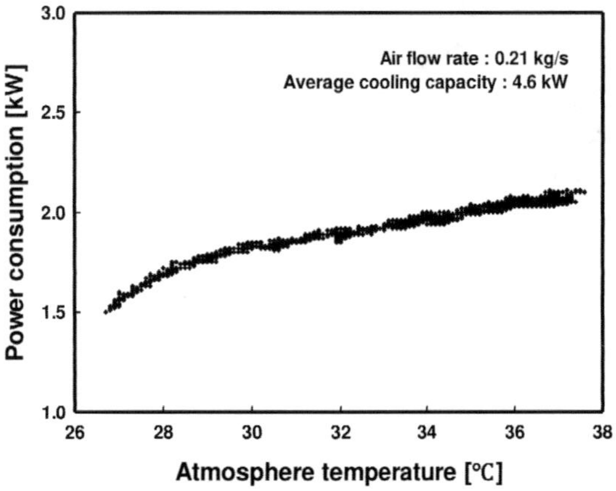

Figure 3. Power consumption according to the change of atmosphere temperature.

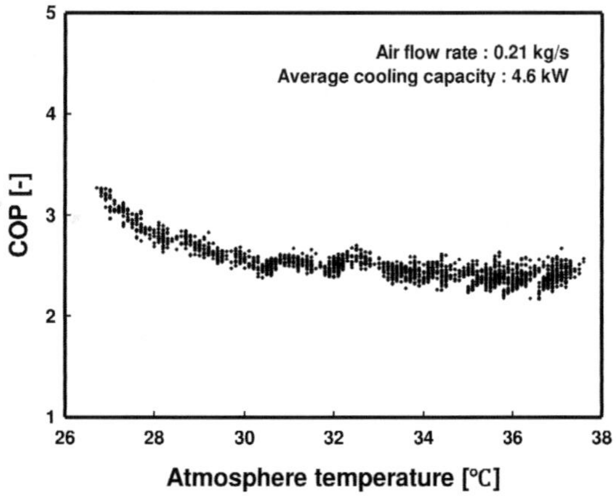

Figure 4. Coefficient of performance according to the change of atmosphere temperature.

Figure 5-8 are graphs of the performance characteristics of the experimental apparatus concerning table 1.

Figure 5 shows the temperature of condensing, discharge air according to spray pump on/off. When the condensate-water spray pump operates, condensation temperature decreased up to 6.2 ℃, and the blowout air temperature of evaporator had no change. The condensation temperature is lowered through the heat exchange with condensate-water.

Figure 6 is a graph comparing the power consumption of the device and measured by installing a power meter. The average power consumption reduced about 0.16 kW when spraying the condensate-water on the condenser. Reduction of the compression ratio leads to the reduction of power consumption of the compressor. Figure 5 shows a result of decreasing compression ratio due to the decrement of condensation temperature.

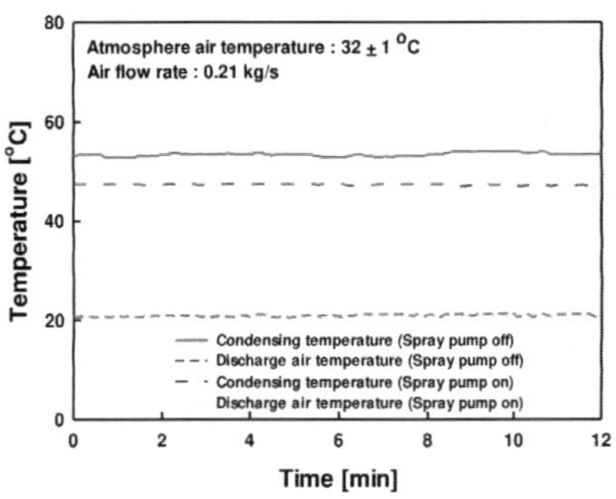

Figure 5. Condensing temperature and discharge air temperature according to change of time.

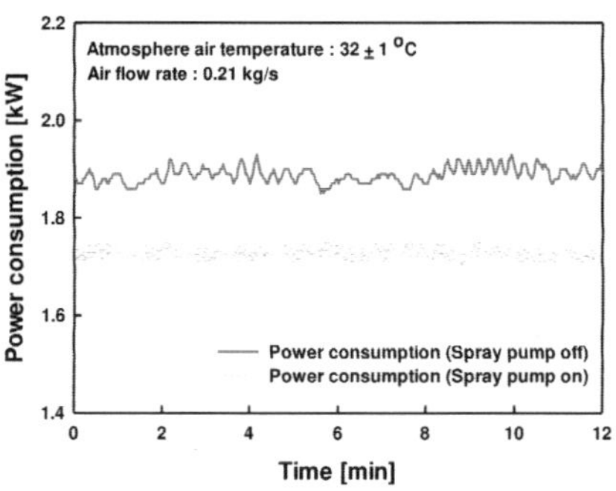

Figure 6. Power consumption according to change of time.

Figure 7 shows the change in cooling capacity according to the spray pump on/off. The cooling capacity increased by 0.2 kW compared to before the condensate-water injection. The increase in cooling capacity is thought to be because of the reduction of the condensation temperature of the cycle due to the injection of condensate-water and the increase of cooling capacity because of the increase of refrigerant circulation.

Figure 8 is a graph comparing the device coefficients of performance according to the condensate-water injection. COP was increased by 16% due to the reduction of the power consumption in figure 6 and the increase in cooling capacity in figure 7.

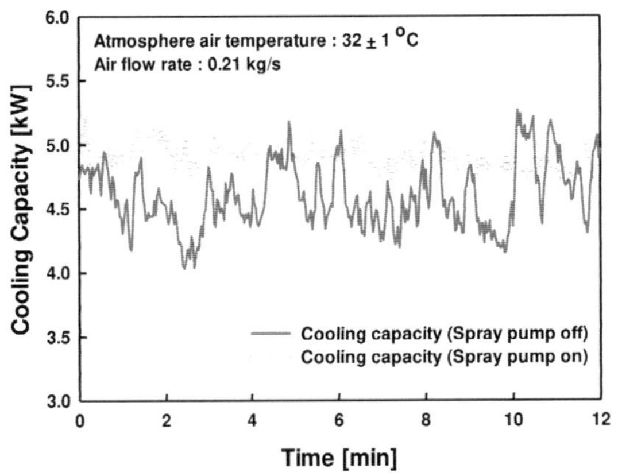

Figure 7. Cooling capacity according to change of time.

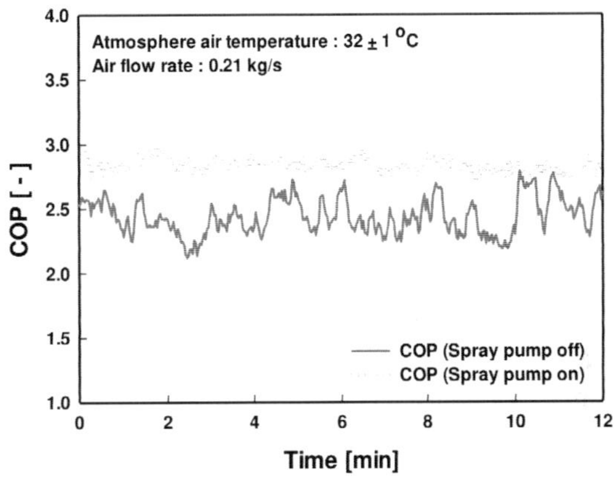

Figure 8. Coefficient of performance according to change of time.

4. Conclusions

In this study, the condensate-water generated from the evaporator of the portable air conditioner was injected to upper part of the condenser, and the following experiments were carried out to compare the condensation temperature, power consumption, cooling capacity and COP according to the presence or absence of spraying.

The condensation temperature was dropped by 6.2 °C under the experimental conditions when spraying condensate-water, and the compression ratio of the cycle was lowered, resulting in a 0.16 kW reduction in power consumption. Also, the cooling capacity was increased by 0.2 kW due to the increase of refrigerant circulation amount. The COP of the device showed a maximum of 16% higher compared to the non-operating spray pump experimental conditions. In the future, to improve the convenience of portable air conditioners research and experiment will be conducted.

References

[1] Khalil A 1993 Dehumidification of atmospheric air as a potential source of fresh water *Desalinatin* **34** pp 587-596

[2] Mahvi A H, Alipour V and Rezaei L 2013 Atmospheric moisture condensation to water recovery by home air conditioners *Am. J. Appl. Sci.* **10(8)** pp 917-923

[3] Algarni S, Saleel C A and Mujeebu M A 2018 Air conditioning condensate recovery and applications - Current developments and challenges ahead *Sustainable Cities and Society* **37** pp 263-274

[4] Akram W, et al 2018 *Recycling of condensed water from an air conditioning unit* (Rajshahi University of Engineering & Technology Press, Rajshahi)

IOP Conf. Series: Materials Science and Engineering **675** (2019) 012043 doi:10.1088/1757-899X/675/1/012043

[5] Brayant J A and Ahmend T 2008 Condensate water collection for an industrial building in Doha, Qatar: An opportunity for water sustainability *Proc. 16th Symposium on Improving Building Systems in Hot and Humid Climates* (Plano, Texas, USA)
[6] Nethaji N, Mohideen T and Nethaji M 2019 Energy conservation in room air conditioner unit by recovering cold energy from condensate *International Journal of refrigeration* **104** pp 95-102
[7] Ardita I N and Subagia I W A 2017 The application of condensate water as an additional cooling media intermittently in condenser of a split air conditioning *Proc. 2nd International Joint Conference on Science and Technology* (Bali, Indonesia)
[8] Psy-Chart-SAREK Ver. 2.0 2019 Society of Air conditioning and Refrigerating Engineers of Korea

Acknowledgments
Following are results of a study on the "Leaders in INdustry-university Cooperation +" Project, supported by the Ministry of Education and National Research Foundation of Korea

Authors' background

Name	*Prefix*	*Research Field*	*Email*
Jung-In YOON	Full Professor	Energy Systems	yoonji@pknu.ac.kr
Chang-Hyo SON	Associate Professor	Refrigeration Systems	sonch@pknu.ac.kr
Choon-Geun MOON	Lecture	Refrigeration Systems	drmchg@gmail.com
Joon-Young KONG	Master Student	Energy Systems	vvkjyvv@naver.com
Ji-Hoon YOON	Master Student	Energy Systems	neverzksch@naver.com

Composite materials for Hi-Tech printed circuit boards

A M Medvedev

Moscow Aviation Institute (National Research University) 125993, Moscow, Russia

E-mail: medvedevam@bk.ru

Abstract. The development of printed circuit board technology is subject to the general trend of electronics development - an increase in functionality and performance. This requires printed circuit boards to increase the assembly density of electronic components and interconnections and reduction of constructive delays in transmission lines of information. In turn, this requires the use of base materials with good dimensional stability and reduced values of dielectric permittivity and dielectric losses. Operability of gigabit electronics largely depends on the state of the surface of the printed conductors, since at high frequencies it is affected by the skin effect. Producers of basic materials, trying to meet these requirements, remain in the field of composite materials, so as not to change the structure of the existing production, based on the use of basic technologies of printed circuit boards. This article is based on the presentation of the properties of new materials from the catalogs of the companies Panasonic, Hitachi (Japan) and Isola (Germany).

Key words: Avionics, Printed Circuits Boards, Composite Materials, Dielectric, Scin-Effect, PTFE, LTCC-ceramics.

1 Introduction

Printed circuit boards (PCB) are the physical basis of all electronic devices. In the production of PCB are used more than 200 operations and numerous basic and auxiliary materials [1].

Usually in the formation of the prime cost of printed circuit boards (PCB), the cost of basic materials is in the first lines. New materials with improved properties are obviously more expensive than usual. From the standpoint of pricing policy, is the use of expensive materials justified? It turned out that - yes, it is justified: if ordinary multilayer printed circuit boards have the structure of prime cost 46%, then complex multi-layer PCBs - about 15%. Reducing the share of the cost of materials in the price of multi-layer PCBs is achieved not only by redistributing the share of the costs of manufacturing operations of complex multi-layer PCBs, but also (principally) by significant yield enhancement mainly due to the high quality of materials.

Traditionally it is believed that materials Hi-Tech PCB for devices gigabit electronics and microwave electronics is mainly based on polytetrafluoroethylene (PTFE). Several displaced accents aside Low Temperature Co-Fired Ceramic (LTCC-ceramics). And do not take into account the possibility of the use of modern epoxy-glass compositions (EGC) specially designed for microwave use. However, if you approach the choice of materials, using multivariate criteria, you can in the microwave range to find a place for modern foil EGC, widely used in the manufacture of printed circuits [1, 2].

The basic technology of PCB focused on use of the EGC. In this regard, two questions often asked, “Will a circuit board made of EGC laminate work for my High Speed or Hi-Tech design?” and, “If EGC isn’t OK, where do I turn?” The answers to these questions are specific to each individual circuit and can only be answered through analysis of the characteristics of FR4 verses other material choices, relative to the demands of the circuit’s noise budget.

This article will outline the material parameters that affect circuit performance and give some direction about how to decide when to choose the correct material for any specific application. The objective of this article to justify the use of epoxy glass compositions for Hi-Tech PCB and show features of their application.

2 Materials and Methods

One of the major goals of this article is to provide information that allows designers and engineers to select a material for every PCB application that optimizes both performance and cost of Hi-Tech PCB. As the fundamental building block for printed circuits, base materials must meet the needs of the printed circuit board (PCB) manufacturer, the circuit assembler, and the original equipment manufacturer (OEM). A balance of properties must be achieved that satisfies each member of the supply chain. In some cases, the desires of one member of the supply chain conflict with another. For example, the need for improved electrical performance by the OEM, or improved thermal performance by the assembler, may necessitate the use of resin systems that require longer multilayer press cycles or less productive drilling processes, or both [3].

Other trends are also driving the need for greater performance. These include:

- Circuit densification
- Higher circuit operating frequencies

Today for the manufacture of Hi-Tech PCB provided a limited number of materials: polytetrafluoroethylene (PTFE, Teflon-RTFE, Rogers is a ceramic filling RTFE), LTCC ceramics and new foiled glass-epoxide composites. Of course, from the viewpoint of dielectric properties the best MICROWAVE-material is PTFE: its dielectric permeability, loss tangent and the lowest water absorption from all variety of solid dielectrics. However, technological difficulties processing make use it only in very exceptional cases, where its unique microwave properties take precedence over constraints its use in production [4].

The same applies to ceramics: the complexity of the formation of these products are cause a relatively limited range of use. This primarily refers to her unpredictably large shrinkage (up to 15% along layers and up to 30% in transversal-direction) that for the larger sizes of mounting substrates is unacceptable. Moreover, the production of circuit boards of LTCC ceramics — very specific production, is fundamentally different from the widespread production of printed circuits on the underlying technology that uses foiled dielectrics EGC. The foiled glass-epoxide dielectrics are best materials for a wide and mass production of electronics. But the possibility of their use in the microwave range cause undeserved doubt that today cannot be categorical. If you approach the choice of materials of substrates from the standpoint of the formation of lines of communication is the main function of the printed circuits for fast digital electronics — that can distinguish two main parameters determining their performance — attenuation (loss) signals in communication lines and signal propagation speed [5].

It is known that the loss of lines proportional to the permittivity and dielectric loss angle tangent [6]. Power loss is measured in lines evaluated as

$P = U^2 \omega C \text{ tg } \delta$,

where U is voltage on the line

C— the line capacitance,

tg δ is the tangent of angle losses.

Power losses in 1 cm^3 of the dielectric in a homogeneous field (e) equal:

$p = E^2 \omega \, \varepsilon_r \text{ tg } \delta$

where ε_r is the relative permittivity.

$L = \text{tg } \delta \, \varepsilon_r$ called dielectric loss factor. This parameter determines the degree of loss of signal in the communication lines.

Second discuss the speeds at which signals propagate on a PCB interconnect. As you know, in a vacuum the speed of propagation of the signal (speed of light) is 299 000 km/s, which corresponds to the delay in the signal line 3.3 ns/m. The speeds of signals on a PCB is less than that in air or vacuum by the square root of ε_r. If $\varepsilon_r = 4$ (like for EGC material types), then the speed of signals on a transmission line is 6-7 ns/m.

Now you can compare a variety of basic materials for these and another important criteria (table 1) [1,7].

Table 1. Base materials for printed circuits.

Type of Materials	Dielectric Constant, ε_r	Loss Tangent, tg δ	Signal Speed, nc/m	Coefficient of Loss, $\varepsilon_r \cdot$tg δ	Manufacturability (ball)	Thermal conductivity, W/m K
Vacuum	1.0	-	3.3	-	-	
PTFE	1.8 — 2,2	0.001	4.7	0.002	6	0.3-0.6
LTCC-ceramic	10 — 12 (20**)	0.006	10.4 — 11.4 (15)	0.07 (0.12)	4	2-4
Epoxy-glass	3.3 — 3.5***	0.0015	5.8- 6,0	0.005	1	0.3-0.4

As a result, we will consider in detail the main properties of Epoxy-glass-composition with respect to Hi-Tech PCB.

3 Results

Epoxy-glass-composition are characterized by a large number of parameters, we consider only those that related to the topic of the article.

Binders - Polymer System

Binders (resin) are the weakest link in composite materials. As a rule, they determine such important parameters as the heat resistance of the multi-layer PCBs and the dimensional stability of the layers of the multi-layer PCBs, which is necessary for the exact combination of elements of interconnections in the three-dimensional structure of the multi-layer PCBs.

Heat resistance is needed to ensure the stability of boards to the group heating of boards for soldering of surface-mounted components. When heating reaches the soldering temperatures, the metallization of the holes experiences large stresses, which are created due to the expansion of the base of printed circuit boards along the Z-axis. With insufficient plasticity of copper galvanic deposition, the metallization of the holes under the action of an expanding dielectric deforms to the point of destruction [8]. Heat resistance in this case is characterized by the glass-transition temperature, above which the deformation of the base of the board is particularly high (Fig. 1). Domestic materials, developed 50 years ago in relation to manual installation, are not suitable for group soldering today, since their glass transition temperature (Tg) does not exceed 105 °C. Today's materials have a Tg of not more than 140 °C. New materials are very diverse and can have Tg up to 220 ° C (see Figure 1 and Table 2) [9].

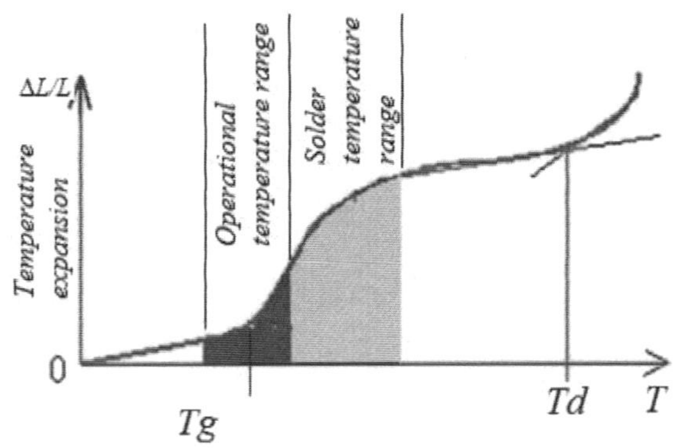

Figure 1. Dependence of the temperature expansion of the composite base of the printed circuit board along the Z axis: T_g is the glass-transition temperature, T_d is the temperature at which the polymer begins to decompose.

Table 2. Basic properties of binders of various firms - suppliers of basic materials of the printed circuit board [10]

Parameter	Method	Unit of measurement	Parameter estimator	HITACHI MCL-BE-67G	Panasonic MC-100MS/EX	Isola DE-104
Glass-transition temperature, T_g, °C	Thermo-gravimetric analysis	°C	Better more	140	137	135
	Thermo-mechanical analysis		Better more	220	147	Not norm
Thermal-expansion coefficient	along the *X*-axis, 30-120°C	ppm	Better less	15	15	16
	along the *Y*-axis, 30-120°C	ppm	Better less	17	15	16
	along the *Z*-axis below T_g	ppm	Better less	40	65	70
	along the Z axis above Tg	ppm	Very critical	170	210	250
Moisture absorption	E-24/50+D-24/23	%	Better less Very critical	0.02	0.1	0.8

The moisture absorption is especially important parameter for microwave devices: the relative dielectric constant of water $\varepsilon_r = 81$ can affect the performance of communication lines of multi-layer PCBs. In addition, if the printed circuit boards are not dried before soldering (it is possible), intensively evaporating moisture at soldering temperatures inflates the composite dielectric base, which can lead to irreversible delamination of the boards.

Glass - glass fiber material

Electrically insulating alkali-free glass based on aluminoboronsilicate has traditionally been used as glass filaments. It was well drawn in the thread and maintained the processes of formation of yarn and fabric without destruction. However, recently it has ceased to meet the requirements of the formation of high-frequency electronic devices: its relative dielectric permittivity is too high $\varepsilon_r = 9$. In the composition with a binder ($\varepsilon_r = 3.2$), its total dielectric permittivity ranges from 5 to 6, depending on the resin content [7]. To reduce the dielectric permittivity (ε_r) and the dielectric loss tangent (tgδ) it is proposed to use a different glass composition - E-glass (boron-free aluminosilicate with light additions of oxides of alkaline-earth metals) with a dielectric permittivity $\varepsilon_r = 6$ and tgδ ≤ 0.004. In combination with the binder properties of such composite at gigahertz frequencies: $\varepsilon_r = 3.5$–4.0, tgδ ≤ 0.002 [12].

Weave of glass fiber

Usually glass threads are twisted into yarn, from which the fabric is woven. The thickness of the yarn density diversifies the range of glass fabrics (Table 3).

Table 3. Parameters of impregnated fiberglass (prepreg) [4]

Designation of glass fiber	Resin content, %	Typical thickness, μm	Frequency, GHz			
			1	20	40	50
1027	75	49	3.2	3.2	3.2	3.2

1035	70	60	3.3	3.3	3.2	3.2
1080	64	76	3.4	3.4	3.4	3.4
1078	75	118	3.4	3.4	3.4	3.4
3313	54	98	3.6	3.6	3.6	3.6
2116	56	132	3.6	3.6	3.6	3.6

But it must be borne in mind that in the process of pressing the material during the transition of the resin of binder from stage B to stage C [5], the resin undergoes polymerization shrinkage and during cooling – thermal shrinkage. The stresses arising at the same time are fixed by the pressed foil and weaving of glass fiber. Then they realize themselves by shrinking (changing linear dimensions) after etching of the pattern (removing part of the foil). Twisted yarn and weaving of glass fiber as springs take on some of the stresses and entail a certain amount of shrinkage of the material of the layers of multi-layer PCBs.

To prevent this phenomenon, the glass threads do not twist into yarn and get a flat weave, as shown in figure 2

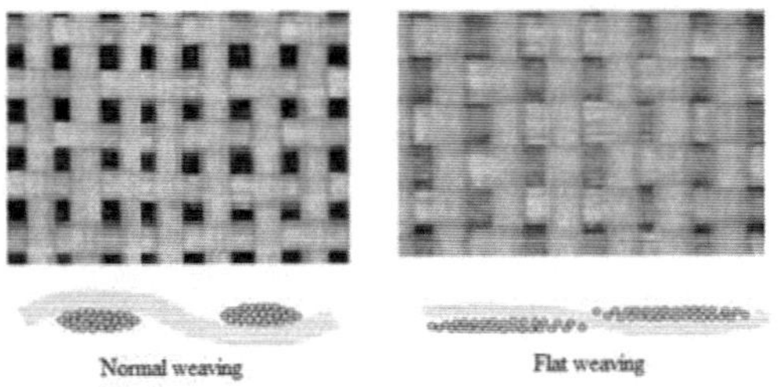

Figure 2. Types of glass fiber weaving.

There is another advantage of flat weaving: dielectric uniformity for communication lines.

Migration

For the long-term operability of electronic devices, it is very important to ensure the complete absence of electrochemical migration processes on the surface and in the volume of the composite material of the processes of electrochemical migration, alignment of current paths between different potential circuits (Figure 3). For their formation, a combination of two factors is necessary: moisture and voltage. Moisture-protective lacquers are more or less successfully protected from surface moisture. But since moisture has a unique property of penetrating everywhere, it finds the ability to condense in the capillaries of composite materials. In this sense, a weak point in composite dielectrics is the surface of glass fiber in adhesion with a binder. When this adhesion is weak, thin capillaries are formed, in which moisture condenses even in conditions close to normal. For mobile devices, the source of moisture can be the evaporation of the human body, condensation when bringing cold devices into a warm room, etc. Automotive electronics, avionics are devices constantly under stress from moisture. So one of the conditions for equipment failure is moisture, which naturally accompanies the maintenance of electronic devices. The only deliverance from such degradation processes of isolation is the use of materials in which there are no micro pores and micro capillaries.
A phenomenon that characterizes the stability of composite dielectrics to electro migration processes is called Conductive Anodic Filament (CAF) in international standards [1, 9] (Fig. 3).

Composite materials in which conditions are created for good impregnation of glass yarn and good adhesion of the binder to glass show good results for long-term resistance to the effects of a moist environment.

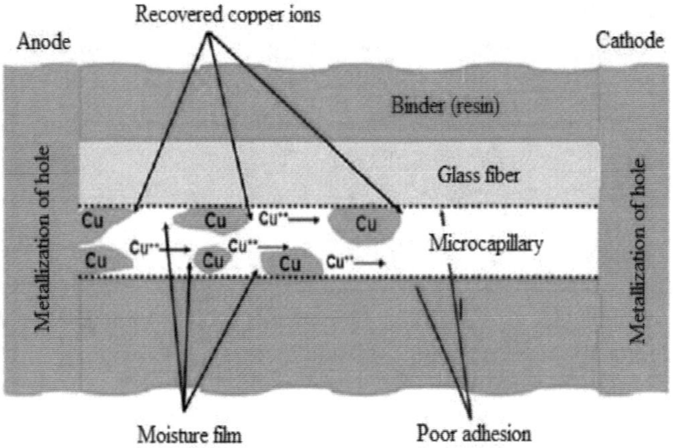

Figure 3. The mechanism of formation of conductive paths in the volume of printed circuit boards - conductive anode filaments.

Foil

Since we are moving farther and farther into the region of high frequencies (high performance of electronic devices), we have to reckon with such a phenomenon as the skin effect - displacement of the conduction of high-frequency signals to the surface region [8]. In table 4, this effect is shown in numbers. It can be seen that the skin effect is already significant when working at gigahertz frequencies typical of today.

Table 4. An approximate representation of the skin effect [13]

Frequency	Thickness of conductive layer, μm
10 kHz	660
100 kHz	210
1 MHz	65
10 MHz	21
100 MHz	6.6
1 GHz	2.1
10 GHz	0.7

It can be seen that at high frequencies the size of the skin effect is comparable with the irregularities (roughness) of the foil (Figure 4). This affects the worse for signal propagation. Obviously, to create a foil without any roughness is very difficult. Moreover, the roughness creates the best conditions for ensuring of adhesion of the foil to the dielectric. Nevertheless, with reference to materials for microwave devices, the roughness of the foil is normalized. According to IPC 4562 [14, 15], the maximum values of the height of the copper foil profile are listed in table 5.

Table 5. Foil roughness standards for microwave device materials according to IPC 4562 [14]

Profile type	Maximum profile height, μm
Standard profile (S)	not indicated
Low profile (L)	10.2
Very low profile (V)	5.1

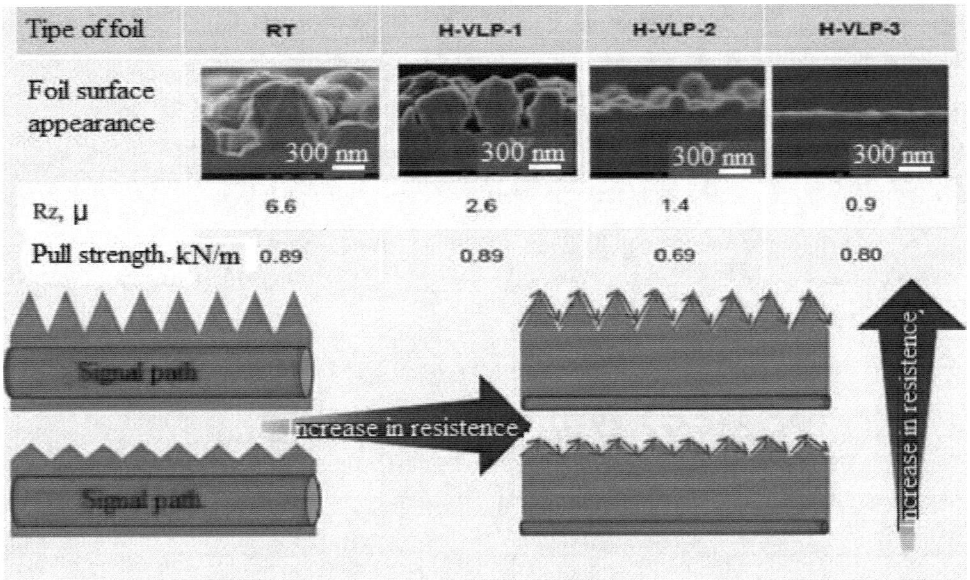

Figure 4. Influence of the skin effect on signal propagation.

4 Discussion

The results show that under certain conditions, the use of epoxy-glass composition acceptable for printed circuit boards, which are made on the underlying technology. However, the designers has to contend with elevated losses and delay signals in these materials. As a rule, they are offset by a decrease in size of boards by increasing density layout of electronic components and interconnections between them on the PCB. This is consistent with the General tendencies of development of technology of modern electronics.

Conclusion

Electronic devices increasingly require the use of Hi-Tech PCB, in particular for microwave range that requires the use of appropriate materials.

Recent developments and proposals for the composite materials market with improved dielectric properties in the direction of microwave allow PCB manufacturers to remain within the framework of basic technologies, i.e. without significant restructuring of production.

The use of high quality, but more expensive, materials reduces production costs, which makes the final product cheaper.

References

[1] Coombs C F 2008, *Handbook Printed Circuits* McGraw-Hill. URL: https://mirror.thelifeofkenneth.com/lib/electronics_archive/Printed_Circuits_Handbook_6Th_Ed_Malestrom.pdf

[2] IPC-PCB Technology Trend 2018. URL: http://www.ipc.org/ContentPage.aspx?pageid=PCB-Technology-Trends-2018

[3] Isola introduces ultra-low loss materials for 100 gigabit ethernet applications. URL: https://www.microwavejournal.com/articles/22539-isola-introduces-ultra-low-loss-materials-for-100-gigabit-ethernet-applications

[4] Medvedev A 2017 *Electronics – Science. Technology. Business* **5(165)** pp 184-7. URL: www.electronics.ru

[5] Tsunooka T, et al 2003 *J. Eur. Ceram. Soc.*, **23(14)** 2573

IOP Conf. Series: Materials Science and Engineering **675** (2019) 012044 doi:10.1088/1757-899X/675/1/012044

[6] IPC-2141 Controlled Impedance Circuit Boards and High Speed Logic Design. URL: http://www.ipc.org/toc/ipc-2141a.pdf

[7] Low Loss Dielectric Stock Materials 2016. URL: https://www.cumingmicrowave.com/products/low-loss-dielectric-stock-materials.html

[8] Medvedev A 2016 *Instruments and Experimental Techniques* **59** pp 879-881

[9] IPC-TM-650, Method 2.6.25A. URL: https://standards.globalspec.com/std/732824/tm-650-2-6-25

[10] Mittal A 2019 Differential Pairs in PCB Transmission Lines: Common Mode Signals *Sierra Circuits Inc* URL: https://www.protoexpress.com/blog/differential-pairs-in-pcb-transmission-lines-common-mode-signals/

[11] Coonrod J and Horn A F 2011 Understanding Dielectric Constant for Microwave PCB Materials. URL: https://www.scribd.com/document/135869348/Understanding-Dielectric-Constant-for-Microwave-PCB-Materials

[12] Atams A A, Atamas N A and Bulavin L A 2005 *J. Mol. Liq.* **120** pp 15– 17

[13] Reynolds J W, LaFrance P A, Rautio J C and Horn A F 2010 Effect of conductor profile on the insertion loss, propagation constant, and dispersion in thin high frequency transmission lines *DesignCon.* **2010**

[14] IPC 4562 Metal Foil for Printed Wiring Applications. URL: https://standards.globalspec.com/std/896578/ipc-4562.

[15] Douglas Brook Skin Effect 2010 (UltraCAD Design Inc., USA)

IOP Conf. Series: Materials Science and Engineering **675** (2019) 012045 doi:10.1088/1757-899X/675/1/012045

Effective sound-absorbing material for transport technological machines of railway transport

I Popova[1], L Leonteva[1], I Danilov[2]

[1]Samara State University of Railway Transport, Saratov, Russian Federation
[2]Peoples' Friendship University of Russia, Moscow, Russian Federation

Email: impopova@mail.ru

Abstract. The article discusses the types and features of the operation of transport technological machines of the railways of the Russian Federation on the example of the most frequently used track machines, and presents methods for noise reducing on the railways of the Russian Federation. The most effective method to reduce the noise arising during the operation of track machines is identified - this is the installation of a steel soundproofing hood (casing) over the electric motor. The calculation of the effectiveness of sound-absorbing materials used for cladding steel bonnet is given. According to the results of the calculation, the material «asbestos felt» turned out to be the most effective, but the other materials under consideration, and in particular, «glass wool», «mineral wool» and «construction felt», turned out to be also applicable to the lining of the steel bonnet, since the calculations revealed the applicable level of sound absorption, which is less than a decibel in terms of noise reduction to the material «asbestos felt».

1. Introduction

In the modern mobile world, the importance of railways is increasing. Only in the first quarter of 2019, passenger traffic amounted to 23.4 million people in long-distance traffic, and 235.8 million people in suburban traffic, in turn, passenger traffic was 24.8 billion passenger-kilometers. As for the freight, the importance of railways is also difficult to overestimate, since the beginning of 2019, the railways transported 318.7 million tons, and the freight turnover amounted to 651.8 billion tons of tariff tons-kilometers. Moreover, in comparison with 2018, a positive increase in both passenger and freight traffic is seen. In the first quarter of 2019 alone, compared with 2018, passenger turnover increased by 3.5%, while freight turnover increased by 2.5% [1].

For the efficiency of the passenger and the freight traffic, the Strategy for the Development of Railway Transport was adopted, which was approved by the decree of the Government of the Russian Federation, according to which by 2030 it is planned to upgrade 23 thousand locomotives, launch 1 million freight cars and more than 23 thousand passenger cars, an increase in motor-car rolling stock by 25 thousand units, it is planned to lay 16 thousand kilometers of new railways, 659 kilometers of which are in the northwestern and central federal districts, 579 kilometers are in the northwestern and Volga federal districts, 210 kilometers are in the northwestern and the Urals Federal District, it is also planned to lay new railways in the southern, north-Caucasian, Siberian and Far Eastern federal districts [2].

The implementation of this strategy, as well as the maintenance of railways, is impossible without transport technological machines. On the railroad, they are called track machines, which function is to prepare a place for laying a new railroad bed, to carry out straightening, ballasting the canvas, to ensure the replacement of sleepers and to clean the canvas from rubble, snow, industrial and domestic waste. Traveling machines are also needed for welding and grinding of rails, star collecting bases, diagnostics of the state of the track, alignment of the track and compaction of the ballast layer [3-6].

Serious importance in the issues of noise insulation is given in the works of many foreign scientists. For example: Zhang X., Jonasson Hans, Busch A., Nugent R., Peng Sun, ZhanchengGuo.

Zhang X and Jonasson Hans G considered the direction of railway noise sources in their works. Busch A and Nugent R E studied the reduced coefficients of insertion loss and absorption of the railway noise barrier, and in particular compared the field measurements and forecasts. Peng Sun and ZhanchengGuo studied the preparation of steel slag of porous sound-absorbing material using coal powder as a blowing material [7-9]. The topic of the absorption studied and Russian scientists, Sychov A., Sychov M., and Rusanov E. presented to the scientific community a method of producing Youmanage protective foam concrete for use in railway transport and Pultznerová A. structured modification of items and equipment for noise control of Railways [10-11].

The purpose of the article is to identify the effective sound-absorbing material that is use to process the inside of the steel bonnet to reduce noise level from the transport technological machines of the Railways of the Russian Federation.

2. Materials and methods

The object of the article is to consider the types and features of the operation of transport technological machines of railways of the Russian Federation, as well as to identify the most effective method for reducing noise arising during the operation of track machines.

The most commonly used railway track machines are the RTM-02 tamping machine, the RTMA4-K tamping machine, the MPD-2 motor platform, the KDE-163 and KDE-253 railway cranes. Tamping machine RTM-02 is use for compaction and supply of ballast under the sleepers and is a self-propelled vehicle on the track. The main components of RTM-02 are diesel engine with transmission, mounted on the chassis frame, compressor, tamping heads with tamper collars, a mechanism for pressing rail heads, wheelset drive mechanisms, tamper wheels and a machine lifting mechanism allowing to remove it from the railway track. This machine uses a worm mechanism with screw movement of links, the advantage of which is the accuracy of the rolled products, since the number of links in the kinematic chain is reduced in this mechanism, as a result, the amount of lateral clearances and the mismatch between the positions of the input and output links are reduced. The ballast is compacted by an eccentric vibration mechanism, a method of horizontal vibration compression, which reduces vibration.

For high-quality ballast compaction, it is necessary to eliminate the vibration of the rails; for this, the machine is equipped with two pressure mechanisms with a pneumatic drive. The configuration of RTM-02 is 16 sleeper, the transport speed of this machine is on average 25.3 km / h, and the speed of movement from the sleeper to the sleeper can vary from 2.2 to 13.5 km / h. The performance of the sleepers per hour is 350-400 units. RTM-02 is used both at the stations and on the hauls with an average lifting and overhauling of the railway track. The tamping machine RTM-02 works in a semi-automatic mode from a remote control panel, the machine is also equipped with a hand brake and a hydraulic brake with a foot and remote control [12-15].

In addition to the RTM-02 tamping machine, a more modified tacking machine RTMA-4K is often used. Its main difference is the presence of automatic control, which allows you to completely eliminate the harmful effects of noise and vibration on the operator. Productivity and technical characteristics practically coincide with RTM-02, but RTMA-4K is lighter, this machine weighs 8 tons, whereas the weight of RTM-02 is 14 tons. Due to the lower weight, RT MA-4K (figure 1) can be used both on wooden and on reinforced concrete sleepers.

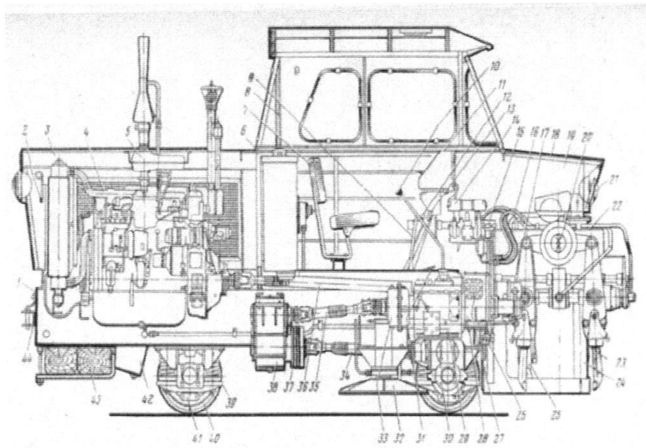

Figure 1. Tamper RTMA-4K (1 - frame; 2 - curtain; 3 - hood; 4 - the engine; 5 - fuel tank; 6 - hydraulic system tank; 7-seat; 8 - cab; 9 - tamping lever; 10 - gear lever; 11 - control panel; 12 - lock the tamping unit; 13 - parking brake lever; 14 - hydraulic control tamping units; 15 pressure switch; 16 - hydraulic pump GP-46; 17 - hydraulic pump GP-10; 18-cap; 19 portal; 20 - unbalance shaft; 21 - headlights; 22-tamping unit; 23 - knockout; 24-guide tamping unit; 25 - hydrodiline of deepening of the tamping block; 26 - distribution valve; 27 - driveshaft; 28 - drive box; 29-box gear changes; 30 - driving wheel pair; 31 - brake and clutch pedals; 32 - lifting and turning device; 33-coupling; parking brake; 34, 35, 36 - driveshafts; 38 - transfer case; 39 - suspension pair of wheels; 40 - not driving wheel pair; 41 - brake; 42 - hydraulic tank; 43 half sleeper; 44 - coupling device).

Another necessary vehicle on the railway is the self-propelled motor platform MPD-2, used in the construction, repair and maintenance of the railway track of 1.435 mm gauge (European gauge) and 1.520 mm gauge (Russian gauge). The machine is actively use when carrying out shunting works, laying new and dismantling the old railroad tracks, in addition, the MPD-2 is equipped with a fire detection and extinguishing system for power plants. Movement control MPD-2 is carried out from the driver's cab, which, using a hydraulic drive, can be transferred from the transport position to the working position and back. The driver's cab is designed to meet the basic ergonomic requirements and is line with sound-absorbing materials, and in particular ABS plastic and noise insulation [16-18].

ABS plastic is a polymer material, one of the most accessible and common, also called a thermoplastic resin. This material has a high impact resistance and elasticity. One of the most important characteristics for cabin facing is heat resistance, which is 103 ^{0}C. The advantages of using ABS plastic are its non-toxicity and a wide range of operating temperatures from -40 to +90 ^{0}C.

Shumoizol is a soundproof material that combines the qualities of thermal insulation and waterproofing. The shumoizol consists of two layers - the base and the bitumen layer and is a non-woven fabric produced in rolls. The operating temperature range of the noise insulation is from -25 to +85 ° C [19].

Continuing consideration of the MPD-2 motor platform (figure 2), it should be noted that its main purpose is to tug the rail and sleeper links in the process of disassembling or laying the track, as well as performing shunting operations within the star collection bases. Inside the frame, under the platform floor, the MPD-2 has two power units consisting of a diesel engine, a generator, a water and oil cooler, which are mounted on an intermediate frame and attached to the frame of the platform with shock absorbers.

IOP Conf. Series: Materials Science and Engineering **675** (2019) 012045 doi:10.1088/1757-899X/675/1/012045

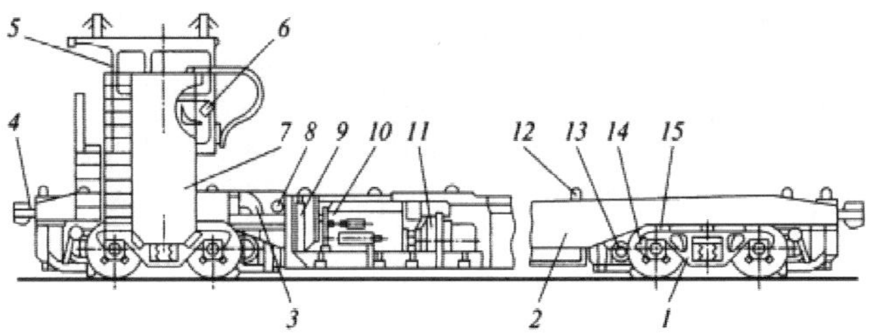

Figure 2. Motor platform MPD-2 (1 - biaxial traction cart; 2 - frame; 3 - winch; 4 - automatic coupler with absorbing device; 5 - control cabin; 6 - control panel; 7 - Portal stand; 8 - compressor; 9 - water and oil radiators; 10 - diesel; 11 - generator; 12 - roller conveyor; 13 - traction motor; 14 - axial gearbox; 15 - wheel pair).

The MPD-2 platform is distinguished by a large carrying capacity of 60 tons, a large traction force from 90 kN to 125 kN and a large traction force of a winch of 58.9 kN; while the train consists of 100 km / h, the weight of the MPD-2 platform is 47 tons.

Railway cranes KDE-163 and KDE-253 (figure 3) are also most often use among railroad track machines. Their purpose is to mechanize loading, unloading, installation and construction works. The KDE-163 m KDE-253 consist of a self-propelled four-axle platform, a swing frame, a diesel generator, working mechanisms, a control cabin, a boom, boom extensions (in the case of KDE-253) and load handling devices.

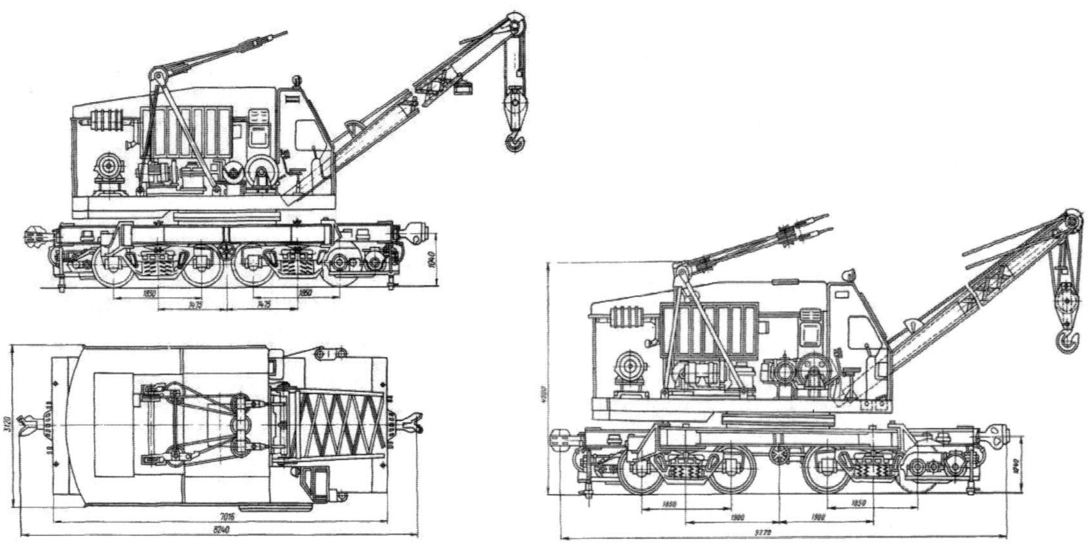

Figure 3. Railway cranes KDE-163 (a), KDE-253 (b).

The KDE-163 and KDE-253 cranes are distinguished, by the carrying capacity, the KDE-163 has a loading capacity of 16 tons and the KDE-253 has a loading capacity of 25 tons, and the crane KDE-253 is equipped with an additional extension of the boom and boom departure. This crane is 20 meters, while the KDE-163 is equipped with a boom of 15 meters, there are small differences in dimensions, the other mechanisms are almost the same.

On the example of the most frequently used railway cars, one should consider the problems of noise and methods for reducing the noise arising from the operation of railway cars.

In octave bands with geometric mean frequencies of 31.5; 63; 125; 250; 500; 1000; 2000; 4000; 8000 Hz (table 1) sound level during operation is:

IOP Conf. Series: Materials Science and Engineering **675** (2019) 012045 doi:10.1088/1757-899X/675/1/012045

- RTM-02 - 108 dBA;
- RTMA-4K - 108 dBA;
- MPD-2 - 93 dBA;
- KDE-163 - 110 dBA;
- KDE-253 - 80 dBA

Table 1. Acoustic characteristics of traveling machine noise sources.

Traveling Cars	Sound pressure levels, dB, in octave bands with geometric average frequencies, Hz									Soundlevel, dBA
	31.5	63	125	250	500	1000	2000	4000	8000	
RTM-02	87	92	96	99	103	105	101	96	88	108
RTMA-4K	96	97	99	101	103	104	102	94	87	108
MPD-2	-	92	90	86	87	90	86	75	65	93
KDE-163	-	92	101	100	102	106	108	98	46	110
KDE-253	79	89	84	80	79	77	71	61	92	80

The average noise level in the city during the day is 50-60 dB, at night this value is reduced to 10 dB, it should be noted that serious hearing problems can start at a sound pressure level exceeding 150 dB [20]. Thus, the noise arising from the operation of traveling machines is not critical, but still significantly exceeds the average level familiar to human hearing.

There are several methods for reducing noise on a railway:

- installation between the subgrade and gravel ballast of polymer pads, reducing noise up to 15 dBA;
- planting forest line along a railroad reducing noise up to 20 dBA;
- acoustic grinding of rails, in which the effect of reducing noise at the source reaches 7-9 dBA;
- vibration damping pads on the rail neck, allowing to reduce noise at the source to 4-5 dBA;
- application of vibration-absorbing mastic to the neck of the rail, carriage and wheels, in which the effect of reducing the noise at the source is up to 7-8 dBA;
- installation of acoustic screens, reducing noise to 12-15 dBA;
- soundproof glazing reducing noise up to 20-30 dBA;
- excavations and embankments that reduce noise up to 8-16 dBA [21-25].

The main source of noise in railway transport is the railway transport itself, and the largest sources of noise are, previously discussed, traveling machines with sufficiently high generator power, therefore, in addition to the noise reduction methods listed above, it is proposed to install a soundproof hood (casing) over the electric motor (figure 4).

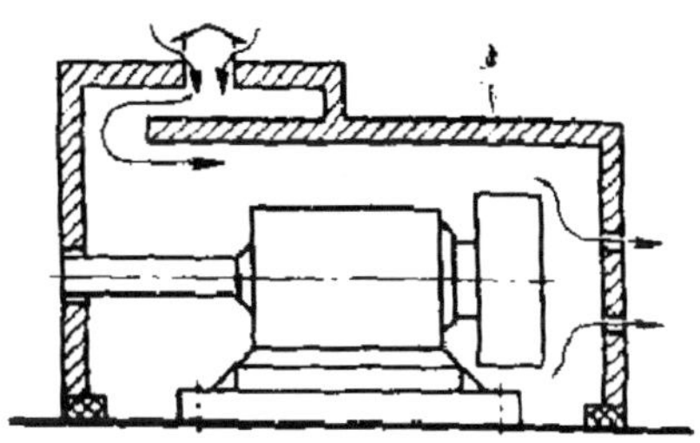

Figure 4. The Sound insulation hood (casing) over the motor.

The basis for making the body of the hood is steel with a thickness of 3-5 mm and a density of 7.8·103 kg/m^3. When cladding a steel bonnet, various sound-absorbing materials can be used, in particular, glass blocks 30 mm thick, asbestos felt 10 mm thick, mineral wool 25 mm thick, and construction felt 25 mm thick.

At each geometric mean frequency, the sound insulation capacity of the hood walls, dB, is determined by the formula:

$$P = 20\lg(G \cdot f) - 47.5 \quad (1)$$

where G is the weight of the hood (kg);
f - geometric mean frequency (Hz);
47.5 - constant value.

The total sound insulation capacity of the hood walls, dB, at each geometric mean frequency, is determined by the formula:

$$P_k = P + 10\lg \cdot \frac{A_2}{A_1} \quad (2)$$

where A_1 and A_2 are equivalent absorption areas before and after bonnet lining.

In turn, the equivalent sound absorption area before and after bonnet lining is calculated by the formula:

$$A = \alpha \cdot S \quad (3)$$

where α is the sound absorption coefficient of the material. α is taken from SN 2.2.4 / 2.1.8.562-96, S-area, covered with sound-absorbing materials of the inner walls of the hood, m^2.

3. Results and discussion

When the calculations were carried out, it was found that when installing a steel bonnet on RTMA-4K with a «glass fiber» sound-absorbing material, the maximum noise reduction is possible by 49 dB.The noise reduction level is 49 dB (with a geometric mean frequency of 125 Hz), 42 dB (with a geometric mean frequency of 250 Hz), 33 dB (with a geometric mean frequency of 500 Hz), 25 dB (with a geometric mean frequency of 1000 Hz).

Steel hood with sound-absorbing material «asbestos felt» reduces noise by: 49 dB (at geometric average frequency of 125 Hz), 42 dB (at geometric average frequency of 250 Hz), 34 dB (with geometric average frequency of 500 Hz), 30 dB (with geometric average frequency of 1000 Hz), 22 dB (with a mean geometric frequency of 2000 Hz).

Steel hood with sound-absorbing material «mineral wool» reduces noise by: 44 dB (at geometric average frequency of 125 Hz), 38 dB (at geometric average frequency of 250 Hz), 31 dB (at geometric average frequency of 500 Hz), 25 dB (at geometric average frequency of 1000 Hz).

Steel bonnet with sound-absorbing material «construction felt» reduces noise by: 45 dB (at geometric average frequency of 125 Hz), 40 dB (at geometric average frequency of 250 Hz), 32 dB (with geometric average frequency of 500 Hz), 17 dB (at geometric average frequency of 1000 Hz), 5 dB (with a mean geometric frequency of 2000 Hz).

Also, calculations of noise reduction were carried out when installing a steel bonnet over the electric motor of the tamping and punching machine RTM-02, the MPD-2 motor platform, the railway crane KDE-163 and KDE-253. The results of the calculations turned out to be identical with the above presented calculation of noise reduction of the tamping and punching machine RTMA-4K.

4. Conclusion

According to the results of the calculations, it was found that the steel hood with sound-absorbing material «asbestos felt» reduces noise more efficiently than the hoods with other sound-absorbing materials, therefore, to effectively reduce noise, you should install a steel hood with sound-absorbing material «asbestos felt». It should be noted that the steel hood with sound-absorbing material «asbestos felt» reduces noise more efficiently than other materials by only 5 dB, therefore sound-absorbing materials such as «glass wool», «mineral wool» and «construction felt» are also applicable and effective.

References

[1] *Russian Railways* URL: https://www.rzd.ru (accessed 26 May 2019)
[2] *The strategy of the development of railway transport until 2030* 2008 (Moscow: Government of the Russian Federation) p 174 URL: https://www.mintrans.ru/documents/2/859
[3] Kolarzh S A and Evdokimov N R 2018 System of control of parameters of functioning of track machines *Mechanics of the XXI* century **17** pp 93-99 URL: https://elibrary.ru/item.asp?id=36295954
[4] Kukin P P, Lapin V L and Ponomarev N L 2009 Safety of technological processes and production (labor protection) (Moscow: Higher school) p 335 URL: https://nashol.com/2015112287408/bezopasnost-jiznedeyatelnosti-bezopasnost-tehnologicheskih-processov-i-proizvodstv-ohrana-truda-kukin-p-p-lapin-v-l-ponomarev-n-l-2007.html
[5] *Instruction on the movement of trains and shunting work on the Railways of the Russian Federation* 2012 (Moscow: Ministry of Transportation of the Russian Federation) p 175 URL: https://search.rsl.ru/ru/record/01004835124
[6] Popovich M V, Volkovoinov B G and Atamanyuk A V 2008 Ensuring the stability of the railway track tracktrack machines after deep cleaning of the ballast layer *Transport of the Russian Federation* **6(19)** pp 48-51 URL: https://elibrary.ru/item.asp?id=11906782
[7] Zhang X and Jonasson Hans G 2006 Directivity of railway noise sources *Journal of Sound and Vibration* **293** pp 995-1006 URL: https://link.springer.com/chapter/10.1007/978-4-431-53927-8_17
[8] Busch A and Nugent R E 2003 A reduced-scale railway noise barrier's insertion loss and absorption coefficients: comparison of field measurements and predictions *Journal of Sound and* Vibration **267** pp 749-759 URL: https://ru.scribd.com/document/232051498/1-s2-0-S0022460X03007387-main
[9] Peng S and Zhancheng G 2015 Preparation of steel slag porous sound-absorbing material using coal powder as pore former *Journal of Environmental Sciences* **36** pp 67-75 URL: https://www.researchgate.net/publication/279736896_Preparation_of_steel_slag_porous_sound-absorbing_material_using_coal_powder_as_pore_former
[10] Sychova A, Sychov M and Rusanova E 2017 A method of obtaining geonoiseprotective foam concrete for use on railway transport *Procedia Engineering* **189** pp 681-687 URL: https://www.sciencepubco.com/index.php/ijet/article/view/23029
[11] Pultznerová A and Ižvolt L 2014 Structural modifications, elements and equipment for railway noise reduction*Procedia Engineering***91** pp 274-279 URL: https://www.researchgate.net/publication/275540809_Structural_Modifications_Elements_and_Equipmn_for_Railway_Noise_Reduction
[12] *Instructions on the device, laying, maintenance and repair of the jointless way* 2016 (Moscow: Open joint stock company Russian Railways) p 171 URL: https://docplan.ru/Index2/1/4294847/4294847268.htm
[13] Glebova E V 2005 Industrial sanitation and occupational health (Moscow: Higher school) pp 91-95 URL: https://www.twirpx.com/file/1020649/
[14] Losev G G 2008 Improving the productivity and quality of universal tamping track machines *Construction and road machines* **12** pp 49-51 URL: https://elibrary.ru/item.asp?id=11712505
[15] Moiseenko V L, Zhigar V I, Pismenay N V and Moiseenko V V 2019 Mobile technological complex for restoration of the destroyed railway track *Mining mechanics and mechanical engineering* **1** pp. 51-59 URL: https://elibrary.ru/item.asp?id=37034333
[16] Funtikov G A 2013 Ecological safety of the mechanized complexes of track cars and ways of its increase *Science and education transport* **1** pp 270-273 URL: https://elibrary.ru/item.asp?id=21808966

[17] Bulaev V G 2016 Environmental safety of track machines *Transport of Ural* **3 (50)** pp 70-74 URL: https://elibrary.ru/item.asp?id=27185224
[18] Khabarov V N, Vershinsky A V andMokin D G 2007 Vibration diagnostics of bearings in axle-box assemblies *The path and track facilities* **5** pp 21-22 URL: https://elibrary.ru/item.asp?id=12894476
[19] Vetrova A M, Klenovich O G, Badrizlova M P, Sineokov A P and Zhdanova O G 2006 Polymer composite materials for repair purposes *Glues, Sealants, Technology* **1** pp 27-29 URL: https://elibrary.ru/item.asp?id=12176358
[20] Sanitary standards SN 2.2.4 / 2.1.8.562-96 *Noise at workplaces, in residential, public buildings and in residential areas* (Moscow: Ministry of health of Russian Federation) p 8 URL: https://base.garant.ru/4174553/
[21] State Standard 33325-2015 *Noise. Methods for calculating the level of external noise emitted by rail* (Moscow: Standardinform) p 28 URL: https://base.garant.ru/71384188/
[22] State Standard 33329-2015 *Acoustic screens for railway transport. Technical requirements* (Moscow: Standardinform) p 20 URL: http://docs.cntd.ru/document/1200124234
[23] State Standard 33328-2015 *Acoustic screens for railway transport. Control methods* (Moscow: Standardinform) p 15 URL: http://docs.cntd.ru/document/1200124233
[24] State Standard 31296.1-2005 *Noise. Description, measurement and assessment of noise on the ground. Part 1. Basic quantities and estimation procedures* (Moscow: Standardinform) p 23 URL: http://docs.cntd.ru/document/1200046350
[25] State Standard 31296.2-2006 *Noise. Description, measurement and assessment of noise on the ground. Part 2. Determination of sound pressure levels* (Moscow: Standardinform) p 35 URL: http://docs.cntd.ru/document/1200062369

Authors'background

Name	*Prefix*	*Research Field*	*Email*	*Personal website*
Irina Popova	Associate Professor	Transport	impopova@mail.ru	-
LyudmilaLeonteva	Associate Professor	Chemistry,labour protection, life safety, ecology	leonteva-saratov@ya.ru	-
Igor Danilov	Full Professor	Transport	danilov-ik@rudn.ru	-

Effect of fixing the pass-through turning tool in the tool holder on the roughness and surface macro deviations during turning

D Allenov[*], K Deinova, V Konoplev and A Korzin

Peoples' Friendship University of Russia, Moscow, Russia

E-mail: allenov_dg@pfur.ru

Abstract. A study of the effect of the tool overhang length from the tool holder and cutting forces corresponding to different sizes of the removed layer on the formation of the quality of the surface layer of parts during turning is presented. Linear displacements arising in a stress-strain state and roughness parameters were used as estimated parameters. It is revealed that the quality of the surface layer improves with a decrease in vibration in the zone of contact between the workpiece and the tool.

Keywords: surface roughness, depth of cut, cutting force, tool overhang, static analysis.

1. Introduction

Currently, there are various methods of manufacturing machine parts and mechanisms, among which turning takes one of the main places. The main parameter of the geometric accuracy of product is the roughness – a set of irregularities forming the microrelief of the surface of the product because of plastic deformation of the workpiece during its processing due to friction and vibration of the tool, tearing material particles from the surface, irregularities of the cutting edges of the tool [1, 2]. The process of formation of the surface layer is influenced by a combination of simultaneously acting factors, the main of which can include processing methods, type and state of tooling, physicomechanical characteristics of the workpieces being machined, cutting modes (feed, depth and cutting speed, spindle rotation speed, length of the tool overhang of the tool holder, cutting power and cutting force), geometric abilities of the cutting plates of the tool, their cutting properties, rigidity technological system, the ability to absorb the energy of vibration (damping ability), the presence and quality of lubricating and cooling technological means and much more [1-6, 8]. Machining attachments, cutting modes, workpiece and tool material are specified by the technological process, but tool overhang is set arbitrarily. Therefore, there is a need to research the effect of tool overhang on the quality of the surface layer and macro deviation of the surface during processing. Among the quality parameters of the surface layer, roughness has a predominant effect on such performance properties of details as: wear resistance of rubbing surfaces, fatigue strength, corrosion resistance, therefore the study of this parameter is of great importance in assessing the quality of the surface of the product. [7].

2. Materials and methods

The paper investigates the influence of the depth of cut and the tool overhang on the quality of the surface layer of the part. The surface roughness parameters Ra and Rz were measured by the contact method [9]: with the model 130 profilometer. A right lathe straight-turning side-facing tool was used as a cutting tool for external turning with a mechanical fastener of the PCLNR2525M12 brand with replaceable inserts made of T15K6 hard alloy [10]. The three-dimensional, solid-state model was prepared in the student version of "Autodesk Inventor Professional" (Figure 1).

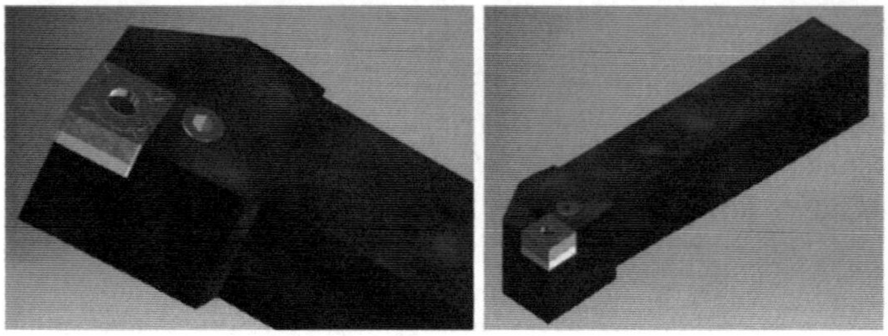

Figure 1. External turning tool PCLNR2525M12.

Steel cylindrical rolled products with a diameter of D = 40 mm (steel grade - St. 30) were used as workpieces. The processing of the workpieces was carried out on a 16K20 turning-cutting machine. The parts to be machined were mounted in a three-cam self-centering chuck with preloading of the rear center. This method of installation of machined parts on the machine allows you to reduce the negative impact on the quality of the surface layer of the vibration that occurs in the area of contact between the tool and the workpiece. In the tool holder, the tool was secured with two bolts. Workpiece processing was carried out at three different depths of cut $t_1 = 0.4$ mm, $t_2 = 1$ mm and $t_3 = 1.5$ mm. The feed rate and spindle speed did not change during the whole experiment and were equal to s = 0.1 mm / rev and n = 1000 rpm, respectively. Actual cutting speed was:

$$V_\phi = \frac{\pi D n}{1000} = 126 m/min \quad (1)$$

Chip formation when cutting materials is carried out under the action of cutting force P, which is usually laid out on the following coordinate axes: tangential P_z, radial P_y and axial P_x [11, 12]. These components are interconnected by these relations: $P_y=(0.25\text{-}0.5)P_z$, $P_x=(0.1\text{-}0.25)P_z$. Numerous studies have established that in most cases the cutting force is approximately equal to the tangential component: $P_z \approx 0.9$ P, that is why practical calculations were made according to the force Pz [11]. Cutting force Pz was determined by the formula [11]:

$$N = 10C_p \cdot t^x \cdot s^y \cdot V_\phi^n \cdot K_p \quad (2)$$

In our case, the calculation was carried out for the material of the working part of the tool - T15K6 hard alloy, the type of treatment - external longitudinal turning and boring. The correction factor K_p, which takes into account the actual cutting conditions, is the product of a number of constant factors:

$$K_p = K_{mp} K_{\varphi p} K_{\gamma p} K_{\lambda p} K_{rp} = 0.846 \quad (3)$$

According to the obtained values of the tangential force and the corresponding cutting modes (depth and cutting speed, as well as the feed rate), the cutting power was determined:

$$N = \frac{P_z V_\phi}{1020 \cdot 60} \quad (4)$$

The obtained values of the tangential component of the cutting force and power were recorded in Table 1:

Table 1 Dependence of tangential force and power on the depth of cut

No.	1	2	4	5	6
1	Depth of cut	t	0.4	1	1.5
2	Tangential component of cutting force	P_z	874	2185	3277
3	Cutting power	N	1.8	4.5	6.7

IOP Conf. Series: Materials Science and Engineering **675** (2019) 012046 doi:10.1088/1757-899X/675/1/012046

The power of the main drive motor is numerically equal to 10 kW. The calculated power at different depths of cut correspond to the given technical characteristics.

3. Results

3.1. The study of the deviation of the tool using computer simulation

Using the KOMPAS-3D and APM FEM three-dimensional modeling systems, a static calculation was made for five overhangs (the length of the first overhang was 35 mm, the others were made in 10 mm increments). The tool, whose dimensions of tool holder block are 25x25 mm, was fastened in a rotary tool holder with two bolts with a diameter of 16 mm.

Temperature equal to 20 degrees Celsius was applied to all surfaces of the created three-dimensional model of the tool under investigation. The tool was fastened in a rotary tool holder with two bolts, therefore, for the analysis, three surfaces were specified as matching (Fig. 2). When attaching the tool, the upper edge of tool holder block with two bolts is in contact (the distance from the working part of the cutter to the first bolt was 38 mm), and the lower edge - with the tool holder itself (the distance from the tool head is 30 mm). The scale of the image of forces that indicated the direction of interaction is numerically equal to 0.1.

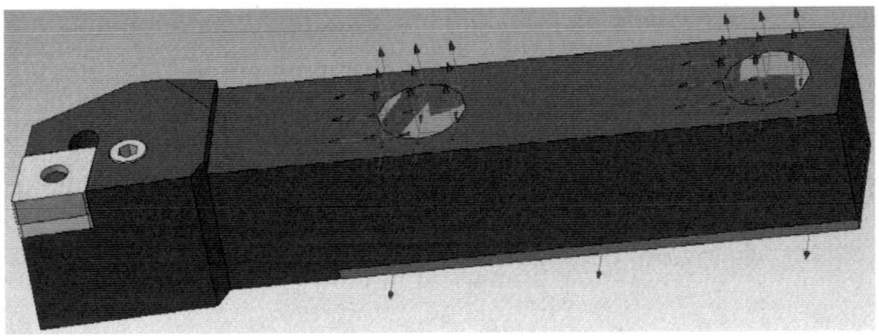

Figure 2. Installation of fixtures.

To indicate the cutting force, a distributed specific force was applied to the tip of the tool. As the magnitude of the loading vector, the cutting forces were selected, shown in Table 1. Steel was set as the sample material.

The last step in creating a model was: the assignment of matching surfaces and the creation of a finite element mesh. The tool model was divided into four nodal tetrahedra. Operation parameters were set from practical considerations: the maximum length of the element side is 0.5, the thickening coefficient on the surface is 1, and the underpressure factor in a volume is 1.5. The resulting number of finite elements of the partition was equal to 5861, nodes - 1896.

In order to identify the relationship between the length of the tool overhang and the depth of cut from the tool deviation from the tool holder, a static calculation was performed. The number of elements selected by the system was 6235 (1896 knots), degrees of freedom - 5688 (443) [13]. It is necessary to pay attention to the increase in the number of selected elements of the system: compared with the stage of creating the finite element mesh, the number of finite elements has increased by 374 units. This is due to the fact that the created computational three-dimensional model, consisting of complex non-uniform geometric transitions, was subjected to additional adaptive splitting. In addition, during the analysis, an automatic calculation of displacements was performed, at which the stiffness matrix was formed and the results for loading were calculated.

The result of the analysis of the 5th sample with the greatest cutting force is presented in Figure 3.

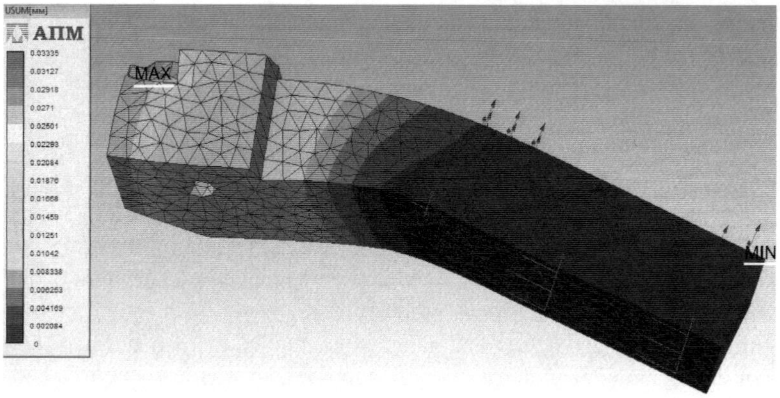

Figure 3. The figure of deformed design of the tool with 5th overhang and the depth of cut of 1.5 mm wound.

The given maps are constructed in the form of iso-areas (the number of iso-levels is 16) on the basis of the deformed structure, while the values at the nodes are averaged. The scale factor (displacement scaling factor for rendering the deformed structure) is 100. For rods, plates, and volume elements, the total linear displacement parameters (USUM) were selected.

The dependences of the obtained linear displacements on the size of the removed layer from the workpiece and the length of the tool overhang from the tool holder are presented in Table 2.

Table 2. Table 2. Dependence of tangential force and power on cutting depth

tool position	depth of cut t (mm)	cutting force P (N)	linear displacement (mm)
	0.4	874	0.003497
1 tool position	1.0	2185	0.008553
	1.5	3277	0.012990
	0.4	874	0.003657
2 tool position	1.0	2185	0.008960
	1.5	3277	0.013710
	0.4	874	0.004710
3 tool position	1.0	2185	0.011660
	1.5	3277	0.017660
	0.4	874	0.006663
4 tool position	1.0	2185	0.016500
	1.5	3277	0.024980
	0.4	874	0.008895
5 tool position	1.0	2185	0.022240
	1.5	3277	0.033350

Static analysis of the stress-strain state of the tool at different tool overhangs and cutting forces showed that linear displacements are directly dependent on the tool overhang and the depth of the layer being removed. Consequently, the tool was subjected to the greatest linear deformation equal to 0.03335 mm at the depth of cut of 1.5 mm, which corresponds to a cutting force of 3277 N, and the fifth tool overhang (Figure 3).

Additionally, when displaying the results of the calculation, the deformed and non-deformed state of the tool were visualized to demonstrate noticeable linear displacements. The scale factor is chosen to be 100 in order to more clearly show the deflection of the tool that appears under the action of stresses.

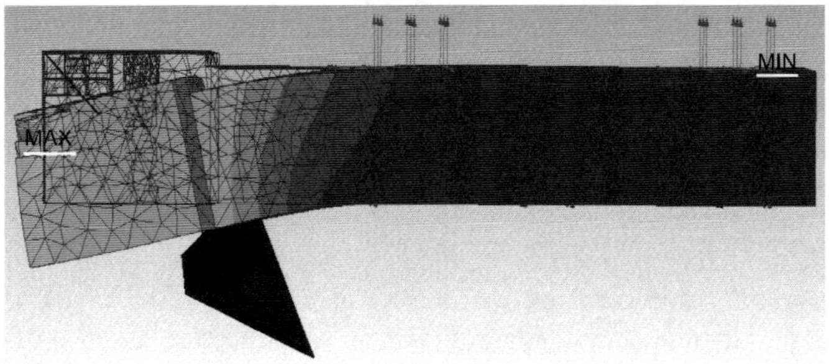

Figure 4. The figure of tool design before and after deformation.

The relationship between the linear displacements that occur during the cutting process, the tool overhang length and the depth of the layer being removed is shown in Figure 5.

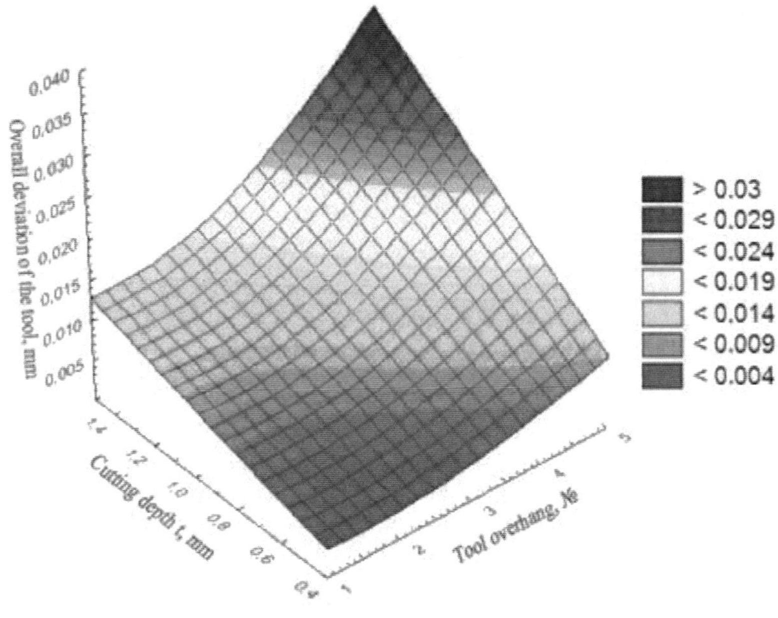

Figure 5. Graph of the deviation of the tool from its overhang and depth of cut.

The difference between the maximum and minimum values of the instrument deviations at t = 0.4 mm is 0.005 mm, at t = 1 mm - 0.015 mm, and at t = 1.5 mm - 0.016 mm.

Such a distribution of deviations by depth of cut is related to the position of the tool in the tool holder: when the first tool over hang occurred, the tool holder block touched the tool holder with its entire edge, which had a slight effect on the tool deviation.

It can be seen from the graph that for each of the considered lengths of tool overhang from the tool holder, the total deviations increase with increasing depth of cut. However, when the first instrument over hang occurred, the linear displacements changed insignificantly with an increase in the size of the removed layer: the difference in linear displacements at t = 1 mm and t = 0.4 mm is approximately equal to the difference in linear displacements at t = 1.5 mm and t = 1 mm (this difference is 0.005 mm). The situation is different from the maximum tool overhang. The difference of linear displacements at t = 1 mm and t = 0.4 mm is 0.015 mm, and at t = 1.5 mm and t = 1 mm the tool was subjected to less linear deformation (0.006 mm). This is explained by the desire for steady-state elastic deformation, which is associated with the mechanical characteristics of the material. In other words,

IOP Conf. Series: Materials Science and Engineering **675** (2019) 012046 doi:10.1088/1757-899X/675/1/012046

when a certain value of linear displacements was reached, the tool began to resist the loads acting on it in order to prevent the onset of plastic deformation of its structure.

According to the obtained data, we can conclude that the length of the tool overhang and the depth of the layer removed from the workpiece significantly affect the tool deflection, that is, as the tool overhang length and cutting depth increase, the tool deviations in the tool holder increase.

3.2. Investigation of the influence of the tool overhang on the surface roughness

To confirm the direct relationship between the quality of the surface layer and cutting conditions by contact, the roughness parameters were measured using a profilometer. Graphs of roughness R_a and R_z on tool overhang and depth of cut are presented in figure 6.

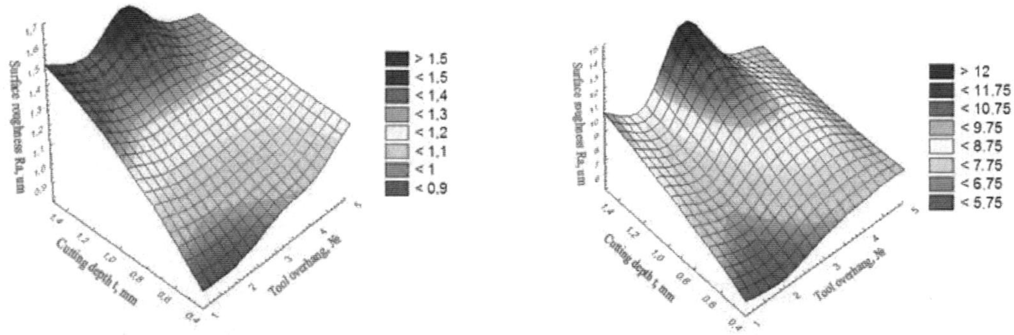

a b

Figure 6. Graphs of roughness parameters R_a and R_z depending on tool overhang and depth of cut: a - for roughness parameter R_a; b - for the roughness parameter R_z.

According to the data obtained, it can be concluded that the tool overhang significantly affects the surface roughness when machining steel grade 30. At the same time, with an increase in the length of the tool overhang from 0 to 50 mm, the surface roughness parameters increase by 30-40% (Figure 6, a). The visible sharp increase in roughness at a depth of cut of 1.5 mm with tool overhang 1 and 3 is due to the fact that when processing workpieces a drain ribbon-like confused chips occurred, which, when removed from the cutting zone, wrapped the treated surface, thereby scratching it [14]. The effect of the depth of cut on the surface roughness is directly proportional to the dependence, which is confirmed in [1, 15]. Similar dependences are observed for Rz. Since this parameter is used to normalize irregularities much less frequently, a more detailed assessment of the dependencies for it was not carried out.

To eliminate the effect of drainage chips on the surface roughness, intermittent cutting was used, that is, the alternation of the cutting process (chip formation) and idling: the chips were crushed into small and easily removable components. This cutting method allowed us to obtain roughness values that are more suitable for plotting (Figure 7).

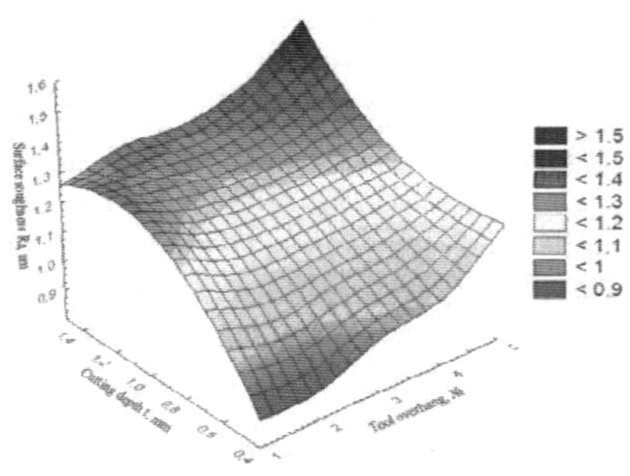

Figure 7. Graph of the roughness parameter R_a versus tool overhang and depth of cut with intermittent cutting.

This graph shows that the ratio of the maximum value of roughness to the minimum with a depth of cut of 0.4 mm is 1.3; with a depth of cut of 1 mm - 1.1; with a depth of cut 1.5 mm - 1.2. These relationships show that the effect of tool overhang on surface roughness at different depths of cut is uneven. Since, in conducting the study, the geometrical parameters of the cutting tool (for example, the cutter angles and the radius of the cutting wedge radius), the cutting conditions and the materials of the workpiece and the turning plate did not change, the main influence on the formation of irregularities was caused by the vibration that occurs during machining in the area of the workpiece and tool. At the same time, in the course of numerous experiments it was found that the surface roughness improves with decreasing vibration, which in turn will be the lower, the greater the cutting speed will be, the cross-sectional size of the cutting tool, the rigidity of the technological system and less depth of cut, feed, and overhang [1, 4 , 6]. According to the data obtained, it can be concluded that with an increase in the length of the tool overhang, the amplitude of oscillations increases, which increases the surface roughness.

4. Discussion

During turning in the zone of contact of the cutting wedge and the work surface, vibrations are formed under the force of friction shavings on the cutter and the cutter on the workpiece. With an increasing of tool overhang, the rigidity of the technological system «machine – device – tool – detail» decreases, as a result the amplitude of the oscillations increases and negatively affects the quality of the surface. Increased vibration in the cutting zone also leads to a reduced service life of the cutting tool due to the increased wear of the cutting wedge [16, 17].

From the above it follows that during turning it is necessary to set the minimum tool overhang. In case of surface treatment at a distance, using a cutter with a long holder, it is necessary to set reduced cutting mode, or to perform processing in multiple passes.

Conclusions

Thus, it should be noted that the quality of the surface layer improves with a decrease in vibration in the zone of contact between the workpiece and the tool. In turn, the emergence of vibrations is influenced by numerous factors, such as, for example, the depth of cut and the tool overhang. With an increase in the size of the layer removed from the workpiece and the length of the tool overhang, an increase in the roughness parameters is observed, and, consequently, the quality of the surface layer deteriorates. In addition, it should be noted that the increase in the values of the above-mentioned cutting conditions adversely affect the occurrence of linear displacements (deviations of the tool in the tool holder). In other words, the strength and wear resistance of the tool, as well as the quality of machining of the parts directly depend on the depth of cut and the overhang of the tool.

References

[1] Suslov A G 2008 *Inzheneriya poverkhnosti detaley* [*Part Surface Engineering*] (Moscow: Publishing House Mashinostroenie) 320 p (in Russian)

[2] Merchant M E 1945 *Journal of Applied Physics* **16** pp 267–275 https://doi.org/10.1063/1.1707586

[3] Bez'yazychnyy V F 2000 Vliyaniye kachestva poverkhnostnogo sloya posle mekhanicheskoy obrabotki na ekspluatatsionnyye svoystva detaley mashin [The influence of the quality of the surface layer after machining on the performance properties of machine parts] *Engineering Journal* **4** pp 9-16 (in Russian)

[4] Kozochkin M, Allenov D and Andryushchenko I 2019 *Proc. 4th International Conference on Industrial Engineering 2018* https://doi.org/10.1007/978-3-319-95630-5_143

[5] Allenov D G 2016 Issledovaniye vliyaniya iznosa rezhushchey kromki instrumenta na chistotu poverkhnosti [Study of the effect of wear of the cutting edge of the tool on the surface cleanliness] *Mechanical Engineering* **4(16)** pp 12-16 (in Russian)

[6] Thomas M, et al 1996 *Computers & Industrial Engineering* **31(3–4)** 637–644 https://doi.org/10.1016/s0360-8352(96)00235-5
[7] Allenov D G, et al 2018 Issledovaniye vliyaniya kachestva poverkhnostnogo sloya na ekspluatatsionnyye kharakteristiki detali [The study of the influence of the quality of the surface layer on the performance characteristics of the part] *Scientific and Technical Bulletin of the Volga region* **6** pp 30-34 (in Russian)
[8] Rogov V A and Gorbani S 2015 Influence of the handle design on a composite lathe cutter *Russian Engineering Research* https://doi.org/10.3103/s1068798x15010220
[9] Markova T V and Kryzhanovskaya I M 2006 *Sherokhovatost poverkhnosti* [*Surface Roughness*] (St. Petersburg: Polytechnic University Press) 32 p (in Russian)
[10] Lim T, et al 2001 *International Journal of Production Research* **39(6)** pp 1239–1256
[11] Dalsky A M, et al 2003 *Spravochnik tekhnologa-mashinostroitelya* [*Reference Book Technologist-Mechanical Engineering*] (Moscow: Publishing House Mashinostroenie) 912 p (in Russian)
[12] Parfeneva I E 2012 *Tekhnologiya konstruktsionnykh materialov. Obshchaya kharakteristika obrabotki rezaniyem* [*Technology of construction materials. General characteristics of machining*] (Moscow: Publishing House Mashinostroenie) 454 p (in Russian)
[13] Ambati R and Yuan H 2011 *International Journal of Advanced Manufacturing Technology* **53(1–4)** 313–323 https://doi.org/10.1007/s00170-010-2818-9
[14] Wolf A M 1973 *Rezaniye metallov* [*Metal cutting*] (Leningrad: Publishing House Mashinostroenie) 496 p (in Russian)
[15] Korsakov V S 1961 *Tochnost mekhanicheskoy obrabotki* [*Accuracy of machining*] (Moscow, Publishing House Mashgiz) 397 p (in Russian)
[16] Allenov D, Kozochkin M and Andryushchenko I 2017 *MATEC Web of Conferences* **129** https://doi.org/10.1051/matecconf/201712901032
[17] Chen J C and Chen W L 1999 *Journal of Intelligent Manufacturing* **10(2)** pp 187–197 https://doi.org/10.1023/A:1008980821787

IOP Conf. Series: Materials Science and Engineering **675** (2019) 012047 doi:10.1088/1757-899X/675/1/012047

Optimization of work of urban passenger transport

I Popova[1*], E Abdulina[1], Yu Danilov[2] and I Danilov[2]

[1] Samara State University of Railway Transport, Saratov, Russia
[2] Peoples' Friendship University of Russia (RUDN University), Moscow, Russia

Email: impopova@mail.ru

Abstract. Recommendations for improving the organization of the logistics system of urban passenger traffic based on the formation of adaptive properties proposed. Definitions of organization and lack of organization of the system, target entropy, orderliness, disorder are considered. The calculation of these indicators allows us to conclude how the company is adapted to the external environment. An important feature of all production systems is the focus of their functioning. If such a system is functioning, it means, first of all, that it solves a finite set of tasks to ensure the adaptability of the urban passenger traffic management system to the flow of passengers and with a certain accuracy the goals of organization and management, of which there is also a finite set. The authors come to a fair conclusion that the lack of organization of functioning is a consequence of the influence of the external environment on the control object, while organization is a quantitative measure of the ability of the system to adapt to the effects of environmental factors. The indicator of organization can be proposed as a criterion that assesses the efficiency of the urban passenger transport system. The worse the internal structures are organized, the less efficiently the control system works and the more mobile and uncertain the external environment, the greater the discrepancy between the goal and the final result.

1. Introduction

The problems of development of logistic methods and their use in the field of transportation dedicated a number of works – B. A. Anikin [1], L. B. Mirotin [2], V.V. Shcherbakov [3], D. Bowersox, D. Kloss [4], S.Keller [5].

The object of the study is the system of urban passenger traffic as part of passenger traffic, as well as providing them with information flows, flows of material, financial and labor resources.

The subject of the study is the organizational, economic, administrative relationships and relationships of participants in the system of urban passenger transport [6].

The aim of the study is to develop methodological approaches to the harmonization of urban passenger transport processes using the strategy of logistics outsourcing.

Research problem -to assess the level of organization of the logistics system of urban passenger transport and the rationale for the transformation of the existing system of urban passenger transport in the logistics system in the light of solving public transport problems [7].

At the same time, the specific application of logistics tools to the passenger transport market has not yet been adequately reflected. Meanwhile, it seems that this issue requires a special approach, since the organization of passenger traffic and the quality of providing various categories of consumers with transport services depend on taking into account the specifics of transport logistics in the field of passenger transportation, which makes the scientific search for organizational and economic solutions for the logistics of this sphere very important [8].

For traditional approaches to the organization of urban passenger traffic, there is a tendency to increase the density of the route network, observe the minimum interval of vehicle traffic and locate stopping points on routes through lines of approximately the same length. Explicitly or implicitly, there is a desire to cover the territory of the city with a route network of passenger transport as much as possible [9].

In passenger traffic, as well as in the traditional understanding of logistics, the requirement to minimize the total costs with an agreed level of quality of services provided remains in force [10]. The formation of the urban passenger transportation system using logistic principles should begin with the identifying the transport needs of city residents.

The logistic approach to the organization of passenger traffic involves the fullest possible consideration of the passengers' requirements who are consumers of urban transport services. An important factor in the quality of transport service is the speed of the message, which directly determines the time spent by the passenger on the trip [11].

2. Materials and methods

Numerous studies have established that the speed of the message is associated by power dependence with the length of the haul. The greater the distance between the stopping points, the higher the speed of the message, the less time the passenger spends on the trip.

The number of stopping points on the route affects the quality of passenger service in an ambiguous manner: if there are few stops, the span length increases and the speed of the message increases. But it may cause an increase in the time the passenger approaches the stop.

The desire to reduce the passenger's approach time to a stop and an increase in the number of stopping points leads in some cases to the fact that the speed of a message in certain sections of the route becomes close to the speed of a pedestrian. This is especially true for urban electric transport, the route of which goes through regulated or unregulated intersections.

An excessive number of stops and an underestimated length of the haul leads to the excessive consumption of fuel and energy, increased vehicles wear and their premature failure, fatigue of drivers, and reduced road safety. Situations of transport services for city dwellers are different, the methods for their resolving should be also different [11].

3. Results and discussion

It should be ensured that vehicles provide not only communication between different areas of the city among themselves, but first of all satisfy the real, having clear temporal and spatial characteristics of the transport needs of city residents. Thereby, one of the basic principles of logistic systems: priority satisfaction of customer requirements will be implemented. Thus, the passenger transportation system of such a large city as Saratov must be considered on the basis of the logistic concept [12].

Under ideal conditions, with an optimal organizational structure, efficient management and the absence of negative impacts from the external environment, the goal of organizing an efficient system for transporting passengers by urban transport will be achieved.

In fact, the goal is not fully achieved due to non-compliance with conditions. The worse the internal structures are organized, the less efficiently the control system works and the more mobile and uncertain the external environment, the greater the discrepancy between the goal and the result obtained [12].

That is, the less the compliance of internal qualities with the urban passenger transport models with the characteristics of the passenger flow, the less it will be adapted to the objective conditions of work, the greater will be the difference between the goal and the result obtained. Quantitatively, this is a variation and consequently, the level of fitness of the organization's models for the transport of passengers to passenger traffic, can be assessed using the level of organization of the system's functioning [13].

Organization is a function of simpler system characteristics: orderliness and complexity.

In practice, when the goal is set, it is important to know not «the absolute» organization of functioning, but disorganization, since it is the operational information about the quality of the organization and the function of the elementary concept of disorder.

Disorder is a measure of the difference of any selected parameter, for example, x_j with respect to the standard of order x_{et}, which tends to zero at $x_j \to x_{et}$ [13].

Since in the general case, the center of order is not a point, but a certain area, which is called a quasi-ordered zone, the disorder is considered in relation to the boundaries of this zone $f(x_j)$.

Then the model of disorder $\overline{Y_j}$ will look like:

$$\overline{Y_j} = |x_j - x_{et}| - f(x_j), \quad (1)$$

where $\overline{Y_j} \geq 0$.

In a more general form, operating with a vector quantity, one can proceed to the expression:

$$\overline{Y} = q_j - f(x_j), \quad (2)$$

where q_j is the error vector module.

The complexity of the system is a function of either the number of elements included in the system, or their diversity. The complexity of the disorganization enters indirectly, that is through the number of considered parameters, possible situations and time intervals.

According to the definition, you can write the following symbolic expression for lack of organization [13]:

$$\overline{O} = \bigcup^{K} \alpha_k \bigcup^{I} S_i \bigcup^{J} P_j \varphi(Y_j), \quad (3)$$

where $\bigcup$ is the symbol of generalization of the characteristics of disorder, respectively, for the j-situations of i-elements and K-time intervals;

$\varphi(Y_j)$ is the function with the help of which weighting of disorder is weighted by the materiality of its manifestation in relation to a certain indicator of the functioning of the system.

In the assumption of the additive nature of the lack of organization by the elements and by the time based on the formula 3, the general expression for lack of organization would be [13]:

$$\overline{O} = \sum_{k=1}^{K} \alpha_k \sum_{i=1}^{I} S_i \sum_{j=1}^{j} P_j \psi\left(P_{\overline{y}}\right), \quad (4)$$

where ψ is the cast function.

The disorder parameter $P_{\overline{y}}$ is found in the expression:

$$P_{\overline{y}} = \omega_j Y_j + c, \quad (5)$$

where ω_j is the weight, taking into account the difference $\overline{Y}$ in the j-th situation compared with other situations;

c is the coefficient, the meaning of which is found from the condition $\psi\left(P_{\overline{y}}\right) = 0$, with $\overline{y_i} = 0$.

The property of the disorganization of the functioning of H_k with the logarithmic function ψ, for the j- situation, has the form:

$$H_k = \sum_{j=1}^{J} P_j \log_a \{[q_j - f(x_j)]\omega_j + 1\}. \quad (6)$$

Such disorganization is conventionally called the target entropy.

Thus, as a result of mathematical transformations, two open spaces are obtained that for the I-th position can be represented graphically (Figure 1).

The first space is the space of the original (scatter, misalignment), which characterizes the amount of disorder, and the second – strictly corresponding to the image space and the value of H_X. These two spaces are similar, i.e. each H_X value in the image space (target entropy) will have its own $H_{X(q)}$ value in the original space.

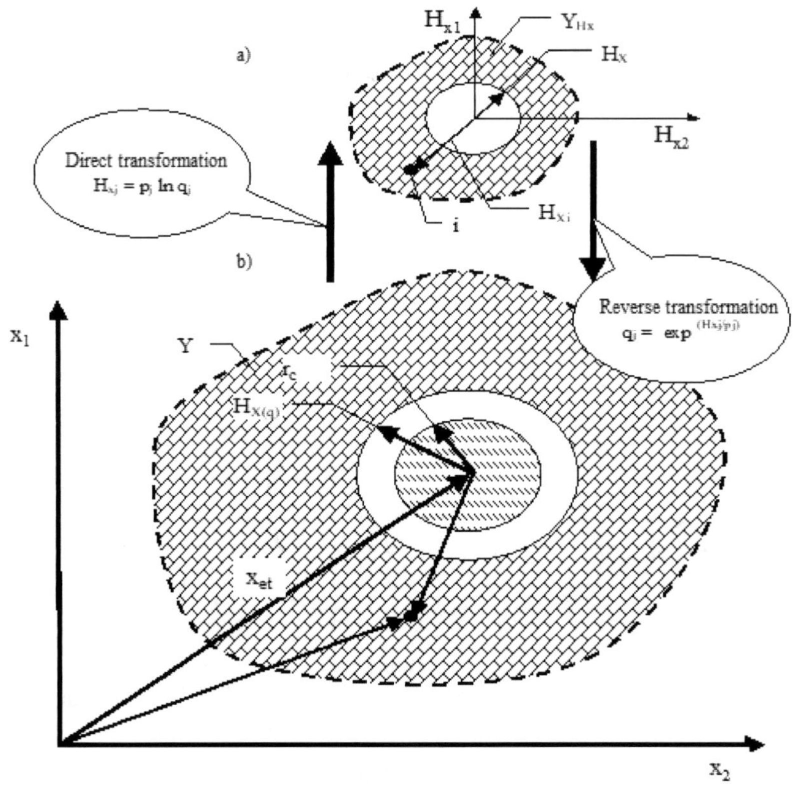

Figure 1. Representation of the relationship between the original and image space (a – space of the original (spread in x_1, x_2 coordinates), b – image space (entropy) in coordinates H_{X1}, H_{X2}, Y – spread area, r_c – the radius of the target area, $H_{X(q)}$ – the average radius of the scatter, q_i – the module of the vector of the error, Y_{H_X} – the area of the target entropy, H_X – the target entropy, the corresponding, $H_{X(q)}$ in the space of dispersion, H_{Xi} – the entropy of the i-th state of the system corresponding to q_i in the scatter space)

The target entropy has the following properties:

– $H_X = 0$, if all $q_j \le f(x_j)$, $P_j = 0$, if all $q_j - f(x_j) > 0$. In this case, the system is fully adapted to the traffic flow;

– for the whole ensemble with $P_j > 0$, and if $\lfloor q_j - f(x_j) \rfloor \to 0$, then $H_X \to 0$, then the system is adapted, if $\lfloor q_j - f(x_j) \rfloor \to \infty$, then $H_X \to \infty$ in the limiting case is not fully adapted;

– when $P_j = \dfrac{1}{j \cdot H_x} = H_{\max}$ is found from the expression:

$$H_{\max} = \log_a \left\{ \max_j \lfloor q_j - f(x_j) \rfloor \omega_j + 1 \right\} \tag{7}$$

In this case, the model of management of the system of transportation of passengers and the flow of passengers are fully not adapted to each other.

Hereinafter, the entropy having been found in accordance with expression 7 will be called the maximum entropy [13].

Despite the fact that the disorganization of the form (6) was conventionally called the target entropy, its consistency under certain conditions can be shown in relation to the classical Shannon entropy. If we take as a parameter the prior probability $\frac{1}{P_j}$ as the standard of order $P_{et}=1$ and substitute these meanings into expression (1), assuming $f(x_j)=0$, then the absolute disorder in the j situation:

$$\overline{Y_j}=\frac{1}{P_j}-1. \tag{8}$$

Obviously, when $P_j\to 0$ $\overline{Y_j}\to\infty$, and when $P_j\to 1$ $\overline{Y_j}\to 0$, which does not contradict the definition of disorder.

Substituting (8) into (5) and assuming that $\omega_j=1$:

$$P_{\bar{y}}=\frac{1}{P_j}. \tag{9}$$

Then we take the logarithmic character ψ and substitute (9) into the expression (4) for the j-situation. As a result, we obtain the well-known expression for Shannon's entropy:

$$\overline{O}=H=-\sum_{j=1}^{m}P_j\log P_j \tag{10}$$

The solution to the problem of evaluating the effectiveness of the functioning of subsystems of urban passenger transport consists primarily in the choice of evaluation criteria. The target entropy fully satisfies the requirements of the universality of the reflection of the specifics of the problem under consideration, consistency with previous theories and dynamism. However, it is more convenient to use relative indicators [14-16].

The increment of organization ΔQ_j in a symbolic form, which is expressed through a change in the system's lack of organization under the influence of the organization's means, looks like this:

$$\Delta O_i=-\Delta O_i=\Delta O_{id}-O_{ip}, \tag{11}$$

where $\Delta O_{id}, O_{ip}$ is the lack of organization in relation to the i-th goal, respectively, before and after the action of the means of the organization.

Then the relative increment of organization due to the action of the organization's funds is called the lack of organization in the functioning of the system and is derived from the expression:

$$O_i=\frac{(O_{id}-O_{in})}{O_{id}}. \tag{12}$$

Taking the logarithmic form of the disorganization function, it is possible to present the relative increment of organization O_i by increasing the adaptive properties of the system and, taking into account expressions 6 and 7, as a function of the system's target and maximum entropy [17]:

$$O_i=\frac{(H_{i\max}-H_{ic})}{H_{i\max}}=1-\frac{H_{ic}}{H_{i\max}}, \tag{13}$$

where H_{ic}, $H_{i\max}$ is respectively, the target and maximum entropy for the i-th target.

Functioning organization of the system has a number of useful properties:

- the scope of this indicator is limited to the top and bottom;
- organization$\rightarrow$ 1, if the probability of achieving the goal $\rightarrow$ 1;
- organization$\rightarrow$0, if the probability of achieving the goal $\rightarrow$0.

Therefore, the target entropy can be considered as a mathematical model of the process of harmonization of urban passenger traffic:

– the measure of disorder is used as an indicator of not achieving the goal;

– the indicator of failure to achieve the goal, taking into account the probability of occurrence of mismatch X_j with respect to the border of the quasi-ordered zone $f(x_j)$, caused by insufficient adaptation of the enterprise to external conditions, is taken as a measure of disorganization of functioning [18].

In the functioning of the system of urban passenger traffic, the compromise of its constituent elements is achieved on the basis of centralized management of its constituent elements, achieved on the basis of centralized management by municipal and federal authorities. Structurally, the logistics information system can be represented as the following components: a database and a bank of models. The functioning of the information system allows to get an idea of the status of the rank of passengers and to adjust the work of the transport's to ensure accurate compliance with the needs of passengers [18].

Conclusion

Thus, the lack of organization of functioning is a consequence of the influence of the external environment on the control object, while organization is a quantitative measure of the ability of the system to adapt to the influence of environmental factors and can be proposed as a criterion for the efficiency of the urban passenger transportation system.

Thus, the target entropy can be considered as a mathematical model of harmonization of passenger work. Depending on the level of organization of the system, we will be able to assess how it is adapted to the external environment.

Scientific novelty - improving the organization of the logistics system of urban passenger traffic on the basis of the formation of adaptive properties through the implementation of the strategy of logistics outsourcing.

References

[1] Anikin B A et al 2015 *Fundamentals of Logistics: Transport infrastructure Logistics* (Publishing House Prospekt, Moscow) p 406

[2] Mirotin L B and Pokrovsky A K 2016 Logistics and resources *Journal of transport* **4** pp 15-17

[3] Shcherbakov V V and Ivanova D P 2013 Positioning of logistics innovation *Innovative activity* **3** pp 98-102

[4] Bowersox D J and Closs D J 2008 *Logistical management The Integrated Supply Chain Process* (Publishing House Olimp-Business, Moscow) p 640

[5] Keller S B 2001 Logistics outsourcing – a management guide *Transportation Journal* **2** pp 52-58

[6] Islam M R, Hadiuzzaman M, Hossain S, Banik R, Hasnat M M and Musabbir S R 2016 Bus service quality prediction and attribute ranking: a neural network approach Public transport 2 pp 295-313

[7] Chefranova O V and Zhigulsky V I 2015 Controlling and monitoring in the implementation of policies and strategies for the development of the transport and road complex *Nauchnaya Mysl'* [*Scientific Thought*] **2** pp 144–147

[8] Oziomek J and Rogowski A 2018 Improvement of regularity of urban public transport lines by means of intervals synchronization *Transport Problems* **13** pp 91–102

[9] Palach S 2014 Trip coordination in municipal passenger transport *Transport Problems* **9** pp 111–117

[10] Vitvitsky E E and Khoroshilova E S 2015 On the manifestation of the random nature of the transport process in the functioning of the delivery motor transportation system in the conditions of the city *Nauchnaya Mysl'* [*Scientific Thought*] **2** pp 178–181

[11] Yarmen M and Sumaedi S 2016 Perceived service quality of youth public transport passengers *Transport Problems* **11** pp 99–111

[12] Rybicka I, Droździel P, Stopka O and Ľupták V 2018 Methodology to propose a regional transport organization within specific integrated transport system: a case study *Transport Problems* **13** pp 115–125

[13] Klochkov V N 1999 *Adaptation and competitiveness of road transport systems in market conditions* (St. Petersburg State University of Engineering and Economics Press, St. Petersburg) p 215

[14] Morchadze T and Rusadze N 2018 Ways to address the challenges in passenger traffic within the urban transport systems *Transport Problems* **13** pp 65–67

[15] Olivkova I 2008 Public transport in Czech Republic Ways to address the challenges in passenger traffic within the urban transport systems *Transport Problems* **3** pp 53–58

[16] Mindur L and Hajdul M 2013 The concept of organizing transport and logistics processes, taking into account the economic, social and environmental aspects *Transport Problems* **8** pp 121–128

[17] Stock J R and Lambert D M 2005 *Strategic logistics management* (Publishing House INFRA-M, Moscow) p 797

[18] Wardlow D, Wood D F, Murphy P R and Johnson J C 1999 *Contemporary Logistics* (Pearson Education, New Jersey) p 608

Authors' background

Name	*Prefix*	*Research Field*	*Email*	*Personal website*
Irina Popova	Associate Professor	Transport	impopova@mail.ru	-
Elena Abdulina	Master Student	Transport	lenacool141@rambler.ru	-
Yuri Danilov	Assistant Professor	Transport	dankostart@mail.ru	-
Igor Danilov	Full Professor	Transport	danilov-ik@rudn.ru	-

IOP Conf. Series: Materials Science and Engineering **675** (2019) 012048 doi:10.1088/1757-899X/675/1/012048

Thermodynamic modeling of a heat pump unit as part of a cogeneration turbine operating in ventilation mode

K Souri*, Y A Antipov and K V Shkarin

Peoples' Friendship University of Russia (RUDN University), Moscow, Russia

E-mail: souri7794@gmail.com

Abstract. Currently, much attention is paid to improving the efficiency of energy generating facilities through the use of secondary energy resources. This article presented a cogeneration turbine in the general energy scheme with a heat pump installation. During the winter period, due to the increased consumption of heat energy, heat-generating turbines, by closing the regulating diaphragm, are switched to full steam extraction mode from the low-pressure flow part to heat the supply water. In this case, the last stages of the turbine are transferred to the ventilation mode with the release of heat due to the vortex flows on the blades of the low pressure part. In order to obtain additional heat energy for network water, the article considers the option of using secondary energy resources with the help of a heat pump unit operating in a steam turbine plant scheme. All necessary calculations of thermodynamic modeling were carried out according to the EES heat pump, the program working as part of a cogeneration turbine, and the results were obtained of increasing the efficiency of using secondary energy resources with various refrigerants.

1. Introduction

Modern cogeneration steam turbines are designed to generate both electrical and thermal energy. However, in winter, in order to maximize the generation of thermal energy, practically all the steam from the flowing part of the low pressure cylinder to the fully closed regulating diaphragm is sent to the mains water heaters, and the turbine stages located behind the low-pressure part are forced to work in the range of ventilation modes. However, due to the reduction of steam consumption to the leakage through the gaps closed by the control diaphragm in the flowing part of low pressure, eddy currents of secondary flows are formed, which lead to the heating of the last stage working vanes to unacceptably high temperature [1]. The relevance of the topic is to increase work efficiency and reduce losses in the cycle of cogenerating turbine plants. Therefore, the development of new ways to improve the efficiency of modern cogeneration turbines without reducing their reliability, maneuverability and available power.

2. Materials and methods

At working blades of the last stage of the CHP turbine T-180 / 210-12.8, operating in the range of ventilation mode, the temperature reaches 240 ^{0}C [2]. At the same time, the reduced steam flow through the closed regulating diaphragm is 0.072. To remove heat generated as a result of ventilation losses in the stages of the low-pressure part, in practice, three main methods are usually used: a ventilation pass of steam through specially designed slots closed by a regulating diaphragm; application of a special cooling steam in the steps of the low-pressure part; injection of condensate into the space of the outlet pipe [3, 4].

However, these methods do not make it possible to use the heat released as a result of the ventilation losses, directly for heating the network water. To solve the problem, the authors propose to select the heat of superheated steam at the outlet from the last stage and use the heat pump plant to direct it to the heating of the network water (figure 1) [5, 6].

A compression type of the heat pump comprises a series-connected evaporator 1, a compressor 2, a heater 3 and a valve 4 connected to the inlet of the evaporator 1. The evaporator 1 is structurally located along the periphery of the flow part of the outlet branch pipe 5 of the low-pressure part. This arrangement makes it possible to organize the heat removal from the vortex flow of superheated steam passing in the peripheral zone of the outlet branch pipe 5. The heater 3 is connected to the regular heater 6 of the network water.

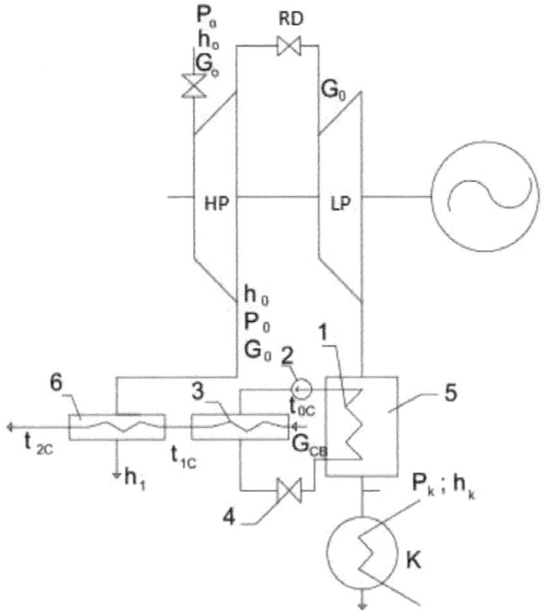

Figure 1. Fragment of the thermal scheme of cogeneration steam turbines plant with a heat pump system.

3. Results

As an example, a heat calculation was performed for the heat pump plant of the T-180 / 210-12.8 steam turbine unit built into the scheme, which operates in the CHP mode [7]. In [8], page 171 of the presented results for measuring the distribution of steam through the height of the working blade of the last stage of this turbine operating in the range of the purge ventilation mode, it is shown that the maximum relative steam flow rate ($\bar{q} = 0.8$) is concentrated on the peripheral part of the blade , which is 20% of its height, when the pressure in the condenser is P_k=3.75 kPa, the vapor velocity at the outlet from the peripheral zone of the blade is 158 m / s. In this case, at such a rate, the vapor flows in the annular space of the evaporator 1, which is designed as a two-row, corridor-type heat exchanger located across the direction of steam movement. In order to increase the heat transfer coefficient, the outer surface of the evaporator tubes is made in the form of annular ribs. The R-123, R-113 and R-718 refrigerants, capable of operating over a wide range of temperatures, were chosen as the working fluids in calculating the heat pump plant. Calculation of the coefficient of heat transfer α1 for boiling refrigerant in the cylindrical tubes of the evaporator 1 was carried out according to Lobuntsov's formula [9]:

$$\alpha_2 = b\left[\frac{\lambda^2}{\nu\delta T_s}\right]^{1/3} q^{2/3} \tag{1}$$

where $b = 0.075\left[1 + 10\left(\frac{\rho''}{\rho' - \rho''}\right)^{2/3}\right]$

$\boldsymbol{\lambda}$ – is the coefficient of thermal conductivity of the refrigerant, $\boldsymbol{\nu}$ – is the coefficient of kinematic viscosity of the refrigerant, T_s – is the saturation temperature of the refrigerant, $\boldsymbol{\rho}'$ and $\boldsymbol{\rho}''$ – are the densities of the liquid and vapor phases of the refrigerant, respectively. Calculation of the heat transfer coefficient α_2 from the vapor to the outer surface of the evaporator tubes 1 was carried out according to the criterion equation [10]:

$$Nu = 0.2Re^{0.65}Pr_l^{0.33}\left(\frac{Pr_l}{Pr_{ct}}\right)^{0.25} \tag{2}$$

where $Nu = \frac{\alpha_1 d_1}{\lambda_1}$

d_l – is the external diameter of the tube, $\boldsymbol{\lambda}_1$ – is the thermal conductivity of the vapor,

$Re = \frac{wd_1}{\nu_1}$

w – the vapor velocity, $\boldsymbol{\nu}_1$ – the kinematic viscosity of the vapor, Pr_l and Pr_{ct} – the Prandtl number at steam and wall temperatures, respectively.

The heat flux from the superheated steam vortex flow in the low-pressure part can be determined by the formula [11]:

$$Q = Gc_p(T_{\Pi 1} - T_{\Pi 2}) \tag{3}$$

where G – is the steam flow rate in the low-pressure part, c_P – the heat capacity of the steam, $T_{\Pi 1}$, and $T_{\Pi 2}$ – are the vapor temperatures respectively at the inlet and outlet of the evaporator 1. On the other hand, this heat flux was directed to evaporate the refrigerant in the evaporator, and it is equal to

$$Q = K(T_{\Pi 1} - T_{\phi})F \tag{4}$$

where T_{Φ} – the refrigerant temperature at the evaporator outlet, F – the area of the heat exchange surface of the evaporator 1, K – the heat transfer coefficient of the evaporator.

$$k = \frac{1}{\frac{1}{\alpha_1}\frac{F_{TP}}{F_P} + \frac{\delta_1}{\lambda_{TP}} + \frac{1}{\alpha_2}} \tag{5}$$

where F_{TP} and F_P – the areas of the outer surface of the tube and the surface of the edges of these tubes respectively, $\boldsymbol{\delta}_1$ – the thickness of the tube wall, $\boldsymbol{\lambda}_{TP}$ – the coefficient of thermal conductivity of the tube. Calculation of the heat pump was carried out according to the method described in [12].

The conversion coefficient *COP* is found from the formula

$$COP = \frac{Q}{N_{COM}} \tag{6}$$

where *Ncom* – the power of the compressor, ΔQ = Q – *Ncom*.

4. Discussion

A program was created in this study [13] that simulates the operation of a heat pump unit operating in the heat recovery scheme selected from an outlet nozzle of a heat and power generation turbine operating in ventilation mode.

This program allows to optimize the operation of the heat pump unit for the Rankine cycle with different refrigerants (R123, R113, R718, etc.) [14] and it also allows to consider various designs of the evaporator and condenser for the operation of the heat pump unit at different temperatures and pressures of the refrigerant. In particular, according to the created program, the conversion factors *COP* and ΔQ were calculated [15][16] at the boiling temperature of the refrigerant from 40 ° C to 70 ° C. The results of the calculations are shown in figures 2, 3, 4, 5, 6 and 7.

IOP Conf. Series: Materials Science and Engineering **675** (2019) 012048 doi:10.1088/1757-899X/675/1/012048

Thus, the heat losses of the heat turbine T-180 / 210-12.8, operating in the ventilation mode, are 4 MW and can be returned for additional heating of the network water by a heat pump. At the same time, as the temperature of saturated vapor, Ts increases from 40 ° C to 70 ° C for R718, the heat pump conversion coefficient *COP* increases from 37 to 571. But on the other hand, the amount of heat taken from the evaporator decreases from 4470 kW to 3840 kW. A further increase in Ts can lead to an increase in the overall dimensions of the heater 3 since the return network water enters the heating circuit with a temperature of 75 ° C. In this case, the temperature decreases to 5 °C.

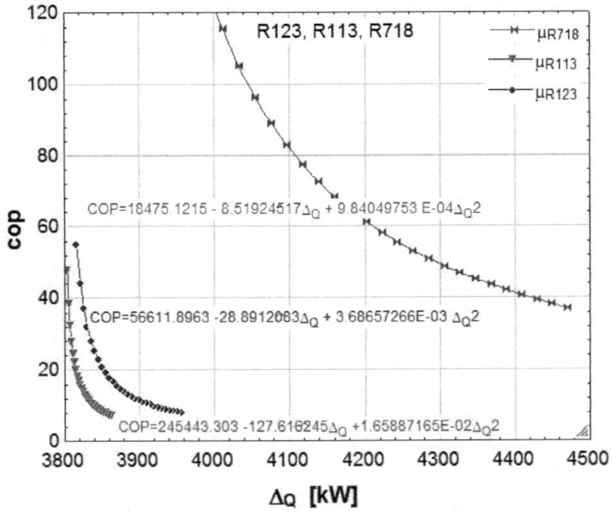

Figure 2. Coefficient of performance for different temperatures of the refrigerants (R123, R113 and R718).

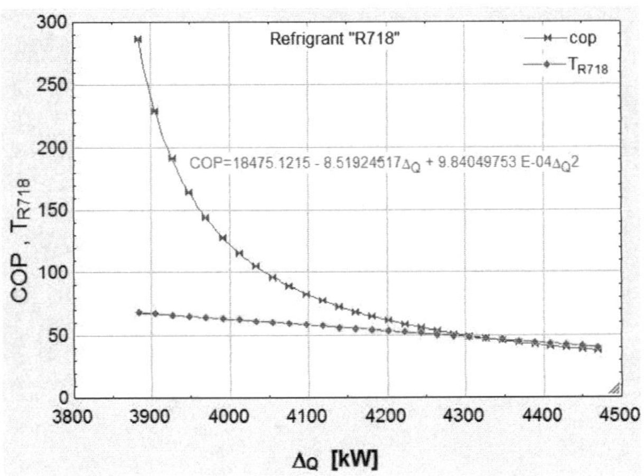

Figure 3. Relation between coefficient of performance (COP) and different temperatures of the refrigerant R718.

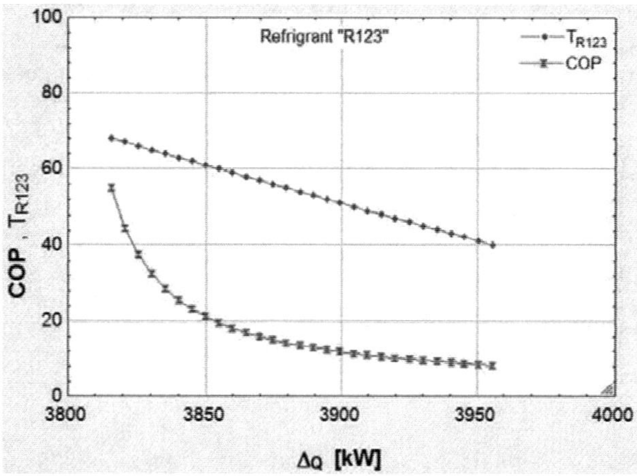

Figure 4. Relation between coefficient of performance (COP) and different temperatures of the refrigerant R123.

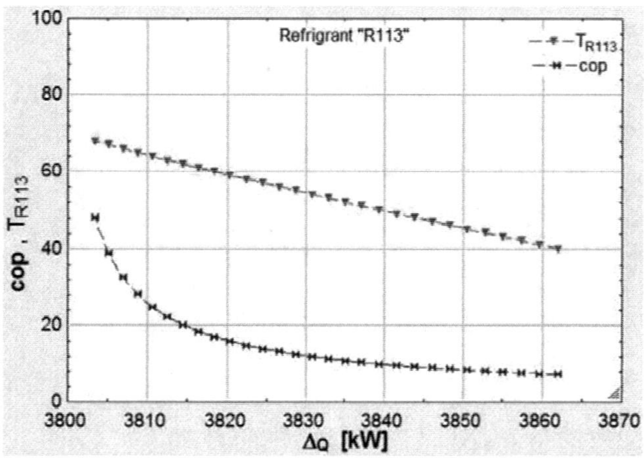

Figure 5. Relation between coefficient of performance (COP) and different temperatures of the refrigerant R113.

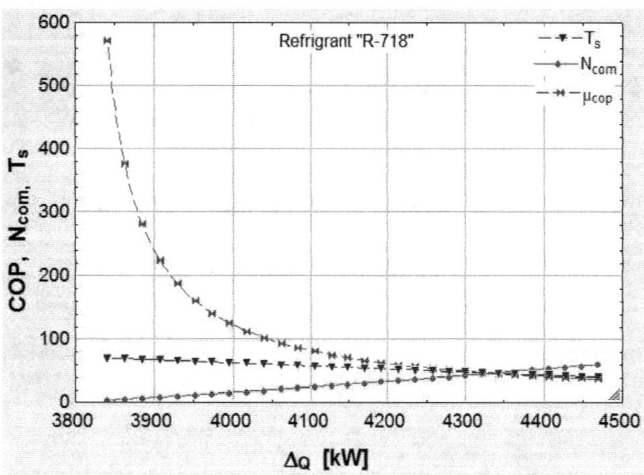

Figure 6. Relation between coefficient of performance (COP) and different temperatures of the refrigerant R718 and compressor power (*Ncom*).

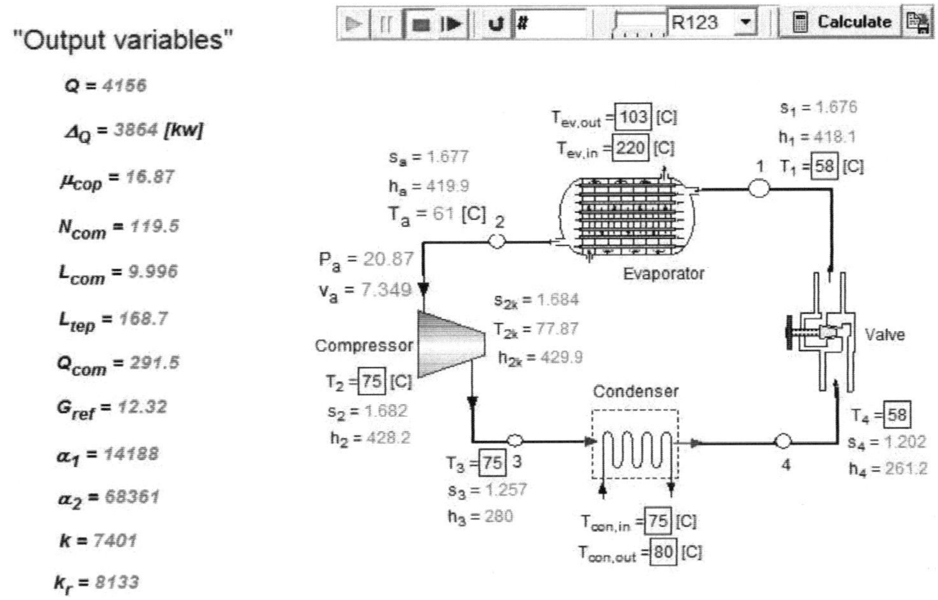

Figure 7. Visualization of the thermodynamic modeling scheme.

5. Conclusions

1. The use of the heat pump plant in the thermal scheme of the turbine T-180 / 201-12.8, operating in the ventilation mode, allows additionally to obtain commercial heat up to 4 MW.
2. As a working fluid in the heat pump plant, it is more promising to use R-718(water) refrigerant, which has a significantly higher conversion factor than R-123 and R_113 for the same initial parameters.
3. The location of the evaporator 1 at the periphery of the exhaust pipe will practically not affect the trajectory of the steam flow in it and will not lead to a significant decrease in the efficiency of the steam-turbine unit operating in the condensation mode.
4. As a result of thermodynamic modeling, dependences $\Delta Q = f$ (COP) were obtained for various refrigerants.

References

[1] Shapiro G A and Efros E I 1982 *Combined motor mode Issues of increasing the maneuverability of thermal and nuclear power plants, Gorlovka,* p. 72 – 73 [in Russian]
[2] Efros E A 1998 *Efficiency and reliability of powerful heat-generating turbines and ways to improve them* , Kirov, p. 351 [in Russian]
[3] Balabanovich V K, Levchenko B L, Makhovko Yu E, Chizh V A, Skoda N I , Soroko E V and Mikhailovsky V V 1986 The method of cooling the last stages of heat turbines on low-flow conditions *Bulletin* **30**
[4] Ivanov C H and Khaimov V A 1989 New cooling system for low-pressure cylinder turbine T-250 / 300-240 *Therm. Eng.* **6** pp.64–66
[5] Souri K and Antipov Y A 2018 *Cogeneration Steam Turbine Plant Bulletin* **34**
[6] Batenin V M, Datsenko V V, Zeigarnik Y A, Kosoi A S and Sinkevich M V 2016 Thermodynamic evaluation of the possibility to increase cogeneration turbine efficiency by using a heat pump operating with steam *Therm. Eng.* **63** 1–6

IOP Conf. Series: Materials Science and Engineering **675** (2019) 012048 doi:10.1088/1757-899X/675/1/012048

[7] Batenin V M, Zeigarnik Y A, Kosoi A S, Datsenko V V and Sinkevich M V 2015 Experimentally studying of axial compressor operation with steam *Therm. Eng.* **62** 906–10
[8] Petukhova V S and Shikova V K, Book 1987 *Heat Exchangers* (Moscow: Energoatomizdat) p.560
[9] Tsvetkov F F, Kerimov R V and Velichko V I. Book 2010, *Heat and mass transfer* , Moscow Publishing house MEI p.93
[10] Konyaeva L P, Makarova G Kh and Shonurova V N, Book 1972, *Methods and dependencies for the theoretical calculation of heat exchange and hydraulic resistance of heat exchange equipment of nuclear power plants* Moscow p.125
[11] Jiri Pitron 2015 Mathematical model of heat pump Electric Power Engineering(EPE) pp 636-640
[12] Shatalov I K, Book 2009 *Heat pump installations driven by heat engines,* Moscow Publishing house RUDN p.112
[13] Souri K, Antipov Y A and Vallejo Maldonado P R 2019 The program Thermodynamic modeling of a heat pump installation with a visual design, Moscow FIPS
[14] Sorin D. Ana Maria B. and Florin B. 2017 The up-to-date heat pump–combined heat and power solution for the complate utilization of the low enthalpy geothermal water potential *International Journal of Energy and Environmental Engineering* **8** pp 189-196
[15] Mastny P 2011 *Obnovitelne elektricke energie* Czech p 254
[16] Gabor B and Karoly P 2014 Energetic examination of air water heat pump for modeling purposes *Int. J. Eng.* **4** pp 131-134

Magnetization modes modelling and analysis for controlled reactors with a rotating field

E I Zabudskiy* and G I Balandina

Peoples' Friendship University of Russia (RUDN University), Moscow, Russia

E-mail: zabudskiy-ei@rudn.ru

Abstract. Electromagnetic reactors are used to carry out practical tasks of automatic control of power lines operation modes, distribution grids and energy supply systems of the industrial enterprises. The reactors' characteristics significantly depend on the harmonic's manifestation magnetic field saturation, which is different in the magnetization modes of reactors: forced, free and symmetrical. For the purpose of their analysis, a generalized mathematical model of the cross-sectional magnetic circuits of static ferromagnetic devices with rotating magnetic field has been developed on the base of nonlinear magnetic circuits theories. It was found that the best mode is symmetric magnetization, characterized by the following advantages: an increased range of reactive power control, enhanced stabilizing effect on the controlled current, reduced losses in steel, increased speed and the absence of "shaking" vibrations of the reactor magnetic core. It is shown that the free magnetization mode is close in its merits to the symmetric magnetization mode.

1. Introduction

In the second decade of the twentieth century, the outlines of one of the problems of high-voltage power transmission technology were outlined, consisting in the fact that at large distances the capacitive conductivity of the line starts to have a significant effect and the capacitive current significantly decreases, which reduces the line capacity. This phenomenon was accompanied by an unacceptable increase in voltage.

In the early 40-ies of XX century R. Rüdenberg (Germany) proposed to solve this problem using a strong magnetic saturation of electrical steel, pointing to the need to eliminate the higher harmonic. R. Rüdenberg's ideas were developed in the works of E. Friedlander (GEC, England). GEC company has manufactured and installed more than 50 ferromagnetic reactors at power from tens to hundreds of Mvar in various countries [1-3].

In Russia, a group of organizations led by M.A. Bryantsev mastered the production of a series of high-voltage controllable reactor 25, 32, 63, 100, 180 MV·A for the 500 kV grids. The series of arc-quenching magnetically controlled reactors with oil cooling with a capacity from 190 to 1520 kV·A and a voltage 6-10 kV were developed and massively introduced [4-6].

As a means of regulating reactive power, reactors are necessary to control the modes of electric power systems in order to solve problems: compensation of excess charging power of power lines (transmission lines) and increase their capacity, limit switching overvoltages and short-circuit currents, reduce voltage fluctuations, rational distribution of voltage and current, etc [7-9].

Controlled reactor (CR) is a static energy nonlinear device, which operation is based on the phenomenon of electromagnetic induction. The active part of the reactor contains one or more windings and a magnetic core. The reactive power it consumes is regulated by changing the saturation of the magnetic circuit. The operating range of magnetic induction values is located behind the "knee" of the magnetization curve.

This is due to the appearance of harmonics of the magnetic field saturation, which affect the characteristics of the reactors.

There are three modes of magnetization, which are determined by the manifestation of these harmonics: the forced magnetization (FM) mode in which harmonics are present in the induction of the magnetic field; the mode of free magnetization by the 2nd harmonic (MH2) in which this harmonics are manifested mainly in the magnetic field strength; the mode of symmetric magnetization (SM) in which, when magnetizing the magnetic circuit, even harmonics are absent in the induction and in the magnetic field strength [3, 8, 10].

The purpose of this research was to analyze the listed magnetization modes, their influence on the properties of the reactors and ascertain the best mode of magnetization, a positive effect on the characteristics of the reactors.

2. Method of research

The process of researching devices should be preceded by the implementation of the model-algorithm-program triad, so the device is replaced with its model, which is then analyzed using PC experimentation using computational logic algorithms.

The mathematical model, based on the theory of nonlinear magnetic circuits, takes into account the discrete structure of the tooth-slot layer, the non-magnetic gap between the stator and rotor, the spatial-temporal spectrum of saturation harmonics, and the MMF values of the reactor windings. The derivation of a mathematical model of the magnetization regimes of reactors in general form is given in [8,3,10]. For the analysis of the free magnetization mode by the 2nd harmonic, the model is presented in the form (1), and for the symmetric magnetization mode – (2). On the basis of the developed generalized mathematical model of the phenomena arising at magnetization of a magnetic circuit, the best modes of magnetization are established.

3. Magnetization modes analysis

3.1. Forced magnetization mode

The forced magnetization (FM) mode under the higher harmonics by the saturation of the magnetic field is considered as a mode in which these harmonics appear in the induction of the magnetic field but are practically absent in the field strength.

In order to explore the influence of the field higher harmonics on the operating modes of a controllable reactor (CR), the graphs, reflected the mutual influence of the main harmonics of a rotating field and a bias field are given. The nature of the change in the amplitudes of higher harmonics of saturation from various factors are also to be deduced. For that, the results of PC calculations of a nonlinear transcendental equation that simulates a magnetic circuit of a reactor operating in various magnetization modes, including the FM mode, are used [8].

Later in the document there are the calculated curves for the reactor, which correspond to rational values of the quantities defined in [6]: the ratio of the transverse geometric dimensions of the magnetic circuit $L^*_{za} = 0.22$, $d^*_a = 0.6$; the number of pole pairs of the three-phase working winding $p_1 = 1$; the relative magnitude of the nonmagnetic gap $\delta^* = 50$.

Dependencies $F^*_1 = f(F^*_{0a})$ at $B^*_1 = const$, named IO curves, are shown in figure 1. They reflect the relationship between the amplitude of the magnetomotive force (MMF) of the three-phase winding F^*_1 and MMF of the control winding F^*_{0a}, which is practically linear in the working range.

Figure 2 displays the dependence $F^*_{0a} = f(\delta^*)$ at $B^*_1 = const$ and $F^*_1 = const$. As can be seen from the graph, with an increase in the gap δ^* for the given B^*_1 и F^*_1 the MMF value of the control ring winding necessary to ensure a certain level of MMF F^*_1, decreases. This is natural, since an increase in the gap in itself leads to an increase in the MMF F^*_1. However, it is not advisable to increase the gap in the reactor in order to reduce the F^*_{0a} at $F^*_1 = const$ and $B^*_1 = const$ or in order to reduce the MMF F^*_1 at $F^*_{0a} = const$ and $B^*_1 = const$, since this drastically limits adjustment capability of the reactor.

Figure 3*a* shows the curves $B^*_1 = f(F^*_1)$ at $F^*_{0a} = const$, which should be used in the reactor calculation. It's seen from the figure, that the reactor working inductions should be taken from the part of the magnetization curve within $6 \leq B^*_1 \leq 7$. More precise recommendations on the choice of B^*_1 are given in [3, 8].

IOP Conf. Series: Materials Science and Engineering **675** (2019) 012049 doi:10.1088/1757-899X/675/1/012049

The curved $B_1^* = f(F_{0a}^*)$ at $F_1^* = const$, shown in figure 3*b* are conveniently used for the calculation when the reactor is switched on with a load [3].

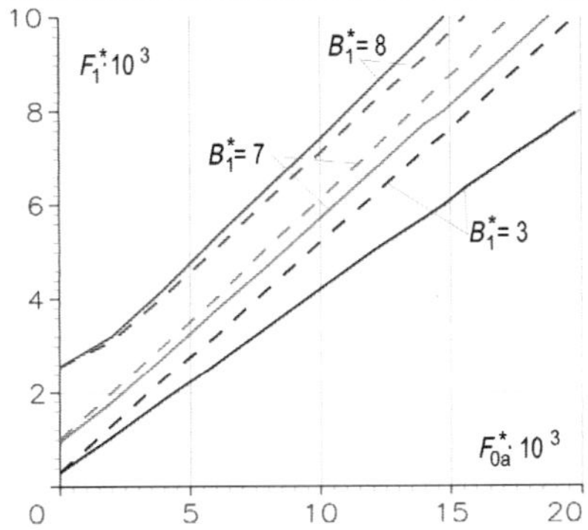

Figure 1. The dependence of the amplitude of the MMF F_1^* on the MMF F_{0a}^* with $\delta^* = 50$, $L_{za}^* = 0.22$, $d_a^* = 0.6$, $p_1 = 1$ in the modes: FM – solid lines and MH2 – dotted.

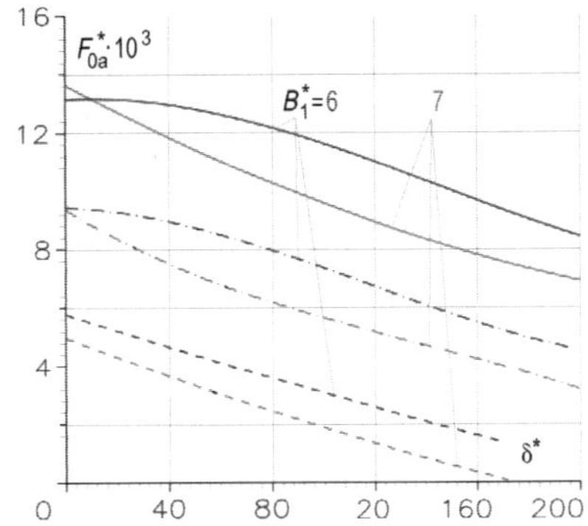

Figure 2. The dependence of the MMF F_{0a}^* on the size of the gap δ^*: the amplitude of the MMF $F_1^* = 7 \cdot 10^3$ (solid lines); $5 \cdot 10^3$ (dot-dashed); $3 \cdot 10^3$ (dotted).

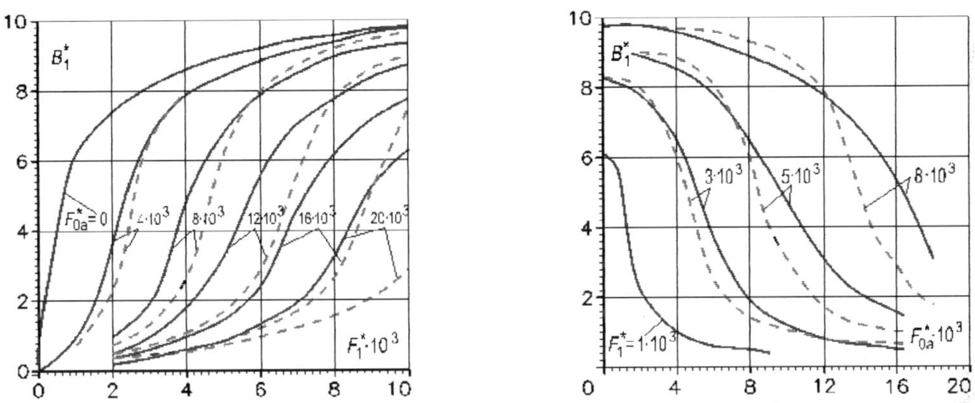

a *b*

Figure 3. The dependence of the amplitude B_1^* 1st induction harmonic in yokes in the modes: FM – solid lines, MH2 – dotted; *a* – on the amplitude F_1^* of the 1st MMF harmonic at $F_{0a}^* = const$; *b* – on the MMF value F_{0a}^* at $F_1^* = const$.

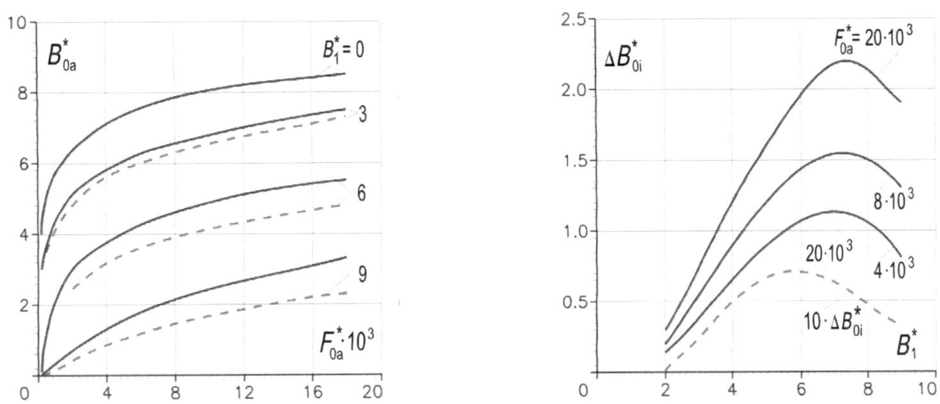

a *b*

Figure 4. The dependence of the induction value B_{0a}^* on MMF F_{0a}^* at $B_1^* = const$ (*a*) and ΔB_{0i}^* on the amplitude B_1^* at $F_{0a}^* = const$ (*b*) in the modes: FM – solid lines, MH2 – dotted.

The curves, shown on the figure 4, illustrate a decrease in the induction of a constant field with an increase in the amplitude of the main harmonic of induction B_1^* (*a*) and a change in the constant induction component ΔB_{0i}^* in the non-magnetized yoke (*b*) [8].

Next, we find out the dependence of the amplitudes of higher harmonics of saturation on the transverse geometry of the magnetic circuit and other factors.

Figure 5*a,b* shows dependences of the 2nd and 3rd harmonics of induction in the function F_{0a}^* and B_1^* for a reactor with $p_1 = 1, L_{za}^* = 0.22, d_a^* = 0.6$ and $\delta^* = 50$. The second harmonic essentially only grows with increasing bias (figure 5*a*).

With a constant magnetization and variations B_1^* the 2nd harmonic of induction has a maximum at certain values of B_1^* (figure 5*b*), which correspond to the operating inductions of the reactor. At the same time, it does not affect the shape of the controlled current curve, since having the number of pole pairs $2p_1 = 2$, the 2nd harmonic induces in the sides of the coils of a three-phase six-zone 2-pole winding EMF equal in magnitude but opposite in sign. The third harmonic for large values of B_1^*, when the steel is saturated with a rotating field, slightly depends on the degree of bias in the constant field.

The effect on the CR properties of odd harmonics of induction is small, since they have an insignificant value at $B_1^* = 6...7$ and the magnetization of the yokes. In particular, the 3rd harmonic with $B_1^* = 7$ and no bias (idle mode) is $B_3^* = 0.38$, or 5.15% of the first harmonic; at $B_1^* = 7$ and $F_{0a}^* = 8000$ $B_3^* = -0.1$, or 1.43% (figure 5*b*).

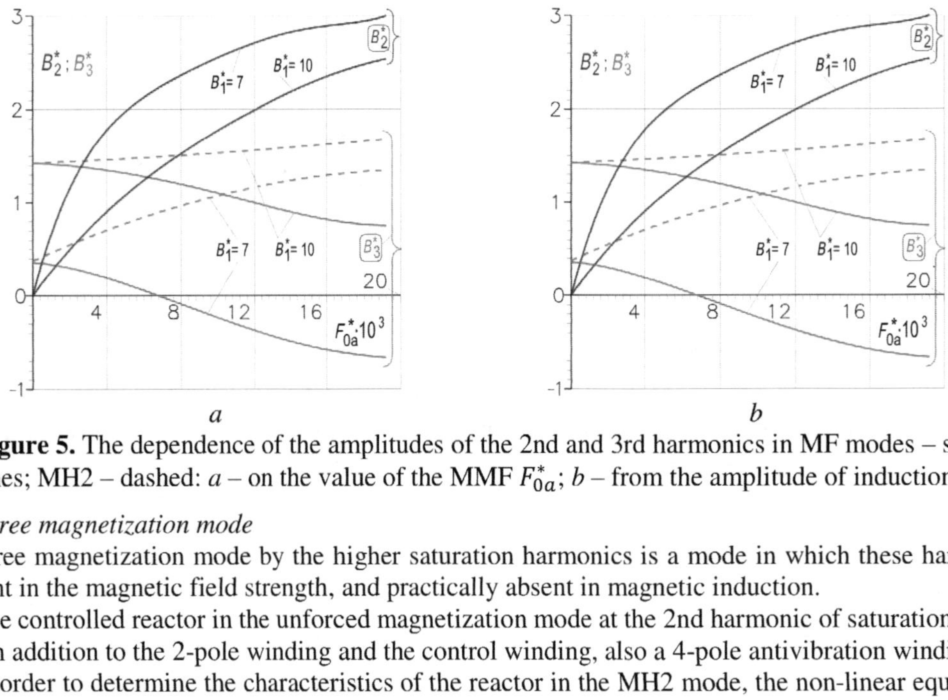

Figure 5. The dependence of the amplitudes of the 2nd and 3rd harmonics in MF modes – solid lines; MH2 – dashed: *a* – on the value of the MMF F_{0a}^*; *b* – from the amplitude of induction B_1^*.

3.2. Free magnetization mode

The free magnetization mode by the higher saturation harmonics is a mode in which these harmonics present in the magnetic field strength, and practically absent in magnetic induction.

The controlled reactor in the unforced magnetization mode at the 2nd harmonic of saturation (UM2) has, in addition to the 2-pole winding and the control winding, also a 4-pole antivibration winding.

In order to determine the characteristics of the reactor in the MH2 mode, the non-linear equation of the magnetic circuit was calculated on a PC [8].

Taking into account the assumptions made in [8] and the reactor design, the nonlinear equation for the mode being analyzed at time $t = 0$ will take the following form:

$$p_1 F_1^* \sin p_1\gamma - 2p_1 F_2^* \cos 2p_1\gamma + \frac{F_{0a}^*}{\pi} + 2\sinh(B_{0a}^* + A_1^*) +$$
$$+2k_s \sinh(\Delta B_{0i}^* + A_1^*) + 2L_{za}^* d_a^* p_1^2 A_2^* \cosh d_a^* p_1 A_3^* + 2\delta^* p_1^2 A_2^* = 0, \quad (1)$$

where A_1^*, A_2^*, A_3^* are respectively expressions

$$A_1^* = -B_1^* \sin p_1\gamma - B_3^* \sin 3p_1\gamma + B_4^* \cos 4p_1\gamma - B_5^* \sin 5p_1\gamma + +B_6^* \cos 6p_1\gamma - B_7^* \sin 7p_1\gamma;$$

$$A_2^* = -B_1^* \sin p_1\gamma - 9B_3^* \sin 3p_1\gamma + 16B_4^* \cos 4p_1\gamma - 25B_5^* \sin 5p_1\gamma + \\ +36B_6^* \cos 6p_1\gamma - 49B_7^* \sin 7p_1\gamma;$$

$$A_3^* = B_1^* \cos p_1\gamma + 3B_3^* \cos 3p_1\gamma + 4B_4^* \sin 4p_1\gamma + 5B_5^* \cos 5p_1\gamma + +6B_6^* \sin 6p_1\gamma +7B_7^* \cos 7p_1\gamma.$$

There are nine unknowns in the equation: $F_1^*, F_2^*, B_{0a}^*, \Delta B_{0i}^*, B_3^*, B_4^*, B_5^*, B_6^*, B_7^*$.

The calculation was carried out for the rational geometric relations defined in [3] $d_a^* = 0.6$; $L_{za}^* = 0.22$ with the number of pole pairs of the working winding $p_1 = 1$. When calculating the relative length of the gap δ^* is assumed to be 50.

In figure 1, 3 and 4*a* dotted lines (MH2 mode) represent the dependencies:

1) $F_1^* = f(F_{0a}^*)$, $B_1^* = const$; 2) $B_1^* = f(F_1^*)$, $F_{0a}^* = const$;

3) $B_1^* = f(F_{0a}^*)$, $F_1^* = const$; 4) $B_{0a}^* = f(F_{0a}^*)$, $B_1^* = const$.

From figure 1 and 3*a*, it follows that the control range in the MH2 mode with operating inductions $B_1^* = 6\ldots7$ is greater than in the FM mode. To achieve the same value of the MMF F_1^* in MH2 mode, a smaller value of the MMF F_{0a}^* is required than in the FM mode (figure 3*b*). Finally, the $B_{0a}^* = f(F_{0a}^*)$ curves of the analyzed mode are lower than the FM mode curves (figure 4*a*). This is explained by the decrease in the additional constant components of the magnetic induction ΔB_{0a}^* and ΔB_{0i}^*, caused by "magnetic rectification" [8]. Figure 4*b* shows the dependence $\Delta B_{0i}^* = f(B_1^*)$ at $F_{0a}^* = const$. A decrease in the induction constant components in the yokes leads to some reduction in the loss in steel. The third harmonic in the MH2 mode as compared to the FM mode does not change its phase (figure 5).

The graphs of the amplitude change of the MMF of the anti-vibration winding $F_2^* = f(F_{0a}^*)$ at $B_1^* = const$ and $F_2^* = f(B_1^*)$ at $F_{0a}^* = const$, which are used in the design of the reactor, are shown in figure 6. With working inductions $B_1^* = 6\ldots7$ and $F_{0a}^* = 12000$ the MMF value F_2^* is approximately 25% of the MDS of the working winding F_1^*.

Summing up the analysis of the free magnetization mode on the 2nd harmonic, we note that the latter favorably differs from the forced magnetization mode by a significant decrease in the vibrations of the magnetic circuit and an extended range of operating current control, as evidenced experimental data [3].

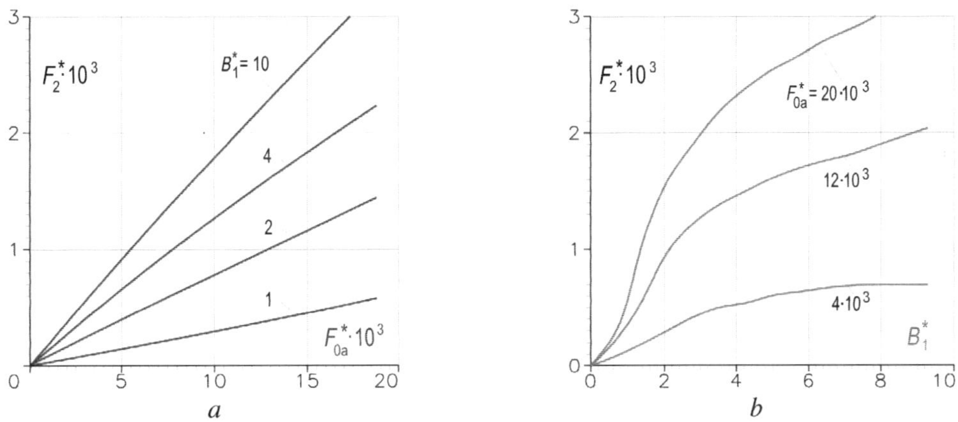

Figure 6. The dependence of the amplitude of the MMF F_2^* in MH2 mode:
a – from the value of the MMF F_{0a}^*; *b* – from the amplitude of induction B_1^*.

3.3. Symmetric magnetization mode

The symmetric magnetization mode (SM) is a mode, in which, with magnetic bias, even saturation harmonics practically do not appear either in induction or in the magnetic field.

For study of the operation of the reactor in the SN mode, we use the results of the calculation of the nonlinear magnetic circuit equation on a PC [8]. Considering the assumptions made in [8] and the reactor design, see (figure 13), this equation for the time $t = 0$ is written as

$$p_1 F_1^* \sin p_1\gamma + \frac{F_{0a}^*}{\pi} + \frac{F_{0i}^*}{\pi} + 2\sinh(B_{0a}^* + A_1^*) + 2k_s \sinh(B_{0i}^* + A_1^*) + \\ +2L_{za}^* d_a^* p_1^2 A_2^* \cosh d_a^* p_1 A_3^* + 2L_{zi}^* d_i^* p_1^2 A_2^* \cosh d_i^* p_1 A_3^* + 2\delta^* p_1^2 A_2^* = 0, \qquad (2)$$

where A_1^*, A_2^*, A_3^* are respectively expressions

$$A_1^* = -B_1^* \sin p_1\gamma + B_2^* \cos 2p_1\gamma - B_3^* \sin 3p_1\gamma + B_4^* \cos 4p_1\gamma - B_5^* \sin 5p_1\gamma + \\ +B_6^* \cos 6p_1\gamma - B_7^* \sin 7p_1\gamma;$$

$$A_2^* = -B_1^* \sin p_1\gamma + 4B_2^* \cos 2p_1\gamma - 9B_3^* \sin 3p_1\gamma + +16B_4^* \cos 4p_1\gamma - 25B_5^* \sin 5p_1\gamma + \\ +36B_6^* \cos 6p_1\gamma - 49B_7^* \sin 7p_1\gamma;$$

$$A_3^* = B_1^* \cos p_1\gamma + 2B_4^* \sin 2p_1\gamma + +3B_3^* \cos 3p_1\gamma + 4B_4^* \sin 4p_1\gamma + 5B_5^* \cos 5p_1\gamma + \\ +6B_6^* \sin 6p_1\gamma + 7B_7^* \cos 7p_1\gamma.$$

There are nine unknowns in the equation (2): $F_1^*, B_{0a}^*, B_{0i}^*, B_2^*, B_3^*, B_4^*, B_5^*, B_6^*, B_7^*$.

The equation is solved at the number of pole pairs of the working winding $p_1 = 1$. The stator and rotor correspond to coefficients $L_{za}^* = 0.22$; $d_a^* = 0,6$ and $L_{zi}^* = 0.11$; $d_i^* = 0.6$.

The ratio of the radii of the average circles of yokes is $k_s = 0.475$. The length of the gap is $\delta^* = 50$.

To illustrate the fact of the absence of even harmonics with symmetric magnetization the figure 7 shows dependences $B_{2,3,4}^* = f(F_{0i}^*, B_1^*)$ at $F_{0a}^* = 8000$. The first quadrant corresponds to the consonant action of the MMFs of the control windings F_{0a}^* and F_{0i}^* with respect to the contour of the rotating field. Therefore, the even harmonics of the same name, generated respectively by the outer and inner yokes, coincide in phase, and the amplitudes of the resulting even harmonics have the greatest value. With a decrease in the MMF, F_{0i}^* the resulting even harmonics also decrease [3].

Since there are no even harmonics with symmetric magnetization, there is no "magnetic rectification" in the yokes [8]. Therefore, in the SM mode, the magnitudes of the constant magnetic inductions B_{0a}^* and B_{0i}^* take their natural values, determined respectively by the MMF F_{0a}^* and F_{0i}^*, the amplitude B_1^* and the magnetic state of the system.

The curves $B_{0a}^*, B_{0i}^* = f(F_{0i}^*, B_1^*)$ at $F_{0a}^* = 8000$ are shown on the figure 8. If $B_1^* = 0$ and $F_{0i}^* = var$, then the induction value $B_{0a}^* = 7.8$ remains constant, and the value B_{0i}^* changes, and the graph $B_{0i}^* = f(F_{0i}^*)$ is located in the 1st and 3rd quadrants symmetrically relative to the origin of the axes of coordinates. When $B_1^* > 0$ and $F_{0i}^* = var$ in the general case even harmonics appear in the field and "magnetic rectification" occurs in the yokes – these are additional constant induction components ΔB_{0a}^* and ΔB_{0i}^*, due to the imposition of even and odd saturation harmonics in them. With variations of B_1^* and F_{0i}^*, the saturation harmonics change, and hence the values ΔB_{0a}^* and ΔB_{0i}^*. Therefore, with a change in the MMF F_{0i}^* within $-8000 < F_{0i}^* \leq 10000$, the induction B_{0a}^* decreases with the transition from the 1st quadrant to the 2nd. When $F_{0i}^* = -8000$ there are no even harmonics in the field, and $\Delta B_{0a}^* = 0$.

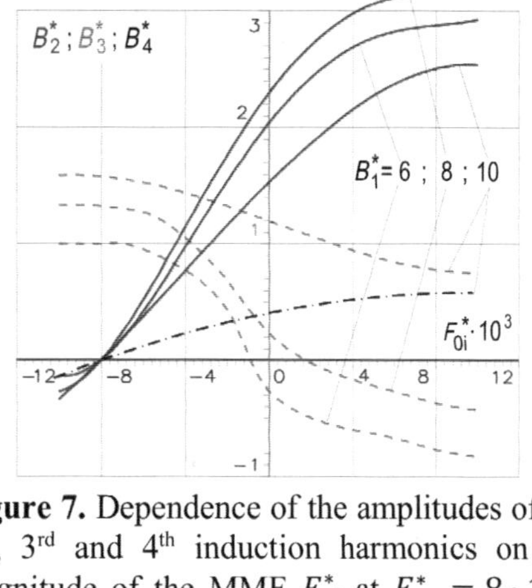

Figure 7. Dependence of the amplitudes of the 2nd, 3rd and 4th induction harmonics on the magnitude of the MMF F_{0i}^* at $F_{0a}^* = 8 \cdot 10^3$: B_2^* – solid lines; B_3^* – dotted; B_4^* – dash-dotted.

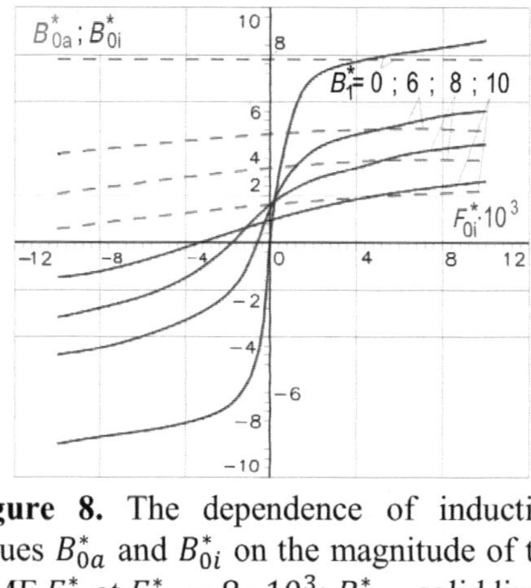

Figure 8. The dependence of induction values B_{0a}^* and B_{0i}^* on the magnitude of the MMF F_{0i}^* at $F_{0a}^* = 8 \cdot 10^3$: B_{0i}^* – solid lines; B_{0a}^* – dash-dotted.

For the values of MMF $|F_{0i}^*| > 8000$ (2nd quadrant) the phase of the resulting even harmonics and the sign ΔB_{0a}^* change to opposite ones. As for the $B_{0i}^* = f(F_{0i}^*, B_1^* \neq 0)$ curves at $F_{0a}^* = 8000$, their symmetry relative to the origin becomes broken due to the "magnetic rectification". When MMF

$F_{0i}^* = 0$, the presence of a constant induction in the inner yoke is determined only by "magnetic rectification", $B_{0i}^* = \Delta B_{0i}^*$. In order to avoid a constant induction in the inner yoke, it is necessary to apply some negative value of the MMF F_{0i}^*. For example, for $B_1^* = 8$ and $F_{0a}^* = 8000$ the induction $B_{0i}^* = 0$, if $F_{0i}^* = -1750$. The difference between the induction B_{0i}^* at $B_1^* = const$, $F_{0i}^* = F_{0a}^* = 8000$ and $F_{0i}^* = -F_{0a}^* = -8000$ is caused by "magnetic rectification".

An interesting fact is that the symmetric magnetization mode affects the 3rd harmonic of magnetic induction qualitatively the same as the mode of free magnetization by the 2nd harmonic (figure 5 and 9). In [3], experimental analogs of the calculated dependences shown in figure 9*b*.

Dependencies $B_{0a}^*, B_{0i}^* = f(F_{0a}^* = |-F_{0i}^*|, B_1^*)$ in SM mode, when $F_{0a}^* = |-F_{0i}^*|$, and in FM mode, are shown in figure 10. Curves must be used when calculating the loss in steel. As follows from figure 10, induction B_{0a}^* and B_{0i}^* in the SM mode are always less than those in the FM mode. Therefore, the yoke losses in the SM mode will be less than in the FM mode.

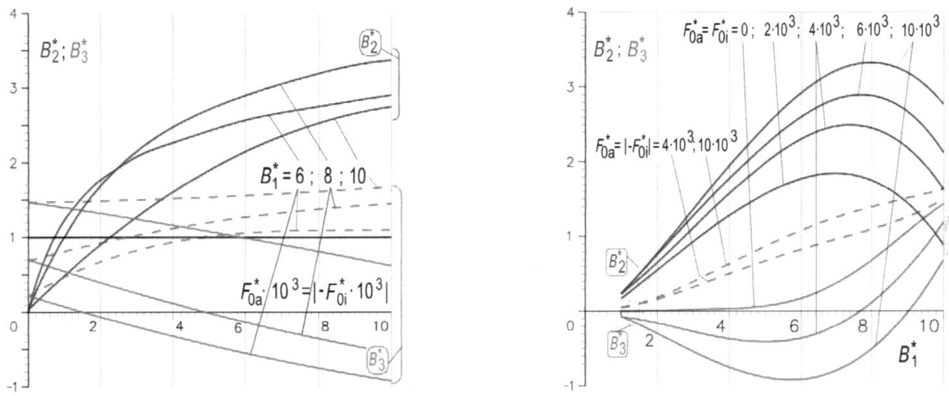

a *b*

Figure 9. The dependence of the amplitudes of the 2nd and 3rd harmonics in FM mode – solid lines; SM mode – dotted: *a* – on of MMF control windings; *b* – on of the induction amplitude B_1^*.

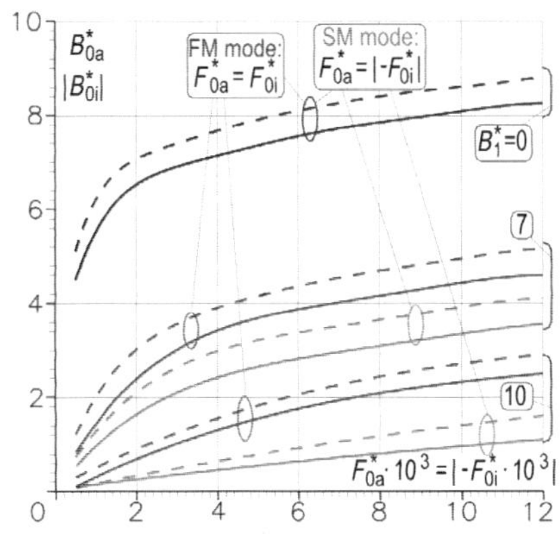

Figure 10. The dependence of the induction values B_{0a}^* and B_{0i}^* on the magnitude of the MMF control windings in the FM and SM modes: B_{0a}^* – solid lines; B_{0i}^* – dotted.

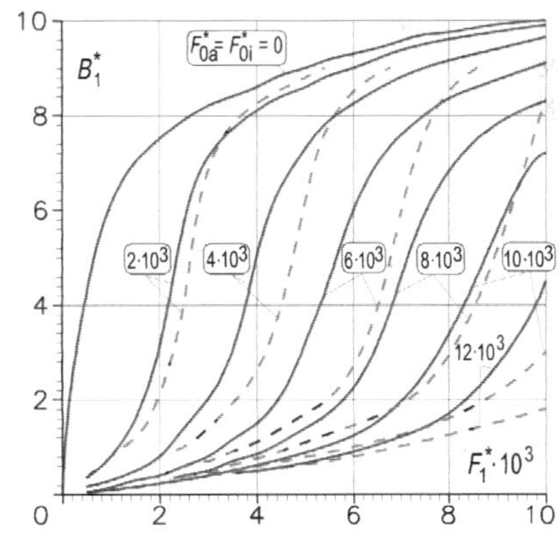

Figure 11. Dependence of amplitude B_1^* 1st harmonic of induction on amplitude F_1^* MMF: in FM mode – solid lines, $F_{0a}^* = F_{0i}^*$; in SM mode – dotted, $F_{0a}^* = |-F_{0i}^*|$.

The dependencies $B_1^* = f(F_1^*, F_{0a}^* = |-F_{0i}^*|)$, illustrating the increase in the range of current control in the SM mode, are shown on the figure 11. These graphs used in calculation of the reactor.

The mode of SM favorably differs from the mode of FM by elimination of the "shaking" vibrations, by the extended control range of the operating current, reduced losses in steel, enhanced stabilizing effect over current and increased speed, which is confirmed by the calculation and experiment [3].

The photo (figure 12) shows a prototype of a reactor, which was used study of the dependence spectrum of the induction of a rotating field in forced and symmetric magnetization modes on the amplitude of the first harmonic of field and control windings.

The photograph (figure 13) shows reactor with symmetrical magnetization, intended for use as a regulating element of the plasmatron power source air plasma cutting of metals [3].

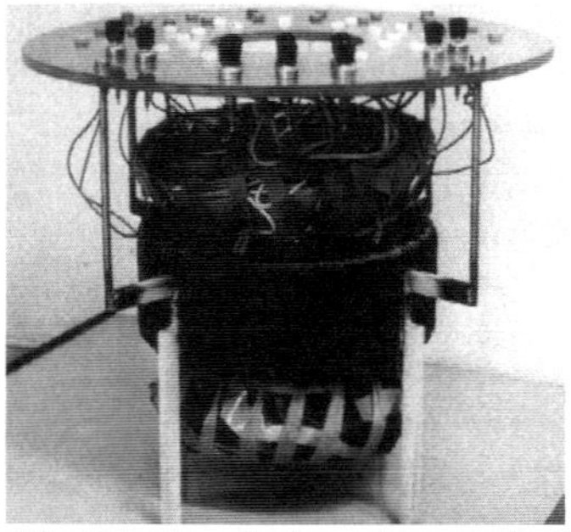

Figure 12. Reactor designed to study the field spectrum and magnetization modes.

Figure 13. Reactor is made of stator plates of the stator from the engine VAO14-4.

Conclusions

- It is shown that the optimal, in terms of maximum current gain, is a reactor in the symmetric magnetization mode, which has a three-phase 2-pole working winding and transverse geometry of the magnetic circuit, corresponding to the stator plates of standard 4-pole AC electrical machines.
- It was found that the reactor in the symmetric magnetization mode favorably distinguishes the following qualities: increased control range of reactive power, enhanced stabilizing effect on current, reduced steel losses, increased speed and the absence of "shaking" vibrations of the magnetic circuit.

References

[1] Alexandrov G N 2002 Fast controllable reactor of transformer type 420 kV, 50 MV *Electrical Technology Russia* **3**
[2] Belyayev A N, Evdokunin G A, Smolovik S V and Chudnyi V S 2009 Use controlled transverse compensation devices in 500 kV transit transmission lines *Electrical Technology Russia* **2** pp 2-13 URL: https://elibrary.ru/item.asp?id=11700350
[3] Zabudskiy E I 2003 Combined adjustable electromagnetic reactors (Moscow: Publishing House Energoatomizdat) p 436
[4] Bryantsev A M 2003 Power reactors controlled by bias magnetization – as an element of the electroenergy system *Electrical Technology Russia* **1**
[5] Dolgopolov A G 2014 Controlled shunt reactors. The principle of operation, construction, relay protection and automation (Moscow: Publishing House Energia) 120 p
[6] Bryantsev A M, Makletsova E E *et al* 2003 Shunting reactors controlled by bias magnetization for (35-500)-kV grids *Electrical Technology Russia* **74(1)** pp 4-12
[7] Dolgopolov A G, Sokolov S E 2012 Controlled reactors. Technology review *Electrical Engineering News* **75(3)**
[8] Zabudskiy E I 2018 Mathematical modeling of controlled electromagnetic reactors (Moscow: Publishing House Megapolis) p 356
[9] Bengtsson C, Gajic Z, Khorami M 2012 Dynamic compensation of reactive power by variable shunt reactors: Control strategies and algorithms *CIGRE 2012* **C1-303**
[10] Zabudskiy E I 2016 Magnetization modes of controlled electro-magnetic reactors *Proc. XVI Int. Conf. on Electromechanics, Electrotechnology, Electromaterials and Components* (Moscow, Russia) pp 171–172

Acknowledgments
The publication has been prepared with the support of the "RUDN University Program 5-100".

IOP Conf. Series: Materials Science and Engineering **675** (2019) 012049 doi:10.1088/1757-899X/675/1/012049

Authors' background

Name	*Prefix*	*Research Field*	*E-mail*	*Personal web-site*
Zabudskiy E I	Full Professor	adjustable electromagnetic facilities for electrical power engineering control systems; mathematical modelling of facilities "anatomy" and "physiology" based on Maxwell's field theory, theory of nonlinear circuits and computer technologies; microprocessor systems for electrical power engineering controlled reactor, rotating field, cross-sectional magnetic circuit, forced magnetization, free magnetization, symmetrical magnetization	zabudskiy-ei@rudn.ru	http://zabudsky.ru
Balandina G I	Associate Professor	control system synthesis, energy facilities automatic control systems, genetic algorithms	balandina-gi@rudn.ru	-

IOP Conf. Series: Materials Science and Engineering **675** (2019) 012050 doi:10.1088/1757-899X/675/1/012050

The study of fuel mixtures with palm oil

P P Oshchepkov[1], S V Smirnov[1], I A Zaev[2] and A A Vorobyev[1]

[1] Peoples' Friendship University of Russia (RUDN University), Moscow, Russia
[2] LLC "Kintekh Lab", Moscow, Russia

E-mail: oshchepkov_pp@pfur.ru

Abstract. The search for alternative fuels for heat engines is an urgent task for scientists around the world. For diesel engines, vegetable oils seem to be the most promising. The use of a partial addition of vegetable oils to diesel fuel allows us to reduce the consumption of hydrocarbons and not cause significant changes in the design of the engine, as well as affect the composition of the exhaust gases. For countries producing palm oil, including Nigeria, this method is economically viable. This article presents the results of computational studies of the features of the combustion process of biodiesel fuel with additives of palm oil and hydrogen peroxide and the effect of additives on the ignition delay period and the combustion process. The calculations have shown the effectiveness of using hydrogen peroxide as an additive to the biofuel under consideration to reduce the ignition delay period and, consequently, improve the power and economic indicators.

Keywords: diesel, alternative fuels, biofuel, biodiesel, palm oil, ignition delay period, hydrogen peroxide additives.

1. Introduction

Transport plays a huge role (over 55% [1]) in forming demand on oil and liquid biofuels worldwide. The stock of hydrocarbons reduces, that is why in some countries the research of alternative fuels for internal combustion engines are held. According to the forecast of energy development worldwide [1] demand on liquid fuels increase in Africa is expected – on 73%: from 183 mtoe in 2015 to 316 mtoe in 2040. Alternative fuels are expected to become 20% of overall energy demand [1] by 2040.

Ethanol and biodiesel – are two most common biofuels [2]. Biodiesel fuel could be produced from a wide specter of vegetable oils. Interesting reviews of biodiesel fuels production give useful information about typical content of original source of fat acid [3-9].

Vegetable oils are a perspective source of alternative fuels. There are more than 350 cultures, such as sunflower, soy, seeds of cotton, rapeseed and peanuts which are potential alternative fuels for diesel engine. [10] the actual usage of vegetable oils in diesel engines meets certain difficulties due to the difference of physical and chemical properties between vegetable fats and petroleum fuels. Disadvantages of those fuels in comparison with diesel fuels are higher viscosity (2-10 times), bad dispersion and auto oxidation [11], bad low-temperature characteristics, low cetane number, high ignition temperature, increased coking ability. In addition to this, calorific value is 7-10% lower than diesel fuels because of the oxygen compounds. That is why vegetable oils are usually used as fuels after refining or in a mix with diesel or alcohol fuels.

Lately, developing countries, such as Malaysia, Indonesia, Thailand, Nigeria pay great attention to the production of biofuels from internal, renewable resources.

Nigeria is in the TOP-10 of palm oil producing countries. Car transport in Nigeria implements the main amount of cargo and passenger transportation [12]. Agriculture occupies 65% of the population. According to *The Global Petroleum Club*, palm oil has the highest productivity (among plants) in kilograms and liters of oil received from hectare.

Application of the palm oil as a supplement to the main fuel in transport sector in agriculture of Nigeria, allows reaching significant economy of fuel, reduce transport costs of petroleum and decrease the amount of flue gases, raising environmental friendliness of diesel transport used in agricultural areas of the country.

Indonesia and Malaysia are world leaders in palm oil production. Scientific research is held in those countries in order to use palm oil as the alternative fuel for diesel engines.

Having physico-chemical properties of vegetable oils compared [13], it is possible to say that palm oil in a number of characteristics (heat of combustion, stoichiometric ratio, cetane number etc.) is the closest to the traditional diesel fuel. At the same time palm oil has some serious differences in characteristics from traditional petroleum fuel (kinematic viscosity, freezing temperature) and that makes it impossible to use pure palm oil in diesel engines without changing the construction of the engine and fuel system. Therefore, mostly acceptable way is using diesel fuel with supplement of palm oil in diesel engines without significant construction changes.

Power and economical rates of engine's work mainly are defined by burning process, therefore, features of biofuel burning process need to be examined basing on palm oil and determine the influence of the hydrogen peroxide and ethanol supplements under consideration on the process. On the basis of this research develop method of biodiesel fuel burning control based on palm oil.

2. Materials and methods

There are two processes depending on fuel's reactional abilities, which need to be pointed out: mixture of air and fuel auto-ignition moment and process of laminar flame diffusion.

Next processes take place while fuel burns inside diesel engine:

1. Fuel injection and spraying
2. Evaporation of the first portions of fuel with generation of homogeneous mixture of air and fuel
3. Auto-ignition of the evaporated fuel in generated mixture of air and fuel
4. Evaporation of the fuel left and its combustion in diffusion-limited mode (turbulent diffusion of fresh air outside torch's line)

During the research of reactional ability of diesel fuel with supplements based on palm oil (further it will be called biodiesel) and its mixtures with peroxides comparative calculations both auto-ignition characteristics (ignition delay) and laminar flame characteristics (laminar flame speed) are of interest.

Auto-ignition and diffusion of the laminar flame are determined by the speeds of chemical reactions. Appropriate chemical model of burning should be chosen for fuel combustion.

Chemical model of biodiesel fuel burning (list of the substances with their thermodynamic and thermophysical properties, list of reactions with approximations correlations between speed constants and pressure/temperature) is suggested in the work [14] for a wide range of fuel components, including biodiesel fuel based on palm oil. Verification of this mechanism took place basing on a great amount of experimental data of auto-ignition delay and laminar flame diffusion speed in air and fuel mixtures. Therefore, this model is used in this research to evaluate burning characteristics of the biodiesel fuel and biodiesel fuel with supplements in comparison with diesel fuel.

The analysis of the accepted model [14] shows that it includes such components as hydrogen peroxide H_2O_2 and ethanol C_2H_5OH, also intermediate products and their dissociation reactions, interactions with oxygen from the air, radicals. Therefore, this model can be applied for analysis of the influence of hydrogen peroxide and ethanol on auto-ignition and laminar burning of biodiesel fuel based on palm oil.

Diesel fuel has complicated composition, for calculations n-heptane based model fuel is used and it is chosen so its cetane number would match diesel's. Its burning model was verified on a number of experiments and it is also taken from [14], which is versatile for a whole group of hydrocarbon fuels.

Next issues are to be solved in order to evaluate burning properties of biodiesel fuel and influence of hydrogen peroxide and ethanol supplements on the burning process:

1. Match characteristics of biodiesel fuel based on palm oil and model diesel fuel based on n-heptane, including ignition delay and laminar flame speed for distinctive diesel engine working modes basing on the kinetic calculations.

IOP Conf. Series: Materials Science and Engineering **675** (2019) 012050 doi:10.1088/1757-899X/675/1/012050

2. Study increasing of reactional ability of biodiesel fuel based on palm oil in relation with volume fraction of hydrogen peroxide and ethanol supplements (gas phase) in comparison with original biodiesel and diesel fuels.

3. Results and discussion.

A series of calculations and studies of model diesel fuel, pure diesel fuel base on palm oil and with supplements of hydrogen peroxide (mole ratio from 0 to 0.1 relating to biodiesel fuel) and ethanol (mole ratio from 0 to 0.1 relating to biodiesel fuel) auto-ignition characteristics took place in order to solve existing problems.

Results of the ignition delay calculations of the model biodiesel fuel based on palm oil without supplements and biodiesel with hydrogen peroxide supplements are shown in figure 1.

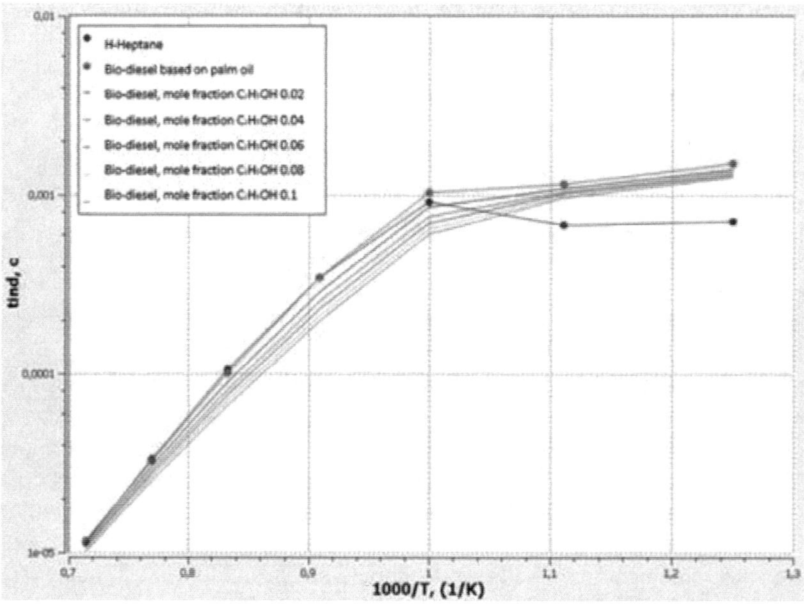

Figure 1. Ignition delay of stoichiometric mixture of biodiesel fuel (based on palm oil) and air (pure and with hydrogen peroxide supplements), diesel fuel and air, biodiesel fuel with supplements of hydrogen peroxide: initial pressure 40 bars, initial temperature 800-1400 K.

Results of the ignition delay calculations of the model biodiesel fuel based on palm oil without supplements and biodiesel with ethanol supplements are shown in figure 2. All calculations were made under next conditions [15]:

- Pressure in the system 40 bars;
- Initial temperature from 800 K up to 1400 K;
- Stoichiometric air-fuel mixture;
- Hydrogen peroxide and ethanol supplements from 1 to 0.1 moles on 1 mole of fuel.

Diesel and biodiesel fuel based on palm oil ignite almost at the same time at high temperatures, over 1000 K. Main differences could be noticed at initial temperatures lower than 1000 K, which are mostly common for the diesel engine auto-ignition conditions. Biodiesel fuel ignites significantly later (differences in auto-ignition delay may reach 100% with initial temperature 800 K). Hydrogen peroxide supplements in 0.02 to 0.1 moles range on 1 mole of fuel can significantly change reactional ability of biodiesel fuel based on palm oil. With initial temperatures at 1000 K - 1100 K and hydrogen peroxide supplements from 0.02 moles/(moles of fuel) ignition delay reduces on 0.2 ms and becomes lower than diesel fuel. In the range of temperatures from 800 K – 1000 K hydrogen peroxide supplements from 0.1 moles/(moles of fuel) allow to reduce the ignition delay on 0.15 ms, which brings biodiesel based on palm oil noticeably closer to the ignition delay parameters of diesel fuel.

IOP Conf. Series: Materials Science and Engineering **675** (2019) 012050 doi:10.1088/1757-899X/675/1/012050

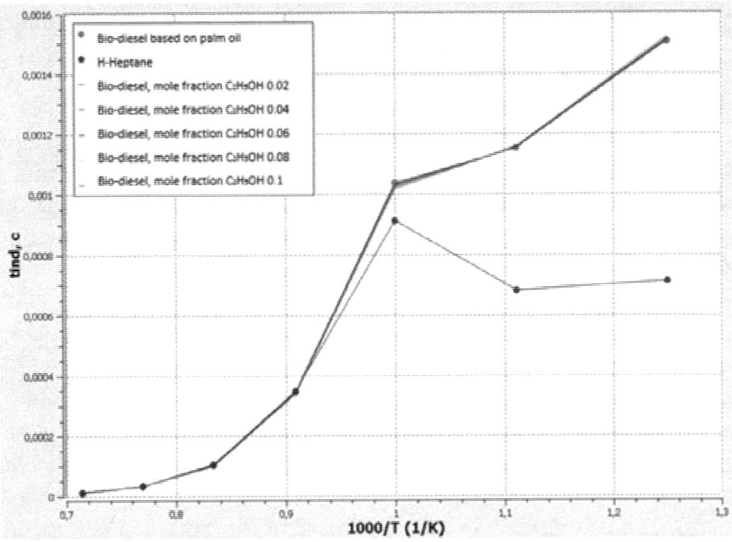

Figure 2. Ignition delay of stoichiometric mixture of biodiesel fuel (based on palm oil) and air (pure and with ethanol supplements), diesel fuel and air, biodiesel fuel with supplements of hydrogen peroxide: initial pressure 40 bars, initial temperature 800-1400 K.

Ethanol supplements to biodiesel based on palm oil do not give us considerable increase of the reactional ability of the fuel being studied (Figure 2).

A series of calculations took place nest to research characteristics of laminar burning of homogeneous mixtures of biodiesel fuel based on palm oil and air with hydrogen peroxide (mole ratio from 0 to 0.1 relating to biodiesel fuel) and ethanol (mole ratio from 0 to 0.1 relating to biodiesel fuel) supplements. Reliance of calculated speed of laminar burning with hydrogen peroxide and ethanol supplements concentration is presented in figure 3. Values of laminar flame in biodiesel mixture fuel/air and diesel fuel (n-heptane)/air are taken as reference levels. All calculations were made under next conditions [15]:

- Stoichiometric mixture fuel-air;
- Fresh mixture temperature 300 K;
- Pressure 1 bar;
- Hydrogen peroxide and ethanol supplements 0-0.1 mole/(moles of fuel).

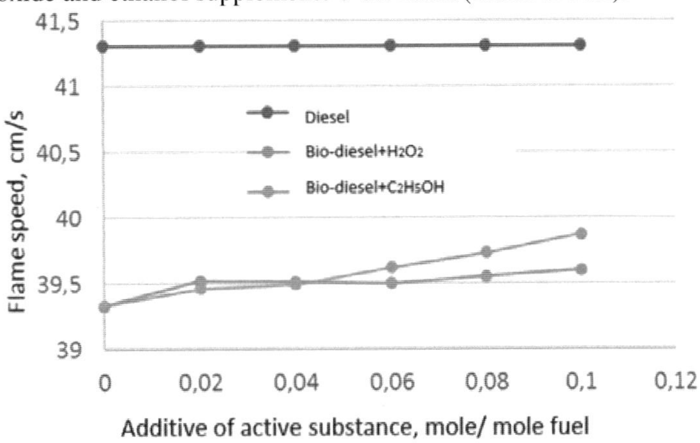

Figure 3. Diffusion speed of laminar flame in stoichiometric mixture of biodiesel fuel (based on palm oil) and air (pure and with ethanol supplements), diesel fuel and air. Pressure 1 bar, initial temperature 300 K.

Hydrogen peroxide and ethanol supplements to biodiesel fuel based on palm oil do not change significantly laminar flame speed in stoichiometric fuel-air mixture (Figure 3) and laminar burning process cannot be substantially accelerated with considered supplements.

4. Conclusions

Even 0.02 moles of hydrogen peroxide supplements on 1 mole of fuel (or 0.0025 g of peroxide on 11 grams of fuel, approximately it is 2 ml of peroxide on 1 liter of fuel) provided the reduction of ignition delay to the numbers comparably lower than diesel. The highest effect hydrogen peroxide in such small concentration may cause on fuel ignition at temperatures 1000 K – 1100 K and make initially less reactional biodiesel fuel more reactional than diesel fuel. Inside low temperature area the efficiency of the supplements to the fuel reduces and concentration of hydrogen peroxide may be increased.

Presented results of the calculations, dependence of the ignition delay from the initial temperature and concentration of the supplement (hydrogen peroxide) permit to suggest the next method of biodiesel fuel based on palm oil burning process control. Having set initial temperature of auto-ignition, it is possible to determine the exact amount of hydrogen peroxide needed to provide auto-ignition delay equal or close to biodiesel fuel with the same conditions. Fuel injection system should supply calculated values of hydrogen peroxide relating to engine working mode in future.

References

[1] Makarov A A, et al 2016 *Energy development forecast in Russia and worldwide* (Publishing House INEI RAN-AC, Moscow)

[2] Aziz A A, Said M F, Awang M A and Said A 2006 The Effects of Neutralized Palm Oil Methyl Esters (NPOME) on Performance and Emission of a Direct Injection Diesel Engine *Proc. 1st International Conference on Natural Resources Engineering and Technology* (Putrajay, Malaysia)

[3] Maa F and Hanna M A 1999 *Bioresour. Technol.* **70(1)** pp 1–15

[4] Leung D Y C, Wu X and Leung M K H 2010 *Appl. Energy* **87** pp 1083–1095

[5] Singh S P and Singh D 2010 *Renew. Sustain. Energy Rev.* **14** pp 200–216

[6] Lin L, et al 2011 *Appl. Energy* **88** pp 1020–1031

[7] Grana R, et al 2012 *Energy* **43** pp 124–139

[8] Grana R, et al 2012 *Combust. Flame* **159** pp 2280–2294

[9] Dayma G, Togbe C and Dagaut P 2009 *Energy Fuels* **23** pp 4254–4268

[10] Rodjanakid K-O and Charoenphonphanich C 2004 Performance of an Engine using Biodiesel from Refined Palm Oil Strearin and Biodiesel from Crude Coconut Oil *Proc. Joint International Conference on Sustainable Energy and Environment* (Hua Hin, Thailand)

[11] May C Y, et al 2005 Palm Diesel: Green and Renewable Fuel from Palm Oil *Palm Oil Developments* **23** pp 3-7

[12] Shumilov K 2004 Transport infrastructure of Nigerian Republic *Foreign military review* **7** pp 18-22

[13] Gusakov S V 2008 *Application possibilities of alternative fuels and renewable energy sources ibn diesel engines: studying manual* (Publishing House of the RUDN University, Moscow)

[14] Saggese C, et al 2013 A lumped approach to the kinetic modeling of pyrolysis and combustion of biodiesel fuels *Proceedings of the Combustion Institute* **34** pp 427–434

[15] Kavtaradze R Z 2008 *Theory of piston engines* (Publishing House of the Bauman Moscow State Technical University, Moscow)

Relationship between hardness and optical properties of diamond-like carbon coatings

C V Suasnavas[1*], M O Makeev[1,2], A S Osipkov[1,2], N B Solano[1], A E Shupenev[2] and P A Mikhalev[1,2]

[1] Peoples' Friendship University of Russia (RUDN University), Moscow, Russia
[2] Bauman Moscow State Technical University, Moscow, Russia

E-mail: carlos.suasnavaslagos@gmail.com

Abstract. Correlation between hardness and optical properties of diamond-like carbon (DLC) coatings is considered. Various methods for the characterization of mechanical, chemical and optical properties of DLC coatings are used; however, some of these methods are destructive, and others are non-destructive. It was found that optical properties of DLC coatings are proportional to their hardness. The mathematical expression that allows to calculate the hardness of a DLC coating according to its refractive index is proposed. Therefore, it is possible to avoid the use of destructive methods (such as nano-indentation technique) to characterize DLC coatings.

Keywords: diamond-like carbon coating, functional properties, optical properties, non-destructive methods, hardness, refractive index, correlation.

1. Introduction

The DLC (diamond-like carbon) coatings are widely used in different fields of science, technology and industry [1-9]. The spectrum is of such magnitude that involves applications in biomedicine, goes through coatings for optical elements, and comes to contribute in the automotive industry, citing a few of the countless applications. Regardless of the application to which the DLC coating is intended, it will always be essential to measure and evaluate the properties of the obtained DLC film to ensure its benefits.

When characterizing functional properties of DLC coatings, e.g., wear resistance, adhesion force, hardness, etc., these properties and the film itself tend to be degraded because the characterization methods are often destructive. On the other hand, when it is required to characterize structural and physical-chemical properties, non-destructive methods are used. In such a way that nowadays it tends to develop mathematical models that interrelate functional, structural, and physicochemical variables of DLC coatings, which at the end will allow to determine the functional properties with high precision and using only non-destructive methods of characterization. In table 1 a summary of typical DLC characterization methods is presented, emphasizing their destructive or non-destructive character.

Table 1. Typical DLC Characterization Methods.

Non-destructive		**Destructive**	
Methods	Characteristics	Methods	Characteristics
Raman Spectroscopy	sp^3/sp^2 bonding ratio, residual stress	Nanoindentation	Young's modulus, hardness, fracture toughness, adhesion force, residual stress
Ellipsometry	optical constants, thickness	Electron Energy-Loss Spectroscopy	sp^3 fraction

Scanning Electron Microscopy (SEM)	topography, composition, defectiveness	Atomic Force Microscopy (AFM)	thickness, roughness

In a second instance, establishing relationships among diverse characteristics, determining the influence of certain characteristics on others, or arriving to conclude that there is a dependence on one characteristic as a function of others, gives the opportunity of a broad understanding about the behavior of this singular amorphous carbon nanostructure, and consequently, it is transformed into valuable information, e.g., in optimization processes during deposition of DLC coatings.

A third argument, no less important, is of quantum conception. The double-slit experiment makes it clear that there is a great uncertainty when making measurements at the nanometer scale, at the quantum scale; the simple fact of making a measurement, however subtle it might be (even non-destructive), inevitably introduces a disturbance to the system, therefore, the measurement is not entirely neat. In short, the less it is measured, the less uncertainty regarding the values obtained.

2. Analysis and discussion

Finding an expression that relates a tribological property with any optical property of a DLC film, would constitute a significant advance since it is firmly aligned in the direction of the pursued objective. Regarding this approach, there is a work developed by a Japanese research team, that claims to have found a proportional relationship between hardness and the refractive index [10]. Next, details about this work are required.

Reference 10 provides a series of useful data and conclusions. The production method is a PVD: Ion beam deposition. This method generally has 2 variables, which can be adjusted or exchanged independently, to obtain a DLC film with certain characteristics. These variables or deposition parameters are the voltage applied to the substrate and the gas source. In table 2, a set of data referring to values of hardness, refractive index, extinction coefficient, hydrogen content, density, and thickness of 4 DLC films deposited on a silicon substrate is observed. The substrate voltage varies between 1 kV and 3 kV, while the gas source used is benzene for 3 of the cases, and cyclohexane for one of them.

Table 2. Characteristics of DLC films obtained by Ion beam deposition [10].

	Deposition parameters		**DLC film type**	**Hardness (GPa)**		**Refractive index (550 nm)**		**Extinction coefficient (550 nm)**	
	Substrate voltage	Source gas		Measurement Method	Magnitude	Measurement Method	Magnitude	Measurement Method	Magnitude
Ion beam deposition	1 kV	C_6H_6	ta-C:H	Nanoindentation	32.3	Ellipsometry	2.43	Ellipsometry	0.33
	2 kV	C_6H_6	ta-C:H		25.8		2.25		0.44
	2 kV	C_6H_{12}	ta-C:H		21.2		2.14		0.41
	3 kV	C_6H_6	a-C:H		23.6		2.17		0.61
	Deposition parameters		**DLC film type**	**Hydrogen content (%)**		**Density (g/cm³)**		**Thickness (nm)**	
	Substrate voltage	Source gas		Measurement Method	Magnitude	Measurement Method	Magnitude	Measurement Method	Magnitude
Ion beam deposition	1 kV	C_6H_6	ta-C:H	Hydrogen Forward Scatterin	23.2	X-Ray Reflectometry	2.3	Ellipsometry	875
	2 kV	C_6H_6	ta-C:H		22.7		2.17		550

2 kV	C_6H_{12}	ta-C:H	25.0	2.09	1077
3 kV	C_6H_6	a-C:H	20.5	2.13	-

Regarding the voltage applied, according to table 2, the following can be asserted:

- The hardness of the nanostructure decreases as the voltage applied to the substrate raises. This relationship is met only if the energy peak (100 eV), leading to the formation of sp^3 bonds, has been exceeded [9]. According to reference 10, the voltage value that produces an energy of around 100 eV in the carbon cations is estimated at 1 kV.
- The extinction coefficient of the nanostructure increases as the voltage applied to the substrate is raised. Unlike the refractive index, which decreases with the growth in the applied voltage.
- By raising the voltage applied to the substrate (above 1 kV), nanostructures of a lower density are achieved.
- The hydrogen content in the nanostructure begins to decrease as the voltage applied to the substrate raises.
- When performing the analysis of the DLC film thickness, as well as the thickness of surface roughness, no clear pattern is observed, and therefore, there is no clear influence of the substrate voltage on the final achievement of a certain thickness. In other words, the substrate voltage alone does not allow to control the DLC film thickness and surface roughness. Figure 1 summarizes the found tendencies when the substrate voltage raises.

The fact that hardness decreases, while substrate voltage raises, is related to the energy acquired by the carbon cations that impact on the DLC film. Increasing the voltage in the substrate results in a greater acceleration in the cations and, therefore, the carbon cations acquire a higher energy. However, to talk about hardness means to talk about a high density of sp^3 bonds in the nanostructure, and according to reference 9, C^+ with energy of 100 eV generate the greatest number of sp^3 links, that is, the peak of performance is reached at this value. In this way, although energy is required to achieve the task, an energy margin must not be exceeded because sp^3 bonds are lost.

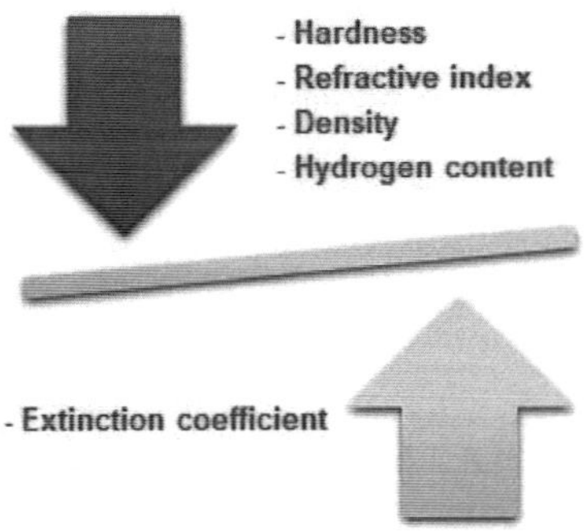

Figure 1. Tendencies of different DLC characteristics when rising substrate voltage.

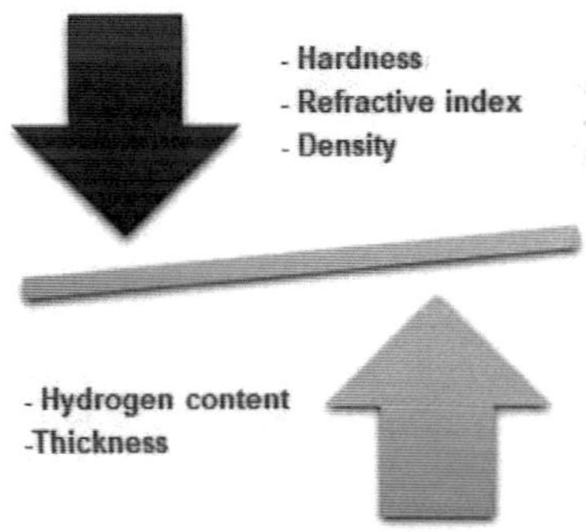

Figure 2. Tendencies of different DLC characteristics when C_6H_{12} is used as gas source.

Analyzing now the type of gas source used in the deposition, it can be concluded that the utilization of one or another substance has a fundamental effect on the hydrogen content of the nanostructure. According to reference 10 and table 2, replacing benzene (C_6H_6) with cyclohexane (C_6H_{12}), and keeping constant the substrate voltage, produces a notorious increase in the hydrogen content of the DLC film. Also, using C_6H_{12} as gas source, it's possible to appreciate a reduction in hardness, refractive index, and density, as compared with the case where C_6H_6 is employed. The situation of extinction coefficient is different, because despite using another type of gas source, the values are practically the same. Finally, a notable change in DLC film thickness is observed, the use of cyclohexane generates films with twice the original thickness. Figure 2 summarizes the found tendencies when C_6H_{12} is used as gas source.

Looking closely at figures 1 and 2, there are clearly 3 properties that go hand in hand: hardness, refractive index and density. Despite the different conditions and parameters of deposition used, the mentioned characteristics always follow the same pattern or trend, in this case the 3 characteristics decrease in their respective values. This situation already becomes an existence sign of a certain relationship between this trio of characteristics.

The Japanese team of scientists decided to continue with the research on the properties of amorphous carbon nanostructures and collected a set of data from a total of 46 samples of DLC, between different laboratories and industries dedicated to the production of this nanostructure. The acquired data corresponded to the hardness and refractive index of the DLC films. It should be noted that among these 46 samples there is a broad spectrum about the method of deposition used, and of course, about the type of DLC film obtained. By joining this group of data and making a correlation between them, the result was amazing: there is a proportional relationship between hardness and refractive index. Figure 3 presents the correlation graph constructed by the research group, which in addition to the 46 samples, includes the 4 samples developed in their laboratory.

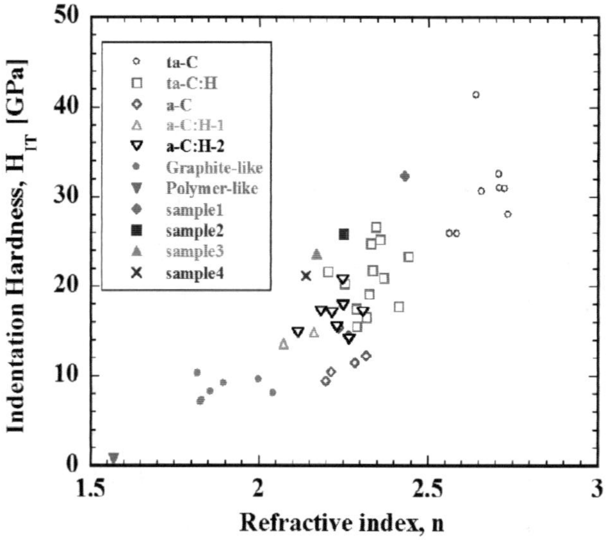

Figure 3. Correlation between the refractive index (550 nm) and hardness of several DLC films [10].

The result is quite encouraging, and since they have worked with different types of DLC films, could be hypothesized that there is a proportional relationship between hardness and refractive index regardless of the deposition method from which the DLC film comes. It would be extraordinary because would grant a degree of universality to the model that it is intended to develop.

Unfortunately, in the referenced work the raw data of the 46 samples are not published, only the data concerning the 4 samples developed in their own laboratory are known. However, as part of the present analysis and with the aim of verifying this relationship, a graph with the known data has been constructed. Linear regression has been applied to adjust a line to the set of points. Figure 4 shows the graph "Refractive Index vs Hardness" for the 4 DLC samples.

As shown in figure 4, it is a strong linear relationship, its correlation index (0.9873) corroborates this. It has been determined, in addition, the equation of the line that would allow to calculate the hardness values of the DLC film as a function of the refractive index (Equation 1).

$$H = 36.388\, n - 56.057 \quad (1)$$

where H – hardness in units of GPa; n – refractive index.

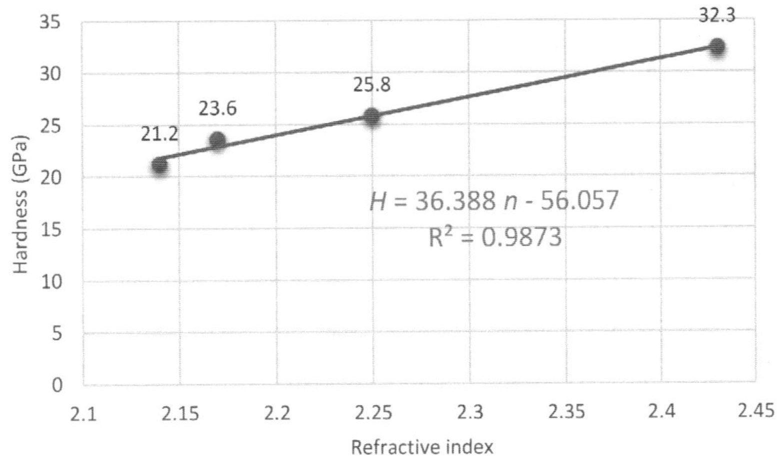

Figure 4. Correlation between the refractive index (550 nm) and hardness of 4 DLC samples.

Using the obtained expression, the hardness values are calculated according to the measured refractive index. Table 3 shows the results of this calculation and the respective relative error. The largest error is less than 3%.

Table 3. Comparative table between measured and calculated hardness.

Refractive index	Measured Hardness (GPa)	Calculated Hardness (GPa)	Relative error
2.43	32.3	32.366	0.20%
2.25	25.8	25.816	0.06%
2.17	23.6	22.905	2.95%
2.14	21.2	21.813	2.89%

The error margin is certainly acceptable, thus supporting the hypothesis of an apparent linear relationship between hardness and refractive index. Of course, the amount of data is insufficient; but nevertheless, it's clear that in order to validate this hypothesis, a considerable set of data resulting from the own experimentation is required.

There are several scientific references that despite using different methods of deposition, come to determine the same close relationship between the content of sp^2 or sp^3 bonds (implicitly material hardness) and the main optical properties: refractive index and extinction coefficient. The conclusion is the following: the increase in sp^2 content from the graphitic structure can be considered to have a large effect on the increase in extinction coefficient, while the increase in C-C sp^3 content can be considered to affect the increase in refractive index [10-13]. Figure 5 summarizes this conclusion. That is, the harder the DLC film is, the higher the refractive index; likewise, as the hardness of the film decreases, it is the extinction coefficient that acquires relevance.

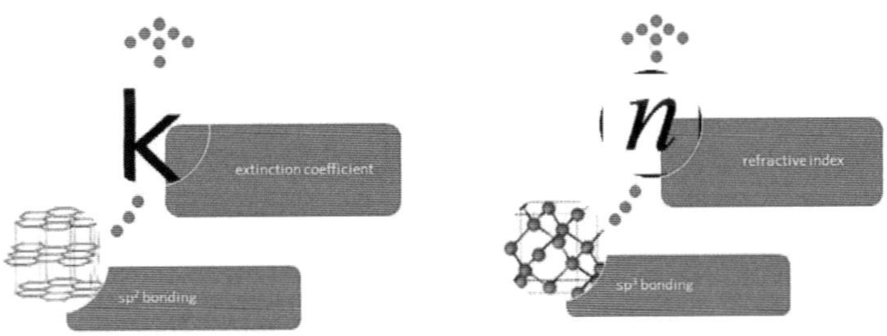

Figure 5. Relationship between optical constants and sp^2/sp^3 bonding ratio.

3. Experimental results

At the moment only 2 samples with DLC coating have been obtained. The technologies used were Cathodic Arc Deposition (Arc-PVD) and Pulsed Laser Deposition (PLD). Hardness and refractive index values were measured using nanoindentation and ellipsometry, respectively. The registered values are shown in table 4.

Table 4. Hardness, refractive index, and sp^3 fraction values of DLC films produced in our laboratory.

	Hardness (GPa)	Refractive index (2μm)	sp^3 fraction %
Sample 1 (Arc-PVD)	14.89	2.12	13
Sample 2 (PLD)	26.48	2.31	51

As a first approximation, given the scarcity of data, it has been decided to ubicate these results within the graph provided by reference 10. As shown in figure 6, the obtained values are within the region that allows to assume a directly proportional relationship between hardness and refractive index.

If a line that passes through the 2 points obtained in the experiment was drawn, the equation of this line would differ in comparison to that obtained for the 4 samples of reference 10. This situation is decisive to conclude that the mathematical relationships depend strictly on the method of deposition that is used. However, there is a defined region by which the different values of hardness and refractive index are distributed. When a general line is drawn through this region, there is an evident consonance with the line obtained for the 4 samples that come from the same method of deposition. That is, both equations belong to the same family of parallel lines. This match is shown in figure 7.

The general region trend is described by the equation (2).

$$H = 36.388\,n - 64.771 \tag{2}$$

where H – hardness in units of GPa; n – refractive index.

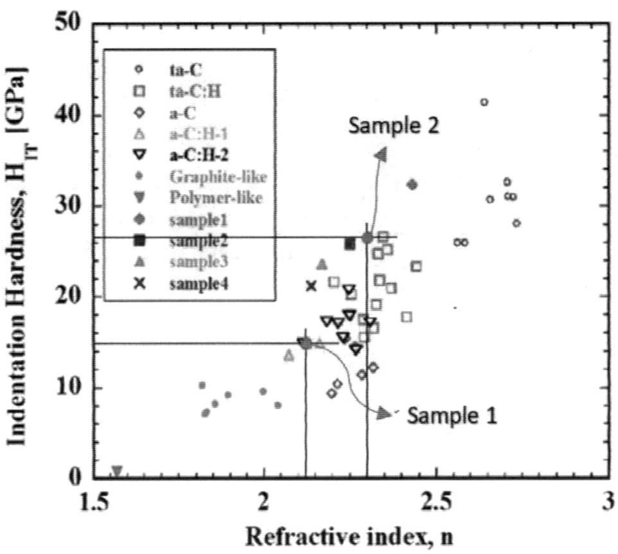

Figure 6. Location of the obtained values in the reference scheme.

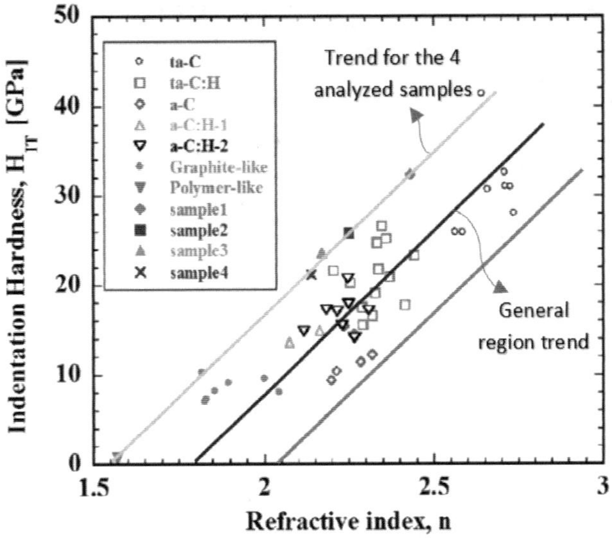

Figure 7. General trend for a set of DLC data.

Conclusions

1. There is a directly proportional linear relationship between the hardness of DLC films and their refractive index.

2. Regardless of the deposition parameters, but if the DLC films are obtained through the same deposition method, it will be possible to determine an equation that describes the relationship between hardness and refractive index.

3. In the graph "refractive index vs. hardness" there is a certain region where all the values corresponding to different types of DLC films are distributed, and this distribution has a completely linear trend.

4. It is possible to indirectly determine variables of tribological type without incurring the use of destructive methods of characterization.

References

[1] Hauert R 2008 *Tribology of diamond-like carbon films: fundamentals and applications* ed C Donnet and A Erdemir (New York: Springer Science & Business Media) pp 494–500

[2] Bewilogua K and Hofmann D 2014 History of diamond-like carbon films–from first experiments to worldwide applications *Surface and Coatings Technology* **242** 214–25

[3] Grill A 2003 Diamond-like carbon coatings as biocompatible materials–an overview *Diamond and related materials* **12(2)** 166–70

[4] Makeev M, Zhukova E, Mikhalev P, Osipkov A and Mironov Y 2015 Physical chemical and protective properties of the diamond-like carbon coatings synthesized from separated plasma of electric arc *Proc. 5th International Workshop on Computer Science and Engineering: Information Processing and Control Engineering 2015* (Chengdu, China) 255–9

[5] Piazza F, Grambole D, Schneider D, Casiraghi C, Ferrari A and Robertson J 2005 Protective diamond-like carbon coatings for future optical storage disks *Diamond and related materials* **14(3-7)** 994–9

[6] Choi W, Kim K, Yi J and Hong B 2008 Diamond-like carbon protective anti-reflection coating for Si solar cell *Materials Letters* **62(4-5)** 577–80

[7] Kutsay O, Gontar A, Novikov N, Dub S, Tkach V, Gorshtein B and Mozkova O 2001 Diamond-like carbon films in multilayered interference coatings for IR optical elements *Diamond and related materials* **10(9-10)** 1846–9

[8] Mironov Y, Stepanov R, Osipkov A, Mironova A, Makeev M, Mikhalev P and Sedih N 2015 Optical and mechanical properties of diamond-like carbon coatings deposited by filtered cathodic vacuum arc deposition *Proc. 5th International Workshop on Computer Science and Engineering: Information Processing and Control Engineering* (Chengdu, China) 295–300

[9] Robertson J 2002 Diamond-like amorphous carbon *Materials science and engineering: R: Reports* **37(4-6)** 129–281

[10] Hiratsuka M, Nakamori H, Kogo Y, Sakurai M, Ohtake N and Saitoh H 2013 Correlation between optical properties and hardness of diamond-like carbon films *Journal of Solid Mechanics and Materials Engineering* **7(2)** 187–98

[11] Smietana M, Szmidt J, Korwin-Pawlowski M, Miller N and Elmustafa A 2008 Influence of RF PACVD process parameters of diamond-like carbon films on optical properties and nano-hardness of the films *Diamond and related materials* **17(7-10)** 1655–9

[12] Lemoine P, Quinn J, Maguire P, Zhao J and McLaughlin J 2007 Intrinsic mechanical properties of ultra-thin amorphous carbon layers *Applied surface science* **253(14)** 6165–75

[13] Su C, Lin C, Chang C, Hung H and Lin T 2006 Mechanical and optical properties of diamond-like carbon thin films deposited by low temperature process *Thin Solid Films* **498(1-2)** 220–3

Accounting for physically nonlinear deformation of the shell under flat loading based on the finite element method

A Sh Dzhabrailov*, Yu V Klochkov and A P Nikolaev

Volgograd State Agrarian University, Volgograd, Russia

E-mail: arsen82@yandex.ru

Abstract. On the basis of the theory of small elastic plastic deformations obtained at the loading step in the curvilinear coordinate system of the relationship between stress increments and strain increments, which are implemented in the matrix of shell stiffness under flat loading In this paper, algorithms for obtaining physically nonlinear relations at the loading step in finite element calculations of shells under flat loading are developed.On the basis of relations of deformation theory of plasticity the equations of connection between increments of stresses and increments of deformations at flat loading in elastic plastic stage of work are received. The coefficients of proportionality are obtained when using the strain diagram of the material. The found defining relations are implemented in the algorithm of forming the stiffness matrix of the finite element of the shell under flat loading. Specific examples show the effectiveness of the developed algorithms for the calculation of the shell with a flat loading beyond the elasticity limits.

Keywords:shell, stiffness matrix, deformation, physical nonlinearity, finite element

1.Introduction

The theory of shells today is at a fairly high level of development [1-6]. Technological progress in the national economy determines the use of shell structures of increasingly complex configuration. The implementation of solving equations of shell theory in practice remains a rather difficult task, in the solution of which the most effective are numerical methods, such as the finite element method [7-10]. Accounting for physical nonlinearity is one of the fundamental problems in the calculation of shell structures. Its solution will allow more efficient use of the strength properties of the materials used in the design of new construction projects, thereby expanding the range of applications of analytical calculations. Among the set of approaches used in finite element calculations of shells taking into account the physical nonlinearity [11-14], the following are distinguished: the preparation of nonlinear equilibrium equations with respect to the initial metric and their solution using an iterative procedure; the method of sequential step loading with a reference metric defined in the initial not deformed state. Each of these approaches has its advantages and disadvantages and none of them can be considered universal. In this paper, the method of step loading is used in the finite element implementation of the calculation of shells of rotation, as it allows to obtain solutions under significant loads. The defining equations at the loading step are obtained by differentiating the basic relations of the deformation theory of plasticity.

2. Materials and Methods

Basic relation. Consider a shell structure in the form of an elliptical cylinder under flat loading. The radius-vector of the middle surface, in this case, is determined by the formula

$$\mathbf{R}^{0}=\mathrm{b}\,sin(t)\,\mathbf{i}+\mathrm{c}\cdot\cos(t)\,\mathbf{k}. \tag{1}$$

The unit vector of the local basis tangent to the meridian can be obtained by differentiating (1) by the global coordinates *s*

$$\mathbf{a}_1^o=\mathbf{R}_{,s}^o=b\cdot\cos(t)\cdot t_{,s}\,\mathbf{i}-c\cdot\sin(t)\cdot t_{,s}\,\mathbf{k}. \quad (2)$$

The unit vector of the normal to the middle surface of an elliptic cylinder is determined by the vector product

$$\mathbf{a}_n^o=\mathbf{a}_1^o\times\mathbf{j}=c\cdot\sin(t)\cdot t_{,s}\mathbf{i}+b\cdot\cos(t)\cdot t_{,s}\mathbf{k}. \quad (3)$$

By differentiating (2) and (3) we find the derivatives of the vectors of the local basis

$$\mathbf{a}_{1,s}^o= b\cdot\left(\cos(t)\cdot t_{,ss}-\sin(t)(t_{,s})^2\right)\mathbf{i}-c\cdot\left(\cos(t)\cdot(t_{,s})^2+\sin(t)t_{,ss}\right)\mathbf{k}, \quad (4)$$

$$\mathbf{a}_{n,s}^o=c\cdot\left(\cos(t)\cdot(t_{,s})^2+\sin(t)t_{,ss}\right)\mathbf{i} + b\cdot\left(\cos(t)\cdot t_{,ss}-\sin(t)(t_{,s})^2\right)\mathbf{k}.$$

For convenience, the expression (2)-(4) write in matrix form

$$\{\mathbf{a}^o\}=[m^o]\{\mathbf{i}\};\ \{\mathbf{a}_{,s}^o\}=[m]\{\mathbf{i}\}, \quad (5)$$

where $\{\mathbf{a}^o\}^T=\{\mathbf{a}_1^o\mathbf{a}_n^o\}$; $\{\mathbf{a}_{,s}^o\}^T=\{\mathbf{a}_{1,s}^o\mathbf{a}_{n,s}^o\}$; $\{\mathbf{i}\}^T=\{\mathbf{ijk}\}$.

On the basis of (5) we obtain the dependence

$$\{\mathbf{i}\}=[m^o]^{-1}\{\mathbf{a}^o\}. \quad (6)$$

Then the derivatives of the vectors of the local basis, taking into account (6), we express through the global coordinates in the original basis

$$\{\mathbf{a}_{,s}^o\}=[m][m^o]^{-1}\{\mathbf{a}^o\}=[n]\{\mathbf{a}^o\}. \quad (7)$$

The position of the midpoint of the surface during deformation is determined by equality

$$\mathbf{R}=\mathbf{R}^o+\mathbf{v}. \quad (8)$$

The displacement vector of a point in the middle surface **v**, part (8), is determined through the nodal unknowns in the following way

$$\mathbf{v}=u\mathbf{a}_1^0+w\mathbf{a}_n^0. \quad (9)$$

By performing the operation of differentiation (9) on the curvilinear coordinate s, taking into account (7) it is possible to obtain derivatives of the displacement vector of the point of the middle surface of the shell

$$\mathbf{v}_{,s}=t_1^1\mathbf{a}_1^o+t_1\mathbf{a}_n^o;\ \mathbf{v}_{,ss}=t_{11}^1\mathbf{a}_1^o+t_{11}\mathbf{a}_n^o. \quad (10)$$

Having differentiated the expression (8) by the global coordinate s, we determine the vectors of the inner point of the middle surface in the deformed state

$$\mathbf{a}_1=\mathbf{R}_{,s}=(\mathbf{R}^o+\mathbf{v})_{,s}=\mathbf{a}_1^o+t_1^1\mathbf{a}_1^o+t_1\mathbf{a}_n^o. \quad (11)$$

The unit normal vector in the deformed state is defined as follows

$$\mathbf{a}_n=\mathbf{a}_1\times\mathbf{j}. \quad (12)$$

The radius vectors of the point of an arbitrary layer of the shell in the initial and deformed states can be represented by dependencies

$$\mathbf{R}^{o\zeta}=\mathbf{R}^o+\zeta\mathbf{a}_n^o;\ \ \mathbf{R}^{\zeta}=\mathbf{R}^{o\zeta}+\mathbf{V}. \quad (13)$$

The displacement vector **V** of the point of an arbitrary layer of the shell in the deformed state is determined on the basis of the hypothesis of direct normals

$$\mathbf{V} = \mathbf{v} + \zeta(\mathbf{a} - \mathbf{a}^o). \quad (14)$$

Curvatures and deformations in an arbitrary layer of the shell spaced from the middle surface at a distance ζ are determined by the formulas of continuum mechanics

$$\varepsilon_{\alpha\beta}^{\zeta}=\left(g_{\alpha\beta}-g_{\alpha\beta}^{0}\right)/2, \quad (15)$$

where $g_{\alpha\beta}$ and $g^0_{\alpha\beta}$ are components of the metric tensor in the deformed and initial states determined by the corresponding scalar products of covariant vectors of the basis of points in an arbitrary shell layer

The expression (15) for plane loading will have the form

$$\varepsilon^{\zeta}_{11}=\mathbf{g}^0_1\mathbf{V}_{,s};\varepsilon^{\zeta}_{22}=\varepsilon^{\zeta}_{12}=0\ , \tag{16}$$

where $\mathbf{g}^0_1=\mathbf{R}^{0\zeta}_{,s}$ and the vector $\mathbf{V}$ is determined by differentiation (14), taking into account (10),(11) and (12).

Finally get

$$\varepsilon^{\zeta}_{11}=t^1_1+\zeta t_{11} \tag{17}$$

where $t^1_1=u_{,s};\ t_{11}=-w_{,ss}$.

Binding ratio of the deformation theory When assembling the plasticity matrix included in the structure of the stiffness matrix, it is necessary to establish the relationship between the strains and stresses at the loading step. For this purpose, the second hypothesis of the deformation theory of plasticity was used in this article, according to which the components of the strain deviator are related to the components of the stress deviator by the ratio

$$\frac{\varepsilon^{\xi}_{11}-\varepsilon_0}{\sigma_{11}-\sigma_0}=\frac{\varepsilon^{\xi}_{22}-\varepsilon_0}{\sigma_{22}-\sigma_0}=\frac{\varepsilon^{\xi}_{33}-\varepsilon_0}{-\sigma_0}=\frac{\varepsilon^{\xi}_{12}}{2\sigma_{12}}=\gamma \tag{18}$$

where the function γ is determined by the ratio of the strain intensity ε_i to the intensity of stress (σ_i) $\gamma=\frac{3}{2}\frac{\varepsilon_i}{\sigma_i}$,and ε_0 and σ_0- the average linear strain and stress, respectively.

Using the data from the tensile test results, it is possible to determine the relationship between the stress intensity and the strain intensity at the point of an arbitrary shell layer

$$\sigma_i = \sigma;\ \varepsilon_i = \varepsilon - \frac{1-2\mu}{E}\sigma, \tag{19}$$

where μ-ratio of Poisson.

On the basis of (19) we can construct a stress-strain diagram. The nonlinear section of the deformation curve in this paper is approximated as a function

$$\sigma_i=a\varepsilon_i^2+b\varepsilon_i+c \tag{20}$$

where the coefficients a,b and c were determined on the basis of experimental data of the curve (19).

The dependence (18) in the case of flat loading will take the form

$$\frac{-\varepsilon^{\xi}_{11}-\varepsilon^{\xi}_{33}}{\sigma_{11}}=\frac{2\varepsilon^{\xi}_{33}-\varepsilon^{\xi}_{11}}{-\sigma_{11}}=\frac{2\varepsilon^{\xi}_{11}-\varepsilon^{\xi}_{33}}{2\sigma_{11}}=\gamma. \tag{21}$$

The increments of the strain tensor are determined by equality

$$\Delta\varepsilon^{\xi}_{\alpha\beta}=\frac{\partial\varepsilon^{\xi}_{\alpha\beta}}{\partial\sigma_{11}}\Delta\sigma_{11}. \tag{22}$$

To determine the differential operators$\partial\varepsilon^{\xi}_{\alpha\beta}/\partial\sigma_{\rho\tau}$, included in (22), it is necessary to express the strain of (21) explicitly

$$\varepsilon^{\xi}_{11} = -\frac{\gamma\sigma_{11}}{3};\ \varepsilon^{\xi}_{33} = -\frac{2\gamma\sigma_{11}}{3}. \tag{23}$$

By performing the differentiation (23) we obtain

$$\frac{\partial\varepsilon^{\xi}_{11}}{\partial\sigma_{11}}=-\frac{2}{E_c}-2\sigma_{11}\left(\frac{1}{E_k}-\frac{1}{E_c}\right)\frac{1}{\sigma_i};\ \frac{\partial\varepsilon^{\xi}_{33}}{\partial\sigma_{11}}=-\frac{1}{E_c}-\sigma_{11}\left(\frac{1}{E_k}-\frac{1}{E_c}\right)\frac{1}{\sigma_i}. \tag{24}$$

IOP Conf. Series: Materials Science and Engineering **675** (2019) 012052 doi:10.1088/1757-899X/675/1/012052

where E_k and E_c - the tangent and the secant modulus deformation diagram.

Finally, the expression (22) can be written in matrix form

$$\left\{\Delta\varepsilon_{\alpha\beta}^{\zeta}\right\}=[D]\{\Delta\sigma_{\alpha\beta}\}, \tag{25}$$

where $[D]^{-1}$-matrix plasticity$\left[C_{pl}\right]$.

Discretization element and stiffness matrix at the loading step. A one-dimensional finite element was chosen to implement the numerical integration procedure. This fragment of the shell, bounded by two parallel planes, is approximated to a linear element with the coordinates of the nodes η=-1 and η=1, where η is the local coordinate used for the convenience of numerical integration (figure 1).

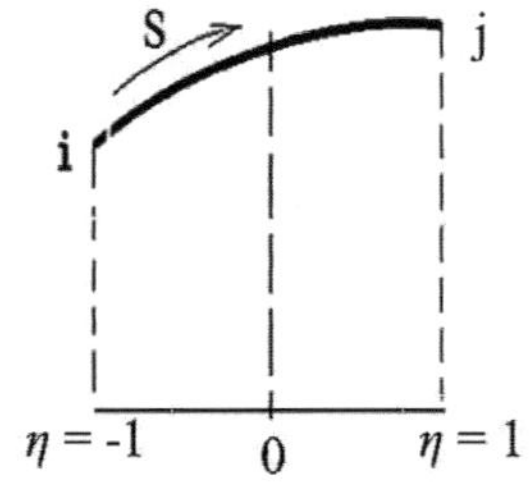

Figure1. The discretization element at the loading step.

In local coordinates the global coordinates of the shell are determined by the dependencies

$$s=\frac{(1-\eta)}{2}s^i+\frac{(1+\eta)}{2}s^j;\ t=\frac{(1-\eta)}{2}t^i+\frac{(1+\eta)}{2}t^j. \tag{26}$$

In the case of finite element analysis of the shell, the components of the displacement vector and their derivatives were chosen as nodal unknowns at the loading step

$$\{U\}^T=\{\{u\}^T\{w\}^T\} \tag{27}$$

where$\{u\}^T=\left\{u^i u^j u,_\eta^i u,_\eta^j\right\},\{w\}^T=\left\{w^i w^j w,_\eta^i w,_\eta^j\right\}$.

The relationship between the derivatives in the local and global coordinate systems is determined by dependencies

$$\frac{\partial q}{\partial \eta}=\frac{\partial q}{\partial s}\frac{\partial s}{\partial \eta};\quad \frac{\partial^2 w}{\partial \eta^2}=\frac{\partial^2 w}{\partial s^2}\left(\frac{\partial s}{\partial \eta}\right)^2, \tag{28}$$

where q is the meridional (u) or normal (w) components of the displacement vector at the loading step.

The meridional and normal displacements of the inner point of the finite element at the loading step are expressed through columns of nodal unknowns as follows

$$u=\{\varphi\}^T\{u\}, w=\{\varphi\}^T\{w\}. \tag{29}$$

The functions of the form included in (29) matrix-rows are determined by Hermite polynomials of the third degree by corresponding expressions

$$\{\varphi\}=\{h_1 h_2 h_3 h_4\}^T \tag{30}$$

By differentiating (29) it is possible to obtain increment derivatives at the loading step in the global coordinate system

$$u,_s=\left\{\varphi,_\eta\right\}^T\frac{\partial\eta}{\partial s}\{u\},\quad w,_s=\left\{\varphi,_\eta\right\}^T\frac{\partial\eta}{\partial s}\{w^n\},\quad w,_{ss}=\left\{\varphi,_{\eta\eta}\right\}^T\left(\frac{\partial\eta}{\partial s}\right)^2\{w\} \tag{31}$$

where $\frac{\partial\eta}{\partial s}=\frac{2}{(s^i\text{-}s^j)}$.

The stiffness matrix and the column of nodal loads of a one-dimensional finite element at the loading step are formed on the basis of the equality of the external and internal forces on the possible displacement [15-17]

$$\int_V \left\{\Delta\varepsilon_{\alpha\beta}^{\zeta}\right\}^T (\{\sigma_{\alpha\beta}\}+\{\Delta\sigma_{\alpha\beta}\})dV=\int_F \{\Delta U\}^T (\{P\}+\{\Delta P\})dF. \qquad (32)$$

In an arbitrary layer of the shell of the increment of deformation can be defined as the matrix product

$$\left\{\Delta\varepsilon_{\alpha\beta}^{\zeta}\right\} = [G][B]\{\Delta U_y^{л}\}, \qquad (33)$$

where matrices *[G]* and *[B]* are formed on the basis of (17).

The stress component increments at the loading step can be determined using the ratio

$$\{\Delta\sigma_{\alpha\beta}\} = [C_{pl}]\left\{\Delta\varepsilon_{\alpha\beta}^{\zeta}\right\}, \qquad (34)$$

where $[C_{pl}]$ - matrix plasticity.

The string of increments of the components of the displacement vector at the loading step can be expressed in a generalized column of increments of the nodal variable parameters of the finite element

$$\{\Delta U\} = [A]\{\Delta U_y^{л}\}, \qquad (35)$$

where [A] is a diagonal matrix containing polynomials of the third degree.

Equality (32), taking into account (33)-(35) can be written in matrix form as follows

$$[K]\{\Delta U_y^{г}\} = \{F\} + \{f\}, \qquad (36)$$

where $[K] = [T]^T \int_V \{B\}^T\{\Gamma\}^T[C_{pl}][\Gamma][B]\, dV$ is the stiffness matrix of a quadrangular finite element at the loading step; $[F]^T=[T]^T \int_F \{A\}^T\{\Delta P\}dF$ is the column of nodal loads at the loading step;$[f]^T=[T]^T\left(\int_F \{A\}^T\{P\}dF - \int_V \{B\}^T\{\Gamma\}^T\{\sigma\}dV\right)$ - Newton-Raphson correction.

3. Results

As an example, the problem of determining the stress-strain state of an arbitrary shell in the form of a fragment of an elliptical cylinder was solved (figure 2) loaded by internal pressure of intensity q. The following initial data were taken: q=0.18 MPa, μ=0.32; E=7.5 · 10^5 MPa; shell thickness t=0.005 m; the first parameter of the cylinder b=1 m ; the second parameter of the cylinder c=0.5 m. Shell in the reference section has a rigid clamping, and the right edge is free.

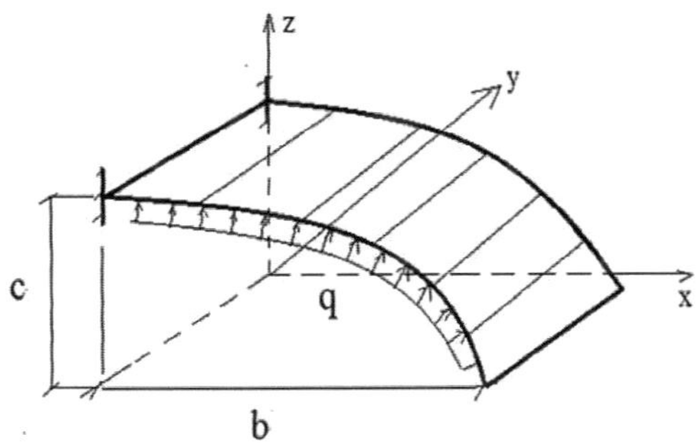

Figure2. Fragment of an elliptical cylinder.

The calculation results are presented in table 1, which shows the numerical values of the meridional stresses on the inner, outer and middle surfaces of the shell, as well as in the fibers of the reference section.

Table 1. The results of calculations of stresses across the shell thickness by varying the number of steps of loading in the reference section

No.	1	2	3	4	5	6	7

1	Stress, MPa	Steps of loading (n=30)	Coordinate on the shell thickness, ζ	Steps of loading (n=70)	Coordinate on the shell thickness, ζ	Steps of loading (n=100)	Coordinate on the shell thickness, ζ
2	σ_{22}	2428.6	$-\zeta/2$	2431.14	$-\zeta/2$	2431.70	$-\zeta/2$
3	σ_{22}	1900.0	$-\zeta/3$	1906.98	$-\zeta/3$	1908.55	$-\zeta/3$
4	σ_{22}	965.82	$-\zeta/6$	969.36	$-\zeta/6$	970.15	$-\zeta/6$
5	σ_{22}	31.63	0	31.73	0	31.75	0
6	σ_{22}	-902.55	$\zeta/6$	-905.89	$\zeta/6$	-906.64	$\zeta/6$
7	σ_{22}	-1836.7	$\zeta/3$	-1843.51	$\zeta/3$	-1845.04	$\zeta/3$
8	σ_{22}	-2413.3	$\zeta/2$	-2415.86	$\zeta/2$	-2416.41	$\zeta/2$

4. Discussion

When varying the number of finite elements, it was found their optimal to achieve a satisfactory convergence of the computational process quantitative value equal to 50.

To find an analytical solution, the calculation scheme presented in figure 3 was used

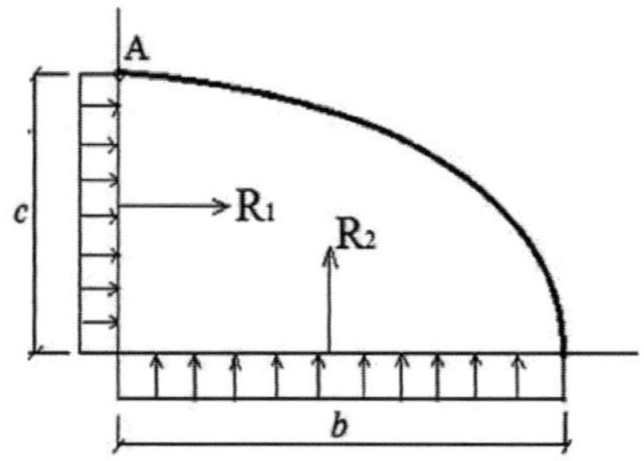

Figure3. Design model.

The total bending moment relative to point A (figure 3) is found as follows:

$$M_A = M_1 + M_2 = \frac{qc^2}{2} + \frac{qb^2}{2} = 112.5(\text{Н·м}). \quad (37)$$

To find the calculated value of the bending moment in the reference section, we present schematically the data in table 1 with the number of loading steps equal to 100 in the form of a diagram (Figure 4).

The calculated moment in the reference section is found as the sum of all moments to the neutral axis:

$$M_c = \sum M_i \ (i=\overline{1\ldots12}) = 109.98(\text{Н·м}) \quad (38)$$

As can be seen from table 1, the increase in the number of loading steps does not lead to a significant change in the numerical values of the controlled parameters of the stress-strain state. In addition, the error of calculation of the bending moment in the reference section was 2.24% compared to the calculated analytically.

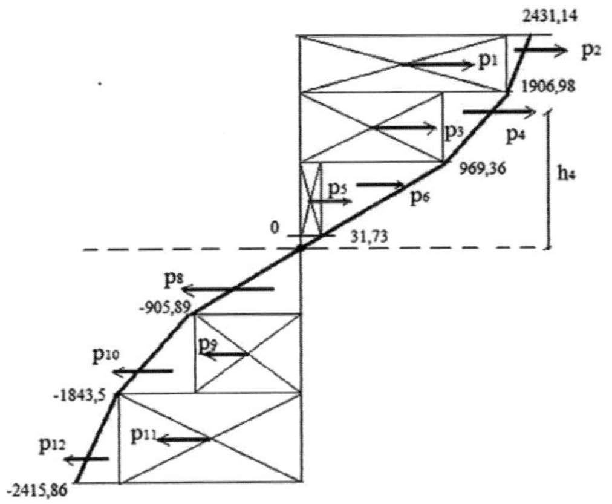

Figure4. Plot of meridionalstresses in the reference section along the shell thickness, MPa.

Conclusions

Thus, it can be concluded that the developed algorithm for taking into account possible plastic deformations in the calculations of arbitrary shells under flat loading has shown sufficient efficiency and can be recommended in the design engineering practice for the analysis of the stress-strain state of such structures. The resulting defining equations at the loading step by differentiating the relations of the deformation theory of plasticity between stresses and strains allow for the finite element implementation to take into account at the j-th loading step the residual equal to the difference between the work of internal and external forces for the previous loading steps.

Acknowledgements

The study was conducted with the support of the grant of the Russian Fund of Federal Investigations and the Administration of Volgograd region №. 18-41-340007r_a.

References

[1] Karpov V V, Ignatev O V and Semenov A A 2017 The stress-strain state of ribbed shell structures *Magazine of Civil Engineering* **74(6)** pp 147-160

[2] Ignatiev V A 2015 Calculation of plane frames with large displacement of nodes by the finite element method in the form of the classical mixed method *Journal of Construction and reconstruction* **2(58)** pp12-19

[3] Kozlov V A 2018 Stress-strain of elements of bridge structures with varying thickness of walls along the length *Russian Journal of Building Construction and Architecture* **1(37)** pp 67-80

[4] Kayumov R A 2017 Postbuckling behavior of compressed rods in an elastic medium *Mechanics of Solids* **52(5)** pp 575-580

[5] Pyatikrestovsky K P and Sokolov B S 2018 Nonlinear analysis of statically ideterminate wooden structures and optimization of cross section dimensions of dome ribs *Internatinal Jurnal for Computational Civil and Structural Engineering* **14(4)** pp 130-139

[6] Badriev I B, Paimushin V N 2017 Refined models of contact interaction of a thin plate with postioned on both sides deformable foundations *Lobachevskii Jurnal of Mathematics* **38(5)** pp 779-793

[7] Bate K Yu 2010 *Numerical methods* (Moscow: Physic. and mathemat. lit.) p 1022

[8] Agapov V P, Golovanov R O and Aidemirov K R 2017 Calculation of load bearing capacity of prestressed reinforsced concrete trusses by the finite element method *IOP Conference Series* : Earth and Environmental Science **90** 01 2018

[9] Agapov V P, Golovanov R O 2018 Comparative analysis of the simplest finite elements of plates in bending *Advances in Inteligent Systems and Computing* **692** pp 1009-1016

IOP Conf. Series: Materials Science and Engineering **675** (2019) 012052 doi:10.1088/1757-899X/675/1/012052

[10] Klochkov Y V, Nikolaev A P and Dzhabrailov A Sh 2013 Finite element analysis of axisymmetrically loaded shells of rotation with branching meridian under elastic-plastic deformation*Journal of Structural Mechanics of Engineering Constructions and Buildings* **3** pp 50-56

[11] Wriggers P and Hudobivnik B 2017 Alow order virtual element formulation for finite elasto-plastic deformation *Computer Methods in Applied Mechanics and Engineering* **53(8)** pp 123-129

[12] Korelc J and Stupkiewicz S 2014 Closed-form matrix exponential and its application in finite strain plasticity International Jurnal for Numerical Methods in Engineering **98** pp 960-987

[13] Souza Neto E A, Peric D and Owen D R J 2008 *Computational Methods for Plasticity, Theory and Applications.* Chichester: Wiley

[14] Miehe C, Welschinger F and Aldakheel F 2014 Variational gradient plasticity at finite strains *Computer Methods in Applied Mechanics and Engineering* **268** pp 704-734

[15] Zheleznov L P, Kabanov V V and Boiko D V 2018 Nonlinear Deformation and Stability of Discrete-Reinforced Elliptical Cylindrical Composite Shells under Torsion and Internal Pressure *Russian Aeronautics* **61(2)** pp 175-182

[16] Matveev A D and Grishanov A N 2018 Multigrid finite elements in the calculations of multilayer cylindrical shells *Siberian Journal of Science and Technology* **19(1)** pp 27-36

[17] Nguyen N and Waas A M 2016 Nonlinear, finite deformation, finite element analysis Zeitschrift fur angewandte Mathematik and Physik **67:35**

IOP Conf. Series: Materials Science and Engineering **675** (2019) 012053 doi:10.1088/1757-899X/675/1/012053

Finite-element analysis of thin-walled shells under various parameterization options of their surfaces

Yu V Klochkov*, A P Nikolaev, T A Sobolevskaya and A S Andreev

Volgograd State Agrarian University, Volgograd, Russia

Email: klotchkov@bk.ru

Abstract. The methods of specifying the middle surface of an elliptical cylinder in curvilinear coordinate systems are described. An algorithm for discretization of an elliptic cylinder by high-precision quadrangular finite elements with a set of nodal variable parameters, which includes components of the displacement vector, as well as their partial derivatives of the first and second orders, is described. Nodal unknowns in global and local coordinate systems are described. Two types of interpolation procedure are presented: vector interpolation of displacement fields, scalar interpolation. Interpolation expressions for the components of the displacement vector and their first and second derivatives are obtained using the vector version of the interpolation procedure.

1 Introduction

Thin shell structures are widely used in many areas of industry: construction, engineering, aircraft, shipbuilding and so on. Among the variety of forms of shells used the most common are cylindrical shells. Since in real engineering practice it is difficult to keep the perfectly round shape of the cross section of the cylinder, the vast majority of cylindrical shells can be considered as elliptical cylinders.

The study of the stress-strain state of cylindrical shells is currently unthinkable without the use of numerical methods of calculation [1-7] with the use of modern computer technology. When creating computational algorithms, the problem of parameterization of the middle surface of an elliptical cylinder inevitably arises. This article discusses two options for parameterization using the angular coordinate and the ellipse parameter, which is the cross section of the cylindrical shell. When using the numerical finite element method (FEM) in shell calculations [8-13], approximation of the required quantities as components of scalar fields is used. As a result, each desired value depends on the nodal values of only the same value, which leads to the problem of accounting for the displacement of the finite element as a solid. Therefore, in this paper we use the vector approximation of the required quantities, allowing to take into account the displacement of the finite element as a solid. The middle surface of an elliptic cylinder is modeled by a set of high-precision quadrangular discretization

2 Materials and Methods

An elliptical cylinder in a Cartesian coordinate system, depending on the orientation in space, can be given by one of the following formulas

$$\frac{y^2}{b^2}+\frac{z^2}{c^2}=1;\quad \frac{x^2}{a^2}+\frac{z^2}{c^2}=1;\quad \frac{x^2}{a^2}+\frac{y^2}{b^2}=1. \tag{1}$$

If we take the first of the formulas (1) for certainty, the radius vector describing the surface of an elliptical cylinder can be written as

$$\vec{R}^0 = x\vec{i} + y\vec{j} + c\sqrt{1-(y/b)^2}\,\vec{k}. \tag{2}$$

Using the differentiation operation (2) x, y we can find expressions for tangent vectors of the local basis

$$\vec{a}_1^0 = \vec{R}_{,x}^0 = \vec{i}; \; \vec{a}_2^0 = \vec{R}_{,y}^0 = \vec{j} - \frac{cy}{b^2\sqrt{1-(y/b)^2}}\vec{k}. \quad (3)$$

The use of formulas (3) is limited by the need to comply with the conditions

$$1-(y/b)^2 > 0. \quad (4)$$

Therefore, it is proposed to use the following formula for specifying the radius vector for an elliptical cylinder

$$\vec{R}^0 = x\vec{i} + r(\theta)\sin\theta\vec{j} + r(\theta)\cos\theta\vec{k}, \quad (5)$$

where θ - is the angular coordinate measured from the vertical counterclockwise in a plane perpendicular to the axis Ox.

The function $r(\theta)$ included in (5) is defined by the expression

$$r(\theta) = \frac{bc}{\sqrt{(c\sin\theta)^2 + (b\cos\theta)^2}}. \quad (6)$$

Performing the operation of differentiation by x and θ over (5), it is possible to obtain the basis vectors located in the tangential plane to the surface of the elliptical cylinder

$$\vec{a}_1^0 = \vec{R}_{,x}^0 = \vec{i};$$

$$\vec{a}_2^0 = \vec{R}_{,\theta}^0 = (r_{,\theta}\sin\theta + r(\theta)\cos\theta)\vec{j} + (r_{,\theta}\cos\theta - r(\theta)\sin\theta)\vec{k}, \quad (7)$$

where $r_{,\theta} = \partial r(\theta)/\partial\theta$.

The normal ort to the cylinder surface can be determined by the vector product

$$\vec{a}^0 = \frac{\vec{a}_1^0 \times \vec{a}_2^0}{\sqrt{a^0}} = \frac{(r(\theta)\sin\theta - r_{,\theta}\cos\theta)\vec{j} + (r_{,\theta}\sin\theta + r(\theta)\cos\theta)\vec{k}}{\sqrt{a^0}}. \quad (8)$$

The incoming (8) determinant of the metric tensor is determined by scalar products (7)

$$a^0 = (\vec{a}_1^0 \cdot \vec{a}_1^0)(\vec{a}_2^0 \cdot \vec{a}_2^0) - (\vec{a}_1^0 \cdot \vec{a}_2^0)^2 = r_{,0}^2 + (r(\theta))^2. \quad (9)$$

On the basis of (7), (8) it is possible to combine the direct and the inverse matrix based on

$$\{\vec{a}^0\} = [l^0]\{\vec{i}\}, \; \{\vec{i}\} = [l^0]^{-1}\{\vec{a}^0\}, \quad (10)$$

where $\{\vec{a}^0\}^T = \{\vec{a}_1^0 \vec{a}_2^0 \vec{a}^0\}$, $\{\vec{i}\}^T = \{\vec{i}\vec{j}\vec{k}\}$.

From (10) we can obtain matrix expressions for the derivatives of the basis vectors of the surface points of an elliptic cylinder

$$\{\vec{a}_{,x}^0\} = [0]\{\vec{i}\}, \; \{\vec{a}_{,\theta}^0\} = [l_{,\theta}^0]\{\vec{i}\} = [l_{,\theta}^0][l^0]^{-1}\{\vec{a}^0\} = [n^0]\{\vec{a}^0\}. \quad (11)$$

In the well-known monograph [14] it is proposed to use parametric dependences as the most compact ones to define surfaces.

Therefore, in the present work, for the elliptic cylinder, it is also possible to use the formula of the radius-vector that contains the parameter t

$$\vec{R}^0 = x\vec{i} + b\sin t\vec{j} + c\cos t\vec{k}, \tag{12}$$

where b and c - the length of the semi-axes of the ellipse, which is the cross section of the cylinder plane perpendicular to the axis Ox, t - the parameter of the ellipse, counted from the vertical axis counterclockwise.

Performing the operation of differentiation over (12) x and t it is possible to obtain the vectors of the local basis lying in the contact plane to the surface of the elliptic cylinder

$$\vec{a}_1^0 = \vec{R}_{,x}^0 = \vec{i}; \;\; \vec{a}_2^0 = \vec{R}_{,t}^0 = b\cos t\vec{j} - c\sin t\vec{k}. \tag{13}$$

Ort normal to the cylinder surface is determined according to (8)

$$\vec{a}^0 = \vec{a}_1^0 \times \vec{a}_2^0 / \sqrt{a^0} = \left(c\sin t\vec{j} + b\cos t\vec{k}\right) / \sqrt{a^0}. \tag{14}$$

From (13) and (14) it is possible to combine the direct and the inverse matrix based on

$$\{\vec{a}^0\} = [b^0]\{\vec{i}\}, \;\; \{\vec{i}\} = [b^0]^{-1}\{\vec{a}^0\} \tag{15}$$

Differentiating (13), (14) by x and t it is possible to obtain derivatives of vectors of the local basis of the surface point of an elliptic cylinder

$$\vec{a}_{1,x}^0 = 0; \;\; \vec{a}_{1,t}^0 = 0; \;\; \vec{a}_{2,x}^0 = 0; \;\; \vec{a}_{2,t}^0 = -b\sin t\vec{j} - c\cos t\vec{k};$$

$$\vec{a}_{,x}^0 = 0; \;\; \vec{a}_{,t}^0 = \left(\frac{c\cos t}{\sqrt{a^0}} - \frac{c\sin t a_{,t}^0}{2\left(a^0\right)^{3/2}}\right)\vec{j} - \left(\frac{b\sin t}{\sqrt{a^0}} + \frac{b\cos t a_{,t}^0}{2\left(a^0\right)^{3/2}}\right)\vec{k},$$

which can be represented in matrix form

$$\{\vec{a}_{,x}^0\} = [0]\{\vec{i}\}, \;\; \{\vec{a}_{,t}^0\} = [b_{,t}^0]\{\vec{i}\} = [b_{,t}^0][b^0]^{-1}\{\vec{a}^0\} = [d^0]\{\vec{a}^0\} \tag{16}$$

Cauchy relations for an elliptic cylinder can be obtained from the relation of continuum mechanics [15]

$$\varepsilon_{\alpha\beta}^{\zeta} = \left(g_{\alpha\beta} - \overset{0}{g}_{\alpha\beta}\right)/2, \tag{17}$$

where $\varepsilon_{\alpha\beta}^{\zeta}$ - deformation at a point located from the middle surface at a distance ζ.

The covariant components of the metric tensor of the initial $\overset{0}{g}_{\alpha\beta}$ and deformed states $g_{\alpha\beta}$ included in (17) are determined by scalar products

$$\overset{0}{g}_{\alpha\beta} = \vec{g}_{\alpha}^0 \cdot \vec{g}_{\beta}^0; \;\; g_{\alpha\beta} = \vec{g}_{\alpha} \cdot \vec{g}_{\beta}. \tag{18}$$

To calculate the vectors of the basis of the initial and deformed states $\vec{g}_{\alpha}$, it is necessary to perform the operation of differentiation of the corresponding radius vectors

$$\vec{g}_{\alpha}^0 = \vec{R}_{,\alpha}^{0\varsigma} = \left(\vec{R}^0 + \varsigma\vec{a}^0\right)_{,\alpha}; \;\; \vec{g}_{\alpha} = \vec{R}_{,\alpha}^{\varsigma} = \left(\vec{R}^{0\varsigma} + \vec{V}\right)_{,\alpha}, \tag{19}$$

where the lower index α successively takes the values x, θ or t.

The incoming (19) vector of displacement of a point located from the middle surface at a distance ς is determined on the basis of the hypothesis of a straight normal by the formula

$$\vec{V} = \vec{v} + \varsigma\left(\vec{a} - \vec{a}^0\right), \tag{20}$$

where $\vec{a}$ is the ort of the normal to the surface of the cylinder is in a deformed state.

The displacement vector of the point of the middle surface of the elliptical cylinder entering (20) and its derivatives along the global curvilinear coordinates can be represented by components related to the local basis

$$\vec{v} = v^{\rho}\vec{a}^{0}_{\rho} + v\vec{a}^{0};\ \vec{v}_{,\alpha} = t^{\rho}_{\alpha}\vec{a}^{0}_{\rho} + t_{\alpha}\vec{a}^{0};\ \vec{v}_{,\alpha\beta} = t^{\rho}_{\alpha\beta}\vec{a}^{0}_{\rho} + t_{\alpha\beta}\vec{a}^{0}, \tag{21}$$

where v^1, v^2 - tangential, a v - normal component of the displacement vector, t^{ρ}_{α} and t_{α} - components of the first and $t^{\rho}_{\alpha\beta}$ and $t_{\alpha\beta}$ - second derivatives of the displacement vector, here Greek indices α, β and ρ consistently take the values 1, 2.

The discretization element of an elliptic cylinder selects a quadrangular fragment of its middle surface with nodes $i,\ j,\ k,\ l$ and located at the vertices of the quadrangle. Given that the Cauchy relations used in the theory of thin shells contain the second derivatives of the normal component, the components of the displacement vector, as well as their first and second derivatives in the global x, θ or t and local coordinate systems $-1 \le \xi,\ \eta \le 1$ were chosen as nodal unknowns

$$\underset{1\times72}{\{U^{G}_{y}\}^{T}} = \left\{ \underset{1\times24}{\{v^{1G}_{y}\}^{T}}\ \underset{1\times24}{\{v^{2G}_{y}\}^{T}}\ \underset{1\times24}{\{v^{G}_{y}\}^{T}} \right\}; \tag{22}$$

$$\underset{1\times72}{\{U^{L}_{y}\}^{T}} = \left\{ \underset{1\times24}{\{v^{1L}_{y}\}^{T}}\ \underset{1\times24}{\{v^{2L}_{y}\}^{T}}\ \underset{1\times24}{\{v^{L}_{y}\}^{T}} \right\}, \tag{23}$$

where

$$\underset{1\times24}{\{q^{G}_{y}\}^{T}} = \{q^{i} q^{j} q^{k} q^{l} q^{i}_{,x} q^{j}_{,x} q^{k}_{,x} q^{l}_{,x} q^{i}_{,\lambda} q^{j}_{,\lambda} q^{k}_{,\lambda} q^{l}_{,\lambda} q^{i}_{,xx} \cdots q^{l}_{,xx} q^{i}_{,\lambda\lambda} \cdots q^{l}_{,\lambda\lambda} q^{i}_{,x\lambda} \cdots q^{l}_{,x\lambda}\};$$

$$\underset{1\times24}{\{q^{L}_{y}\}^{T}} = \{q^{i} q^{j} q^{k} q^{l} q^{i}_{,\xi} q^{j}_{,\xi} q^{k}_{,\xi} q^{l}_{,\xi} q^{i}_{,\eta} q^{j}_{,\eta} q^{k}_{,\eta} q^{l}_{,\eta} q^{i}_{,\xi\xi} \cdots q^{l}_{,\xi\xi} q^{i}_{,\eta\eta} \cdots q^{l}_{,\eta\eta} q^{i}_{,\xi\xi} \cdots q^{l}_{,\xi\xi}\}.$$

Here takes λ the value of θ or t, and q is understood as a component of the displacement vector v^1, v^2 or v.

3 Results

Standard for FEM is the interpolation procedure of each component of the displacement vector through its own nodal values [16]

$$q = \underset{1\times24}{\{\varphi\}^{T}} \underset{24\times1}{\{q^{L}_{y}\}}, \tag{24}$$

where $\{\varphi\}^{T}$ - matrix-a string of form functions composed of the product of Hermite polynomials of the fifth degree.

The standard interpolation procedure (24) is quite correct when using a Cartesian coordinate system. In the SSS analysis of ellipsoidal shells, it is more convenient to use curvilinear coordinate systems as the most natural for curved surfaces. However, the application of interpolation of the form (24) to the calculation of ellipsoidal shells can lead to significant calculation errors, since the interpolation expressions (20) do not contain the parameters of the used curvilinear coordinate system. Therefore, this article implements the vector form of the interpolation procedure

$$\vec{v} = \underset{1\times24}{\{\varphi\}^{T}} \underset{24\times1}{\{\vec{v}^{L}_{y}\}} = \underset{1\times24}{\{\varphi\}^{T}} \underset{24\times24}{[T]} \underset{24\times1}{\{\vec{v}^{G}_{y}\}}, \tag{25}$$

where

$$\underset{1\times 24}{\left\{\vec{v}_y^G\right\}^T} = \left\{\vec{v}^i \vec{v}^j \vec{v}^k \vec{v}^l \vec{v}^i_{,\xi} \vec{v}^j_{,\xi} \vec{v}^k_{,\xi} \vec{v}^l_{,\xi} \vec{v}^i_{,\eta} \vec{v}^j_{,\eta} \vec{v}^k_{,\eta} \vec{v}^l_{,\eta} \vec{v}^i_{,\xi\xi} .. \vec{v}^l_{,\xi\xi} \vec{v}^i_{,\eta\eta} .. \vec{v}^l_{,\eta\eta} \vec{v}^i_{,\xi\eta} .. \vec{v}^l_{,\xi\eta}\right\} \quad (26)$$

$$\underset{1\times 24}{\left\{\vec{v}_y^G\right\}^T} = \left\{\vec{v}^i \vec{v}^j \vec{v}^k \vec{v}^l \vec{v}^i_{,x} \vec{v}^j_{,x} \vec{v}^k_{,x} \vec{v}^l_{,x} \vec{v}^i_{,\lambda} \vec{v}^j_{,\lambda} \vec{v}^k_{,\lambda} \vec{v}^l_{,\lambda} \vec{v}^i_{,xx} .. \vec{v}^l_{,xx} \vec{v}^i_{,\lambda\lambda} .. \vec{v}^l_{,\lambda\lambda} \vec{v}^i_{,x\lambda} .. \vec{v}^l_{,x\lambda}\right\} \quad (27)$$

Here $[T]$ is the transition matrix from (26) to (27).

The displacement vectors of the nodes of a quadrangular finite element and their derivatives taking into account (21) can be represented by the components of the nodal vectors of the local basis

$$\vec{v}^m = v^{m\rho}\vec{a}_\rho^{0m} + v^m \vec{a}^{0m};\quad \vec{v}^m_{,\alpha} = t_\alpha^{m\rho}\vec{a}_\rho^{0m} + t_\alpha^m \vec{a}^{0m};\quad \vec{v}^m_{,\alpha\beta} = t_{\alpha\beta}^{m\rho}\vec{a}_\rho^{0m} + t_{\alpha\beta}^m \vec{a}^{0m}, \quad (28)$$

where $m = i,\ j,\ k,\ l.$

A column of vector nodal unknowns $\left\{\vec{v}_y^G\right\}$ in the global coordinate system based on (28) can be represented by a matrix product

$$\underset{24\times 1}{\left\{\vec{v}_y^G\right\}} = \underset{24\times 72}{\left\lfloor\vec{A}\right\rfloor} \underset{72\times 1}{\left\{t_y\right\}} = \underset{24\times 72}{\left\lfloor\vec{A}\right\rfloor} \underset{72\times 72}{\left[P_R\right]} \underset{72\times 1}{\left\{U_y^G\right\}}, \quad (29)$$

where

$$\left\{t_y\right\}^T = \left\{v^{1i} v^{2i} v^i v^{1j} v^{2j} v^j ... v^l t_1^{1i} t_1^{2i} t_1^i t_1^{1j} t_1^{2j} t_1^j .. t_1^l t_2^{1i} t_2^{2i} t_2^i t_2^{1j} t_2^{2j} t_2^j .. t_2^l t_{11}^{1i} t_{11}^{2i} t_{11}^i .. t_{11}^l t_{22}^{1i} t_{22}^{2i} t_{22}^i .. t_{22}^l t_{12}^{1i} t_{12}^{2i} t_{12}^i .. t_{12}^l\right\},$$

$[P_R]$ - matrix of transition from post $\left\{t_y\right\}$ to post $\left\{U_y^G\right\}$ (22).

The interpolation expression (25) taking into account (28) will take the form

$$\vec{v} = \underset{1\times 24}{\left\{\chi\right\}^T} \underset{24\times 72}{\left\lfloor\vec{A}\right\rfloor} \underset{72\times 72}{\left[P_R\right]} \underset{72\times 1}{\left\{U_y^G\right\}}, \quad (30)$$

where $\underset{1\times 24}{\left\{\chi\right\}^T} = \underset{1\times 24}{\left\{\varphi\right\}^T} \underset{24\times 24}{\left[T\right]}.$

The input to $\left\lfloor\vec{A}\right\rfloor$ the nodal vectors of local bases $\vec{a}_1^{0m}$, $\vec{a}_2^{0m}$, $\vec{a}^{0m}$ on the basis of (15) can be expressed in terms of the vectors of the local basis of a point belonging to the inner region of a quadrangular finite element, for example $\lambda = t$

$$\left\{\vec{a}^{0m}\right\} = \left[b^{0m}\right]\left\{\vec{i}\right\} = \left[b^{0m}\right]\left[b^0\right]^{-1}\left\{\vec{a}^0\right\} = \left[c^{0m}\right]\left\{\vec{a}^0\right\}. \quad (31)$$

The ratio (30) taking into account (31) can be presented in the following form

$$\underset{1\times 3}{\left\{\vec{a}^0\right\}^T} \underset{3\times 1}{\begin{Bmatrix} v^1 \\ v^2 \\ v \end{Bmatrix}} = \underset{1\times 3}{\left\{\vec{a}^0\right\}^T} \underset{3\times 72}{\left[B\right]} \underset{72\times 72}{\left[P_R\right]} \underset{72\times 1}{\left\{U_y^G\right\}}, \quad (32)$$

where $\underset{3\times 72}{\left[B\right]} = \left[\chi_1 \underset{3\times 3}{\left[c^{0i}\right]^T} \middle| \chi_2 \underset{3\times 3}{\left[c^{0j}\right]^T} \middle| \chi_3 \underset{3\times 3}{\left[c^{0k}\right]^T} \middle| \chi_4 \underset{3\times 3}{\left[c^{0l}\right]^T} \middle| \chi_5 \underset{3\times 3}{\left[c^{0i}\right]^T} ... \middle| \chi_{24} \underset{3\times 3}{\left[c^{0l}\right]^T}\right].$

From (32) it is possible to obtain the required interpolation expressions for the components of the displacement vector using the vector variant of the interpolation procedure

$$v^1 = \left\{\chi_1 \left\{c_{11}^{0i} c_{21}^{0i} c_{31}^{0i}\right\} \middle| \chi_2 \left\{c_{11}^{0j} c_{21}^{0j} c_{31}^{0j}\right\} \middle| \chi_3 \left\{c_{11}^{0k} c_{21}^{0k} c_{31}^{0k}\right\} \middle| \chi_4 \left\{c_{11}^{0l} c_{21}^{0l} c_{31}^{0l}\right\} \middle| \chi_5 \left\{c_{11}^{0i} c_{21}^{0i} c_{31}^{0i}\right\} .. \middle| \chi_{24} \left\{c_{11}^{0l} c_{21}^{0l} c_{31}^{0l}\right\}\right\}\left\{W_y\right\};$$

$$v^2 = \left\{ \chi_1 \left\{ c_{12}^{0i} c_{22}^{0i} c_{32}^{0i} \right\} \middle| \chi_2 \left\{ c_{12}^{0j} c_{22}^{0j} c_{32}^{0j} \right\} \middle| \chi_3 \left\{ c_{12}^{0k} c_{22}^{0k} c_{32}^{0k} \right\} \middle| \chi_4 \left\{ c_{12}^{0l} c_{22}^{0l} c_{32}^{0l} \right\} \middle| \chi_5 \left\{ c_{12}^{0i} c_{22}^{0i} c_{32}^{0i} \right\} .. \middle| \chi_{24} \left\{ c_{12}^{0l} c_{22}^{0l} c_{32}^{0l} \right\} \right\} \{W_y\}, \quad (33)$$

$$v = \left\{ \chi_1 \left\{ c_{13}^{0i} c_{23}^{0i} c_{33}^{0i} \right\} \middle| \chi_2 \left\{ c_{13}^{0j} c_{23}^{0j} c_{33}^{0j} \right\} \middle| \chi_3 \left\{ c_{13}^{0k} c_{23}^{0k} c_{33}^{0k} \right\} \middle| \chi_4 \left\{ c_{13}^{0l} c_{23}^{0l} c_{33}^{0l} \right\} \middle| \chi_5 \left\{ c_{13}^{0i} c_{23}^{0i} c_{33}^{0i} \right\} .. \middle| \chi_{24} \left\{ c_{13}^{0l} c_{23}^{0l} c_{33}^{0l} \right\} \right\} \{W_y\},$$

where $\{W_y\} = [P_R]\{U_y^G\}$

To obtain the interpolation of the dependencies for the component of derivative of the displacement vector procedure must be repeated differentiation of (25) x and t

$$\vec{v}_{,\alpha} = \left(\{\varphi_{,\xi}\}^T \cdot \xi_{,\alpha} + \{\varphi_{,\eta}\}^T \cdot \eta_{,\alpha} \right) [T] \{\vec{v}_y^G\} = \{\chi^\alpha\}^T [\vec{A}][P_R]\{U_y^G\},$$

$$\vec{v}_{,\alpha\beta} = \left(\{\varphi_{,\xi\xi}\}^T \xi_{,\alpha}\xi_{,\beta} + \{\varphi_{,\xi\eta}\}^T \left(\xi_{,\alpha}\eta_{,\beta} + \xi_{,\beta}\eta_{,\alpha} \right) + \{\varphi_{,\eta\eta}\}^T \eta_{,\alpha}\eta_{,\beta} + \right.$$

$$\left. + \{\varphi_{,\xi}\}^T \xi_{,\alpha\beta} + \{\varphi_{,\eta}\}^T \eta_{,\alpha\beta} \right) [T] \{\vec{v}_y^G\} = \{\chi^{\alpha\beta}\}^T [\vec{A}][P_R]\{U_y^G\} \quad (34)$$

By performing over (34) transformations (31) and (32), you can write interpolation dependencies for the components of the first and second derivatives of the displacement vector

$$\begin{Bmatrix} t_\alpha^1 \\ t_\alpha^2 \\ t_\alpha \end{Bmatrix} = \left[\chi_1^\alpha [c^{0i}]^T \middle| \chi_2^\alpha [c^{0j}]^T \middle| \chi_3^\alpha [c^{0k}]^T \middle| \chi_4^\alpha [c^{0l}]^T \middle| \chi_5^\alpha [c^{0i}]^T \ldots \middle| \chi_{24}^\alpha [c^{0l}]^T \right] \{W_y\},$$

$$\begin{Bmatrix} t_{\alpha\beta}^1 \\ t_{\alpha\beta}^2 \\ t_{\alpha\beta} \end{Bmatrix} = \left[\chi_1^{\alpha\beta} [c^{0i}]^T \middle| \chi_2^{\alpha\beta} [c^{0j}]^T \middle| \chi_3^{\alpha\beta} [c^{0k}]^T \middle| \chi_4^{\alpha\beta} [c^{0l}]^T \middle| \chi_5^{\alpha\beta} [c^{0i}]^T \ldots \middle| \chi_{24}^{\alpha\beta} [c^{0l}]^T \right] \{W_y\} \quad (35)$$

4 Discussion

Comparing (24) and (33), (35), we can note the fundamental difference between them. In the conventional interpolation procedure, the individual component of the displacement vector of the point of the inner region of the finite element depends on the nodal values of the same component and its derivatives, and the interpolation expression (24) does not contain the parameters of the curvilinear coordinate system used. When using the vector method of the interpolation procedure, the individual component of the displacement vector depends on the complete set of nodal values of all three components and their derivatives, and the interpolation dependences include the parameters of the curvilinear coordinate systems used.

The further procedure of forming the stiffness matrix and the column of nodal forces is performed in the manner generally accepted in FEM [17-21].

Conclusions

The analysis of the developed algorithms for obtaining stiffness matrices of high-precision finite quadrangular finite elements shows that the proposed version of the representation of the middle surface of the shell in the form of an elliptic cylinder using a radius (5) depending on the angular coordinate is the most appropriate. In addition, the use of vector approximation of the desired values in the algorithm of formation of the stiffness matrix of a quadrangular finite element more correctly reflects the geometric meaning of the problem.

Acknowledgements

The study was carried out with the financial support of the Russian Foundation for Basic Research and the Administration of the Volgograd Region in the framework of the research project No. 18-41-343003 r_mol_a.

References

[1] Zheleznov L P, Kabanov V V and Boiko D V 2014 Nonlinear deformation and stability of discretely reinforced elliptical cylindrical shells under transverse bending and internal pressure *Russian Aeronautics* **57(2)** pp 118-126

[2] Kayumov R A, Shakirzyanov F R and Gavryushin S S 2014 Modeling of the deformation process and evaluation of the bearing capacity of the soil – thin-walled structure *Proceedings of higher educational institutions. Engineering* **6** pp 20-24

[3] Krivoshapko S N and Galishnikova V V 2017 *Architectural and construction structures: textbook for academic undergraduate* (Moscow: Yurayt Publishing House) p 476

[4] Belostotsky A M, Akimov P A, Afanasyeva I N and Kaytukov T B 2018 Contemporary problems of numerical modelling of unique structures and buildings *International Journal for Computational Civil and Structural Engineering* **13(2)** pp 9-34

[5] Galishnikova V V and Pahl P J 2018 Constrained construction of planar delaunay triangulations without flipping *Construction mechanics of engineering structures and structures* **14(2)** pp 154-174

[6] Mirsaidov M M, Abdikarimov R A, Vatin N I, Zhgutov V M, Khodzhaev D A and Normuminov B A 2018 Nonlinear parametric oscillations of viscoelastic plate of variable thickness *Magazine of Civil Engineering* **82 (6)** pp 112-126

[7] Badriev I B, Paimushin V N 2017 Refined models of contact interaction of a thin plate with positioned on both sides deformable foundations *Lobachevskii Journal of Mathematics* **38(5)** pp 779-793

[8] Levyakov S V and Kuznetsov V V 2011 Application of triangular element invariants for geometrically nonlinear analysis of functionally graded shells *Computational Mechanics* **48(4)** pp 499-513

[9] Javili A, Mc Bride A, Steinmann P and Reddy B D 2014 A unified computational framework for bulk andsurface elasticity theory: a curvilinear-coordinate based finite element methodology *Computational Mechanics* **54(3)** pp 745-762

[10] Hanslo P, Larson M G and Larson F 2015 Tangential differential calculus and the finite element modeling of a large deformation elastic membrane problem *Computational Mechanics* **56(1)** pp 87-95

[11] Ren Hui 2015 Fast and robust full-guad-rature triangular elements for thin plates/ shells, with large deformations and large rotations *ASME Journal of Computational and Nonlinear Dynamics* **10(5)** pp 051018/1-051018/13

[12] Nguyen N and Waas A 2016 Nonlinear, finite deformation, finite element analysise *ZAMP. Z. Angew. math. and Phys* **67(9)** pp 35/1-35/24

[13] Paznanova S L, Vasilev G P, Dineva P S and Manolis G D 2016 Dynamic analysis of nanoheterogeneities in a finite-sized solidby boundary and finite element methods *International Journal of Solids and Structures* **80** pp 1-18

[14] Krivoshapko S N and Ivanov V N 2010 *Encyclopedia of analytical surfaces* (Moscow: Publishing House "LIBROKOM") p 560

[15] Klochkov J V, Nikolaev A P and Vakhnina O V 2011 Numerical analysis of stress-strain state of thin shells when using a triangular finite element with Lagrange multipliers *Proceedings of the lower Volga agrodiversity complex: Science and higher professional education* **3(23)** pp 186-192

[16] Storozhuk E A and Yatsura A V 2017 Analytical-numerical solution of static problems for noncircular cylindrical shells of variable thickness *International Applied Mechanics* **53(3)** pp 313-325

[17] Karpov V V, Ignatev O V and Semenov A A 2017 The stress-strain state of ribbed shell structures *Magazine of Civil Engineering* **74 (6)** pp 147-160

[18] Dzhabrailov A Sh, Klochkov Yu V and Nikolaev A P 2009 The finite element analysis of shells of revolution with a branching meridian *Russian Aeronautics* **52(1)** pp 22-29

IOP Conf. Series: Materials Science and Engineering **675** (2019) 012053 doi:10.1088/1757-899X/675/1/012053

[19] Agapov V P and Aidemirov K R 2016 Calculation of farms by the finite element method taking into account geometric nonlinearity *Industrial and civil engineering* **11** pp 4-7
[20] Lalin V, Rybakov V and Sergey A 2014 The finite elements for design of frame of thin-walled beams *Applied Mechanics and Materials* **578** pp 858-863
[21] Klochkov Y V, Nikolaev A P, Kiseleva T A and Marchenko S S 2016 Comparative analysis of the results of finite element calculations based on an ellipsoidal shell *Journal of Machinery Manufacture and Reliability* **45(4)** pp 328-336

IOP Conf. Series: Materials Science and Engineering **675** (2019) 012054 doi:10.1088/1757-899X/675/1/012054

Mathematical model of building envelope element insolation

N P Kopytov, E A Mityushov, N E Misyura* and S A Berestova

Institute of Fundamental Education Ural Federal University, Ekaterinburg, Russia

E-mail: n_misura@mail.ru

Abstract. In this paper, the problem of estimation of building envelope oriented element insolation by direct sunlight is solved using mathematical model of Earth's rotation. Quaternions are used as mathematical tool for description of rotation. The model allows to obtain automatically estimation of building element insolation by direct sunlight for given latitude and given time interval (month, week, day, etc.). The model takes into account schedule of change of day and night, change of direct sunlight angle on a given element of the surface caused by sun motion over horizon (changes of its height and azimuth during time of day). Distinctive feature of the model is simplicity of program-algorithmic implementation due to using description of rotation by means of quaternions.

1. Introduction

The level of solar insolation of building envelopes plays essential role in modern architectural design. The solar insolation factor is important for designing the forms of individual buildings, designing of sets and ensembles of buildings, as well as micro districts in cities as a whole. It requires the development of tools for estimating the level of solar insolation for building envelopes. Today, there are two main reasons for the necessity of such tools having in the design of building forms. On the one hand, it is necessary to provide sufficient inflow of sunlight, or vice versa, the level of shading for office, industrial and residential rooms of buildings in order to create comfortable conditions for their inhabitants [1]. On the other hand, it is necessary to take into account the development of renewable energy sources based on photovoltaic panels embedded in building envelopes (facades and roofs) – Building Integrated Photovoltaic (BIPV) [2-5]. For such technologies, the problem of optimal integration of PV elements into the building envelope, their orientation in accordance with insolation level are especially important, since their effectiveness directly depends on it. It is necessary to describe role and area of optimal application of BIPV technologies, which, according to the World Energy Council, is one of priority areas for the development of solar power in general [6]. The main reason for this is the intersection of trends in urbanization and trends in the development of electric power systems. As for urbanization, it is necessary to note the rapid development of cities, population growth and increasing of population density, it leads to construction of a large number of high buildings, increasing of buildings energy consumption. In some developed countries, the share of energy consumption by buildings is 35-40% of total energy consumption [6]. As for electrical power industry, one of the trends of development of power supply systems, in general, is the active usage of renewable energy sources, in particular, solar panels. The advantages of using solar generation to supplement existing power supply systems are: no need to build infrastructure for delivery of energy fuel (gas, diesel, coal, etc.); environmental friendliness, that is especially important in the conditions of modern city; constant reduction of cost of solar panels and components for systems with solar panels (controllers, inverters, energy storages); local generation, which increases the reliability of power supply for critical systems of buildings during power outages from external electrical grids, decreasing of load on the external power lines. The disadvantages of solar generation include instability (the level of solar irradiation on the surface of the Earth is not constant), large area that is needed to install panels. The first disadvantage can be partially overcome by the using of electricity storage devices; the second one is very difficult to be overcome in cities, because of absence and high cost of land areas in cities for installation of arrays of solar panels. One of the approaches that can help to overcome these contradictions in urban conditions is integration of solar panels into the building envelopes - roof and facades - that is using of BIPV technologies. Aspects and information about the active development of BIPV technologies outlined above are supported by large number of various scientific works, performed using a variety of theoretical and experimental research methods, individual reviews [7], reports of international specialized agencies and committees [6, 8]. Separately, works can be noted that are dedicated to the development of specialized software for assessing insolation, not only in scope of individual separate buildings, but also in scope of entire ensembles and micro districts of cities [9, 10]. Additionally, it is necessary to note works about optimality of shapes of PV surfaces themselves [11]. Good illustrations and photos of buildings with BIPV are presented in [7, 11]. In addition, active work dedicated to regulation of BIPV technologies are described in [8]. All facts and number of papers speak about stability of this trend. As mentioned above, for BIPV technologies, the optimal orientation of solar panel arrays is particularly important from the point of view of direct sunlight falling on them, since the maximum generation of electricity, in the general case and for most solar panel types, is carried out when the sun's rays fall on the elements at 90° angle. In order to solve the problem of optimal integration of PV elements into the building envelope, mathematical models and algorithms are needed that give values of estimation of solar insolation. In this paper, an original method for estimating the insolation of the oriented element of the building envelope by direct sunlight is proposed. The core part of proposed method is quaternion model of Earth's rotation.

2. Quaternion model of Earth's rotation

In order solve problem of insolation we need to build model of Earth's rotation. Some simplifications can be taken due to problem has character of estimation. As base for model let us consider the following points:

1. Earth is considered as a ball;
2. Center of mass of Earth moves uniformly over circular orbit around the Sun, making full turn during 8760 hours;
3. Axis of Earth has a constant nutation angle 23.44 degrees;
4. Earth makes a full turn around its axis (self-rotation) during $24\dfrac{365}{366} \approx 23.9344$ hours, which corresponds to the complete revolution of the Earth around its axis relative to distant stars (It is «Sidereal day» in this model);
5. Earth period of rotation relative to Sun is 24 hours (period between two Sun culminations at one meridian);
6. Beginning of time in model is moment of the summer solstice for Greenwich (prime) meridian.

Let us consider the orbital coordinate system $O\varsigma\eta\xi$ with origins in Earth center of mass центре масс and following directions of axes: axis $O\varsigma$ is directed opposite to Sun orbit movement, axis $O\eta$ is directed opposite to direction to Sun, asis $O\xi$ is perpendicular to ecliptic plane (figure 1). Introduced system $O\varsigma\eta\xi$ is right-handed.

In this coordinate system Earth has regular precession with constant angular velocity ω_ψ, self-rotation with constant angular velocity ω_φ, constant angle of nutation:

$$\omega_\psi = -\frac{2\pi}{365\times 24}\ {rad}/{h},\quad \omega_\phi = \frac{2\pi}{24\times(365/366)}\ {rad}/{h},\quad \vartheta_0 = \frac{\pi}{180}\times 23.44\ rad\ .$$

Sign of minus in expression of ω_ψ is related to direction chosen axes of coordinate system.

For determination of orientation and position of objects on Earth surface we will use unit quaternion

$$\mathrm{q}(t) = \{q_0(t), q_1(t), q_2(t), q_3(t)\},$$

where $q_0^2(t) + q_1^2(t) + q_2^2(t) + q_3^2(t) = 1$.

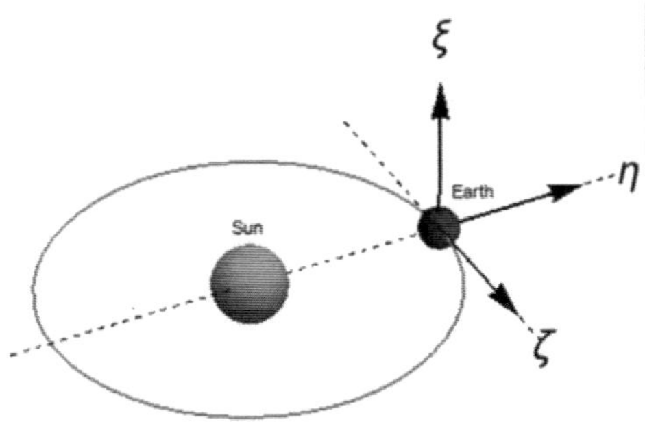

Figure 1. Illustration of orientation of introduced orbital coordinate system.

Taking into account precession movement of Earth in selected coordinate system $O\varsigma\eta\xi$ components of quaternion of orientation are determined as [12]:

$$q_0(t) = \cos\frac{\vartheta_0}{2}\cos\frac{(\omega_\psi + \omega_\varphi)\,t}{2}, \quad q_1(t) = \sin\frac{\vartheta_0}{2}\cos\frac{(\omega_\psi - \omega_\varphi)\,t}{2},$$

$$q_2(t) = \sin\frac{\vartheta_0}{2}\sin\frac{(\omega_\psi - \omega_\varphi)\,t}{2}, \quad q_3(t) = \cos\frac{\vartheta_0}{2}\sin\frac{(\omega_\psi + \omega_\varphi)\,t}{2}.$$

Changing of position of any point related to Earth can be found using following rotation transformation:

$$r(t) = R(t)r_0,$$

where r_0 and $r(t)$ - radius-vectors of initial ant current positions of point in orbital coordinate system, and $R(t)$ - matrix of rotation that is determined by components of quaternion:

$$R(t) = \begin{pmatrix} 1-2q_2^2(t)-2q_3^2(t) & -2q_0(t)q_3(t)+2q_1(t)q_2(t) & 2q_0(t)q_2(t)+2q_1(t)q_3(t) \\ 2q_0(t)q_3(t)+2q_1(t)q_2(t) & 1-2q_1^2(t)-2q_3^2(t) & -2q_0(t)q_1(t)+2q_3(t)q_2(t) \\ -2q_0(t)q_2(t)+2q_1(t)q_3(t) & 2q_0(t)q_1(t)+2q_3(t)q_2(t) & 1-2q_2^2(t)-2q_1^2(t) \end{pmatrix}.$$

3. Main orientations and schedule of changing of day and night

Let's take into consideration $a(t)$ - vector of Earth axis (in direction from center of Earth to geographical North Pole), movement of this axis is determined by equation

$$a(t) = R(t)\begin{pmatrix} 0 \\ 0 \\ 1 \end{pmatrix}.$$

Let us consider angle ϑ between normal vector to Earth's surface at given latitude λ of Greenwich (prime) meridian and Earth's axis

$$\vartheta = \frac{\pi}{2} - \frac{\pi}{180} \times \lambda,$$

here $\lambda > 0$ for Northern Hemisphere and $\lambda < 0$ for Southern Hemisphere.

Normal vector $n_\lambda(t)$ to Earth's surface at given latitude λ of Greenwich (prime) meridian is determined by equation

$$n_\lambda(t) = R(t)Q\begin{pmatrix}0\\0\\1\end{pmatrix},$$

where $Q = \begin{pmatrix}1 & 0 & 0\\ 0 & \cos\vartheta & -\sin\vartheta\\ 0 & \sin\vartheta & \cos\vartheta\end{pmatrix}$ - direction cosines matrix.

In this model we will consider direction to Sun sl to be constant for all points of Earth's surface (it is simplification but it is frequently used by reason of long distance between Earth and Sun and size of Sun). Vector sl according to selected coordinate system is determined as

$$sl = \begin{pmatrix}0\\-1\\0\end{pmatrix}.$$

Visualization of positions and orientations of main vectors in selected orbital coordinate system at time zero point (summer solstice) is presented in figure 2.

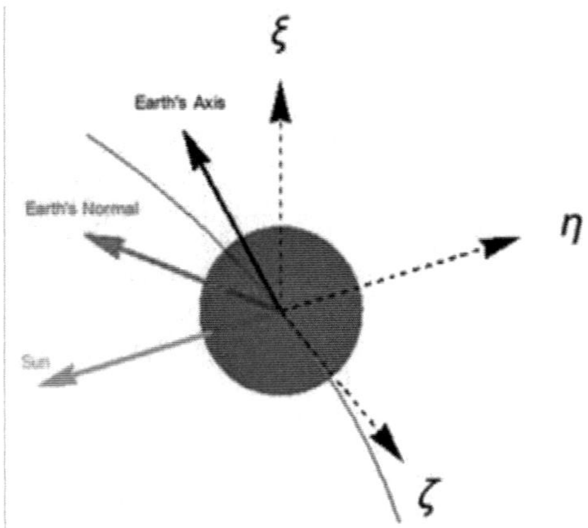

Figure 2. Visualization of main vectors (Earth's axis, normal vector at given latitude 60 degrees, vector directed to Sun) at summer solstice and orbital coordinate system.

The entered parameters and operations on them allow to obtain various information related to the movement of Earth. For example, the schedule of day and night, as well as the time of sunrise and sunset, can be determined by angle between the normal vector at a given point on surface of Earth $n_\lambda(t)$ and vector of direction to the Sun sl. In case if $c_1(t) = \langle n_\lambda(t), sl\rangle > 0$, there is day in this point on Earth surface, and if $c_1(t) = \langle n_\lambda(t), sl\rangle < 0$, then night. When $c_1(t) = \langle n_\lambda(t), sl\rangle = 0$, it corresponds to event of sunrise or sunset and depends on sign of expression of $\langle n_\lambda(t), sl(t)\rangle$ before.

4. Determination of insolation of building envelop element

Let us consider the local coordinate system $Oxyz$, that is constantly rigidly related to Earth surface in location of considered object (for example, considered building). Axis Ox is directed to East, axis Oy is directed to North, axis Oz is directed by normal vector to Earth surface (direction of axis Oz is the same as vector $n_\lambda(t)$). Orientations of local coordinate system $Oxyz$ and orbital coordinate system are demonstrated in figure 3.

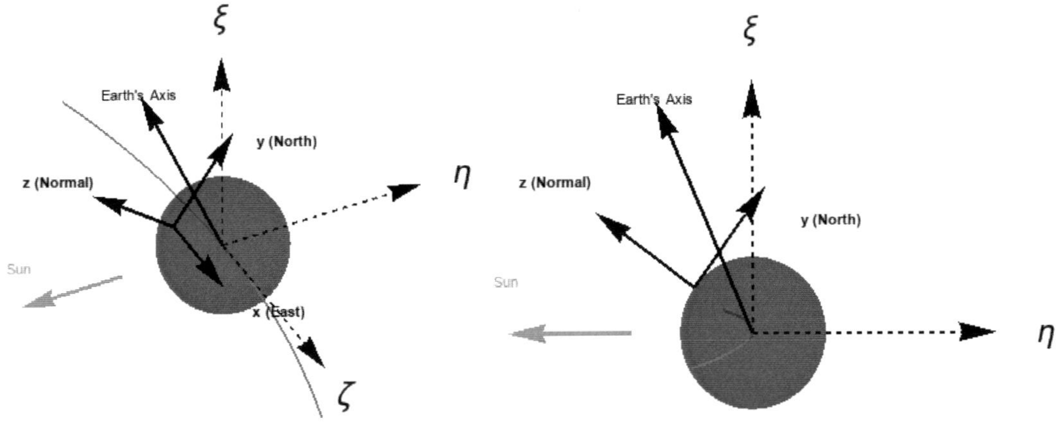

Figure 3. Local and orbital coordinate systems.

Let us consider normal vector to element of building envelope in local coordinate system:

$$n = \begin{pmatrix} n_x \\ n_y \\ n_z \end{pmatrix}.$$

Coordinates of this vector in orbital coordinate system are determined by transformation

$$\begin{pmatrix} n_\varsigma \\ n_\eta \\ n_\xi \end{pmatrix} = Q \begin{pmatrix} n_x \\ n_y \\ n_z \end{pmatrix},$$

where Q is direction cosines matrix and it depends on given latitude (see above), in this case it is applied to go from local to orbital coordinate system.

The major part of insolation in clear sky day corresponds to direct sun irradiation [13]. Direct insolation of element of building envelope is determined by the variable geometric factor in form of cosine $c_2(t)$ of angle between the unit vector to Sun and unit vector normal of element of building envelope. This can be expressed in orbital coordinate system by expression

$$c_2(t) = \left\langle R(t) \begin{pmatrix} n_\varsigma \\ n_\eta \\ n_\xi \end{pmatrix}, sl \right\rangle.$$

In order to solve comparative problems of estimation of insolation by direct sunlight, it is necessary to introduce a characteristic that takes into account changing of angle of incidence of direct sunlight on oriented element of building envelope over time. It should be taken into account that in some cases when $c_2(t) > 0$, $c_1(t)$ may be even less than 0, i.e. Sun is lower than the horizon (for example, at winter solstice).

Taking into account discussion above we propose to use integral geometrical characteristic of insolation of oriented building envelop element in form of

$$I = \frac{1}{4}\int_{t_1}^{t_2}\left(1+\frac{c_1(t)}{\sqrt{(c_1(t))^2}}\right)\left(1+\frac{c_2(t)}{\sqrt{(c_2(t))^2}}\right)c_2(t)dt\,,$$

that describes insolation of selected unite element by direct sunlight on time interval (t_1, t_2). Here $c_1(t)$ - cosine of angle between normal vector of Earth's surface and direction to Sun, it takes positive value when Sun is above the horizon; $c_2(t)$ - cosine of angle between normal vector of oriented element and direction to Sun, it takes positive value when angle between normal vector of oriented element and direction to the Sun is less than 90;

expression $\frac{1}{2}\left(1+\frac{c_1(t)}{\sqrt{(c_1(t))^2}}\right)$ - gives 1 when angle between normal vector of Earth's surface and direction to Sun is less than 90 degrees, that is Sun is above the horizon, and gives 0 when it is night (it is a kind of «day indicator»);

expression $\frac{1}{2}\left(1+\frac{c_2(t)}{\sqrt{(c_2(t))^2}}\right)$ - gives 1 when angle between normal vector of selected oriented element and direction to Sun is less than 90 degrees.

5. Results of numerical simulation and interpretation

In order to demonstrate work of proposed model and algorithm let us chose two conditional unit elements of building envelope – element of vertical facade oriented strictly in direction to geographical South Pole, and element of roof installed at angle of 45 degrees to surface of Earth and adjacent to facade under consideration (figure 4).

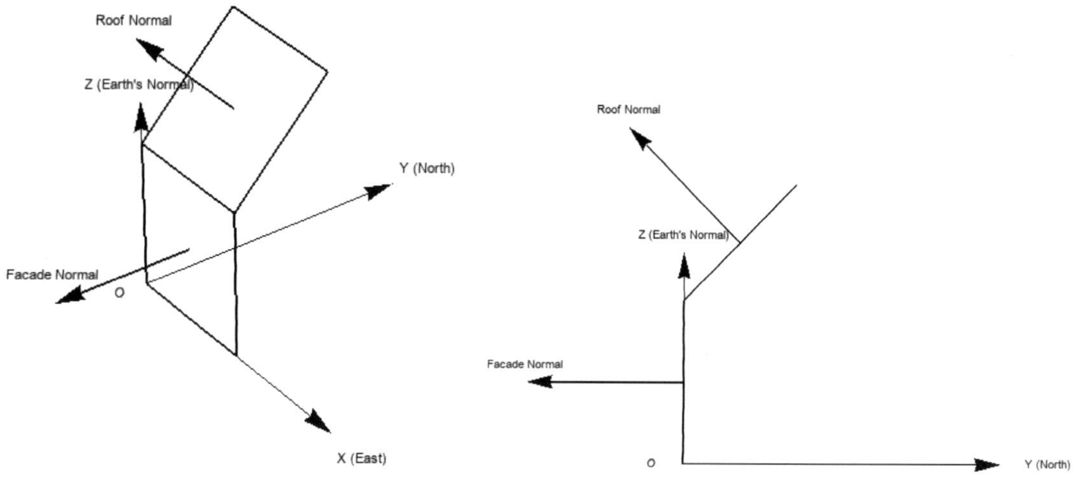

Figure 4. Illustration of orientations of vertical element of facade element of element of roof and their normal vectors in local coordinate system.

In local coordinate system $Oxyz$ normal vectors of selected elements are

$$n_{Fasade}^{Oxyz} = \begin{pmatrix} 0 \\ -1 \\ 0 \end{pmatrix}, n_{Roof}^{Oxyz} = \begin{pmatrix} 0 \\ -\sqrt{2}/2 \\ \sqrt{2}/2 \end{pmatrix}.$$

In orbital coordinate system they are

$$n_{Fasade}^{O\varsigma\eta\xi} = Qn_{Fasade}^{Oxyz}, \; n_{Roof}^{O\varsigma\eta\xi} = Qn_{Roof}^{Oxyz}.$$

Illustration of orientations of vertical element of facade element of element of roof and their normal vectors in local coordinate system $Oxyz$ is presented in figure 4. Visualization of the same vectors in orbital coordinate system $O\varsigma\eta\xi$ is presented in figure 5.

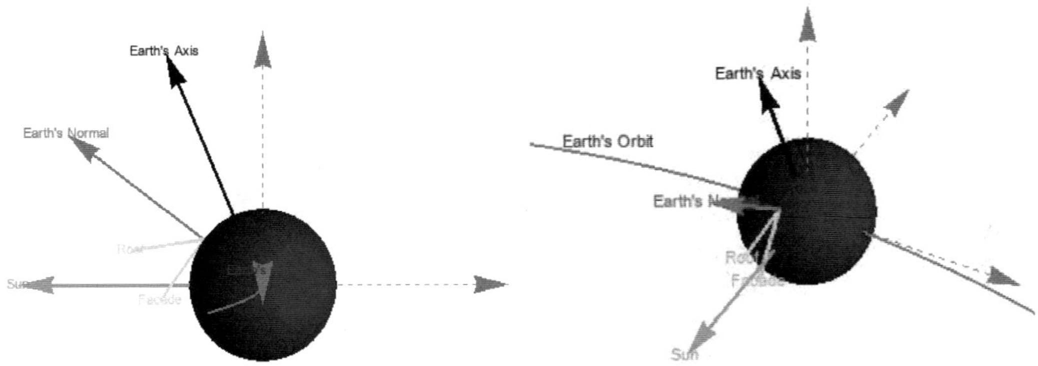

Figure 5. Illustration of orientations of normal vectors of vertical facade element and roof element in orbital coordinate system.

Now it is possible to calculate values of insolation integral I for facade and roof elements during intervals of four days - day of summer solstice (0 day), 91 days later (close to the autumn equinox), 182 days later (close to winter solstice), 274 days later (close to the spring equinox). Location is set at latitude of 60 degrees in Northern Hemisphere. The interval of integration is 24 hours. The results of numerical simulation are shown in table 1.

Table 1. Results of numerical simulation.

	0 day	**+ 91 days**	**+ 182 days**	**+ 274 days**
Facade	3.8756	6.6042	5.1038	6.6042
Roof	8.0534	7.3828	3.9032	7.3828

The insolation values for +91 days point and for +274 days point are equal, which is explained by the fact that planet occupies diametrically opposite positions in orbit according to the model (365-91 = 274). Insolation of facade in winter (+182 days point) is higher than insolation of roof, it is explained by lower height of Sun above the horizon in winter.

6. Conclusions

Proposed kinematic model of Earth rotation based on quaternions and proposed dimensionless characteristic I, that takes into account geometrical factor of direct sunlight inclination within selected time interval, allow to solve wide range of problems of comparative estimations. The proposed model is quite simple for implementation in mathematical simulation packages, such as MathCAD or Wolfram Mathematica, for example. The vector determination of positions of points on Earth's surface, orientations of main axes and elements of building envelope open wide opportunities for various visualization and solving of other related tasks.

IOP Conf. Series: Materials Science and Engineering **675** (2019) 012054 doi:10.1088/1757-899X/675/1/012054

References

[1] Ivanova S 2011 Estimation of solar radiation for buildings with complex architectural layouts, *Proceeding of the 5th International Conference SOLARIS 2011* (Brno University of Technology, Czech Republic) pp 104-109

[2] Yang H and Lu L 2007 The Optimum Tilt Angles and Orientations of PV Claddings for Building-Integrated Photovoltaic (BIPV) Applications *Journal of Solar Energy Engineering* **129(2)** pp 253-255

[3] Waseef A A 2014 Tilt and orientation: a preference factor among photovoltaic roof systems, *International Journal of Innovative Research in Science, Engineering and Technology* **3(3)** pp 10750-10758

[4] Meggers F, Aviv D, Charpentier V, Teitelbaum E, Ainslie1 A and Adriaenssens S 2017 Co-optimization of solar tracking for shading and photovoltaic energy conversion *Proceedings of the 15th IBPSA Conference 2017* (San Francisco, USA) pp 2224-2231

[5] Salem T and Kinab E 2015 Analysis of Building-Integrated Photovoltaic Systems: A Case Study of Commercial Buildings under Mediterranean Climate *Procedia Engineering* **118** pp 538–545

[6] World Energy Resources: 2013 Survay URL: https://www.worldenergy.org/wp-content/uploads/2013/09/Complete_WER_2013_Survey.pdf

[7] Zhang T, Wang M and Yang H 2018 A Review of the Energy Performance and Life-Cycle Assessment of Building-Integrated Photovoltaic (BIPV) Systems *Energies* **11(11)**: 34p

[8] International definition of "BIPV" 2018 Report IEA-PVPS T15-04 of the International Energy Agency - Photovoltaic Power Systems Programme URL: http://www.iea-pvps.org/fileadmin/dam/public/report/national/IEA-PVPS_Task_15_Report_C0_International_definitions_of_BIPV_hrw_180823_03.pdf

[9] Good C S, Lobaccaro G and Harklau S 2014 Optimization of solar energy potential for buildings in urban areas – a Norwegian case study *Energy Procedia* **58** pp 166-171

[10] Jaugsch F, Löwner M O 2016 Estimation of solar energy on vertical 3D building walls on city quarter scale, *Proceeding of the 11th International 3D Geoinfo Conference 2016* (Athens, Greece) pp 135-143

[11] Cheng S 2009 Curved Photovoltaic Surface Optimization for BIPV: An Evolutionary Approach Based on Solar Radiation Simulation MSc Thesis (London: University of London) URL: http://discovery.ucl.ac.uk/18982/1/18982.pdf

[12] Kovalev M and Gorr G V 2014 Danilyuk Application of Rodrigues–Hamilton parameters in investigation of precessional motion of a rigid body with a fixed point Proceedings of IAMM NAS of Ukraine pp 93-101 URL: http://iamm.su/upload/iblock/fcd/93_101.pdf

[13] Naraghi M H 2009 Optimum solar panel tilt angle for maximum annual irradiation *ASME International Mechanical Engineering Congress and Exposition 2009* (Lake Buena Vista, Florida, USA) pp 1-10 URL: https://www.researchgate.net/publication/267592248_Optimum_Solar_Panel_Tilt_Angle_for_Maximum_Annual_Irradiation

IOP Conf. Series: Materials Science and Engineering **675** (2019) 012055 doi:10.1088/1757-899X/675/1/012055

Stress and strain intensity factors within angle area of the boundary

L U Frishter

National Research University Moscow State University of Civil Engineering, Moscow, Russian Federation

Email: lfrishter@mail.ru

Abstract. Stress and strain state of the designs and structures within the area (zone) with angle cutouts, boundary cuts are characterized by emergence of the stress concentration areas and requires assessment of strength and reliability of facilities that is an actual task of the engineering practice. The theoretical study of stress and strain state within angle cutouts of the boundary area is reduced to research of singular solutions for the elastic problem with power singularity. Moreover, definition of stress or strain concentration in the non-regular point of the area boundary loses its significance. The stress and strain state within angle cutout top of the plane area boundary, which is recorded with use of intensity factor, is considered in the article. Two approaches to obtaining of formula for displacements and stresses within non-regular point of the plane area boundary with use of the stress and strain intensity factor are given. A difference in formula of stresses and displacement obtained for the limit values of stress and strain determine the practical significance of the work during experiments and determination of critical values for stress and strain.

1. Introduction

The stress-strain state of designs and structures with a geometrically non-linear border shape - cutouts, boundary cuts, is characterized by the emergence of the stress concentration areas, deformations with significant quantities and their gradients.

The solution of the physically linear problem of the theory of elasticity in the geometrically linearized formulation of the boundary conditions is determined by infinite stresses, strains and their gradients at the apex of an ideal cut or angular cut of the boundary of the area [1-7]. The concept of stress concentration at an irregular point of the area boundary loses its meaning.

The study of the stress-strain state of designs and structures in the area of an irregular point of the boundary is characterized by solving a homogeneous boundary value problem of the theory of elasticity [1–4, 8–14].

The power feature of the stress-strain state is due to the geometry of the shape of the boundary of the area — cutout, boundary cut, the order of which depends on the eigenvalues of the homogeneous elastic problem [1-2, 7-8, 15-16]. The eigenvalues of the homogeneous boundary-value problem depend on the shape of the boundary, the type of boundary conditions, the mechanical characteristics of the material of the area, and have many values [2, 7-10, 16].

The proper (singular) solutions of homogeneous boundary value problem are determined with an accuracy of unknown arbitrary constants.

Excluding the irregular point itself and its small neighborhood, it is possible to characterize the stress-strain state in a small area of the top of the cut as an asymptotic solution of an elastic homogeneous problem, depending on the unknown constants (intensity coefficients).

In fracture mechanics for an ideal mathematical section, singular stresses are characterized by stress intensity factors [4, 8, 9, 15].

In this work, the stress-strain state in the corner cut-out area is characterized by stress intensity factors and strain intensity factors that do not match.

The purpose of this research work is to develop a general approach for determining stress intensity factors and coefficients of intensity of deformations as stress value limits and deformations within an angular cut of the area boundary.

This research work considers two approaches to obtaining expressions for displacements and stresses in the vicinity of the apex of the cut-out of the flat area boundary using the stress intensity factors and the strain intensity factors. The expressions of stresses and displacements recorded with the help of stress intensity factors and strain intensity factors determine the practical significance of the study work when conducting experiments and determining critical values of stresses and strains.

The task of studying the intensity factors of stresses and strains is relevant for the study of the stress-strain state of designs and structures with geometric nonlinear shape of the boundaries.

2. Method of research

2.1. Problem Setting. Case of Plane Strain

The stress and strain state of composite structures within the areas of element mating under action of forced strains rupturing along the element contact line (surface) is characterized by the feature of power type. Stress and strain state within non-regular point of the area boundary under action of forced strains, particularly, temperature ones is determined by solving the uniform elastic boundary value problem. The concentrated efforts at the top of the cut are not considered. Solving of the uniform elastic problem within the area of non-regular point on special line is reduced [1, 2, 3] to solving the plane strain problem: $\sigma_x \neq 0$, $\sigma_y \neq 0$, $\tau_{xy} \neq 0$, $\tau_{xz} = \tau_{yz} = 0$, $\frac{\partial}{\partial z} \approx 0$, $\sigma_{zz} = \nu(\sigma_{xx} + \sigma_{yy})$ and anti-plane strain problem: $\varepsilon_{xz} \neq 0;$ $\varepsilon_{yz} \neq 0;$ $W \neq 0;$ $\frac{\partial}{\partial z} \approx 0;$ $n_z = 0$, $\varepsilon_x = \varepsilon_y = \varepsilon_z = \varepsilon_{xy} = 0,$ $U = V = 0$. Solving of the plane problems in the polar coordinate system is sought by variable separation method [3].

The uniform elastic boundary value problem for plane V-shaped area with symmetrical opening 2α within angle boundary point – top of the angle cutout with random opening is considered [1-5, 15-16]. Solving of the plane strain problem in displacements in polar coordinate system shall be sought as follows:

$$\mathrm{u}_r(r,\theta) = r^{\lambda} f(\theta), \qquad \mathrm{u}_\theta(r,\theta) = r^{\lambda} g(\theta) \tag{1}$$

where $f(\theta),$ $g(\theta)$ – unknown function of angle θ to be determined, λ – unknown parameter. Substituting ratios (1) in Lame equation, formula for displacements and stresses within non-regular point of the area boundary are obtained [1-5, 7, 9, 15]:

$$\begin{aligned}
&\mathrm{r}^{-\lambda}\,\mathrm{u}_r = A\cos\left[(1+\lambda)\theta\right] + B\sin\left[(1+\lambda)\theta\right] + C\cos\left[(1-\lambda)\theta\right] + D\sin\left[(1-\lambda)\theta\right],\\
&\mathrm{r}^{-\lambda}\,\mathrm{u}_\theta = B\cos\left[(1+\lambda)\theta\right] - A\sin\left[(1+\lambda)\theta\right] + \nu_2\, D\cos\left[(1-\lambda)\theta\right] - \nu_2\, C\sin\left[(1-\lambda)\theta\right],\\
&\mu^{-1}\mathrm{r}^{1-\lambda}\,\sigma_\theta = -2\lambda A\cos\left[(1+\lambda)\theta\right] - 2\lambda B\sin\left[(1+\lambda)\theta\right] -\\
&\qquad -(1+\lambda)(1-\nu_2)\,C\cos\left[(1-\lambda)\theta\right] - (1+\lambda)(1-\nu_2)D\sin\left[(1-\lambda)\theta\right],\\
&\mu^{-1}\mathrm{r}^{1-\lambda}\,\tau_{r\theta} = -2\lambda A\sin\left[(1+\lambda)\theta\right] + 2\lambda B\cos\left[(1+\lambda)\theta\right] -\\
&\qquad -(1-\lambda)(1-\nu_2)\,C\sin\left[(1-\lambda)\theta\right] + (1-\lambda)(1-\nu_2)D\cos\left[(1-\lambda)\theta\right],\\
&\mu^{-1}\mathrm{r}^{1-\lambda}\sigma_r = 2\lambda\{(A\cos\left[(1+\lambda)\theta\right] + B\sin\left[(1+\lambda)\theta\right] +\\
&\qquad + \frac{3-\lambda}{k-\lambda} C\cos\left[(1-\lambda)\theta\right] + \frac{3-\lambda}{k-\lambda}\mathrm{D}\sin\left[(1-\lambda)\theta\right]\},
\end{aligned} \tag{2}$$

where $k = 3 - 4\nu$, $\nu_2 = \dfrac{3 + \lambda - 4\nu}{3 - \lambda - 4\nu}$, factors A, B, C, D – random constants to be determined; λ – proper values of uniform boundary problem, generally, complex numbers, ν, μ – Poisson ratio and shear modulus accordingly.

Meeting the uniform boundary conditions: $\sigma_\theta = \tau_{r\theta} = 0$ at $\theta = \pm\alpha$, two uniform linear equation systems are obtained in regard to unknown constants A, B, C, D, which solving determines the ratios for the following factors:

$$A = \frac{(1-\lambda)\sin[(1-\lambda)\alpha]}{(k-\lambda)\sin[(1+\lambda)\alpha]} C, \qquad B = \frac{(1+\lambda)\sin[(1-\lambda)\alpha]}{(k-\lambda)\sin[(1+\lambda)\alpha]} D. \tag{3}$$

The stress and strain state (2) is obtained with limit values of stresses within non-regular point of the boundary called stress intensity factors similar to stress intensity factors in mechanics of damage [3, 8, 12-13].

2.2. Area Stress Intensity Factors

To determine factors A and C in formula (2) for stresses σ_θ, go to the limit at $\theta = 0$:

$$\lim_{r \to 0} \mu^{-1} r^{1-\lambda} \sigma_{\theta,\theta=0} = -2A\lambda - (1+\lambda)(1-\nu_2)C. \tag{4}$$

Specify

$$K_I^\sigma = \lim_{r \to 0} \mu^{-1} r^{1-\lambda} \sigma_{\theta,\theta=0}. \tag{5}$$

Considering ratios (3) for A and C, symbols (4), (5) we obtain unknown factors in formula (2) as follows:

$$A = \frac{(1-\lambda)\sin[(1-\lambda)\alpha]}{2\lambda[(\lambda-1)\sin[(1-\lambda)\alpha] + (\lambda+1)\sin[(1+\lambda)\alpha]]} K_I^\sigma, \tag{6}$$

$$C = \frac{(k-\lambda)\sin[(1+\lambda)\alpha]}{2\lambda[(\lambda-1)\sin[(1-\lambda)\alpha] + (\lambda+1)\sin[(1+\lambda)\alpha]]} K_I^\sigma. \tag{7}$$

To determine factors B and D in formula (2) for stresses $\tau_{r\theta}$, go to the limit at $\theta = 0$:

$$\lim_{r \to 0} \mu^{-1} r^{1-\lambda} \tau_{r\theta,\theta=0} = 2\lambda B + (1-\lambda)(1-\nu_2)D. \tag{8}$$

Specify

$$K_{II}^\sigma = \lim_{r \to 0} \mu^{-1} r^{1-\lambda} \tau_{r\theta,\theta=0}. \tag{9}$$

Considering ratios (3) for B and D, symbols (8), (9) we obtain unknown factors in formula (2) as follows:

$$B = \frac{(1+\lambda)\sin[(1-\lambda)\alpha]}{2\lambda[(\lambda+1)\sin[(1-\lambda)\alpha] - (1-\lambda)\sin[(1+\lambda)\alpha]]} K_{II}^\sigma, \tag{10}$$

$$D = \frac{(k-\lambda)\sin[(1+\lambda)\alpha]}{2\lambda[(\lambda+1)\sin[(1-\lambda)\alpha] - (1-\lambda)\sin[(1+\lambda)\alpha]]} K_{II}^\sigma. \tag{11}$$

Factors

$$K_I^\sigma = \lim_{r \to 0} \mu^{-1} r^{1-\lambda} \sigma_{\theta,\theta=0}, \quad K_{II}^\sigma = \lim_{r \to 0} \mu^{-1} r^{1-\lambda} \tau_{r\theta,\theta=0}$$

by analogy with symbols taken in the mechanics of damage [1, 2, 13–16] shall be determined as stress intensity factors, where

$$\lambda = \min\left|\operatorname{Re}\lambda_i\right|.$$

Considering the detected factors (6), (7), (10), (11) and symbols of limit stress values as stress intensity factors in form of (5), (9), initial stress and strain state within non-regular point of the area boundary (2) shall be recorded as a function of these factors.

2.3. Area Strain Intensity Factors

Specify formula for strains

$$r^{1-\lambda}\varepsilon_\theta = -A\lambda\cos\left[(1+\lambda)\theta\right] - B\lambda\sin\left[(1+\lambda)\theta\right] + \\ + \frac{(1+\lambda-4\nu)}{k-\nu}C\lambda\cos\left[(1-\lambda)\theta\right] + \frac{(1+\lambda-4\nu)}{k-\nu}D\lambda\sin\left[(1-\lambda)\theta\right], \quad (12)$$

To determine factors A and C in formula (3) for strains ε_θ, go to the limit at $\theta = 0$:

$$\lim_{r\to 0} r^{1-\lambda}\varepsilon_{\theta,\theta=0} = \lambda[-A + \frac{(1+\lambda-4\nu)}{k-\nu}C]. \quad (13)$$

Specify

$$K_I^\varepsilon = \lim_{r\to 0} r^{1-\lambda}\varepsilon_{\theta,\theta=0}. \quad (14)$$

Considering ratios (3) for A and C, symbols (13), (14) we obtain unknown factors in formula (2) as follows:

$$C = \frac{(k-\lambda)\sin\left[(1+\lambda)\alpha\right]}{\lambda\left((1+\lambda-4\nu)\sin\left[(1+\lambda)\alpha\right] - (1-\lambda)\sin\left[(1-\lambda)\alpha\right]\right)}K_I^\varepsilon, \quad (15)$$

$$A = \frac{(1-\lambda)\sin\left[(1-\lambda)\alpha\right]}{\lambda\left((1+\lambda-4\nu)\sin\left[(1+\lambda)\alpha\right] - (1-\lambda)\sin\left[(1-\lambda)\alpha\right]\right)}K_I^\varepsilon. \quad (16)$$

To determine factors B and D in the formula for strains $\varepsilon_{r\theta}$, go to the limit at $\theta = 0$:

$$r^{1-\lambda}\varepsilon_{r\theta} = -2\lambda A\sin\left[(1+\lambda)\theta\right] + 2\lambda B\cos\left[(1+\lambda)\theta\right] - \\ -(1-\lambda)(1-\nu_2)C\sin\left[(1-\lambda)\theta\right] + (1-\lambda)(1-\nu_2)D\cos\left[(1-\lambda)\theta\right]. \quad (17)$$

$$\lim_{r\to 0} r^{1-\lambda}\varepsilon_{r\theta,\theta=0} = 2\lambda B + (1-\lambda)(1-\nu_2)D. \quad (18)$$

Specify

$$K_{II}^\varepsilon = \lim_{r\to 0} r^{1-\lambda}\varepsilon_{r\theta,\theta=0}. \quad (19)$$

Considering ratio

$$\varepsilon_{r\theta} = \mu^{-1}\tau_{r\theta}$$

we obtain

$$K_{II}^\varepsilon = \lim_{r\to 0} r^{1-\lambda}\varepsilon_{r\theta,\theta=0} = \lim_{r\to 0}\mu^{-1}r^{1-\lambda}\tau_{r\theta,\theta=0} = K_{II}^\sigma, \text{ i.e. } K_{II}^\sigma = K_{II}^\varepsilon,$$

thus factors B and D have the same form (10), (11), i.e.

$$B = \frac{(1+\lambda)\sin\left[(1-\lambda)\alpha\right]}{2\lambda\left[(\lambda+1)\sin\left[(1-\lambda)\alpha\right] - (1-\lambda)\sin\left[(1+\lambda)\alpha\right]\right]}K_{II}^\varepsilon, \quad (20)$$

$$D = \frac{(k\text{-}\lambda)\sin\left[(1+\lambda)\alpha\right]}{2\lambda\left[(\lambda+1)\sin\left[(1-\lambda)\alpha\right] - (1-\lambda)\sin\left[(1+\lambda)\alpha\right]\right]}K_{II}^\varepsilon. \quad (21)$$

Factors

$$\mathrm{K}_I^{\varepsilon} = \lim_{r\to 0} r^{1-\lambda}\,\varepsilon_{\theta,\theta=0},\quad \mathrm{K}_{II}^{\varepsilon} = \lim_{r\to 0} r^{1-\lambda}\varepsilon_{r\theta,\theta=0}$$

by analogy with symbols taken in the mechanics of damage [3, 8, 13–15] shall be determined as strain intensity factors, where

$$\lambda = \min\left|\mathrm{Re}\,\lambda_i\right|.$$

Considering the detected factors (15), (16), (20), (21) and symbols of limit strain values as strain intensity factors in form of (14), (19), initial stress and strain state within non-regular point of the area boundary (2) shall be recorded as a function of these factors.

For the anti-plane strain problem solving within the non-regular point of the area boundary shall be recorded similarly as follows:

$$\begin{aligned} &w = r^{\lambda_1}[D_1 \sin\lambda_1\theta + D_2\cos\lambda_1\theta], \\ &\mu^{-1}r^{1-\lambda_1}\tau_{xz} = \lambda_1\{-D_1\sin[(1-\lambda_1)\theta] + D_2\cos[(1-\lambda_1)\theta]\}, \\ &\mu^{-1}r^{1-\lambda_1}\tau_{yz} = \lambda_1\{D_1\cos[(1-\lambda_1)\theta] + D_2\sin[(1-\lambda_1)\theta]\}, \end{aligned} \tag{22}$$

where D_1, D_2 – random constants to be determined, where λ_1 – proper values of uniform boundary problem.

Specifying

$$\mathrm{K}_{III}^{1} = \lim_{r\to 0}\mu^{-1}r^{1-\lambda_1}\tau_{yz,\theta=0},\quad \mathrm{K}_{III}^{2} = \lim_{r\to 0}\mu^{-1}r^{1-\lambda_1}\tau_{xz,\theta=0}, \tag{23}$$

Factors D_1 and D_2 are recorded, moreover, the stress and strain intensity factors numerically coincide.

3. Results

The procedure for determination of stress and strain state (2) within non-regular point of the plane area boundary is performed. The unknown factors are determined with use of limit stress values as stress intensity factors (5), (9) in the form of (6), (7), (10), (11) and with use of limit strain values as strain intensity factors (14), (19) in the form of (15), (16), (20), (21).

4. Discussion

Difference in stress and strain intensity factors for normal stresses and linear strains is detected. The intensity factors for shear stresses and angular strains coincide. In the main case of stress and strain state within the cut top of the area boundary one cannot speak about proportionality of stress and strain intensity factors that shall be considered during experimental determination of their values.

Conclusion

The novelty of the obtained results consists in the development of a general approach for determining stress intensity factors and strain intensity factors, which do not coincide in general case. The stress and strain state within the angle cut outs of the designs and structure boundary recorded in the form of limit values for stress and strain can be used in future for determination of stress and strain intensity factors that is an independent problem of mechanics of solids.

References

[1] Timoshenko S P and Goodier J 1975 *Elasticity Theory* (Moscow: Nauka) p 576
[2] Parton V Z and Perlin P I 1981 *Methods of Mathematical Elasticity Theory* (Moscow: Nauka) p 688
[3] Kondratyev V A 1967 Boundary Problems for Elliptical Equations within the Areas with Conical and Angular Points *Works of Moscow Mathematical Society* **16** pp 209–292
[4] Cherepanov G P 1974 *Mechanics of Brittle Fracture* (Moscow: Nauka) 1974 p 640
[5] Khesin G L et al 1975 *Photoelastic Method vol. 3* (Moscow: Stroyizdat) p 311
[6] Vardanyan G.S., Frishter L.Yu. 2007 *Int. J. Comput. Civ. Struct. Eng.* **3**, p 75–81

[7] Frishter L 2017 Evaluations of the solution to the homogeneous plane problem of the theory of elasticity in the neighborhood of an irregular boundary point *MATEC Web of Conferences* **117** 00047
[8] Kuliev V D 2005 *Singular Boundary Problems* (Moscow: Nauka) p 719
[9] Razumovsky I A 2007 *Interferential Optical Method of Mechanics of Strained Solid Body* (Moscow: Publishing House of the Bauman Moscow State Technical University) p 240
[10] Frishter L 2016 The analysis of the methods of research the influence of nonuniform local stress on a crack resistance of protective coating *MATEC Web of Conferences* **86** 04022
[11] Xu L R, Kuai H and Sengupta S 2004 *Exp. Mech.* **44** pp 608–615
[12] Xu L R, Kuai H and Sengupta S 2004 *Exp. Mech.* **44** pp 616–621
[13] Yao X F, Yeh H Y and Xu W 2006 *Int. J. Solid Struct.* **43** pp 1189–1200
[14] Matveenko V P, Nakarjakova T O, Sevodina N V and Shardakov I N 2008 Stress singularity at the top of homogeneous and composite cones with different boundary conditions *J. Math. Mech.* **72** pp 477–484
[15] Frishter L Yu 2017 Photoelasticity-based study of stress-strain state in the area of the plain domain boundary cut-out area vertex *Advances in Intelligent Systems and Computing* **692** pp 836–844
[16] Pestrenin V M, Pestrenina I V and Landik L V 2013 *Tomsk State University Journal (Mathematics and Mechanics)* **4** pp 80–87

IOP Conf. Series: Materials Science and Engineering **675** (2019) 012056 doi:10.1088/1757-899X/675/1/012056

Defining of nonlocal damping model parameters based on composite beam dynamic behaviour numerical simulation results

E S Shepitko[1*] and V N Sidorov[1,2,3]

[1]Russian University of Transport (MIIT), Moscow, Russia
[2]Peoples' Friendship University of Russia (RUDN University), Moscow, Russia
[3]Perm National Research Polytechnic University, Perm, Russia

E-mail: shepitko-es@mail.ru

Abstract. The nonlocal model of internal damping is considered in this paper using the Euler-Bernoulli beam with fixed ends as an example. Equation of beam motion considering nonlocal damping is solved by Galerkin method to develop the model. The required number ofeigenmodes is obtained for the beam under an instantly applied distributed load. The influenceof nonlocal damping model parameters variation on the beam vibration process simulation results is shown. The beam is considered under a periodic deterministic distributed load. Defining of nonlocal damping model parameters is carried out using results of numerical three-dimensional finite element simulation. The parameters are defined for the glass fibre reinforced plastic beam with the fixed ends under the instantly applied distributed load. The parameters obtained for two different kernel functions and corresponding standard errors are compared. An opportunity of one-dimensional beam model use for the design of composite elements is shown and justified.

1. Introduction

Various composite materials find appropriate place in the contemporary construction industry. High strength under the different loads, low weight, high fire resistance could be named among the main advantages of such materials. For successful design of composite elements it is necessary to take into account that composite materials have inhomogeneous structure: physical properties of their matrix and reinforcement are significantly different. Generally, to provide required precision of calculations the composite beam elements behaviour is simulated using detailed 3D finite element modelling. This approach allows assigning material different material properties in different directions depending on reinforcement orientation.

However, in some cases one-dimensional beam models can be more convenient than detailed 3D. For example on the initial design stage when multiple variants should be compared and the best variant should be chosen. To use one-dimensional model of composite beam and not to sacrifice the calculation precision special hypotheses that approximate the orthotropic material properties are needed. The nonlocal hypothesis can be used to approximate the damping properties of composite material.

Nonlocal damping model is based on principals of nonlocal mechanics. [1] Damping in the certain point of structure with longitudinal coordinate x1 assumed to be dependent not only on local value of strain rate at this point, but also on the values of strain rates at the neighboring points. The bigger is the distance between the two points the lower influence one of them has on the other [3].

In this paper the nonlocal damping problem is solved on the example of glass fiber reinforced plastic beam with the fixed ends. The method of nonlocal model parameters defining is also shown on the beam of the same type.

2. Materials and methods

2.1. Nonlocal damping model for the beam with fixed ends

The Kelvin-Voigt material model is commonly used in engineering practice to describe the damping process:

$$\sigma = E\varepsilon + \gamma E\dot{\varepsilon}, \tag{1}$$

where σ, ε – normal stress and axial strain, $\dot{\varepsilon}$ – rate of strain change, E – Young modulus, γ – damping ratio.

Dot indicates the time derivative.

Consider nonlocal damping equation (1) transforms to [3]:

$$\sigma(x,t) = E\left[\varepsilon(x,t) + \gamma\int_0^l C_v(|x-\theta|)\dot{\varepsilon}(\theta,t)d\theta\right]. \tag{2}$$

here $C_v(|x-\theta|)$ – the kernel function, which characterize the nonlocal damping. The $C_v(|x-\theta|)$ function responds to normalization requirement, which is:

$$\int_{-\infty}^{\infty} C_v(|x-\theta|)d\theta = 1. \tag{3}$$

The kernel function can have rectangle shape, triangle shape, exponential shape or error function shape. [4,8] These shapes are just mathematical approximation and none of them is correlated with the internal structure of the material. Therefore none of them has an advantage before another. In this paper two types of kernel function are applied and compared:

a) Exponential kernel function:

$$C_v(|x-\theta|) = \frac{\mu}{2}\cdot e^{-\mu|x-\theta|}. \tag{4}$$

b) Error kernel function:

$$C_v(|x-\theta|) = \frac{\mu}{\sqrt{2\pi}}\cdot e^{\frac{-\mu^2(x-\theta)^2}{2}}. \tag{5}$$

Here μ is the parameter that characterizes the influence distance in the nonlocal damping model x and θ are the longitudinal coordinate.

The equilibrium equation for an elementary part of the beam is:

$$\frac{\partial^2 M(x,t)}{\partial x^2} = m\frac{\partial^2 w(x,t)}{\partial t^2} - q(x,t), \tag{6}$$

where $w(x,t)$ – beam deflection, m – distributed mass, $q(x,t)$ – distributed load.

Considering that plane sections remain plain the bending moment expression is:

$$M(x,t) = -EI\left[\frac{\partial^2 w(x,t)}{\partial x^2} + \gamma\int_0^l C_v(|x-\theta|)\frac{\partial^3 w(\theta,t)}{\partial\theta^2\partial t}d\theta\right]. \tag{7}$$

Here *EI* is beam bending stiffness.

Substitute the second derivative of the moment expression to the left part of the equation (5), and obtain the expression regarding the deflection function $w(\theta,t)$:

$$\frac{\partial^2 w(x,t)}{\partial t^2} + \frac{EI}{m}\left[\frac{\partial^4 w(x,t)}{\partial x^4} + \gamma\frac{\partial^2}{\partial x^2}\int_0^l C_v(|x-\theta|)\frac{\partial^3 w(\theta,t)}{\partial\theta^2\partial t}d\theta\right] = \frac{q(x,t)}{m}. \tag{8}$$

Solution of this equation should satisfy the boundary conditions for $x=0$ and $x=l$.

Function $w(x,t)$ is presented as normal modes of beam vibrations:

$$w(x,t)=\sum_{i=1}^{n} f_i(t)V_i(x). \tag{9}$$

Here $f_i(t)$ – generalized displacements, $V_i(x)$ – coordinate functions.

For the beam with fixed ends the coordinate functions are [5]:

$$V_i(x)=(shk_i l-\sin k_i l)(chk_i x-\cos k_i x)-(chk_i l-\cos k_i l)(shk_i x-\sin k_i x). \tag{10}$$

Here k_i is calculated as the *i*-root of the characteristic equation $ch(kl)\cos(kl)=1$.

To obtain the generalized displacements $f_i(\tau)$ the Galerkin method is used. The resulting system of the integral-differential equations with exponential function as kernel function is:

$$\begin{aligned}&\ddot{f}_j(\tau)+\left(\frac{k_j}{k_1}\right)^4 f_j(\tau)+\frac{2\in}{a_j k_1^4 l^4}\sum_{i=1}^{m}\int_0^1 V_j^{``}(z)\int_0^1 \frac{\mu l}{2}e^{-\mu l|z-y|}V_i^{``}(y)dydz\dot{f}_i(\tau)\\&=\frac{l^4}{k_1^4 a_j}\int_0^1 \frac{q(y,\tau)}{EI}V_j(y)dy,\end{aligned} \tag{11}$$

where $k_j^4=m\omega_j^2/EI$, ω_j – natural frequency, $a_j=\int_0^l V_j^2(x)dx$, z and y – dimensionless coordinates - $z=x/l$, $y=\theta/l$, τ – dimensionless time $\tau=\omega_1 t$, $\omega_1^2=Elk_1^4/m$, $2\in=\gamma\omega_1$, ω_1 – minimum natural frequency.

The $f_i(\tau)$ values are obtained by solving the equation system by numerical Runge-Kutta method of 4th order [6]that applied in MATLAB, and beam deflections - using expression (10).

3. Results

3.1. Influence of nonlocal damping consideration on the computer modelling results

First of all it is necessary to determine the number of vibration modes that need to be considered to obtain results of required precision [9]. For this purpose the beam with the fixed is considered under the instantly applied distributed load $q=1000t$. The beam is 10 m long and has bending stiffness $EI=9.82\cdot 10^5 t\cdot m^2$. The deflection value is picked up in the middle point of the beam at the moment when the vibrations are totally damped and the deflection value is equal to static one.

As it is shown in table 1, when 5 and 7 eigenmodes are hold the results are equal up to the third sign.

Table 1. Deflection in the middle point of the beam corresponding to the number of eigenmodes.

Number of Eigen modes	Deflection, m
1	-0.0268
3	-0,0264
5	-0,0265
7	-0,0265

In case of periodic or stochastic load the number of eigenmodes should be taken according to the load frequency.

To figure out how variation of the influence distance μ, which is the main parameter of nonlocal model, affects the results of the beam vibration process computer modelling, the beam is loaded with the periodic distributed load:

$$q=A\sin(\omega t), \tag{12}$$

where A – load amplitude, ω – load frequency, t – time.

The influence distance μ shows the level of nonlocal properties in material. The lower is μ the higher is the level. High μ values make the model close to local (figure 1).

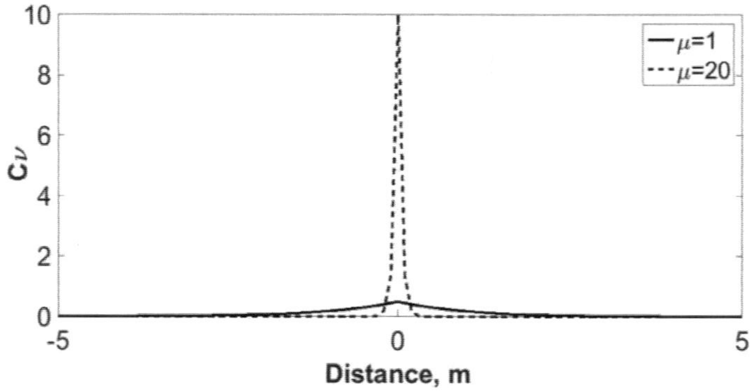

Figure 1. Exponential kernel function for different values of μ.

The vibration process of the beam under the periodic load (12) and $\mu = 0.2$ *1/m* is shown on figure 2.

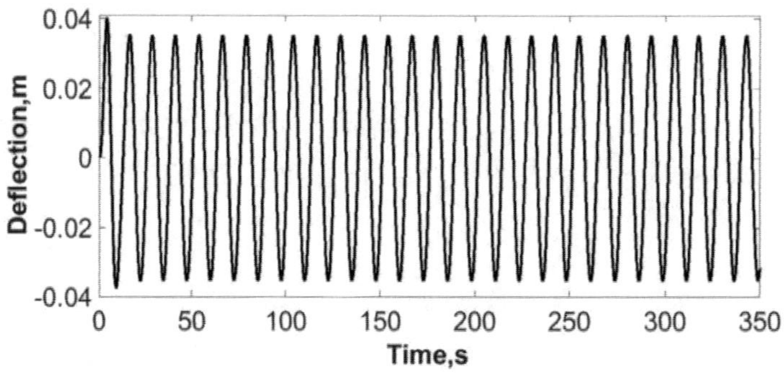

Figure 2. Time history of deflection in the middle of the beam for μ=0.2 *1/m*.

The deflection amplitudes for different μ values are shown in table 2.

Table 2. Deflection amplitudes for different influence distance values.

Influence distance μ, 1/m	Deflection amplitude, m
0.2	0.0350
0.5	0.0304
1	0.0237

These amplitudes decrease as μ gets bigger, thus model with higher nonlocal properties gives the bigger deflection amplitude.

3.2. Defining of nonlocal model properties

The influence distance for the material has to be determined to make nonlocal model applicable for practical needs. For this purpose experimental or numerical simulation data can be used.

Consider GFRP beam with the fixed ends made of orthotropic thermoset vinyl ester class 1 FRP under instantly applied distributed load.

The characteristics of the material obtained experimentally in [7,2,10] are presented in table 3.

These characteristics were assigned to a 3D beam created in Abaqus SIMULIA. The beam is 6 m long and has 0.2x0.3m rectangular cross-section. The time history of vibrations under the instantly applied load was obtained for this beam and imported to MATLAB. The difference between Abaqus results obtained with 3D orthotropic material model and the results obtained with one-dimensional beam where local Kelvin-Voigt model is used to describe damping is shown on figure 3. To make the damping model close to Kelvin-Voigt μ was assumed equal to 100.

Table 3. Characteristics of thermoset vinyl ester class 1 FRP

Characteristics	Value
Young modulus (longwise)	17.2 GPA
Young modulus (crosswise)	12.2 GPA
Poisson ration (longwise)	0.32
Poisson ration (crosswise)	0.15
Density	1.9 Mg/m3
Damping ratio	0.042

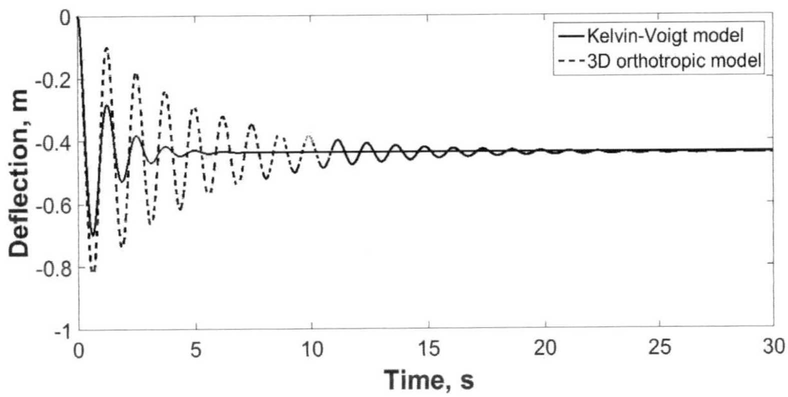

Figure 3. Deflection of 6m long beam (0.2x0.3m) using local damping model in comparison to numerical simulation data.

Standard error for deflection can be calculated as follows:

$$\sigma = \sqrt{\frac{\sum_{i=1}^{N}(y_i - f_i(\mu))}{N}}. \tag{13}$$

Here y_i is 3D orthotropic data value at a time step i, $f_i(\mu)$ – 1D model data value at a time step i, N – number of time steps.

For Kelvin-Voight model the standard error is 16.7% of the deflection value at time step of 25 seconds.

For nonlocal-model the special script was created in MATLAB that allows defining μ by the method of least squares. The kernel function was assumed exponential. In this case the results look as in figure 4.

IOP Conf. Series: Materials Science and Engineering **675** (2019) 012056 doi:10.1088/1757-899X/675/1/012056

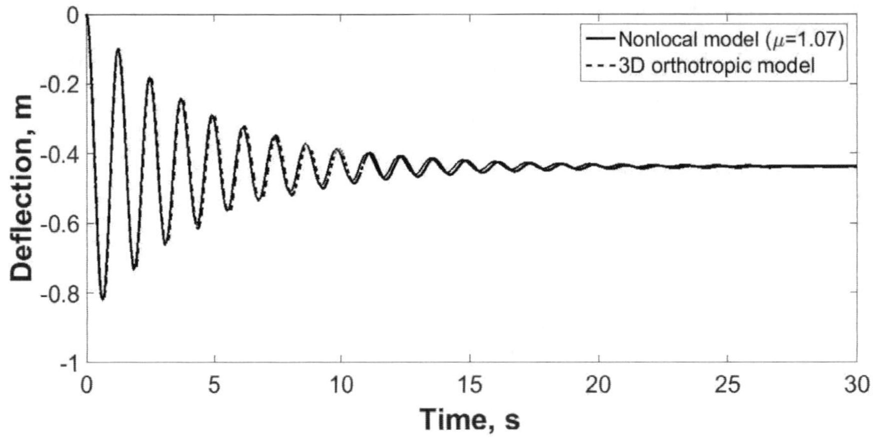

Figure 4. Deflection of 6m long beam (0.2x0.3m) using nonlocal damping model in comparison to numerical simulation data.

For nonlocal model with parameters defined by method of least squares the standard error is 3.1% of deflection at time step of 25 seconds.

The influence distance is a material property. Thus, the nonlocal model with defined μ should work for any beam made of the same material. Two more beams with different geometry were modelled in 3D orthotropic and 1D nonlocal formulation to compare their vibration processes. It was assumed that the material is the same, so the influence distance does not change. The influence distance and standard errors for all three beams are presented in table 4, along with the influence distance and standard errors obtained using error kernel function.

Table 4. Comparison of 3D orthotropic and 1D nonlocal models results

Kernel functiion	Defined influence distance μ, 1/m	Beam length and X-section	Standard error, %
Exponential kernel function	1.07	6 m, 0.2x0.3	3.1
		7 m, 0.2x0.3	3.2
		7 m, 0.3x0.4	4
Error kernel function	0.96	6 m, 0.2x0.3	3.1
		7 m, 0.2x0.3	3.2
		7 m, 0.3x0.4	4

The standard error stays less than 5% for beams with changed geometry, which means that nonlocal damping model can be used to simulate the beam vibration process with one-dimensional beam model. There is almost no difference between the results obtained by exponential and error kernel function. Both of them are just mathematical approximation and none of them is better than the other. To decrease the error it is necessary to carry out kernel function that will be based on the internal structure of composite material.

4. Discussion

The review of different studies of composite elements design has shown that their dynamic behaviour is often simulated using detailed 3D finite element modelling which requires significant amount of time and efforts. To apply one-dimensional models for the composite beam design it is possible to use the nonlocal damping model to describe the internal friction process in composite material. It is possible to change the characteristics of the beam vibrations flexibly by varying the parameters of nonlocal damping model.

The nonlocal damping model with properties which are defined using the method of least squares properly fits the numerical simulation data. The relative error says less then 4%.

Comparison of two different kernel functions – exponential and error function – has shown almost no difference between results precision. Working out the kernel function based on the internal structure of material will be an object of further research.

Conclusion

The nonlocal damping model that can be used to simulate the composite beams dynamic behaviour is presented in this paper. The number of eigenmodes needed to obtain required calculation precision is determined and the influence of the nonlocal damping parameters on the beam vibrations is analysed.

It is shown in the paper that increasing the level of nonlocal properties in material (decreasing the influence distance μ) leads to increasing of deflection amplitude.

The nonlocal damping model presented in this paper is more flexible in comparison to traditional local models. This allows modelling dynamic behaviour of beams that made of anisotropic and orthotropic materials in one-dimensional statement. This might be rewarding on the initial design stages when it is required to check and reject a significant amount of structure variants.

References

[1] Banks H T and Inman D J 1991 On damping Mechanisms in Beams *Journal of Applied Mechanics* **58(3)** pp 716–723

[2] Landherr J. C. Dynamic analysis of a FRP deployable box beam. Master of Applied Science Thesis. Kingston: Queen's University, 2008

[3] Lei Y, Friswell M I and Adhikari S 2006 A Galerkin Method for Distributed Systems with Non-local Damping *Int. Journal of Solids and Structures* **43** pp 3381–3400

[4] Russell D L 1992 On Mathematical Models for the Elastic Beam with Frequency-Proportional Damping *Banks, H.T. (Ed.), Control and Estimation in Distributed Parameter Systems* (Philadelphia: SIAM) pp 125–169

[5] Filippov A P 1970 *Dynamics of Deformable Systems*. (Moscow: Mashinostroyenie) p 734

[6] Kalitkin N N 2011 *Numerical Methods: Study Course – 2nd edition* (St. Petersburg: BHV-Peterburg p 592

[7] Lim R A 2016 *Structural monitoring of a 10m fibre reinforced polymer bridge subjected to severe damage* (Kingston: Queen's University)

[8] Potapov V D 2012 Stability of Rods under Stochastic Loading Considering Nonlocal Damping *Problems of Machinery and reliability* **4** pp 25–31

[9] Fyodorov V S, Sidorov V N and Shepitko E S 2018 Nonlocal damping consideration for the computer modelling of linear and nonlinear systems vibrations under the stochastic loads *IOP Conf. Series: Materials Science and Engineering* **456** 012040

[10] Xie A. Development of an FRP Deployable Bridge, Master of Applied Science Thesis, Department of Civil Engineering, Royal Military College of Canada, 2007

IOP Conf. Series: Materials Science and Engineering **675** (2019) 012056 doi:10.1088/1757-899X/675/1/012056

Authors' background

Name	*Prefix*	*Research Field*	*Email*	*Personal website*
Vladimir N. Sidorov	Full Professor	Structural mechanics, mathematical modeling, numerical methods	sidorov.vladimir@gmail.com	
Elena S. Shepitko	PhD Candidate	Structural dynamics, nonlocal mechanics, numerical simulation	shepitko-es@mail.ru	

IOP Conf. Series: Materials Science and Engineering **675** (2019) 012057 doi:10.1088/1757-899X/675/1/012057

Wavelet-based numerical and semianalytical methods of computational local structural analysis

P A Akimov[1,2,3,4*], A M Belostosky[2,3,4], T B Kaytukov[1], L S Lyakhovich[3] and M L Mozgaleva[5]

[1] Russian Academy of Architecture and Construction Sciences, 24, ul. Bolshaya Dmitrovka, Moscow, 107031, Russia
[2] Department of Civil Engineering, Peoples' Friendship University of Russia, 6, Miklukho-Maklaya str., Moscow, 117198, Russia
[3] Department of Structural Mechanics, Tomsk State University of Architecture and Building, 2, Solyanaya sq., Tomsk, 634003, Russia
[4] Research & Development Centre StaDyO, office 810, 18, 3ya Ulitsa Yamskogo Polya, Moscow, 125040, Russia
[5] Department of Applied Mathematics, National Research Moscow State University of Civil Engineering, 26, Yaroslavskoe Shosse, Moscow, 129337, Russia

Email: akimov@raasn.ru

Abstract. Numerical or semianalytical solution of problems of structural mechanics of high dimensionality is computationally costly process in many cases. However, structural engineer does not normally face the task of obtaining a solution of the problem with high accuracy at all points of the considered domain occupied by the structure. As a rule, all subdomains that are potentially dangerous in terms of structural strength are well known in advance. Operational and variational formulations of boundary problems of structural mechanics with the use of method of extended domain are presented. After corresponding (finite element or finite difference) discretization and passage to governing equations considering problems are transformed to a multilevel space with the use of multilevel wavelet transform (discrete Haar basis is used). Special algorithms of averaging are presented.

Keywords: numerical methods, semianalytical methods, local structural analysis, wavelet-based methods, boundary problem, reduction.

1. Introduction

Since the 90s of the last century, wavelet analysis has found its application in such areas of structural mechanics and solid mechanics as structural dynamics, structural health monitoring, detection of defects (fracture zones, cracks, etc.), probabilistic mechanics, data visualization, identification systems in civil engineering and others. We should mention here research works of A. Barinka, T. Barsch [1], P. Charton, A. Cohen [2], S. Dahlke [3], W. Dahmen [4], S. Dumont [5], S. Jaffard [6], A. Kunoth [1], A.J. Kurdila [7], F. Lebon [5], P. Monasse, P. Oswald [8], V. Perrier [9], S. Prossdorf [10], S. Qian [11], R. Schneider [10], K. Urban [1] and J. Weiss [11].

Among conventional numerical methods (finite element method (FEM), boundary element method (BEM), Galerkin method, collocation methods) wavelets are used, as a rule, for approximation of unknown functions. The development of mathematical models, algorithms and program systems for performing wavelet analysis of complex oscillations for some distributed systems are presented in research works of M.V. Zhigalov [12], A.V. Krysko [12-15], V.A. Krysko [12-15] and V.V. Soldatova. The averaging method for elliptic differential equations based on wavelet transform and FEM for predicting effective properties and analyzing averaged solutions of equations for composites with known structure and component properties was developed by S.P. Kopysov and Yu.A. Sagdeeva

[16]. Wavelet-based methods were used in research works of O.V. Mkrtychev, G.A. Ginchvelashvili and A.A. Reshetova within the study of nonstationary dynamic processes, modeling of accelerogram of earthquakes in the form of a nonstationary random process, development of nonlinear dynamic methods of structural analysis [17].

Historically, the first and most ineffective from the computational point of view is the use of a sufficiently detailed approximating mesh (grid) over the entire domain occupied by the considering structure. On the one hand, such an approach provides a high-precision solution in the selected local subdomains, occupied by the structure, on the other hand, it is associated with an unreasonably large amount of essentially computational work.

The second technique, which is widely recommended in the literature dealing with numerical methods (FEM) is the use of non-uniform approximating meshes, condensing in selected local subdomains. At the same time, usually there is no rigorous mathematical justification for the construction of such meshes (there are no rules and principles of the corresponding thickening). Some alternative is the use of finite elements with a special form function (for instance, high order polynomial).

The third technique, which widely used in modern practice of structural analysis includes "cutting out" the selected subdomain, specifying corresponding boundary conditions, based on the results of more rough analysis or on any a priori estimates. Thus, we have so-called fragmentation method developed by A. B. Zolotov and O.V. Garden for solution of problems of structural mechanics. As an alternative, we can consider using the technique of the method of super-elements.

The fourth technique is the use of the Monte-Carlo method, which has shown its effectiveness in finding local solutions even in multidimensional problems of high dimensionality, but which has, nevertheless, a major drawback in the form of low convergence rate.

This paper continues the series of papers of P.A. Akimov [18-21], D.N. Alekseev, M. Aslami [18,19], T.B. Kaytukov [19], M.L. Mozgaleva and A.B. Zolotov, devoted to the development of wavelet-based numerical methods of local structural analysis. In this case, the theoretical foundation for construction of local approximation meshes is the representation of solutions of the corresponding boundary problems of structural mechanics through the fundamental functions of differential operators with the subsequent use of the Saint-Venant principle [22]. Behavior of the fundamental function (including the decay rate) predetermines the nature of the enlargement of the mesh. In case of the absence of a decrease in the fundamental function, the construction of local approximating meshes is inexpedient.

2. Materials and methods

2.1. Basic formulas of fast discrete Haar transforms and averaging for one-dimensional problems

2.1.1. Algorithms of fast direct and inverse discrete Haar transforms. Let $\omega = \{ x : a \le x \le b \}$ be considering one-dimensional domain, where x is coordinate, a is lower limit of interval; b is upper limit of interval. Domain ω can be divided into $(n-1)$ equal parts, where $n = 2^M$ and M is the number of levels of the Haar basis [23-27]. Coordinates of mesh nodes are defined by formulas.

$$x_i = a + (i-1)h, \ i = 1, 2, ..., n ; \quad h = (b-a)/(n-1) . \tag{1}$$

Haar mesh functions are defined by formulas (N_p is the number of Haar functions at level p):

$$\psi_1^M(i) = \alpha_M^{-1}, \ \ i = 1, 2, ..., n ; \tag{2}$$

$$\psi_j^p(i) = \alpha_p^{-1} \begin{cases} 1, \ \ 2^{p+1}(j-1) < i \le 2^p(2j-1) \\ -1, \ \ 2^p(2j-1) < i \le 2^{p+1} j \\ 0, \ \ i \le 2^{p+1}(j-1) \ \bigcup \ i > 2^{p+1} j, \end{cases} \quad i = 1, 2, ..., n, \ \ 0 \le p < M; \tag{3}$$

$$N_p = \begin{cases} n/2^{p+1}, \ 0 \le p < M \\ 1, \ p = M; \end{cases} \qquad \alpha_p = \begin{cases} \sqrt{2^{p+1}}, \ 0 \le p < M \\ \sqrt{2^M} = \sqrt{n}, \ p = M. \end{cases} \tag{4}$$

Let $f(i)$ be arbitrary mesh function. Then we have

$$f(i) = \sum_{p=0}^{M} \sum_{j=1}^{N_p} v_j^p \psi_j^p(i), \quad v_j^p = \sum_{i=1}^{n} f(i) \psi_j^p(i), \ j = 1, 2, ..., N_p, \ p = 0, 1, ..., M, \tag{5}$$

where v_j^p, $j = 1, 2, ..., N_p$, $p = 0, 1, ..., M$ are wavelet (Haar) expansion coefficients.

Let us consider algorithm of fast direct discrete wavelet (Haar) transform. We have

$$u_j^0 = f(j), \ j = 1, 2, ..., n; \quad \alpha_0 = \sqrt{2} \tag{6}$$

and for all $p = 0, 1, ..., M-1$, $j = 1, 2, ..., N_p$ we can use the following formulas:

$$v_j^p = \alpha_p^{-1}(u_{2j-1}^p - u_{2j}^p); \ u_j^{p+1} = u_{2j-1}^p + u_{2j}^p; \ \alpha_{p+1} = \sqrt{2}\alpha_p; \tag{7}$$

$$\alpha_M = \sqrt{n}; \ v_1^M = \alpha_M^{-1} u_1^M, \tag{8}$$

where u_j^p, $j = 1, 2, ..., N_p$, $p = 0, 1, ..., M$ are so-called auxiliary quantities.

Let us also consider algorithm of fast inverse discrete wavelet (Haar) transform. We have

$$\alpha_M = \sqrt{n}; \ \alpha_{M-1} = \sqrt{n}; \ u_1^M = \alpha_M^{-1} v_1^M \tag{9}$$

and for all $p = M-1, M-2, ..., 0$, $i = 1, 2, ..., N_p$ we can use the following formulas:

$$j = [(i+1)/2]; \ z = (-1)^{i+1}; \ u_j^p = \alpha_p^{-1} z v_j^p + u_j^{p+1}; \ \alpha_{p-1} = \alpha_p / \sqrt{2}. \tag{10}$$

Thus, we have

$$f(i) = u_i^0, \ i = 1, 2, ..., n. \tag{11}$$

2.1.2. Algorithm of averaging. The number of unknowns can be decreased with the use of special procedure of reduction of problems without significant loss of accuracy. Corresponding errors in local solutions are relatively small. Let us assume that it is necessary to use averaging at some level and q. Thus, for all $p = 0, 1, ..., q$ and $j = 1, 2, ..., N_p$ we can suppose

$$(Du^p)_{2j-1} \approx (Du^p)_{2j} \approx (D\tilde{u}^p)_{2j-1}, \ v_{2j-1}^p = v_{2j}^p, \ j = 1, 2, ..., N_{p+1}, \tag{12}$$

$$\tilde{u}_{2j-1}^p = (u_{2j-1}^p + u_{2j}^p)/2; \ (D\tilde{u}^p)_{2j-1} = (\tilde{u}_{2j}^p - \tilde{u}_{2j-1}^p)/(2^{p+1}h); \tag{13}$$

Formulas of averaging have the following form

$$v_{2j-1}^p = v_{2j}^p = \beta v_j^{p+1}, \ j = 1, 2, ..., N_{p+1}; \ \beta = 1/(2\sqrt{2}). \tag{14}$$

2.2. Basic formulas of fast discrete Haar transforms and averaging for two-dimensional problems

2.2.1. Algorithms of fast direct and inverse discrete Haar transforms. Let ω be considering two-dimensional rectangular domain, $\omega = \{ (x_1, x_2) : 0 \le x_1 \le l_1, \ 0 \le x_2 \le l_2 \}$, where x_1 and x_2 are corresponding coordinates; l_1 and l_2 are dimensions along x_1 and x_2 (length of domain along x_1

and length of domain along x_2). Domain ω can be divided into $(n-1)$ equal parts along coordinate x_1 and into $(n-1)$ equal parts along x_2 ($h_1 = l_1/(n-1)$, $h_2 = l_2/(n-1)$), where $n = 2^M$, M is the number of levels of the Haar basis. Coordinates of mesh nodes are defined by formulas

$$x_{1,i} = (i_1 - 1)h_1,\ i_1 = 1, 2, ..., n;\quad x_{2,i} = (i_2 - 1)h_2,\ i_2 = 1, 2, ..., n. \tag{15}$$

Haar mesh functions $\psi^p_{s_1,s_2,j_1,j_2}(i_1,i_2)$, $p = 1,2,...,M$, $j_1, j_2 = 1,2,...,N_p$, $s_1, s_2 = 0,1$ (except $s_1 = s_2 = 0$) are defined by formulas:

$$\psi^p_{s_1,s_2,j_1,j_2}(i_1,i_2) = \alpha_p^{-1}\begin{cases}(-1)^{k_1 s_1 + k_2 s_2}, & \bigcap_{q=1}^{2}\bigcup_{k_q=0}^{1}\left(\begin{array}{l}2^{p+1}(j_q - 1 + \frac{k_q}{2}) < i_q \wedge \\ i_q \le 2^{p+1}(j_q - \frac{1}{2} + \frac{k_q}{2})\end{array}\right) \\ 0, & \text{in other cases;}\end{cases}\quad \psi^M_{0,0,1,1}(i_1,i_2) = \alpha_M^{-1}; \tag{16}$$

$$N_p = \begin{cases} n/2^{p+1}, 0 \le p < M \\ 1,\ p = M; \end{cases}\qquad \alpha_p = \begin{cases} 2^{p+1},\ 0 \le p < M \\ 2^M = n,\ p = M. \end{cases} \tag{17}$$

Let $f(i_1,i_2)$ be an arbitrary mesh function and $v^p_{1,0,j_1,j_2}, v^p_{0,1,j_1,j_2}, v^p_{1,1,j_1,j_2}$, $j_1, j_2 = 1,2,...,N_p$, $p = 1, 2, ..., M$ are wavelet (Haar) expansion coefficients. Then we have

$$\begin{aligned} f(i_1,i_2) = v^M_{0,0,1,1}\psi^M_{0,0,1,1} + \sum_{p=0}^{M-1}\sum_{j_1=1}^{N_p}\sum_{j_2=1}^{N_p}(v^p_{1,0,j_1,j_2}\psi^p_{1,0,j_1,j_2}(i_1,i_2) + \\ + v^p_{0,1,j_1,j_2}\psi^p_{0,1,j_1,j_2}(i_1,i_2) + v^p_{1,1,j_1,j_2}\psi^p_{1,1,j_1,j_2}(i_1,i_2)); \end{aligned} \tag{18}$$

$$v^p_{s_1,s_2,j_1,j_2} = \sum_{i_1=1}^{N}\sum_{i_2=1}^{N} f(i_1,i_2)\psi^p_{s_1,s_2,j_1,j_2}(i_1,i_2). \tag{19}$$

Let us consider algorithm of fast direct discrete wavelet (Haar) transform. We have

$$u^0_{j_1,j_2} = f(j_1,j_2),\ j_1 = 1, 2, ..., n,\ j_2 = 1, 2, ..., n;\ \alpha_0 = 2. \tag{20}$$

and for all $p = 0, 1, ..., M-1$, $j_1, j_2 = 0, 1, ..., N_p$, $s_1, s_2 = 0, 1$ (except $s_1 = s_2 = 0$)) we can use:

$$z_1 = (-1)^{s_1},\ z_2 = (-1)^{s_2};\ \alpha_{p+1} = 2 \cdot \alpha_p; \tag{21}$$

$$v^p_{s_1,s_2,j_1,j_2} = \alpha_p^{-1}(u^p_{2j_1-1,2j_2-1} + z_1 u^p_{2j_1,2j_2-1} + z_2 u^p_{2j_1-1,2j_2} + z_1 z_2 u^p_{2j_1,2j_2}); \tag{22}$$

$$u^{p+1}_{j_1,j_2} = u^p_{2j_1-1,2j_2-1} + u^p_{2j_1,2j_2-1} + u^p_{2j_1-1,2j_2} + u^p_{2j_1,2j_2}; \tag{23}$$

$$\alpha_M = n;\ v^M_{0,0,1,1} = \alpha_M^{-1} u^M_{1,1}, \tag{24}$$

where $u^p_{j_1,j_2}$, $j_1, j_2 = 1, 2, ..., N_p$, $p = 1, 2, ..., M$ are so-called auxiliary quantities.
Let us also consider algorithm of fast inverse discrete wavelet (Haar) transform. We have

$$\alpha_M = n;\ \alpha_{M-1} = n;\ u^M_{1,1} = \alpha_M^{-1} v^M_{0,0,1,1}. \tag{25}$$

and for all $p = M-1, M-2, ..., 0, i_1, i_2 = 1, 2, ..., N_p$ we can use the following formulas:

$$j_1 = [(i_1 + 1)/2]; \quad j_2 = [(i_2 + 1)/2]; \quad z_1 = (-1)^{i_1+1}; \quad z_2 = (-1)^{i_2+1}; \quad \alpha_{p-1} = \alpha_p / 2; \tag{26}$$

$$u^p_{i_1,i_2} = \alpha_p^{-1}(z_1 v^p_{1,0,j_1,j_2} + z_2 v^p_{0,1,j_1,j_2} + z_1 z_2 v^p_{1,1,j_1,j_2}) + u^{p+1}_{j_1,j_2}. \tag{27}$$

Thus, we have

$$f(i_1,i_2) = u^0_{i_1,i_2}, \quad i_1 = 0,1,\dots,n, \quad i_2 = 0,1,\dots,n. \tag{28}$$

2.2.2. Algorithm of averaging. Let us assume that it is necessary to make averaging at some level q. For all $p = 1,2,\dots,q$, $j_1, j_2 = 1,2,\dots,N_p$, $s_1, s_2 = 0,1$ (except $s_1 = s_2 = 0$) we can suppose

$$(D_1 u^p)_{2j_1-1,2j_2-1} = (D_1 u^p)_{2j_1-1,2j_2} = (D_1 u^p)_{2j_1,2j_2-1} = (D_1 u^p)_{2j_1,2j_2} \approx (D_1 \tilde{u}^p)_{2j_1-1,2j_2-1}; \tag{29}$$

$$(D_2 u^p)_{2j_1-1,2j_2-1} = (D_2 u^p)_{2j_1-1,2j_2} = (D_2 u^p)_{2j_1,2j_2-1} = (D_2 u^p)_{2j_1,2j_2} \approx (D_2 \tilde{u}^p)_{2j_1-1,2j_2-1}; \tag{30}$$

$$(D_2^+ D_1^+ u^p)_{2j_1-1,2j_2-1} = (D_2^+ D_1^+ u^p)_{2j_1-1,2j_2} = (D_2^+ D_1^+ u^p)_{2j_1,2j_2-1} = (D_2^+ D_1^+ u^p)_{2j_1,2j_2} \approx (D_2^+ D_1^+ \tilde{u}^p)_{2j_1-1,2j_2-1}; \tag{31}$$

$$v^p_{s_1,s_2,2j_1-1,2j_2-1} = v^p_{s_1,s_2,2j_1,2j_2-1} = v^p_{s_1,s_2,2j_1-1,2j_2} = v^p_{s_1,s_2,2j_1,2j_2}, \tag{32}$$

$$\tilde{u}^p_{j_1,j_2} = (u^p_{j_1,j_2} + u^p_{j_1+1,j_2} + u^p_{j_1,j_2+1} + u^p_{j_1+1,j_2+1})/4; \tag{33}$$

$$(D_1^+ u^p)_{j_1,j_2} = (u^p_{j_1+1,j_2} - u^p_{j_1,j_2})/(2^p h); \quad (D_2^+ u^p)_{j_1,j_2} = (u^p_{j_1,j_2+1} - u^p_{j_1,j_2})/(2^p h); \tag{34}$$

$$(T_1^+ u^p)_{j_1,j_2} = u^p_{j_1+1,j_2} + u^p_{j_1,j_2}; \quad (T_2^+ u^p)_{j_1,j_2} = u^p_{j_1,j_2+1} + u^p_{j_1,j_2}; \tag{35}$$

$$D_1 = 0.5 \cdot T_2^+ D_1^+; \quad D_2 = 0.5 \cdot T_1^+ D_2^+. \tag{36}$$

Formulas of averaging have the following form

$$v^p_{1,0,2j_1-1,2j_2-1} = v^p_{1,0,2j_1,2j_2-1} = v^p_{1,0,2j_1-1,2j_2} = v^p_{1,0,2j_1,2j_2} = \beta_{1,0} v^{p+1}_{1,0,j_1,j_2}, \quad j_1, j_2 = 1,2,\dots,N_{p+1}; \tag{37}$$

$$v^p_{0,1,2j_1-1,2j_2-1} = v^p_{0,1,2j_1,2j_2-1} = v^p_{0,1,2j_1-1,2j_2} = v^p_{0,1,2j_1,2j_2} = \beta_{0,1} v^{p+1}_{0,1,j_1,j_2}, \quad j_1, j_2 = 1,2,\dots,N_{p+1}; \tag{38}$$

$$v^p_{1,1,2j_1-1,2j_2-1} = v^p_{1,1,2j_1,2j_2-1} = v^p_{1,1,2j_1-1,2j_2} = v^p_{1,1,2j_1,2j_2} = \beta_{1,1} v^{p+1}_{1,1,j_1,j_2}, \quad j_1, j_2 = 1,2,\dots,N_{p+1}; \tag{39}$$

$$\beta_{1,0} = 0.25; \quad \beta_{0,1} = 0.25; \quad \beta_{1,1} = 0.125. \tag{40}$$

2.3. About fast discrete Haar transforms and averaging for three-dimensional problems

This most cumbersome case is described in [28]. Corresponding algorithms of fast direct and inverse discrete Haar transforms and algorithm of averaging are similar to those described earlier.

2.4. Intro to multilevel wavelet-based numerical method of local structural analysis

In accordance with the method of extended domain [29], the domain Ω, occupied by considering structure, is embordered by extended one ω of arbitrary shape, particularly elementary. Operational formulation of the problem in extended domain ω normally has the form

$$Lu = F, \tag{41}$$

where L is the operator of considering boundary problem with allowance for boundary conditions; u is the unknown function; F is the given right-side function.
Variational formulation of the problem can be obtained directly from operational formulation:

$$\Phi(u) = 0.5 \cdot (Lu, u) - (F, u)\,. \tag{42}$$

Solution of (42) is the critical point of (41). (f, g) denotes dot product of functions f and g.
Obviously, it can be shown that discrete (numerical) formulation of the problem has the form:

$$A\bar{u} = \bar{f}\,, \tag{43}$$

where $A = \{a_{i,j}\}_{i,\,j=1,2,\ldots,n_{gl}}$ is the finite difference (finite element) approximation of L; $\bar{u} = [\,u_1 \;\; u_2 \;\; \ldots \;\; u_{n_{gl}}\,]^T$ is the unknown mesh function; $\bar{f} = [\,f_1 \;\; f_2 \;\; \ldots \;\; f_{n_{gl}}\,]^T$ is the given right-side mesh function; n_{gl} is dimension of considering problem.
Major peculiarities of specified method of basis (local) variations include universality and full adaptability to computer implementation. Particularly for linear problems we can use formulas

$$a_{i,j} = \Phi(\bar{e}^{(i)} + \bar{e}^{(j)}) - \Phi(\bar{e}^{(i)}) - \Phi(\bar{e}^{(j)}) + \Phi(\bar{0})]; \quad f_i = 0.5 \cdot [\Phi(\bar{e}^{(i)}) - \Phi(-\bar{e}^{(i)})], \tag{44}$$

$$\bar{e}^{(i)} = [\,e_1^{(i)} \;\; e_2^{(i)} \;\; \ldots \;\; e_{n_{gl}}^{(i)}\,]^T, \;\; i = 1, 2, \ldots, n_{gl}; \;\; e_j^{(i)} = \delta_{i,j}, \;\; j = 1, 2, \ldots n_{gl}; \tag{45}$$

$\bar{e}^{(i)}$, $i = 1, 2, \ldots, n_{gl}$ are basis mesh vectors; $\bar{0}$ is the null function; $\delta_{i,j}$ is the Kronecker delta.
Obviously, we can write the following wavelet-based (Haar-based) formulation of the problem:

$$\Phi(\bar{u}) = 0.5 \cdot (A\bar{u}, \bar{u}) - (\bar{f}, \bar{u}) = 0.5 \cdot (LQ\bar{v}, Q\bar{v}) - (\bar{f}, Q\bar{v}) = 0.5 \cdot (Q^* LQ\bar{v}, \bar{v}) - (Q^* \bar{f}, \bar{v}), \tag{46}$$

where Q is so-called transition matrix consisting from Haar basis vectors, located in rows; $\bar{v}$ is vector of Haar expansion coefficients of the vector $\bar{u}$. Thus, expression (46) can be rewritten in the form:

$$\tilde{\Phi}(\bar{v}) = 0.5 \cdot (Q^* LQ\bar{v}, \bar{v}) - (Q^* \bar{f}, \bar{v})\,. \tag{47}$$

Corresponding operational formulation of the considering problem has the form

$$\tilde{L}\bar{v} = \tilde{\bar{f}}\,, \quad \tilde{L} = Q^* LQ; \quad \tilde{\bar{f}} = Q^* \bar{f}\,. \tag{48}$$

Further reduction of the problem is based on the averaging algorithm specified above.

2.5. Intro to multilevel wavelet-based semianalytical method of local structural analysis

The objects of the multilevel wavelet-based semianalytical (discrete-continual) method are structures with piecewise constancy of physical and geometrical parameters in one dimension. This dimension or direction is called “basic direction”. As a result so-called wavelet-based discrete-continual design model is introduced. Wavelet approximation of extended domain along non-basic directions is introduces. However, along the basic direction problem remains continual. Analytical solution is apparently preferable in all aspects for providing high-precision structural analysis. It allows structural engineer considering boundary effects. Thus, we can correctly solve problems of structural mechanics when some components of corresponding solutions are rapidly varying functions. Due to dramatic changes of these components inside of mesh elements their rate of change can’t be correctly and adequately considered within conventional numerical methods. Analytical methods of solution are extremely effective here. Another feature of the proposing multilevel wavelet-based semianalytical method of local structural analysis is the absence of limitations on lengths of structures. It can be shown, that resultant multipoint boundary problem after corresponding reduction has the form [30-32]

$$\begin{cases} \bar{y}' = A_k \bar{y} + \bar{f}_k, \quad x \in (x_k^b, x_{k+1}^b), \quad k = 1, ..., n_k - 1 \\ B_k^- \bar{y}(x_k^b - 0) + B_k^+ \bar{y}(x_k^b + 0) = \bar{g}_k^- + \bar{g}_k^+, \quad k = 2, ..., n_k - 1 \\ B_1^+ \bar{y}(x_1^b + 0) + B_{n_k}^- \bar{y}(x_{n_k}^b - 0) = \bar{g}_1^+ + \bar{g}_{n_k}^-, \end{cases} \tag{49}$$

where $x_k^b = x_{3,k}^b$, $k = 1, ..., n_k$ are coordinates of boundary points; A_k, $k = 1, 2, ..., n_k - 1$ are matrices of constant coefficients of order n; B_k^-, B_k^+, $k = 2, ..., n_k - 1$ and $B_1^+, B_{n_k}^-$ are matrices of boundary conditions of order n at point x_k^b; $\bar{g}_k^-, \bar{g}_k^+$, $k = 2, ..., n_k - 1$ and $\bar{g}_1^+, \bar{g}_{n_k}^-$ are right-side vectors of boundary conditions at point x_k^b; $\bar{y} = \bar{y}(x) = [\, y_1(x) \;\; y_2(x) \;\; ... \;\; y_n(x)\,]^T$ is the unknown vector function; $\bar{y}^{(1)} = \bar{y}^{(1)}(x) = d\bar{y}/dx$; $\bar{f}_k = \bar{f}_k(x) = [f_{k,1}(x) \;\; f_{k,2}(x) ... f_{k,n}(x)\,]^T, k = 1,2,...,n_k - 1$ are right-side vector functions.

Correct solution of considering multipoint boundary problem of structural analysis is rather complicated due to several factors. In the first place we should mention boundary effects and immense number of differential equations. Thus, stiff systems are under consideration. In the second place matrices of coefficients of considering type of systems have eigenvalues of opposite signs and corresponding Jordan matrices are not diagonal. It should be emphasized, that correct method of solution of multipoint boundary problems for systems of ordinary differential equations with piecewise constant coefficients in structural analysis has been already developed. This method overcomes all specified difficulties. Besides, we should mention the following advantages of the method: universality, full adaptability to computer implementation, computational stability, optimal conditionality of resultant systems, special algorithm of partial Jordan decomposition of matrix of coefficient, eliminating necessity of computing of root vectors.

3. Results

As is known, development, research and verification of correct mathematical models and methods of structural mechanics are the most important aspects of ensuring safety of buildings and complexes. Generally, in practice global solution in the whole domain is not required in many cases. On the contrary it is normally necessary to obtain high-precision solution of the problem in a certain set of prescribed (selected) subdomains. Wavelet-based numerical and semianalytical methods of corresponding computational local structural analysis are presented. These methods allow reducing dimensionalities of problems. Moreover high-precision results in selected subdomains are provided. Wavelets provide effective and powerful tool for such objectives.

4. Discussion

Results of the distinctive paper is certainly significant because so-called multilevel (local and global (conventional)) structural analysis is required for solution of various technical problems. On the one hand, defects and collapses are most commonly local (subdomains with edge effects deserve special attention in this regard because corresponding significant stresses can lead to destruction), but on the other hand, corresponding load-bearing capacity is determined by the global behavior of structure. Besides, it is associated with the condition of limit equilibrium. After expansion of the solution with the use of local wavelet basis corresponding components are considered at each level of the basis. Reduction of the problem with the use of original algorithm of averaging allows obtaining high accuracy solution of the problem in selected domains with an acceptable reduced size. This localization can be imposed to any selected subdomain by choosing an optimum reduction matrix.

5. Conclusion

Thus, considering wavelet-based multilevel numerical and semianalytical methods of local structural analysis are promising, practically and theoretically significant. Besides, semianalytical formulations are contemporary mathematical models which currently becoming available for computer realization.

Comparison between conventional FEM and wavelet-based numerical and semianalytical methods of local structural analysis must allow to conclude that the localization of the problem provide high-precision results for selected domains even in high level of reduction in wavelet coefficients. However, results of such local analysis may be unacceptable in the other (unselected) subdomains. Analysis of the behavior of the fundamental functions of boundary problems must be also under consideration in order to ensure the correct choice of reducing parameters.

6. References

[1] Barsch T, Kunoth A and Urban K 1997 *Wavelet Analysis and Its Applications* **6** 383-412
[2] Cohen A, Tiplica T and Kobi A 2016 *Control Engineering Practice* **49** 129-138
[3] Karbalaali H, Javaherian A, Dahlke S and Torabi S 2017 *Journal of Applied Geophysics* **146** 67-79
[4] Dahmen P 2001 *Journal of Computational and Applied Mathematics* **128(1-2)** 133-185
[5] Dumont S and Lebon F 1996 *Computers & Structures* **61(1)** 55-65
[6] Esser C and Jaffard S 2018 *Advances in Mathematics* **328** 928-958
[7] Ko J, Kurdila A J and Pilant M S 1997 *Computer Methods in Applied Mechanics and Engineering* **146(1-2)** 1-17
[8] Ko J, Kurdila A J and Oswald P 1997 *Wavelet Analysis and Its Applications* **6** 413-438
[9] Clausel M, Oberlin T and Perrier V 2015 *Applied and Computational Harmonic Analysis* **39(3)** 2015 450-486
[10] Dahmen W, Prossdorf S and Schneider R 1994 *Wavelet Analysis and Its Applications* **5** 385-424
[11] Qian S and Weiss J 1993 *Applied Mathematics Letters* **6(1)** 47-52
[12] Awrejcewicz J, Krysko A V, Pavlov S P, Zhigalov M V and Krysko V A 2017 *Mechanical Systems and Signal Processing* **93** 415-430
[13] Awrejcewicz J, Krysko A V, Saltykova O A, Vetsel S S and Krysko V A 2018 *Communications in Nonlinear Science and Numerical Simulation* **62** 339-351
[14] Krysko A V, Awrejcewicz J, Saltykova O A, Vetsel S S and Krysko V A 2016 *Chaos, Solitons & Fractals* **91** 622-638
[15] Krysko V A, Awrejcewicz J, Dobriyan V, Papkova I V and Krysko V A *Journal of Sound and Vibration* **446** 374-386
[16] Kopysov S and Sagdeeva Yu A 2007 *Russian Mathematics* **51(7)** 76-79
[17] Mkrtychev O V, Dzhinchvelashvili G A, Busalova M S 2014 *Procedia Engineering* **91** 54-57
[18] Aslami M and Akimov P A 2016 *Thin-Walled Structures* **98(B)** 392-402
[19] Aslami M, Akimov P A and Kaytukov T B 2015 *Procedia Engineering* **111** 57-64
[20] Akimov P A and Negrozov O A 2016 *Procedia Engineering* **153** 24-31
[21] Negrozov O A, Akimov P A and Lantsova I Yu 2016 *Procedia Engineering* **153** 926-932
[22] Horgan C O and Carlsson L A 2018 *Comprehensive Composite Materials II* **7** 38-55
[23] Arbabi S, Nazari A and Darvishi M T 2017 *Applied Mathematics and Computation* **292** 33-46
[24] Babaaghaie A and Maleknejad K 2017 *Journal of Computational and Applied Mathematics* **317** 643-651
[25] Dai Q and Cao Q 2018 *Mechanical Systems and Signal Processing* **105** 200-213
[26] Dai Q, Cao Q and Chen Y 2018 *Applied Mathematical Modelling* **57** 603-613
[27] Prakash B, Setia A and Alapatt D 2017 *Journal of Computation Science* **22** 109-118
[28] Mozgaleva M L and Akimov P A 2015 *Procedia Engineering* **111** 569-574
[29] Akimov P A and Mozgaleva M L 2014 *Applied Mechanics and Materials* **580-583** 2898-2902
[30] Akimov P A 2012 *Applied Mechanics and Materials* **204-208** 4502-4505
[31] Akimov P A, Mozgaleva M L, Aslami M and Negrozov O A 2014 *Procedia Engineering* **91** 8-13
[32] Aslami M and Akimov P A 2016 *Archives of Civil and Mechanical Engineering* **16(4)** 668-677

Authors' background

Name	*Prefix*	*Research Field*	*Email*	*Personal website*
Pavel A. Akimov	Prof.	Mathematical modelling	pavel.akimov@gmail.com	
Alexander M. Belostotsky	Prof.	Mathematical modelling	amb@stadyo.ru	
Taymuraz B. Kaytukov	Dr.	Mathematical modelling	tkaytukov@gmail.com	
Leonid S. Lyakhovich	Prof.	Solid mechanics	lsl@sibmail.com	
Marina L. Mozgaleva	Prof.	Mathematical modelling	marina.mozgaleva@gmail.com	

Spectral determination of novel methacryloyl guanidine compounds in water solutions

V A Tikhomirov, D I Mendeleev, V V Kurenkov, M R Menyashev, N A Sivov and V A Gerasin*

Topchiev Institute of Petrochemical Synthesis, Russian Academy of Sciences (TIPS RAS), Moscow, Russia

E-mail: gerasin@ips.ac.ru

Abstract. For production and application of novel biocidal guanidine polymer compounds techniques for monitoring the toxic monomer and oligomer components in polymer aqueous solutions are required. A direct spectroscopic method is offered for determination of methacryloyl guanidine salts, polymer and copolymer in water solutions at native pH. The impact of unsubstituted guanidine and methacrylate guanidine contaminants is also discussed. The spectroscopic method offered is simple, rapid and effective for monitoring contamination of polymethacryloylguanidines with toxic monomer components during synthesis of polymers and copolymers from different monomer salts, during dialysis and other manufacturing stages of biocidal polymers.

1. Introduction

Synthesis of novel guanidine compounds (both macromolecular and low molecular weight) is of particular importance due to them being highly potent in biocide and antiseptic applications [1–3]. Guanidine polymers possess general biocidal properties versus a wide spectrum of both gram-positive and gram-negative bacteria [4–6], and are widely used as a scaffold in the design of specific bioactive compounds with low cytotoxicity and, therefore, human- and livestock-safe [6].

During the last 10 years a series of novel substituted guanidine compounds were synthesized TIPS RAS, their high biocidal activity proven both individually and as a part of polymer composite materials [5, 7]. Substituted guanidines, particularly methacryloyl-guanidines, surpass unsibstituted guanidine in biocidal activity [8, 9]. Due to promising prospects of their practical usage, the question of their determination and identification in solutions is indispensable during industrial and commercial introduction. As the toxicity of biocidal guanidine compounds varies profoundly with their molecular weight (monomers are generally more toxic than polymers), it is essential to develop techniques for monitoring and controlling the low molecular weight components in polymer products concurrently with their synthesis.

Usually spectral determination of guanidine compounds is carried out via IR-range guanidine group characteristic bands, which do not change on polymerization [10]; for methacryloylguanidines, double bond opening on polymerization is clearly and quantitatively identifiable [5]. In UV-vis spectroscopy, guanidine attenuation happens in near UV range (200–300 nm) [11]. Thus, for quantitative determination of guanidines and guanidine groups in arginine, colored complex formation with naphthol is usually employed (Sakaguchi reaction, [12]), which is a finicky sophisticated technique and due to some curcumstances [13] is unfit for determination of guanidine monomers in the presence of their polymers in aqueous solutions.

For PHMG, a colored complex of polymer with Eosin Y at pH 2.5, absorbing at 540 nm, is employed for its determination in water [14]. The latter is proposed for creating samplers for detecting PHMG aerosol presence in air [15]. Solution separation could be achieved in general case by

IOP Conf. Series: Materials Science and Engineering **675** (2019) 012058 doi:10.1088/1757-899X/675/1/012058

electrophoresis, which was proven for mixtures of PHMG, its monomer and guanidine hydrochloride (GHC) [16], and is potentially applicable for novel substances.

Some substituted guanidines demonstrate a red shift in their attenuation bands, allowing for direct quantitative determination in mid UV. Ab initio quantum chemical calculations prove the red shifts to be connected with both increase in the size of conjugated system and ionized complex formation of guanidinium cation with organic counterion. Guanidinium cation π-π* transition band lies at about 160 nm [17], while for phenilguanidinium and 1-naphtylguaniinium at 190 and 220 nm, accordingly [18]. Complex formation shifts the n-π* transition band; phenilguanidinium maximum in the presence of complexing counterions is shifted from 210 nm to up to 20 nm (230 nm, F^-) [18]. In view of the above, methacryloylguanidines in aqueous solution should have attenuation bands in the UV-vis range, sensitive to their degree of polymerization and monomer counterions.

The aim of this article is to obtain and analyze UV-vis spectra of novel methacryloyl substituted guanidine compounds in the 190–250 nm range. Special attention is paid to mixed aqueous solutions, at natural pH typical for their synthesis and manufacturing. We believe these results to be useful for analytical and express monitoring methods in production of guanidine-based biocidal additives.

2. Materials and methods

The objects of study were novel methacryloyl substituted guanidine compounds (table 1, table 2), synthesized and characterized at TIPS RAS. Polymers were fully separated and purified by filtering the reaction mass, followed by precipitation into acetone, multiple leaching with same, and dialysis on "Spectrapor MWCO 3500" membranes to remove <3000 Da fraction. Products' structure and composition were verified by 1H NMR spectroscopy immediately afterwards. The syntheses are described in detail in [5, 19].

Table 1. Low molecular weight guanidine compounds studied.

Compound (**designation**)	Chemical formula	MW (g/mol)[a]	pH[b]
Guanidine hydrochloride (**GHC**)	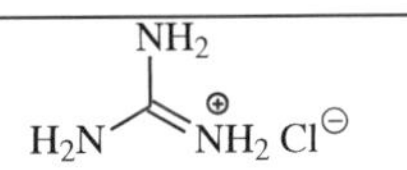	95.5	6.4
Methacrylate guanidine (**MAG**)	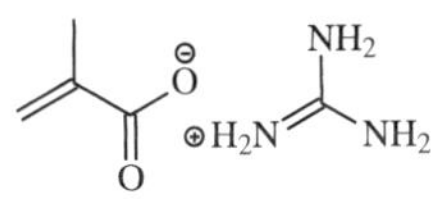	145.2	6.2
Methacryloylguanidine hydrochloride (**MGHC**)	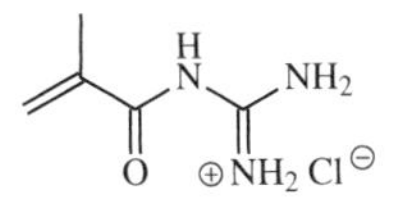	163.6	6.1
Methacryloylguanidine trifluoroacetate (**MGTFA**)	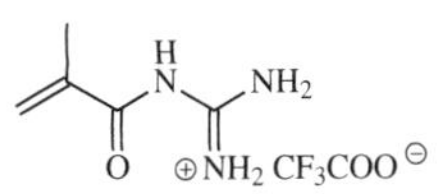	241.2	5.6
Methacryloylguanidine acetate (**MGAc**)	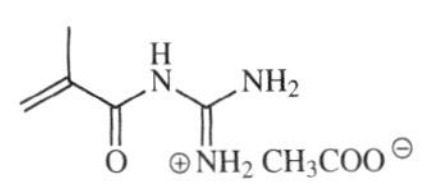	187.2	6.1

[a] molecular weight

[b] pH of 1.14 mmol/l aqueos solution at 20°C

Preconditioned *in vacuo* at 34 °C for 50 hours, the solid powder samples were kept in a dessicator due to their hygroscopic behavior. Fresh solutions of monomers and polymer with concentration 2.85±0.05 mmol/l were prepared and diluted to working concentrations with bidistilled water using standard analytical techniques. Measurements were carried out on a split-beam Specord M40 spectrometer (Germany), in 2 cm quartz cells at 20±1 °C in the range of 188–1100 nm to a precision of 0.001. The total concentration of mixed solutions was equalized at 30 μmol/l concentration by the absorption in right isosbestic point (233 nm). pH was measured at 20±1 °C on an ANION 7000 pH-meter (Russia) with a glass–silver electrode pair.

Table 2. High molecular weight guanidine compounds studied.

Compound (**designation**)	Chemical formula	pH[c]
Polymer of MAG (**PolyMAG**)	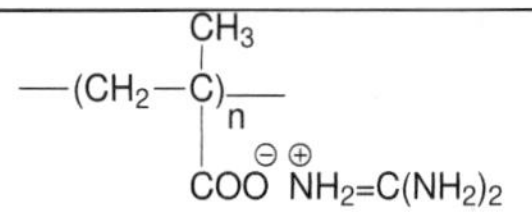	7.5
Polymer of MGHC (**PolyMGHC**)	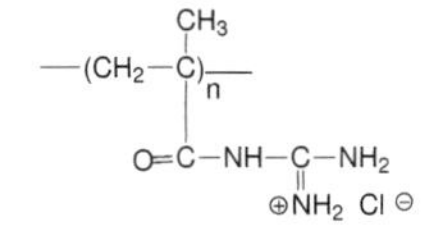	3.9
Copolymer of MGHC with diallyldimethylammonium chloride 78:22 mol. % (**coPolyMGHC**)	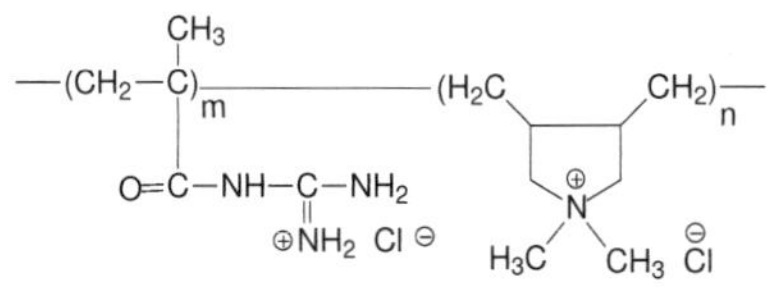	3.7

[c] pH of 1.14 mmol/l aqueos solution at 20°C

3. Results and discussion

The spectra of low-molecular weight guanidine compounds are presented on figure 1.

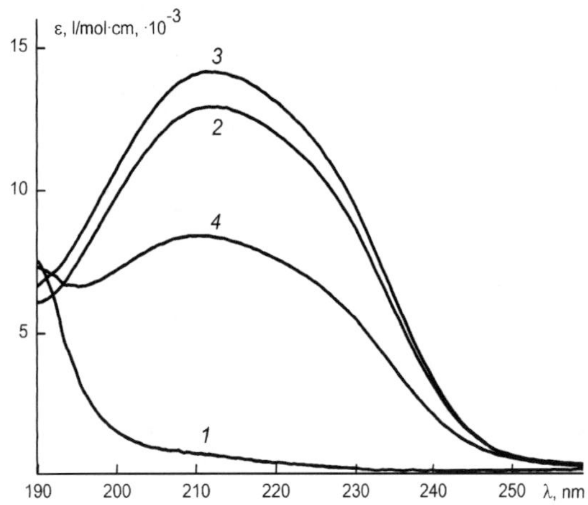

Figure 1. Electronic absorption spectra of low molecular weight guanidine compounds in bidistilled water at natural pH:
1 — GHC;
2 — MGHC;
3 — MGTFA;
4 — MGAc;
concentration 30 μmol/l.

GHC solutions a show an absorption maximum at 190 nm with ε = 7.6·103 l/mol·cm (figure 1, curve 1). In aqueous solutions GHC exists as a guanidinium cation (pKa=13.5 [11]) with absorption band in the far UV range; the detected maximum lies at the shoulder of the absorption band and at the boundary of the measurement range, so the intensity dependence is linear only at concentrations of 16–30 μmol/l. Furthermore, in aqua guanidine is prone to oxidizing to urea (and subsequently to ammonium carbonate) with the same band position and ε ≈ 1·103 l/mol·cm [20], leading to a decrease in measured concentration compared to freshly prepared solutions. This also hinders the direct quantitative measurement of GHC in aqueous solution.

A direct spectropolarimetric technique is described for measuring unsubstituted GHC and urea in dialyzed aqueous protein solutions [21], where the first hindrance is curcumvented by measuring the attenuance at the inflection point (in the range 190–230 nm), allowing for quantitative determination in concentrations from 6 μmol/l in distilled water and from 120 μmol/l in phosphate buffer.

Nevertheless, applying the same treatment to our spectra did not enhance neither the limit of detection nor the precision of measurement.

The introduction of auxochromic methacryloyl group into guanidinium cation leads to a red shift of attenuation band to a maximum of 210 nm (figure 1, curves 2–4). Molar extinction coefficients are similar for MGHC and MGTFA, and is slightly lower for MGAc. As the cation structure for those compounds is the same, it can be accounted for by the difference in counterions: MGHC and MGTFA are strong acid salts (pKa = –10 and pKa = 0.23, correspondingly), while MGAc is the salt of weak acetic acid (pKa = 4.76). Neither GHC nor MGHC in this concentrations shift the absorption maxima of anionic dye Fast Green FCF, which forms complexes with arginine moieties in proteins.

The polymerization of MGHC with –C=C– bond of methacryloyl group being opened results in π-π^* band shifting to a 201 nm position in polymer (figure 2, curve 2).

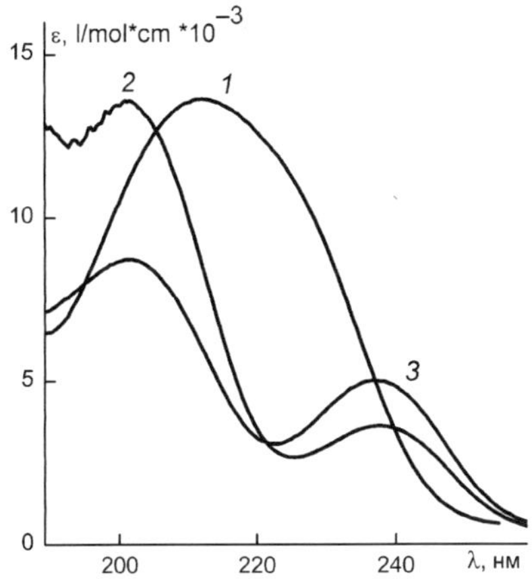

Figure 2. Electronic absorption spectra of methacryloylguanidine compounds in bidistilled water at natural pH:
1 — MGHC;
2 — PolyMGHC;
3 — coPolyMGHC;
concentration 30 μmol/l.

PolyMGHC quantitative measurement is possible both via 201 nm band and via n-π^* band at 238 nm, or at the saddle point (223 nm). For both PolyMGHC and coPolyMGHC, the maximum at 238 nm coincides with the right isosbestic point at 238 nm (figure 2). As neither diallyldimethylammonium, nor its oligomer blocks have an absorption band at λ>200 nm [22], it is possible to recalculate the attenuance of coPolyMGHC to MGHC monomer equivalent for similar determination in mixed monomer/copolymer solution.

CoPolyMGHC spectra have the same band positions as PolyMGHC (201 and 238 nm), but their relative intensity is different; furthermore, for solutions diluted to less than 10 μmol/l, the relative intensities also change, compared to higher concentrations. This could be attributed to different methacryloyl block conformation depending on polymer chain structure in solution, and by general supramolecular structure differences; the implication to spectral measurement is that for those concentrations, the 238 nm band is better suitable.

No attenuance was detected which could be attributed to colloidal scattering in those conditions and concentrations; additional care should be taken, however, on measurements in higher ionic strength solutions and at higher polymer concentrations.

These factors allow for quantitative determination of isolated MGHC, PolyMGHC and coPolyMHC in aqueous solutions (figure 3).

Band characteristics, ranges of detection and linear regression parameters are presented in table 3. For mixed solutions, the use of absorption at the right shoulder of 238 nm band or a total peak decomposition is necessitated for polymer, and the 223 nm for monomer, due to close proximity of absorption bands and their high half-width (25–30 nm), and a non-additive increase in absorption in sub-205 nm range.

Thus, direct spectrophotometry allows for effective determination of methacryloylguanidine monomers, polymer, and copolymer in unbuffered aqueous solution at natural pH without prior component separation, but lacks in GHC contaminant detection. The spectral measurement range rules out the usage of default glass or plastic transmission cells, and a split-beam geometry for the spectrometer is preferred to minimize the water attenuation.

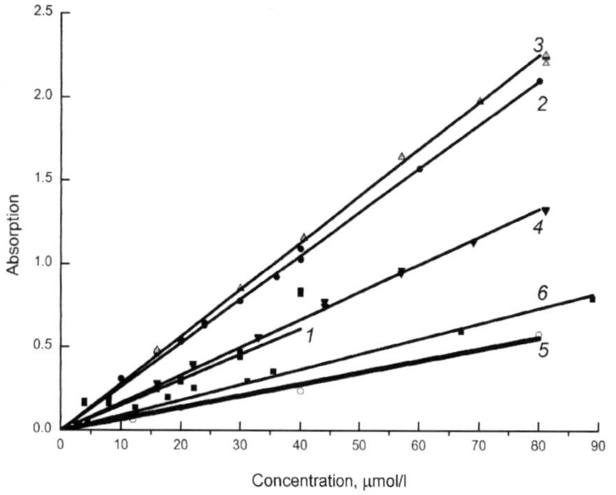

Figure 3. Calibration curves:
1 — GHC;
2 — MGHC;
3 — MGTFA;
4 — MGAc;
5 — PolyMGHC (238 nm);
6 — coPolyMGHC (238 nm).
Measurement uncertainty is within the size of symbols.

Table 3. Attenuation bands and molar extinction coefficients of guanidine compounds[d].

Compound	λ_{max} (nm)	$\varepsilon\cdot10^{-3}$ (l/mol·cm)	Linear range of detection (μmol/l)	Linear regression $A=k\cdot C$ parameters	
				$k\cdot10^{-3}$ (l/mol)	Regression coefficient r^2
GHC	190	7.6±0.2	16–30	15.2±0.2	0.999
MGHC	212	13.1±0.1	4–80	26.2±0.2	0.998
MGTFA	212	14.3±0.1	4–80	28.2±0.2	0.999
MGAc	210	8.5±0.1	4–80	16.6±0.2	0.999
PolyMGHC	201	13.5±0.3	4–40	27.0±0.2	0.985
	238	3.6±0.1	4–80	6.9±0.2	0.993
coPolyMGHC	201	8.9±0.1	4–90	17.7±0.3	0.999
	238	5.3±0.2	12–120	9.2±0.2	0.994

[d] For all substances studied, the limit of determination is sufficiently low (<1 μmol/l)

One additional contaminant to consider is methacrylate guanidine, which is a byproduct of methacryloyl guanidine synthesis, capable of polymerization in the same condition as methacryloylguanidine. To estimate the possibility of analytical monitoring this contaminant in MGHC/PolyMGHC aqueous solutions, additional spectra were taken (figure 4).

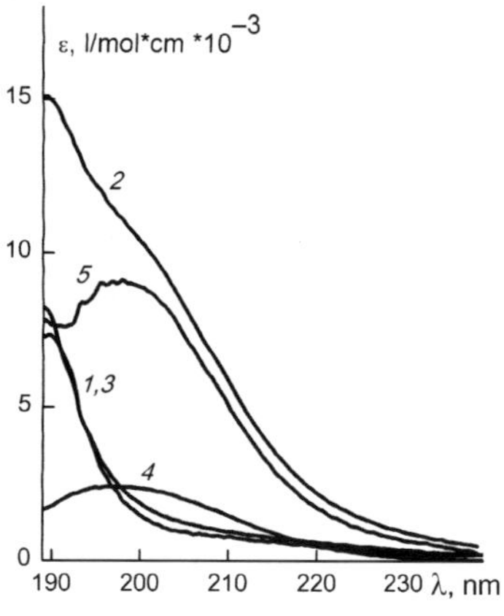

Figure 4. Electronic absorption spectra of guanidine compounds and contaminants in bidistilled water at natural pH:
1 — GHC;
2 — MAG;
3 — PolyMAG;
4 — methacrylic acid;
5 — difference spectrum of MAG and GHC;
concentration 30 μmol/l.

On methacrylate guanidine dissociation in water, unsubstituted guanidinium cation and methacrylic acid anion are produced. MAG solution spectrum amounts to a superposition of guanidinium and methacrylate ion attenuation bands (figure 4, curves 1, 2, 5), as the band positions in difference spectrum (figure 4, curve 5) are identical to those in the spectrum of methacrylate anion (figure 4, curve 4), and the difference in intensity of extinction is explained by methacrylic acid partial dissociation in measurement conditions (pKa = 4.46). On polymerization of MAG the methacrylic group double bond opens, so the aqueous solution of PolyMAG (curve 3) has the same spectrum as guanidinium.

MAG attenuation bands are essentially non-linear vs concentration, and so is the case for pure methacrylic acid without guanidinium cations [23]). Furthermore, both methacrylic acid and its salts are unstable in aqueous solution and the bands coincide with the bands of other components and contaminants. Thus, they could not be used for direct quantitative determination, and monitoring MAG in the reaction mixture or product should be done in other ways.

4. Conclusion

Novel monomer and polymer methacryloylguanidine compounds were spectroscopically characterized for the first time in the UV-vis range. Attenuation bands and molar extinction coefficients are determined for them in aqueous solutions, allowing for quantitative determination in 4–80 µmol/l range. UV-vis spectroscopy is effective for monitoring comtamination of polymethacryloylguanidines with monomers during polymer and copolymer synthesis from different monomer salts at natural pH, dialysis and other biocidal polymer manufacturing stages.

Acknowledgments

This work was supported by the Ministry of Education and Science of the Russian Federation within agreement № 14.607.21.0135, code «2015-14-579-0173-012», project №RFMEFI60715X0135.

References

[1] Rogalsky S, Bardeau J F, Wu H, Lyoshina L, Bulko O, Tarasyuk O, Makhno S, Cherniavska T, Kyselov Y and Koo J H 2016 *J. Mater. Sci.* **51** 7716–30
[2] Carmona-Ribeiro A M and de Melo Carrasco L D 2013 *Int. J. Mol. Sci.* **14** 9906–46.
[3] Berlinck R G S, Burtoloso A C B, Trindade-Silva A E, Romminger S, Morais R P, Bandeira K and Mizuno C M 2010 *Nat. Prod. Rep.* **27** 1871–907
[4] Kukharenko O, Bardeau J F, Zaets I, Ovcharenko L, Tarasyuk O, Porhyn S, Mischenko I, Vovk A, Rogalsky S and Kozyrovska N 2014 *Eur. Polym. J.* **60** 247–54.
[5] Menyashev M R Synthesis of guanidine containing polymers and development of guanidine polymer-based composite materials. PhD thesis. 2017 (Moscow: TIPS RAS). (***in Russian***)
[6] Sączewski F and Balewski Ł 2013 *Expert Opinion on Therapeutic Patents.* **23** 965–95
[7] Antipov E E, Gerasin V A, Antipova L A and Sivov N A 2006 Patent RU 2424797 C1
[8] Timofeeva L M and Kleshcheva N A 2011 *Appl. Microbiol. Biotechnol.* **89** 475–92
[9] Tanzer J M, Slee A M and Kamay B A 1977 *Antimicrob. Agents Chemother.* **12** 721–9
[10] Angell C L, Sheppard N, Yamaguchi A, Shimanouchi T, Miyazawa T and Mizushima S 1957 *Trans. Faraday Soc.* **53** 589–600
[11] Tables of Physical & Chemical Constants, 16th ed. 3.8.7 UV-visible spectroscopy. 2005 (Kaye & Laby Online. Version 1.0)
[12] Weber C J 1930 *J. Biol. Chem.* **86** 217–22
[13] Jorpes E, Thoren S 1932 *Biochem. J.* **26** 1504–06
[14] Monakhova Y B, Kuballa T, Leitz J and Lachenmeier D W 2011 Int. J. Anal. Chem. **2011** 704795
[15] Choi S H, Park S K, Kang H J, Lee K and Kwon J H 2015 *Bull. Korean Chem. Soc.* **36** 1819–23
[16] Rudnev A V and Dzherayan T G 2006 *J. Anal. Chem.* **61** 1002–5
[17] Antol I, Glasovac Z, Crespo-Otero R and Barbatti M 2014 *J. Chem. Phys.* **141** 1–11

[18] Antol I, Glasovac Z, Margetic D, Crespo-Otero R and Barbatti M 2016 *J. Phys. Chem.* A. **120** 7088–100
[19] Menyashev M R, Martynenko A I, Popova N I, Kleshcheva N A and Sivov N A 2016 *Polym. Sci. Ser.* B **58** 556–63
[20] Filutowicz Z, Lukaszewski K and Pieszynski K 2004 *Journal of Medical Informatics & Technologies* **8** 105–110
[21] Fuertes M A, Pérez J M and Alonso C J 2004 *Biochem. Bioph. Methods.* **59** 209–16
[22] Mwangi I W, Ngila J C, Ndungu P and Msagati T A M 2013 *Water, Air and Soil Pollution.* **224** 1638.
[23] Ungnade H E and Ortega I 1951 *J. Amer. Chem. Soc.* **73** 1564–7

Authors' background

Name	*Prefix*	*Research Field*	*Email*	*Personal website*
Tikhomirov Vsevolod Andreevich	PhD candidate	Optical spectroscopy	hasn200@mail.ru	–
Mendeleev Dmitry Ivanovich	PhD candidate	Spectroscopy, materials science	dmen@ips.ac.ru	–
Kurenkov Viktor Vladislavovich	PhD candidate	Materials science	viktorkur@yandex.ru	–
Menyashev Marat Ravilievich	Assistant Professor	Polymer chemistry, materials science	menyashev.marat@gmail.com	–
Sivov Nikolay Aleksandrovich	Associate Professor	Organic chemistry, polymer chemistry	sivov@ips.ac.ru	–
Gerasin Viktor Anatolievich	Associate Professor	Materials science, physical chemistry	gerasin@ips.ac.ru	https://istina.msu.ru/profile/gerasin/

IOP Conf. Series: Materials Science and Engineering **675** (2019) 012059 doi:10.1088/1757-899X/675/1/012059

Nonlinear deformation of a solid body on the basis of flow theory and realization of FEM in mixed formulation

N A Gureeva*, R Z Kiseleva and A P Nikolaev

Volgograd State Agrarian University, Volgograd, Russia

Email: natalya-gureeva@yandex.ru

Abstract. In the mixed formulation, an algorithm for the formation of a hexahedral finite element deformation matrix at the loading step to determine the stress-strain state of a solid body beyond the elastic limit based on the theory of plastic flow is developed in a curvilinear coordinate system. Stresses and displacements are taken as nodal unknowns. The approximation of the required values of the inner point of the finite element through the nodal unknowns was carried out by trilinear functions. To obtain the deformation matrix of the hexahedral finite element, a mixed functional on the equality of the real and possible works of external and internal forces with the replacement of the actual work of internal forces by the difference of the total and additional work of internal forces at the loading step is used.

1 Introduction

The stress-strain state (SSS) theory of solids developed in the present time enough [1-5]. The complexity of the differential equations obtained in determining SSS of solids does not allow to obtain analytical solutions. In order to find the main deformation parameters, numerical methods are actively used, the main of which is the FEM. The most common finite element method in the formulation of the displacement method [6-10], rarely used in a mixed formulation [6, 11]. The unknown functions determined by the method of displacements are obtained continuous only at the nodal points of the finite element, and at its boundaries are not continuous. In the case of a mixed method, the found unknown functions have continuity both at the nodes and at the boundaries of the elements. Therefore, the use of the FEM in a mixed formulation is preferable.

The problems of reducing the material consumption of the designed engineering structures for various purposes dictate the need to consider the processes of deformation of solids beyond the elastic limit. For this purpose, in this paper, an algorithm for obtaining the defining relations at the loading step is developed for use in the formation of the deformation matrix of a bulk hexahedral finite element in the mixed formulation of the finite element method (FEM). As a research tool, an algorithm for obtaining relations between stress increments and strain increments is developed in a curvilinear coordinate system based on the hypothesis of the theory of plastic flow on the proportionality of the components of the plastic strain increment tensor to the components of the stress deviator. The obtained relations are used in the algorithm for the development of the deformation matrix of a three-dimensional hexagonal finite element in a mixed formulation.

2 Materials and Methods

The geometry of a rigid body. The position of an arbitrary point of the M body in the coordinate system xOz is described by the radius vector

$$\vec{R} = x^k\left(\theta^m\right)\vec{i}_k;\; k,m = 1,2,3, \tag{1}$$

where x^k - cartesian coordinates of the point M ; θ^m - curvilinear coordinates of the point M; $\vec{i}_k$ - orts of the cartesian coordinate system.

The vectors of the local basis of a point M are defined by the expression

$$\vec{g}_i = \vec{R},_i . \tag{2}$$

The basis vectors (2) and their derivatives can be represented by a matrix expression

$$\underset{3\times1}{\{\vec{g}\}} = \underset{3\times3}{[s]}\underset{3\times1}{\{\vec{i}\}};\quad \underset{3\times1}{\{\vec{i}\}} = \underset{3\times3}{[s]^{-1}}\underset{3\times1}{\{\vec{g}\}};\quad \underset{3\times1}{\{\vec{g}_{,i}\}} = \underset{3\times3}{[m_i]}\underset{3\times1}{\{\vec{g}\}};\quad i = 1,2,3, \tag{3}$$

where $\underset{1\times3}{\{\vec{g}\}^T} = \{\vec{g}_1\ \vec{g}_2\ \vec{g}_3\}$; $\underset{1\times3}{\{\vec{i}\}^T} = \{\vec{i}_1\ \vec{i}_2\ \vec{i}_3\}$; $\underset{1\times3}{\{\vec{g}_{,3}\}^T} = \{\vec{g}_{i,3}\ \vec{g}_{i,3}\ \vec{g}_{i,3}\}$.

Based on the displacements from the deformation at loading step. When a rigid body is deformed, its arbitrary point is considered in three states: initial M , deformed after j loading steps M^j (displacement vector V) and deformed after $(j+1)$ loading step M^{j+1} (displacement vector $\vec{w}$).

Displacement vectors V are $\vec{w}$ expressed by the components in the basis of the M^t initial state point in the form

$$\vec{V} = v^i\vec{g},\quad \vec{w} = w^i\vec{g}_i \tag{4}$$

Their derivatives are defined by expressions

$$\vec{V},_i = v,_i^k\,\vec{g}_k + v^k\vec{g}_k,_i = f_i^k\vec{g}_k;\quad \vec{w},_i = w,_i^k\,\vec{g}_k + w^k\vec{g}_k,_i = \alpha_i^k\vec{g}_k, \tag{5}$$

where f_i^k - functions of components v^i and their derivatives; α_i^k - functions component w^i and their derivatives.

Deformations and their increments at the points M^j, M^{j+1} and after j the $(j+1)$ loading steps are determined as the differences between the metric tensors of the initial and deformed states of the body and are written in the form [4], which, taking into account (4) and (5), are written in matrix form

$$\underset{6\times1}{\{\Delta\varepsilon\}} = \underset{6\times3}{[L]}\underset{3\times1}{\{v\}};\quad \underset{6\times1}{\{\Delta\varepsilon\}} = \underset{6\times3}{[L]}\underset{3\times1}{\{w\}}, \tag{6}$$

where $\underset{1\times6}{\{\varepsilon\}^T} = \{\varepsilon_{11}\ \varepsilon_{22}\ \varepsilon_{33}\ 2\varepsilon_{12}\ 2\varepsilon_{13}\ 2\varepsilon_{23}\}$ - matrix - line deformation; $[L]$ - the matrix of differential operators; $\underset{1\times3}{\{v\}^T} = \{v^1\ v^2\ v^3\}$ - matrix-line displacements of a point M^j; $\underset{1\times6}{\{\Delta\varepsilon\}^T} = \{\Delta\varepsilon_{11}\ \Delta\varepsilon_{22}\ \Delta\varepsilon_{33}\ 2\Delta\varepsilon_{12}\ 2\Delta\varepsilon_{13}\ 2\Delta\varepsilon_{23}\}$ - matrix - string deformations; $\underset{1\times3}{\{w\}^T} = \{w^1\ w^2\ w^3\}$ - the matrix-line displacements of the point M^{j+1}.

Relations of the theory of plastic flow. The total strain increments are determined by the sum of the elastic and plastic strain increments

$$\Delta\varepsilon_{ij} = \Delta\varepsilon_{ij}^e + \Delta\varepsilon_{ij}^p, \tag{7}$$

where $\Delta\varepsilon_{ij}$ – the total increment of deformation; $\Delta\varepsilon_{ij}^e$ – the increments of elastic deformation; $\Delta\varepsilon_{ij}^p$ - the increment of plastic deformation.

The increment of elastic strain using the stress increment is expressed by the Hooke's law in matrix form

$$\underset{6\times1}{\{\Delta\varepsilon^e\}} = \underset{6\times6}{[C]}\underset{6\times1}{\{\Delta\sigma\}}, \tag{8}$$

where $\underset{1\times6}{\{\Delta\varepsilon\}}^T = \{\Delta\varepsilon_{11}^e\ \Delta\varepsilon_{22}^e\ \Delta\varepsilon_{33}^e\ 2\Delta\varepsilon_{12}^e\ 2\Delta\varepsilon_{13}^e\ 2\Delta\varepsilon_{23}^e\}$ - matrix-row increments of elastic deformation; $\underset{1\times6}{\{\Delta\sigma\}}^T = \{\Delta\sigma_{11}\ \Delta\sigma_{22}\ \Delta\sigma_{33}\ \Delta\sigma_{12}\ \Delta\sigma_{13}\ \Delta\sigma_{23}\}$ - the matrix-the row of increments of stress.

The increments of plastic deformations at the loading step are determined on the basis of the hypothesis of the theory of plastic flow about the proportional dependence of increments of plastic deformations on the components of the stress deviator, which is written in the form

$$\Delta\varepsilon_{ij}^p = k\left(\sigma_{ij} - \frac{1}{3}P(\sigma)g_{ij}\right), \tag{9}$$

where k is the coefficient of proportionality; $P(\sigma) = \sigma^{mn}g_{mn}$ - is the first invariant of the stress tensor.

The coefficient of proportionality in the expression (9) is the formula [3]

$$k = \frac{3}{2}\frac{\varepsilon_i^{\Delta p}}{\sigma_i}, \tag{10}$$

where $\varepsilon_i^{\Delta p}$ is the intensity of increments of plastic deformations; σ_i – the intensity of stresses.

At small steps of loading the intensity of increments of plastic deformations and increments of their intensity can be taken equal to

$$\varepsilon_i^{\Delta p} = \Delta\varepsilon_i^p, \tag{11}$$

where $\Delta\varepsilon_i^p$ – the increment of the intensity of plastic deformations.

And we can also assume that the increment of the intensity of total deformations $\Delta\varepsilon_i$ is equal to the sum of the increments of the intensities of elastic $\Delta\varepsilon_i^e$ and plastic $\Delta\varepsilon_i^p$ deformations

$$\Delta\varepsilon_i = \Delta\varepsilon_i^e + \Delta\varepsilon_i^p, \tag{12}$$

The increment of the intensity of elastic deformations is according to the formula

$$\Delta\varepsilon_i^e = \frac{\Delta\sigma_i}{E_н}, \tag{13}$$

where $E_н$ is the deformation module of the initial section of the deformation diagram.

At elastic-plastic deformation at a loading step the relation takes place

$$\Delta\sigma_i = E_k\Delta\varepsilon_i. \tag{14}$$

where E_k is the tangent module of the deformation diagram.

The increment of the intensity of plastic deformations is determined from the expression (13)

$$\Delta\varepsilon_i^p = \Delta\varepsilon_i - \Delta\varepsilon_i^e. \tag{15}$$

The intensity of the stress is the formula

$$\sigma_i = \sqrt{3S^{ij}S_{ij}}\ , \tag{16}$$

where $S_{ij} = \sigma_{ij} - \frac{1}{3}P(\sigma)g_{ij}$ is the stress deviator.

At the loading step, the stress intensity increment is determined using the derived expression (16).

The step of loading the dependencies between the total increments of strain increments of stress taking into account (8) and (9) can be written in matrix form

$$\underset{6\times1}{\{\Delta\varepsilon\}} = \underset{6\times6}{\left[C^n\right]}\underset{6\times1}{\{\Delta\sigma\}}. \tag{17}$$

3 Results and discussion

Finite element deformation matrix at the loading step. The finite element is selected in the form of a hexagon, with nodes i, j, k, l, m, n, p, h. The nodal unknowns are taken as increments of displacements and increments of stresses. To perform numerical integration, an arbitrary hexagon is mapped to a rectangular cube with local coordinates ξ, η, ζ, varying from -1 to 1. The global coordinates of an x^1, x^2, x^3 arbitrary point of the hexahedronal finite element, depending on their nodal values, are expressed by relations

$$x^i = \underset{1\times8}{\{\phi(\xi,\eta,\zeta)\}^T}\underset{8\times1}{\{x^i{}_y\}};\quad i = 1,2,3\,, \tag{18}$$

where $\underset{1\times8}{\{\varphi(\zeta,\eta,\xi)\}^T} = \{\varphi_1\ \varphi_2\ \varphi_3\ \varphi_4\ \varphi_5\ \varphi_6\ \varphi_7\ \varphi_8\}$ - matrix -a string of form functions;

$\underset{1\times8}{\{x^i{}_y\}^T} = \{x^{ii}\ x^{ij}\ x^{ik}\ x^{il}\ x^{im}\ x^{i1n}\ x^{ip}\ x^{ih}\}$ - matrix - a string of nodal values of the coordinates of the finite element.

The displacements of an arbitrary point of M^t a finite element are approximated by the displacements of its nodal points by functions (18) and are represented in matrix form

$$\underset{3\times1}{\{v\}} = \underset{3\times24}{[A]}\underset{24\times1}{\{v_y\}}, \tag{19}$$

where $\underset{1\times24}{\{v_y\}^T} = \left\{\underset{1\times8}{\{v^1_y\}^T}\ \underset{1\times8}{\{v^2_y\}^T}\ \underset{1\times8}{\{v^3_y\}^T}\right\}.$

The relations between deformations and displacements, increments of deformations and displacements at the loading step (6) taking into account (19) are written in the form

$$\underset{6\times1}{\{\varepsilon\}} = \underset{6\times3}{[L]}\underset{3\times24}{[A]}\underset{24\times1}{\{v_y\}} = \underset{6\times24}{[B]}\underset{24\times1}{\{v_y\}};\quad \underset{6\times1}{\{\Delta\varepsilon\}} = \underset{6\times3}{[L]}\underset{3\times24}{[A]}\underset{24\times1}{\{w_y\}} = \underset{6\times24}{[B]}\underset{24\times1}{\{w_y\}}, \tag{20}$$

where $\underset{1\times24}{\{w_y\}^T} = \left\{\underset{1\times8}{\{w^1_y\}^T}\ \underset{1\times8}{\{w^2_y\}^T}\ \underset{1\times8}{\{w^3_y\}^T}\right\}.;\ \underset{1\times8}{\{w^i{}_y\}^T} = \{w^{ii}\ w^{ij}\ w^{ik}\ w^{il}\ w^{im}\ w^{\backslash in}\ w^{ip}\ w^{ih}\}.$

The stresses at an arbitrary point of a finite element are approximated through the stress components in its nodal stresses by expressions

$$\sigma_{ij} = \underset{1\times8}{\{\phi(\xi,\eta,\zeta)\}^T}\underset{8\times1}{\{\sigma_{ijy}\}};\quad i = 1,2,3;\quad j = 1,2,3, \tag{21}$$

where $\underset{1\times 8}{\left\{\sigma_{ij\,y}\right\}^T} = \left\{\sigma_{ij}^i\ \sigma_{ij}^j\ \sigma_{ij}^k\ \sigma_{ij}^l\ \sigma_{ij}^m\ \sigma_{ij}^n\ \sigma_{ij}^p\ \sigma_{ij}^h\right\}$; - matrix – stress lines in the nodes of the finite element.

The approximation of stress increments at an arbitrary point of a finite element through its nodal stresses and their increments is performed in the same way as the expressions (21) and is written in matrix form

$$\underset{6\times 1}{\{\sigma\}} = \underset{6\times 48}{[G]}\underset{48\times 1}{\{\sigma_y\}},\quad \underset{6\times 1}{\{\Delta\sigma\}} = \underset{6\times 48}{[G]}\underset{48\times 1}{\{\Delta\sigma_y\}}, \tag{22}$$

where $\underset{1\times 6}{\{\sigma\}^T} = \{\sigma_{11}\ \sigma_{22}\ \sigma_{33}\ \sigma_{12}\ \sigma_{13}\ \sigma_{23}\}$;

$$\underset{1\times 48}{\{\sigma_y\}^T} = \left\{\underset{1\times 8}{\{\sigma_{11\,y}\}^T}\ \underset{1\times 8}{\{\sigma_{22\,y}\}^T}\ \underset{1\times 8}{\{\sigma_{33\,y}\}^T}\ \underset{1\times 8}{\{\sigma_{12\,y}\}^T}\ \underset{1\times 8}{\{\sigma_{13\,y}\}^T}\ \underset{1\times 8}{\{\sigma_{23\,y}\}^T}\right\};$$

$$\underset{1\times 6}{\{\Delta\sigma\}^T} = \{\Delta\sigma_{11}\ \Delta\sigma_{22}\ \Delta\sigma_{33}\ \Delta\sigma_{12}\ \Delta\sigma_{13}\ \Delta\sigma_{23}\};$$

$$\underset{1\times 48}{\{\Delta\sigma_y\}^T} = \left\{\{\Delta\sigma_{11\,y}\}^T\ \{\Delta\sigma_{22\,y}\}^T \ldots\ldots \{\Delta\sigma_{13\,y}\}^T\ \{\Delta\sigma_{23\,y}\}^T\right\}$$

The functional expressing the equality of possible and actual works of internal and external forces at the loading step for an arbitrary stress state of the shell of rotation is written in the form

$$\begin{aligned} \Pi \equiv & \int_V \underset{1\times 6}{\{\Delta\sigma\}^T} \underset{6\times 3}{[L]} \underset{3\times 1}{\{w\}} dV - \frac{1}{2} \underset{1\times 6}{\{\Delta\sigma\}^T} \underset{6\times 6}{\left[C^{\Pi}\right]} \underset{6\times 1}{\{\Delta\sigma\}} dV - \\ & -\frac{1}{2}\int_S \underset{1\times 3}{\{w\}^T} \underset{3\times 1}{\{\Delta q\}} dS - \int_S \underset{1\times 3}{\{v\}^T} \underset{3\times 1}{\{q\}} dS + \int_V \underset{1\times 6}{\{\sigma\}^T} \underset{6\times 1}{\{\Delta\varepsilon\}} dV, \end{aligned} \tag{23}$$

where $\underset{1\times 3}{\{q\}^T} = \{q_1 q_2 q_3\}$; $\underset{1\times 3}{\{\Delta q\}^T} = \{\Delta q_1 \Delta q_2 \Delta q_3\}$ - matrix – row components of the total external loads and their increments, respectively.

Taking into account (17), (21) and (22) the functional (23) for the finite element will take the form

$$\begin{aligned} \Pi \equiv & \underset{1\times 48}{\{\Delta\sigma_y\}^T} \int_V \underset{48\times 6}{[G]^T} \underset{6\times 24}{[B]} dV \underset{24\times 1}{\{w_y\}} - \frac{1}{2} \underset{1\times 48}{\{\Delta\sigma_y\}^T} \int_V \underset{48\times 6}{[G]^T} \underset{6\times 6}{\left[C^{\Pi}\right]^T} \underset{6\times 48}{[G]} dV \underset{48\times 1}{\{\Delta\sigma_y\}} - \\ & -\frac{1}{2} \underset{1\times 24}{\{w_y\}^T} \int_S \underset{24\times 3}{[A]^T} \underset{3\times 1}{\{\Delta q\}} dS - \underset{1\times 24}{\{w_y\}^T} \left(\int_S \underset{24\times 3}{[A]^T} \underset{3\times 1}{\{q\}} dS - \int_V \underset{24\times 6}{[B]^T} \underset{6\times 1}{\{\sigma\}} dV \right). \end{aligned} \tag{24}$$

After minimization of the functional (24) by nodal unknowns, the systems of equations $\{w_y\}^T$ are $\{\Delta\sigma_y\}^T$ obtained

$$\begin{aligned} & \frac{\partial \Pi}{\partial (\Delta\sigma_y)^T} \equiv \underset{48\times 24}{[Q]} \underset{24\times 1}{\{w_y\}} - \underset{48\times 48}{[H]} \underset{48\times 1}{\{\Delta\sigma_y\}} = 0; \\ & \frac{\partial \Pi}{\partial (w_y)^T} \equiv \underset{24\times 48}{[Q]} \underset{48\times 1}{\{\Delta\sigma_y\}} - \underset{24\times 1}{\{\Delta f_q\}} - \underset{24\times 1}{\{R\}} = 0, \end{aligned} \tag{25}$$

where $\underset{48\times 48}{[H]} = \int_V \underset{48\times 6}{[G]^T} \underset{6\times 6}{\left[C^{\Pi}\right]^T} \underset{6\times 48}{[G]} dV$; $\underset{28\times 24}{[Q]} = \int_V \underset{48\times 6}{[G]^T} \underset{6\times 24}{[B]} dV$;

$\underset{24\times1}{\{\Delta f_q\}} = \int_S \underset{24\times3}{[A]^T} \underset{3\times1}{\{\Delta q\}} dS$; $\underset{124\times1}{\{R\}} = \int_S \underset{124\times3}{[A]^T} \underset{3\times1}{\{q\}} dS - \int_V \underset{24\times6}{[B]^T} \underset{6\times1}{\{\sigma\}} dV$ - Raphson's residual.

Systems of equations (25) are presented in the traditional form for the finite element method

$$\underset{72\times72}{[K]}\underset{72\times1}{\{z_y\}} = \underset{72\times1}{\{f_y\}} \tag{26}$$

where $\underset{72\times72}{[K]} = \begin{bmatrix} -\underset{48\times48}{[H]} & \underset{48\times24}{[Q]} \\ \underset{24\times48}{[Q]^T} & \underset{24\times24}{[0]} \end{bmatrix}$ - is the deformation matrix of the finite element;

$\underset{1\times72}{\{z_y\}^T} = \left\{ \underset{1\times48}{\{\Delta\sigma_y\}^T} \underset{1\times24}{\{w_y\}} \right\}$ - vector of nodal unknowns of the finite element;

$\underset{1\times72}{\{f_y\}^T} = \left\{ \underset{1\times48}{\{0\}^T} \left\{ \underset{1\times24}{\{\Delta f_q\}^T} + \underset{1\times24}{\{R\}^T} \right\} \right\}$ - vector of nodal loads.

The formation of the deformation matrix of the entire structure is performed using the traditional FEM procedure.

The resulting deformation matrix of the hexahedral finite element in the mixed formulation has a size of (72x72), which is smaller than the size of the stiffness matrix of the same finite element in the formulation of the displacement method (96x96). In addition, the advantage of the algorithm for obtaining the finite element stiffness matrix in a mixed formulation is to provide stress compatibility between elements not only at nodal points, but also on the faces of adjacent elements.

Conclusion
The mixed functional used to obtain the deformation matrix of a bulk finite element allows us to take into account the residual at the loading step as a difference in the numerical values of the external and internal forces for the previous loading steps, as well as to take into account the boundary stress conditions.

Acknowledgements
The study was carried out with the financial support of the Russian Foundation for Basic Research and the Administration of the Volgograd Region in the framework of the research project No. 18-41-340007 r_ a.

References
[1] Novozhilov V V 1962 *Theory of thin shells* (Leningrad: Sudpromgiz) p 432
[2] Sedov L I 1976 *Continuum Mechanics Vol 1* (Moscow: Nauka) p 536
[3] Grigolyuk E I and Lopanitzin E A 2004 *Finite deflections, stability and supercritical behavior of thin shallow shells* (Moscow: Moscow State Technological University MAMI) p 162
[4] Petrov V V 2017 *Nonlinear incremental construction mechanics* (Moscow: Infra-Engineering) p 479
[5] Petrov V V 2013 Calculation of flexible inhomogeneous shallow shells made of a physically nonlinear material *Collection of scientific works of VolGAS* pp 373 - 377
[6] Badriev I B, Makarov M V and Paymushin V N 2012 Solvability of physically and geometrically nonlinear problem of three-layer plates with transversally soft filler *Izv. Higher Educational. Mathematics* 2 pp 171-183
[7] Malinin N N 1975 *Applied theory of plasticity and creep* (Moscow: Mechanical Engineering) p 399

[8] Bathe K J 2010 *Finite element Methods* (Moscow: Fizmatlit) p 1022

[9] Golovanov A I, Tuleneva O N and Shigabutdinov A F 2006 *Finite element Method in statics and dynamics of thin-walled structures* (Moscow: Fizmatlit) p 391

[10] Klochkov Yu V, Nikolaev A P, Vakhnina O V and Kiseleva T A 2016 Analysis of the stress-strain state of the fuselage fragment in the form of a thin shell using a triangular finite element with Lagrange multipliers *Izv. Higher Educational. Aviation* **3** pp 20–26.

[11] Gureeva N A 2007 Eight-node volumetric finite element of the shell of rotation with unknown stresses and displacements in the nodes *Izv. Higher Educational. Construction* **4** pp 33–39

[12] Ignatiev V A and Ignatiev A V 2006 Mixed form of FEM in problems of construction mechanics *Construction mechanics and calculation of structures* **1** pp 59-64.

[13] Myoung-Gyu Lee and Chung-Souk Han 2012 An explicit final element approach with paten projection technique for strain gradient plasticity formulas *Computational Mechanics* **2** pp 171-183

[14] Beirao L, Veiga Da, Brezzi F, and Marini L D 2013 Virtual Elements for linear elasticity problems *SIAM Journal on Numerical Analysis* **2** pp 794-812.

[15] Gain A L, Talischi C and Paulino G H 2014 On the Virtual Element Method for three-dimensional linear elasticity problems on arbitrary polyhedral meshes *Computer Methods in Applied Mechanics and Engineering* pp 132-160.

[16] Beirao L, Lovadina C and Mora D 2015 A Virtual Element Method for elastic and inelastic problems on polytope meshes *Computer Methods in Applied Mechanics and Engineering* pp 327-346.

[17] Aldakheel F, Hudobivnik B and Wriggers P 2018 A Virtual element formulation for phase-field modeling of ductile fracture *Computational Engineering.*

[18] Artioli E, Veiga L B D, Lovadina C and Sacco E. 2017 Arbitrary order 2d virtual elements for polygonal meshes: Part ii, inelastic problem *Computational Mechanics* pp 643-657.

Application of a triangular finite element with Lagrange correction factors in calculations of thin shells of ellipsoidal type

Yu V Klochkov[1*], A P Nikolaev[1], O V Vakhnina[1], T A Sobolevskaya[1] and M Yu Klochkov[2]

[1]Volgograd State Agrarian University, Volgograd, Russia
[2]Lomonosov Moscow state University, Moscow, Russia

Email: klotchkov@bk.ru

Abstract. The study of the stress-strain state of a thin shell of an ellipsoidal type was carried out. The finite element method was used as a research tool using the triangular fragment of the middle surface as the discretization element. The nodal variable parameters of the triangular element of discretization selected the components of the displacement vector, their first and second derivatives with respect to global curvilinear coordinates. Fifth-degree polynomials were used as polynomial functions. To improve the consistency of the triangular element of discretization on the boundaries of adjacent elements, the Lagrange correction factors in the nodes entered in the midpoints of the sides of the triangular element were additionally used as unknowns. The analysis of the results of the calculation of the thin shell of the ellipsoidal type showed high efficiency of using the modified triangular finite element.

1. Introduction

Thin-walled elements of many building and engineering constructions have the shape of a triaxial ellipsoid or its fragments. The study of the stress-strain state (SSS) of a thin shell of ellipsoidal type requires the preliminary solution of the problem of parametrization of its middle surface. In monograph [1], it is proposed to set the surface of a triaxial ellipsoid using parametric dependencies. However, when using numerical calculation methods [2-8], it is preferable to use parameters that have convenient geometric interpretation for fixing, for example, the coordinates of the discrete grid points on the ellipsoid surface. To this end, this article proposes to use the formula of the radius vector, which is a function of two global curvilinear coordinates: the axial coordinate and the angle, measured from the vertical axis parallel to the axis in a plane perpendicular to the axis. As a tool for investigating the SSS of a triaxial ellipsoid, it is proposed to apply a numerical finite element method (FEM) using the triangular fragment of the middle surface as a discretization element with high-order interpolation polynomials. To improve the consistency of the triangular discretization element at the boundaries of adjacent elements, it is proposed to additionally use Lagrange multipliers at the nodes entered in the middle of the sides of the triangular element as unknown.

2. Materials and methods

2.1. The geometry of a triaxial ellipsoid

The radius vector defining the surface of a triaxial ellipsoid in a cartesian coordinate system is determined by the expression

$$\vec{R}^0 = x\vec{i} + y\vec{j} + c\sqrt{1-(x/a)^2-(y/b)^2}\,\vec{k}, \qquad (1)$$

where a, b, c - the lengths of the semiaxes of the triaxial ellipsoid.

Vectors of the local basis tangent to the surface of the ellipsoid can be obtained by differentiating (1) with respect to x, and y

$$\vec{a}_1^0 = \vec{R}_{,x}^0 = \vec{i} - \frac{cx}{a^2\sqrt{1-(x/a)^2-(y/b)^2}}\vec{k};$$
$$\vec{a}_2^0 = \vec{R}_{,y}^0 = \vec{j} - \frac{cy}{b^2\sqrt{1-(x/a)^2-(y/b)^2}}\vec{k}. \tag{2}$$

It is not quite convenient to use formulas (2), since it is necessary to observe the condition

$$1-(x/a)^2-(y/b)^2 > 0. \tag{3}$$

Therefore, this paper proposes the use of the following expression for the radius vector of a triaxial ellipsoid

$$\vec{R}^0 = x\vec{i} + r(\theta)\sin\theta\vec{j} + r(\theta)\cos\theta\vec{k}, \tag{4}$$

where θ - is angle measured from the vertical counterclockwise in a plane perpendicular to the axis Ox.

The function $r(x,\theta)$ included in (4) is determined by the following formula

$$r(x,\theta) = \frac{bc\sqrt{1-(x/a)^2}}{\sqrt{b^2\cos^2\theta + c^2\sin^2\theta}}. \tag{5}$$

The tangential vectors of the local basis of an arbitrary point on the surface of the ellipsoid can be determined by differentiating (4) with respect by x and θ

$$\vec{a}_1^0 = \vec{i} + r_{,x}\sin\theta\,\vec{j} + r_{,x}\cos\theta\,\vec{k};$$
$$\vec{a}_2^0 = (r_{,\theta}\sin\theta + r\cos\theta)\vec{j} + (r_{,\theta}\cos\theta - r\sin\theta)\vec{k}, \tag{6}$$

where $r_{,x} = \partial r(x,\theta)/\partial x = \frac{-x}{\sqrt{1-(x/a)^2}} \cdot \frac{bc}{\sqrt{b^2\cos^2\theta + c^2\sin^2\theta}};$

$r_{,\theta} = \partial r(x,\theta)/\partial\theta = \frac{-bc\sqrt{1-(x/a)^2}\sin 2\theta(c^2-b^2)}{2(b^2\cos^2\theta + c^2\sin^2\theta)^{3/2}};\quad r \equiv r(x,\theta).$

Orth normal to the surface of the ellipsoid is determined by the vector product

$$\vec{a}^0 = \vec{a}_1^0 \times \vec{a}_2^0/\sqrt{a^0} = \left[(-r_{,x}\cdot r)\vec{i} + (r\sin\theta - r_{,\theta}\cos\theta)\vec{j} + (r_{,\theta}\sin\theta + r\cos\theta)\vec{k}\right]/\sqrt{a^0}, \tag{7}$$

where $a^0 = (\vec{a}_1^0\cdot\vec{a}_1^0)(\vec{a}_2^0\cdot\vec{a}_2^0) - (\vec{a}_1^0\cdot\vec{a}_2^0)^2$ - determinant of the metric tensor.

Based on (6) and (7), we can form a direct and inverse matrix dependencies

$$\underset{3\times1}{\{\vec{a}^0\}} = \underset{3\times3}{[d^0]}\underset{3\times1}{\{\vec{i}\}},\quad \{\vec{i}\} = [d^0]^{-1}\{\vec{a}^0\}, \tag{8}$$

where $\underset{1\times3}{\{\vec{a}^0\}^T} = \{\vec{a}_1^0\ \vec{a}_2^0\ \vec{a}^0\};\quad \underset{1\times3}{\{\vec{i}\}^T} = \{\vec{i}\ \vec{j}\ \vec{k}\}$

Differentiation (6) and (7) by x and θ it is possible to obtain derivatives of the vectors of the local basis of an arbitrary point on the surface of the ellipsoid

$$\vec{a}^0_{1,x} = r_{,xx}\sin\theta\,\vec{j} + r_{,xx}\cos\theta\,\vec{k}; \qquad \vec{a}^0_{2,x} = \left(r_{,\theta x}\sin\theta + r_{,x}\cos\theta\right)\vec{j} + \left(r_{,\theta x}\cos\theta - r_{,x}\sin\theta\right)\vec{k};$$

$$\vec{a}^0_{,x} = \vec{i}\left(\left(-r_{,xx}\cdot r - \left(r_{,x}\right)^2\right)\frac{1}{\sqrt{a^0}} - \left(r_{,x}\cdot r\right)\left(\frac{1}{\sqrt{a^0}}\right)_{,x}\right) +$$

$$+\vec{j}\left(\left(r_{,x}\sin\theta - r_{,\theta x}\cos\theta\right)\frac{1}{\sqrt{a^0}} + \left(r\sin\theta - r_{,\theta}\cos\theta\right)\left(\frac{1}{\sqrt{a^0}}\right)_{,x}\right) +$$

$$+\vec{k}\left(\left(r_{,\theta x}\sin\theta + r_{,x}\cos\theta\right)\frac{1}{\sqrt{a^0}} + \left(r_{,\theta}\sin\theta + r\cos\theta\right)\left(\frac{1}{\sqrt{a^0}}\right)_{,x}\right);$$

$$\vec{a}^0_{1,\theta} = \left(r_{,x\theta}\sin\theta + r_{,x}\cos\theta\right)\vec{j} + \left(r_{,x\theta}\cos\theta - r_{,x}\sin\theta\right)\vec{k};$$

$$\vec{a}^0_{2,\theta} = \left(r_{,\theta\theta}\sin\theta + 2r_{,\theta}\cos\theta - r\sin\theta\right)\vec{j} + \left(r_{,\theta\theta}\cos\theta - 2r_{,\theta}\sin\theta - r\cos\theta\right)\vec{k};$$

$$\vec{a}^0_{,\theta} = \vec{i}\left(\left(-r_{,x\theta}\cdot r - r_{,x}\cdot r_{,\theta}\right)\frac{1}{\sqrt{a^0}} - r_{,x}\cdot r\left(\frac{1}{\sqrt{a^0}}\right)_{,\theta}\right) +$$

$$+\vec{j}\left(\left(2r_{,\theta}\sin\theta + r\cos\theta - r_{,\theta\theta}\cos\theta\right)\frac{1}{\sqrt{a^0}} + \left(r\sin\theta - r_{,\theta}\cos\theta\right)\left(\frac{1}{\sqrt{a^0}}\right)_{,\theta}\right) +$$

$$+\vec{k}\left(\left(r_{,\theta\theta}\sin\theta + 2r_{,\theta}\cos\theta - r\sin\theta\right)\frac{1}{\sqrt{a^0}} + \left(r_{,\theta}\sin\theta + r\cos\theta\right)\left(\frac{1}{\sqrt{a^0}}\right)_{,\theta}\right). \tag{9}$$

Derivatives of the vectors of the local basis (9) can be represented in a matrix form

$$\underset{3\times1}{\left\{\vec{a}^0_{,x}\right\}} = \underset{3\times3}{\left[d^0_x\right]}\underset{3\times1}{\left\{\vec{i}\right\}}, \quad \underset{3\times1}{\left\{\vec{a}^0_{,\theta}\right\}} = \underset{3\times3}{\left[d^0_\theta\right]}\underset{3\times1}{\left\{\vec{i}\right\}}, \tag{10}$$

where $\underset{1\times3}{\left\{\vec{a}^0_{,x}\right\}^T} = \left\{\vec{a}^0_{1,x}\ \vec{a}^0_{2,x}\ \vec{a}^0_{,x}\right\}$; $\underset{1\times3}{\left\{\vec{a}^0_{,\theta}\right\}^T} = \left\{\vec{a}^0_{1,\theta}\ \vec{a}^0_{2,\theta}\ \vec{a}^0_{,\theta}\right\}$

Taking into account (8) from (10), it is possible to obtain matrix relations that allow us to represent the derivatives of the vectors of the local basis in the local basis of the same point on the surface of the ellipsoid

$$\begin{gathered}\underset{3\times1}{\left\{\vec{a}^0_{,x}\right\}} = \left[d^0_x\right]\left[d^0\right]^{-1}\left\{\vec{a}^0\right\} = \left[m^0\right]\left\{\vec{a}^0\right\}, \\ \underset{3\times1}{\left\{\vec{a}^0_{,\theta}\right\}} = \left[d^0_\theta\right]\left[d^0\right]^{-1}\left\{\vec{a}^0\right\} = \left[n^0\right]\left\{\vec{a}^0\right\}.\end{gathered} \tag{11}$$

In the process of deformation point M^0 the middle surface of the ellipsoidal shell will move to a points M, a point $M^{0\zeta}$, distance from the point M^0 along the distance normal ζ, take a new position M^ζ. Point moves M^0 and $M^{0\zeta}$ in new positions M and M^ζ will be characterized by displacement vectors $\vec{v}$ and $\vec{V}$ respectively. Vector of the displacement of points M^0 and its derivatives by x and θ determined by the expressions

$$\vec{v} = u\vec{a}^0_1 + v\vec{a}^0_2 + w\vec{a}^0; \quad \vec{v}_{,x} = t^1_x\vec{a}^0_1 + t^2_x\vec{a}^0_2 + t_x\vec{a}^0; \quad \vec{v}_{,\theta} = t^1_\theta\vec{a}^0_1 + t^2_\theta\vec{a}^0_2 + t_\theta\vec{a}^0, \tag{12}$$

IOP Conf. Series: Materials Science and Engineering **675** (2019) 012060 doi:10.1088/1757-899X/675/1/012060

where u, v - tangential, w - normal components of the displacement vector; $t_x^1, t_x^2, t_x, t_\theta^1, t_\theta^2, t_\theta$ - polynomials containing the first derived components of the displacement vector.

Point displacement vector $M^{0\zeta}$ based on the hypothesis of a direct normal, can be determined by the formula

$$\vec{V} = \vec{v} + \zeta\left(\vec{a} - \vec{a}^0\right), \tag{13}$$

where $\vec{a}$ - ort normal to the surface of the ellipsoid in the deformed state.

Radius vectors defining points $M^{0\zeta}$, M and M^ζ determined by the appropriate expressions

$$\vec{R}^{0\zeta} = \vec{R}^0 + \zeta\vec{a}^0; \quad \vec{R} = \vec{R}^0 + \vec{v}; \quad \vec{R}^\zeta = \vec{R}^{0\zeta} + \vec{V}. \tag{14}$$

Differentiating (14) by x and θ can get the corresponding basis vectors

$$\vec{g}_x^0 = \vec{R}_{,x}^{0\zeta}; \quad \vec{g}_\theta^0 = \vec{R}_{,\theta}^{0\zeta}; \quad \vec{g}_x = \vec{R}_{,x}^\zeta; \quad \vec{g}_\theta = \vec{R}_{,\theta}^\zeta, \tag{15}$$

scalar products which determine the components of the metric tensor of the original and deformed states

$$g_{xx}^0 = \vec{g}_x^0 \cdot \vec{g}_x^0; \quad g_{\theta\theta}^0 = \vec{g}_\theta^0 \cdot \vec{g}_\theta^0; \quad g_{xx} = \vec{g}_x \cdot \vec{g}_x; \quad g_{\theta\theta} = \vec{g}_\theta \cdot \vec{g}_\theta; \quad g_{x\theta}^0 = \vec{g}_x^0 \cdot \vec{g}_\theta^0; \quad g_{x\theta} = \vec{g}_x \cdot \vec{g}_\theta. \tag{16}$$

Deformations at the point of the ellipsoidal shell spaced at a distance from the middle surface ζ, calculated using continuum mechanics formulas

$$\varepsilon_{xx}^\zeta = \left(g_{xx} - g_{xx}^0\right)/2; \quad \varepsilon_{\theta\theta}^\zeta = \left(g_{\theta\theta} - g_{\theta\theta}^0\right)/2; \quad \varepsilon_{x\theta}^\zeta = \left(g_{x\theta} - g_{x\theta}^0\right)/2. \tag{17}$$

2.2. Triangle sampling element

The median surface of the ellipsoidal shell is approximated by an ensemble of triangular discretization elements with nodes i, j, k. For convenience of numerical integration over the area, a triangular fragment of the median surface of the ellipsoid is mapped onto a right-angled triangle with unit lengths in the local coordinate system $0 \le \xi, \eta \le 1$.

Considering that the deformations (17) are functions of the components of the displacement vector, as well as their first and second derivatives by x and θ in addition to the displacement vector components, the first and second derivatives with respect to the global x, θ or local ξ, η coordinates

$$\underset{1\times54}{\left\{U_y^G\right\}^T} = \left\{ \underset{1\times18}{\left\{u_y^G\right\}^T} \underset{1\times18}{\left\{v_y^G\right\}^T} \underset{1\times18}{\left\{w_y^G\right\}^T} \right\}; \tag{18}$$

$$\underset{1\times54}{\left\{U_y^L\right\}^T} = \left\{ \underset{1\times18}{\left\{u_y^L\right\}^T} \underset{1\times18}{\left\{v_y^L\right\}^T} \underset{1\times18}{\left\{w_y^L\right\}^T} \right\}, \tag{19}$$

where

$$\underset{1\times18}{\left\{q_y^G\right\}^T} = \left\{q^i q^j q^k q_{,x}^i q_{,x}^j q_{,x}^k q_{,\theta}^i q_{,\theta}^j q_{,\theta}^k q_{,xx}^i q_{,xx}^j q_{,xx}^k q_{,\theta\theta}^i q_{,\theta\theta}^j q_{,\theta\theta}^k q_{,x\theta}^i q_{,x\theta}^j q_{,x\theta}^k\right\};$$

$$\underset{1\times18}{\left\{q_y^L\right\}^T} = \left\{q^i q^j q^k q_{,\xi}^i q_{,\xi}^j q_{,\xi}^k q_{,\eta}^i q_{,\eta}^j q_{,\eta}^k q_{,\xi\xi}^i q_{,\xi\xi}^j q_{,\xi\xi}^k q_{,\eta\eta}^i q_{,\eta\eta}^j q_{,\eta\eta}^k q_{,\xi\xi}^i q_{,\xi\xi}^j q_{,\xi\xi}^k\right\},$$

here q is understood as a component of the displacement vector u, v or w.

As polynomial functions of a triangular discretization element, it is proposed to use fifth-degree full polynomials

$$q = \underset{1\times 18}{\{\varphi\}^T} \underset{18\times 1}{\{q_y^L\}}. \tag{20}$$

Differentiating (20) by x and θ you can get the first and second derivatives of the displacement vector components

$$\begin{aligned}
q_{,x} &= \left(\{\varphi_{,\xi}\}^T \xi_{,x} + \{\varphi_{,\eta}\}^T \eta_{,x}\right)\{q_y^L\};\\
q_{,\theta} &= \left(\{\varphi_{,\xi}\}^T \xi_{,\theta} + \{\varphi_{,\eta}\}^T \eta_{,\theta}\right)\{q_y^L\};\\
q_{,xx} &= \left(\{\varphi_{,\xi\xi}\}^T (\xi_{,x})^2 + 2\{\varphi_{,\xi\eta}\}^T \xi_{,x}\eta_{,x} + \{\varphi_{,\eta\eta}\}^T (\eta_{,x})^2\right)\{q_y^L\};\\
q_{,\theta\theta} &= \left(\{\varphi_{,\xi\xi}\}^T (\xi_{,\theta})^2 + 2\{\varphi_{,\xi\eta}\}^T \xi_{,\theta}\eta_{,\theta} + \{\varphi_{,\eta\eta}\}^T (\eta_{,\theta})^2\right)\{q_y^L\};\\
q_{,x\theta} &= \left(\{\varphi_{,\xi\xi}\}^T \xi_{,x}\xi_{,\theta} + \{\varphi_{,\eta\eta}\}^T \eta_{,x}\eta_{,\theta} + \{\varphi_{,\xi\eta}\}^T (\xi_{,\theta}\eta_{,x} + \xi_{,x}\eta_{,\theta})\right)\{q_y^L\}.
\end{aligned} \tag{21}$$

3. Results

To improve the compatibility of the considered element of discretization, it is proposed to introduce three additional nodes 1, 2 and 3, one in each of the midpoints. If in additional nodes we calculate the derivatives of the normal component of the displacement vector along the normals to the sides of the triangular elements, then in the values of these derivatives, calculated from the side of the neighboring elements, there will be a jump, which is caused by the incompatibility of this element of discretization. To eliminate the noted deficiency, you can enter Lagrange multipliers as nodal unknown additional nodes λ_l $(l = 1,2,3)$. Then the following equality holds

$$\left(\frac{\partial w_l}{\partial S_{nl}} + \frac{\partial w_l'}{\partial S_{nl}'}\right)\lambda_l = 0, \tag{22}$$

where the bar indicates the values related to the adjacent sample element.

For a single triangular finite element, you can write the following expression

$$\frac{\partial w_1}{\partial S_{n1}}\lambda_1 + \frac{\partial w_2}{\partial S_{n2}}\lambda_2 + \frac{\partial w_3}{\partial S_{n3}}\lambda_3 = 0. \tag{23}$$

Derivatives $\frac{\partial w_l}{\partial S_{nl}}$ can be expressed in terms of derivatives of the normal component of the displacement vector in global coordinates x and θ

$$\frac{\partial w_l}{\partial S_{nl}} = \frac{\partial w_l}{\partial x}\cos\alpha_l + \frac{\partial w_l}{\partial \theta}\cos\beta_l, \tag{24}$$

where α_l and β_l - angles between the normals to the side of the triangular discretization element and the tangential vectors of the local basis $\vec{a}_1^0$, $\vec{a}_2^0$ in additional node l.

Cosines of corners α_l and β_l can be calculated by the formulas

$$\cos\alpha_l = \frac{\vec{a}_{1l}^0 \cdot \vec{n}_l}{|\vec{a}_{1l}^0| \cdot |\vec{n}_l|}; \quad \cos\beta_l = \frac{\vec{a}_{2l}^0 \cdot \vec{n}_l}{|\vec{a}_{2l}^0| \cdot |\vec{n}_l|}. \tag{25}$$

Included in (25) the normal vectors to the sides of the triangular element in the additional nodes can be determined by the vector product

$$\vec{n}_l = \vec{a}_l^0 \times \vec{t}_l, \tag{26}$$

where $\vec{a}_l^0$ - ort normal to the surface of the ellipsoid in the additional node l; $\vec{t}_l$ - vector tangent to the side of a triangular discretization element, defined by the formula

$$\vec{t}_l = \left(\vec{R}^0_{,x}\right)_l + \left(\vec{R}^0_{,\theta}\right)_l \left(\partial\theta/\partial x\right)_l . \tag{27}$$

Derivatives of the normal components of the displacement vector along the normals to the sides of the triangular element in the additional nodes, taking into account (24), can be expressed in terms of a column of nodal unknowns $\frac{\partial w_l}{\partial S_{nl}} = \left(\left(\{\varphi_{,\xi}\}^T \cdot \xi_{,x} + \{\varphi_{,\eta}\}^T \cdot \eta_{,x}\right)_l \cos\alpha_l + \left(\{\varphi_{,\xi}\}^T \cdot \xi_{,\theta} + \{\varphi_{,\eta}\}^T \cdot \eta_{,\theta}\right)_l \cos\beta_l\right)\{w_y^L\}$ or in matrix form

$$\frac{\partial w_l}{\partial S_{nl}} = \underset{1\times54}{\{\chi_l\}^T}\ \underset{54\times54}{[T]}\ \underset{54\times1}{\{U_y^G\}} \tag{28}$$

where $[T]$ - transition matrix from the nodal unknowns column in the local coordinate system $\{U_y^L\}$ to the column of unknowns in the global coordinate system $\{U_y^G\}$.

Relation (23) with regard to (28) can be represented in the matrix form

$$\underset{1\times3}{\{\lambda\}^T} \underset{3\times54}{\begin{bmatrix} \{\chi_1\}^T[T] \\ \{\chi_2\}^T[T] \\ \{\chi_3\}^T[T] \end{bmatrix}} \underset{54\times1}{\{U_y^G\}} = \underset{1\times3}{\{\lambda\}^T} \underset{3\times54}{[B]} \underset{54\times1}{\{U_y^G\}} = 0. \tag{29}$$

The Lagrange functional for the triangular element of discretization, taking into account (29), will be written as

$$\Phi_L = \int_V \{\varepsilon^{\varsigma}_{\alpha\beta}\}^T \{\sigma^{\alpha\beta}\}^T dV - \int_F \{U\}^T \{P\} dF + \{\lambda\}^T [B]\{U_y^G\}, \tag{30}$$

where $\{\varepsilon^{\varsigma}_{\alpha\beta}\}^T = \{\varepsilon_{11}\ \varepsilon_{22}\ 2\varepsilon_{12}\}$; $\{\sigma^{\alpha\beta}\}^T = \{\sigma^{11}\ \sigma^{22}\ \sigma^{12}\}$ - matrix rows of deformations and stresses; at a point at a distance from the middle surface ζ ; $\{U\}^T = \{u\ v\ w\}$- line component of the displacement vector of the middle surface point; $\{P\}^T = \{P_{11}\ P_{22}\ P_{12}\}$ - external surface load string .

Minimizing the functionality (30) by $\{U_y^G\}^T$ and $\{\lambda\}^T$ can get a system of equations, from which you then assemble the stiffness matrix $[K]$ and nodal force column $\{R\}$ triangular bin

$$\underset{57\times57}{[K]} = \begin{bmatrix} \underset{54\times54}{[K']} & \underset{54\times3}{[B]^T} \\ \underset{3\times54}{[B]} & \underset{3\times3}{[0]} \end{bmatrix}; \qquad \underset{1\times57}{\{R\}^T} = \left\{ \underset{1\times54}{\{R'\}^T}\ \underset{1\times3}{\{000\}^T} \right\}, \tag{31}$$

where $[K']$and $\{R'\}$- stiffness matrix and nodal force column of a standard triangular element without Lagrange multipliers λ_l .

Further construction of the stiffness matrix and the column of nodal forces for the calculation of the ellipsoidal shell is performed in the standard way for FEM [9–19].

4. Discussion

As an example, the SSS of a triaxial ellipsoid loaded with an internal intensity pressure was investigated q. Due to the presence of symmetry planes, 1/8 of the shell was calculated.

The calculations were carried out in two variants: in the first variant, a triangular element without Lagrange multipliers with a stiffness matrix was used as a discretization element 54×54; in the second variant, a triangular discretization element with Lagrange multipliers with a stiffness matrix was used 57×57. The results of the variant calculation showed significant advantages of the second version of the calculation in the accuracy of finite-element solutions.

5. Conclusion

Based on the analysis of the calculation results, we can conclude that the triangular discretization element with Lagrange multipliers 57×57 has the advantage over the standard triangular finite element with the size of the stiffness matrix 54×54 in the problems of analyzing the SSS of thin shells of ellipsoidal type. A promising direction of development of the developed algorithm is the use of the interpolation procedure for the components of the displacement vector as components of vector fields [20].

Acknowledgements

The study was carried out with the financial support of the Russian Foundation for Basic Research and the Administration of the Volgograd Region in the framework of the research project No. 18-41-340007 r_ a.

References

[1] Krivoshapko S N and Ivanov V N 2010 *Encyclopedia of analytical surfaces* (Moscow: Publishing house "LIBROKOM") p 560

[2] Storozhuk E A and Komarchuk S M 2018 Stress Distribution Near a Circular Hole in a Flexible Orthotropic Cylindrical Shell of Elliptical Cross-Section *International Applied Mechanics* **54(6)** pp 687-694

[3] Storozhuk E A, Chernyshenko I S and Yatsura A V 2018 Stress–Strain State Near a Hole in a Shear-Compliant Composite Cylindrical Shell with Elliptical Cross-Section *International Applied Mechanics* **54(5)** pp 559-567

[4] Badriev I B, Paimushin V N 2017 Refined models of contact interaction of a thin plate with positioned on both sides deformable foundations *Lobachevskii Journal of Mathematics* **38 (5)** pp 779-793

[5] Pyatikrestovsky K P and Sokolov B S 2018 Nonlinear analysis of statically indeterminate wooden structures and optimization of cross section dimensions of dome ribs *International Journal for Computational Civil and Structural Engineering* **14(4)** pp 130-139

[6] Kayumov R A 2017 Postbuckling behavior of compressed rods in an elastic medium *Mechanics of Solids* **52(5)** pp 575-580

[7] Kozlov V A 2018 Stress-strain of elements of bridge structures with a varying thickness of walls along the length *Russian Journal of Building Construction and Architecture* **1(37)** pp 67-80

[8] Karpov V V, Ignatev O V and Semenov A A 2017 The stress-strain state of ribbed shell structures *Magazine of Civil Engineering* **74 (6)** pp 147-160

[9] Agapov V P and Golovanov R O 2018 Comparative analysis of the simplest finite elements of plates in bending *Advances in Intelligent Systems and Computing* **692** pp 1009-1016

[10] Solodovnikov A S and Sheshenin S V 2017 Numerical study of strength properties for a composite material with short reinforcing fibers *Moscow University Mechanics Bulletin* **72(4)** pp 94-100

[11] Agapov V P, Golovanov R O and Aidemirov K R 2017 Calculation of load-bearing capacity of prestressed reinforced concrete trusses by the finite element method *IOP Conference Series: Earth and Environmental Science* **90** 012018

[12] Zheleznov L P, Kabanov V V and Boiko D V 2014 Nonlinear deformation and stability of discretely reinforced elliptical cylindrical shells under transverse bending and internal pressure *Russian Aeronautics* **57(2)** pp 118-126

[13] Galishnikova V V and Pahl P Ja 2018 Constrained construction of planar delaunay triangulations without flipping *Construction mechanics of engineering structures and structures* 14(2) pp 154-174

[14] Ignatyev A V, Ignatyev V A and Onischenko E V 2016 Analysis of systems with unilateral constraints through the finite element method in the form of a classical mixed method *Procedia Engineering* **150** pp 1754-1759

[15] Zheleznov L P, Kabanov V V and Boiko D V 2018 Nonlinear Deformation and Stability of Discrete-Reinforced Elliptical Cylindrical Composite Shells under Torsion and Internal Pressure *Russian Aeronautics* **61(2)** pp 175-182

[16] Kositsyn S B, Fedorov V S and Akulich V Yu 2018 Geotechnical projection of the influence of the construction of the designed metropolitene tunnel by the method of shield passage on the sedimentation of the earth's surface *Russian Journal of Building Construction and Architecture* **1(37)** pp 81-91

[17] Matveev A D and Grishanov A N 2018 Multigrid finite elements in the calculations of multilayer cylindrical shells *Siberian Journal of Science and Technology* **19 (1)** pp 27-36

[18] Nguyen N and Waas A M 2016 Nonlinear, finite deformation, finite element analysis *Zeitschrift für angewandte Mathematik und Physik* **67:35**

[19] Hanslo Peter, Larson Mats G and Larson Fredrik 2015 Tangential differential calculus and the finite element modeling of a large deformation elastic membrane problem *Comput. Mech* 56(1) pp 87-95

[20] Klochkov Y V, Nikolaev A P and Vakhnina O V 2016 Calculation of rotation shells using finite triangular elements with Lagrange multipliers in variative approximation of displacements *Journal of Machinery Manufacture and Reliability* **45(1)** pp 51-58

IOP Conf. Series: Materials Science and Engineering **675** (2019) 012061 doi:10.1088/1757-899X/675/1/012061

Control method of high-precision soft docking of moving objects using low-accurate values of relative speed and distance

I A Mukhametzyanov, O I Chekmareva*

Peoples' Friendship University of Russia (RUDN University), Moscow, Russia.

E-mail: spacefrogling@mail.ru

Abstract. An analytical expression is constructed for the main vector of control forces, providing the pursuit of an unpredictably moving object with the purpose of shockless high-precision docking with it or its shockless capture in a finite period of time using low-accurate values of the measured relative velocity and distance. The constructed control ensures the pursuit of the goal on the principle of proportional navigation, brings the pursuing point in a finite time to the "initial position", which provides the possibility of reducing the distance to the target, and then ensures the subsequent sliding movement of a point from the "initial position" along the line of discontinuity to the origin of the phase coordinate system. To solve the problem with large relative measurement errors, it was proposed to construct a continuous part of the control force, which provides the process of sliding movement of the representative point along the line of discontinuity, as a function, depending on the current values of the relative velocity and distance. The article also analyzes the issue related to the expansion of the class of control forces by introducing some positive parameter in the expression for the continuous component of the control force. The choice of this parameter affects the duration of the rapprochement of the pursued and pursuing points.

Keywords: control forces, shockless, high-precision docking, finite time, quasi-speed.

1. Introduction

At this stage of development of modern civilization, the total automation of all management processes takes place in technology and other areas of human activity. A major obstacle here is the presence of unpredictable random disturbances. In particular, in technology, this leads to the appearance of shock loads at the moment of contact, dangerous for both controlled and pursued objects. Therefore, an extremely important task is the selection of such a control algorithm that allows you to perform the task of docking the objects or capturing the targets, avoiding an impact, in a finite period of time. Every year, a large number of works devoted to the problem of automatic control in conditions of uncertainty appear. We list only a small part of them as an example. In [1], a control scheme for a hybrid robot to capture a non-cooperative target in space is proposed. The study [2] sets up a class of absolutely continuous robust controllers which seem to be effective in counteracting both uncertain dynamics and unbounded disturbances. In [3], an adaptive control scheme is constructed which guarantees the asymptotic satisfaction of the control objective in the presence of bounded parametric uncertainties. In [4], a robust adaptive tracking controller is proposed to control the nonholonomic mobile manipulator. In [5], robust task-space motion control of a mobile manipulator using a nonlinear control with an uncertainty estimator is constructed. The work [6] deals with orbit and attitude control of a chaser spacecraft when approaching in the vicinity of the target spacecraft; then the impact of the robot manipulator dynamics on the attitude motion and the associated control effort to keep the attitude stable during the manipulator's operation is analyzed. In [7], the control of flexible-joint robotic manipulators while avoiding actuator saturation is investigated. The paper [8] proposes a finite time synergetic control scheme which is synthesized with synergetic theory and a terminal attractor technique for controlling robot manipulators; this control scheme has the characteristics of finite time convergence and chattering free phenomena. In [9], a voltage-based sliding mode control is presented to control the position of rigid-link flexible-joint serial robot manipulator in the presence of structured and unstructured uncertainties. The study [10] investigates a novel optimal adaptive radial basis function neural network control for a class of multiple-input-multiple-output nonlinear robot manipulators with uncertain dynamics. In [11] a control algorithm of the movement of robotic manipulator by the method of Lyapunov function with a significant influence of the disturbing moments in operating conditions is obtained. The work [12] presents a robust interaction control approach for underwater vehicle manipulator systems. The paper [13] investigates the task-space prescribed performance tracking control problem of free-floating space manipulators with kinematic and dynamic uncertainty. In [14], a model reference adaptive state-dependent Riccati equation control of nonlinear uncertain systems is proposed and applied for regulation and tracking of free-floating space manipulator. The work [15] demonstrates the capability of the model free controller to be able to command a highly nonlinear and uncertain system. Neither of the above studies (the list of which can be continued) does not aim to avoid an impact at the moment of contact with the target. In [1], it was noted that the problem of shock loads exists and they are caused by errors in measuring. In [16], an attempt was made to construct the control of a collision-free contact of two bodies, but none of applicable in practice algorithm was constructed. Thus, in none of the well-known studies the problem is solved in the most general formulation in which it is posed and solved in this work.

The aim of the research is to build the most autonomous ("self-tuning") control that allows to avoid the appearance of the impact contact forces and to achieve the target in finite time in conditions of uncertainty. In this formulation, all the problems arising in the control process are completely solved.

The results can be used to build control systems for the berthing of ships; landing of aircraft; landing, docking or berthing of spacecraft. Also, this control can be adapted for robotic manipulators with the purpose of shockless capture of various objects, including space debris.

2. Materials and methods

2.1 Formulation of the problem

Consider a controlled body with the center of mass at point C. The task is to construct an analytical expression for the main vector of control forces, providing the pursuit of an unpredictably moving object with the purpose of shockless high-precision docking with it or its shockless capture in a finite period of time using low-accurate values of the measured relative velocity and distance.

The terms "high-precision" and "low-accurate" are estimated by the value of the relative error of meters.

2.2 The vector of control force, providing movement on the principle of proportional navigation

We construct the force $\bar{F}$, providing the movement of point C after point O according to the principle of proportional navigation in the form

$$\bar{F} = m\tilde{b}(\bar{\omega} \times \bar{V}), \tag{1}$$

where m and $\bar{V}$ – the mass and absolute velocity of point C, $\bar{\omega}$ – absolute angular velocity of the line of sight $\overline{CO}$, $\tilde{b}$ is a positive coefficient of proportionality.

It is assumed that the following conditions are to be met:

$$(\bar{s} \cdot \dot{\bar{s}}) < 0, \quad |\bar{V}| > |\bar{V}_0|, \tag{2}$$

where $\bar{V}_0$ – the absolute speed of the point O, $\bar{s} = \overline{CO}$.

Note that $\bar{s}$ is a law of the relative motion of the point C in the moving coordinate system with the origin at the point O, one axis of which is directed along the vector $\overline{CO}$, the other axis $\dot{\bar{s}}$ is orthogonal to the vector $\overline{CO}$.

2.3 Bringing the haunting point to the "initial position"

We define the component $\tilde{R}$ of the main vector of control forces, which ensures the shockless bringing of the point C to the point O in a finite period of time.

Note that to solve this problem, it is necessary to ensure that the conditions (2) are met. This is due to the fact that at the initial moment of time $t = 0$ the first of the conditions (2) may be unfulfilled. In this case, the point C must be brought to its initial position in a finite period of time, where this condition is met.

We show that as an initial point we can use any point of the line

$$\dot{s} + \mu s = 0 \tag{3}$$

in the phase plane $(\bar{s} \cdot \dot{\bar{s}})$, where s and $\dot{s}$ are projections of the vectors $\bar{s}$ and $\dot{\bar{s}}$ onto $\overline{CO}$, $\mu = \text{const} > 0$.

From (3) it follows that the first condition (2) is guaranteed by the choice of μ in the form

$$\mu = -\frac{\dot{s}(0)}{s(0)} > 0.$$

Therefore, the first condition from (2) in the case of

$$\dot{s}(0) + \mu s(0) \neq 0$$

can be achieved by vanishing the value

$$\tilde{V} = \dot{s}(t) + \mu s(t) \tag{4}$$

in a finite period of time.

This problem can be solved as follows. We use the basic law of dynamics to describe relative motion of the point C in the projection on the $\overline{CO}$ axis:

$$m\ddot{s}(t) = R + \tilde{U}, \tag{5}$$

where R is the projection of a part of the main vector of control forces, $\tilde{U}$ is the projection of the main vector of non-control forces, including inertial forces, on the same axis $\overline{CO}$.

Choose R in the form

$$R = -R^0 \text{sign}\tilde{V}, \tag{6}$$

where R^0 is a positive constant or piece-wise constant function satisfying the condition

$$R^0 \geq \left|\tilde{U} + m\mu\dot{s}\right| + \delta, \quad \delta = \text{const} > 0, \tag{7}$$

the choice of the magnitude of which in each specific case is carried out according to the principle of feedback on quasi-accelerations (see, for example, [17,18]).

Now, substituting (6) into (5) and taking into account the time-differentiated equality (4), we get

$$m\frac{\mathrm{d}\tilde{V}}{\mathrm{d}t} = -R^0 \text{sign}\tilde{V} + \tilde{U} + m\mu\dot{s}, \tag{8}$$

where R^0 satisfies the condition (7).

Multiplying (8) by $\tilde{V}$, we get

$$m\tilde{V}\frac{\mathrm{d}\tilde{V}}{\mathrm{d}t} = -R^0\tilde{V}\text{sign}\tilde{V} + \left(\tilde{U} + m\mu\dot{s}\right)\tilde{V}. \tag{9}$$

With the condition (7), we have

$$\frac{1}{2}\frac{\mathrm{d}\left(m\tilde{V}^2\right)}{\mathrm{d}t} \leq -\delta\left|\tilde{V}\right|,$$

and

$$\frac{\mathrm{d}\tilde{V}^2}{\left|\tilde{V}\right|} \leq -\frac{2\delta}{m}\mathrm{d}t. \tag{10}$$

Integrating (10) and taking into account $\left|\tilde{V}\right| = \sqrt{\tilde{V}^2}$, we get

$$2\left|\tilde{V}\right|\Big|_{|\tilde{V}(0)|}^{0} \leq -\frac{2\delta}{m}t\Big|_{0}^{\tilde{t}},$$

where $\tilde{t}$ is the instant of time when $\tilde{V}$ vanishes.

Hence,

$$\tilde{t} \leq \frac{m|\tilde{V}(0)|}{\delta},$$

where $m|\tilde{V}(0)|/\delta$ is the maximum possible value of the time interval for bringing of the representative point $(s, \dot{s})$ on a straight line (3).

2.4 Bringing the representative point to the position $s = 0, V = 0$

Starting from the moment $t = \tilde{t}$ we organize the process of sliding movement of the representative point along the line of discontinuity of the form

$$V(\tau) - \frac{V(\tilde{t})}{\sqrt{|s(\tilde{t})|}}\sqrt{|s(\tau)|} = 0, \quad (11)$$

where $\tau = t - \tilde{t}$. To do this, as the quasi-speed, we choose

$$\tilde{V}' = V(\tau) - \frac{V(\tilde{t})}{\sqrt{|s(\tilde{t})|}}\sqrt{|s(\tau)|}. \quad (12)$$

We represent the control force R as the sum of the continuous component

$$R_1 = \frac{mV^2}{2s}$$

and the piece-wise constant component R' of the form

$$R' = -\tilde{R}^0 \mathrm{sign}\tilde{V}', \quad (13)$$

where $\tilde{R}^0$ satisfies the condition

$$\tilde{R}^0 \geq |\tilde{U}| + \tilde{\delta}, \qquad \tilde{\delta} = \mathrm{const} > 0. \quad (14)$$

In each specific case the choice of the magnitude of $\tilde{R}^0$ is carried out according to the principle of feedback on quasi-accelerations (see, for example, [17,18]).

Now we substitute $R = R_1 + R'$ into (5). Taking into account (12) and (14) we get

$$m\frac{\mathrm{d}\tilde{V}'}{\mathrm{d}t} = -\tilde{R}^0 \mathrm{sign}\tilde{V}' + \tilde{U}. \quad (15)$$

Multiplying (15) by $\tilde{V}'$, we get

$$m\tilde{V}'\frac{\mathrm{d}\tilde{V}'}{\mathrm{d}t} = -\tilde{R}^0\tilde{V}'\mathrm{sign}\tilde{V}' + \tilde{U}\tilde{V}'. \quad (16)$$

With the condition (14) we have

$$\frac{1}{2}m\frac{\mathrm{d}(\tilde{V}')^2}{\mathrm{d}t} \leq -\tilde{\delta}|\tilde{V}'|,$$

and

$$\frac{\mathrm{d}(\tilde{V}')^2}{|\tilde{V}'|} \leq -\frac{2\tilde{\delta}}{m}\mathrm{d}t. \quad (17)$$

Integrating (17) and taking into account $|\tilde{V}'| = \sqrt{(\tilde{V}')^2}$, we get

$$2|\tilde{V}'|\Big|_{|\tilde{V}'(\tau_0)|}^{0} \leq -\frac{2\tilde{\delta}}{m}t\Big|_{\tau_0}^{\tilde{\tau}}, \qquad \tau_0 = \tilde{t},$$

where $\tilde{\tau}$ is the moment in time when $\tilde{V}'$ vanishes.

From here

$$\tilde{\tau} \leq \frac{m|\tilde{V}'(\tilde{t})|}{\tilde{\delta}}.$$

Consequently, the maximum possible value of the moment of time at which $\tilde{V}'$ vanishes does not exceed the value

$$\max\tilde{\tau} = \frac{m|\tilde{V}'(\tilde{t})|}{\tilde{\delta}}.$$

IOP Conf. Series: Materials Science and Engineering **675** (2019) 012061 doi:10.1088/1757-899X/675/1/012061

So, controlling in the time interval $t \in [0, \tilde{t}]$ by the force of the form (6), then in the interval $t > \tilde{t}$ by the force $R_1 + R'$, we get the desired control force of the point C directed along line of sight $\overline{CO}$. For $0 \le t \le \tilde{t}$, the control has the form

$$R = -R^0 \mathrm{sign}\tilde{V},$$

where

$$R^0 \ge \left|\tilde{U} + m\mu\dot{s}\right| + \delta, \quad \delta = \mathrm{const} > 0.$$

When $\tau = t - \tilde{t}$, the control force is $R = R_1 + R'$, where

$$R_1 = \frac{mV^2}{2s}, \quad R' = -\tilde{R}^0 \mathrm{sign}\tilde{V}'.$$

Here

$$\tilde{R}^0 \ge \left|\tilde{U}\right| + \tilde{\delta}, \quad \tilde{\delta} = \mathrm{const} > 0,$$

$$\tilde{V}' = V(\tau) - \frac{V(\tilde{t})}{\sqrt{|s(\tilde{t})|}}\sqrt{|s(\tau)|}, \quad \tilde{V}(t) = \dot{s}(t) + \mu s(t).$$

A general view of the control force $\tilde{R}'$ can be represented as

$$\tilde{R}' = R\ \mathrm{sign}\tilde{V}^2 + (R_1 + R')(1 - \mathrm{sign}\tilde{V}^2). \quad (18)$$

Note that $\mathrm{sign}\tilde{V}^2 = 0$ for $\tilde{V} = 0$; and for $\tilde{V} \ne 0$, the value of this expression is equal to one. The first term $R\ \mathrm{sign}\tilde{V}^2$ on the right side of (18) is completely excluded from the control for all $t > \tilde{t}$, so for $t > \tilde{t}$ $R = R_1 + R'$.

Thus, the problem is solved by using the force $\bar{F}$ in the form (1), providing the pursuit mode of the point C of the controlled body according to the principle of proportional navigation, and the force $\tilde{R}'$ of the form (18), which ensures the shockless capture of the point O of the pursued body in a finite period of time.

It should be particularly noted that the second term of the control force $\tilde{R}'$ allows for high-precision soft docking of moving objects (pursuing and pursued bodies) at low-accurate values of the relative speed and distance between the points C and O, used in the construction of the force $\tilde{R}'$. This conclusion is confirmed by the formula (11), since at the time $\tilde{\tau}$ when $\tilde{V}'$ vanishes, takes place

$$V(\tilde{\tau}) - \frac{V(\tilde{t})}{\sqrt{|s(\tilde{t})|}}\sqrt{|s(\tilde{\tau})|} \equiv 0. \quad (19)$$

It follows that the origin of the phase coordinate system is the only point where the equalities $V = 0$, $s = 0$ are satisfied for any initial values of V_0 and s_0 (even low-accurate) that satisfy the condition (2). Note that the condition (2) is satisfied in two regions of the phase plane out of four formed by the reference axes s and $\dot{s}$. In one of these areas $s < 0$, $\dot{s} > 0$, and in the second $s > 0$, $\dot{s} < 0$.

Due to the fact that earlier when a line of discontinuity was chosen in the form (11) this choice was not justified, we consider it appropriate here to look into this question. To do this, we return to the equation (5). When $R = mV^2/2s$, $\tilde{U} = 0$ we get:

$$m\frac{\mathrm{d}V}{\mathrm{d}t} = \frac{mV^2}{2s}.$$

From here we have

$$\frac{\mathrm{d}V}{\mathrm{d}t} = \frac{V^2}{2s}. \quad (20)$$

The equation (20) is an equation with separable variables s and V and is easily integrated using the expressions

$$s = |s|\mathrm{sign}\ s, \quad \mathrm{d}s = (\mathrm{sign}\ s)\mathrm{d}|s|.$$

The general solution of the equation (20) is

$$|s| = \left[\frac{V_0}{2\sqrt{|s_0|}}t + \sqrt{|s_0|}(\mathrm{sign}\ s_0)\right]^2, \quad (21)$$

where $V_0 = V(\tilde{t})$, $s_0 = s(\tilde{t})$. From here we get the following expression for s:

$$s = \left[\frac{V_0}{2\sqrt{|s_0|}}t + \sqrt{|s_0|}(\text{sign } s_0)\right]^2 \text{sign } s. \tag{22}$$

Differentiating s by t, we have

$$V = \frac{V_0}{\sqrt{|s_0|}}\left[\frac{V_0}{2\sqrt{|s_0|}}t + \sqrt{|s_0|}(\text{sign } s_0)\right] \text{sign } s. \tag{23}$$

Considering

$$\sqrt{|s|}(\text{sign } s) = \frac{V_0}{2\sqrt{|s_0|}}t + \sqrt{|s_0|}(\text{sign } s_0), \tag{24}$$

from (23) we get

$$V = \frac{V_0}{\sqrt{|s_0|}}\sqrt{|s|}. \tag{25}$$

This expression for $t > \tilde{t}$ is the equation of the discontinuity line (11), and the quasi-velocity $\tilde{V}'$ has the form (12). Therefore, the sum of the forces $\bar{F}$ of the form (1) and $\tilde{R}'$ of the form (18) completely solves the posed problem.

2.5 Expansion of the class of control forces

Now let us choose the force of control R in (5) in the form

$$R = \frac{mV^2}{bs}, \tag{26}$$

where b is any positive constant.

Then the equation (20) is expressed as

$$\frac{\mathrm{d}V}{\mathrm{d}t} = \frac{V^2}{bs}. \tag{27}$$

This equation, as well as the equation (20), is an equation with separable variables and integrates without much difficulty. The general solution of the equation (27) is

$$s = \left[\frac{b-1}{b}\frac{V_0}{\sqrt[b]{|s_0|}}t + {s_0}^{(b-1)/b}(\text{sign } s_0)\right]^{b/(b-1)} \text{sign } s. \tag{28}$$

Differentiating s by t, we get

$$V = \frac{V_0}{\sqrt[b]{|s_0|}}\left[\frac{b-1}{b}\frac{V_0}{\sqrt[b]{|s_0|}}t + {s_0}^{(b-1)/b}(\text{sign } s_0)\right]^{1/(b-1)} \text{sign } s. \tag{29}$$

Hence, for $s = 0$, we have $t = \tilde{\tau}$, where

$$\tilde{\tau} = -\frac{bs_0}{(b-1)V_0}. \tag{30}$$

Therefore, under the condition (2), $\tilde{\tau} > 0$ exists for all even positive integers $b > 1$. In particular, for $b = 2$, $\tilde{\tau}_2 = 2|s_0|/|V_0|$ holds. Note that for $b \to \infty$, the value of $\tilde{\tau}_\infty = |s_0|/|V_0|$ is two times less than the value of $\tilde{\tau}_2$.

Consequently, with an increase in the value of b, the time of bringing the representative point to the point $s = 0$, $V = 0$ decreases, but not more than twice as compared with the case $b = 2$.

3. Results and discussion

3.1 Results

An analytical expression is constructed for the main vector of control forces, providing the pursuit of an unpredictably moving object with the purpose of shockless high-precision docking with it or its shockless capture in a finite period of time using low-accurate values of the measured relative velocity and distance. The terms “high-precision” and “low-accurate” are estimated by the value of the relative error of meters.

The constructed control ensures the pursuit of the target on the principle of proportional navigation, brings the pursuing point in a finite time to the "initial position", which provides the possibility of reducing the distance to the target, and then ensures the subsequent sliding movement of a point from the "initial position" along the line of discontinuity to the origin of the phase coordinate system. To solve the problem with large relative measurement errors, it was proposed to construct a continuous part of the control force, which provides the process of sliding movement of the representative point along the line of discontinuity, as a function, depending on the current values of the relative velocity and distance. The article also analyzes the issue related to the expansion of the class of control forces by introducing some positive parameter in the expression for the continuous component of the control force. The choice of this parameter affects the duration of the rapprochement of the pursued and pursuing points.

3.2 Discussion

In previously published papers by the authors, when brining the point C to the point O, the constant control force $R_1 = mV_0^2/2s_0$ was used. The discontinuity line was an arc of a parabola

$$V^2(t) = V_0^2\left(1 - \frac{s(t)}{s_0}\right).$$

This shows that the velocity $V(t)$ (for $V_0 \neq 0$) vanishes at the time point $\tilde{\tau}$ when the equality $s(\tilde{\tau}) = s_0$ is reached. Consequently, the equilibrium point was located on the axis of reference s at a distance s_0 from the point from which the rapprochement began.

If this distance were measured with ideal accuracy, then at the moment of time $\tilde{\tau}$ the representative point would be at the origin of the phase coordinate system. But in practice, this distance could be measured with an error, so the representative point at $V = 0$ would not be at the point $s = 0$, but at some distance from it (depending on the size of the measurement error).

This article proposes to make part of the control force R_1 changing as

$$R_1 = \frac{mV^2(t)}{2s(t)},$$

where $V(t)$ and $s(t)$ are the current values of V and s. In this case, the angular coefficient μ of the straight (3) becomes a variable.

In this case, it is necessary to observe condition of positivity of μ, then the straight line (3) does not lose the property of the mutually opposition of the signs of s and $\dot{s}$ in all its points, only slightly changing the angle of inclination to the axis of reference s.

Thereby the values of $\tilde{t}$ and $\tilde{\tau}$ will also vary somewhat, remaining finite positive values. Thus, in the case of choosing R_1 in the form proposed in this article, when using inaccurate (with significant relative errors) values of s and $\dot{s}$, the representative point $(s, \dot{s})$ comes to a point equilibrium O for a finite period of time, as well as in the case of using precisely measured s and $\dot{s}$.

In this case, while the representative point approaches the point O, the absolute values of s and $\dot{s}$ becomes more and more accurate. At the moment of time $t = \tilde{\tau}$ they become absolutely accurate if the meters of s and $\dot{s}$ have no stagnation zone.

If there is a zone of stagnation, then the distance between the points C and O at $t = \tilde{\tau}$ will be commensurate with its size. Note that the size of the stagnation zone is measured in small relative values, and, therefore, does not have much practical value.

4. Conclusions

This article proposes the most autonomous ("self-tuning") control, which does not require human intervention and allows to avoid the appearance of the impact contact forces and to achieve the target in finite time in conditions of uncertainty.

The results can be used to build control systems for the berthing of ships; landing of aircraft; landing, docking or berthing of spacecraft; also for shockless capture of various objects (including space debris) by manipulators and robots.

5. References

[1] Jun He, Haichao Zheng, Feng Gao and Haibo Zhang 2019 *Mechanism and Machine Theory* **140** pp 83–103
[2] Galicki M 2015 *Automatica* **51** pp 49–54
[3] Bartolini G, Ferrara A and Utkin V I 1995 *Automatica* **31(5)** pp 769–773
[4] Jinzhu Peng, Jie Yu and Jie Wang 2014 *ISA Transactions* **53(4)** pp 1035–1043
[5] Mishra S, Londhe P S, Mohan S, Vishvakarma S K and Patre B M 2018 *Computers & Electrical Engineering* **67** pp 729–740
[6] Ijar M Da Fonseca, Luiz C S Goes, Narumi Seito, Mayara K. da Silva Duarte and Élcio Jeronimo de Oliveira 2017 *Acta Astronautica* **137** pp 490–497
[7] Caverly R J, Zlotnik D E, Bridgeman L J and Forbes J R 2014 *Robotics and Computer-Integrated Manufacturing* **30(6)** pp 658–666
[8] Chi-Hua Liu and Ming-Ying Hsiao 2012 *Computers & Mathematics with Applications* **64(5)** pp 1163–1169
[9] Saeed Zaare, Soltanpour M R and Moattari M 2019 *Robotics and Autonomous Systems* **118** pp 204–219
[10] Runxian Yang, Chenguang Yang, Mou Chen and Andy SK Annamalai 2017 *Neurocomputing* **234** pp 107–115
[11] Bragina A A, Shcherbakov V P and Shiryaev V I 2018 *IFAC-PapersOnLine* **51(32)** pp 298–303
[12] Heshmati-Alamdari S, Bechlioulis C P, Karras G C, Nikou A, Dimarogonas D V and Kyriakopoulos K J 2018 *Annual Reviews in Control* **46** pp 315–325
[13] Zhi-Gang Zhou, Yong-An Zhang and Di Zhou 2017 *Aerospace Science and Technology* **71** pp 568–579
[14] Saeed Rafee Nekoo 2019 *Aerospace Science and Technology* **84** pp 348–360
[15] Abouaïssa H and Chouraqui S 2019 *Journal of Computational Science* **31** pp 6–16
[16] Matyukhin V I 2010 *Journal of Applied Mathematics and Mechanics* **74(5)** pp 599–610
[17] Mukhametzyanov I A and Chekmaryova O I 2014 Process control of unstressed docking of plurality of moving objects in an ordered time points *Discrete and Continuous Models and Applied Computational Science* **4** pp 106–111 URL: http://journals.rudn.ru/miph/article/view/8582
[18] Mukhametzyanov I A 2014 The principle of feedback on the quasi-accelerations for unstressed stabilization in finite time of given manifolds of mechanical and generalized systems *Discrete and Continuous Models and Applied Computational Science* **3** pp 107–114 URL: http://journals.rudn.ru/miph/article/view/8236

Efficiency of a thermal machine with final processing time

Yu G Rudoy*, O I Chekmareva

Peoples' Friendship University of Russia (RUDN University), Moscow, Russia

Email: rudikar@mail.ru

Abstract. The Novikov and Curzon-Ahlborn approaches which lead to the improving of the classical Carnot effectiveness expression in the direction of its diminishing were studied in this paper. This effect is due to the account for finite time of non-isothermic processes of the heat exchange which lead to the non-zero values of the thermal machine output power. Obvious expressions for these values are found in both analytic and graphical forms.

Keywords: thermal machine, finite time action, effectiveness.

1. Introduction

1.1. General concept of a heat engine and its efficiency

The concept of a heat engine and its efficiency (often referred as efficiency coefficient) are the basis of the classical or equilibrium phenomenological thermodynamics created in the first half of the 19th century by Carnot, Clapeyron, Kelvin and Clausius [1]. As is known, the simplest prototype of a heat engine consists of three physical objects: two thermostats with fixed temperatures (a "heater" with a temperature T_1 and a "refrigerator" with a temperature T_2, and always $T_1 > T_2 > 0$, all inequalities are strict!), as well as of working fluid, capable of heat exchange with both thermostats (Q_1 and Q_2, respectively) and performing mechanical work W on the external environment.

All values Q_1, Q_2 and W are algebraic (with the condition that $|Q_1| > |Q_2|$), and by virtue of the first law of thermodynamics for any closed cycle of a heat engine, the equality $W = Q_1 + Q_2$ takes place. The signs of all these values depend on the direction of the process performed by the heat engine: direct ($W > 0$, $Q_1 > 0$, $Q_2 < 0$) in the mode of a heat engine or reverse ($W < 0$, $Q_1 < 0$, $Q_2 > 0$) in one of the modes: heat pump or refrigerator [2].

Accordingly, the efficiency of a heat engine in each of these sub-modes is determined by different dimensionless ratios of Q_1, Q_2 and W, but from both physical and technical points of view, the efficiency $\eta = W/Q_1$, sometimes called the Carnot's C function, is traditionally considered to be the most interesting which has an obvious lower bound ($C > 0$). By proven by Carnot universality of the action of a heat engine (in the sense of independence of efficiency on the type of working body), function C depends only on temperatures T_1 and T_2, and so that C increases with increasing temperature difference between T_1 and T_2, which (and this turns out to be quite significant!) is measured either by the difference $T_1 - T_2 > 0$, or by the ratio $T_1/T_2 > 1$.

Within the framework of Carnot's physical representations (i.e., without using the first law of thermodynamics, according to which $W = Q_1 + Q_2$), the dimensionless function $C(T_1, T_2)$ has the form $(T_1 - T_2)/T'$, where T' is some finite (arbitrary, but fixed) temperature value. It is clear that the efficiency in the formulation by Carnot has no limit from above, which, of course, is not confirmed by real observations of heat engines. Later (1849), Clausius took into account the fact $W = Q_1 + Q_2$, so that $\eta = 1 - |Q_2|/Q_1$; since, by definition, $0 < |Q_2|/Q_1 < 1$, the Carnot function is bounded both from below and above, so $0 < C(T_1, T_2) < 1$. Clausius found the explicit form of this function by setting the ratio $|Q_2|/Q_1 = T_2/T_1$, therefore

$$\eta_C = C(T_1, T_2) = 1 - T_2/T_1.$$

1.2. The difficulties with physical meaning of Carnot–Clausius formulation

This Carnot–Clausius result for efficiency is given in all thermodynamic textbooks as a model of the logically consistent derivation of an important physical result. However, for almost all known heat engines, the indicated efficiency value turns out to be too high - the real efficiency is usually 0.25 to 0.40 of the Carnot's efficiency. Therefore, in thermodynamics, there has long been a need to obtain a more realistic upper limit for the efficiency of heat engines. Of course, it was clear to Carnot himself that for any real heat engine with possible useless leaks of heat (besides its transformation into work) the inequality $\eta < \eta_C$ holds; the same is true for friction losses during the performance of mechanical work by the working fluid. In addition, Kelvin [3] as far back as 1853 drew attention to the fact that the heat exchange process between thermostats and the working fluid cannot, strictly speaking, be purely isothermal, as is assumed in the ideal Carnot cycle, but quantitative estimates of the effect of this fact on the value of efficiency were received neither then nor long after that.

Indeed, Kelvin's observation could open up a new stage in the study of a heat engine and its efficiency, but the real need for this actually arose only after a century (!) in the 40s–50s of the 20th century due to the needs of a completely new type of heat engines, namely, nuclear power plants (in the terminology of the time – nuclear reactors). The reason for the partial revision of the Carno–Clausius theory was primarily that within the framework of this theory any heat engine has zero output power $P = 0$, since with nonzero work $W > 0$, the time t of its execution is very large ($t \to \infty$), so that $P = W/t \to 0$.

On the other hand, it is the very large (or even infinite) time of performing any thermodynamic process (except, possibly, adiabatic) in the cycle of a heat engine is a necessary condition for the quasi-equilibrium state of such a process. The rejection of this condition and, consequently, the transition to a finite process time leads to far-reaching consequences. First of all, it is necessary to abandon the assumption about the exact coincidence of the temperature of the working fluid and each of the two thermostats (on the corresponding isothermal sections of the Carnot cycle).

Moreover, taking into account the non-equilibrium and, therefore, irreversible process of thermal conductivity, the condition $Q_1/T_1 + Q_2/T_2 = 0$, which describes the preservation of the "external" entropy, established by Clausius when analyzing the Carnot equilibrium cycle, will be violated. With a small deviation from equilibrium, the term $\Delta S > 0$ should be added to the left-hand part of the Clausius equation, describing the production of "internal" entropy and always strictly positive due to the second law of thermodynamics. Thus, generally speaking, the entire entropy balance should be recalculated based on non-equilibrium thermodynamics (a linear approximation is sufficient).

In the context of the efficiency problem, it is significant that violation of the Clausius equality also simultaneously leads to a violation of the expression for the efficiency of the Carnot cycle based on this equality (see above). This raises the question of obtaining a modified expression for both efficiency η and nonzero power of heat engine $P > 0$. It is clear from the beginning that for a real heat engine operating over a Carnot-type cycle (but with a finite duration), the theoretical upper limit η for efficiency will be below the Carnot limit η_C. An important question is whether the expression for η will preserve the valuable property of universality inherent for the expression for η_C, which depends only on the bath temperatures T_1 and T_2, but not on the properties of the working fluid or its boundaries with thermostats. Moreover, in view of the refusal of the isothermal condition of the processes of production and release of heat by the working fluid to thermostats (or baths), the question arises of choosing the average temperature T_0 of the working fluid in the interval between the bath values T_1 and T_2. Thus, the value T_0 can play the role of a parameter optimizing the operation of a heat engine (in the sense of maximizing its average mechanical or electrical power P).

1.3. The short history of the search for more adequate physical meaning

As far as the authors know, for the first time this set of questions was raised by compatriots of Carnot, French physicists and engineers, in the late 40s - early 50s of the last century in connection with the problem of optimizing the design of atomic reactors. The first prototype of the *Zoé* reactor on uranium oxide with heavy water as a moderator was launched on December 15, 1948 and developed only a few kilowatts of electrical power. The next generation reactor (Saclay, end of October 1952) already gave from 1,000 to 1,500 kW of power. A detailed description of all the issues related here is given in the report [4] of J. Yvon in August 1956.

One of the decisive factors of such a significant increase in power was precisely the selection of the *optimal* average temperature T_0 of the working fluid (heat-transfer fluid). In [4] it was stated that for this purpose it should be $T_0 = \sqrt{T_1 T_2}$, and the corresponding efficiency is $\eta = 1 - \sqrt{T_2/T_1}$, but no calculations were given (probably, for reasons of secrecy). An analysis of the literature indicates that another French physicist, a well-known thermodynamic expert P. Chambadal, who published a small book on atomic energy [5] in 1958 and later (in 1963) – rather a thorough analysis of the efficiency of a heat engine of any nature [6].

However, much earlier (in 1957-58), the famous Soviet thermal physicist I.I. Novikov published in the open press two papers [7, 8], where he gave a mathematically not quite strict, but physically quite convincing justification of the formulas for T_0 and η. Only after nearly 20 years (1975), two American physicists Curzon and Ahlborn [6] offered a fairly consistent derivation of formulas for T_0 and η, which has since been simplified, generalized and improved many times [10]. We note, however, that in article [6] for some reason none of the above-mentioned sources is mentioned at all.

Later in the physical literature, a whole scientific field appeared, known as finite-time thermodynamics. A review of its state in the period of greatest development (the 80s) can be found in the review [11], and the current state (combined with the theory of optimization of heat engines) is fully reflected in the monograph [12].

The aim of this paper is twofold: first, to give a short review of the original attempts to account for the finite time corrections of the thermal machine efficiency and, also, to give mostly appropriate derivation of this efficiency on the grounds of variational methods.

2. Materials and method of research

2.1 Novikov's calculation for the efficiency of a nuclear power plant

In 1957, the Soviet thermal physicist I.I. Novikov published an article entitled "Efficiency of atomic energy installation" in which he gave a physically quite convincing justification of the formulas for the temperature of the working fluid and the efficiency of a heat engine. In brief, this rationale was as follows.

Coming to the expression for the actual useful work produced in the installation [7, 8]:

$$L_{\text{act}} = B\left(T_a - T_p - a\right)\left[1 - \frac{T_0}{T_p}(1 + A)\right],$$

where A and B are constant values depending on the degree of irreversibility of the thermodynamic cycle, T_0 is the ambient temperature (i.e., of the atmosphere), T_p is the average temperature of the working fluid in the cycle at the heat supply section, T_a is the reactor temperature, a is a positive value, representing the average value of the temperature difference between the primary refrigerator and the working substance (as a rule, some kind of fluid) in the process of heat supply to the latter.

From the maximum condition of the function L_{act}, we find the optimal temperature of the working fluid T': $T' = \sqrt{(T_a - a)T_0(1 + A)}$. If the irreversibility of processes in the thermodynamic cycle can be neglected, i.e. put $A = 0$, then $T' = \sqrt{(T_a - a)T_0}$. The value of a, small in comparison with T_a, in some cases can also be neglected, then $T' = \sqrt{T_a/T_0}$. This is the approximate formula for calculating the optimal average temperature of heat supply to the working fluid in the cycle.

Using this formula, it is easy to determine the value of the effective efficiency of the nuclear power plant. The effective efficiency is the ratio of the produced useful work to the amount of heat released in the reactor Q, i.e.

$$\eta_c = (1 - a)\left[1 - \left(\frac{T_0(1 + A)}{T_a - a}\right)^{1/2}\right].$$

IOP Conf. Series: Materials Science and Engineering **675** (2019) 012062 doi:10.1088/1757-899X/675/1/012062

If we ignore the heat leaks and the irreversibility of the thermodynamic cycle and neglect a, then the following simple formula is obtained for the coefficient of efficiency:

$$\eta_c = 1 - \sqrt{T_0/T_a}.$$

Later in the articles [7,8], the author makes a very important explanation that these formulas do not depend on the type of primary coolant used, i.e. can be used for various types of nuclear power plants.

Thus, I.I. Novikov was the first to give a mathematically incompletely rigorous, but physically completely convincing justification of the formulas for the optimum temperature of the working fluid and efficiency.

2.2 Derivation by Curzon and Ahlborn

Only 20 years later, in 1975, two American physicists Curzon and Ahlborn offered a fairly consistent derivation of these formulas, which has since been repeatedly improved and simplified.

Considering the heat flows through the reservoir containing the working fluid proportional to the temperature difference between the reservoir walls, we can write for the isothermal expansion and contraction stage, respectively:

$$F_1 = \alpha(T_1 - T_{1w}), \quad F_2 = \beta(T_{2w} - T_2),$$

where F_1 and F_2 are heat flows, T_{1w} and T_{2w} are the temperature of the working fluid, T_1 is the heater temperature, T_2 is the refrigerator temperature, α and β are heat transfer coefficients.

Now we can write the expressions for the transferred amount of heat for each stage, respectively:

$$W_1 = F_1 t_1 = \alpha t_1(T_1 - T_{1w}), \quad W_2 = F_2 t_2 = \beta t_2(T_{2w} - T_2),$$

where t_1 and t_2 are the corresponding intervals of the processes.

For the adiabatic stages, we take into account the absence of heat fluxes due to their reversibility:

$$W_1/T_{1w} = W_2/T_{2w}.$$

The power of the heat engine is determined by the expression:

$$P = \frac{W_1 - W_2}{(t_1 + t_2)\gamma},$$

where $(\gamma - 1)(t_1 + t_2)$ is the time required for performing adiabatic cycles.

Using the above ratios, we arrive at the following expression:

$$P = \frac{\alpha\beta xy(T_1 - T_2 - x - y)}{y[\beta T_1 y + \alpha T_2 x + xy(\alpha - \beta)]},$$

where $x = T_1 - T_{1w}$, $y = T_{2w} - T_2$.

The power P is maximal at x and y values satisfying the conditions:

$$\frac{\partial P}{\partial x} = 0, \quad \frac{\partial P}{\partial y} = 0,$$

i.e.

$$\beta T_1 y(T_1 - T_2 - x - y) = x[\beta T_1 y + \alpha T_2 x + xy(\alpha - \beta)],$$
$$\alpha T_2 y(T_1 - T_2 - x - y) = y[\beta T_1 y + \alpha T_2 x + xy(\alpha - \beta)].$$

It follows that:

$$y = \left(\frac{\alpha T_2}{\beta T_1}\right)^{1/2} x.$$

We introduce the notation $\mu = x/T_1$ and make a quadratic equation with respect to μ:

$$\left(1 - \frac{\alpha}{\beta}\right)\mu^2 - 2\left[\left(\frac{\alpha T_2}{\beta T_1}\right)^{1/2} + 1\right]\mu + \left(1 - \frac{T_2}{T_1}\right) = 0.$$

We can find its solution that has a physical meaning:

$$\mu = \frac{x}{T_1} = \frac{1 - \sqrt{T_2/T_1}}{1 + \sqrt{\alpha/\beta}}.$$

Using the previous calculations, you can rewrite it as follows:

$$\frac{y}{T_2} = \frac{\sqrt{T_1/T_2} - 1}{1 + \sqrt{\beta/\alpha}}.$$

Then:

$$\eta' = \frac{W_1 - W_2}{W_1} = 1 - \frac{T_{2w}}{T_{1w}} = 1 - \frac{T_2 + y}{T_1 - x} = 1 - \sqrt{T_2/T_1}.$$

Like Novikov, the authors of [9] make an important observation that the result obtained depends only on the temperature of the heat carriers.

2.3 Derivation of efficiency expression by means of Lagrange multipliers method

This derivation, being simpler and more general than the two above, also helps to find a compromise necessary to maximize the output power of the thermal installation.

During the first stage of the cycle between t_0 and $t_1 = t_0 + \tau_1$, the working fluid has a temperature $\theta_1(t)$, receiving heat Q_1 from the heater at a temperature T_1. We assume that the heat transfer process occurs according to the Fourier law, therefore the expression for the heat flux $\varphi_1(t)$ from the heater to the working body is $\varphi_1(t) = C_1[T_1 - \theta_1(t)]$, where C_1 is assumed to be known. During the second stage between t_1 and $t_2 = t_1 + \tau_a$, the temperature adiabatically drops from $\theta_1(t)$ to $\theta_2(t)$. During the third stage between t_2 and $t_3 = t_2 + \tau_2$, the working fluid has a temperature of $\theta_2(t)$, giving off heat $|Q_2|$ to the refrigerator, which has a temperature of T_2 and $\varphi_2(t) = C_2[\theta_2(t) - T_2]$. Finally, between t_3 and $t_4 = t_3 + \tau_a'$ there is a return to the initial state with the temperature increasing from $\theta_2(t_3)$ to $\theta_1(t_4) = \theta_1(t_0)$. The total cycle duration is $\tau = t_4 + t_0 = \tau_1 + \tau_a + \tau_2 + \tau_a'$. The work done by the heat engine during the cycle is the difference between the amount of heat

$$Q_1 = \int_{t_0}^{t_1} \varphi_1(t)dt$$

obtained from the heater, and

$$|Q_2| = \int_{t_2}^{t_3} \varphi_2(t)dt$$

given to the refrigerator. Thus, the output power:

$$P = \frac{1}{\tau}\left\{C_1 \int_{t_0}^{t_1} [T_1 - \theta_1(t)]dt - C_2 \int_{t_2}^{t_3} [\theta_2(t) - T_2]dt\right\}.$$

The above cycle is closed, so the entropy change in the time interval between t_0 and t_4 is zero:

$$\Delta S = C_1 \int_{t_0}^{t_1} \frac{T_1 - \theta_1(t)}{\theta_1(t)} dt - C_2 \int_{t_2}^{t_3} \frac{\theta_2(t) - T_2}{\theta_2(t)} dt = 0.$$

We need to maximize P as a function of temperatures $\theta_1(t)$, $\theta_2(t)$ and durations τ_1, τ_a, τ_2, τ_a' for given values of T_1, T_2, C_1, C_2, τ with $\Delta S = 0$. Introducing the Lagrange multiplier λ/τ and writing down that $P - \lambda\Delta S/\tau$ is constant with respect to $\theta_1(t)$ and $\theta_2(t)$, we get:

$$\theta_1(t) = \sqrt{\lambda T_1}, \quad \theta_2(t) = \sqrt{\lambda T_2}. \tag{1}$$

The optimal cycle for a thermal installation is a Carnot cycle with constant temperatures $\theta_1(t) \equiv \theta_1$ and $\theta_2(t) \equiv \theta_2$. Maximizing P with respect to τ_a and τ_a', we obtain $\tau_a/\tau \approx 0$, $\tau_a'/\tau \approx 0$: the adiabatic steps should be as short as possible. This allows you to write that $P - \lambda\Delta S/\tau$ is invariable relative to $\tau_1/\tau = 1 - \tau_2/\tau$, which gives:

$$C_1(T_1 - \theta_1)(1 - \lambda/\theta_1) + C_2(\theta_2 - T_2)(1 - \lambda/\theta_2).$$

Or using (1):

$$\sqrt{\lambda} = \frac{\sqrt{C_1 T_1} + \sqrt{C_2 T_2}}{\sqrt{C_1} + \sqrt{C_2}}.$$

From the condition $\Delta S = 0$, we get the optimal duration of isothermal steps of the cycle:

$$\frac{\tau_1}{\tau} = \frac{\sqrt{C_2}}{\sqrt{C_1}+\sqrt{C_2}}, \quad \frac{\tau_2}{\tau} = \frac{\sqrt{C_1}}{\sqrt{C_1}+\sqrt{C_2}}.$$

Maximum power is

$$P_{\max} = \frac{C_1 C_2}{\left(\sqrt{C_1}+\sqrt{C_2}\right)^2}(T_1 - T_2).$$

This power is equivalent at best to 1/4 (attained with $C_1 = C_2$) heat flux $C(T_1 - T_2)$, which could be transferred from the heater to the refrigerator with a coefficient C equivalent to $\sqrt{C_1 C_2}$. Finally, the efficiency value obtained from $\Delta S = Q_1/\theta_1 - |Q_2|/\theta_2 = 0$:

$$\eta = 1 - \frac{|Q_2|}{Q_1} = 1 - \frac{\theta_2}{\theta_1} = 1 - \sqrt{T_2/T_1}.$$

As you can see, the formula obtained using the Lagrange multipliers method is identical to that obtained earlier by other methods, and still remains independent of the characteristics of the working body.

At the isothermal expansion section, the gas receives from the heater the amount of heat Q_1, and at the isothermal compression section it gives the amount of heat $|Q_2|$, therefore, a fraction of the heat which is converted into useful work is equal to $W = Q_1 - |Q_2|$.

Entering the auxiliary designation $q = |Q_2|/Q_1$, we write the formula of efficiency and power:

$$\eta = 1 - \frac{|Q_2|}{Q_1} = 1 - q, \qquad q > 0. \tag{2}$$

$$P = \frac{W}{\tau} = \frac{Q_1 - |Q_2|}{\tau} = \left(\frac{Q_1 - |Q_2|}{|Q_2|}\right)\frac{Q_1}{\tau} = \frac{\eta Q_1}{\tau},$$

where τ is the cycle time. From here

$$W = \eta Q_1. \tag{3}$$

Using the second law of thermodynamics, we write the condition for the equality of the total entropy change to zero during the entire cycle:

$$\frac{Q_1}{T_1} - \frac{|Q_2|}{T_2} = 0.$$

Consequently,

$$\eta = 1 - \frac{T_2}{T_1}. \tag{4}$$

The resulting formula of Carnot's efficiency does not have a root that would appear if we took into account the process time.

Now we use the second law of thermodynamics for working fluid temperatures θ_1 and θ_2:

$$\frac{Q_1}{\theta_1} - \frac{|Q_2|}{\theta_2} = 0.$$

From here

$$\frac{\theta_1}{\theta_2} = \frac{Q_1}{|Q_2|} = \frac{1}{q}, \tag{5}$$

and the efficiency formula is

$$\eta = 1 - \frac{\theta_2}{\theta_1} = 1 - q. \tag{6}$$

Denoting the Carnot efficiency (4) as η_c and subtracting (6) from (4), we get:

$$\eta_c - \eta = q - \frac{T_2}{T_1}. \tag{7}$$

Similarly, we suppose (as in the conclusion with the help of additional Lagrange multipliers) that the heat transfer process during the first stage of the cycle occurs according to the Fourier law, therefore the heat flux $\varphi_1(t)$ from the heater to the working fluid has the form $\varphi_1(t) = C_1[T_1 -$

$\theta_1(t)]$, where C_1 is assumed to be known. During the third stage, the working fluid has a temperature $\theta_2(t)$, giving off heat $|Q_2|$ to a refrigerator having a temperature T_2 and $\varphi_2(t) = C_2[\theta_2(t) - T_2]$. Then, since $Q_1 = \int_{t_0}^{t_1} \varphi_1(t)dt$, and $|Q_2| = \int_{t_2}^{t_3} \varphi_2(t)dt$:

$$Q_1 = C_1 \int_{t_0}^{t_1}(T_1 - \theta_1)dt = C_1\tau_1(T_1 - \theta_1) = C_1\tau_1T_1\left(1 - \frac{\theta_1}{T_1}\right), \tag{8}$$

$$|Q_2| = C_2 \int_{t_2}^{t_3}(\theta_2 - T_2)dt = C_2\tau_2(\theta_2 - T_2) = C_2\tau_2T_2\left(\frac{\theta_2}{T_2} - 1\right). \tag{9}$$

Expressing from (5) $|Q_2| = qQ_1$ and $\theta_2 = q\theta_1$, and also using (8) and (9), we get:

$$\theta_1 = \frac{qC_1\tau_1T_1 + C_2\tau_2T_2}{q(C_1\tau_1 + C_2\tau_2)}.$$

Substituting this expression into (8), we get:

$$Q_1 = \frac{C_1C_2\tau_1\tau_2T_1(q - T_2/T_1)}{q(C_1\tau_1 + C_2\tau_2)}.$$

Using (7) and (2) and denoting $C_1C_2\tau_1\tau_2/(C_1\tau_1 + C_2\tau_2)$ as C', we write the expression for Q_1 in the form:

$$Q_1 = C'T_1\frac{\eta_c - \eta}{1 - \eta}. \tag{10}$$

Substituting (10) into (3), we get:

$$P = \frac{W}{\tau} = \frac{\eta Q_1}{\tau} = \frac{C'}{\tau}T_1\eta\frac{\eta_c - \eta}{1 - \eta}. \tag{11}$$

Combining C' and $\tau = \tau_1 + \tau_2$ into one constant C, we finally write:

$$P = CT_1\eta\frac{\eta_c - \eta}{1 - \eta}. \tag{12}$$

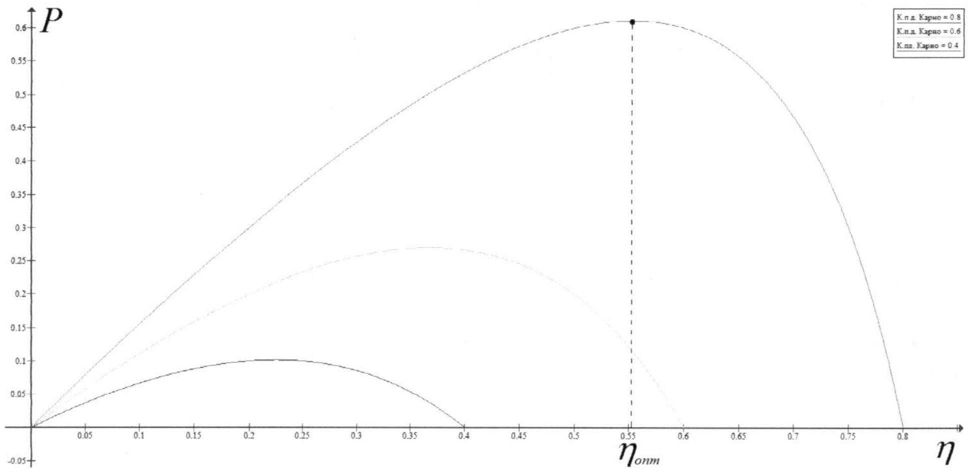

Figure 1. Dependence graph (12) for η_c, equal to 0.4, 0.6, 0.8.

Let's consider the function (12). Differentiation gives the following expression for the first derivative:

$$P' = CT_1\frac{\eta^2 - 2\eta + \eta_c}{(1 - \eta)^2}.$$

Equating it to zero, we get the following expression for the critical point:

$$\eta = 1 - \sqrt{1 - \eta_c}.$$

We next obtain the second derivative of the function (12):

$$P'' = 2CT_1\frac{(\eta_c - 1)}{(1 - \eta)^3}. \tag{13}$$

Since η and η_c are less than 1, expression (13) is negative, which allows us to say that this critical point is the maximum of function (12), which was to be proved.

3. Results and discussion

The new formula for the real efficiency of a heat engine

$$\eta = 1 - \sqrt{1 - \eta_c} = 1 - \sqrt{T_2/T_1},$$

meeting the maximum power condition (for given heat transfer parameters during thermal contact of the working fluid and thermostats), fully corresponds to the result obtained for the first time by Novikov (see Section 2). It is clear that for any values of T_1 and T_2 the value of η is obviously less than η_c, (e.g., in lowest approximations $\eta \approx (1/2)\eta_c$+(1/8) $)\eta_c{}^2$) which, in turn, is obviously less than one (see figure), and at which the heat engine has zero power.

Simple calculations show that at a given temperature difference, the values of η_c correspond much better to the actually observed efficiency values than the theoretically correct, but almost strongly overestimated value of η_c. This is clearly shown by the table below from [9].

Power Plant	T_c (°C)	T_h (°C)	η (Carnot)	η (Endoreversible)	η (Observed)
West Thurrock (UK) coal-fired power plant	25	565	0.64	0.40	0.36
CANDU (Canada) nuclear power plant	25	300	0.48	0.28	0.30
Larderello (Italy) geothermal power plant	80	250	0.33	0.178	0.16

4. Conclusions

Derivation of the formula of efficiency by means of the method of Lagrange multipliers is simpler and more general than the well-known approaches of Novikov and Curzon & Alborne and also allows to select the average temperature of the working fluid that maximizes the mechanical power of the power plant. The next step is to see how this conclusion may be extended on more complex types of thermal machines – for example those with thermal radiation as the working body.

5. References

[1] Gelfer Ya M 1969 *The History and Methodology of Thermodynamics and Statistical Physics* (Moscow: Publishing House Vysshaya Shkola) p 475

[2] Rudoy Yu G 2011 The effectiveness of the thermal machine in the regimes of heat engine, heat pump and refrigerator *Physics in Higher Education* **17(4)** pp 126–33 URL: http://pinhe.lebedev.ru/tom17n4.en.htm#The_Effectiveness_of_the_Thermal_Machine

[3] Thomson W 1853 *Philosophical Magazine* **5(30)** https://doi.org/10.1080/14786445308562743

[4] Yvon J 1955 La pile de Saclay. Experience acquise en deux ans sur le transfert de chaleur par gaze comprime *Communication du C.K.A. à la Conférence de Genève* p 36

[5] Chambadal P 1957 *Les Centrales Nucleaires* (Paris: Armand Colin) p 58

[6] Chambadal P 1963 *Evolution et Applications du Concept d'Entropie* (Paris: Dunod) p 280

[7] Novikov I I 1957 Efficiency of atomic energy installation *Atomnaya Energiya* **3(11)** pp 409-412

[8] Novikov I I 1958 Efficiency of atomic energy installation (a review) *J. of Nuclear Energy II* **7** pp 125–128

[9] Curzon F L and Ahlborn B 1975 Efficiency of a Carnot engine at maximum power output *American J. of Phys.* **43** pp 22–24

[10] Van den Broeck C 2005 Thermodynamic efficiency at maximum power *Phys. Rev. Letters* **95** 190602 (1-3)

[11] Andresen B, Salamon P and Berry R S 1984 Thermodynamics in finite time *Phys. Today* **37(9)** pp 62–70

[12] Mironova V A, Amelkin S A, Tsirlin A M 2000 *Mathematical Methods of Thermodynamics at Finite Time* (Moscow: Publishing House Khimia) p 384

Fullerenes as solitons in chiral model of graphene

Yu P Rybakov* **and M Umar**

Peoples' Friendship University of Russia (RUDN University), Moscow, Russia

Email: rybakov-yup@rudn.ru

Abstract. Taking into account the sp^2- hybridization effect for valence electrons in carbon atoms, we introduce a unitary matrix U as an order parameter and suggest a scalar chiral model of graphene for the description of graphene-like configurations. Using the well-known hedgehog ansatz for modeling Fullerene C_{60}, we estimate the polarizability of a single Fullerene specie in a uniform external electric field. Finally, we use this result for calculating the parameters of the van der Waals potential for the two interacting Fullerene species.

1 Introduction

Since the discovery of mono-atomic carbon layers called graphenes [1, 2] this material attracted deep interest of researchers due to its extraordinary properties concerning magnetism, stiffness and high electric and thermal conductivity [3, 4]. The interesting connection was revealed with other graphene-based materials: Fullerenes [5] and carbon nano-tubes [6]. A very simple explanation of these unusual properties of graphene was suggested in [7], where the idea of massless Dirac-like excitations of honeycomb carbon lattice was discussed, the latter one being considered as a superposition of two triangular sub-lattices. Some phenomenological development of this idea was realized in [8] and [9]. In the first paper [8] the simplest scalar chiral model of graphene excitations was considered, spin and magnetic excitations being studied in [9].

The objective of the current study was to show the applicability of the scalar chiral model for the description of electromagnetic properties of Fullerene configurations. In particular, we calculated the electric polarizability of a single Fullerene specie and also estimated the parameters of the interaction potential for the two Fullerene species at large distance between them.

The concept behind this research is the following. As is well known, the carbon atom possesses of four valence electrons in the so-called hybridized sp^2- states, the one of them being "free" in graphene lattice and all others forming strong covalent bonds with the neighbors. It appears natural to introduce scalar a_0 and 3-vector $\vec{a}$ fields corresponding to the $s-$orbital and the $p-$orbital states of the "free" electron respectively. These two fields can be combined into the unitary matrix $U \in SU(2)$ considered as the order parameter of the model in question:

$$U = a_0\tau_0 + i\,\vec{a}\cdot\vec{\tau}, \tag{1}$$

where τ_0 is the unit $2\times 2-$matrix and $\vec{\tau}$ are the three Pauli matrices, with the $SU(2)-$condition

$$a_0^2 + \vec{a}^2 = 1 \tag{2}$$

being imposed. It is convenient to construct, via differentiating the chiral field U, the so-called left chiral current

$$l_\mu = U^+\partial_\mu U, \quad \mu = 0, 1, 2, 3, \tag{3}$$

the index μ denoting the derivatives with respect to the time $x^0 = ct$ and the space coordinates x^i, $i = 1, 2, 3$. The Lagrangian density reads

$$L = -\frac{I}{4} tr\,(l_\mu l^\mu) - \frac{\lambda^2}{2} \vec{a}^2 \tag{4}$$

and corresponds to the sigma-model approach in the field theory with a small mass term. Here the constant model parameters I and λ are introduced. Comparing the Lagrangian density (4) with that of the Landau – Lifshits theory corresponding to the quasi-classical long-wave approximation to the Heisenberg magnetic model, one can interpret the parameter I in (4) as the exchange energy between the atoms (per spacing). As was shown in [8], the model (4) admits the kink-like or the domain-wall configuration

$$U = \exp\,(i\tau_1\Theta), \quad \Theta = \Theta(z), \tag{5}$$

corresponding to the ideal graphene plane with the normal oriented along the $z-$axis. In this case

$$\Theta(z) = 2\arctan\exp(-z/\ell)\,,$$

where the characteristic length appears:

$$\ell = \frac{\sqrt{I}}{\lambda}. \tag{6}$$

The model (4) admits also the two-dimensional hedgehog configuration [10]

$$U = \exp\,(i\sigma\Theta), \quad \Theta = \Theta(\rho), \tag{7}$$

where

$$\sigma = (\tau_1 x + \tau_2 y)/\rho, \quad \rho^2 = x^2 + y^2.$$

The configuration (7) corresponds to the infinite carbon nano-tube oriented along the $z-$axis. In the sequel we consider the three-dimensional hedgehog configurations describing Fullerene species.

2 Materials and Methods

As a material we consider Fullerene species described by spherically-symmetric soliton-like configurations. For approximating these configurations we use trial functions found via the variational method.

2.1 Fullerene configurations

It is worth-while to stress that three-dimensional configurations should meet the scale-invariance Hobart – Derrick requirements [11, 12]. To this end, the Lagrangian density (4) should be modified through including the Skyrme – Faddeev term [13, 14]:

$$L = -\frac{I}{4} tr(l_\mu l^\mu) - \frac{\lambda^2}{2} \vec{a}^2 + \frac{\varepsilon^2}{16} \left([l_\mu, l_\nu][l^\mu, l^\nu]\right), \tag{8}$$

with ε being some new dimensionless coupling constant. For the description of C_{60} Fullerenes we suggest to use the well-known hedgehog ansatz

$$U = \exp[i\tau\Theta(r)], \quad \tau = (\vec{\tau}\,\vec{n}), \quad \vec{n} = \vec{r}\,/\,r, \tag{9}$$

corresponding to the nucleon structure in the Skyrme model [13], where the baryon charge was identified with the unit topological invariant of the degree type:

$$Q = \deg(S^3 \to S^3) = 1, \quad \Theta(0) = \pi, \quad \Theta(\infty) = 0. \tag{10}$$

Inserting (9) into (3), one finds the space components of the left chiral current:

$$l_j = i\frac{x_j}{r}\Theta'\tau + \frac{i}{2r}\sin 2\Theta\left(\tau_j - \frac{x_j}{r}\tau\right) + i\sin^2\Theta\frac{x_k}{r^2}\varepsilon_{kjl}\tau_l. \tag{11}$$

Using (11), one can reduce the Lagrangian density (8) to the following form:

$$L = -\frac{I}{2}\left(\Theta'^2 + \frac{2}{r^2}\sin^2\Theta\right) - \frac{\lambda^2}{2}\sin^2\Theta - \varepsilon^2\frac{\sin^2\Theta}{r^2}\left(\frac{\sin^2\Theta}{r^2} + 2\Theta'^2\right). \tag{12}$$

The configuration with the unit topological charge satisfying boundary conditions (10), can be approximated by the trial function

$$\tan(\Theta/2) = \exp\left(-\eta - k(1+e^{\eta})\right), \quad \eta = \ln(r/r_0), \quad k = r_0/\ell. \tag{13}$$

Identifying this configuration with that of C_{60} Fullerene [15], one finds $r_0 = 3.55\,\mathring{A}$.

3 Results

3.1 Electromagnetic properties of Fullerenes

The interaction with the electromagnetic field can be introduced into the chiral model through the gauge invariance principle. To this aim, we consider the gauge transform:

$$U' = VUV^{-1}, \quad V = \exp(i\gamma\tau_3), \quad A'_\mu = A_\mu + \partial_\mu\Lambda, \quad L'_\mu = VL_\mu V^{-1}, \tag{14}$$

combined with extending the derivative:

$$D_\mu U = \partial_\mu U - ie_0 A_\mu[\Gamma_e, U], \quad L_\mu = U^+ D_\mu U, \tag{15}$$

where e_0 and Γ_e stand for the electromagnetic coupling constant and the charge operator respectively. Finally, the gauge invariance holds, if $\Gamma_e = \tau_3$, $\gamma = e_0\Lambda$.

Let us now estimate the electric polarizability α of a single Fullerene specie in a uniform electric field with the intensity E_0. In the simplest dipole approximation one can put:

$$A_0 = -E_0 r\cos\vartheta + \frac{p_0 r\cos\vartheta}{r^3 + r_0^3}, \tag{16}$$

where $p_0 = \alpha E_0$ is the induced dipole moment of the Fullerene specie and the spherical coordinates r, ϑ are used. Exploiting (16), one can calculate the part of the electromagnetic energy depending on p_0:

$$W_{\text{em}} = -\frac{p_0}{3}E_0 + \frac{p_0^2}{3r_0^3}. \tag{17}$$

Minimizing (17) with respect to p_0, one easily obtains

$$p_0 = \frac{1}{2}r_0^3 E_0 = \alpha E_0, \quad \alpha = \frac{1}{2}r_0^3. \tag{18}$$

3.2 Interaction of two Fullerene species

Let us now consider the two Fullerene species with the centers O_1, O_2, $O_1O_2 = R$, where $R >> r_0$. Our goal will be the calculation of the induced dipole moments and the van der Waals interaction potential. In this case one can put in the first approximation

$$U = U_1 U_2, \quad U_j = \cos\Theta_j + i\sin\Theta_j(\vec{\tau}\,\vec{R}_j)/R_j, \quad \Theta_j = \Theta(R_j), \tag{19}$$

where $\vec{R}_1 = \vec{r} + \vec{e}_z R/2$, $\vec{R}_2 = \vec{r} - \vec{e}_z R/2$ and $\Theta(R_j)$ corresponds to (13). In the static case the electromagnetic part of the Lagrangian reads:

$$L_{\text{em}} = -\frac{I}{4}trL_0^2 + \frac{1}{8\pi}(\nabla A_0)^2 - j_0 A_0 - \frac{\nu_0}{4}A_0^4, \tag{20}$$

where

$$j_0 = \frac{1}{4\pi}div\left(\vec{E}_1 + \vec{E}_2\right)$$

and the nonlinear term of the Mie type is added, with ν_0 being some new coupling constant. Inserting (19) into (15), one gets

$$-\frac{1}{2}trL_0^2 = e_0^2 A_0^2\, tr\left[1 - (U_1^+\tau_3 U_1)(U_2\tau_3 U_2^+)\right] \approx e_0^2 A_0^2 \rho^2\left(\frac{\sin^2\Theta_1}{R_1^2} + \frac{\sin^2\Theta_2}{R_2^2}\right). \tag{21}$$

Let us estimate the electromagnetic energy W_{em} of the system. To this end, we use the dipole deformation of the Fullerene species:

$$A_0 = p\frac{(z + R/2)R_1^3}{\left(R_1^2 + r_0^2\right)^3} + p\frac{(z - R/2)R_2^3}{\left(R_2^2 + r_0^2\right)^3} \equiv A_0^{(1)} + A_0^{(2)}. \tag{22}$$

It should be underlined that the proper electrostatic part of W_{em} is small due to the Maxwell equations and therefore

$$W_{\text{em}} = 2\pi \int_0^{\infty} \rho d\rho \int_{-\infty}^{\infty} dz L_{em} \approx W_1 + W_2, \tag{23}$$

where the following main contributions are separated:

$$W_1 = 8\pi I e_0^2 \int \rho d\rho \int dz \left(A_0^{(1)}\right)^2 \rho^2 \frac{\sin^2 \Theta_1}{R_1^2}, \quad W_2 = -\pi \nu_0 \int\int \rho d\rho dz I e_0^2 \left(A_0^{(1)}\right)^4. \tag{24}$$

Inserting (22) into (24), one finds

$$W_1 + W_2 \approx \frac{\pi p^2 I e_0^2}{15(1+k) r_0} - 0.17 \frac{\pi \nu_0 p^4}{r_0^5}. \tag{25}$$

Minimizing the energy (25) with respect to p^2, one gets

$$p^2 \approx \frac{I e_0^2 r_0^4}{5\nu_0 (1+k)}. \tag{26}$$

Now we are in a position to estimate the interaction potential between the two Fullerene species. To this end, in view of (18), we can calculate the interaction force between two effective dipoles:

$$\vec{F}_{12} = (\vec{p}_1 \nabla_2) \vec{E}_2 \approx r_0^3 (\vec{E}_2 \nabla_2) \vec{E}_2 / 2 = -\nabla_2 V, \tag{27}$$

where the potential V coincides with that of van der Waals:

$$V = -\frac{r_0^3}{4} E_2^2 = -\frac{r_0^3 p^2}{R^6}. \tag{28}$$

Here it was taken into account that

$$E_2 = E_z = \frac{2p}{R^3}.$$

Finally, inserting (26) into (28), one finds

$$V = -\frac{I e_0^2 r_0^7}{5\nu_0 (1+k) R^6}. \tag{29}$$

Identifying the potential (29) with that of Girifalco [16, 17, 18, 19]:

$$V = -\frac{\sigma}{R^6}, \quad \sigma = 2^9 r_0^6 7.5 \times 10^{-21} J,$$

one can estimate the value of the constant v_0:

$$v_0 = \frac{I e_0^2 r_0^7}{5(1+k)\sigma} = 1.7 \times 10^{11} (Jm)^{-1},$$

where it was taken into account that $I = 1.3\, nN,\quad k = 2.67,\quad e_0 = e/(\hbar c)$.

4 Discussion

For the description of Fullerene electromagnetic properties we used the method of consecutive approximations. First, we estimated the electric dipole moment

$$p_0 = \frac{1}{2} r_0^3 E_0 = \alpha E_0$$

induced in a single Fullerene specie under the influence of the external electric field E_0. As a result, we found the electric polarizability α, which proved to be proportional to the volume of the Fullerene specie. This fact was taken into account for estimating the interaction potential V between the two separated Fullerene species. This potential proved to coincide with that of van der Waals [cf. (27) and (28)].

It is worth-while to stress that this result follows from minimizing the electric energy of the two interacting Fullerene species in supposition that the special nonlinear term of the Mie type was added to the Lagrangian [cf. (20)]. Comparing the form of the interaction potential V with that of Girifalco [16], it was possible to estimate the Mie constant v_0 in (20).

Conclusions

1. Within the scope of the suggested earlier scalar chiral model of graphene we constructed trial spherically-symmetric Fullerene configurations corresponding to the “hedgehog” ansatz or the Skyrmions in nuclear physics.
2. Using the gauge invariance principle, we included in the model the interaction with the electromagnetic field via extending the derivatives.
3. Using the trial Fullerene configurations, we estimated the electric polarizability of a single Fullerene specie in a uniform electrostatic field.
4. Minimizing the electromagnetic part of the energy for the case of the two Fullerene species at large distance between them, we confirmed the validity of the van der Waals interaction potential.

Acknowledgements

This publication has been prepared with the support of the RUDN University Program 5-100.

References

[1] Novoselov K S et al 2004 Electric field effect in atomically thin carbon films *Science* **306** 666
[2] Geim A K 2009 Graphene: status and prospects *Science* **324** 1530
[3] Lee C et al 2008 Measurement of elastic properties and intrinsic strength of monolayer graphene *Science* **321** 385
[4] Bolotin K I et al 2008 Ultrahigh electron mobility in suspended graphene *Solid State Communications* **146** 351
[5] Lu X and Chen Z 2005 Curved *pi*-conjugation aromaticity, and the related chemistry of small fullerenes (C_{60}) and single-walled carbon nanotubes *Chemical Review* **105** 3643

[6] Terrones M 2003 Science and technology of the twenty-first century: synthesis, properties, and applications of carbon nanotubes *Annual Review of Materials Research* **33** 419
[7] Semenoff G W 1984 Condensed-matter simulation of a three-dimensional anomaly *Physical Review Letters* **53** 2449
[8] Rybakov Yu P 2012 On chiral model of graphene *Solid State Phenomena* **190** 59
[9] Rybakov Yu P 2015 Spin excitations in chiral model of graphene *Solid State Phenomena* **233-234** 16
[10] Rybakov Yu P 2018 Magnetic excitations of carbon nanotubes in chiral model of graphene *European Physical Journal Web of Conferences* **185** 11007
[11] Hobart R H 1963 On the instability of a class of unitary field models *Proceedings of the Physical Society* **82** (part 2) 201
[12] Derrick G H 1964 Comments on nonlinear wave equation as a model for elementary particles *Journal of Mathematical Physics* **5** 1252
[13] Skyrme T H R 1962 A unified field theory of mesons and baryons *Nuclear Physics* **31** 556
[14] Faddeev L D 1973 Gauge invariant model of electromagnetic and weak interactions of leptons *Reports of the USSR Academy of Sciences* **210** 807
[15] Dresselhaus M S, Dresselhaus G and Eklund P C 1995 *Science of Fullerenes and Carbon Nanotubes* (San Diego: Academic)
[16] Girifalco L A 1991 Interaction potential for carbon C_{60} molecules *The Journal of Physical Chemistry* **95** 5370
[17] Girifalco L A 1992 Molecular properties of C_{60} in the gas and solid phases *The Journal of Physical Chemistry* **96** 858
[18] Zubov V I, Zubov I V 2009 Influence of heat vacancies on thermodynamic properties and stability of Fullerenes *Dynamics of Complex Systems* **3** 52
[19] Bahaa Khedr M, Osman S M and Al Bussaidi M S 2009 Surface tension, shear viscosity and isothermal compressibility of liquid C_{60} along the liquid–vapor coexistence *Molecular Physics* **107** 1355

IOP Conf. Series: Materials Science and Engineering **675** (2019) 012064 doi:10.1088/1757-899X/675/1/012064

Generalizing Darcy's law for filtration radial flows through porous media

Yu P Rybakov*, N V Semenova and J S Safarov

Peoples' Friendship University of Russia (RUDN University), Moscow, Russia

Email: rybakov-yup@rudn.ru

Abstract. We study the filtration process in porous media and compare the filtration coefficients for two possible geometries of flows: cylindrical and radial ones. Solving balance equations for impurity concentration in the simplest axially-symmetric stationary case, one finds the radial filtration process more effective. Therefore, we restrict our attention to the radial filters with non-homogeneous grain filling. First, we describe the transverse diffusion effect and then suggest the generalization of the Darcy's filtration law, its dynamical origin being stressed. Using the perturbation method, we find the structure of the Stokes stream function for some particular choices of the porosity.

1 Introduction

We study the hydrodynamics of flow in a porous medium modelling the grain filling in filters [1 - 9]. Using the lattice approximation, we derive the structure of the fluid current and obtain the transverse diffusion coefficient *D*, which proves to be proportional to the diameter *d* of the grains as constituents of the medium [10 - 14]. However, we begin with some remarks concerning the geometry of flows in porous media.

The objective of the current study was threefold. First, we intended to choose the appropriate geometry for the filtration process, the goal being the improvement of its effectiveness. Beside this, on the base of the lattice approximation, we intended to take into account a very important transverse diffusion effect, which was typical for fluid flows in porous media. Finally, we planned to find some arguments in favour of generalizing the standard Darcy filtration law.

1.1 Comparing filtration coefficients for two flow geometries

Let us first estimate the so-called filtration coefficient, which is given by the ratio of the impurity concentration n for the outgoing flow to that for the incoming one. Taking into account that n satisfies the balance equation of the form

$$\partial_t n + div(n\vec{u}) = -\beta n,$$

where $\vec{u}$ stands for the velocity of the fluid and β – for the absorption coefficient of the porous medium, one gets for the stationary process the relation:

$$div\,(n\vec{u}) = -\beta n. \tag{1}$$

For the rude estimation of the solution to (1) in the case of a cylindrical filter, one can approximate the velocity by its constant vertical part u_0 and obtain the following solution for the concentration $n(z)$:

$$n(z) = n_0 \exp\left(-\frac{\beta z}{u_0}\right). \tag{2}$$

Consider now a radial filter, with the flow velocity being radial. Taking into account that for the radial flow the velocity appears to be inverse proportional to the radial coordinate ρ :

$$u(\rho) = u_0 \frac{\rho_0}{\rho}, \quad \rho \geq \rho_0,$$

where $u_0 \rho_0$ stands for the constant of integration, one easily finds the evident form of the stationary balance equation (1):

$$div\,(n\vec{u}) \approx \frac{1}{\rho} \partial_\rho (\rho n u) = u_0 \frac{\rho_0}{\rho} \partial_\rho n = -\beta n. \quad (3)$$

Its solution has the form

$$n(\rho) = n_0 \exp\left[\frac{\beta}{2\rho_0 u_0} (\rho_0^2 - \rho^2) \right]. \quad (4)$$

Comparing the formulas (2) and (4), giving the expressions for the filtration coefficient, one can see that for $\rho >> \rho_0$ the radial filter proves to be more effective than the cylindrical one. Taking into account this conclusion, from now on we concentrate on the radial flow in porous media.

2 Materials and Methods

We consider as a material the porous medium in the filter and to model the grain filling we use the lattice approximation method for representing the stationary mass balance equation. In the limiting case of the zero lattice spacing we obtain the differential equation of the second order corresponding to the transverse diffusion. To solve the hydrodynamic equations of motion for axially-symmetric radial flows, we introduce the Stokes stream function. However, for solving the full system of equations we use the perturbation method and exploit the corresponding two small parameters. The first one concerns the ratio of the grain diameter *d* to the radius *a* of the filter, the second one being related with the small deformation of the Darcy filtration coefficient.

3 Results

3.1 Lattice approximation for mass balance equation in porous media: transverse diffusion

We use the cylindrical coordinates ρ, φ, z, with φ being the azimuth angle. Let us first consider the discrete variant of the mass conservation equation and number the lattice vertices by the indices i, j (transverse to the flow) and k (along the flow), the corresponding cylindrical coordinates being φ, z, and ρ, respectively. Let us denote the local radial stream of the fluid by

$$G_{ijk} = \Delta S_k u_{ijk},$$

where u_{ijk} stands for the velocity of the flow and ΔS_k is the area of the gap between the grains. It means that

$$\Delta S_k = \rho_k \Delta\varphi \Delta z \, S_k,$$

where S_k denotes the porosity of the medium. Therefore, the local conservation law reads

$$G_{ijk}= r_{k-1}\, G_{ijk-1}+p_{k-1}\,(G_{i-1\,jk-1}+G_{i+1\,jk-1})$$
$$+\,q_{k-1}\,(G_{ij-1k-1}+G_{ij+1k-1}),$$

where the branching coefficients *r, p, q* are introduced. Supposing that $p = 0$ due to the axial symmetry of the flow, let us write the mass conservation equation

$$\sum_{ij} G_{ijk} = \sum_{ij} G_{ijk-1}\,, \tag{5}$$

which immediately implies the constraint on the branching coefficients:

$$r_k + 2q_k = 1. \tag{6}$$

Identifying now the lattice spacing with the diameter *d* of the grain, it can be proved through (5) and (6) that in the continuous limit the following differential equation is valid:

$$div\,(rS\,\vec{u}) = d\,\partial_z(S\,q\,\partial_z u_\rho). \tag{7}$$

One can rewrite the equation (7) in the form of the stationary conservation law:

$$div\,\vec{j} = 0,$$

where the components of the current $\vec{j}$ in cylindrical coordinates read:

$$j_z = S(\rho)u - D(\rho)S(\rho)\partial_z w, \quad j_\rho = r(\rho)S(\rho)\,w, \tag{8}$$

and the transverse diffusion coefficient is introduced:

$$D(\rho) = q(\rho)d(\rho). \tag{9}$$

Here we use the following denotations for the components of the velocity:

$$u_\rho = w,\;\; u_z = u,$$

the density of the fluid being unity.

Now it is worth - while to stress that the effect of the transverse diffusion in porous media is widely discussed in literature [11, 12].

3.2 The hydrodynamics of the radial flow in porous media: generalizing Darcy's law

To find the profiles of the velocity $\vec{u}$ and the pressure *P*, it is necessary to solve also the Euler equation, with the force density $\vec{f}$ including the gravity acceleration $\vec{g}$ and the Darcy force

$$\vec{f}_D = -k_D\vec{u}.$$

In the first approximation, the Darcy coefficient k_D appears to be constant: $k_D = k_0 = const$, but in general, it should be some function of the velocity and the pressure. We suggest a generalization of the Darcy law [7, 15 - 18] by including in k_D the invariant

$$I = (\vec{u}\nabla)P \tag{10}$$

via the simplest linear form:

$$k_D = k_0 + k \cdot I.$$

The well-known Darcy's law of filtration was found empirically in 1856 by the French engineer H. Darcy [7, 15] who specialized in constructing fountains. According to this law, the velocity of the fluid flow in a porous medium appears to be proportional to the pressure gradient and depends on the structure of grains as constituent elements of the medium. Later various modifications of the Darcy law were suggested, with the correction terms containing some degrees of the velocity, which was considered as a formal small parameter [16]. As an alternative to such an activity, we suggest a dynamical approach based on the structure of the fundamental Euler equation, which contains the velocity and the pressure:

$$(\vec{u}\nabla)\vec{u} + \nabla P = \vec{g} - k_D\vec{u}. \tag{11}$$

Some deviations from the standard Darcy law were observed in [20, 21]. We intend to analyze the dependence of the filtration process on the coefficient k^*. First, we write down the mass balance equation (7) and the Euler equation (11) in cylindrical coordinates:

$$(w\partial_\rho + u\partial_z)u + \partial_z P = g - (k_0 + k^* I)u, \tag{12}$$

$$(w\partial_\rho + u\partial_z)w + \partial_\rho P = -\,(k_0 + k^* I)w, \tag{13}$$

where the invariant I reads

$$I = (w\partial_\rho + u\partial_z)P.$$

Multiplying the equations (12) and (13) by u and w respectively and adding the results, one easily finds for the invariant I the following expression:

$$I = \left[1 + k^*(u^2 + w^2)\right]^{-1}\left[gu - k_0 u^2 - \frac{1}{2}(w\partial_\rho + u\partial_z)(u^2 + w^2)\right]. \tag{14}$$

Using (14), one can eliminate the pressure P between the equations (12) and (13), the resulting equation for u and w being:

$$\begin{gathered}\partial_\rho\left[(w\partial_\rho + u\partial_z)u + (k_0 + k^* I)u\right] = \\ \partial_z\left[(w\partial_\rho + u\partial_z)w + (k_0 + k^* I)w\right].\end{gathered} \tag{15}$$

Let us now introduce the stream function Ψ through the substitution

$$\rho S\, r\, w = -\partial_z\Psi, \quad \rho S\,(u - D\partial_z w) = \partial_\rho\Psi, \tag{16}$$

the equation (7) being satisfied identically. Taking into account that

$$u << w, \tag{17}$$

one can use the perturbation method by putting

$$\Psi = \psi_0 + \psi, \quad u = \tilde{u}, \quad w = \overline{w}(\rho) + \tilde{w}, \quad \psi_0 >> \psi, \tag{18}$$

where it is supposed that

$$\rho S\, r\,\overline{w} = -\partial_z \psi_0, \quad \partial_\rho \psi_0 = 0. \tag{19}$$

In view of (19), one finds

$$\psi_0 = -C_0 z, \quad \overline{w} = C_0 (\rho S r)^{-1}, \tag{20}$$

with C_0 being the constant of integration.

Due to the definition (16) of the stream function, one gets

$$\widetilde{w} = -(\rho S r)^{-1} \partial_z \psi, \quad u = (\rho S)^{-1} \partial_\rho \psi + D \partial_z \widetilde{w}. \tag{21}$$

Inserting (21) into (15), one obtains the final equation for the stream function ψ, which can be linearized with respect to $u << w$ and solved by separating variables:

$$\psi = R(\rho) Z(z). \tag{22}$$

3.3 Structure of the stream function: perturbation approach

Now we suppose that the radial filter in question has the external radius $a >> d$, the internal radius ρ_0 and several plates (layers) of the height 2ℓ. Therefore, the boundary condition for the fluid velocity reads:

$$u(\rho, z = \pm\ell) = 0, \quad -\ell \le z \le \ell, \quad \rho_0 \le \rho \le a. \tag{23}$$

It is worth - while to notice that in our case the following two small parameters can be introduced:

$$\varepsilon = k^* w^2 << 1, \quad \mu = d / a << 1.$$

This fact permits one, in the first approximation, to neglect the terms in (15) containing the gravity acceleration g and the diffusion coefficient D. Taking into account that the asymmetry of $\widetilde{w}$ in z depends on g, one can suppose that

$$\widetilde{w}(-z) = \widetilde{w}(z), \quad \psi(-z) = -\psi(z). \tag{24}$$

In view of the boundary condition (23), the property (24) implies the simplest structure of $Z(z)$:

$$Z(z) = \sin(\lambda_n z), \quad \lambda_n = n\pi / \ell, \quad n = 1, 2, 3, \ldots \tag{25}$$

Inserting (25), (21) and (22) into (15), one obtains in the first approximation the following equation for the function $R(\rho)$:

$$\left[\overline{w} \left(\frac{R'}{\rho S} \right)' + \frac{k_0 R'}{\rho S} \right]' = \lambda_n^2 \left[\left(\frac{\overline{w} R}{\rho S r} \right)' + \frac{k_0 R}{\rho S r} \right]. \tag{26}$$

The equation (26) can be simplified if one supposes that the function $r(\rho)$ weakly depends on ρ:

$$r(\rho) \approx r_0 = const.$$

IOP Conf. Series: Materials Science and Engineering **675** (2019) 012064 doi:10.1088/1757-899X/675/1/012064

Using the denotations:

$$\lambda_0 = \lambda_n^2 / r_0, \quad \sigma_0 = k_0 r_0 / C_0, \quad p = \rho S \tag{27}$$

and introducing the new dependent variable

$$X(\rho) = (R'/p)' - \lambda_0 R / p, \tag{28}$$

one derives from (26) the following equation for $X(\rho)$:

$$(X/p)' + \sigma_0 X = 0. \tag{29}$$

The equation (29) has the evident solution:

$$X(\rho) = pY_0 \exp\left(-\sigma_0 \int_{\rho_0}^{\rho} p\, d\rho\right), \tag{30}$$

with Y_0 being the constant of integration. Inserting (30) into (28), one finds the non-homogeneous equation for the function $R(\rho)$, which can be solved in some particular cases.

3.3.1 The case of constant porosity $S = S_0 = const$

In this case $p = S_0 \rho$ and from (30) one gets

$$X = pY_0 \exp\left[-\gamma(\rho^2 - \rho_0^2)\right], \quad \gamma = \sigma_0 S_0 / 2. \tag{31}$$

Inserting (31) into (28), one finds the following equation for $R(\rho)$:

$$\rho^{-1}(\rho^{-1}R')' - \lambda_0 \rho^{-2} R = S_0^2 Y_0 \exp\left[-\gamma(\rho^2 - \rho_0^2)\right] \tag{32}$$

Taking into account that the homogeneous part of the equation (32) admits the solution

$$R_0 = \rho K_1(k_n \rho), \quad k_n^2 = \lambda_0 = \lambda_n^2 / r_0,$$

where K_1 is the Macdonald function of the first order (or the modified Bessel function) [19], one can represent the radial function $R(\rho) \equiv R_n(\rho)$ in the form:

$$R_n(\rho) = \rho K_1(k_n \rho) U_n(\rho). \tag{33}$$

Using the substitution (33) and taking into account the equation for $K_n(z)$:

$$K_n'' + z^{-1} K_n' = (1 + n^2 z^{-2}) K_n,$$

one can deduce from (32) the corresponding equation for the function $U_n(\rho)$:

$$(\rho K_1^2 U_n')' = Y_0 S_0^2 \rho^2 K_1 \exp\left[-\gamma(\rho^2 - \rho_0^2)\right] \tag{34}$$

It should be noted that solution to (34) can be found by simple double integration.

Finally, the resulting stream function takes the form

$$\psi(\rho, z) = \sum_{n=1}^{\infty} U_n(\rho)\rho K_1(k_n\rho)\sin(n\pi z/\ell), \tag{35}$$

and the velocity components can be easily found through (21).

It is worth - while to underline that the functions $U_n(\rho)$ contain three constants of integration, which can be found through fixing the boundary values of the functions $w, u, \partial_\rho w$ at the central radius $\rho = \rho_0$.

3.3.2 The case of the porosity $S = S_1/\rho$, $S_1 = const$

In this case $p = \rho S = S_1$ and the radial function $R(\rho)$ satisfies the equation

$$R'' - \lambda_0 R = Y_0 \exp[-\sigma_0 S_1(\rho - \rho_0)]. \tag{36}$$

The corresponding substitution reads

$$R_n(\rho) = U_n(\rho)\exp(-k_n\rho),$$

and the equation for $U_n(\rho)$ takes the simple form:

$$\left[U_n' \exp(-2k_n\rho)\right]' = Y_0 \exp(-k_n\rho)\exp[-\sigma_0 S_1(\rho - \rho_0)].$$

3.3.3 The case of the porosity $S = S_2/\rho^2$, $S_2 = const$

In this case $p = \rho S = S_2/\rho$ and the equation (29) has the following solution:

$$X(\rho) = X_0(\rho_0/\rho)^{\sigma+1}, \quad \sigma = \sigma_0 S_2, \quad X_0 = const. \tag{37}$$

Inserting (37) into (28), one finds the equation for $R_n(\rho)$:

$$\rho R_n'' + R_n' - k_n^2 \rho R_n = S_2 X_0(\rho_0/\rho)^{\sigma+1}. \tag{38}$$

The equation (38) can be solved through the substitution

$$R_n = U_n(\xi)K_0(\xi), \quad \xi = k_n\rho, \tag{39}$$

where K_0 stands for the Macdonald function of the zero order and $U_n(\xi)$ satisfies the equation

$$(\xi K_0^2 U_n')' = X_0 k_n^{\sigma} K_0(\rho_0/\xi)^{\sigma+1}. \tag{40}$$

The solution U_n to the equation (40) can be found by simple double integration.

4 Discussion

For analyzing the structure of the stationary axially-symmetric mass balance equation we used the lattice approximation method and introduced the porosity *S* and the branching coefficients *r, p, q*. In the continuous limit we obtained the differential equation of the second order describing the transverse diffusion effect, the corresponding diffusion coefficient (9) being proportional to the vertical branching coefficient q and the diameter of the grain d.

The next step concerned generalizing the Darcy force in the Euler equation through including the natural invariant (10) into the Darcy coefficient. The special structure of the Darcy force permits one to exclude the pressure from the Euler equation and obtain the main equation (15) for the velocity components. The equation (15) can be solved via the perturbation method, if one neglects the vertical velocity component. The structure of the corresponding Stokes stream function can be found by separating variables. In the first approximation, if one neglects the terms containing the gravity acceleration, the Stokes function has the following remarkable property: it appears to be an odd function with respect to the vertical coordinate z. It should be noticed that the Stokes function can be found analytically or expressed through the Macdonald functions for some special choices of the porosity.

Conclusions

1. We studied the hydrodynamics of radial flows in non-homogeneous porous media and showed that the equations of motion should be modified to take into account the effect of the transverse diffusion and that of generalizing the Darcy's law.
2. Due to introducing the Stokes stream function the special perturbation approach was elaborated for solving the modified equations.
3. For some particular choices of the porosity structure $S(\rho)$ the analytical solutions for the stream function were obtained.

Acknowledgements
This publication has been prepared with the support of the RUDN University Program 5-100.

References

[1] Bear J 1988 *Dynamics of Fluids in Porous Media* (Mineola: Dover Publications)
[2] Cheremisinoff N P and Azbel D S 1998 *Liquid Filtration* (Boston: Butterworth-Heinemann)
[3] Dullien F A L 2012 *Porous Media: Fluid Transport and Pore Structure* (San Diego: Academic Press)
[4] Kim S and Karila S J 1991 *Microhydrodynamics: Principles and Selected Applications* (Boston: Butterworth-Heinemann)
[5] Pinder G F and Gray W G 2008 *Essentials of Multiphase Flow and Transport in Porous Media* (New York: John Wiley & Sons)
[6] Sahimi M 2011 *Flow and Transport in Porous Media and Fractional Rock* (New York: John Wiley & Sons)
[7] Sheidegger A E 1960 *The Physics of Flow Through Porous Media* (New York: MacMillan)
[8] Sheidegger A E 1964 Statistical hydrodynamics in porous media *Journal of Applied Physics* **25** 997
[9] Vafai K 2015 *Handbook of Porous Media* (Boca Raton: CRC Press, Taylor & Francis Group)
[10] Fara H D, Sheidegger A E 1961 Statistical geometry of porous media *Journal of Geophysical Research* **66** 3279
[11] Harlemaii D R F, Rumer R R 1963 Longitudinal and lateral dispersion in an isotropic porous medium *Journal of Fluid Mechanics* **16** 385
[12] Josselin de Jong G 1958 Longitudinal and transverse diffusion in granular deposits *Transactions of American Geophysical Union* **39** 67
[13] Saffman P G 1959 A theory of dispersion in a porous medium *Journal of Fluid Mechanics* **6** 321
[14] Sheidegger A E 1961 General theory of dispersion in porous media *Journal of Geophysical Research* **66** 3273
[15] Darcy H 1856 *Les Fontaines Publiques de la Ville de Dijon* (Paris: Dalmont)

[16] Polubarinova - Kochina P Ya 1960 *Theory of Ground Water Movement* (Princeton: Princeton University Press)
[17] Whitaker S 1966 The equations of motion in porous media *Chemical Engineering Science* **21** 291
[18] Whitaker S 1986 Flow in porous media I: A theoretical derivation of Darcy's law *Transport in Porous Media* **1** 3
[19] Whittaker E T and Watson G N 1927 *A Course of Modern Analysis Vol 1, 2* (London: Cambridge University Press)
[20] Olsen H W 1965 Deviations from Darcy's law in saturated clays *Proc. Soil Sci. Soc. Amer.* **29** 135
[21] Firdaouss M, Guermond J-L, Le Quéré P 1997 Nonlinear corrections to Darcy's law at low Reynolds numbers *J. Fluid Mech.* **343** 331

IOP Conf. Series: Materials Science and Engineering **675** (2019) 012065 doi:10.1088/1757-899X/675/1/012065

The choice of dynamic variables in quantum theory

N V Samsonenko[1*], F Ndahayo[2], A Usman Manga[3] and M V Semin[1]

[1] Peoples' Friendship University of Russia (RUDN University), Moscow, Russia
[2] University of Rwanda, Butare, Rwanda
[3] University of Abdou Mumuni, Niamey, Niger

E-mail: nsamson@bk.ru

Abstract. The influence of various choices of dynamic variables in quantum theory on the interpretation of measurement results and experiments are discussed. Dynamic variables in the generally accepted quantum mechanics have no visual sense, which in turn leads to the loss of the image of the studied phenomena and processes of the microworld. Thus, quantum mechanics forces us to abandon the possibility of even mentally imagining the conditions that determine the behavior of an individual system in a quantum region. Also, various representations of the wave function and their use in solving dynamic equations are shown. By now, quantum mechanics is not a complete theory. It resembles classical (Newtonian) mechanics, which explains how movement occurs as a result of the action of certain forces, but does not say what kind of forces it is. Thus, before the advent of basic theories, the application of quantum mechanics to specific physical problems always implied the need to build certain guesses about the nature of these forces.

1. Introduction

Studies have shown that the existing large number of interpretations of quantum mechanics is partly due to the different choice of generalized dynamic variables for describing the same physical system. Very often, by definition, these variables have no visual physical meaning, which makes it difficult to interpret a theory [1]. For example, by now it is already well known that the same elementary particle can be described by both a reducible and an irreducible representation of the Lorentz group.

The clearest example of this is the elegant reduction [2] of the Klein-Gordon equation for a scalar particle (s = 0)

$$(\in + m^2)\varphi = 0 \qquad [\in \text{ is the d'Alembert operator}] \tag{1}$$

to the Petiau-Duffin-Kemmer equation [3-5], which is usually written in the form of the Dirac equation (but with matrices β_μ instead of matrices γ_μ). In our work we use the Pauli metric (+1, +1, +1, +1), $x_\mu = x^\mu = (x_1, x_2, x_3, x_4 = ict)$, in which "co" and "contr" are variant components of vectors do not differ and all the γ - matrices are Hermitian $\left(\gamma_\mu{}^+ = \gamma_\mu\right)$ [6]:

$$\gamma_k = -i\beta\alpha_k \; (k = 1,2,3), \quad \gamma_4 = \beta, \quad \gamma_5 = -\rho_1 = \gamma_1\gamma_2\gamma_3\gamma_4 = i\alpha_1\alpha_2\alpha_3.$$

$$\left(\beta^\mu \frac{\partial}{\partial x^\mu} + m\right)\psi = 0\,. \tag{2}$$

Here, the 5-component wave function realizes the reducible representation of the Lorentz group and consists of a scalar and a 4-vector:

$$\psi = \begin{pmatrix} \varphi \\ \varphi_\mu \end{pmatrix}. \tag{3}$$

In this case, the matrices β satisfy the Petiau-Duffn-Kemmer algebra (23).

In addition, it is obvious that any linear equation for a complex wave function can be written (separating the real and imaginary parts) as a system of equations only for real functions, but which will be related to each other by the normalization condition. There are such (equivalent to the usual) formulations of quantum mechanics without the explicit use of complexity.

The objective of the current study was to explain that the use of complex wave functions is one of the convenient ways to describe quantum systems, and not their obligatory attribute, as is often considered. The use of complex dynamic variables increases the number of degrees of freedom of a physical system. For example, in quantum field theory, this allows one to simply describe the charges of particles, assuming that real fields do not have charges.

2. Materials and methods of research

It is well known that the wave function of a particle in quantum mechanics is determined up to a unitary transformation. However, there are still "dark" spots.

As is known, the Dirac equation

$$\left(\gamma^{\mu}\frac{\partial}{\partial x^{\mu}}+m_0\right)\psi_D=0\,, \tag{4}$$

describes particles with left and right helicities, which is evident from the identical representation of the wave function in the form:

$$\psi_D\equiv\frac{1}{2}(1+\gamma_5)\psi+\frac{1}{2}(1-\gamma_5)\psi=\psi_L+\psi_R\,. \tag{5}$$

In the standard model, only the left-handed states of massless leptons with exact γ_5 - invariance (chiral symmetry) enter the seed lepton sector:

$$\gamma_5\psi_L=\psi_L\,. \tag{6}$$

In 1952, Corben [6] proposed to add a pseudoscalar mass to the Dirac equation:

$$\left(\gamma^{\mu}\frac{\partial}{\partial x^{\mu}}+m_0+im_c\gamma_5\right)\psi_C=0\,. \tag{7}$$

When considering specific problems, the Dirac-Corben equation (7) led to the same predictions [7] as the Dirac equation. Later [8] it was shown that there is a unitary transformation

$$\psi_C=U\psi_D, \tag{8}$$

$$U=\frac{1}{\sqrt{2}}\sqrt{1+\frac{m_0}{m}}\left[1+i\rho_1\frac{m_c}{m}\left(1+\frac{m_0}{m}\right)^{-1}\right], \tag{9}$$

which reduces the Dirac-Corben equation to the Dirac equation, in which the mass term takes the form

$$m=\sqrt{m_0^2+m_c^2}\,. \tag{10}$$

That is, the square of the mass can take on different values:

$$m^2=m_0^2+m_c^2\;\Rightarrow\;\begin{cases}>0 \text{ for bradyons } (v<c),\\ =0 \quad \text{for luxons} \quad (v=c),\\ <0 \text{ for tachyons } (v>c).\end{cases}$$

All operators are converted by the usual rule:

$$O_c=UO_DU^{-1}\,. \tag{11}$$

Using the notation $\frac{m_0}{m}=cos\theta,\ \ \frac{m_c}{m}=sin\theta$, the unitary operator (9) can be written in a more compact form [9]:

$$U = \exp\left(-i\gamma_5 \frac{\theta}{2}\right). \tag{12}$$

Let us make several important observations regarding the Dirac-Corben equation:

1) If the scalar mass is zero ($m_0 = 0$) from (7), we obtain the equation

$$\left(\gamma^\mu \frac{\partial}{\partial x^\mu} + i m_c \gamma_5\right)\psi = 0\,, \tag{13}$$

in which the real mass $m_c > 0$ according to (10) in the experiments will be observed as the usual scalar mass m_0 in the Dirac equation:

$$m_0 = \sqrt{m_c{}^2}\,. \tag{14}$$

2) If $m_0 = 0$ and $m_c = i\mu$ (tachyon), then in this case the transition operator to the Dirac equation is no longer unitary and the physical equivalence between equations (4) and (7) no longer exists.

3) When $m_0 \neq 0$ and $m_c = i\mu$ for the effective mass, we get

$$m = \sqrt{m_0{}^2 - \mu^2}\,, \tag{15}$$

and a particle can be a tachyon ($v > c$) with $m_0 < \mu$, a luxon ($v = c$) with $m_0 = \mu$ and a bradion ($v < c$) with $m_0 > \mu$.

4) When $m_c = i m_0$, we obtain from (7) the equation

$$\left[\gamma^\mu \frac{\partial}{\partial x^\mu} + m_0(1-\gamma_5)\right]\psi = 0\,. \tag{16}$$

This equation is γ_5 - invariant for right-handed neutrinos with an effective zero mass

$$m = \sqrt{m_0{}^2 - \mu^2} = 0. \tag{17}$$

The number m_0 has the meaning of the eigenvalue of the operator of the projection of the spin on the left chiral states (with the right helicity), since

$$m_c(1-\gamma_5)\psi_L = 0, \qquad m_c(1-\gamma_5)\psi_R = m_c(1-\gamma_5)\psi. \tag{18}$$

Consider the problem of relativistic quantum equations from a more general point of view. It is known [10] that by extracting the square root from the Klein-Gordon operator $\left(\sqrt{\Box + m^2}\right)$, one can obtain infinitely many equations of the Dirac type, but with different matrices:

$$\left(\gamma^\mu \frac{\partial}{\partial x^\mu} + m\right)\psi = 0\,, \tag{19}$$

$$\left(\beta^\mu \frac{\partial}{\partial x^\mu} + m\right)\psi = 0\,, \tag{20}$$

$$\left(u^\mu \frac{\partial}{\partial x^\mu} + m\right)\varphi = 0\,. \tag{21}$$

Here the matrices γ^μ, β^μ, u^μ satisfy the Dirac-Clifford algebra

$$\gamma^\mu\gamma^\nu + \gamma^\nu\gamma^\mu = 2\delta^{\mu\nu}, \tag{22}$$

Petiau-Duffin-Kemmer algebra

$$\beta^\mu\beta^\nu\beta^\alpha + \beta^\alpha\beta^\nu\beta^\mu = \delta^{\mu\nu}\beta^\alpha + \delta^{\alpha\nu}\beta^\mu \tag{23}$$

and abelian algebra

$$u^\mu u^\nu - u^\nu u^\mu = 0\,. \tag{24}$$

In the last equation, u^μ is the 4-particle velocity vector.

The relativistic covariance of equations (19), (20), (21) is either obvious (equation (21)), or simply proved. We emphasize that there are infinitely many equations of each type (19), (20), (21), if you do not limit the rank of the Lorentz group representations to which the wave functions ψ, φ belong, but for practical purposes most often (but not always [11], [12]) equations only with matrices of lowest rank are used.

But even algebras (22), (23), (24) do not exhaust all possibilities. Consider this on the example of Maxwell's equation.

Since the time of Majorana [13], dozens of formulations of these equations have been known in the form of (19) and (20), including when mapping the field strengths $\vec{E}$ and $\vec{H}$ (electromagnetic field tensor $F_{\mu\nu}$) to spaces of spinors and quaternions.

We introduce real non-Hermitian matrices $\vec{\sigma} = (\sigma_1, \sigma_2, \sigma_3)$ of rank 3, which are the generators of the group of three-dimensional rotations SO(3) [14]

$$\sigma_3 = \begin{pmatrix} 0 & 1 & 0 \\ -1 & 0 & 0 \\ 0 & 0 & 0 \end{pmatrix}, \quad \sigma_2 = \begin{pmatrix} 0 & 0 & -1 \\ 0 & 0 & 0 \\ 1 & 0 & 0 \end{pmatrix}, \quad \sigma_1 = \begin{pmatrix} 0 & 0 & 0 \\ 0 & 0 & 1 \\ 0 & -1 & 0 \end{pmatrix}, \tag{25}$$

satisfying commutation relations

$$\sigma_1\sigma_2 - \sigma_2\sigma_1 = \sigma_3\,. \tag{26}$$

Using the obvious identity $(\vec{\sigma}\cdot\vec{\nabla})\vec{a} = rot\vec{a}$ Maxwell's equation in vacuum we write in the form:

$$\frac{\partial\vec{E}}{\partial t} - (\vec{\sigma}\cdot\vec{\nabla})\vec{H} = 0, \tag{27.a}$$

$$\frac{\partial\vec{H}}{\partial t} - (\vec{\sigma}\cdot\vec{\nabla})\vec{E} = 0. \tag{27.b}$$

Multiplying equation (27.b) by "i" and, adding two equations, we get:

$$\vec{F} = \vec{E} + i\vec{H}, \qquad \left[\frac{\partial}{\partial t} + i(\vec{\sigma}\cdot\vec{\nabla})\right]\vec{F} = 0. \tag{28}$$

Here we have introduced, following Majoran [13], a complex vector $\vec{F}$, satisfying the Cartan isotropy condition:

$$|\vec{F}|^2 = \vec{F}^*\cdot\vec{F} = \vec{E}^2 + \vec{H}^2 \neq 0, \qquad \vec{F}^2 = \vec{F}\cdot\vec{F} = \vec{E}^2 - \vec{H}^2 + 2i(\vec{E}\cdot\vec{H}) = 0. \tag{29}$$

Let us check whether the quadratic equation satisfies the Dalamber wave equation, as is the case with the Dirac equation for each of the 4 complex (or 8 real) components of the wave function separately. To do this, multiply (28) from the left by the adjoint operator in square brackets:

$$\left[\frac{\partial}{\partial t} - i(\vec{\sigma}\cdot\vec{\nabla})\right]\left[\frac{\partial}{\partial t} + i(\vec{\sigma}\cdot\vec{\nabla})\right]\vec{F} = 0, \tag{30}$$

$$\left[\frac{\partial^2}{\partial t^2} + (\vec{\sigma}\cdot\vec{\nabla})(\vec{\sigma}\cdot\vec{\nabla})\right]\vec{F} = 0\,. \tag{31}$$

Considering the relation $(\vec{\sigma}\cdot\vec{\nabla})(\vec{\sigma}\cdot\vec{\nabla})\vec{F} = \vec{\nabla}(\vec{\nabla}\cdot\vec{F}) - \Delta\vec{F}$, which in conventional notation is equivalent to the formula $rot\,rot\,\vec{F} = grad\,div\,\vec{F} - \Delta\vec{F}$, with the additional condition $div\,\vec{F} = 0$, we obtain the wave equation

$$\left(\Delta - \frac{\partial^2}{\partial t^2}\right)\vec{F} = 0. \tag{32}$$

IOP Conf. Series: Materials Science and Engineering **675** (2019) 012065 doi:10.1088/1757-899X/675/1/012065

Separating the real and imaginary parts in equation (32), we obtain 6 scalar wave equations for the fields $\vec{E}$ and $\vec{H}$:

$$\in \vec{H} = 0, \qquad \in \vec{E} = 0. \tag{33}$$

Thus, Maxwell's equations (27.a) and (27.b) can be written in the Dirac form (19), but using the **3-row real non-Hermitian** matrices $\vec{\sigma}$ in (25), which are completely different from the Majorana matrices. Writing Maxwell's equations in the form (28) has certain advantages in the transition to quantum theory, since the complex vector $\vec{F} = \vec{E} + i\vec{H}$ can be considered as **a wave function**, since the quantity $\vec{F}^* \cdot \vec{F} = \vec{E}^2 + \vec{H}^2$ is the energy density, as

$$\int \vec{F}^* \cdot \vec{F}\, dV = \int (\vec{E}^2 + \vec{H}^2)\, dV = \mathcal{E}\,. \tag{34}$$

After normalization to the total energy $\mathcal{E}$ (dividing both parts of expression (34) by the constant $\mathcal{E}$), the expression under the integral sign can be interpreted as **a probability density**:

$$\int \left(\frac{1}{\sqrt{\mathcal{E}}}\vec{F}^*\right)\left(\frac{1}{\sqrt{\mathcal{E}}}\vec{F}\right) dV = 1\,. \tag{35}$$

Interestingly, the introduced non-Hermitian matrices $\vec{\sigma}$ do not satisfy the commutation relations (22), (23), (24).

If we add the complex 3-vector $\vec{F} = \vec{E} + i\vec{H}$ to the complex 4-vector

$$\psi = \begin{Bmatrix} \vec{H} + i\vec{E} \\ i(\mathcal{H} + i\mathcal{E}) \end{Bmatrix}, \tag{36}$$

where the additional scalar $\mathcal{E}$ and pseudoscalar $\mathcal{H}$ fields are introduced (**invariant** under the Lorentz transformations), then we arrive at a generalized symmetric system of Maxwell equations with electric and magnetic charges, allowing for the existence of longitudinal electric and magnetic fields **in vacuum** [15]. In this case, α - representation of 4×4 matrices were used, satisfying the algebra (22), but **not** the γ - matrices by virtue of the additional relation $\alpha^i\alpha^j = i\alpha^k$. In fact, these are the generators of the rank 4 representation of the SU(2) group, as the matrices ρ_i, σ^k, used by Dirac, also satisfying relations (22) and the additional conditions $\rho^i\rho^j = i\rho^k$, $\sigma^i\sigma^j = i\sigma^k$, $(i, j, k = 1, 2, 3$; cyclic permutation of indexes).

3. Results and discussion

The choice of the isotropic vector $\vec{F} = \vec{E} + i\vec{H}$ as a dynamic variable turns out to be very successful, since it allows describing not only particles with integer spin (photons), but also particles with half-integer spin (for example, neutrinos).

This would seem to contradict the generally accepted opinion that the tensor representations of the Lorentz group (in this particular case, $\boldsymbol{F_{\mu\nu}}$) can describe only particles with integer spin. But it turns out that it is not.

Let us demonstrate this using the Cartan method [16], which uses the mapping of 2-component 3-dimensional complex spinors

$$\xi = \begin{pmatrix} \xi_1 \\ \xi_2 \end{pmatrix} \in C^2 \tag{37}$$

on the space of 3-component 3-dimensional complex isotropic vectors:

$$\vec{F} = \vec{E} + i\vec{H} \in C^3\,. \tag{38}$$

We emphasize once again that the isotropy condition

$$\vec{F}^2 = \vec{E}^2 - 2i\vec{E} \cdot \vec{H} + \vec{H}^2 = 0 \tag{39}$$

means the presence of two scalar non-linear links:

$$\vec{E}^2 - \vec{H}^2 = 0, \quad \vec{E} \cdot \vec{H} = 0 . \tag{40}$$

Herewith

$$\vec{F}^2 \neq |\vec{F}|^2 = \vec{F}^* \cdot \vec{F} = \vec{E}^2 + \vec{H}^2 \neq 0. \tag{41}$$

Using mapping [16]

$$b^0(\xi,\tau) = -(\xi_1\tau_2 - \xi_2\tau_1),$$
$$\vec{b}(\xi,\tau) = \begin{bmatrix} \xi_1\tau_1 - \xi_2\tau_2 \\ i(\xi_1\tau_1 + \xi_2\tau_2) \\ -(\xi_1\tau_2 + \xi_2\tau_1) \end{bmatrix} \tag{42}$$

in the particular case of a single spinor ($\xi \equiv \tau$), assuming $b^0(\xi,\xi) = 0$, $\vec{b}(\xi,\xi) = -i(\vec{E} + i\vec{H})$, the equation for neutrinos in the spinor form

$$p^0\xi = (\vec{p} \cdot \vec{\sigma})\xi \tag{43}$$

takes the form of Maxwell-like equations

$$rot\vec{E} + \frac{\partial \vec{H}}{\partial t} = v_k \vec{\nabla} H_k, \tag{44}$$

$$rot\vec{H} - \frac{\partial \vec{E}}{\partial t} = -v_k \vec{\nabla} E_k. \tag{45}$$

Here, the unit vector $\vec{v}$ is directed along the rate of energy transfer and is determined by the formula:

$$\vec{v} = \frac{\vec{J}}{J_0} = \frac{\vec{E} \times \vec{H}}{\vec{E}^2} . \tag{46}$$

Similarly, one can obtain a system of Maxwell-like equations for electrons [17]. In this last case, instead of two already, it will contain 4 equations of the type (44-45) for two waves (**E**, **H**) and ($\mathbf{E}^1, \mathbf{H}^1$), connected by a mass term.

Here are the main properties of the Cartan map:

1. The Cartan map is a non-linear transformation of dynamic variables (from spinors to isotropic vectors)

$$\psi \leftrightarrow \vec{E} + i\vec{H}, \tag{47}$$

similar to nonlinear canonical transformations in classical mechanics.

2. The spinor equation for neutrinos (43) is the Lagrange equation of the second kind without constraints (a total of 4 independent scalar equations).

3. The equations for the neutrino in the Maxwell-like form (44-45) are the Lagrange equations of the first kind with constraints (the number of independent variables does not change: 6-2 constraints = 4).

4. Despite the seemingly noncovariant form of equations (44-45), it can be proved that they satisfy the principle of relativity, like Maxwell's differential equations written in three-dimensional form.

5. The Cartan map preserves the eigenvalues of the operators of physical quantities.

6. It commutes with the relativistic transformations of the Lorentz group.

7. Despite the choice of the tensor representation of the Lorentz group as the dynamic variables for the neutrino from the equations (42-43), after the second quantization follows a half-integer spin $\frac{1}{2}$.

8. The isotropic complex vector $\vec{F}$ plays the role of the wave function.

9. The use of isotropic vectors to calculate the probabilities and cross sections of various processes [18] shows their high efficiency, especially when taking into account the polarizations of particles.

10. Solutions of Maxwell-like nonlinear equations for neutrinos satisfy the Maxwell system of equations (but not vice versa), which makes it possible to treat the neutrino field as a kind of electromagnetic field with half-integer spin quanta $s = \frac{1}{2}$.

4. Conclusion

Summarizing, we can say that the enormous arbitrariness in the choice of dynamic variables for describing relativistic quantum systems of different spin (and, therefore, a huge amount of interpretation possibilities) allows us to hope that in the end optimal values will be found that will clarify their physical meaning in the most simple way. It may finally be possible to clarify the physical essence of the de Broglie wave, although "the explanation of this meaning will be a difficult task for the extended theory of electromagnetism" [19].

5. References

[1] Samsonenko N V 2018 Interpretation of quantum mechanics 100 years after its creation *Metaphysics* **2(28)** pp 62-65
[2] Petiau G 1980 *Geophysical Prospecting* **28** pp 792-804
[3] Petiau G 1936 PhD Thesis University of Paris (Acad. R. Belg. CI. Sci. Mem. Collect.)
[4] Duffin R J 1938 *Phys. Rev.* **54** 1114
[5] Kemmer N 1939 *Proc. Roy. Soc.* **A173 91**
[6] Corben N C 1952 *Nuovo Cimento* **9(11)** pp 1071-1079
[7] Caldirola P and Gulmanelli P 1952 *Nuova Cimento* **9(9)** pp 834-845
[8] Bellomo E and Loinger A 1952 *Nuovo Cimento* **9(12)** pp 1240-1241
[9] Bergmann O and Baker N 1954 *Nuovo Cimento* **11(2)** pp 203-204
[10] Samsonenko N V, Usman Manga A and Ndahayo F 2014 Extraction of the square root of dalambertian *Proc. of the 50th All-Russian Conference on Problems of Dynamics, Particle Physics, Plasma Physics and Optoelectronics* (Moscow, Russia) pp 3-5
[11] Samsonenko N V 1981 Les equation du premier ordre les plus simples pour des particules de spin "0" *Annales de la Fondation Lois de Broglie* **6(4)** pp 341-346
[12] Samsonenko N V 1981 Une nouvelle equation du premier ordre pour une particule de spin "1" *Annales de la Fondation Lois de Broglie* **6(4)** pp 347-372
[13] Majorana E 1928-1932 *Scientific Manuscripts* (Domus Galilaeana, Pisa) p 101, p 3, p 11, pp 16-17, p 83, p 159
[14] Davydov A S 1973 *Quantum Mechanics* (Publishing House Nauka, Moscow) p 82
[15] Samsonenko N V 1977 To the problem of Maxwell's equations in spinor form *Problems of Quantum Physics* (RUDN University Press, Moscow) pp 23-28
[16] Cartan E 1959 *Theory of spinors* (Publishing House Mir, Moscow)
[17] Samsonenko N V and Bulikunzira S 1994 Tensor description of particles with spin ½ *RUDN Journal of Physics* **2** pp 104-117
[18] Samsonenko N V and Ndahayo F 1998 Tensor formalism of the description of direct and inverse β-decay *Russian Academy of Science ser. Physics* **62(1)** pp 12-20
[19] De Broglie L 1923 Quanta de lumière, diffraction et interferences *Comptes Rendus* **177** p 548

IOP Conf. Series: Materials Science and Engineering **675** (2019) 012066 doi:10.1088/1757-899X/675/1/012066

Shells analysis in orthogonal curvilinear coordinate system with variation-difference method

V N Ivanov and A A Shmeleva*

Peoples' Friendship University of Russia (RUDN University), Moscow, Russia

Email: aritad@tutanota.com

Abstract. The variation-difference method is a convenient numerical method for shells of complex forms. It is enough when only cinematic boundary conditions are satisfied because the method is based on the principle of Lagrange. Another advantage of the variation-difference method is the better opportunity to create computer programs based on it. For shell analysis in orthogonal coordinate system as well as for shell analysis in principal curvatures the system of equations describing stress-strain state can be simplified. In this paper the difference between analysis in orthogonal coordinate system and analysis in principal curvatures of the surface is considered. The main distinction of the analysis of shells in orthogonal curvilinear coordinate system is the necessity of determination of components which include curvature of torsion of coordinate lines. The addition of these components in the equations of the theory of shells for the coordinate system in principal curvatures gives possibility to analyze shells in common orthogonal coordinate system. In this article shell analysis in orthogonal coordinate system is applied to shells based on normal cyclic surfaces.

1. Introduction

The object of this work is the analysis of shells of complex forms with the variation-difference method. The most often used shells are shells of simple forms such as some surfaces of revolution, analytic formulas of which are less complex in comparison with other forms of shells [1-5].Using different types of shells in construction requires studying of geometry of these surfaces that is considered in Encyclopedia [6], the technology of construction as well as methods of analysis.

Analytic methods [7] in most cases cannot be applied for shells of non-canonic forms at all or there are difficulties in this application. That is the reason for which numerical methods are used: the difference method, the finite element method [8, 9] and the variation-difference method.

The disadvantage of the difference method is the necessity to satisfy both cinematic and static boundary conditions. This is not necessary if the finite element method or the variation-difference method, which are based on the principle of Lagrange, are used. The software such as ANSYS, NISA, NASTRAN etc. is based on the finite element method [10-11].

When the variation-difference method is used the construction is covered by the difference mesh and the derivatives in the functional of total energy of deformations are replaced by difference equations. Minimizing the functional of total energy of deformations (the function of nodal displacements) and obtaining a system of algebraic equations nodal displacements are got. Using difference derivatives internal forces and deformations in nods are obtained.

The analysis of the stress-strain state is more precise for shells of complex form if the variation-difference method is used. The finite element method takes insufficient into account the external and internal geometry. The variation-difference method is easier to realize in program. The usage of variation-difference method of calculation of plates and shells was considered in the work [7].

At the Department of Civil Engineering of People's Friendship University of Russia (RUDN University) the software for analysis of shells in principal curvatures on the basis of variation-difference method was developed. The subject of this paper is the analysis of shells in orthogonal curvilinear coordinate system. The objective of this work is to find the way to simplify this analysis. In this paper the necessary changes in formulas of shells analysis in principal curvatures are given and program realization is discussed.

2. Materials and methods

In this paper the variation-difference method is used which is convenient to the analysis of shells based on normal cyclic shells in the orthogonal coordinate system. In this article the common orthogonal curvilinear coordinate system is considered, in which coefficient of the second fundamental form of surfaces M is not equal to zero.

Orthogonal curvilinear coordinate system is a coordinate system, where the curvilinear coordinate lines of the surface u and v form a right angle $\chi = 90°$ and, therefore, the coefficient of the first fundamental form F, which is expressed as a dot product equal zero: $F = (\bar{\rho}_u, \bar{\rho}_v) = |\bar{\rho}_u| \cdot |\bar{\rho}_v| \cdot cos\chi = AB \cdot cos\chi$, where A, B, F are coefficients of the first fundamental form of the surface $\varphi_1 = ds^2 = |d\rho^2| = (d\bar{\rho}d\bar{\rho}) = A^2du^2 + 2Fdudv + B^2dv^2, A^2 = (\bar{\rho}_u, \bar{\rho}_u), B^2 = (\bar{\rho}_v, \bar{\rho}_v)$. When a shell is analyzed in the orthogonal coordinate system the equations which characterize its stress-strain state are simplified.

For the coordinate system, in which the coefficient of the second fundamental form M is different from zero:

$$\varphi_2 = (d^2\bar{\rho} \cdot \bar{m}) = -(d\bar{\rho} \cdot d\bar{m}) = Ldu^2 + 2Mdudv + Ndv^2,$$

$$L = \frac{[\bar{\rho}_u\,\bar{\rho}_v\,\bar{\rho}_{uu}]}{\Sigma}, M = \frac{[\bar{\rho}_u\,\bar{\rho}_v\,\bar{\rho}_{uv}]}{\Sigma} N = \frac{[\bar{\rho}_u\,\bar{\rho}_v\,\bar{\rho}_{vv}]}{\Sigma},$$

where Σ is a discriminant of a surface: $\Sigma = \sqrt{A^2B^2 - F^2}$. It is convenient to analyze normal cyclic surfaces in the orthogonal coordinate system. Coefficients of fundamental form of these surfaces are given in the paper [12]. Some examples of using cyclic shells are considered in the articles [13-15].

3. Results and discussion

The coefficient of the first and the second fundamental forms of the normal cyclic surface in the orthogonal coordinate system are expressed as:

$$A = R'^2 + [s' - Rk_s\cos\omega]^2; F = 0; B = R^2;$$

$$L = \frac{R'T_{33} - T_{13}T_{31}}{\sigma}; \ M = \frac{1}{\sigma}RR'k_s\sin\omega\,; \ N = \frac{R^2[s' - Rk_s(\bar{e}\bar{v})]}{\Sigma} = \frac{RT_{13}}{\sigma},$$

where $T_{13} = s' - Rk_s\,cos\,\omega\,; \ T_{31} = s'k_s\,cos\,\omega - Rk_s^2\cos^2\omega + R''$;

$$T_{33} = s'' - 2k_sR'\cos\omega - Rk_s'cos\omega + Rk_s\chi_s(\bar{e}\,\bar{\beta}); \sigma = \frac{\Sigma}{R} = \sqrt{R'^2 + [s' - Rk_s(\bar{e}\bar{v})]^2}.$$

The stress-strain state of elastic shells in the linear analysis is expressed as a system of equations, which consists of equilibrium equations, which are satisfied, when the functional of total energy of deformations is minimum; geometric equations; Hooke's equations; expressions of potential energy of deformation and work of external forces. For calculation using programming methods it is more convenient to use a vector-matrix formulation.

Let coordinate lines of the surface be signified as $\alpha_1 = u, \alpha_2 = v$, coefficients of the first fundamental form $A = A_1, B = A_2$, components of the vector of displacements u_1, u_2, u_3, components of tangential deformations $\varepsilon_1, \varepsilon_2, \varepsilon_{12}$, components of bending deformations χ_1, χ_2, χ_3, components of tangential forces N_1, N_2, S, bending forces M_1, M_2, H, shearing forces Q_1, Q_2.

Deformations and curvatures are expressed by geometric equations. The system of equations in the orthogonal curvilinear coordinate system is written in the linear theory of shells as:

$$\varepsilon_1 = \frac{1}{A_1}\frac{\partial u_1}{\partial \alpha_1} + \frac{1}{A_1 A_2}\frac{\partial A_1}{\partial \alpha_2}u_2 + k_1 u_3;\ \varepsilon_2 = \frac{1}{A_1 A_2}\frac{\partial A_2}{\partial \alpha_1}u_1 + \frac{1}{A_2}\frac{\partial u_2}{\partial \alpha_2} + k_2 u_3;$$

$$\varepsilon_{12} = -\frac{1}{A_1 A_2}\frac{\partial A_1}{\partial \alpha_2}u_1 + \frac{1}{A_2}\cdot\frac{\partial u_1}{\partial \alpha_2} - \frac{1}{A_1 A_2}\frac{\partial A_2}{\partial \alpha_1}u_2 + \frac{1}{A_1}\cdot\frac{\partial u_2}{\partial \alpha_1} - \mathbf{2k_{12}u_3};$$

$$\chi_1 = \frac{1}{A_1}\frac{\partial \gamma_1}{\partial \alpha_1} + \frac{1}{A_1 A_2}\frac{\partial A_1}{\partial \alpha_2}\gamma_2 - \boldsymbol{k_{12}\omega_3};\ \chi_2 = \frac{1}{A_2}\frac{\partial \gamma_2}{\partial \alpha_2} + \frac{1}{A_1 A_2}\frac{\partial A_2}{\partial \alpha_1}\gamma_1 + \boldsymbol{k_{12}\omega_3};$$

$$\chi_3 = \tau_2 + k_2\omega_1 + \boldsymbol{k_{12}\varepsilon_2} = \tau_1 + k_1\omega_2 + \boldsymbol{k_{12}\varepsilon_1},$$

where

$$\gamma_1 = -\frac{1}{A_1}\frac{\partial u_3}{\partial \alpha_1} + k_1 u_1 - \boldsymbol{k_{12}u_2};\ \gamma_2 = -\frac{1}{A_2}\frac{\partial u_3}{\partial \alpha_2} + k_2 u_2 - \boldsymbol{k_{12}u_1};$$

$$\omega_1 = \frac{1}{A_1}\frac{\partial u_2}{\partial \alpha_1} - \frac{1}{A_1 A_2}\frac{\partial A_1}{\partial \alpha_2}u_1 - \boldsymbol{k_{12}u_3};\ \omega_2 = \frac{1}{A_2}\frac{\partial u_1}{\partial \alpha_2} - \frac{1}{A_1 A_2}\frac{\partial A_2}{\partial \alpha_1}u_2 - \boldsymbol{k_{12}u_3};$$

$$\omega_3 = \frac{1}{2A_1 A_2}\left(\frac{\partial A_2 u_2}{\partial \alpha_1} - \frac{\partial A_1 u_1}{\partial \alpha_2}\right);$$

$$\tau_1 = \frac{1}{A_1}\frac{\partial \gamma_2}{\partial \alpha_1} - \frac{1}{A_1 A_2}\frac{\partial A_1}{\partial \alpha_2}\gamma_1;\ \tau_2 = \frac{1}{A_2}\frac{\partial \gamma_1}{\partial \alpha_2} - \frac{1}{A_1 A_2}\frac{\partial A_2}{\partial \alpha_1}\gamma_2,$$

where γ_1, γ_2 are the angles made by unit vectors $\bar{e}_i$ at the direction to the normal $\bar{m}(\bar{e}_i)$ in the plane $(\bar{e}_i, \bar{m})$.

Unlike the analysis of surfaces in principal curvatures, it is required to calculate the coefficients, which contain k_{12}.

The relations between vectors of internal forces and vectors of tangential and bending strains of a mid-surface are determined according to the Hook's law:

$$\bar{T} = c[N]\bar{\varepsilon};\ \bar{M} = d[N]\bar{\chi};\ [N] = \begin{bmatrix} 1 & \nu & 0 \\ \nu & 1 & 0 \\ 0 & 0 & \frac{1-\nu}{2} \end{bmatrix};\ \mathrm{c} = \frac{Eh}{1-\nu^2};\ d = \frac{Eh^3}{12(1-\nu^2)},$$

where $\bar{T}$ is a vector of tangential internal forces; $\bar{M}$ is a vector of bending internal forces; $[N]$ is a elasticity matrix; c is a tangential rigidity of the shell; d is a flexural rigidity of the shell; E is a modulus of elasticity of a material of the shell; ν is a Poisson's ratio; h is a thickness of the shell.

Tangential and bending potential energies can be expressed in terms of an integral taken over a mid-surface of the shell Ω:

$$U_T = \frac{1}{2}\iint_\Omega \bar{T}^* \bar{\varepsilon} d\Omega = \iint_\Omega (N_1\varepsilon_1 + N_2\varepsilon_2 + 2S\varepsilon_3)\, d\Omega;$$

$$U_u = \frac{1}{2}\iint_\Omega \bar{M}^* \bar{\chi} d\Omega = \iint_\Omega (M_1\chi_1 + M_2\chi_2 + 2H\chi_3)\, d\Omega;\ d\Omega = A_1 A_2 d\alpha_1 d\alpha_2.$$

When the Hooke's law is taken into account, the formulas of potential energy are written as:

$$U_T = \frac{1}{2}\iint_\Omega \bar{T}^* \bar{\varepsilon} d\Omega = \frac{c}{2}\iint_\Omega \bar{\varepsilon}^* [N]\bar{\varepsilon} d;\ U_u = \frac{1}{2}\iint_\Omega \bar{M}^* \bar{\chi} d\Omega = \frac{d}{2}\iint_\Omega \bar{\chi}^* [N]\,\bar{\chi} d\Omega.$$

The index * means that it is a transpose of the matrix.

$$\bar{T} = \begin{Bmatrix} N_{\alpha_1} \\ N_{\alpha_2} \\ S \end{Bmatrix} = \begin{Bmatrix} T_1 \\ T_2 \\ T_3 \end{Bmatrix};\ \bar{M} = \begin{Bmatrix} M_{\alpha_1} \\ M_{\alpha_2} \\ H \end{Bmatrix} = \begin{Bmatrix} M_1 \\ M_2 \\ M_3 \end{Bmatrix};\ \bar{\varepsilon} = \begin{Bmatrix} \varepsilon_{\alpha_1} \\ \varepsilon_{\alpha_2} \\ \varepsilon_{12} \end{Bmatrix} = \begin{Bmatrix} \varepsilon_1 \\ \varepsilon_2 \\ \varepsilon_3 \end{Bmatrix};\ \bar{\chi} = \begin{Bmatrix} \chi_{\alpha_1} \\ \chi_{\alpha_2} \\ 2\chi_{\alpha_1\alpha_2} \end{Bmatrix} = \begin{Bmatrix} \chi_1 \\ \chi_2 \\ \chi_3 \end{Bmatrix},$$

where T_1, T_2 are normal tangential forces, $T_3 = S$ is a shearing tangential force; M_1 and $M_2$are bending moments, $M_3 = M_{12}$ is a twisting moment; ε_1 and ε_2 are tangential deformations, $\varepsilon_3 = \gamma_{12}$ is a bending deformation; χ_1 и χ_2 are changes of principle curvature of a shell, $\chi_3 = 2\chi_{12}$ is a twist of a shell.

The work of external forces can be expressed as:

$$A = \iint_{\Omega} \bar{q}^* \bar{u} d\Omega + \sum \bar{p}_i^* \bar{u}_i ; \; \bar{q} = \begin{Bmatrix} q_1 \\ q_2 \\ q_3 \end{Bmatrix}; \; \bar{p} = \begin{Bmatrix} p_1 \\ p_2 \\ p_3 \end{Bmatrix}; \; \bar{u} = \begin{Bmatrix} u \\ v \\ w \end{Bmatrix} = \begin{Bmatrix} u_1 \\ u_2 \\ u_3 \end{Bmatrix},$$

where $\bar{q}$ is a vector of projections of a continuously distributed load onto the coordinate system of the surface and onto the normal to the surface, $\bar{p}$ is a vector of projections of single loads; $\bar{u}$ is a vector of tangential and normal displacements of the mid-surface of the shell; $\bar{u}_i$ is a vector of these displacements in the point where the single load $\bar{p}_i$ is applied.

The expression of tangential and bending deformation in matrix form is:

$$[\varepsilon] = \begin{bmatrix} \varepsilon_1^1(u_1) & \varepsilon_1^2(u_2) & \varepsilon_1^3(u_3) \\ \varepsilon_2^1(u_1) & \varepsilon_2^2(u_2) & \varepsilon_2^3(u_3) \\ \varepsilon_3^1(u_1) & \varepsilon_3^2(u_2) & \varepsilon_3^3(u_3) \end{bmatrix}; \quad [\chi] = \begin{bmatrix} \chi_1^1(u_1) & \chi_1^2(u_2) & \chi_1^3(u_3) \\ \chi_2^1(u_1) & \chi_2^2(u_2) & \chi_2^3(u_3) \\ \chi_3^1(u_1) & \chi_3^2(u_2) & \chi_3^3(u_3) \end{bmatrix}.$$

In ε_ℓ^k and χ_ℓ^k the index k means the number of the column and the number of a distraction, the index ℓ means the number of the row and a number of a deformation ε_ℓ and χ_ℓ .

$$\bar{\varepsilon} = [\varepsilon]\bar{I}; \; \bar{\chi} = [\chi]\bar{I}, \text{ where } \bar{I} = \begin{Bmatrix} 1 \\ 1 \\ 1 \end{Bmatrix} \text{ is a summarizing vector.}$$

The formulas for the potential energy will be:

$$U_T = \frac{c}{2} \iint_{\Omega} \bar{\varepsilon}^* [N] \bar{\varepsilon} d\Omega = \frac{c}{2} \iint_{\Omega} \bar{I}^* [\varepsilon]^* [N] [\varepsilon] \bar{I} d\Omega; \; U_u = \frac{d}{2} \iint_{\Omega} \bar{\chi}^* [N] \bar{\chi} d\Omega = \iint_{\Omega} \bar{I}^* [\bar{\chi}]^* [N] [\bar{\chi}] \bar{I} d\Omega.$$

The components of the deformation matrix are expressed as:

$$\varepsilon_1^1 = \frac{1}{A_1} \frac{\partial u_1}{\partial \alpha_1}; \; \varepsilon_1^2 = \frac{1}{A_1 A_2} \frac{\partial A_1}{\partial \alpha_2} u_2; \; \varepsilon_1^3 = k_1 u_3; \; \varepsilon_2^1 = \frac{1}{A_1 A_2} \frac{\partial A_2}{\partial \alpha_1} u_1; \; \varepsilon_2^2 = \frac{1}{A_2} \frac{\partial u_2}{\partial \alpha_2}; \; \varepsilon_2^3 = k_2 u_3;$$

$$\varepsilon_3^1 = -\frac{1}{A_1 A_2} \frac{\partial A_1}{\partial \alpha_2} u_1 + \frac{1}{A_2} \cdot \frac{\partial u_1}{\partial \alpha_2}; \; \varepsilon_3^2 = -\frac{1}{A_1 A_2} \frac{\partial A_2}{\partial \alpha_1} u_2 + \frac{1}{A_1} \cdot \frac{\partial u_2}{\partial \alpha_1}; \; \varepsilon_3^3 = \boldsymbol{-2 k_{12} u_3};$$

$$\chi_1^1 = \frac{1}{A_1} \cdot \left(\frac{\partial k_1}{\partial \alpha_1} - \boldsymbol{\frac{k_{12}}{2A_2} \frac{\partial A_1}{\partial \alpha_2}} \right) u_1 + \frac{k_1}{A_1} \cdot \frac{\partial u_1}{\partial \alpha_1} + \boldsymbol{\frac{1}{2} \frac{k_{12}}{A_2} \frac{\partial u_1}{\partial \alpha_2}};$$

$$\chi_1^2 = \left(\boldsymbol{-\frac{1}{A_1} \frac{\partial k_{12}}{\partial \alpha_1}} + \frac{k_2}{A_1 A_2} \frac{\partial A_1}{\partial \alpha_2} - \boldsymbol{\frac{k_{12}}{2A_1 A_2} \frac{\partial A_2}{\partial \alpha_1}} \right) u_2 - \boldsymbol{\frac{3}{2} \frac{k_{12}}{A_1} \frac{\partial u_2}{\partial \alpha_1}};$$

$$\chi_1^3 = \frac{1}{A_1^3} \frac{\partial A_1}{\partial \alpha_1} \frac{\partial u_3}{\partial \alpha_1} - \frac{1}{A_1 A_2^2} \frac{\partial A_1}{\partial \alpha_2} \frac{\partial u_3}{\partial \alpha_2} - \frac{1}{A_1^2} \frac{\partial^2 u_3}{\partial \alpha_1^2};$$

$$\chi_2^1 = \left(\boldsymbol{-\frac{1}{A_2} \frac{\partial k_{12}}{\partial \alpha_2}} + \frac{k_1}{A_1 A_2} \cdot \frac{\partial A_2}{\partial \alpha_1} - \boldsymbol{\frac{k_{12}}{2A_1 A_2} \frac{\partial A_1}{\partial \alpha_2}} \right) u_1 - \boldsymbol{\frac{3}{2} \frac{k_{12}}{A_2} \frac{\partial u_1}{\partial \alpha_2}};$$

$$\chi_2^2 = \left(\frac{1}{A_2} \frac{\partial k_2}{\partial \alpha_2} - \boldsymbol{\frac{k_{12}}{2A_1 A_2} \frac{\partial A_2}{\partial \alpha_1}} \right) u_2 + \boldsymbol{\frac{k_{12}}{2A_1} \frac{\partial u_2}{\partial \alpha_1}} + \frac{k_2}{A_2} \frac{\partial u_2}{\partial \alpha_2};$$

$$\chi_2^3 = -\frac{1}{A_1^2 A_2}\frac{\partial A_2}{\partial \alpha_1}\frac{\partial u_3}{\partial \alpha_1} + \frac{1}{A_2^3}\frac{\partial A_2}{\partial \alpha_2}\frac{\partial u_3}{\partial \alpha_2} - \frac{1}{A_2^2}\frac{\partial^2 u_3}{\partial \alpha_2^2};$$

$$\chi_3^1 = \left(\frac{1}{A_2}\frac{\partial k_1}{\partial \alpha_2} + \boldsymbol{\frac{2k_{12}}{A_1 A_2}\frac{\partial A_2}{\partial \alpha_1}} - \frac{k_2}{A_1 A_2}\frac{\partial A_1}{\partial \alpha_2}\right)u_1 + \frac{k_1}{A_2}\frac{\partial u_1}{\partial \alpha_2};$$

$$\chi_3^2 = \left(-\boldsymbol{\frac{1}{A_2}\frac{\partial k_{12}}{\partial \alpha_2}} - \frac{k_2}{A_1 A_2}\frac{\partial A_2}{\partial \alpha_1}\right)u_2 + \frac{k_2}{A_1}\frac{\partial u_2}{\partial \alpha_1};$$

$$\chi_3^3 = \frac{1}{A_1^2 A_2}\frac{\partial A_1}{\partial \alpha_2}\frac{\partial u_3}{\partial \alpha_1} + \frac{1}{A_1 A_2^2}\frac{\partial A_2}{\partial \alpha_1}\frac{\partial u_3}{\partial \alpha_2} - \frac{1}{A_1 A_2}\frac{\partial^2 u_3}{\partial \alpha_1 \partial \alpha_2}.$$

The obtained components of deformations can be written as:

$$\varepsilon_\ell^k(u_k) = \textstyle\sum_{k=0}^{5} a_{\ell,m}^k\, \partial^m u_k; \text{ and } \chi_\ell^k(u_k) = \sum_{k=0}^{5} b_{\ell,m}^k\, \partial^m u_k,$$

where

$$\partial^0 u_k = u_k;\ \partial^1 u_k = \frac{\partial u_k}{\partial \alpha_1};\ \partial^2 u_k = \frac{\partial u_k}{\partial \alpha_2};\ \partial^3 u_k = \frac{\partial^2 u_k}{\partial \alpha_1^2};\ \partial^4 u_k = \frac{\partial^2 u_k}{\partial \alpha_1 \partial \alpha_2};\ \partial^5 u_k = \frac{\partial^2 u_k}{\partial \alpha_2^2},$$

where $a_{\ell,m}^k, b_{\ell,m}^k$ are coefficients of derivatives of the components of deformations.

Coefficients of tangential deformations for a displacement u_1:

for ε_1^1: $a_{1,0}^1 = 0;\ a_{1,1}^1 = \frac{1}{A_1};\ a_{1,m}^1 = 0$ if $m > 1$; for ε_2^1: $a_{2,0}^1 = \frac{1}{A_1 A_2}\frac{\partial A_2}{\partial \alpha_1};\ a_{2,m}^1 = 0$ if $m > 0$;

for ε_3^1: $a_{3,0}^1 = -\frac{1}{A_1 A_2}\frac{\partial A_1}{\partial \alpha_2};\ a_{3,1}^1 = 0;\ a_{3,2}^1 = \frac{1}{A_2};\ a_{3,m}^1 = 0$ if $m > 2$.

Coefficients of tangential deformations for a displacement u_2:

for ε_1^2: $a_{1,0}^2 = \frac{1}{A_1 A_2}\frac{\partial A_1}{\partial \alpha_2};\ a_{1,m}^2 = 0$ if $m > 0$;

for ε_2^2: $a_{2,0}^2 = 0;\ a_{2,1}^2 = 0;\ a_{2,2}^2 = \frac{1}{A_2};\ a_{2,m}^2 = 0$ if $m > 2$;

for ε_3^2: $a_{3,0}^2 = -\frac{1}{A_1 A_2}\frac{\partial A_2}{\partial \alpha_1};\ a_{3,1}^2 = \frac{1}{A_1};\ a_{3,m}^1 = 0$ if $m > 1$.

Coefficients of tangential deformations for a displacement u_3:

$$a_{1,0}^3 = k_1;\ a_{2,0}^3 = k_2;\ a_{3,0}^3 = \boldsymbol{-2k_{12}};\ a_{1,m}^3 = a_{2,m}^3 = a_{3,m}^3 = 0 \text{ if } m > 0.$$

Coefficients of bending deformations for a displacement u_1:

for χ_1^1: $b_{1,0}^1 = \frac{1}{A_1}\cdot\left(\frac{\partial k_1}{\partial \alpha_1} - \boldsymbol{\frac{k_{12}}{2A_2}\frac{\partial A_1}{\partial \alpha_2}}\right);\ b_{1,1}^1 = \frac{k_1}{A_1}; b_{1,2}^1 = \boldsymbol{\frac{1}{2}\frac{k_{12}}{A_2}};\ b_{1,m}^1 = 0$ if $m > 2$;

for χ_2^1: $b_{2,0}^1 = -\boldsymbol{\frac{1}{A_2}\frac{\partial k_{12}}{\partial \alpha_2}} + \frac{k_1}{A_1 A_2}\cdot\frac{\partial A_2}{\partial \alpha_1} - \boldsymbol{\frac{k_{12}}{2A_1 A_2}\frac{\partial A_1}{\partial \alpha_2}};\ b_{2,1}^1 = 0;\ b_{2,2}^1 = -\boldsymbol{\frac{3}{2}\frac{k_{12}}{A_2}};\ b_{2,m}^1 = 0$ if $m > 2$;

for χ_3^1: $b_{3,0}^1 = \frac{1}{A_2}\frac{\partial k_1}{\partial \alpha_2} + \boldsymbol{\frac{2k_{12}}{A_1 A_2}\frac{\partial A_2}{\partial \alpha_1}} - \frac{k_2}{A_1 A_2}\frac{\partial A_1}{\partial \alpha_2};\ b_{3,1}^1 = 0;\ b_{3,2}^1 = \frac{k_1}{A_2};\ b_{3,m}^1 = 0$ if $m > 2$.

Coefficients of bending deformations for a displacement u_2:

for χ_1^2: $b_{1,0}^2 = -\boldsymbol{\frac{1}{A_1}\frac{\partial k_{12}}{\partial \alpha_1}} + \frac{k_2}{A_1 A_2}\frac{\partial A_1}{\partial \alpha_2} - \boldsymbol{\frac{k_{12}}{2A_1 A_2}\frac{\partial A_2}{\partial \alpha_1}};\ b_{1,1}^2 = -\boldsymbol{\frac{3}{2}\frac{k_{12}}{A_1}};\ b_{1,m}^1 = 0$ if $m > 1$;

for χ_2^2: $b_{2,0}^2 = \frac{1}{A_2}\frac{\partial k_2}{\partial \alpha_2} - \boldsymbol{\frac{k_{12}}{2A_1 A_2}\frac{\partial A_2}{\partial \alpha_1}};\ b_{2,1}^2 = \boldsymbol{\frac{k_{12}}{2A_1}};\ b_{2,2}^2 = \frac{k_2}{A_2};\ b_{2,m}^2 = 0$ if $m > 2$;

for χ_3^2: $b_{3,0}^2 = -\boldsymbol{\frac{1}{A_2}\frac{\partial k_{12}}{\partial \alpha_2}} - \frac{k_2}{A_1 A_2}\frac{\partial A_2}{\partial \alpha_1};\ b_{3,1}^2 = \frac{k_2}{A_1};\ b_{3,m}^2 = 0$ if $m > 1$.

Coefficients of bending deformations for a displacement u_3:

for χ_1^3: $b_{1,0}^3 = 0$; $b_{1,1}^3 = \frac{1}{A_1^3}\frac{\partial A_1}{\partial \alpha_1}$; $b_{1,2}^3 = -\frac{1}{A_1 A_2^2}\frac{\partial A_1}{\partial \alpha_2}$; $b_{1,3}^3 = -\frac{1}{A_1^2}$; $b_{1,m}^1 = 0$ if $m > 3$;

for χ_2^3: $b_{2,0}^3 = 0$; $b_{2,1}^3 = -\frac{1}{A_1^2 A_2}\frac{\partial A_2}{\partial \alpha_1}$; $b_{2,2}^3 = \frac{1}{A_2^3}\frac{\partial A_2}{\partial \alpha_2}$; $b_{2,3}^3 = b_{2,4}^3 = 0$; $b_{2,5}^3 = -\frac{1}{A_2^2}$;

for χ_3^3: $b_{3,0}^3 = 0$; $b_{3,1}^3 = \frac{1}{A_1^2 A_2}\frac{\partial A_1}{\partial \alpha_2}$; $b_{3,2}^3 = \frac{1}{A_1 A_2^2}\frac{\partial A_2}{\partial \alpha_1}$; $b_{3,3}^3 = 0$; $b_{3,4}^3 = -\frac{1}{A_1 A_2}$; $b_{3,5}^3 = 0$.

The matrixes of these coefficients will be:

$$\overline{a^l} = \begin{bmatrix} a_{1,0}^k & a_{1,1}^k & a_{1,2}^k & a_{1,3}^k & a_{1,4}^k & a_{1,5}^k \\ a_{2,0}^k & a_{2,1}^k & a_{2,2}^k & a_{2,3}^k & a_{2,4}^k & a_{2,5}^k \\ a_{3,0}^k & a_{3,1}^k & a_{3,2}^k & a_{3,3}^k & a_{3,4}^k & a_{3,5}^k \end{bmatrix}; \ \overline{b^l} = \begin{bmatrix} b_{1,0}^k & b_{1,1}^k & b_{1,2}^k & b_{1,3}^k & b_{1,4}^k & b_{1,5}^k \\ b_{2,0}^k & b_{2,1}^k & b_{2,2}^k & b_{2,3}^k & b_{2,4}^k & b_{2,5}^k \\ b_{3,0}^k & b_{3,1}^k & b_{3,2}^k & b_{3,3}^k & b_{3,4}^k & b_{3,5}^k \end{bmatrix}.$$

$$\overline{a^1} = \begin{bmatrix} 0 & \frac{1}{A_1} & 0 & 0 & 0 & 0 \\ \frac{1}{A_1 A_2}\frac{\partial A_2}{\partial \alpha_1} & 0 & 0 & 0 & 0 & 0 \\ -\frac{1}{A_1 A_2}\frac{\partial A_1}{\partial \alpha_2} & 0 & \frac{1}{A_2} & 0 & 0 & 0 \end{bmatrix}; \ \overline{a^2} = \begin{bmatrix} \frac{1}{A_1 A_2}\frac{\partial A_1}{\partial \alpha_2} & 0 & 0 & 0 & 0 & 0 \\ 0 & 0 & \frac{1}{A_2} & 0 & 0 & 0 \\ -\frac{1}{A_1 A_2}\frac{\partial A_2}{\partial \alpha_1} & \frac{1}{A_1} & 0 & 0 & 0 & 0 \end{bmatrix}; \ \overline{a^3} = \begin{bmatrix} k_1 & 0 & 0 & 0 & 0 & 0 \\ k_2 & 0 & 0 & 0 & 0 & 0 \\ -2\boldsymbol{k_{12}} & 0 & 0 & 0 & 0 & 0 \end{bmatrix};$$

$$\overline{b^1} = \begin{bmatrix} \frac{1}{A_1}\cdot\left(\frac{\partial k_1}{\partial \alpha_1} - \frac{\boldsymbol{k_{12}}}{\boldsymbol{2A_2}}\frac{\boldsymbol{\partial A_1}}{\boldsymbol{\partial \alpha_2}}\right) & \frac{k_1}{A_1} & \frac{\boldsymbol{1}}{\boldsymbol{2}}\frac{\boldsymbol{k_{12}}}{\boldsymbol{A_2}} & 0 & 0 & 0 \\ -\frac{\boldsymbol{1}}{\boldsymbol{A_2}}\frac{\boldsymbol{\partial k_{12}}}{\boldsymbol{\partial \alpha_2}} + \frac{k_1}{A_1 A_2}\cdot\frac{\partial A_2}{\partial \alpha_1} - \frac{\boldsymbol{k_{12}}}{\boldsymbol{2A_1 A_2}}\frac{\boldsymbol{\partial A_1}}{\boldsymbol{\partial \alpha_2}} & 0 & -\frac{\boldsymbol{3}}{\boldsymbol{2}}\frac{\boldsymbol{k_{12}}}{\boldsymbol{A_2}} & 0 & 0 & 0 \\ \frac{1}{A_2}\frac{\partial k_1}{\partial \alpha_2} + \frac{\boldsymbol{2k_{12}}}{\boldsymbol{A_1 A_2}}\frac{\boldsymbol{\partial A_2}}{\boldsymbol{\partial \alpha_1}} - \frac{k_2}{A_1 A_2}\frac{\partial A_1}{\partial \alpha_2} & 0 & \frac{k_1}{A_2} & 0 & 0 & 0 \end{bmatrix};$$

$$\overline{b^2} = \begin{bmatrix} -\frac{\boldsymbol{1}}{\boldsymbol{A_1}}\frac{\boldsymbol{\partial k_{12}}}{\boldsymbol{\partial \alpha_1}} + \frac{k_2}{A_1 A_2}\frac{\partial A_1}{\partial \alpha_2} - \frac{\boldsymbol{k_{12}}}{\boldsymbol{2A_1 A_2}}\frac{\boldsymbol{\partial A_2}}{\boldsymbol{\partial \alpha_1}} & -\frac{\boldsymbol{3}}{\boldsymbol{2}}\frac{\boldsymbol{k_{12}}}{\boldsymbol{A_1}} & 0 & 0 & 0 & 0 \\ \frac{1}{A_2}\frac{\partial k_2}{\partial \alpha_2} - \frac{\boldsymbol{k_{12}}}{\boldsymbol{2A_1 A_2}}\frac{\boldsymbol{\partial A_2}}{\boldsymbol{\partial \alpha_1}} & \frac{\boldsymbol{k_{12}}}{\boldsymbol{2A_1}} & \frac{k_2}{A_2} & 0 & 0 & 0 \\ -\frac{\boldsymbol{1}}{\boldsymbol{A_2}}\frac{\boldsymbol{\partial k_{12}}}{\boldsymbol{\partial \alpha_2}} - \frac{k_2}{A_1 A_2}\frac{\partial A_2}{\partial \alpha_1} & \frac{k_2}{A_1} & 0 & 0 & 0 & 0 \end{bmatrix};$$

$$\overline{b^3} = \begin{bmatrix} 0 & \frac{1}{A_1^3}\frac{\partial A_1}{\partial \alpha_1} & -\frac{1}{A_1 A_2^2}\frac{\partial A_1}{\partial \alpha_2} & -\frac{1}{A_1^2} & 0 & 0 \\ 0 & -\frac{1}{A_1^2 A_2}\frac{\partial A_2}{\partial \alpha_1} & \frac{1}{A_2^3}\frac{\partial A_2}{\partial \alpha_2} & 0 & 0 & -\frac{1}{A_2^2} \\ 0 & \frac{1}{A_1^2 A_2}\frac{\partial A_1}{\partial \alpha_2} & \frac{1}{A_1 A_2^2}\frac{\partial A_2}{\partial \alpha_1} & 0 & -\frac{1}{A_1 A_2} & 0 \end{bmatrix}.$$

Let us use the vector of derivatives:

$$\bar{\partial}^* = \{\partial^0, \partial^1, \partial^2, \partial^3, \partial^4, \partial^5\} = \{1, \frac{\partial}{\partial \alpha_1}, \frac{\partial}{\partial \alpha_2}, \frac{\partial^2}{\partial \alpha_1^2}, \frac{\partial^2}{\partial \alpha_1 \partial \alpha_2}, \frac{\partial^2}{\partial \alpha_2^2}\}.$$

Then the matrix formulation of the vector of components of deformations can be written as:

$$\bar{\varepsilon}^k(u_k) = \bar{a}^k \bar{\partial} u_k; \ \bar{\chi}^k(u_k) = \bar{b}^k \bar{\partial} u_k, \ \bar{\varepsilon} = \textstyle\sum_{i=1}^3 \bar{a}^k \bar{\partial} u_k; \ \bar{\chi} = \sum_{i=1}^3 \bar{b}^k \bar{\partial} u_k.$$

Using these formulas we will have the expressions of the potential energy of deformation as:

$$U_T = \frac{c}{2}\iint_\Omega \textstyle\sum_{k=1}^3 \sum_{\ell=1}^3 \left(\bar{a}^k \bar{\partial} u_k\right)^* [N] \left(\bar{a}^\ell \bar{\partial} u_\ell\right) d\Omega; \ U_u = \frac{d}{2}\iint_\Omega \sum_{k=1}^3 \sum_{\ell=1}^3 \left(\bar{b}^k \bar{\partial} u_k\right)^* [N] \left(\bar{b}^\ell \bar{\partial} u_\ell\right) d\Omega.$$

Calculating with the variation-difference method includes replacing derivatives in the functional of potential energy with difference equations, and the shell is covered by the mesh of constant or variable step (Figure 1). Variable step is ordinary used, if there are domains where the function is changing fast.

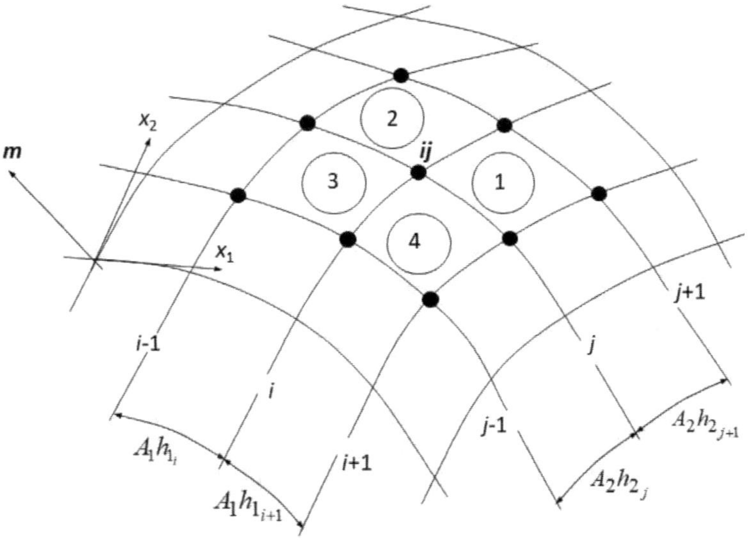

Figure 1. Coordinate mesh at region of nod ij.

Let us consider a coordinate mesh with the axes α_1, α_2. Let the domain of the mesh, which surround the node ij located at the intersection of axes i and j be divided in four quadrants (Fig. 2). There will be four types of difference equations (left and right) for these quadrants. In this case the solution will be more precise. For the first quadrant a step forward in both coordinates is taken (++), for the second quadrant it is a step back in α_1 and a step forward in α_2 (- +), for the third it is a step back in both α_1 and α_2 (- -) and for the fourth it is a step forward in α_1 and a step back in α_2 (+ -).

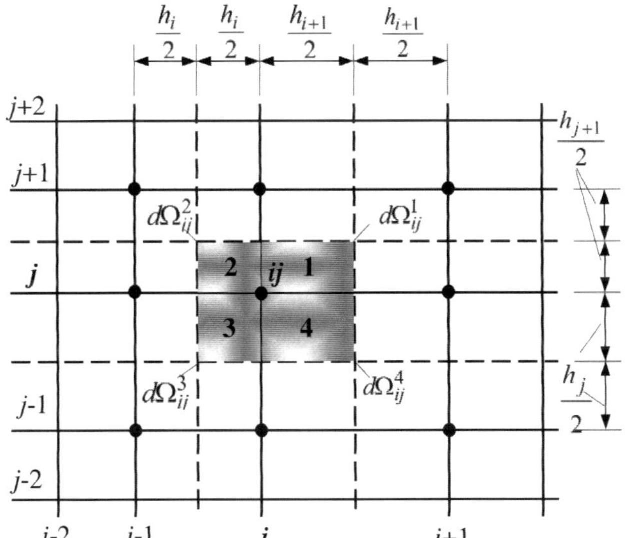

$$d\Omega'_{ij} = A_1^{ij} A_2^{ij} d\tilde{\Omega}$$

Figure 2. Nod points at the *region of nod ij* and subregions of integration

$$d\tilde{\Omega}_{ij}^1 = \frac{h_{i+1}h_{j+1}}{4};$$

$$d\tilde{\Omega}_{ij}^2 = \frac{h_i h_{j+1}}{4}, \quad d\tilde{\Omega}_{ij}^3 = \frac{h_i h_j}{4}$$

Using this division first derivatives are written as:

$$\left.\frac{\partial u_k}{\partial \alpha_1}\right|_+^{ij} = \frac{u_k^{i+1,j} - u_k^{i,j}}{h_{1_{i+1}}}; \quad \left.\frac{\partial u_k}{\partial \alpha_1}\right|_-^{ij} = \frac{u_k^{i,j} - u_k^{i-1,j}}{h_{1_i}}; \quad \left.\frac{\partial u_k}{\partial \alpha_2}\right|_+^{ij} = \frac{u_k^{i,j+1} - u_k^{i,j}}{h_{2_{j+1}}}; \quad \left.\frac{\partial u_k}{\partial \alpha_2}\right|_-^{ij} = \frac{u_k^{i,j} - u_k^{i,j-1}}{h_{2_j}}.$$

Second-order mixed partial derivatives are written as:

$$\left.\frac{\partial^2 u_k}{\partial\alpha_1\partial\alpha_2}\right|_{++}^{ij} = \frac{u_k^{i+1,j+1}-u_k^{i+1,j}-u_k^{i,j+1}+u_k^{i,j}}{h_{1_{i+1}}\cdot h_{2_{j+1}}};\ \left.\frac{\partial^2 u_k}{\partial\alpha_1\partial\alpha_2}\right|_{-+}^{ij} = \frac{u_k^{i,j+1}-u_k^{i,j}-u_k^{i-1,j+1}+u_k^{i-1,j}}{h_{1_i}\cdot h_{2_{j+1}}};$$

$$\left.\frac{\partial^2 u_k}{\partial\alpha_1\partial\alpha_2}\right|_{\pm}^{ij} = \frac{u_k^{i+1,j}-u_k^{i+1,j-1}-u_k^{i,j}+u_k^{i,j-1}}{h_{1_{i+1}}\cdot h_{2_j}};\ \left.\frac{\partial^2 u_k}{\partial\alpha_1\partial\alpha_2}\right|_{--}^{ij} = \frac{u_k^{i,j}-u_k^{i,j-1}-u_k^{i-1,j}+u_k^{i-1,j-1}}{h_{1_i}\cdot h_{2_j}}.$$

Second-order non-mixed partial derivatives based on central difference equations are written as:

$$\left.\frac{\partial^2 u_k}{\partial\alpha_1^2}\right|^{ij} = \frac{2(\beta_1^{ij}u_k^{i-1,j}-(1+\beta_1^{ij})u_k^{i,j}+u_k^{i+1,j}}{\beta_1^{ij}(1+\beta_1^{ij})h_{1_i}^2};\ \left.\frac{\partial^2 u_k}{\partial\alpha_2^2}\right|^{ij} = \frac{2(\beta_2^{ij}u_k^{i,j-1}-(1+\beta_2^{ij})u_k^{i,j}+u_k^{i,j+1}}{\beta_2^{ij}(1+\beta_2^{ij})h_{2_j}^2},$$

where $\beta_1^{ij} = \frac{h_{1_{i+1}}}{h_{1_i}}$; $\beta_2^{ij} = \frac{h_{2_{j+1}}}{h_{2_j}}$; $u_k^{i,j}$ is a displacement of a nod ij with a number k of the displacement; $h_{1_i}, h_{1_{i+1}}$, are steps of the mesh at the left side and at the right side of the nod in α_1, $h_{2_j}, h_{2_{j+1}}$ are analogic steps in α_2.

According to the formulas of derivatives, nine displacements of the nod determine difference derivatives near to this nod. The vector of the displacement in nods is:

$$\{\delta_k^{ij}\}^* = \{u_k^{i-1,j-1}, u_k^{i-1,j}, u_k^{i-1,j+1}, u_k^{i,j-1}, u_k^{i,j}, u_k^{i,j+1}, u_k^{i+1,j-1}, u_k^{i+1,j}, u_k^{i+1,j+1},$$

where $k = 1,2,3$.

After that four types of difference matrix $\bar{d}_t^{ij}$ and the matrix form of the vector of derivatives $\bar{\partial} u_k = \bar{d}_t^{ij}\ \bar{\delta}_k^{ij}$ can be written, where *t* means the number of quadrant near to the nod i, j, $t = 1, 2, 3, 4$. The functional of potential energy of deformation can be written in the difference form as:

$$U_T^{ij} = \frac{c}{2}\iint_\Omega \sum_{k=1}^3 \sum_{\ell=1}^3 (\bar{a}_{ij}{}^k \bar{d}_t^{ij}\ \bar{\delta}_k^{ij})^* [N] (\bar{a}_{ij}{}^\ell \bar{d}_t^{ij} \bar{\delta}_\ell^{ij})\, d\Omega_{ij}^t = \frac{1}{2}(\ \bar{\delta}_k^{ij})^* \bar{r}_T^{ij}\ \bar{\delta}_k^{ij};$$

$$U_u^{ij} = \frac{d}{2}\iint_\Omega \sum_{k=1}^3 \sum_{\ell=1}^3 \left(\bar{b}_{ij}{}^k \bar{d}_t^{ij} \bar{\delta}_k^{ij}\right)^* [N] \left(\bar{b}_{ij}{}^\ell \bar{d}_t^{ij} \bar{\delta}_\ell^{ij}\right) d\Omega_{ij}^t = \frac{1}{2}(\bar{\delta}_k^{ij})^* \bar{r}_u^{ij}\ \bar{\delta}_k^{ij},$$

where $\bar{r}_T^{ij}$ is a matrix of tangential rigidity, $\bar{r}_u^{ij}$ is a matrix of flexural rigidity:

$$\bar{r}_T^{ij} = \frac{c}{2}\iint_\Omega \sum_{t=1}^4 (\bar{a}_{ij}{}^k \bar{d}_t^{ij}\ )^* [N] (\bar{a}_{ij}{}^\ell \bar{d}_t^{ij}\ )\, d\Omega;\ \bar{r}_u^{ij} = \frac{d}{2}\iint_\Omega \sum_{t=1}^4 \left(\bar{b}_{ij}{}^k \bar{d}_t^{ij}\ \right)^* [N] \left(\bar{b}_{ij}{}^\ell \bar{\delta}_\ell^{ij}\right) d\Omega_{ij}^t;$$

$\bar{a}_{ij}{}^k, \bar{b}_{ij}{}^k$ are matrixes for coefficients near this nod i, j; Ω_{ij}^t are subdomains near it.

The functional of potential energy is defined by summarizing nodal integrals. The condition of the minimum of the total potential energy of deformation is $\frac{\partial Э}{\partial Z_k} = \frac{\partial U}{\partial Z_k} - \frac{\partial A}{\partial Z_k} = 0$, that is $\frac{\partial U}{\partial Z_k} = \frac{\partial A}{\partial Z_k}$. Tangential and bending deformations are calculated using difference derivatives, and internal forces are calculated with the formulas of Hook's law. To obtain shearing forces the formulas below are used:

$$Q_1 = \frac{1}{A_1 A_2}\left[\frac{\partial}{\partial\alpha_2}(A_1 M_2) - \frac{\partial A_1}{\partial\alpha_2} M_1 + \frac{1}{A_2}\frac{\partial}{\partial\alpha_1}(A_2^2 H)\right]$$

$$Q_2 = \frac{1}{A_1 A_2}\left[\frac{\partial}{\partial\alpha_1}(A_2 M_1) - \frac{\partial A_2}{\partial\alpha_1} M_2 + \frac{1}{A_1}\frac{\partial}{\partial\alpha_2}(A_1^2 H)\right].$$

Shell analysis in the orthogonal coordinate system is convenient because of the more simple form of the equations than in general form and it gives the opportunity to calculate shells which is impossible to analyze in the coordinate system in principal curvatures. For example, in this coordinate system it is possible to analyze normal cyclic shells, for which in general case generatrix circles are not the lines of principal curvatures. It is true only for tubular surfaces and surfaces of rotation.

The difference between the analysis of surfaces in principal curvatures and their analysis in orthogonal coordinate system is in the geometric equations of deformations and curvatures. In the case of the orthogonal coordinate system the deformation matrix include the components containing k_{12}. With the addition of these components in the analysis it is possible to enlarge the field of application of the software which is developed for analysis of shells in principal curvatures. For the software it is convenient to use the variation-difference method.

Conclusions

Formulas for analysis in the orthogonal curvilinear coordinate system may be obtained through adding to formulas of strains and curvatures in the coordinate system in principal curvatures components which include k_{12}. This allows analysis of structures for which it is inconvenient to use the coordinate system in principal curvatures. Changing of formulas of strains in the program for analysis of shells in the coordinate system in principal curvatures and consideration of the geometry of shells allow analyzing normal cyclic shells in the orthogonal coordinate system.

References

[1] Vasiliev V and Morozov E 2018 *Advanced mechanics of composite materials and structures* Chapter 11 (Amsterdam: Elsevier) pp 761-785

[2] Jasion P 2009 Stability analysis of shells of revolution under pressure conditions *Thin-walled structures* **47** pp 311-317

[3] Krivoshapko S N 2002 Static, vibration and buckling analyses and applications to one-sheet hyperboloidal shells of revolution *Applied Mechanics Reviews* **55** pp 241-270

[4] Yoo Chai H, Lee Sung C 2011 Buckling of thin cylindrical shell elements *Stability of Structures* chapter 9 pp 441-474

[5] Li J, Shi Zh and Liu Lei 2019 A scaled boundary finite element method for static and dynamic analyses of cylindrical shells *Engineering Analysis with Boundary Elements* **98** pp 217-231

[6] Krivoshapko S N and Ivanov V N 2015 *Encyclopedia of analytical surfaces* (New York: Springer) p 752

[7] Ivanov V N and Krivoshapko S N 2010 *Analytical methods of analysis of shells of complex form* (Moscow: RUDN University Press) p 542

[8] Hutton D V 2004 *Fundamentals of finite element analysis* (New York: McGraw-Hill) p 640

[9] Chapelle D and Bathe K-J 2011 *The finite element analysis of shells – Fundamentals* (Berlin: Springer) p 410

[10] Bhavikatti S S 2005 *Finite element analysis* (New Delhi: New Age International Publisher) p 334

[11] Madenci E and Guven I 2018 *The finite element method and applications in engineering using ANSYS* (New York: Springer) p 686

[12] Ivanov V N, Shmeleva A A 2016 Geometry and formation of the thin-walled space shell structures on the base of normal cyclic surfaces *Structural Mechanics of Engineering Constructions and Buildings* **6** pp 3-8

[13] Krivoshapko S N and Bock Hyeng C A 2012 Geometrical research of rare types of cyclic shells *International Journal of Research and Reviews in Applied Sciences* **12(3)** pp 346-359

[14] Krivoshapko S N and Bock Hyeng C A 2012 Static and dynamic analysis of thin walled cyclic shells *International Journal of Modern Engineering Research* **2(5)** pp 3502-3508

[15] Bock Hyeng C A, Yamb E B 2012 Application of cyclic shells in architecture, machine design, and bionics *International Journal of Modern Engineering Research* **2(3)** pp 799-806

Stability of metal cylindrical grid shell structures

A I Siyanov[1], M Rynkovskaya[2], E Tupikova[2*]

[1] Technical Institute (branch) of North-Eastern Federal University, Neryungri, Russia.
[2] Peoples' Friendship University of Russia, Moscow, Russia.

E-mail: emelian-off@yandex.ru

Abstract. Based on the fundamental thin shell theory applied to thin continuous cylindrical shells, a method for calculating the stability of the corresponding grid shell structures is developed. Mathematical modeling is performed and factors of critical state of structures are determined. Critical deformation zones are identified and the patterns of changes in shell geometry under actual loads are established. The dependencies of strength and shape-forming parameters are obtained depending on the material and the number of half-waves of the stability loss shape. The critical load is determined taking into account the possible rigidity of multi-element grids, and the results of numerical studies are presented. A noticeable difference in the values of critical load is shown and the admissible limits of applicability of geometric parameters are found. A significant sensitivity of shells with a rhombic mesh are noted.

Keywords: cylindrical grid shell, critical load, buckling analysis, continual model, discrete model

1. Introduction

The problem of the stability of cylindrical grid shell structures now is a special part of design of reliable roofs for buildings and structures. There are studies by Zhonghao Zhang et al. [1], devoted to systems with reinforcing elements. A detailed analysis of the stability of similar structures was carried out. This study focused on the tension member installation and examined the effect of tension member on buckling load and strength of single layer two-way grid shell, taking account of the units of grid, load distribution, initial imperfection and the initial axial force. However, the main attention is paid to single-layered surfaces. According to the results of recent studies (Raghavan Ramalingam and S. Arul Jayachandran [2]) the stability loss of particular zones influenced by actual loads may occur. Based on the collected data and the design experience, Huihuan Ma et al. [3] conducted experimental and numerical studies of the cylindrical grid shell structures. The complete load–deflection response curve and the buckling mode of the test structure were obtained. FEA models were also established to carry out numerical analysis. Through comparison of the FEA results and the corresponding test results, it is shown that the proposed FEA model can be used to effectively predict the mechanical behavior of the reticulated shell including its buckling load. Obviously, the mesh, type of external loads, and the boundary conditions have a significant effect on the structures behavior [4-9], and the need to follow economic principles further complicates the problem. There are known discrete-continuous methods of calculation using iterative procedures and the results of experimental tests of continuous shells [10]. The developed algorithms contain analytical dependencies and are based on a discrete or continual representation of the surface. The solutions obtained are applied to grid shells of arbitrary geometry and have a unified approach to determining the value of the critical [12]. The shape of the shells is often specified in the first approximation and is remote from the actual operation of the structures. The conducted studies showed a number of factors which had not been taken into account before and considerable margin of bearing capacity [12]. However, cylindrical grids, in comparison with other curvilinear multi-element structures [13], are subjected to substantial curvature of surfaces, have different schemes of geometry variation and the smallest strength parameters of stability. However, in past research using numerical models, shell elements have been widely used to simulate bridges, beams, columns and structural components [14-15], concrete slabs, and steel plates [16]. Therefore, the goal of this paper is to develop a mathematical model to determine the critical load and to investigate the stability of the cylindrical grid shell structures. The finite element method and Lira-SAPR 2018 software are used. The analytical study is included in order to calculate the critical load.

2. Scheme of the model

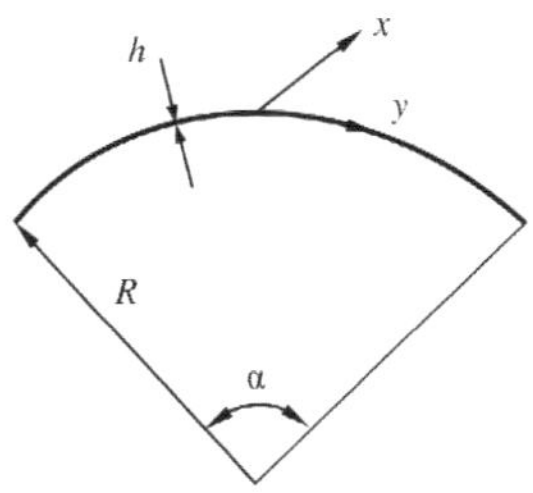

Figure 1. Geometrical parameters

To determine the behavior and the features of grids deformation, a continuum method is used [17]. In accordance with it, geometric and strength parameters are given.

The shape of the surface is described by the radius of curvature R, the angle α and the thickness h, the position of each point being determined by the x and y coordinates along the generator and along the arc direction of the circle, respectively [12] (see figure 1).

The displacements are represented in the form of deflections w. The deviations of the span points of the surface along the external normal to the center of curvature are assumed positive. The material is taken into account by Young modulus E. Flexural cylindrical rigidity and subcritical stress are given by the force parameters D and $p_y = qR/h$, respectively. The load q is uniformly distributed over the surface and applied radially along the outer normal.

3. Continual model analysis

The differential equation of stability [17] is:

$$D\nabla^8 w + \frac{Eh}{R^2}\frac{\partial^4 w}{\partial x^4} + qR\nabla^4\left(\frac{\partial^2 w}{\partial y^2}\right) = 0 \tag{1}$$

Equation (1) includes bending D and axial Eh rigidities. The grid surface of the shell is defined from longitudinal, transverse and diagonal rods with the formation of a diagonal grating (figure 2).

Thin shell theory can be applied to the grid shell structure analysis for the first preliminary calculation. Such analytical model, in which grid reticular structure can be substituted by equivalent continuous shell was developed in [17].

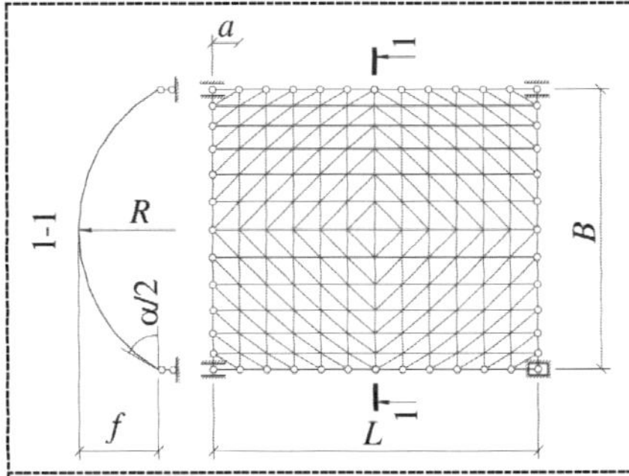

Figure 2. The scheme of a cylindrical grid shell structure

Using the differential equation for continuous shells, the equivalent rigidity relations are written [17]:

$Eh \sim \frac{EA}{a}$; $D \sim \frac{EI}{a}$, where I is moment of inertia, A is cross-sectional area of the structural elements. Being combined with the elastic modulus E these geometrical parameters I and A form the flexural EI and the membrane EA rigidities of the elements respectively.

Then, equation (1) can be written as [17]:

$$\frac{EI}{a}\nabla^8 w + \frac{EA}{aR^2}\frac{\partial^4 w}{\partial x^4} + qR\nabla^4\left(\frac{\partial^2 w}{\partial y^2}\right) = 0 \tag{2}$$

Taking into account the hinged support of the grid on the contour, the boundary conditions are represented in the form:

$$\begin{cases} w\big|_{x=0} = w\big|_{x=L} = 0; \quad \left.\dfrac{\partial^2 w}{\partial x^2}\right|_{x=0} = \left.\dfrac{\partial^2 w}{\partial x^2}\right|_{x=L} = 0 \\ w\big|_{y=0} = w\big|_{y=\alpha R} = 0; \quad \left.\dfrac{\partial^2 w}{\partial y^2}\right|_{y=0} = \left.\dfrac{\partial^2 w}{\partial y^2}\right|_{y=\alpha R} = 0 \end{cases} \tag{3}$$

where L is the structure length along the generator; αR – is the arc length.
According to the proposed conditions (5), the deflection w is defined as a function of $\left(\sin\dfrac{m\pi x}{L}\sin\dfrac{n\pi y}{\alpha R}\right)$.

Here, m is the number of half-waves of the curved surface along the generator; n is the number of half-waves in the direction of the arc. Substituting into the stability equation (2), it is found

$$q_{cr} = \frac{EA}{aR}\left[\frac{I}{A}\left(\frac{\pi\alpha R}{nL^2} + \frac{n\pi}{\alpha R}\right)^2 + \left(\frac{(\alpha R)^2}{R^2 n^2 \pi^2}\right) / \left(1 + \frac{n^2 L^2}{(\alpha R)^2}\right)^2\right] \tag{4}$$

The shell changes its shape along the generator always in the form of one half-wave. Therefore, when determining the critical load q_{cr} in Eq. (4) the quantitative parameter m is omitted and is not used in the calculations.
To carry out analytical preliminary calculation, it is introduced the value of the dimensionless parameter

$$\hat{q} = q_{cr}\frac{aR}{EA}$$

The geometric and strength characteristics are set and the necessary calculations are done.

4. Discrete model (FEA)

The finite element method and building a model in Lira SAPR 2018 software are used. A rectangular grid type with a rod that divides the cell region into two identical triangles is chosen. The geometric stability of the structure is ensured by the hinged support conditions, what is more, at one corner point there is the complete prohibition of displacements (X, Y, Z axes), and at the other three corner points we impose bonds along the axes X, Z. The other contour support nodes are hinged vertically (Z axis). Mechanical properties are given in Table1.

This scheme allows to determine the maximum force in the elements and the biggest vertical displacements of nodes. The basic data for constructing the design scheme of the grid surface is shown in Table 1.

Table 1. Geometric parameters and material characteristics

B (m)	f (m)	R (m)	a (m)	E (kN/m^2)	Steel
36	9	22.5	3.0	2.06×108	S235

The stiffness characteristics of tubular profile rods are assigned based on two sizes of cross-sections of the software assortment (see Table 2).

Table 2. Cross-section parameters of tubular profile rods

System	D (m)	t (m)	A (m^2)	I (m^4)
1	168×10^{-3}	5.5×10^{-3}	28.1×10^{-4}	92.9×10^{-7}
2	140×10^{-3}	4.5×10^{-3}	19.2×10^{-4}	44.2×10^{-7}

The influence of the length L on the magnitude $\hat{q}$ and number of half-wave n of the structural stability loss form is investigated with the L/a ratio varying along the generator in the range from 12 to 48. The correlations of shell deformation under critical load are revealed.

Finally the correlation of dimentionless critical pressure $\hat{q}$ and L/a ratio can be presented graphically. The curves show differential equation solution got by numeric method, the lower eigenvalue critical pressure is obtained. So, these curves present the results of the calculation, based on continual hypothesis.

The points, marked by circles, present the results of FEA calculations of special cases, realized by Lira SAPR2018. The discrete model is realized, the rods are modeled by rod FE. This calculation is provided for verification of the continual model analysis.

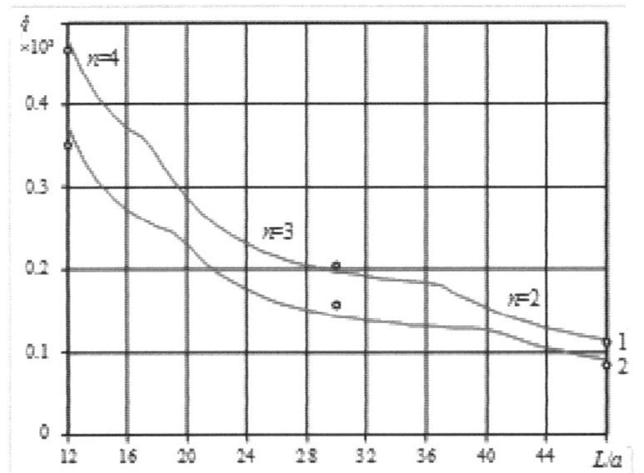

Figure 3. The curves of $\hat{q}$ - L/a ratio dependence (obtained by continual method) and control points of FE calculation

It is obvious that the second system is more vulnerable to the loss of stability than the first one. The rods of the second system have a cross section of 140×4.5 mm, and the critical load got using computer analysis and the obtained by analytical dependence shows almost the same value. For a design with any of the accepted cross-sections of the rods, a decrease in the parameter $\hat{q}$ and a decrease in the number of half-wave n of the shape of the loss of stability are noticed with an increase in the number of cells along the generator. That is, increasing the length in the direction of the generator led to reduction in the critical load. The deformation of the system occurs in the form of one half-wave *m* along the length of the structure and different half-wave *n* along the direction of the arc. The forms of stability loss of the most vulnerable shell with three values of the shape-forming parameter are shown in figures 4(a)-(c).

The increase in the length of the shell is seen to adversely affect the stability of the structure and

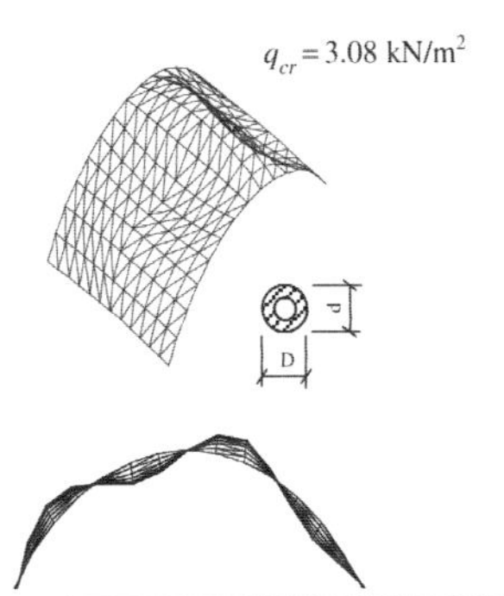

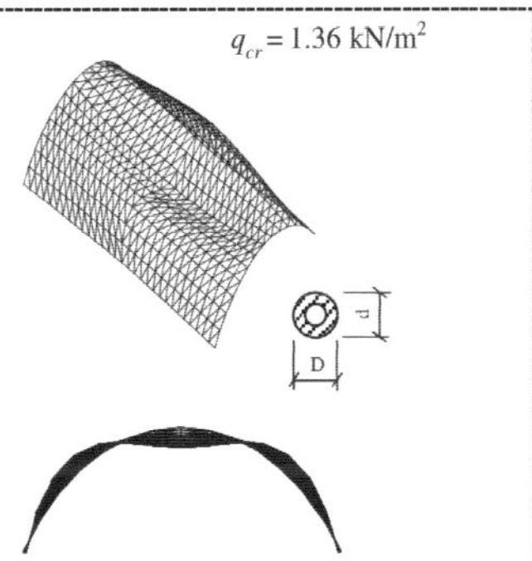

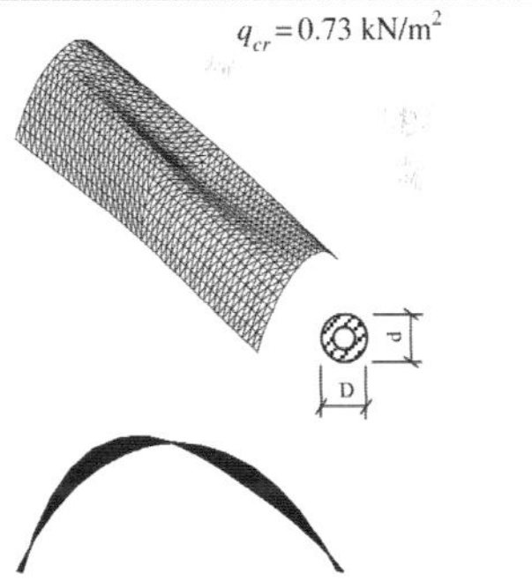

Figure 4 (a). *L*/*a*=12 **(b)** *L*/*a*=30 **(c)** *L*/*a*=48

reduce the magnitude of the critical load. As shown by the calculations using Lira SAPR 2018, software based on FEM, the maximally compressed rod bears a higher critical load. A comparison of the corresponding parameter of the most compressed element with critical loading ($q_{cr.b}$) and the structure as a unit (q_{cr}) for the three shells with the smallest cross-section of the rods is shown in Table 3.

Table 3. Critical loads ratio $q_{cr.b}/q_{cr}$

D (m)	*t* (m)	*L*/*a*=12	*L*/*a*=30	*L*/*a*=48
140×10^{-3}	4.5×10^{-3}	22.97/3.08	10.98/1.36	4.25/0.73

Thus, the larger value of the parameter $q_{cr.b}$ (see Table 3) makes it necessary to use in calculations the value of the lower critical load, at which the shell is already losing its stability.

5. Comparison of grids

The rods arrangement in the cylindrical surface is important. The detailed descriptions and some recommendations for calculation of systems with square and rhombic grids are known. However, various software based on the finite element method is widely used to analyze critical load values. That is why it was decided to create the similar mesh models (see Table 4) and examine them to verify results.

Table 4. Parameters in arc direction and material data

B (m)	f (m)	R (m)	E (kN/m^2)	Steel
41.72	14.07	22.50	2.06 × 108	S235

The different shape of the intersection of the rods allows to obtain the values of the length of the construction with orthorhombic and triangular mesh with the appropriate sizes and cell numbers (see Table 5).

Table 5. Parameters and quantitative indicators along the length

L_r (m)	a_r (m)	L_r/a_r	L_t (m)	a_t (m)	L_t/a_t
63.90	4.26	15	90.0	3.0	30
Subscript: r – orthorhombic; t – triangular					

Here L is the length of the shell; a is the size of the grid (figure 2). Subscript r and t indicate the shape of the cell: r - rhombic; t - triangular Stiffness characteristics are assumed and chosen from the software library (see Table 6).

Table 6. Cross-section parameters of tubular rods

D (m)	t (m)	A (m^2)	I (m^4)
140×10^{-3}	4.5×10^{-3}	19.2×10^{-4}	44×10^{-7}

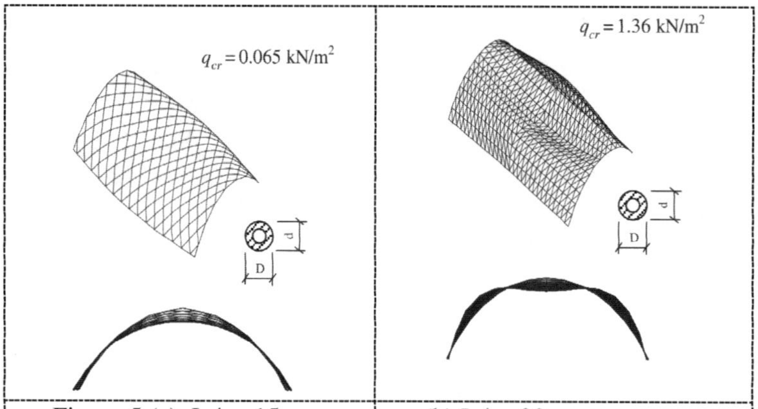

Figure 5 (a). L_r/a_r=15 (b) L_t/a_t=30
The forms of stability loss of shells with possible nets

Modeling is performed for a system with a rhombic and triangular grid. The corresponding shell calculations with different parameters along the length have been made. The values of the critical loads and the loss of stability forms are shown in figures 6(a) and 6(b).

The general condition of the structure with two schemes of rods distribution on a cylindrical surface is revealed. In the absence of longitudinal and transverse ribs, the shell with a rhombic grid is more sensitive and vulnerable. The significant difference in the values of the critical loads q_{cr} and the number of half-wave n of the loss-of-stability form confirms this. It can be explained by the fact that the rhombic cell does not have the required rigidity in its plane and is almost geometrically unstable system. The largest energy reserve is characterized by a triangular grid, which has both bending and membrane stiffness at the same time.

6. Conclusions

Based on thin shell theory, a mathematical model of cylindrical grid shell structure is built and a technique for the stability analysis of this structure is developed. The analysis of the stability of the cylindrical grid shell structures is performed, and zones of the geometry change are determined depending on the geometrical parameters and the number of half-wave n in the direction of the arc of the circle. The necessary components of multi-element grids rigidity are identified, and the possibility to determine the critical load qcr taking into account the exact ratios of the force and shape-forming characteristics. The paper shows the acceptable agreement between the values of the finite element and analytical analysis. A significant decrease in performance and a decrease in the number of half-wave n of the structural stability loss form with an increase in the number of cells along the generator from 12 to 48 is revealed.

The elimination of the maximally compressed rod from the grid system is investigated, and by comparing the parameters $q_{cr.b}/q_{cr}$ the necessity to perform the shell calculation taking into account the lower critical load is revealed. Limit values of B×L dimensions of the grid surface horizontal projection are determined depending on the magnitude of the applied load and different stiffness characteristics of the rods. The range of permissible values of parameters and possible design of the structure is shown. For a given length L and different values of width B, the number of half-waves n varied from 2 to 4. It is found that at B = 35.91 m, the number of half-waves n of stability loss form has reached its limiting value. Based on the results of introducing a fixed length L and changing the width B of the system, the overall dimension condition L>B is recommended. The design of shells with a rhombic mesh is undesirable because of the increased vulnerability and the strong sensitivity of the structure.

7. References

[1] Zhang Z, Ma H, Fujimoto M, and Fu Q 2014 *The Open Civil Engineering Journal* **8** 311-319
[2] Ramalingam R and Jayachandran S A 2015 *J. of Constructional Steel Research* **114** 136-145
[3] Ma H, Fan F, Wen P, Zhang H, and Shen S 2015 *Thin-Walled Structures* **86** 1-9
[4] Sakai Y and Ohsaki M 2018 *Engineering Structures* **169** 55-67
[5] Mesnil Y, Douthe C, Baverel O, and Leger B 2017 *Engineering Structures* **132** 337-348
[6] Ma H, Fan F, Zhong J, and Cao Z 2013 *Thin-Walled Structures* **72** 128-138
[7] Richardson J N, Adriaenssens S, Coelho R F, Bouillard P 2013 *Engineering Structures* **52** 230-239
[8] Hao W and Minger W 2018 *Engineering Structures* **168** 576-588
[9] He Zhou Y X and Liu D 2014 *Thin-Walled Structures* **82** 233-244
[10] Cabrero J M and Bayo E 2005 *Engineering Structures* **27** 1125-1137
[11] Chen P and Kawaguchi M 2006 *Int. J.of Space Structures* **2** 111-118
[12] Gotsulyak E A and Siyanov A I 2004 *International Applied Mechanics* **40** 426-431
[13] Grigorenko Y V, Savula Y G, Mukha I S 2000 *International Applied Mechanics* **36** 979-1000
[14] Rahnavard R, Naghavi M, Abudi M, Suleiman M 2018 *Case Studies in Construction Materials* **8** 476-488
[15] Naghavi M, Rahnavard R, Thomas R J, Malekinejad M 2018 *J. of Building Engineering* **22** 415-428
[16] Rahnavard R, Hassanipour A, Mounesi A 2016 *J. of Constructional Steel Research* **121** 441-456
[17] Volmir A S 1965 *Stability of deformable systems* (NASA AD)

Acknowledgments

This research paper was carried on the program to improve the competitiveness of Technical Institute (branch) of North-Eastern Federal University, Neryungri among the world's leading research and education centers in the 2016-2020. The publication has been prepared with the support of the "RUDN University Program 5-100".

IOP Conf. Series: Materials Science and Engineering **675** (2019) 012067 doi:10.1088/1757-899X/675/1/012067

Authors' background

Name	*Prefix*	*Research Field*	*Email*	*Personal website*
Alexander I. Siyanov	Associate Professor	Steel structures	Siyanov-vntu@mail.ru	
Marina Rynkovskaya	Associate Professor	Shells and plates	Marine_step@mail.ru	
Evgenia Tupikova	Assistant Professor	Shells and plates	Emelian-off@yandex.ru	

Development of methodology for calculation of permissible stresses for evaluation of contact strength of cylindrical gears in case of overload

Yu V Belousov[1,2], S L Shambina[2*] and F V Rekach[2]

[1] Bauman Moscow State Technical University (National Research University), Moscow, Russia
[2] Peoples' Friendship University of Russia (RUDN University), Moscow, Russia

E-mail: shambina_sl@mail.ru

Abstract. The article focuses on the development of methods for determining the allowable stresses to check the contact strength of the spur and helical gears under overload. Short-term overloads in gears occur under the action of maximum moments during the start-up of the device or due to the other reasons. The calculation of these loads is performed to check the static strength of the transmission teeth. The analysis of the limiting state of the material in the working area of the gear teeth was performed using the Pisarenko-Lebedev strength criterion, which is one of the most common strength conditions. The method of calculation of allowable stresses was developed. The calculation of limit stresses for normalized, improved or hardened wheels, as well as for the wheels after hardening by chemical-thermal treatment. The calculation results were compared with the values recommended by the standard. It was shown that the permissible contact stress for helical gears is slightly higher than for straight gears. Practically for all heat-treated steels used in the manufacture of gears, the norm of permissible limit stresses recommended by the standard is higher (in some cases significantly) compared to the stresses required from the conditions of contact strength.

Keywords: permissible stresses, contact area, strength criterion, stress intensity

1. Introduction

Gears are designed to transmit motion with the necessary change of angular velocity and torque by the magnitude and direction [1]. The force from one element of the coupling pair to another is transmitted by means of teeth, sequentially engaging. Gears are mainly used to transmit rotational motion between parallel shafts, shafts with intersecting and crisscrossing axes, to convert rotational motion into translational motion and vice versa.

If there are parallel shafts they use transmission with cylindrical gears, which may be with straight, slanting or herringbone teeth. These transmissions are usually called cylindrical. Among the mechanical transmission they are the most common. They are used in a fairly large range of loads and operating conditions.

The force interaction of the gear teeth is accompanied by the destruction of their working surfaces. The main types of teeth destruction, that are associated with contact stresses and friction, are fatigue chipping, jamming, wear, as well as residual deformation or brittle destruction of the surface layer under overload.

The method of determining the allowable stress for the evaluation of the last two of these types of damage is presented in the State Standard of Russia (STST) 21354-87. The permissible stresses in this case are designed to eliminate the appearance of unacceptable plastic deformations or brittle destruction of the surface layer of the teeth under the action of the ultimate load.

When the maximum torque T_{max} is applied, the contact stress σ_{Nmax} must not exceed the permissible contact stress for the specified service life:

$$\sigma_{N_{max}} \leq \sigma_{NP_{max}} . \qquad (1.1)$$

The stress $\sigma_{N_{max}}$ is determined by the formula:

$$\sigma_N = \sigma_N \sqrt{\frac{T_{max} K_{N_{max}}}{T_N K_N}}, \qquad (1.2)$$

where T_N and K_N are the nominal torque and the corresponding load factor; K_N is the load factor, determined at a load of T_{max}.

Short-term operating maximum loads T_{max} are used to check the static strength of the transmission teeth. These loads (moments) include such which number of cycles during the service life of the transmissio

$$N_{N_{min}} \leq 0{,}03 N_{GN} ,$$

where N_{GH} is the number of loading cycles corresponding to the abscissa of the inflection point of the fatigue curve for contact stresses:

$$N_{GN} = 30(HB)^{2{,}4},$$

here HB is the surface hardness of the transmission wheel.

When $\text{H} \geq 560$ HB (HRC 56)) take $N_{GN} = 12 \cdot 10^7$.

Peak loads are not included in the typical loading conditions, they are indicated separately. If the peak moment T_{peak} is not set, its value is found by the starting torque of the electromotor, by the limiting torque of the safety devices, by the inertial moments arising during sudden braking, etc. [2, 3, 13].

According to STST 21354-87, the permissible contact stress at maximum load, which does not cause residual deformation or brittle destruction of the surface layer $\sigma_{NP_{max}}$, depends on the method of chemical and thermal treatment of the gear and the nature of the hardness change along the depth of the tooth:

- for teeth subjected to normalization, improvement or quenching with low tempering, they are:

$$\sigma_{NP_{max}} = 2{,}8\, \sigma_T;$$

- for teeth subjected to cementation or contour hardening they are:

$$\sigma_{NP_{max}} = 44 H_{HRC_E};$$

- for nitrided teeth they are:

$$\sigma_{HP_{max}} = 3\text{H}_{\text{HV}} .$$

According to STST 21354-87 different criteria were used to determine the permissible stresses: Brinell hardness numbers HB, ones by Rockwell σ_T. The choice of the criterion depends on the kind of chemical-thermal surface treatment of teeth.

This situation makes it difficult to analyze the load capacity of gears. In addition, all the above dependencies are empirical. Methods of accurate calculation of permissible contact stresses at the moment does not exist [6-8]. The authors of this paper tried to develop a uniform method of calculating the allowable stress.

2. Calculation of allowable stresses for plastic materials.

At first it is necessary to give the definition of allowable stresses for the calculation of the strength of the transmission under short-term overload. In cylindrical gears, the initial contact of the teeth occurs along the line. Then the linear contact passes into contact on a narrow platform bounded by two straight lines. With increasing load elastic deformation is being replaced by elastic-plastic one. The greatest shear stresses initially appear at some depth. With a further increase in the load, the plastic deformation comes to the surface, where it initially occurs at the points of the longitudinal axis of symmetry of the contact site. It can be assumed that the laws of the theory of elasticity remain valid until the appearance of plastic deformations on the surface [9-12, 16].

For the quantitative description of the material's limiting state in the contact zone the Pisarenko-Lebedev strength criterion was used. According to this criterion, the limit state of the material is determined by both tangential and normal stresses. The expression for the equivalent stress has the following form:

$$\sigma_{eq} = x\sigma_i + (1-x)\sigma_{lar}\,, \tag{2.1}$$

where $x = \sigma_t/\sigma_c$; σ_t is the ultimate strength of the material under uniaxial tension; σ_c is the absolute value of the compressive strength of the material; σ_i is stress intensity; σ_{lar} is the largest principal stress.

The Pisarenko-Lebedev strength criterion is the most common strength condition for such a case. At $x = 1$, which is typical for a plastic material, the Pisarenko-Lebedev strength criterion switches into the stress intensity criterion. At $x = 0$ for a very brittle material, this criterion coincides with the criterion of the highest normal stresses.

Let's express the intensity of stresses through the main stresses:

$$\sigma_i = \frac{1}{\sqrt{2}}\sqrt{(\sigma_1-\sigma_2)^2 + (\sigma_2-\sigma_3)^2 + (\sigma_3-\sigma_1)^2}\,. \tag{2.2}$$

For toothed gears at the points of the longitudinal axis of symmetry of the contact strip:

$$\sigma_1 = -2\mu p_0\,; \qquad \sigma_2 = \sigma_3 = -p_0, \tag{2.3}$$

where где p_0 – is the highest pressure on the contact surface; μ is the coefficient of transverse deformation of the material.

Solving together the equations (2) and (3), we obtain:

$$\sigma_i = p_0(1-2\mu)\,. \tag{2.4}$$

In [5, 6] it was shown that the contact area of the teeth in a helical gear has the form of a highly elongated ellipse, not a strip. The main stresses will have the greatest value in the center of the contact area [4]. They can be defined by formulas:

$$\sigma_1 = -\left[2\mu + \frac{(1-2\mu)b}{a+b}\right]q_0\,; \quad \sigma_2 = -\left[2\mu + \frac{(1-2\mu)a}{a+b}\right]q_0; \quad \sigma_3 = -q_0\,, \tag{2.5}$$

where q_0 – the maximum pressure at the center of the contact area; σ_1 – tension in the direction of the smaller axis of the ellipse; σ_2 – tension in the direction of the larger axis of the ellipse; σ_3 – tension normal to the contact area; a and b are the size of the greater and lesser axes of the ellipse, respectively.

Substituting the expression (5) in (2), for the oblique transmission we obtain:

$$\sigma_i' = (1-2\mu)q_0\frac{\sqrt{a^2+b^2-ab}}{a+b}\,. \tag{2.6}$$

With the average ratio of the ellipse half-axes $a/b \approx 50$ according to [5] we have:

$$\sigma_i{}' = 0{,}97(1-2\mu)q_0\,. \tag{2.7}$$

The gear limit state will occur when the following conditions are true:

- for spur gears expressions (3) and (4) we will have:

$$\sigma_{eq} = p_0[x(1-4\mu) + 2\mu] = \sigma_{ul}\,, \tag{2.8}$$

- for helical gears taking in the account expressions (5) and (7) we will have:

$$\sigma_{eq} = q_0\left[0{,}97x(1-2\mu) + (1-x)\left(2\mu + (1-2\mu)\frac{b}{a+b}\right)\right] = \sigma_{ul}\,, \tag{2.9}$$

IOP Conf. Series: Materials Science and Engineering **675** (2019) 012068 doi:10.1088/1757-899X/675/1/012068

where σ_{ul} is the value characterizing the limit state of the material (yield strength or ultimate strength). For materials in the plastic state, when $x = 1$, the condition of reaching the limit state is:

- for straight gears:

$$\sigma_{eq} = p_0(1 - 2\mu) = \sigma_T \, , \tag{2.10}$$

- for helical gears:

$$\sigma_{eq}{=}0{,}97\, q_0\, (1 - 2\mu) = \sigma_T \; . \tag{2.11}$$

For structural materials, which after surface hardening by chemical-thermal treatment were prone to brittle fracture $\sigma_{eq} = \sigma_t$.

At short-term overloads for plastic materials (wheels subjected to normalization, improvement or throughout quenching with low tempering) basing on formulas (10) and (11) we obtain expressions for the highest pressure $p_{0_{max}}$ ($q_{0_{max}}$) on the surface of the contact area at the time of the appearance of plastic deformations.

- for straight-toothed gears, they are determined by the following formula:

$$p_{0_{max}} = \frac{\sigma_T}{(1-2\mu)} \, , \tag{2.12}$$

- for helical gears according to the formula:

$$q_{0_{max}} = \frac{\sigma_T}{0{,}97(1-2\mu)} \, . \tag{2.13}$$

By the contact stresses σ_N we mean the greatest pressure at the contact site. Therefore, the permissible contact stress σ_N at maximum load:

- for straight gears:

$$\sigma_{NP_{max}} = p_{0_{cr}} = (2{,}00 \ldots 2{,}78)\sigma_T \, ,$$

- for helical gears:

$$\sigma'_{NP_{max}} = q_{0_{max}} = (2{,}06 \ldots 2{,}87)\sigma_T$$

when the Poisson's ratio μ changes from 0.25 to 0.32.

Standard STST 21354-87 for steels which are usually used for the manufacture of gears and subjected to volumetric heat treatment (normalization, improvement, through hardening with low tempering) recommended the following rule, limiting the value of the permissible limit contact stresses, $\sigma_{NP_{max}} = 2.8\,\sigma_T$. It corresponds only to the extreme upper limit in the obtained dependences. The average values of permissible contact stresses in the obtained dependences are about 15% lower than the recommended ones.

3. Calculation of allowable stresses for materials prone to brittle destruction.

For materials which as a result of hardening heat treatment have a tendency to brittle destruction of the surface layer (for teeth subjected to cementation or contour hardening, for nitrided teeth) permissible contact stress limit:

- for straight gears:

$$\sigma_{NP_{max}} = p_{0_{cr}} = \frac{\sigma_t}{x(1-4\mu)+2\mu} \, , \tag{3.1}$$

- for helical gears:

$$\sigma'_{NP_{max}} = q_{0_{cr}} = \frac{\sigma_t}{0{,}97x(1-2\mu)+(1-x)(2\mu+(1-2\mu)\frac{b}{a+b})} \, . \tag{3.2}$$

For steels subjected to heat treatment, the coefficient characterizing their brittleness $x = 0.4 \ldots 0.78$. For the average values of $x = 0.59$ and the coefficient of transverse deformation $\mu = 0.29$ permissible contact stress limit:

- for straight gears:

$$\sigma_{NP_{max}} = 2.06\sigma_t,$$

- for helical gears:

$$\sigma'_{NP_{max}} = 38.13\ HRC.$$

Tensile strength of steels, which are commonly used for the manufacture of gears and are subjected to improvement and hardening of HDTV along the entire tooth contour and through hardening of HDTV, $\sigma_t \approx 18.33$ HRC. For them, the permissible contact limit stress has the following value:

- for straight gears:

$$\sigma_{NP_{max}} = 37.76\ HRC,$$

- for helical gears:

$$\sigma'_{NP_{max}} = 38.13\ HRC.$$

Tensile strength of steels, which are commonly used for the manufacture of gears and are subjected to cementation and contour hardening $\sigma_t \approx 16.81$ HRC. For them, the permissible contact limit stress has the following value:

- for straight gears:

$$\sigma_{NP_{max}} = 34.63\ HRC,$$

- for helical gears:

$$\sigma'_{NP_{max}} = 34.96\ HRC\ .$$

Tensile strength of steels subjected to nitriding $\sigma_c \approx 18.49\ HRC$. For them, the permissible contact limit stress has the following value:

- for straight gears:

$$\sigma_{NP_{max}} = 38.1\ HRC,$$

- for helical gears:

$$\sigma'_{NP_{max}} = 38.46\ HRC.$$

In the first two cases, the STST 21354-87 recommends the following:

$$\sigma_{NP_{max}} = 40\ HRC.$$

In the third case:

$$\sigma_{NP_{max}} = (30 \ldots 40)\ HRC.$$

4. Conclusion

Analyzing the obtained dependences, it can be noted that the permissible limit contact stress for helical gears is slightly higher than for straight gears. In the manufacture of gears made of steel, which are subjected to volumetric heat treatment (normalization, improvement, through hardening with low tempering), this difference is 3.3%. In other cases, it does not exceed 1%. The difference here is negligible and it can be assumed that the permissible limit contact stress for helical gears coincides with the same for straight teeth.

For steels that undergo volumetric heat treatment (normalization, improvement, through quenching with low tempering), the norm of permissible limit contact stresses recommended by the standard can be almost 30% higher compared to the values calculated from the dependencies obtained in this work.

For steels, which are usually used for the manufacture of gears and are subject to improvement and hardening of HDTV throughout the tooth contour and through hardening of HDTV, the rate of permissible limit contact stresses is almost 6% higher compared to the values calculated from the dependencies obtained in this work.

For steels subjected to cementation and contour hardening, the permissible limit contact stresses are 16% higher than the values calculated from the dependencies obtained in this work.

For nitrided steels, the values of permissible contact stress limits set by the standard are quite consistent with the calculations according to the above dependencies.

However, it should be noted that for nitrided teeth due to the small thickness of the hardened surface layer, it is possible to press the surface layer into the relatively soft core of the tooth under the action of high contact stresses, as well as the occurrence of deep contact destruction. In this case, the values of permissible contact stresses determined by formulas (11) and (12) should be reduced by approximately 1.35 times. Then for nitrided teeth:

- spur gears:

$$\sigma_{NP_{max}} = 28.49\ HRC,$$

- helical gears:

$$\sigma'_{NP_{max}} = 28.76\ HRC\ .$$

These values are below the standard by 18.6%.

References

[1] Shambina S and Virchenko V 2013 Modular variant geometric modelling of complex technical objects *RUDN J. of Eng. Researches* **3** pp 5-8
[2] Andrienko L A, et all 2014 *Detali mashin* (Moscow: Publishing House MGTU im. N. Baumana, Moscow) p 465
[3] Ivanov M N and Finogenov VA 2007 *Detali mashin* (Publishing House Higher Education, Moscow) p 408
[4] Belousov Yu V 2014 Calculation contact pressure on the working surface of the teeth of spur gearing *J. Struct. Mech. of Eng. Constr. and Build.* **2** pp 63-67
[5] Belousov Yu V 2015 Calculation of the stresses on the working surface of the teethes of oblique angle spur gears *J. Struct. Mech. of Eng. Constr. and Build.* **4** pp 35-38
[6] Belousov Yu V 2015 Calculation of the admissible tensile stresses to appreciate contact strength of spur gears at the unfavorable loading *J. Struct. Mech. of Eng. Constr. and Build.* **6** pp 29-32
[7] Kirilovsky V V 2018 Reserves for Improving the Design of Shafts *RUDN J. of Eng. Researches* **19(4)** pp 428-39
[8] Sosnovskii L A, Zhmailik V A and Komissarov V V 2009 Size effect contact fatigue *J. Strength of Mater.* **41(1)** pp 88-94
[9] Lu Xi 2010 Investigation of the region of fatigue crack initiation transmission gear *J. Mater. Sci. and Eng.: A.* **527(6)** pp 1377-82
[10] Hoffman N P and Stolz V 2010 On transient growth of wear pattern properties *J. Wear.* **268(8)** pp 886-92
[11] Korotkin V I, Sukhov D Yu and Kolosova E M 2010 Stress in the base of straight barrel-shaped gearteeth in cylindrical involute transmissions *J. Russian Eng. Res.* **30(10)** pp 967-75
[12] Korotkin V I and Gazzaev D A 2011 Effective contact stress at the surface of barrel-shaped straight teeth on involute gears *J. Russian Eng. Res.* **31(8)** pp 725-30
[13] Chernets M V, Chernets Y O and Yarema R Y 2013 A method for the evaluation of the influence of correction and wear of the teeth of a cylindrical gear on its durability and strength. Part 2. Contact strength *J. Mater. Science* **48(6)** pp 752-56

[14] Babichev D and Storchak M 2015 Synthesis of cylindrical gears with optimum rolling fatigue strength *J. Produc. Eng. Res. and Devel.* **9(1)** pp 87-97

[15] Deithard T 2015 Einteilige wdlzfrdzer vertreiben kosten aus der scrienfertigung *J. WB Werkstatt und Betried* **3** pp 30-32

[16] Shambina S L, Rekach F V and Belousov Y V 2016 On new modification of some strength criteria for anisotropic materials// *J. Key Eng. Mat.* **724** pp 53-57

Acknowledgements

This paper was financially supported by the Ministry of Education and Science of the Russian Federation on the program to improve the competitiveness of Peoples' Friendship University of Russia (RUDN University) among the world's leading research and education centers in the 2016-2020. This publication was prepared with the support of the "RUDN University Program 5-100".

Relaxation of stresses in the elements of reinforced concrete structures

T V Zhdanova* **and E A Larionov**

Department of applied mathematics, Moscow state university of civil engineering, Moscow, Russia

Email: zhdanovatv@mgsu.ru

Abstract. The issues of calculating and forecasting the long-term safety of reinforced concrete structures and structures are related to the dynamics of their stress state and lead to relaxation problems to take into account the redistribution of stress from concrete to reinforcement. In this paper, an approach based on the concept of the strength structure of structural materials is proposed for their solution. According to this concept, in contrast to traditional hypotheses, the material is considered as the combination of its fractions (links, layers, fibers) with statistically distributed strengths. The loading generates the loss of the power resistance capacity by a part of the element fractions which entails a redistribution of stresses on its whole fractions. As a result of this process, a nonlinear dependence of the strain on stress calculated under the assumption of equal strength of all fractions, arises. A significant role in the evolution of stresses in the components of a reinforced concrete element is played by creep of the material. In the linear formulation for a material with a uniform strength, the stresses are determined by solving a linear integral equation coupled with its rheological equation of a mechanical state. Based on the concept of the statistical distribution of the strengths of fractions, a modification of L. Boltzmann's superposition principle, well-known in the linear creep theory, is proposed for its applicability and the non-linear dependence of strain on stress. This allows the derivation of a linear integral equation with respect to the so-called structural stress experienced by the part of the element component capable of force resistance. The desired design stress is determined by solving an algebraic equation relating structural and design stresses. The proposed approach greatly simplifies the establishment of stress estimates in the components of structural elements required in the long-term safety forecast of structures.

1. Introduction

The problems considered in this paper are associated with the phenomenon of creep – an increase in the strain $\sigma(\tau_0)$ generated by stress $\varepsilon(\tau)$ at $\tau > \tau_0$. Over time τ stresses $\sigma(\tau)$ with unchanged deformations $\varepsilon(t_0)$ decrease and this phenomenon is called stress relaxation.

Stress relaxation is a consequence of the development of creep deformations $\varepsilon_{\Pi}(t,t_0)$ in the material. With a constant total strain $\varepsilon(t_0)=\varepsilon_M(t)+\varepsilon_{\Pi}(t,t_0)$, the fraction of the instantaneous elastic-plastic deformation $\varepsilon_M(t)$ decreases and the corresponding stress fraction decreases. Note that in real constructions, the phenomena of creep and relaxation occur simultaneously.

1.1. Linear formulation of the problem

In the linear formulation, the stress $\sigma(\tau)=\sigma(t_0)$ generates a strain

$$\varepsilon(t)=\frac{\sigma(t_0)}{E(t)}+\sigma(t_0)C(t,t_0), \tag{1}$$

here $E(t)$ is the elastic modulus of instantaneous deformations; $C(t,t_0)$ – a measure of simple creep at the moment t when loading at the moment t_0.

According to the superposition principle of L. Boltzmann [1], partial increment of creep strain

$$\varepsilon_{\Pi}(t,\tau_i)=C(t,\tau_i)\cdot\Delta\sigma(\tau_i) \tag{2}$$

is determined only by the magnitude and duration $\Delta\tau_i=\tau-\tau_i$ of the stress $\sigma(\tau)$ increment $\Delta\sigma(\tau_i)$ and does not depend neither on the rest of its partial increments $\Delta\sigma(\tau_j)$, $(j\neq i)$, nor on their duration.

The interdependence of increments $\Delta\varepsilon_{\Pi}(t,\tau_i)$ allows the representation of the increment $\Delta\varepsilon_{\Pi}(t,t_0)$ of creep strain corresponding to the incremental stress

$$\Delta\sigma(t)=\sum_{i=1}^{n}\Delta\sigma(\tau_i) \tag{3}$$

by the sum

$$\Delta\varepsilon_{\Pi}(t,t_0)=\sum_{i=1}^{n}C(t,\tau_i)\Delta\sigma(\tau_i). \tag{4}$$

Passing in the integral sum to the limit, we have

$$\Delta\varepsilon_{\Pi}(t,t_0)=\int_{t_0}^{t}C(t,\tau)d\sigma(\tau), \tag{5}$$

and then, integrating (5) in parts, we get

$$\Delta\varepsilon_{\Pi}(t,t_0)=C(t,t)\sigma(t)-C(t,t_0)\sigma(t_0)-\int_{t_0}^{t}\sigma(\tau)\frac{\partial C(t,\tau)}{\partial\tau}d\tau, \tag{6}$$

here $C(t,t)$ – the so-called short-term creep.

The value $C(t,t)$ is experimentally indefinable and, as in most works on the theory of concrete creep, we assume $C(t,t)=0$. Adding to $\Delta\varepsilon_{\Pi}(t,t_0)$ the stress-induced $\sigma(t_0)$ creep strain $C(t,t_0)\sigma(t_0)$ over the gap $[t,t_0]$, we have

$$\Delta\varepsilon_{\Pi}(t,t_0)=-\int_{t_0}^{t}\sigma(\tau)\frac{\partial C(t,\tau)}{\partial\tau}d\tau. \tag{7}$$

Adding the creep deformation $\varepsilon_{\Pi}(t,t_0)$ with instantaneous deformation $\varepsilon_M(t)=\dfrac{\sigma(t)}{E(t)}$ we arrive at the equality

$$\varepsilon(t)=\frac{\sigma(t)}{E(t)}-\int_{t_0}^{t}\sigma(\tau)\frac{\partial C(t,\tau)}{\partial\tau}d\tau, \tag{8}$$

representing a linear rheological equation of the mechanical state of concrete under prolonged loading of structural elements (beam, column).

According to (8), we obtain the equation

$$\sigma(t)=\varepsilon(t)E(t)+E(t)\int_{t_0}^{t}\sigma(\tau)\frac{\partial C(t,\tau)}{\partial\tau}d\tau. \tag{9}$$

For given deformations $\varepsilon(t)$ and parameters $E(t)$ and $C(t,\tau)$, the stress $\sigma(t)$ is determined by solving a linear integral equation by known methods, for example, using the iteration method.

For mature and old concrete, it is usually assumed that $C(t,\tau)=C_0\left[1-e^{-\gamma(t-\tau)}\right]$, and if aging $C(t,\tau)=\theta(\tau)\left[1-e^{-\gamma(t-\tau)}\right]$ is taken into account, then equation (9) reduces to a linear differential equation of the first and second orders, respectively. Here $\theta(\tau)$ is the aging function, for example [2], $\theta(\tau)=\frac{A}{\tau}+C_0$, and γ is a parameter reflecting the measure of the influence of an open surface on the process of concrete formation. Thus, in the linear formulation of the problem of stress relaxation in a concrete element with its axial loading, they are solved rather simply [3].

Consider for illustration a reinforced concrete beam, which is subject to axial load.

$$N(\tau)=\begin{cases} N_1;\ 2kT \le \tau < (2k+1)T \\ N_2;\ (2k+1)T \le \tau < 2(k+1)T \end{cases}, \tag{10}$$

here $T>0;\ N_1<N_2;\ k=0,1,2,...$.

Efforts N_1 and N_2 generate in concrete in moments $t_1=2kT$ and $t_2=(2k+1)T$ stresses equal to

$$\sigma_1=\frac{N_1}{A_b(1+m\mu)};\ \sigma_2=\frac{N_2}{A_b(1+m\mu)} \tag{11}$$

respectively; $m=\frac{E_a}{E_b};\ \mu=\frac{A_a}{A_b}$.

The area $A(\tau)$ of the normal section of the whole $\sigma(\tau)=\sigma_1$ with fractions at destruction does not change, and therefore

$$S^0(\tau)=1+V\left[\frac{\sigma_1}{R(\tau)}\right]. \tag{12}$$

For mature and old concrete it is believed $C(t,\tau)=C_0\left[1-e^{-\gamma(t-\tau)}\right]$, and the strength $R(\tau)$ is equal to its long-term strength R.

At each time interval we have a linear integral equation

$$\sigma_c(\tau)=\sigma_{ic}+\lambda\int_{t_i}^{t}\sigma_{ic}e^{-\gamma(t-\tau)}d\tau,\ i=1,2, \tag{13}$$

$$\lambda=\frac{\gamma m E_a C_0}{1+m\mu};\ t_1=2kT;\ t_2=(2k+1)T;\ k=0,1,2,..., \tag{14}$$

reduced to a differential equation

$$\frac{d\sigma_{bc}(t)}{dt}+\delta\sigma_{bc}(t)=\sigma_i\gamma;\ \delta=\gamma+\lambda;\ i=1,2. \tag{15}$$

By solving this equation with the initial condition $\sigma(0)=\sigma_i$; $i=1,2$ is the function

$$\overset{*}{\sigma}_{bc}(t)=\sigma_{ic}\left[\frac{\gamma}{\delta}+\frac{\lambda e^{-\delta(t-nT)}}{\delta}\right], \tag{16}$$

$$\sigma_i=\begin{cases}\sigma_1;\ n=2k \\ \sigma_2;\ n=2k+1\end{cases}. \tag{17}$$

The calculated stress in the concrete component is determined from the equation

$$1+V\left[\frac{\sigma_b(t)}{R}\right]^m\left[\frac{\gamma}{\delta}+\frac{\lambda e^{-\delta(t-nT)}}{\delta}\right]=\overset{*}{\sigma}_{bc}(t). \tag{18}$$

According to (18), the stress in the reinforcement is estimated at $t\to\infty$.

1.2. The relaxation of stresses

In a reinforced concrete element, the reduction of stresses $\sigma_b(\tau)$ in the concrete component is accompanied by an increase in the stresses $\sigma_s(\tau)$ in the reinforcement.

It should be noted that over time, the increase $\sigma_s(\tau)$ is significant and can lead to rupture of reinforcement and, as a result, to serious consequences in critical structures (in reinforced concrete shells of reactors, conduits). In this regard, in the long-term forecast of the safety of structures and structures, the problem arises of estimating the maximum possible stresses in the reinforcement.

Consider, for example, the relaxation of stresses in a single-reinforced reinforced concrete bar that is bent by the moment $M(t)$. At a distance h_a from the neutral axis Ox we have

$$\sigma_a(t)=\frac{1}{\mu n_0}\left[\frac{M(t)h_a}{J_b}-\sigma_b(t,h_a)\right], \tag{19}$$

here $\mu=\frac{A_a}{A_b}$; A_a and A_b – the area of normal sections of reinforcement and the concrete part of the timber; $n_0=\frac{J}{J_b}$, where J and J_b are the moments of inertia of the concrete component and the reduced normal section about the axis Ox; $\sigma_b(t,h_a)$ – normal stress in the concrete layer in contact with the reinforcement.

According to the condition of compatibility deformation at h_a

$$\sigma_b(t,h_a)=\tilde{\sigma}_b(t,h_a)+\lambda(t)\int_{t_0}^{t}\sigma_b(\tau,h_a)\frac{\partial C(t,\tau)}{\partial\tau}d\tau, \tag{20}$$

where $\tilde{\sigma}_b(t,h_a)=\frac{M(t)h_a}{J_b(\mu n_0 m(t)+1)}$ is the instantaneous elastic stress; $\lambda(t)=\frac{M(t)E_a(t)}{\mu n_0 m(t)+1}$; $m(t)=\frac{E_a(t)}{E_b(t)}$; $E_a(t)$ – modulus of elastic deformations of reinforcement.

The integral equation is solved by simple iterations with a zero approximation $\sigma_{b_0}(t,h_a)=\hat{\sigma}_b(t,h_a)$.

In the calculations, it is assumed that $E_b(t)=E_b$, $E_a(t)=E_a$, $C_b(t,\tau)=C_{0_b}\left[1-e^{-\gamma(t-\tau)}\right]$ and equation (11) is reduced to a differential equation

$$\frac{d\sigma_b(t,h_a)}{dt}+(\lambda+\gamma)\sigma_b(t,h_a)=\gamma\hat{\sigma}_b(t,h_a)+\frac{d\hat{\sigma}_b(t,h_a)}{dt}, \tag{21}$$

and at a constant bending moment M to the equation

$$\frac{d\sigma_b(t,h_a)}{dt}+(\lambda+\gamma)\sigma_b(t,h_a)=\gamma\hat{\sigma}_b(t,h_a). \tag{22}$$

According to the solutions $\sigma_b(0,h_a)=\hat{\sigma}_b(0,h_a)$ of these equations with the initial condition, the creep of concrete leads to the relaxation of its stresses and, for sufficiently large ones t, are respectively $\sigma_b(t,h_a)$

$$\sigma_b(\infty,h_a)=\frac{\gamma\hat{\sigma}_b(t,h_a)+\gamma\dot{\hat{\sigma}}_b(t,h_a)}{\lambda+\gamma}, \tag{23}$$

$$\sigma_b(\infty,h_a)=\frac{\gamma}{\lambda+\gamma}\hat{\sigma}_b(0,h_a). \tag{24}$$

As a result of a prolonged redistribution of stresses $\hat{\sigma}_b(0,h_a)$ from concrete to reinforcement, its initial stress $\sigma_a(0)=\dfrac{Mh_a n_0 m}{J(\mu n_0 m+1)}$ increases to

$$\sigma_a(\infty)=\frac{1}{\mu n_0}\left[\frac{Mh_a}{J_0}-\frac{\gamma}{\gamma+\lambda}\hat{\sigma}_b(0,h_a)\right]. \quad (25)$$

2. Method of solution for nonlinear formulation of the problem

The increasing load on the structural element entails structural changes, giving rise to a progressive process of discontinuity of its material [4], [5]. A material (concrete, steel, wood, plastic) is considered as a combination of its fractions (links, fibers, layers) with statistically distributed strengths $R_i(\tau)$. The concept of statistical variation of the quantity $R_i(\tau)$, which goes back to [6], was developed in [7]. This concept serves as the basis for modifying L. Boltzmann's superposition principle for its applicability in a nonlinear formulation.

A force $N(\tau)$ increasing to a normal cross section results in the destruction of a part of the fractions, reducing the cross-sectional area A to the area $A(\tau)$ formed by the whole fractions at the moment τ.

Magnitude

$$\sigma_c(\tau)=\frac{N(\tau)}{A(\tau)}, \quad (26)$$

associated with structural damage to the material is structural, and the value

$$\sigma(\tau)=\frac{N(\tau)}{A} \quad (27)$$

calculated normal stress in the element. According to (17) and (18)

$$\sigma_c(\tau)=\frac{A}{A(\tau)}\sigma(\tau), \quad (28)$$

and the function $S^0(\tau)\equiv\dfrac{A}{A(\tau)}$ describes the process of destruction of fractions during force deformation. This process is accompanied by a redistribution of loading $N(\tau)$ on the area $A(\tau)$, because only fractions that are whole at this point have a resistance.

In contrast to the linear formulation, which means equal strength of all links, interdependent increments of creep deformations do not correspond to stepwise increment $\Delta\sigma=\sum_{i=1}^{n}\Delta\sigma(\tau_i)$ – the action $\Delta\sigma(\tau_i)$ at the moment $\tau=\tau_j$ is enhanced by the action $\Delta\sigma(\tau_j)$ that destroys part of the whole fractions by this time [8].

Let us mentally select in the concrete (steel) component of an element V its part V_t consisting of integers in between $[t_0,t]$ fractions. The stress V_t in the current $\tau\in[t_0,t]$ coincides with the stress in the part V_τ of the element consisting of integers at the moment τ of fractions. Thus, the stress in the part V_t is a structural stress $\sigma_c(\tau)$. Since the fractions V_t are equally strong until the partial $\tau=t$ increments $\Delta\varepsilon_{\Pi}(t,\tau_i)=C(t,\tau_i)\cdot\Delta\sigma_c(\tau_i)$ of creep deformations are independent of each other and the stress $\Delta\sigma_c(t)=\sum_{i=1}^{n}\Delta\sigma_c(\tau_i)$ generates an increment

$$\Delta\varepsilon_{\Pi}(t,t)=\sum_{i=1}^{n} C(t,\tau_i)\Delta\sigma_c(\tau_i) \tag{29}$$

creep strain. This allows the repetition of standard operations to derive the rheological equation of the mechanical state in a nonlinear formulation [9], [10]

$$\varepsilon(t)=\frac{\sigma_c(t)}{E(t)}-\int_{t_0}^{t}\sigma_c(\tau)\frac{\partial C(t,\tau)}{\partial\tau}d\tau . \tag{30}$$

Since $\sigma_c(\tau)=S^0(\tau)$, then according to (30)

$$\varepsilon(t)=\frac{S^0(t)\sigma_c(t)}{E(t)}-\int_{t_0}^{t}S^0(\tau)\sigma_c(\tau)\frac{\partial C(t,\tau)}{\partial\tau}d\tau . \tag{31}$$

3. Results

Equality (29) represents a modification of the L. Boltzmann principle of superposition of creep deformations.

In [11], the equation is obtained in a nonlinear formulation

$$\varepsilon(t)=\frac{S_M\left[\frac{\sigma(t)}{R(t)}\right]}{E(t)}-\int_{t_0}^{t}S_{\Pi}\left[\frac{\sigma(\tau)}{R(\tau)}\right]\frac{\partial C(t,\tau)}{\partial\tau}d\tau , \tag{32}$$

where $S_M\left[\frac{\sigma(t)}{R(t)}\right]$ and $S_{\Pi}\left[\frac{\sigma(\tau)}{R(\tau)}\right]$ are nonlinear functions of stresses, respectively, of instantaneous deformations and creep deformations. Since in the physical aspect both types of deformations are generated by a single force factor, these functions coincide and represent nothing more than structural stress $\sigma_c(t)$ and $\sigma_c(\tau)$.

According to $\sigma_c(\tau)=S^0(\tau)\sigma(t)$ a single function of stress is represented as $S[\sigma(\tau)]=S^0(\tau)\sigma(\tau)$. Non-linear function $S^0(\tau)$ in applications is given by the equality

$$S^0(\tau)=1+V\left[\frac{\sigma(\tau)}{R(\tau)}\right]^m , \tag{33}$$

where V and m – empirical parameters; for concrete it is commonly believed $m=4$.

The function $S^0(\tau)$ describing the process of loss of power resistance ability by fractions of $R_i<\sigma(\tau)$ depends on the stress level $\eta(\tau)=\frac{\sigma(\tau)}{R(\tau)}$.

Similar to equation (8) and equation (30) is reduced to a linear integral equation already with respect to structural stress. After determining its solution $\sigma_c^*(t)$, the calculated stress $\sigma^*(t)$ is found by solving the equation

$$S^0(\tau)\sigma(t)=\sigma_c^*(t). \tag{34}$$

In applications [12], according to (33) and (34), the quantity $\sigma^*(t)$ is the root of the equation

$$[\sigma(t)]^{m+1}+\frac{[R(t)]^m}{V}\sigma(t)-\frac{[R(t)]^m}{V}\sigma_c^*(t)=0 . \tag{35}$$

Along with (33), the nonlinear function of stresses $\sigma_c(\tau)=a[\sigma(\tau)]^b$ is used [13], where a and b and are empirical parameters. In this case $a[\sigma^*(t)]^b=\sigma_c{}^*(t)$ and $\sigma^*(t)=\left[\frac{\sigma_c{}^*(t)}{a}\right]^{\frac{1}{b}}$.

3.1. Calculation of stresses

The increasing normal force $N(\tau)$ reduces the area $A_b(\tau)$ and $A_a(\tau)$ the normal sections of the components of the element that are capable of force resistance and increases the calculated stresses $\sigma_b(\tau)$ and $\sigma_a(\tau)$ to structural stresses $\sigma_{b_c}(\tau)$ and $\sigma_{a_c}(\tau)$.

Equations (19) - (22) and estimates (23) - (25) also hold for structural stresses $\sigma_{b_c}(\tau)$ and $\sigma_{a_c}(\tau)$, and estimates for the maximum values of design stresses are determined according to (34) and (35).

4. Different approach to the problem

In [19] the approach to the problem is described according to the equation

$$\varepsilon_b(t)=\frac{\sigma_b(t)}{E_b(t)}+\int_{t_0}^{t}\sigma_b(\tau)\frac{\partial C_b(t,\tau)}{\partial\tau}d\tau+\int_{\sigma(t_0)}^{\max\sigma}f[\sigma(t)]F[T(\sigma,\tau)]d\sigma, \tag{36}$$

here $f[\sigma(t)]=\xi[\sigma(t)]^u$; ξ – small parameter; u – nonlinearity parameter; $F[T(\sigma,\tau)]=F_0\left[1-e^{-\varphi T(\sigma,t)}\right]$; $T(\sigma,t)$ – total duration of the stress σ ; F_0 and φ – empirical parameters. Stress $\sigma_b(t)$ is sought in the form

$$\sigma_b(t)=\sigma_0(t)+\xi\sigma_1(t)+\xi^2\sigma_2(t)+... \tag{37}$$

Poincaré's small parameter method involving the Laplace transform. It should be noted that in this case, the procedure for finding stresses $\sigma_b(t)$ and $\sigma_a(t)$ is difficult and busy.

Some other approaches to this problem are given in the following papers [15 – 17].

5. Conclusion

1. Estimates of the maximum possible (over a long time) stress values in the reinforcement are important in the long-term forecast because its rupture causes serious consequences.
2. The proposed approach based on the strength structure of structural materials (for example, the distribution of the strengths $R_i(\tau)$ of element fractions according to the normal law) is much simpler than the known methods for estimating stresses in the components of elements.

6. References

[1] Bolzman L 1874 *Zur Theorie der elastishen Nfchwirkung* (Berlin: Wiener).
[2] Arutyunyan N Kh 1952 *Some questions of the theory of creep* (Moscow: Gostekhizdat) p 130
[3] Vasilkova N T, Bashkatova M E and Larionov E A 2012 Stress relaxation under axial loading of a reinforced concrete beam *Structural mechanics of engineering construction and building* **1** pp 63-68
[4] Berg O Ya 1962 *The physical basis of the theory of strength of concrete and reinforced concrete* (Moscow: Stroiizdat) p 96
[5] Gvozdev A A 1949 *Calculation of the bearing capacity of structures by the method of limiting equilibrium* (Moscow: Gosstroyizdat) p 280
[6] Wiebull W 1949 A statistical representation of fatigue. *Trans. Roy. Inst. Techn.* **27**
[7] Kholmyansky M M 1997 *Concrete and reinforced concrete: deformability and strength* (Moscow: Stroiizdat) p 576

[8] Bondarenko V M and Larionov E A 2011 The principle of deformation during structural damage to structural elements *Structural mechanics of engineering construction and building* **2** pp 16-22
[9] Larionov E A and Larionov A E 2015 To the theory of nonlinear creep *Structural mechanics and calcul of buildind* **2** pp 58-65
[10] Larionov E A and Larionov A E 2017 To the theory of nonlinear creep. *Structural mechanics of engineering construction and building* **4** pp 35-39
[11] Bondarenko V M and Bondarenko S V 1982 *Engineering methods of the nonlinear theory of reinforced concrete* (Moscow: Stroiizdat) p 278
[12] Larionov E and Zveryaev E 2017 Stress relaxation of constructions elements *MATEC Web of Conferences* **117** 00101
[13] Bondarenko V M 2015 Exposure of effective prestressing of reinforced concrete structures *Structural mechanics of engineering construction and building* **2** pp 41-45
[14] Galustov K Z 2006 *Nonlinear theory of concrete creep and calculation of reinforced concrete structures* (Moscow: Fizmatlit) pp 248
[15] Sidorov V N and Nowak K 2018 Numerical investigation of the long-term work of arches under material creep conditions *IOP Conference Series: Materials Science and Engineering* **456** **(1)** 012007
[16] Zamaliev F S and Zakirov M A 2018 Stress-strain state of a steel-reinforced concrete slab under long-term *Magazine of Civil Engineering* **83** **(7)** pp 12-23
[17] Korotchenko I A, Ivanov E N, Manovitsky S S, Borisova V A, Semenov K V and Barabanshchikov Yu G 2017 Deformation of concrete creep in the thermal stress state calculation of massive concrete and reinforced concrete structures *Magazine of Civil Engineering* **69** **(1)** pp 56-63

Radiative properties of Tm^{3+} doped TeO_2-ZnO-Na_2O-TiO_2 glasses studied using Judd-Ofelt Theory

A Marzuki*, Riyatun, M Larasati, G T Singgih

Department of Physics, Sebelas Maret University, Surakarta 57216 Indonesia

Email: amarzuki@mipa.uns.ac.id

Abstract. New Tm^{3+}-doped tellurite glasses were prepared by melt quenching technique. X-ray Diffraction (XRD) characterization were carried out to confirm the glassy state of the samples. Characterization using UV-VIS-NIR spectrophotometer and FTIR were performed in order to determine the width of the transparent window, optical band gap and bonding structure of the samples. Density and refractive index were measured at room temperature using pycnometer and Brewster's angle technique. Utilizing these data, lasing performance of the samples was predicted. For this purpose, the Judd-Ofelt theory was applied to calculate Judd-Ofelt parameters (Ω_t). From these results several radiative transitions were studied in terms of transition probabilities (A_i), branching ratios (β_i) and radiative lifetime (τ). The predicted results were then compared to that obtained from fluorescence spectra measurement so that quantum efficiency of the samples can be calculated.

Keywords: Tellurite glasses, Judd-Ofelt analysis, Spectroscopic properties, Tm^{3+} doped glasses, Optical absorption.

1. Introduction

Triggered by the need for optical amplifiers to compensate for the propagation loss in the second and third telecommunication windows, Nd^{3+} and Er^{3+} doped glasses have been attracted more attention and researched more than other rare earth ions doped glasses [1, 2,3]. Among the less studied rare earth doped glasses, Tm^{3+} doped glasses have been researched more than others especially due to their capability to provide mid-infrared laser source.

Many papers report emission characteristic at about 1.8 μm resulted from $^3F_4 \rightarrow {}^3H_6$ transition for different glass hosts. Using the product of stimulated emission cross-section (σ) and lifetime (τ) as an important parameter characterizing laser performance of a Tm^{3+} doped glass, Tian et al., [4] who studied Tm^{3+} doped 53.3$BiO_{1.5}$–33.3GeO_2–13.4$NaO_{0.5}$ glasses calculate σ x τ and their results are compared to other glasses such as fluorophosphates glasses [5], tellurite glasses [6], germinate glasses [7] and ZBLAN glasses[8]. Glasses fabricated by Tian, et. al., [4] have σ x τ which are higher than the first three glasses [5, 6, 7] but higher ZBLAN glasses [8] suggesting that Tm^{3+} doped 53.3$BiO_{1.5}$–33.3GeO_2–13.4$NaO_{0.5}$ glasses are potentially used as materials for optical amplifiers.

Other laser transition encounters within the scheme of Tm^{3+} ions energy level diagram is that resulted from $^3H_5 \rightarrow {}^3H_6$ emitting light at about 1.22 μm. This emission is potentially used for optical amplifiers at 1.22 μm. This electronic transition, however, suffers from the narrow gap between $^3H_5$ and $^3F_4$ resulting in $^3H_5$ is easily depopulated [9]. For this purpose, low phonon energy glass host is required. In addition to the above transitions, glass host dependence are also apply for other transition schemes, e.g., $^3F_4 \rightarrow {}^3H_5$, $^3H_4 \rightarrow {}^3H_6$ and $^3H_5 \rightarrow {}^3F_4$. In this study, we applied Judd-Ofelt theory to analyze different composition of Tm_2O_3:TeO_2-ZnO-Na_2CO_3-TiO_2 in order to predict the strength of an electronic transition relative to other

IOP Conf. Series: Materials Science and Engineering **675** (2019) 012070 doi:10.1088/1757-899X/675/1/012070

transitions originated from the same energy level expressed in oscillator strength (f) and branching ratios (β).

2. Experiment

Tm^{3+} doped tellurite glasses having compositions (in mol%): $60TeO_2$-$25ZnO$-$(10\text{-}x)Na_2CO_3$-$5TiO_2$-xTm_2O_3 (with x = 1, 2, 3, 4, 5) were fabricated by melt quenching method. The starting materials in the form of powder were weighed in nitrogen controlled glove box so that a mixture of 10 gram was obtained. After thoroughly mixed, the mixture was transferred to platinum crucible. Melting was carried out in electric furnace at 900^0C for 60 minutes while periodically stirred to get a homogeneous melt. Casting was carried out by pouring the melt into a pre-heated parallel brass mold followed by annealing at 360^0C to remove any residual stressed formed during casting.

Density measurements were carried out at room temperature using pycnometer. Refractive indices were measured at wavelength of 625 nm at room temperature using Brewster angle method. Transmittance within UV-VIS-NIR region was measured using Spectrophotometer Perkin-Elmer UV-VIS-NIR Lambda-25. Based on the data of density, refractive index and absorption spectra; samples were analyzed their lasing performance by applying Judd-Ofelt theory.

3. Results and discussion

3.1. X-Ray diffraction

Six glasses fabricated in this experiment show good transparency (Figure 1). In order to confirm the amorphous nature of the samples, XRD analysis was performed. It can be seen from figure2 that all the XRD patterns show no narrow sharp peak. Instead, broad peaks at 2θ at about 29^0 characterizing the amorphous nature of solids were observed.

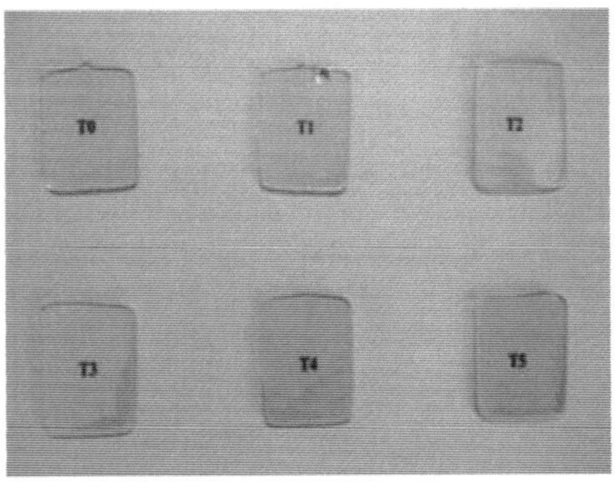

Figure 1. Photograph of six transparent glasses showing that they possess high transparency.

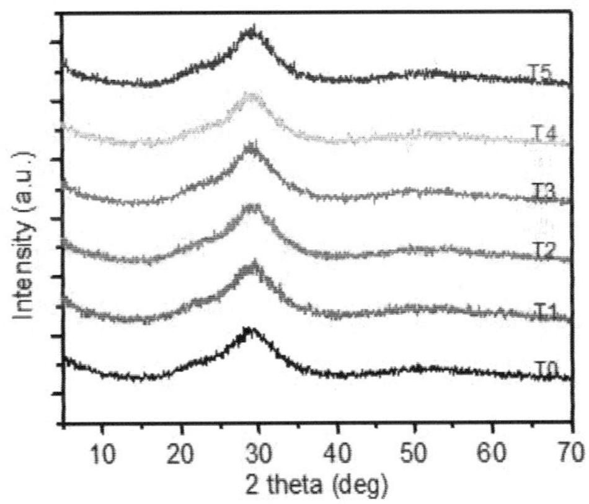

Figure 2. X-Ray Diffraction patterns of samples with different concentration of Tm_2O_3.

3.2. Density and refractive index

Density of all the samples measured at room temperature are shown in figure 3. It shows that density increases with increasing Tm_2O_3 concentration. Adding Tm_2O_3 (MR = 388 smu) to replace Na_2O (MR = 62) leads to increasing average molecular mass of glass. From definition of density, which is mass (m) divided by volume (V), this Tm_2O_3 addition contribute to the increasing m. As can be seen from figure 3, however, this addition causes the molar volume is increased. This suggest that the change in average molecular mass is more dominant in determining the glass density than the change in molar volume.

Figure 4 shows compositional variation of refractive index obtained from both measurement and calculation based on Lorentz-Lorenz equation:

$$n = \sqrt{\frac{2N_A\alpha_m\rho+3\sum x_iM_i}{3\sum x_iM_i-N_A\alpha_m\rho}}, \tag{1}$$

where n is refractive index, N_A is Avogadro number, α_m is molecular polarizability, M_i is molecular weight, x_i is molar fraction and ρ is density.

It is seen that refractive index increases as Tm_2O_3 concentration is increased. From equation (1) it is clearly shown that the value of n is controlled by ρ. In this regard, ρ can be related to electron density.

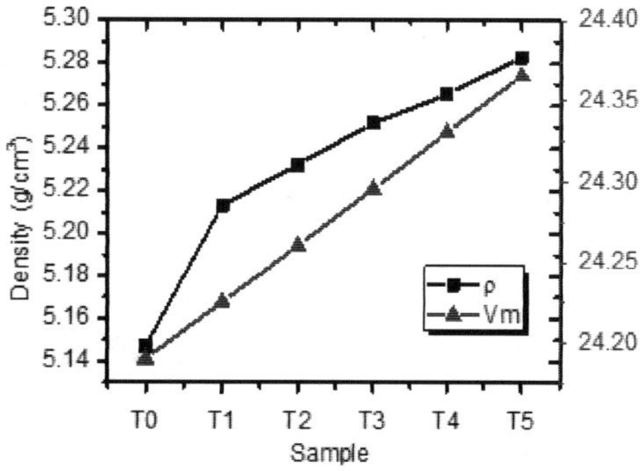

Figure 3. The change in density and molar volume as different concentration of Tm_2O_3 were added into the glass.

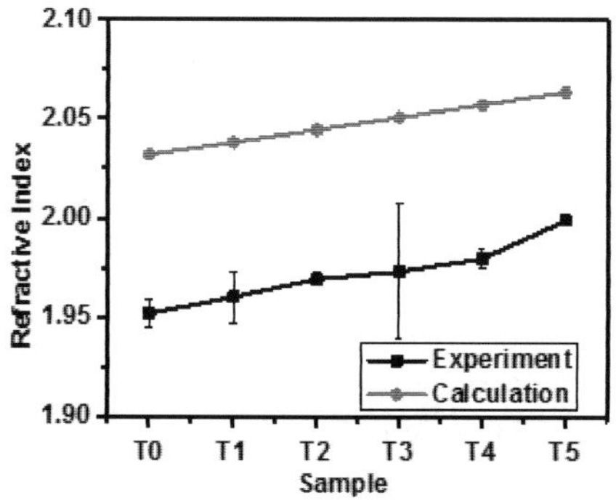

Figure 4. Compositional variation of refractive resulted from experiment and calculation based on Lorentz-Lorenz equation.

3.3. Absorption spectra

Figure 5 shows linear absorption spectra of all samples. Four distinguishable peaks corresponding to f-f electronic transition of Tm^{3+} ions are observed. They are bands corresponding to electronic transitions: $^3H_6 \rightarrow {}^3H_4$ (793 nm), $^3H_6 \rightarrow {}^3F_3$ (687 nm), $^3H_6 \rightarrow {}^3F_2$ (660 nm) and $^3H_6 \rightarrow {}^1G_4$ (468 nm). Assignments were made referring to Carnall et al. [7]. From figure 5, it can also be seen that there is an increase in absorption band intensity with the addition of Tm_2O_3. Adding Tm_2O_3 means supplying more electrons at the ground energy level of Tm^{3+} resulting in more photons are absorbed.

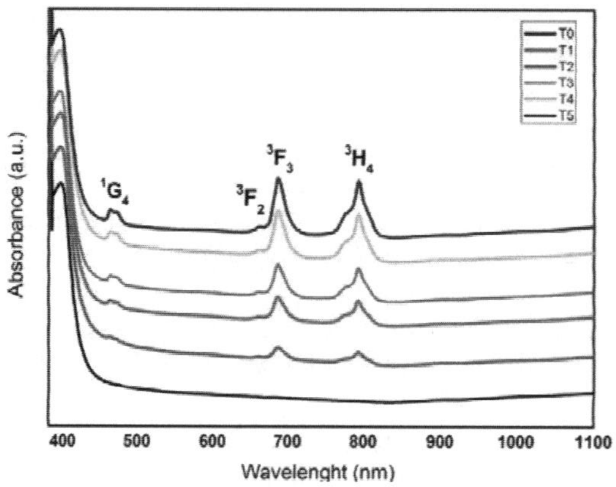

Figure 5. Absorption spectra of the present glasses at different concentration of Tm_2O_3.

In order to evaluate the lasing performance from the data of absorption spectra, Judd-Ofelt theory was applied. All glass samples were evaluated in term of measured oscillator strength (f_{meas}), calculated oscillator strength (f_{calc}), radiative transition probalities (A), lifetime (τ) and branching ratios (β) expressed in the following equations:

$$f_{meas} = \frac{3hcn}{8\pi^3 e^2 N} \frac{9}{(n^2+2)^2} \frac{(2J+1)}{\bar{\lambda}} \int \sigma(\lambda) d\lambda \quad (2)$$

$$f_{calc}(J \rightarrow J') = \sum_{t=2,4,6} \Omega_t \left| \langle (S,L)J \middle\| U^{(t)} \middle\| (S',L')J' \rangle \right|^2 \quad (3)$$

$$A[(S,L)J;(S',L')J'] = \frac{64\pi^4 e^2 n}{3h(2J+1)\bar{\lambda}^3} \left(\frac{n^2+2}{9}\right)^2 \sum_t \Omega_t \left| \langle (S,L)J \middle\| U^{(t)} \middle\| (S',L')J' \rangle \right|^2 \quad (4)$$

$$\beta[(S,L)J;(S',L')J'] = \frac{A[(S,L)J;(S',L')J']}{\sum_{S',L',J'} A[(S,L)J;(S',L')J']} \quad (5)$$

$$\tau_{rad} = \left\{ \sum_{S',L',J'} A\,[(S,L)J;(S',L')J'] \right\}^{-1} \quad (6)$$

where h is Plank's constant, c is velocity of light in vacuum, n is refartive index, N is number of Tm^{3+} per mole, J and J' are total angular momenta of ground state and excited states, respectively, $\Omega_{\lambda=2,4,6}$ are Judd-Ofelt intensity parameters and $||U^{\lambda}||^2$ are elements of doubly reduced matrix operator calculated from the intermediate-coupling approximation.

Since $||U^{\lambda}||^2$ of a rare earth ion is insensitive to its environment, their values do not depend on the glass host. In most work, the value of $||U^{\lambda}||^2$ are taken from Carnall table [10]. The measured oscillator strength (f_{meas}) was obtained from absorption spectra by firstly calculating area under absorption peaks $\int \sigma(\lambda) d\lambda$ as given in equation (2). In order to calculate Ω, A, τ and β, firstly, we have to calculate $\int \sigma(\lambda) d\lambda$, area under the peak of an absorption band. Inserting $||U^{\lambda}||^2$ corresponding to each absorption band into equation (3) and then using Root-mean-square deviation (rms), one can obtain $\Omega_{\lambda=2,4,6}$ (Table 1).

Table 1. Judd-Ofelt intensity parameters ($\Omega_{\lambda=2,4,6}$ x 10^{-20} cm^2) of the present glasses and those taken from literatures.

Samples	Ω_2	Ω_4	Ω_6	Ω_4/Ω_6	Reference
T1	8,88	1,34	1,91	0.70	
T2	8,78	1,80	1,67	1.08	
T3	8,56	1,87	1,69	1.11	Present work
T4	8,14	1,75	1,56	1.12	
T5	8,95	1,88	1,66	1.13	
TNZL: Tm	4,16	1,87	1,14	1.64	[11]
Fluorophosphate	3,01	2,56	1,54	1.66	[4]

It can be seen from table 1 that the magnitudes of $\Omega_{\lambda=2,4,6}$ are similar to those reported values. Except for sample T_1, their magnitude increase following the trend of $\Omega_2 > \Omega_6 > \Omega_4$. The high value of Ω_2 indicates that the present glasses possess high covalence and asymmetry. The values of the spectroscopic factor defined by Ω_4/Ω_6 are also similar to those reported in other papers, which are in the range of 0.65 – 1.46. Once the values of $\Omega_{\lambda=2,4,6}$ are obtained, one can then compute radiative transition rate (A), branching ratios (β) and radiative lifetime (τ) using equation (4-6) and the results for transition originated from $^3H_4$ and $^3H_5$ are tabulated in Table 2. Radiative transition rate (A) in all samples for $^3H_5 \rightarrow {}^3H_6$ are the higher than that for $^3H_5 \rightarrow {}^3F_4$ and so are their branching ratios, which are more than 94% of the emitted light from $^3H_5$ goes to $^3H_6$ resulting in emission at about 1220 nm.

IOP Conf. Series: Materials Science and Engineering **675** (2019) 012070 doi:10.1088/1757-899X/675/1/012070

There are three possible electronic transition originated from $^3H_4$. They are $^3H_4 \rightarrow {}^3H_5$, $^3H_4 \rightarrow {}^3F_4$ and $^3H_4 \rightarrow {}^3H_6$. Among them, emission resulted by transition $^3H_4 \rightarrow {}^3H_6$ resulting in emission at about 800 nm is the strongest ($\beta > 91\%$). This value tends to decrease with increasing Tm_2O_3.

Table 2. *Transition probabilities* (*A*), *branching ratio* (β), dan *radiative lifetime* (τ) of the present glasses at different concentration of Tm_2O_3 for electronic transition originated from $^3H_4$ and $^3H_5$.

Samples	Transitions	Energy (cm^{-1})	A (s^{-1})	, β (%)	τ (ms)
T1	$^3H_5 \rightarrow {}^3F_4$	2167	51,08	5,96	1,17
	$\rightarrow {}^3H_6$	8264	806,07	94,04	
	$^3H_4 \rightarrow {}^3H_5$	4362	51,74	1,19	0,23
	$\rightarrow {}^3F_4$	6529	310,51	7,11	
	$\rightarrow {}^3H_6$	12626	4003,04	91,70	
T2	$^3H_5 \rightarrow {}^3F_4$	2167	50,65	5,93	1,17
	$\rightarrow {}^3H_6$	8264	803,06	94,07	
	$^3H_4 \rightarrow {}^3H_5$	4362	63,70	1,48	0,23
	$\rightarrow {}^3F_4$	6529	315,34	7,32	
	$\rightarrow {}^3H_6$	12626	3927,07	91,20	
	$^3H_5 \rightarrow {}^3F_4$	2167	50,15	5,83	1,16
T3	$\rightarrow {}^3H_6$	8264	810,62	94,17	
	$^3H_4 \rightarrow {}^3H_5$	4362	65,73	1,53	0,23
	$\rightarrow {}^3F_4$	6529	315,24	7,33	
	$\rightarrow {}^3H_6$	12626	3918,44	91,14	
T4	$^3H_5 \rightarrow {}^3F_4$	2167	48,12	5,84	1,21
	$\rightarrow {}^3H_6$	8264	775,98	94,16	
	$^3H_4 \rightarrow {}^3H_5$	4362	63,18	1,54	0,24
	$\rightarrow {}^3F_4$	6529	302,69	7,38	
	$\rightarrow {}^3H_6$	12626	3738,37	91,09	
T5	$^3H_5 \rightarrow {}^3F_4$	2167	54,35	5,97	1,10
	$\rightarrow {}^3H_6$	8264	856,50	94,03	
	$^3H_4 \rightarrow {}^3H_5$	4362	69,21	1,50	0,22
	$\rightarrow {}^3F_4$	6529	338,18	7,33	
	$\rightarrow {}^3H_6$	12626	4203,30	91,16	

4. Conclusions

New tellurte glasses with compositions (in mol%): $60TeO_2$-$25ZnO$-$(10\ -x)Na_2CO_3$-$5TiO_2$-Tm_2O_3 (with x = 1, 2, 3, 4, 5) have been fabricated, characterized and analyzed in order to predict their lasing performance especially light emitted from $^3H_5$ and $^3H_4$. For transition from $^3H_5$, more than 94 % of the emitted light is at about λ = 1220 nm while for the transition from $^3H_4$, about 91 % of the emitted light is at about λ = 800 nm. This domination tend to slightly decrease with the increase of Tm_2O_3 concentration.

Acknowledgements

The authors acknowledge for the financial support from PNBP UNS Research Funding 2019.

References

[1] Dan H K, Qiu J, Zhou D and Wang R 2018 Broadband near-infrared emission and energy transfer in Nd-Bi co-doped transparent silicate glass-ceramics for optical amplifiers *Optical Materials* **85** pp 517-522

[2] Marzuki A, Yunianto M, Riyatun and Purwanto H 2018 Absorption spectra analysis of Er^{3+}-Doped TeO2- ZnO-Bi2O3 Glasses *Key Engineering Materials* **772** pp 85-89

[3] Marzuki A and Fausta D A 2018 Physical and Optical Studies of Bi^{3+}-Modified Erbium Doped Tellurite Glasses *IOP Conf. Series: Materials Science and Engineering* **333** 012015

[4] Tian Y, Xu R, Zhang L, Hu L and Zhang J 2010 1.8 μm emission of highly thulium doped fluorophosphate glasses J. Appl. Phys. **108** 083504

[5] Wang M, et al 2009 Effect of $Al(PO_3)_3$ content on physical, chemical and optical properties of fluorophosphate glasses for 2 μm application *Mater. Chem. Phys.* **114(1)** pp 295-299

[6] Fusari F, et al 2008 Spectroscopic and lasing performance of Tm3+- doped bulk TZN and TZNG tellurite glasses operating around 1.9 μm, *Opt. Express* **16(23)** pp 19146-51

[7] Balda R, Lacha L M, Fernández M J, Fernández-Navarro J M 2005 Optical spectroscopy of Tm^{3+} ions in GeO_2–PbO–Nb_2O_5 glasses *Optical Materials* **27(11)** 1771-75

[8] Özen G, et al 1995 Excited state absorption mechanisms of red to UV and blue conversion luminescence in Tm^{3+} doped fluorophosphate glass *Journal of Luminescence* **63(1-2)** pp 85-96

[9] Huihua X, Gao T, Lan L and Wei C 2011 50GeSe2-25In2Se3-25CsI glass doped with Tm3+, Tm3+/Ho3+ and Tm3+/Er3+ for amplifiers working at 1.22 μm *J. Rare Earth* **29(10)** 920-3

[10] Carnall W T, Fields P R and Rajnak K 1968 Electronic Energy Levels in the Trivalent Lanthanide Aquo Ions. I. Pr^{3+}, Nd^{3+}, Pm^{3+}, Sm^{3+}, Dy^{3+}, Ho^{3+}, Er^{3+}, and Tm^{3+} *J. Chem. Phys.* **49** 4424 https://doi.org/10.1063/1.1669893

[11] Lachheb R, et al 2015 Characterization of Tm^{3+} doped TNZL glass laser material *Journal of Luminescence* **161** pp 281-287

AUTHOR INDEX